PERFORMING DATA ANALYSIS USING IBM SPSS®

PERFORMING DATA ANALYSIS USING IBM SPSS®

LAWRENCE S. MEYERS

Department of Psychology
California State University, Sacramento
Sacramento, California

GLENN C. GAMST

Department of Psychology
University of La Verne
La Verne, California

A. J. GUARINO

Department of Biostatistics
MGH Institute of Health Professions
Boston, Massachusetts

Marc Brock
847-271-4857

WILEY

Published by John Wiley & Sons, Inc., Hoboken, New Jersey.
Published simultaneously in Canada.

Library of Congress Cataloging-in-Publication Data:

Meyers, Lawrence S.

Performing data analysis using IBM SPSS® / Lawrence S. Meyers, Department of Psychology, California State University, Sacramento, Sacramento, CA, Glenn C. Gamst, Department of Psychology, University of La Verne, La Verne, CA, A. J. Guarino, Department of Biostatistics, MGH Institute of Health Professions, Boston, MA.

pages cm

Includes bibliographical references and index.

ISBN 978-1-118-35701-9 (pbk.) – ISBN 978-1-118-51494-8 – ISBN 978-1-118-51492-4 (ePDF) – ISBN 978-1-118-51493-1 (ePub) – ISBN 978-1-118-51490-0 1. Social sciences–Statistical methods–Computer programs. 2. SPSS (Computer file) I. Title.

HA32.M4994 2013

005.5'5–dc23

2013002844

Printed in the United States of America

10 9 8 7 6 5 4

CONTENTS

PREFACE

The IBM SPSS® software package is one of the most widely used statistical applications in academia, business, and government. This book, *Performing Data Analysis Using IBM SPSS*, provides readers with both a gentle introduction to basic statistical computation with the IBM SPSS software package and a portal to the more comprehensive and statistically robust multivariate procedures. This book was written to be a stand-alone resource as well as a supplementary text for both undergraduate introductory and more advanced graduate-level statistics courses.

For most of the chapters, we provide a consistent structure that includes the following:

- *Overview:* This is a brief conceptual introduction that furnishes a set of relevant details for each statistical procedure being covered, including a few useful references that supply additional background information.
- *Numerical Example:* This includes a description of the research problem or question, the name of the data file, a description of the variables and how they are coded, and (often) a screenshot of the IBM SPSS Data View.
- *Analysis Strategy:* When the analysis is performed in stages, or when alternative data processing strategies are available, we include a description of how we have structured our data analysis and explain the rationale for why we have performed the analyses in the way presented in the chapter.
- *Analysis Setup:* This includes how to configure each dialog window with screenshots and is accompanied (within reason) with explanations for why we chose the particular options we utilized.
- *Analysis Output:* This elucidates the major aspects of the statistical output with pertinent screenshots and discussion.

Because of the multiple audience we are attempting to reach with this book, the complexity of the procedures covered varies substantially across the chapters. For example, chapters that cover IBM SPSS basics of data entry and file manipulation, descriptive statistical procedures, correlation, simple linear regression, multiple regression, one-way chi-square, *t* tests, and one and two-way analysis of variance designs are all appropriate topics for first- or second-level statistics and data analysis courses. The remaining chapters, data transformations, assumption violation assessment, reliability analysis, logistic regression, multivariate analysis of variance, survival analysis, multidimensional scaling, cluster analysis, multilevel modeling, exploratory and confirmatory factor analysis, and structural equation modeling, are all important topics that may be suitable for more advanced statistics courses.

There are 66 chapters in this book. They are organized into 19 sections or "Parts." Different authors might organize the chapters in somewhat different ways and present them in a somewhat different order, as there is no fully agreed upon organizational structure for this material. However, except for the chapters presented in the early parts that show readers how to work with IBM SPSS data files, most of the data analysis chapters can be used as a resource on their own, allowing users to work with whatever analysis procedures meet their needs; the order in which users would choose to work with the chapters is really a function of the foundations on which the material is based (e.g., users should undertake structural equation modeling only after acquiring some familiarity with regression techniques and factor analysis).

Part 1, "Getting Started With IBM SPSS®," consists of three chapters that provide the basics of IBM SPSS. Chapter 1 provides an introduction to IBM SPSS, Chapter 2 describes how to enter data, and Chapter 3 demonstrates how to import data from Excel to IBM SPSS.

Part 2, "Obtaining, Editing, and Saving Statistical Output," consists of three chapters that describe ways of manipulating the IBM SPSS statistical output. Chapter 4 conveys how to perform a statistical procedure. Chapter 5 demonstrates how to edit statistical output. Chapter 6 provides information on saving and copying output.

Part 3, "Manipulating Data," contains three chapters that focus on how to organize existing data. Chapter 7 examines the sorting and selecting of cases. Chapter 8 demonstrates how to split a data file. Chapter 9 discusses how to merge cases and variables.

Part 4, "Descriptive Statistics Procedures," consists of three chapters that provide descriptive statistical summary capabilities. Chapter 10 focuses on the analysis of frequency counts for categorical variables. Chapter 11 describes how to compute measures of central tendency and variability. Chapter 12 provides additional options to examine variables in the data file.

Part 5, "Simple Data Transformations," consists of five chapters that demonstrate how to manipulate variables. Chapter 13 describes how to standardize a variable through the creation of z scores. Chapter 14 demonstrates how to recode the values of a variable. Chapter 15 provides a discussion of visual binning used in categorizing data. Chapter 16 demonstrates how to compute a new variable from existing data, and Chapter 17 shows how to transform data into time variables.

Part 6, "Evaluating Score Distribution Assumptions," consists of four chapters that examine the assumptions underlying most of the statistical procedures covered in the book. Chapter 18 focuses on the detection of univariate outliers, and Chapter 19 examines their multivariate counterpart. Chapter 20 focuses on the assessment of normality, and Chapter 21 demonstrates how to remedy assumption violations through data transformation.

Part 7, "Bivariate Correlation," consists of two chapters dealing with correlation. Chapter 22 demonstrates how to perform a Pearson product moment correlation (r), and Chapter 23 depicts how to compute a Spearman rho and Kendall tau-b correlation.

Part 8, "Regressing (Predicting) Quantitative Variables," consists of six chapters dealing with simple and multiple regression and multilevel modeling. Chapter 24 covers simple linear regression. Chapter 25 demonstrates how to center a predictor variable. Chapter 26 covers multiple linear regression. Chapter 27 covers hierarchical linear regression. Chapter 28 describes polynomial (curve estimation) regression. Chapter 29 provides an introduction to multilevel modeling.

Part 9, "Regressing (Predicting) Categorical Variables," consists of three chapters that deal with logistic regression. Chapter 30 covers binary logistic regression. Chapter 31 demonstrates ROC (receiver operator curve) analysis. Chapter 32 examines multinomial logistic regression.

Part 10, "Survival Analysis," consists of three chapters that depict various types of survival analysis. Chapter 33 demonstrates life table analysis. Chapter 34 covers the Kaplan–Meier procedure. Chapter 35 demonstrates the Cox regression procedure.

Part 11, "Reliability as a Gauge of Measurement Quality," is covered in two chapters. Chapter 36 covers reliability analyses related to issues of internal consistency. Chapter 37 covers reliability analyses that focus on inter-rater reliability.

Part 12, "Analysis of Structure," is covered in two chapters and deals with various types of factor analysis. Chapter 38 covers principal components analysis and factor analysis, and Chapter 39 covers confirmatory factor analysis.

Part 13, "Evaluating Causal (Predictive) Models," contains four chapters that deal with model building, as it pertains to mediation analysis and structural equation modeling. Chapter 40 covers simple mediation analysis. Chapters 41 and 42 cover path analysis using multiple regression and Amos, respectively. Chapter 43 provides an introduction to structural equation modeling.

Part 14, "*t* Test," consists of three chapters that cover various types of *t* tests. Chapter 44 demonstrates how to conduct a single sample *t* test. Chapter 45 covers the independent groups *t* test, and Chapter 46 covers the correlated samples *t* test.

Part 15, "Univariate Group Differences: ANOVA and ANCOVA," consists of seven chapters that cover various one- and two-way analyses of variance procedures. Chapter 47 demonstrates the one-way between-subjects ANOVA using the IBM SPSS GLM (general linear model) procedure. Chapter 48 demonstrates a trend analysis using polynomial contrasts. Chapter 49 covers one-way between-subjects ANCOVA (analysis of covariance). Chapter 50 examines two-way between-subjects ANOVA. Chapter 52 covers one-way repeated linear mixed models, and Chapter 53 examines the two-way simple mixed design.

Part 16, "Multivariate Group Differences: MANOVA and Discriminant Function Analysis," covers multivariate analysis of variance (MANOVA) and discriminant function analysis. Chapter 54 examines how to conduct a one-way between-subjects MANOVA. Chapter 55 covers discriminant function analysis, and Chapter 56 describes how to compute a two-way between-subjects MANOVA.

Part 17, "Multidimensional Scaling," consists of two multidimensional scaling chapters. Chapter 57 describes multidimensional scaling using the classic metric approach, while Chapter 58 describes multidimensional scaling using the individual differences scaling approach.

Part 18, "Cluster Analysis," consists of two cluster analysis chapters. Chapter 59 demonstrates hierarchical cluster analysis, while Chapter 60 depicts the *k*-means approach.

Part 19, "Nonparametric Procedures for Analyzing Frequency Data," completes the book and consists of six chapters dealing with nonparametric statistical procedures. Chapter 61 covers the binomial test, and Chapter 62 covers the one-way chi-square test. Chapter 63 demonstrates the two-way chi-square test with observed versus expected frequencies. Chapter 64 demonstrates how to do a risk analysis, and Chapter 65 covers the chi-square layers procedure. Lastly, Chapter 66 demonstrates hierarchical log-linear analysis.

Entering ISBN 9781118357019 at booksupport.wiley.com allows users to access the IBM SPSS data sets that were used in each of the chapters. These files can be downloaded, and users can shadow our analyses of the data on their own computers, assuming that they have the IBM SPSS software on such systems.

Because this book has multiple intended audiences, we recommend several different reading strategies. For the beginning IBM SPSS user, we suggest a very careful reading of Chapters 1 through 9, before moving into Chapters 10, 11, 12, 22, 24, 45, 47, 50, and 62. For the advanced undergraduate student, graduate student, or researcher, the remaining chapters should be pursued as the need arises.

PART 1

GETTING STARTED WITH IBM SPSS®

CHAPTER 1

Introduction to IBM SPSS®

1.1 WHAT IS IBM SPSS?

IBM SPSS is a computer statistical software package. This software can perform many types of data-oriented tasks such as recoding a variable (e.g., "flipping" the values of a reverse-worded survey item). It will perform these tasks for each case in the data set, even if there are tens of thousands of cases (a daunting job to perform by hand). IBM SPSS can also perform a huge range of statistical procedures, ranging from computing simple descriptive statistics such as the mean, standard deviation, and standard error of the mean, through some fundamental procedures such as correlation and linear regression, to a variety of multivariate procedures such as factor analysis, discriminant function analysis, and multidimensional scaling.

SPSS at one time was an acronym for *Statistical Package for the Social Sciences* but it is now treated as just a familiar array of letters. This is just as well, as researchers from a wide array of disciplines, not just those in the social sciences, use this software. Relatively recently, IBM purchased SPSS and beginning with version 19 has officially renamed the software as IBM SPSS.

1.2 BRIEF HISTORY

As described by Gamst, Meyers, and Guarino (2008), in the long-ago days, users did not have the luxury of pointing and clicking but instead actually typed syntax (SPSS computer code) as well as their data onto rectangular computer cards that were then physically read into a very large mainframe computer. Eventually, the cards gave way to computer terminals where users would type their data together with the syntax to structure their analysis via a keyboard and CRT (cathode ray tube) screen. The software finally reached the relatively early personal computers (PCs) in the middle 1980s, and it has gained considerable sophistication over the years.

As the program developed, one aspect has remained consistent: the statistical procedures are still driven by syntax. As we interact with the dialog windows, IBM SPSS is actually converting our actions and selections into its own code (syntax).

Performing Data Analysis Using IBM SPSS®, First Edition.
Lawrence S. Meyers, Glenn C. Gamst, and A. J. Guarino.

FIGURE 1.1
The three types of IBM SPSS files with which we ordinarily work: data files (.sav), output files (.spv), and syntax files (.sps).

1.3 TYPES OF IBM SPSS FILES AND FILE NAME EXTENSIONS

There are three kinds of files with which we ordinarily work when using IBM SPSS: *data files*, *output files*, and *syntax files*. We constantly deal with data and output files; more seasoned users also use syntax files extensively. Each file type has its own file name extension and distinctive icon, as shown in Figure 1.1. We discuss data files in Chapter 2 and output files in Chapter 4 and leave any discussion of syntax files for more specialized applications in some of the later topics covered in the book (e.g., performing simple effects in analysis of variance). The story on each file type in simplified form is as follows:

- *Data File.* This is a spreadsheet containing the data that were collected from the participating entities or *cases* (e.g., students in a university class, patients in a clinic, retail stores in a national chain). In the data file, the variables are represented as columns; cases, as rows. This file type uses the extension **.sav** and its icon shows a grid.
- *Output File.* This file is produced when IBM SPSS has performed the requested statistical analysis (or other operations such as saving the data file). It contains the results of the procedure. This file type uses the extension **.spv** and its icon shows a window with a banner.
- *Syntax File.* This file contains the IBM SPSS computer code (syntax) that drives the analysis. This file type uses the extension **.sps** and its icon shows a window with horizontal lines.

If the extensions do not show on your screen, here is what can be done to show the file extensions. If you are using Windows 7

- select **Control Panel ➔ Folder Options ➔ View Tab**;
- uncheck the checkbox for **Hide extensions for known file types**;
- click **OK**.

Here is what that can be done to show the file extensions in Mac OS X:

- Select **Finder ➔ Preferences ➔ Advanced Tab**.
- Check the checkbox for **Show all filename extensions**.
- Close the window.

CHAPTER 2

Entering Data in IBM SPSS®

2.1 THE STARTING POINT

When opening the IBM SPSS software program, we are presented with the view shown in Figure 2.1. We can navigate to an existent file, run the tutorial, type in data, and so on. By selecting the choice **Type in data** or by selecting **Cancel**, we can reach the IBM SPSS spreadsheet (the **Data View** display). We will select **Cancel**.

2.2 THE TWO TYPES OF DISPLAYS

The spreadsheet that is initially displayed is shown in Figure 2.2. This view, which is the default display, is called by IBM SPSS the **Data View** because it is, quite literally, where we enter and view our data. But as shown in Figure 2.2, it is also possible to display the **Variable View**. Whether we are entering our own data or importing an already constructed data set (as described in Chapter 3), we will need to work in both the **Data View** and the **Variable View** screens. Although we can deal with these screens in any order, we strongly encourage those new to IBM SPSS to begin with the **Variable View** screen when entering a new data set.

2.3 A SAMPLE DATA SET

Figure 2.3 shows a very simple set of fictional results of a research study just to illustrate how to go through the steps of entering data. The variables and their meaning are as follows:

- *ID.* This is an arbitrary identification code associated with each research participant (case). The ID de-identifies participants, thus protecting their anonymity and guaranteeing confidentiality. The ID also allows us to review the original data (which should also contain the identification codes) if data entry questions or errors occurred.
- *Gender.* This indicates the gender of the participant; in Figure 2.3, M stands for male and F stands for female.

Performing Data Analysis Using IBM SPSS®, First Edition.
Lawrence S. Meyers, Glenn C. Gamst, and A. J. Guarino.

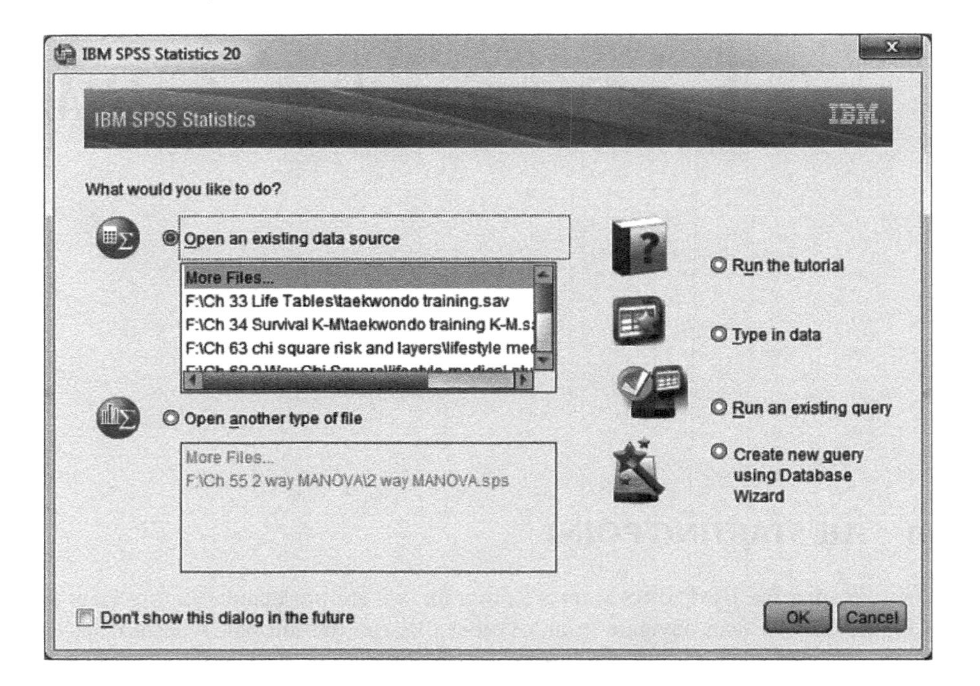

FIGURE 2.1 The screen presented on opening IBM SPSS.

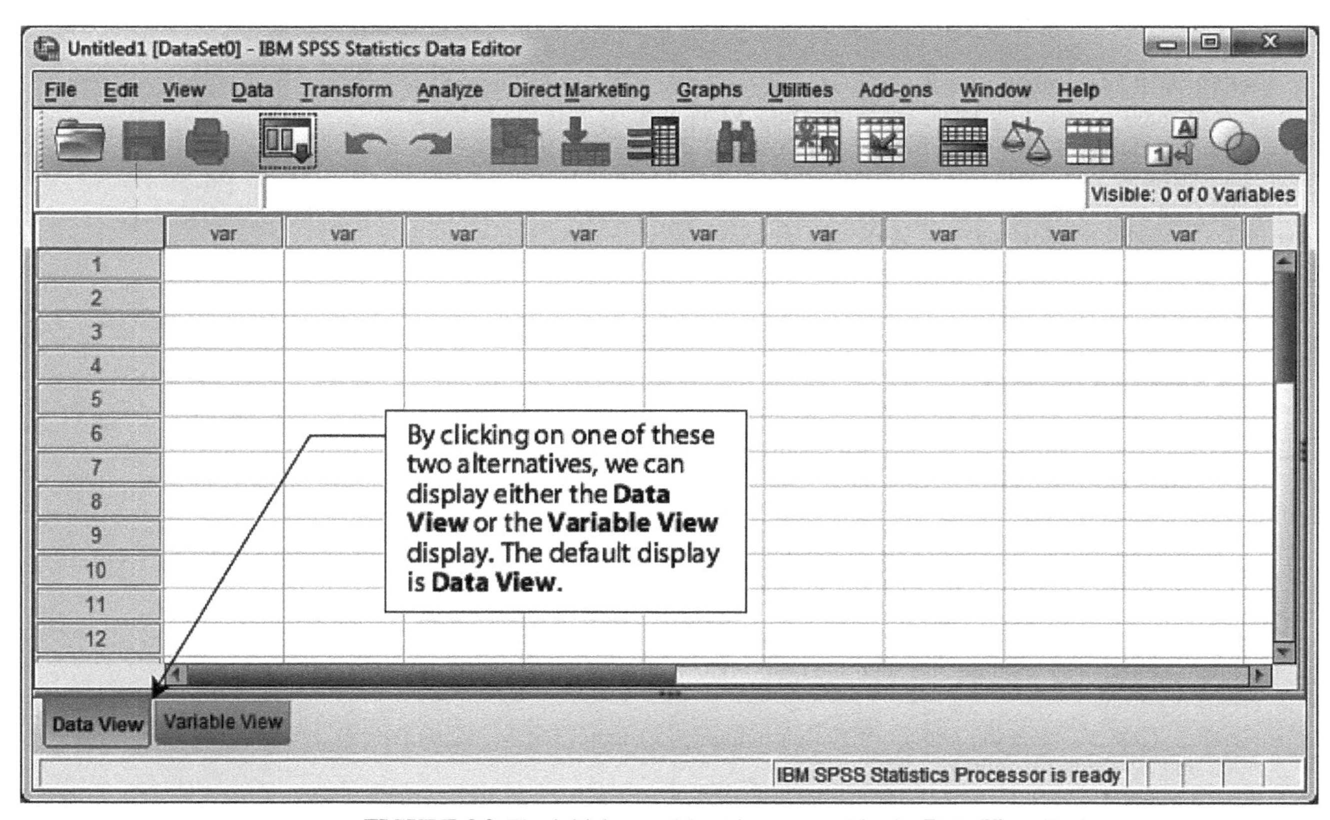

FIGURE 2.2 The initial spreadsheet is presented in the **Data View** display.

ID	Gender	Extraversion	Sales in Thousands of Dollars
1	M	9	50
2	F	3	26
3	F	7	40
4	M	5	38
5	M	6	34
6	F	8	42

FIGURE 2.3
Raw data that are to be entered in IBM SPSS.

- *Extraversion.* A personality characteristic indicating, roughly speaking, the degree to which the person is outgoing. In this study, it is represented by a 10-point scale with 1 indicating *very low* and 10 indicating *very high.*
- *Sales in Thousands of Dollars.* This indicates sales figures for each participant (salesperson) during a given month in a given department of a large retail chain.

2.4 THE VARIABLE VIEW DISPLAY

Selecting **Variable View** at the bottom of the new (blank) spreadsheet gives rise to the screen shown in Figure 2.4. This display is editable and allows us to specify the various properties of the variables in the data file. The columns address the following variable specifications:

- *Name.* A reasonably short but descriptive name of the variable. There is a 64-character maximum for English, but we ordinarily want very much shorter names. No spaces or special characters are allowed but underscores can be used. We suggest using letters and numbers only.

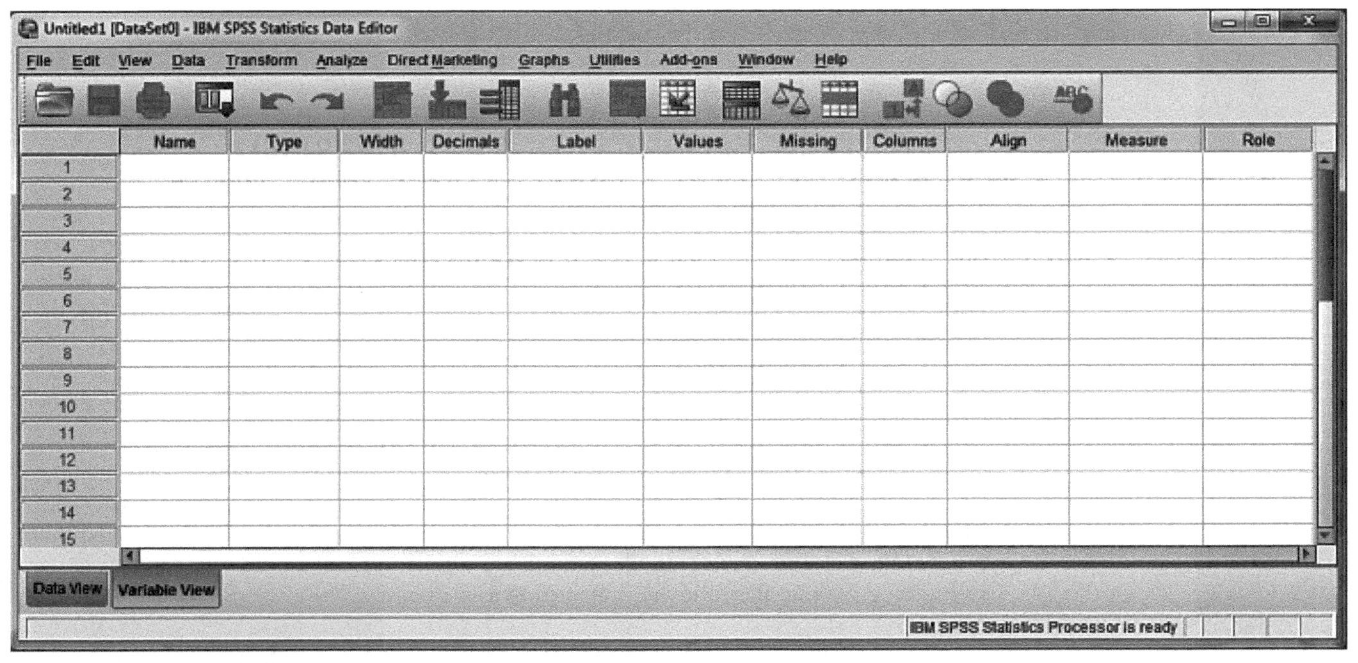

FIGURE 2.4 The **Variable View** display.

- *Type.* There are several types of data (e.g., scientific notation, date, string) that IBM SPSS can read, but we will restrict ourselves in this example to **Numeric** (i.e., regular numbers), which is the default.
- *Width.* This is the number of spaces the data occupy. The default is **8**.
- *Decimals.* This is the number of decimal places shown in the **Data View** for that variable. The default is **2**. Note that IBM SPSS computes values to 16 decimals no matter what; if we ask to see fewer, as most of us do, the displayed values will be rounded to the number of decimals specified here.
- *Label.* A phrase to describe the variable. This is often omitted by researchers if the variable name is sufficiently descriptive.
- *Values.* When entering information on a categorical variable (such as gender), it is appropriate for most of the analyses we cover in the book to use arbitrary numeric codes for the categories. This is the place that we can (and absolutely should) specify labels for each category code.
- *Missing.* We can designate a missing value by either leaving the cell empty in the process of data entry or using an arbitrary numeric code. Arbitrary numeric codes are useful when there are different reasons for defining a value as missing (e.g., the original source is not legible, there is a double answer) permitting us to differentiate why a value may be missing.
- *Columns.* This is the number of spaces the data are allowed to occupy. The default is **8**.
- *Align.* This specifies right, left, or center alignment in the **Data View** display. The default is **Right**.
- *Measure.* This specifies the scale of measurement of the variable. The options are **Scale** (representing approximately interval-level data at a minimum), **Ordinal** (containing only less than, equal to, and greater than information), and **Nominal** (representing categorical data), with the default shown initially as **Unknown**.
- *Role.* There are a variety of roles that can be assigned to variables, but we will restrict ourselves to **Input**, which is the default (and thus permits variables to be placed in all of our analyses).

2.5 ENTERING SPECIFICATIONS IN THE VARIABLE VIEW DISPLAY

In the **Variable View** window, double-click the cell under **Name** in the first row and type in the name of the first variable (ID). Then click in the cell directly below to allow IBM SPSS to fill in its defaults in the first row. This is shown in Figure 2.5. Modify the default specifications as follows:

- Click the **Decimals** cell and change the specification to 0 by clicking on the down toggle.
- Click the **Measure** cell and select **Scale**.

The finished first row for the **ID** variable is shown in Figure 2.6. We have repeated this process for the other three variables, and the completed specifications thus far are shown in Figure 2.7. Note the following (also shown in Figure 2.7):

- We used a short **Name** for the dollar amount of sales (**sales**) and so supplied a **Label** to provide a more complete description of the variable.

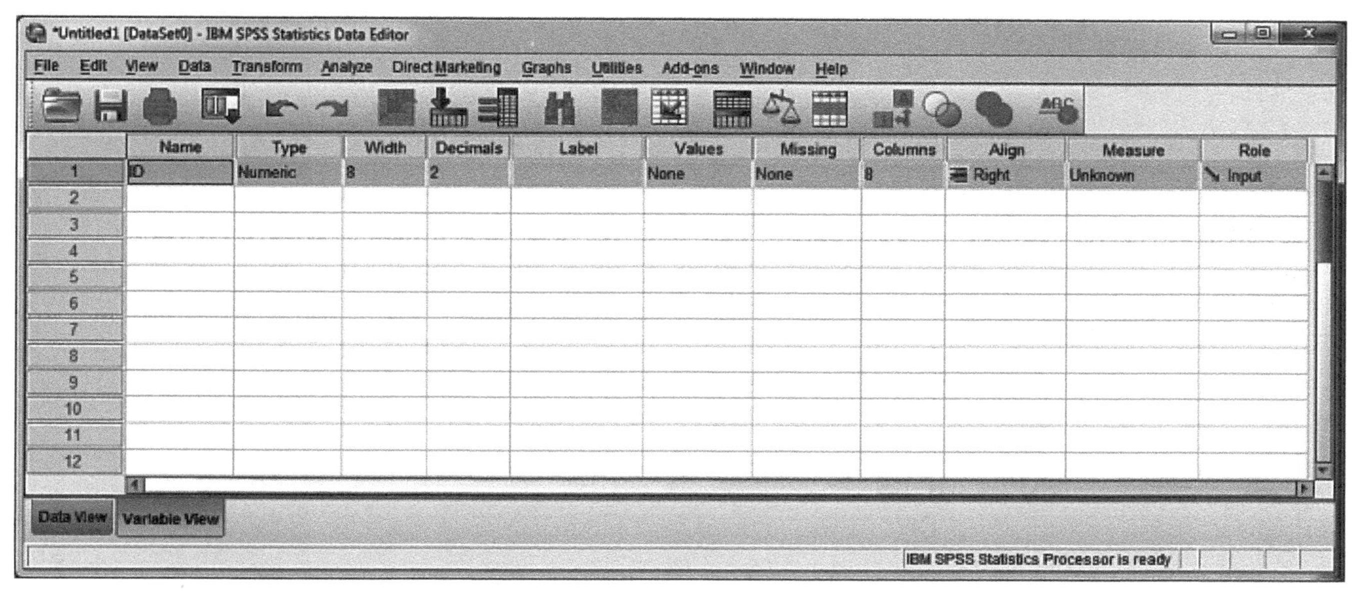

FIGURE 2.5 The **Variable View** display with the first variable typed in and the defaults showing.

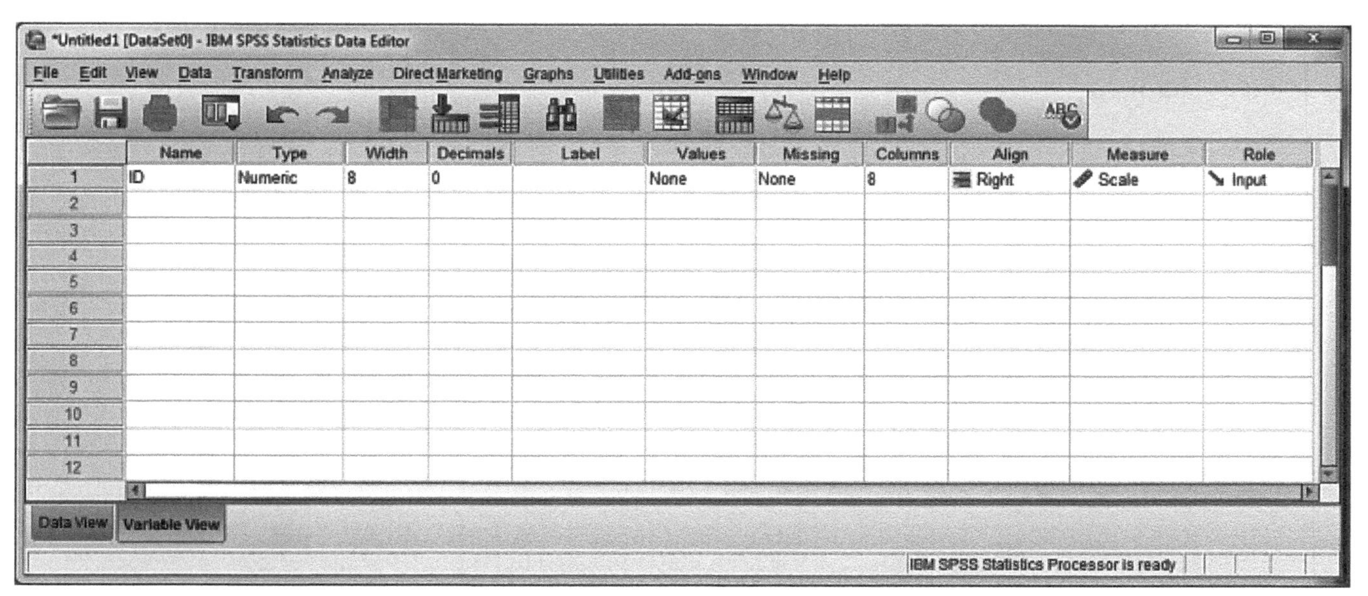

FIGURE 2.6 The **Variable View** display with all of the variables typed in, the defaults modified, and the first step in specifying the gender codes showing.

- The variable **gender** is a categorical (nominal) variable and we have indicated that in the **Measure** column; **extraversion** and **sales** are quantitative variables called by IBM SPSS as **Scale** measures.

The only job remaining is to specify the value labels for the **gender** variable. It is a categorical (nominal) variable and we will use the values (numerical codes) **1** for female and **2** for male. Such codes are arbitrary and idiosyncratic to each researcher and thus need to be specified so that (a) other researchers can understand the data and (b) these value labels can appear in some of our output.

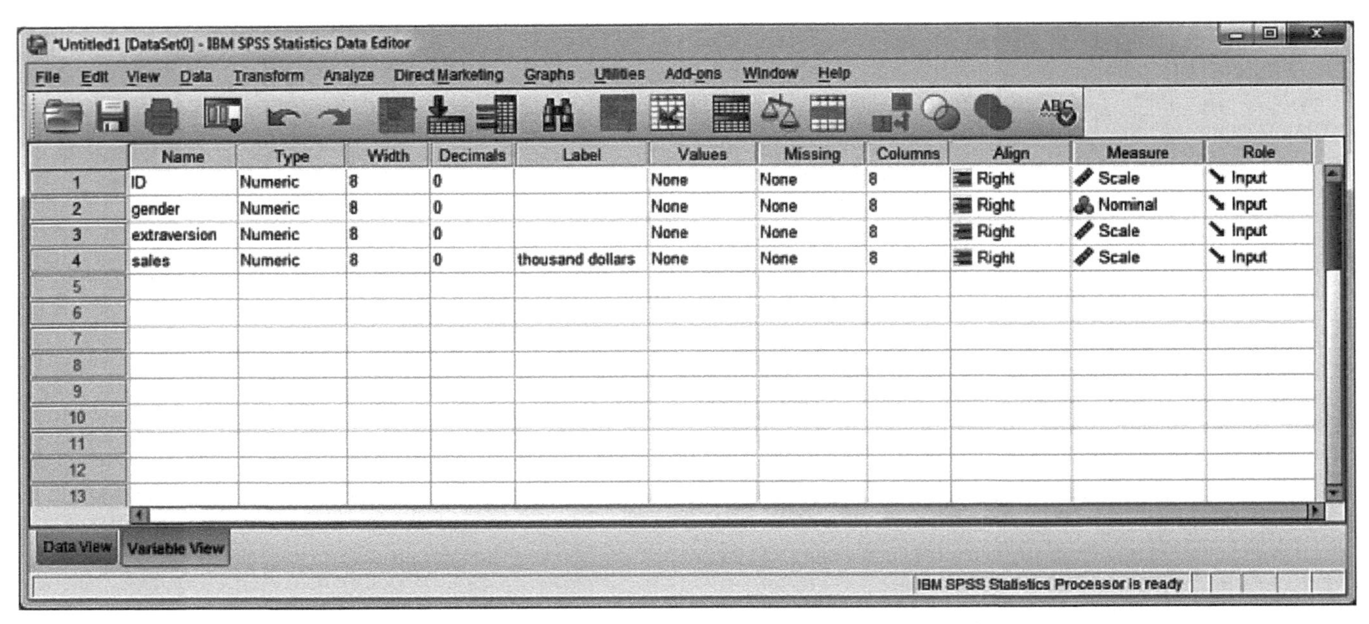

FIGURE 2.7 The variables now have most of their specifications.

To accomplish this, we click the cell under **Values** on the row for **gender**. This produces the **Value Labels** dialog window. Follow these steps as shown in Figure 2.8 to provide the specifications:

- Type **1** in the **Value** panel.
- Type **female** in the **Label** panel.
- Click **Add**.
- Type **2** in the **Value** panel.
- Type **male** in the **Label** panel.
- Click **Add**.

To register the labels with IBM SPSS, we click **OK**. The labels are now contained in the **Values** cell for gender (see Figure 2.9).

2.6 SAVING THE DATA FILE

We now do something that all users should do on a very frequent basis: we will save the data file (and will do so every time we make any modification to it). Select **File → Save As** (or select the **Save File** icon shown in Figure 2.9). This opens a standard file-saving dialog screen in the operating system (e.g., Windows 7, Mac OS X). Navigate to the desired location, name the file in some way that makes sense, and save it. (IBM SPSS will display an acknowledgment of the saving operation—we close that acknowledgment window without saving it.) The name of the file will appear in the banner of the IBM SPSS window; double-clicking the file icon in the directory will directly open the file next time.

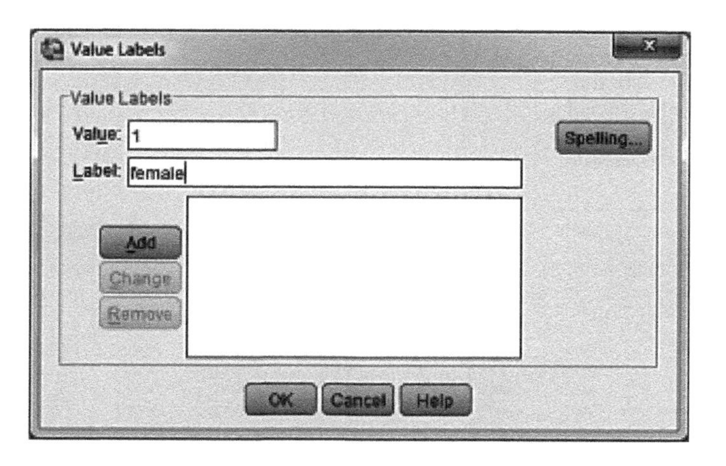

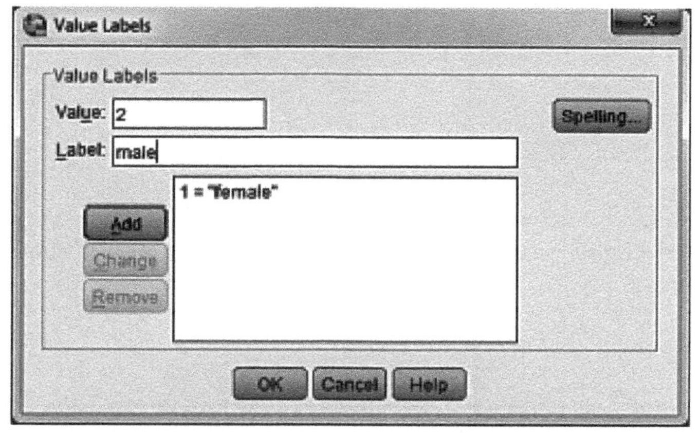

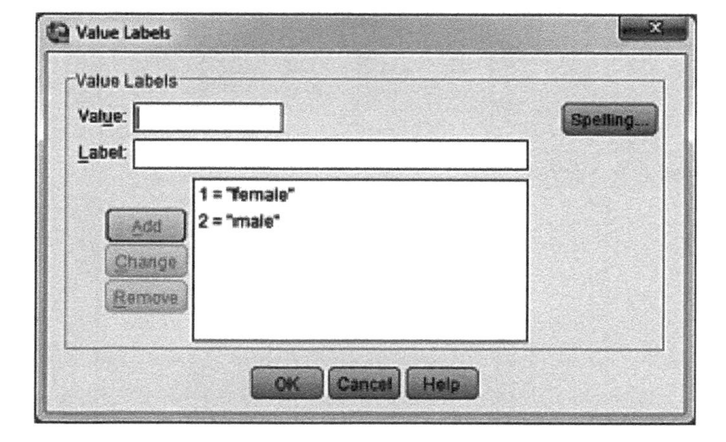

FIGURE 2.8
The **Value Labels** window with the **gender** codes entered.

2.7 ENTERING SPECIFICATIONS IN THE DATA VIEW DISPLAY

With the specifications now in place and the file saved, the variable names are visible after selecting **Data View** and the file name is now shown in the banner (see Figure 2.10). In the **Data View** spreadsheet, we simply enter our numbers, using the arrow keys or tab key to move from cell to cell. Once the data have been entered as shown in Figure 2.10, we again save the data file (clicking the **Save File** icon will overwrite the older file with

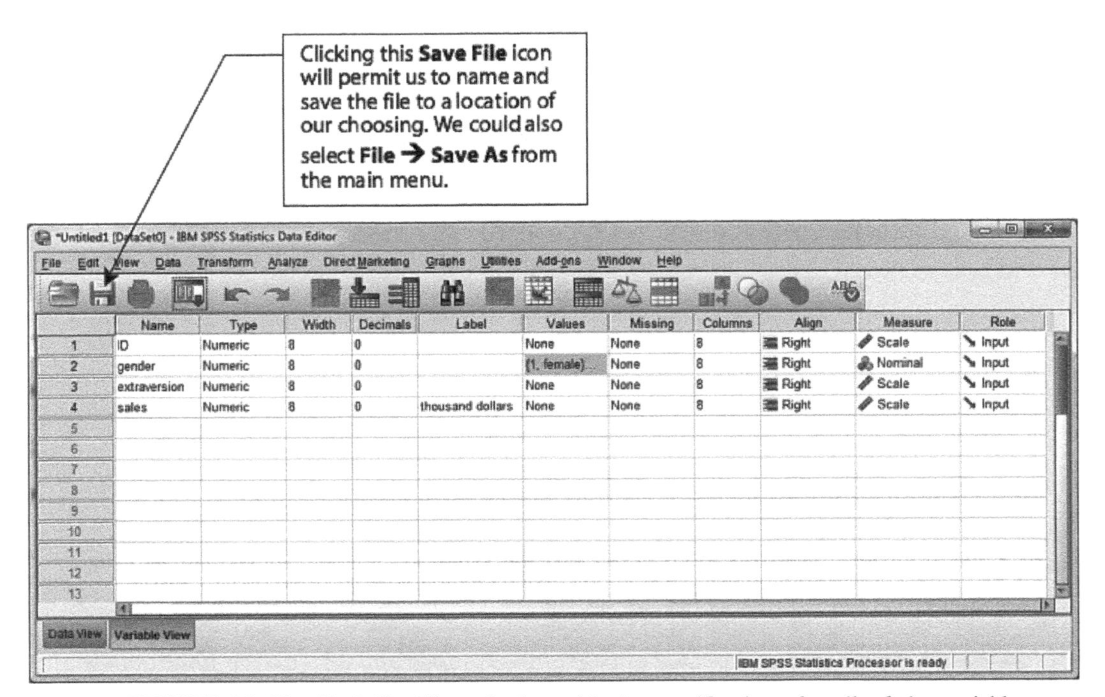

FIGURE 2.9 The **Variable View** display with the specifications for all of the variables now complete.

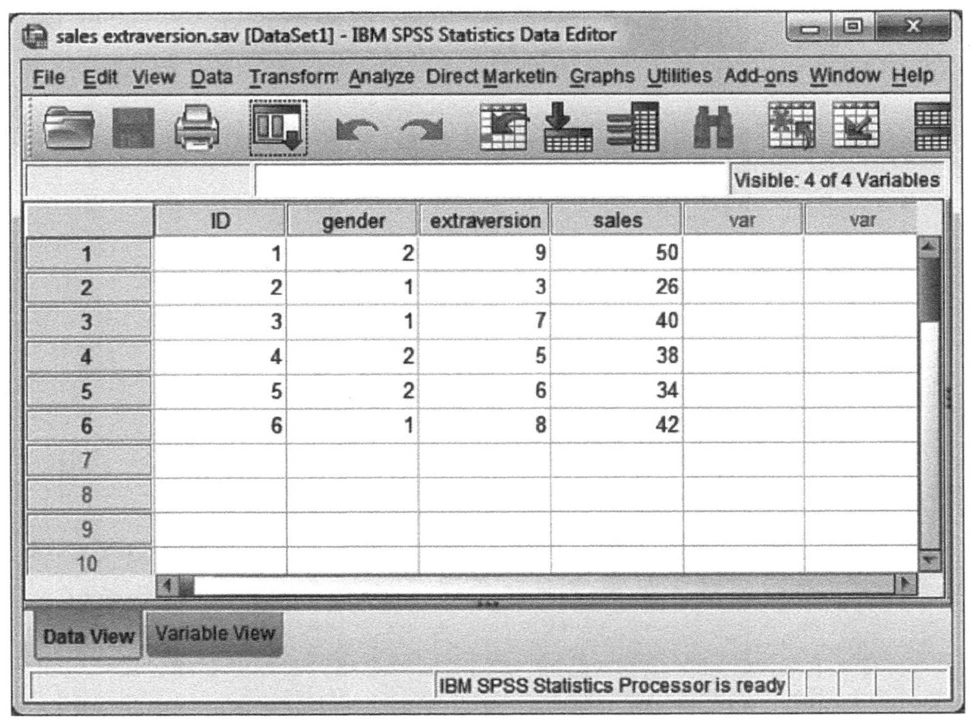

FIGURE 2.10 The **Data View** with the data entered and the file saved.

the modified file). A "dimmed" ("grayed") icon indicates that the current version of the file as shown on the screen is saved. As soon as we make any changes to it, the icon becomes "normal" and signifies that what we see on the screen is not currently saved. In Figure 2.10, the icon is dim indicating that we have not modified the data file since the last time it was saved.

CHAPTER 3

Importing Data from Excel to IBM SPSS®

3.1 THE STARTING POINT

In addition to entering data directly into IBM SPSS, it is possible to bring in (import) data that have been entered in another software spreadsheet. We illustrate this using Excel. The data set used in Chapter 2 was entered into an Excel worksheet and is shown in Figure 3.1. We will import this data set into IBM SPSS.

3.2 THE IMPORTING PROCESS

As described in Section 2.1, open IBM SPSS and reach the **Data View** display of a blank (new) data file. Select **File → Open → Data** to reach the **Open Data** dialog window shown in Figure 3.2. From the **Files of type** drop-down menu, select **Excel**; these files have a base **.xls** extension, some with different extra letters depending on the version of Excel (e.g., **.xlsx**, **.xlsm**). Then navigate through the storage drives to locate the Excel file containing the data (**sales extraversion.xls**). Selecting (clicking) its name in the panel will cause the file name to appear in the **File name** panel. This is also shown in Figure 3.2.

With the name and type of file now identified, select **Open**. This produces the **Opening Excel Data Source** window shown in Figure 3.3. The checkbox for **Read variable names from the first row of data** is already checked as a default. We keep it that way because our Excel spreadsheet contains the variable names in the appropriate SPSS format. If the Excel file variable name is not in the acceptable SPSS format, SPSS will assign VAR0001, VAR0002, and so on to the variables. Clicking **OK** initiates the importing process, the result of which is shown in Figure 3.4.

The importing process has produced an IBM SPSS data file. It is currently untitled, as it has just been created. This file should be saved. Once saved, we would select the **Variable View**, modify the default specifications to match our data (overruling the IBM SPSS defaults), and again save the modified version of the data file. When finished, it would be indistinguishable from the one we created in Chapter 2.

IBM SPSS will always provide an acknowledgment of its actions; acknowledgment of the importing process is shown in Figure 3.5. This is the syntax that mediated the

Performing Data Analysis Using IBM SPSS®, First Edition.
Lawrence S. Meyers, Glenn C. Gamst, and A. J. Guarino.

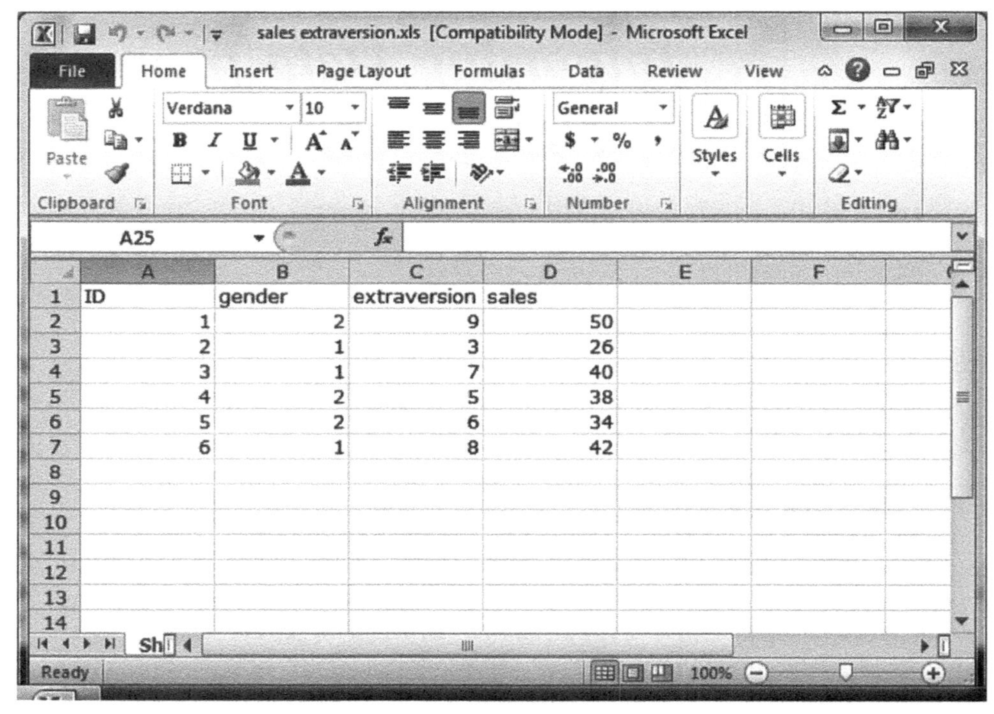

FIGURE 3.1 Excel spreadsheet containing the data set we wish to import.

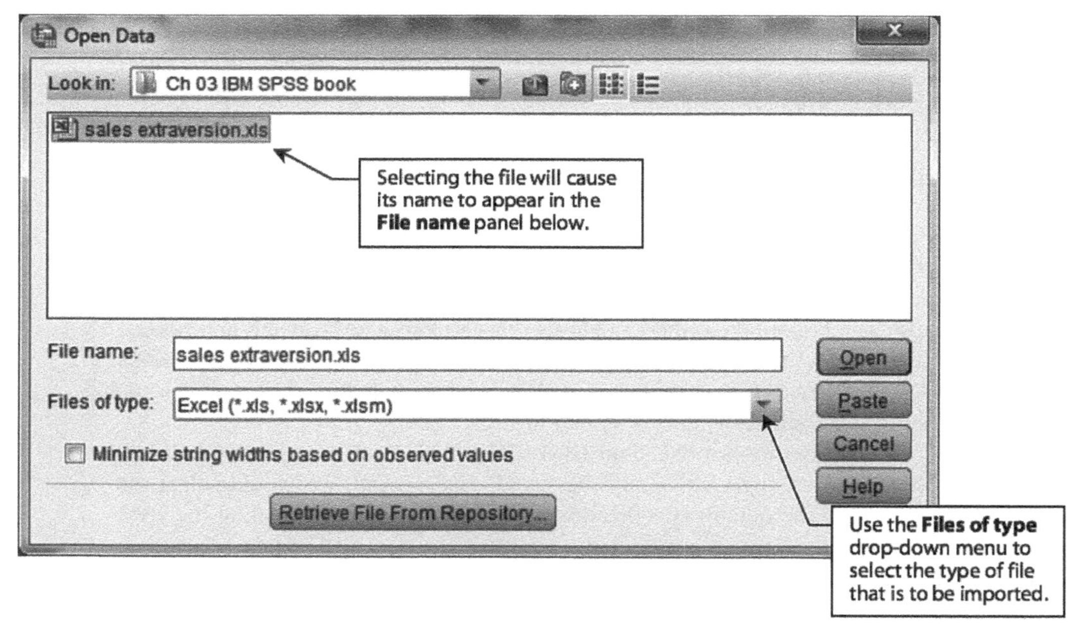

FIGURE 3.2 The **Open Data** dialog window.

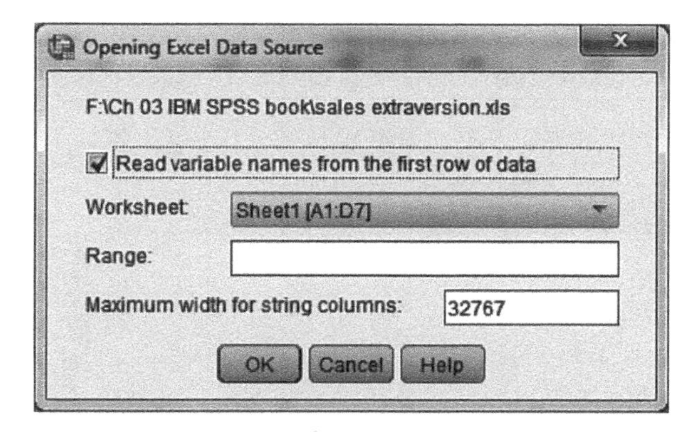

FIGURE 3.3
The **Opening Excel Data Source** dialog window.

*Untitled3 [DataSet3] - IBM SPSS Statistics Data Editor

File Edit View Data Transform Analyze Direct Marketing Graphs Utilities Add-ons Window Help

Visible: 4 of 4 Variables

	ID	gender	extraversion	sales	var	var	var
1	1	2	9	50			
2	2	1	3	26			
3	3	1	7	40			
4	4	2	5	38			
5	5	2	6	34			
6	6	1	8	42			
7							
8							
9							
10							
11							
12							
13							

Data View Variable View

IBM SPSS Statistics Processor is ready

FIGURE 3.4 The imported data file needs to be named and saved.

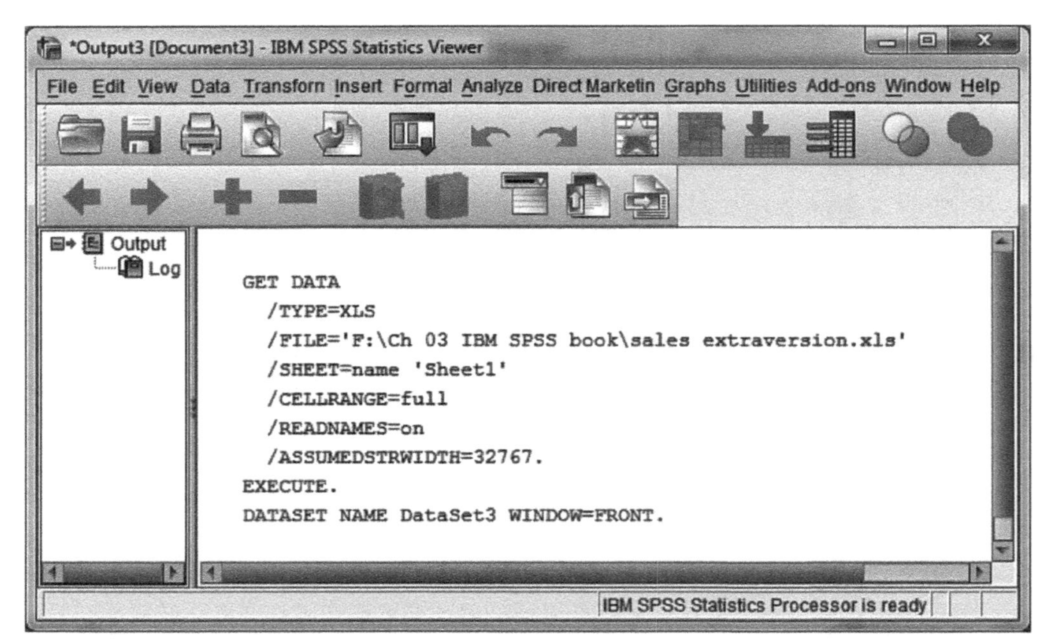

FIGURE 3.5 The acknowledgment of importing need not be saved.

importing of the Excel file; a good deal of specific information is shown, but the general message is to deal with an **.xls** format to be found via the given path and capture all of the data fields with their names. It is sufficient to close that window without saving (there will be a prompt requesting a decision about saving the file when we attempt to close it).

PART 2

OBTAINING, EDITING, AND SAVING STATISTICAL OUTPUT

CHAPTER 4

Performing Statistical Procedures in IBM SPSS®

4.1 OVERVIEW

The **Analyze** menu located in the IBM SPSS main menu bar (see Figure 4.1) gives us access to a range of statistical procedures (e.g., analysis of variance, linear regression), and we will use it extensively, but we will also be working to a certain extent with some of the other menus in the main menu (e.g., **Data**, **Transform**). All of this work will involve interacting with dialog windows and, in most circumstances, obtaining output.

Performing statistical analyses is carried out in two stages: analysis setup and viewing/interpreting the output. Here, we treat these processes in a generic manner, without being concerned about the details of the setup or the interpretation of the output; rather, our purpose is to present the general process of what will be done for most analyses.

4.2 USING DIALOG WINDOWS TO SETUP THE ANALYSIS

In order to set up an analysis, it is necessary to have a data file open, as the analysis will be performed on the active data file. We will use the data file named **sales extraversion** that was created in Chapter 2 to illustrate our analysis setup, and we will call upon the **Bivariate Correlations** procedure simply as a matter of convenience (we could have chosen any other procedure as they are all structured similarly) to illustrate how to perform a statistical analysis.

With the data file as the active window, select from the IBM SPSS menu **Analyze ➔ Correlate ➔ Bivariate** to reach the main **Bivariate Correlations** dialog window shown in Figure 4.2. The main window of a statistical procedure is the one where we identify the variables to be included in the analysis; it is almost always the window that opens when we invoke most procedures.

Although there are many elements contained in these main dialog windows, working with them is pretty straightforward. Following are the major elements of a main dialog window using **Bivariate Correlations** as our medium:

- The banner of the window provides the name of the procedure.

Performing Data Analysis Using IBM SPSS®, First Edition.
Lawrence S. Meyers, Glenn C. Gamst, and A. J. Guarino.

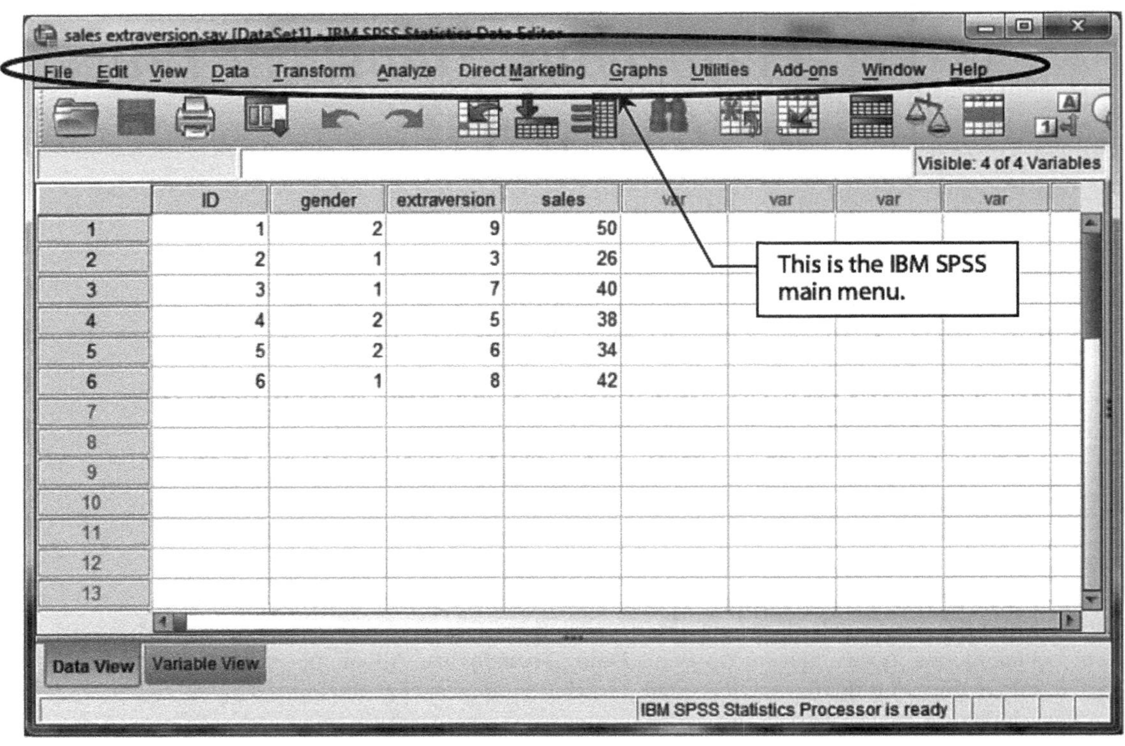

	ID	gender	extraversion	sales	var	var	var	var
1	1	2	9	50				
2	2	1	3	26				
3	3	1	7	40				
4	4	2	5	38				
5	5	2	6	34				
6	6	1	8	42				
7								
8								
9								
10								
11								
12								
13								

FIGURE 4.1 The main menu of IBM SPSS.

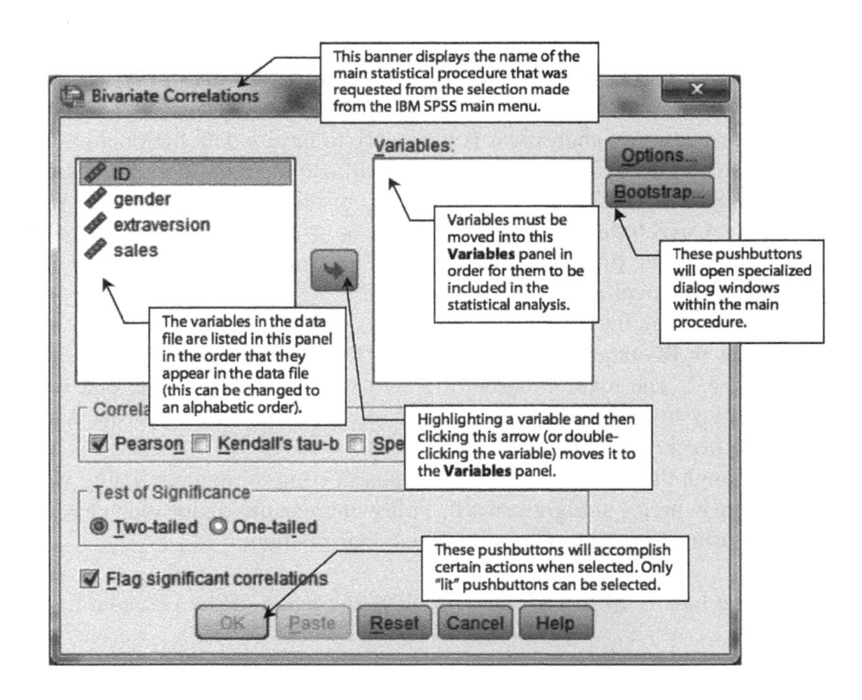

FIGURE 4.2
The main **Bivariate Correlations** dialog window.

- The unnamed panel in the upper left quadrant of the window, which we will call the **Variable List** panel, contains the names of the variables in the data file. They are listed in the order that they appear in the data file but it is possible to have them listed alphabetically (e.g., select **Edit** ➔ **Options** ➔ **General** tab and click **Alphabetical** under **Variable Lists** as shown in Figure 4.3).
- The **Variables** panel identifies the variables that are to be included in the analysis. Variables may be moved into the **Variables** panel either by highlighting them in the **Variable Lists** panel and clicking the arrow button or by double-clicking the name of the variable.
- The pushbuttons on the far right of the main window ordinarily open subordinate dialog screens in order to customize some aspect of the statistical analysis or to instruct IBM SPSS to print certain information in the output.
- The pushbuttons on the bottom of the window enable some action to take place. They must be active to be eligible for selection. For example, the **OK** pushbutton (this enables the analysis to be performed) is not currently available (it is not active) because no variables have yet been moved into the **Variables** panel (there are no variables identified on which an analysis can be performed).

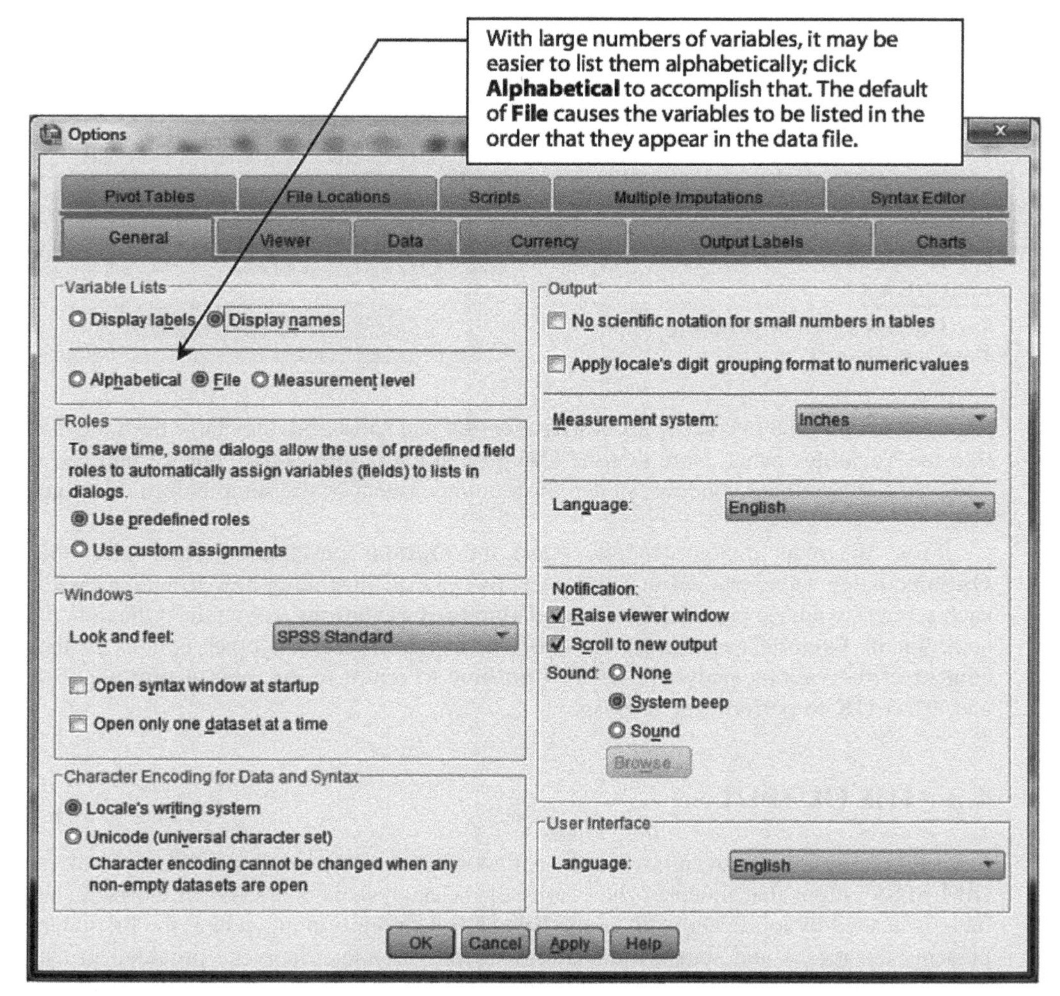

FIGURE 4.3 The **General** tab in editing the IBM SPSS system **Options**.

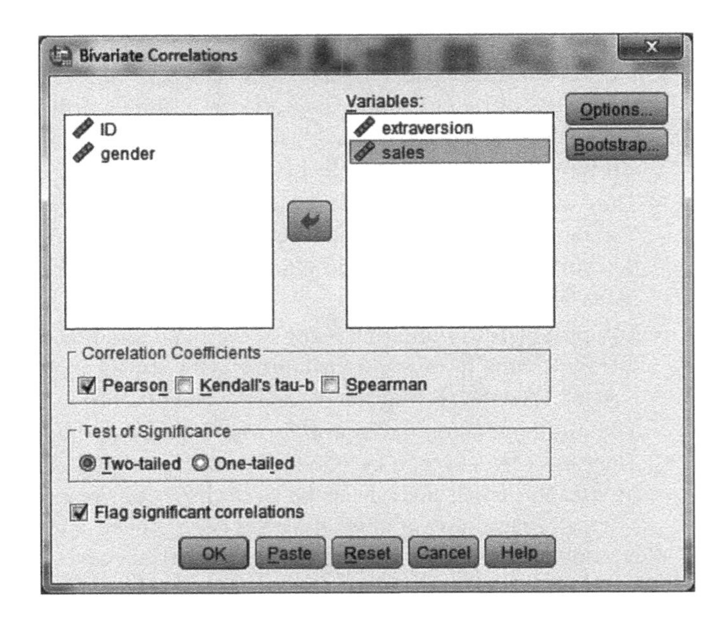

FIGURE 4.4

The main **Bivariate Correlations** dialog window with **extraversion** and **sales** entered into the **Variables** panel.

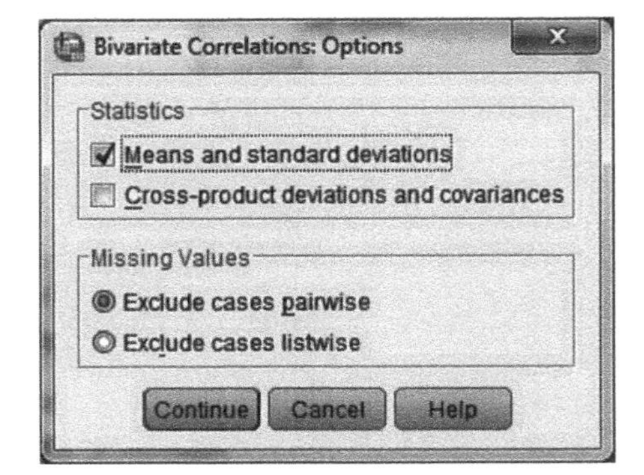

FIGURE 4.5

The **Options** screen of **Bivariate Correlations**.

Figure 4.4 shows the two variables of **extraversion** and **sales** after they have been moved into the **Variables** panel. Note that the **OK** pushbutton is now active, but we will open one of the subordinate windows to demonstrate this aspect of the setup before selecting **OK**.

From the main dialog window, select the **Options** pushbutton. This opens the **Options** dialog screen shown in Figure 4.5. By way of illustrating how to interact with such screens, we have checked **Means and standard deviations** and retained the default selection of **Exclude cases pairwise** (we will explain these and other options in the context of the various analyses). Select **Continue** to return to the main dialog window and select **OK** to perform the analysis.

4.3 THE OUTPUT

Figure 4.6 displays the results (usually called *output*) of the analysis. It is typical in IBM SPSS output that much of the results of the analysis is contained in tables. Each table is headed by a title above it. In the **Bivariate Correlation** procedure, the first table presents the means and standard deviations of the variables; these are provided in the output because we requested in the **Options** dialog screen that these descriptive statistics be displayed.

Descriptive Statistics

	Mean	Std. Deviation	N
extraversion	6.33	2.160	6
sales in thousand dollars	38.33	8.042	6

Correlations

		extraversion	sales in thousand dollars
extraversion	Pearson Correlation	1	.936**
	Sig. (2-tailed)		.006
	N	6	6
sales in thousand dollars	Pearson Correlation	.936**	1
	Sig. (2-tailed)	.006	
	N	6	6

**. Correlation is significant at the 0.01 level (2-tailed).

FIGURE 4.6

Sample output from the **Bivariate Correlations** procedure.

The second table shows the Pearson correlation between extraversion and sales (a rather exaggerated .936). In the **Bivariate Correlation** procedure, IBM SPSS footnotes with asterisks show different probability levels. Here, there is only one correlation. Also shown in the printout is **Sig. (2-tailed)**; this is the exact probability of obtaining a Pearson correlation of .936 based on an *N* of 6 cases assuming that the null hypothesis (maintaining that the population correlation value is zero) is true.

CHAPTER 5

Editing Output

5.1 OVERVIEW

As noted in Section 4.3, much of the output in IBM SPSS® is presented in the form of tables. The tables produced by IBM SPSS in its output are "generic" and may be less well formatted than we would prefer, especially if the table is to be included in a report. However, tables in the output can be edited to a certain extent before copying them to a word processing document or saving them to a PDF (Portable Document Format) file. In this chapter, we illustrate some simple editing tasks that can be done.

To generate an output table so that we can illustrate these editing tasks, we have performed a **Bivariate Correlations** analysis with four variables and have obtained the table of **Descriptive Statistics** shown in Figure 5.1. Assume that we wish to show this table in a presentation and, in that context we determine that (a) the **Std. Deviation** column is disproportionally wide and (b) we would prefer to have the word "Standard" written out rather than being abbreviated.

To edit this table, we double-click on it. The result is shown in Figure 5.2. We can determine that it is editable because of three visual cues:

- It becomes outlined in a dashed line.
- It takes on a red arrow to its left.
- The table title format changes to white font on a black background.

5.2 CHANGING THE WORDING OF A COLUMN HEADING

Double-clicking the column heading **Std. Deviation** permits us to edit it (see Figure 5.3). We then type out the full term (with a hard return after **Standard**) to replace the abbreviation as shown in Figure 5.4.

5.3 CHANGING THE WIDTH OF A COLUMN

We can also adjust the width of the columns. Double-clicking in the table and placing the cursor on one of the vertical border lines of the **Standard Deviation** column gives

Performing Data Analysis Using IBM SPSS®, First Edition.
Lawrence S. Meyers, Glenn C. Gamst, and A. J. Guarino.

Descriptive Statistics

	Mean	Std. Deviation	N
academic	3.6250	.80623	16
family	2.3125	.79320	16
social	3.1250	1.36015	16
personal	3.7500	1.29099	16

FIGURE 5.1

A table of **Descriptive Statistics** obtained from the **Bivariate Correlations** procedure.

Descriptive Statistics

	Mean	Std. Deviation	N
academic	3.6250	.80623	16
family	2.3125	.79320	16
social	3.1250	1.36015	16
personal	3.7500	1.29099	16

FIGURE 5.2

The table is now editable.

Descriptive Statistics

	Mean	Std. Deviation	N
academic	3.6250	.80623	16
family	2.3125	.79320	16
social	3.1250	1.36015	16
personal	3.7500	1.29099	16

FIGURE 5.3

Double-clicking the column heading **Std. Deviation** permits us to edit it.

Descriptive Statistics

	Mean	Standard Deviation	N
academic	3.6250	.80623	16
family	2.3125	.79320	16
social	3.1250	1.36015	16
personal	3.7500	1.29099	16

FIGURE 5.4

The full wording of **Standard Deviation** is now the column heading.

us a double horizontal arrow as shown in Figure 5.5. Clicking and dragging this border to the left allows us to narrow the column as shown in Figure 5.6. Clicking anyplace outside of the table takes it out of edit mode. Having modified the file, we now save it.

5.4 VIEWING MORE DECIMAL VALUES

IBM SPSS carries out its computations to 16 decimal places, but the full range of these decimal values is almost never displayed in the output tables, a strategy that makes a good deal of sense, given the level of measurement precision in our research instrumentation. Nevertheless, the fact that they are not displayed belies the fact that they are there—they are present but hidden from view, as it is a rare occurrence when we wish to see all of that information.

Descriptive Statistics

	Mean	Standard Deviation	N
academic	3.6250	.80623	16
family	2.3125	.79320	16
social	3.1250	1.36015	16
personal	3.7500	1.29099	16

As one feature of table editing, we can drag a column vertical border in or out. Double-click and drag to reposition it.

FIGURE 5.5 In editing mode, it is possible to move the vertical column border in or out.

Descriptive Statistics

	Mean	Standard Deviation	N
academic	3.6250	.80623	16
family	2.3125	.79320	16
social	3.1250	1.36015	16
personal	3.7500	1.29099	16

FIGURE 5.6
The **Standard Deviation** column is now narrower.

Descriptive Statistics

	Mean	Standard Deviation	N
academic	3.6250	.80623	16
family	2.3125	.79320	16
social	3.1250	1.3601470508735443	
personal	3.7500	1.29099	16

FIGURE 5.7
Double-clicking a numerical entry in a table yields the decimal value to 16 places.

To view the full set of decimal values, we double-click the table to place it in edit mode. Then we double-click the entry whose full decimal values we wish to see. This is shown in Figure 5.7 where we have selected the entry for the standard deviation of **social** whose value is displayed in the table as 1.36015. By double-clicking that entry, we can see the full decimal value 1.3601470508735443 that was heretofore (and gratefully) hidden. Note that the ordinarily displayed tabled entry is a properly rounded representation of the full decimal value as computed by IBM SPSS.

It is common that SPSS will print a **sig**. value (the probability of the statistic occurring by chance alone if the null hypothesis is true) to allow us to test the statistical significance of an obtained statistic (e.g., a Pearson r value) against the alpha level we have established. In many of the statistical procedures, the **sig**. value is given to three decimal places.

It is not uncommon for these probability values to be sufficiently low that the number of zero digits well exceeds the three-decimal printing limitations in the IBM SPSS tables. For example, the computed probability might be .000316. This conundrum is resolved by IBM SPSS, sometimes to the dismay of students, by presenting in the output table a **sig**. value of .000.

In presenting the results of any analysis in a report, probability values should never be reported as .000; rather, they should be reported as $p < .001$ (American Psychological

Association, 2009). This is because the probability is never zero, but just a very low value. Double-clicking the displayed value of .000 will yield the longer decimal.

Some probability values may be sufficiently low that IBM SPSS will present them in what is known as exponential notation. For example, the value of .000316 would be displayed as 3.16E−4. This notation is interpreted as follows:

- 3.16 is the base nonzero numeral in the expression.
- E indicates that the value is written in exponential notation.
- The dash is a minus sign directing us to move the decimal to the left, adding zeros as needed.
- 4 is the number of decimal places involved in the move.

Putting all this together, the exponential notation in this instance directs us to move the decimal in 3.16 four places to the left. In order to comply, it is necessary to add three zeros to the left of the 3; thus, the end result is .000316.

5.5 EDITING TEXT IN IBM SPSS OUTPUT FILES

Although much of the information in output files is contained in tables, text is also produced. As an example, text in an output file will document the analysis setup by displaying the underlying syntax. This will be the case even when IBM SPSS provides an acknowledgment that it has carried out an instruction.

The output text can be edited. While we would not wish to change it, we may wish to copy such text to a word processing or other document. Double-clicking on the text gives us editing access to it, and we can, for example, copy and paste it where we wish.

CHAPTER 6

Saving and Copying Output

6.1 OVERVIEW

The statistical analysis of our data is contained in IBM SPSS® output files. These results need to be saved and often need to be copied into reports or other documents. In this chapter, we show how to accomplish these operations.

6.2 SAVING AN OUTPUT FILE AS AN IBM SPSS OUTPUT FILE

In Section 1.3, we indicated that IBM SPSS output files are associated with an **.spv** extension. The standard file-saving routine within IBM SPSS will save the output file in that format. To save a newly generated output file, select **File → Save As** or select the **Save File** icon (shown in Figure 2.9). This opens a standard file-saving dialog screen in the operating system (e.g., Windows 7, Mac OS X). Navigate to the desired location (e.g., a personal flash drive), name the file, and save it.

It is possible to directly open the output file later. Double-clicking the file icon in the directory will open the file provided that there exists on that computer the same or a more recent version of IBM SPSS that created the file. Note that if for some reason we are using a computer that does not have IBM SPSS (e.g., a home computer), or one that contains an earlier version of the software, then trying to open the output file is not possible.

6.3 SAVING AN OUTPUT FILE IN OTHER FORMATS

IBM SPSS allows for an output file to be saved in a myriad of formats including HTML, Excel, PowerPoint, Microsoft Word/RTF, and PDF. We discuss how to do this for PDF, but our description can also be generalized to the other formats.

A PDF document is a type of file that is in Portable Document Format. It is a faithful copy of the original but it is not editable unless it is opened in the full version of Adobe Acrobat or some comparable application. When PDF documents are printed

Performing Data Analysis Using IBM SPSS®, First Edition.
Lawrence S. Meyers, Glenn C. Gamst, and A. J. Guarino.

or viewed on the screen, they mirror what was on the screen originally even though the current computer may not have the fonts that were used in the document; that is what makes them portable—the PDF contains within it all the information necessary for the document to be displayed on the screen or printed.

IBM SPSS is capable of saving a PDF version of the output file. This is an ideal way to view the *full set* of results (view a copy of the output file) when we cannot or choose not to access IBM SPSS.

To instruct IBM SPSS to save an output file as a PDF document, the file should be the active window (click its banner to make sure that it is active). Then select from the main menu **File ➔ Export**. This opens the **Export Output** screen shown in Figure 6.1. Select **Portable Document Format (*.pdf)** from the **File Type** drop-down menu as shown in Figure 6.2.

With the **File Type** specified as PDF, select **Browse** and navigate to the location where the file is to be saved. Name the file. Figure 6.3 shows the result of this browsing and naming process. Then click **Save**. This returns us to the **Export Output** screen. Click **OK** and wait for the creation process to be finished.

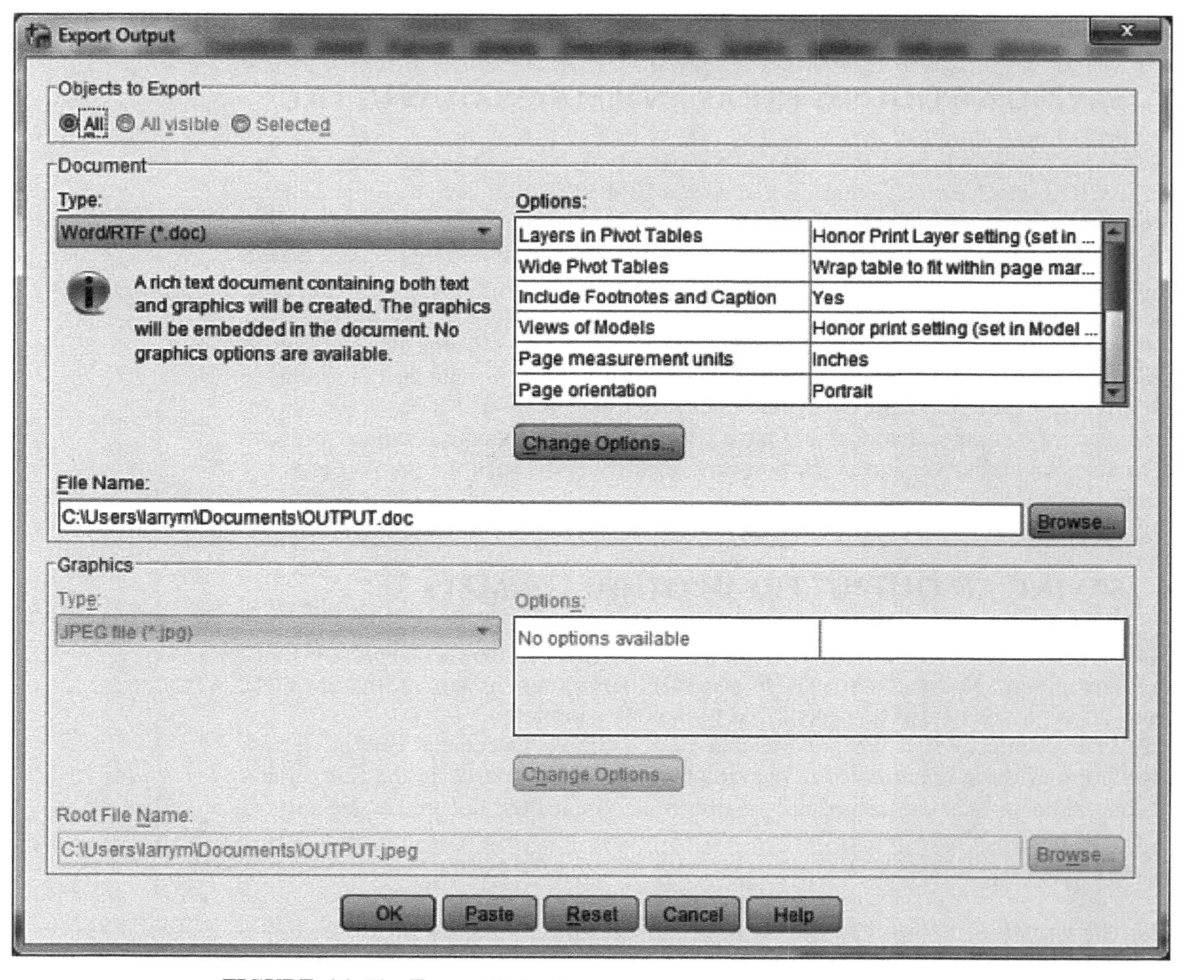

FIGURE 6.1 The **Export Output** screen.

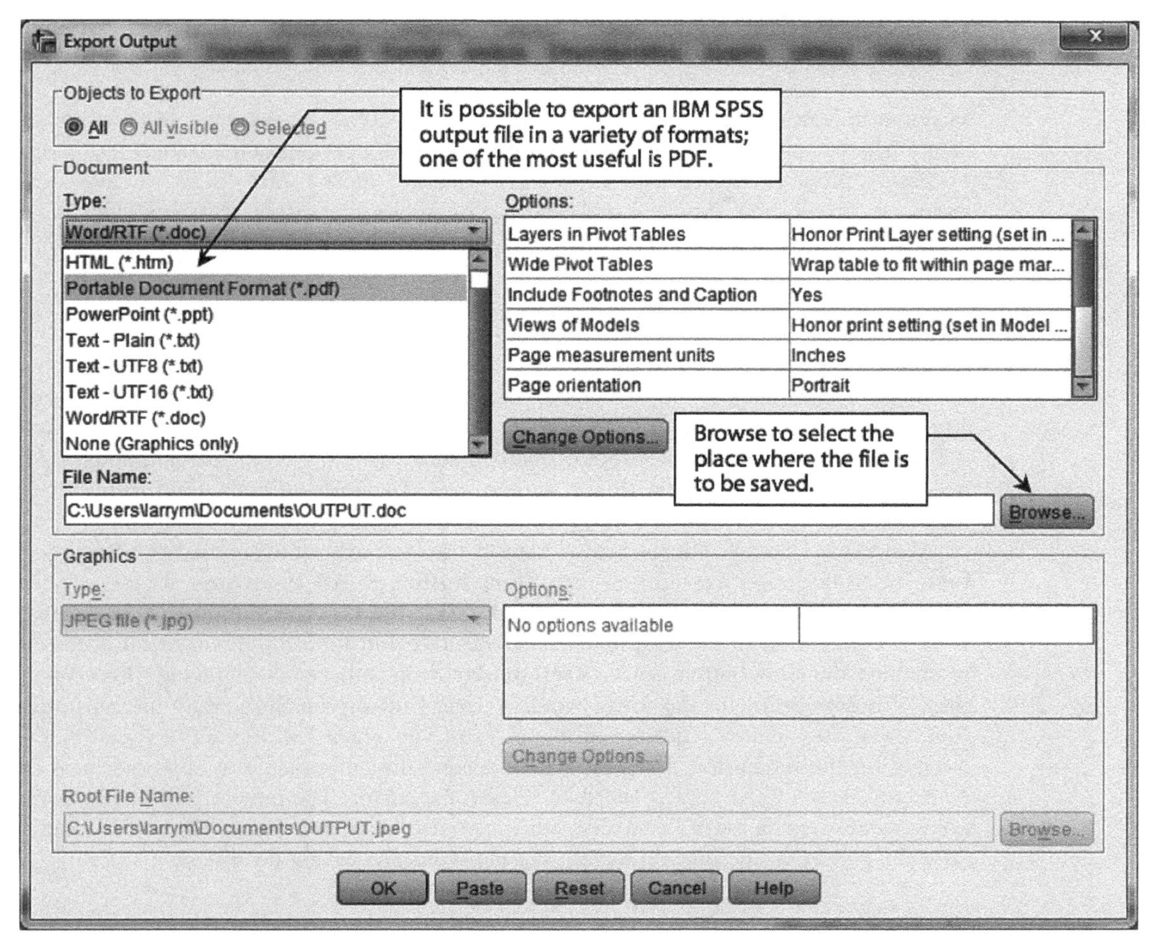

FIGURE 6.2 Some of the choices in the **File Type** drop-down menu on the **Export Output** screen.

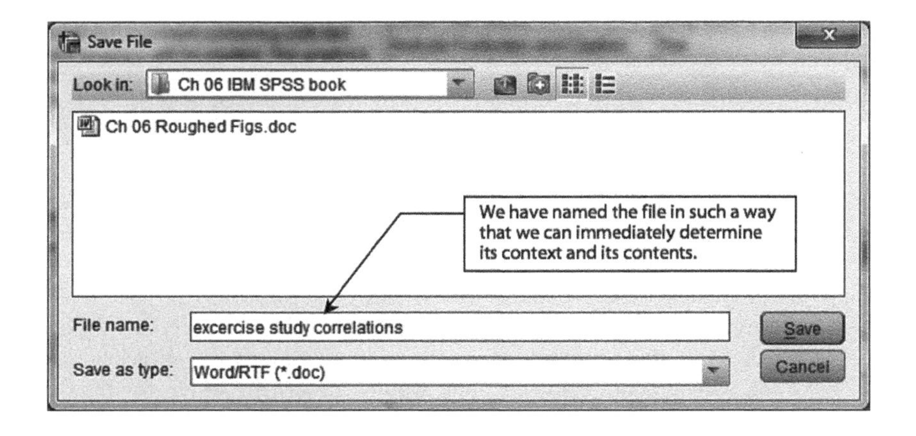

FIGURE 6.3 Name the file in a way that suggests its context and contents.

6.4 USING OPERATING SYSTEM UTILITIES TO COPY AN IBM SPSS TABLE TO A WORD PROCESSING DOCUMENT

As noted in Section 4.3, much of the output in IBM SPSS is presented in the form of tables. It is not uncommon for users to save (copy) only certain output tables to a word processing program (e.g., Microsoft Word). The two most useful forms to save such copies are screenshots (images) and word processing tables. Both PCs and Macs have utility programs as a part of their operating systems that can generate such copies.

Screenshots of the Active Window on a Windows-Based PC

There are utilities in the Windows operating systems that allow users to take screenshots of windows. The quickest way to take a screenshot of the entire screen on most versions of Windows is to tap the **Print Screen** key (on some keyboards this key may be designated as **Prnt Scrn**). To take a screenshot of the one active window on the screen, hold down the **Alt** key while tapping the **Print Screen** key. The screenshot in both instances is placed on the clipboard and may be pasted into any word processing document.

Windows 7 and Windows 8 also have a user-friendly utility called the **Snipping Tool**; it can be found through the path **Start Button ➔ All Programs ➔ Accessories ➔ Snipping Tool**. Open the **Snipping Tool**. Click the **Options** button to confirm that **Always copy snips to the Clipboard** is checked. From the drop-down menu accessed by clicking the **New** button, select **Rectangular Snip**; other choices include **Free-form Snip**, **Window Snip** (for the active window), and **Full-screen Snip**. With the **Snipping Tool** active (its window is visible on the screen), the screen will lighten to signal that it is ready for the screenshot. Drag the cursor around the output table or any other area of the screen that is to be copied and then release the cursor. The picture will be placed in a new window (which can be saved), but it can also be placed on the clipboard. Simply paste the picture in the desired location in the word processing document.

Screenshots of Any Part of the Output on a Mac

For Mac users, it is also possible to take a screenshot of a selected section of anything that appears on the screen, of an open window that is visible on the screen, or of the entire screen. The Mac OS X contains a utility called **Grab** (located in the Utilities folder in the Applications folder). Open **Grab** and select from its **Capture** menu **Selection**, **Window**, or **Screen**. Then follow these guidelines:

- If **Selection** was chosen, click and drag to create a rectangle around the portion of the screen to be captured.
- If **Window** was chosen, click the inside of the window to be captured.
- If **Screen** was chosen, click on the screen.

After making a selection, the picture that has been taken will appear in its own window. Save the screenshot (the Desktop may be the most convenient location) and drag its icon into the word processing document.

An alternative utility that is even more user-friendly is **SnapNDrag**, available as a free download from Yellow Mug Software (http://www.yellowmug.com). Open **SnapNDrag** and select from the pushbuttons in its window **Selection**, **Window**, or **Screen**. Then make the screen capture in the same way as described above for **Grab**. The resulting picture appears in a miniwindow inside the **SnapNDrag** window. Drag the minipicture directly into the word processing document or anyplace else (e.g., an e-mail). It can also be dragged to the Desktop and saved.

6.5 USING THE COPY AND PASTE FUNCTIONS TO COPY AN IBM SPSS OUTPUT TABLE TO A WORD PROCESSING DOCUMENT

It is possible on either Macs or PCs to use the **Copy** and **Paste** commands to place a copy of an output table into a word processing document. Several combinations of **Copy** in the IBM SPSS menu and **Paste** in the Microsoft Word menu will accomplish the job, and we identify one that will work for each platform:

- Move the cursor to a position inside the table that is to be copied from the IBM SPSS output file.
- Click anyplace in the output table; this will produce a "box" around the table.
- In Mac OS X, select in the IBM SPSS menu **Edit ➔ Copy Special**. This will open the **Copy Special** dialog window with several choices all of which are checked. Click **OK**. Open the Word document and set the cursor at the place where the output table is to appear. Select in the Word menu **Edit ➔ Paste Special** and select **Picture** to obtain a screenshot or select **Formatted Text (RTF)** to produce a rich text formatted word processing table.
- In Windows 7, select in the IBM SPSS menu **Edit ➔ Copy Special**. This will open the **Copy Special** dialog window with some of the several choices checked. To produce a word processing table, select only **Rich Text**; to obtain a screenshot of the table, select only **Image**. Click **OK**. Open the Word document and set the cursor at the place where the output table is to appear. Right-click and select the (leftmost) icon for **Keep Source Formatting**; whatever format was chosen in the **Copy Special** dialog window will appear in the word processing document.
- Resize the screenshot as necessary or edit the word processing table for font and size preferences as appropriate.
- Save the Word document.

PART 3

MANIPULATING DATA

CHAPTER 7

Sorting and Selecting Cases

7.1 OVERVIEW

IBM SPSS® has procedures that enable us to display the data file in different ways and to process selected portions of the data. These procedures are quite useful when there are many more cases (rows) in the data file than we can easily view on the screen. We discuss two of these procedures in this chapter.

7.2 SORTING CASES

Cases are arranged in the data file in the order that they were entered. Sometimes researchers may wish to view the data file in a different case order. IBM SPSS provides a very simple procedure to sort cases in either an ascending or a descending order. Figure 7.1 shows a portion of the data file named **Personality** ordered by a variable named **subid** that represents an arbitrary identification code unique to each case.

To sort the data file based on another variable (let us say **age**), from the IBM SPSS main menu select **Data ➔ Sort Cases**. This produces the **Sort Cases** dialog window shown in Figure 7.2. Double-click **age** to move it from the **Variable List** to the **Sort by** panel (see Figure 7.2). **Ascending** is the default option that is checked under **Sort Order**, and we retain it. Click **OK**. The results of this are shown in Figure 7.3. IBM SPSS acknowledges in an output window that it has executed the sorting by age; this output file can be deleted without saving.

7.3 SELECTING CASES

There are occasions when researchers want to focus on a subset of the sample, for example, one of the sexes or ethnicities or those who were coded as exhibiting symptoms of a certain syndrome. It is possible to accomplish this using the **Select Cases** routine. Cases selected by this routine are eligible to be included in any subsequent analysis; cases excluded are not eligible to be included in any subsequent analysis.

Performing Data Analysis Using IBM SPSS®, First Edition.
Lawrence S. Meyers, Glenn C. Gamst, and A. J. Guarino.

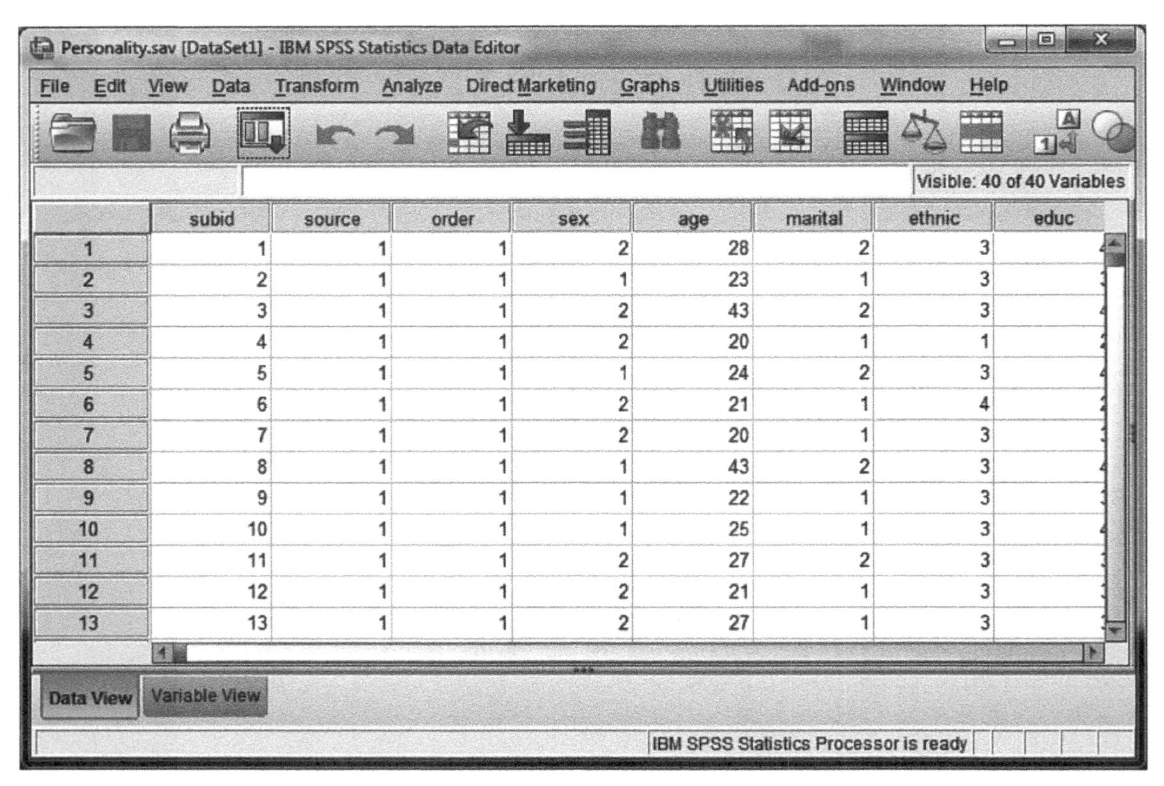

FIGURE 7.1 The original data file.

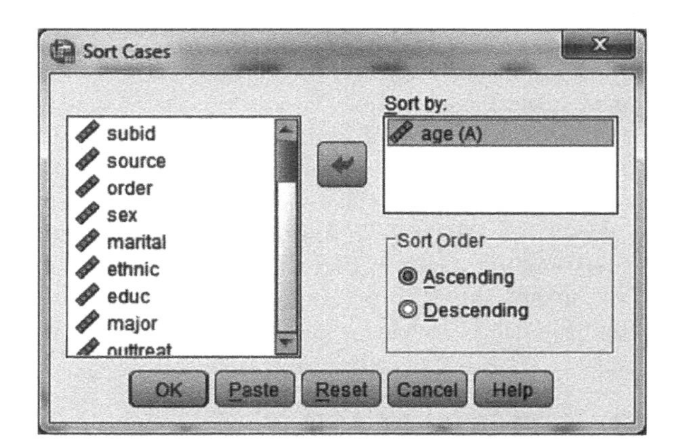

FIGURE 7.2
The **Sort Cases** dialog window with **age** to be sorted in an **Ascending Sort Order**.

We will demonstrate the selection routine with the data file shown in Figure 7.1 by selecting the female cases in the data file. In this particular data file, males are coded as **1**; females, **2**; and missing values, **9**. From the IBM SPSS main menu select **Data → Select Cases**. This produces the **Select Cases** dialog window shown in Figure 7.4. As shown in the figure, we have checked the option **If condition is satisfied**; this condition will be that the case is a female.

Selecting the **if** pushbutton under **If condition is satisfied** opens the **Select Cases If** screen. It is possible to configure a complicated condition to be met, but it also allows us to specify our simple condition. The steps to select females (the end result of which is shown in Figure 7.5) are as follows:

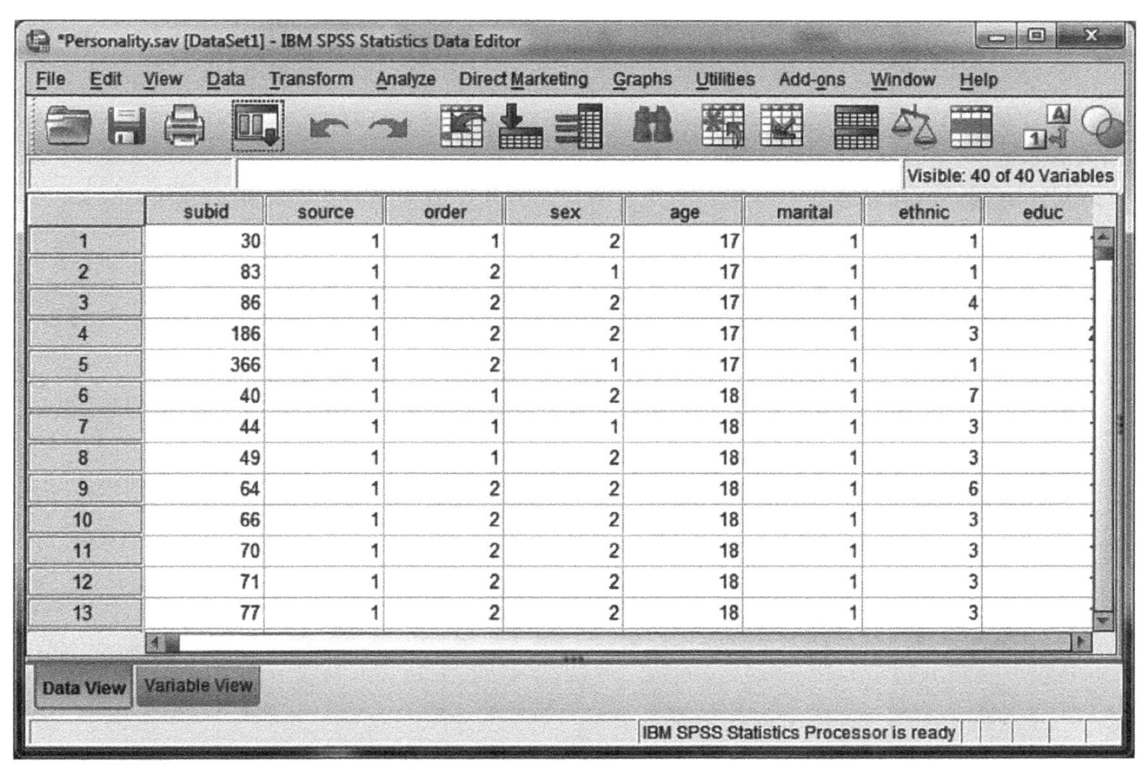

	subid	source	order	sex	age	marital	ethnic	educ
1	30	1	1	2	17	1	1	
2	83	1	2	1	17	1	1	
3	86	1	2	2	17	1	4	
4	186	1	2	2	17	1	3	
5	366	1	2	1	17	1	1	
6	40	1	1	2	18	1	7	
7	44	1	1	1	18	1	3	
8	49	1	1	2	18	1	3	
9	64	1	2	2	18	1	6	
10	66	1	2	2	18	1	3	
11	70	1	2	2	18	1	3	
12	71	1	2	2	18	1	3	
13	77	1	2	2	18	1	3	

FIGURE 7.3 The data file is now sorted by **age**.

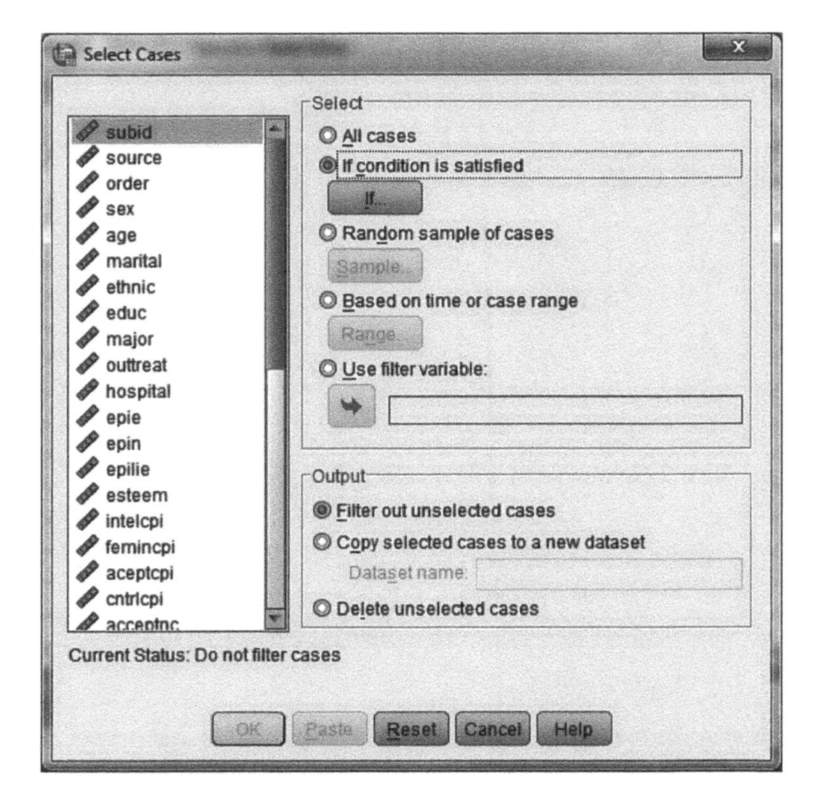

FIGURE 7.4

The **Select Cases** dialog window with **If condition is satisfied** checked.

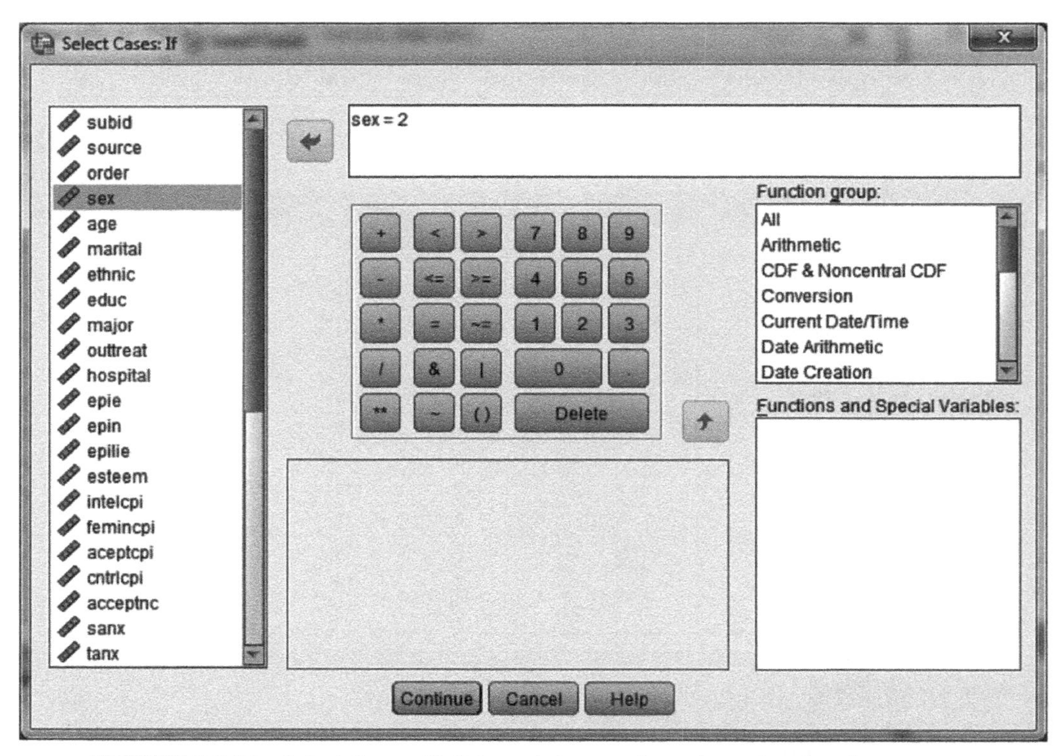

FIGURE 7.5 The **Select Cases: If** dialog window configured to select females.

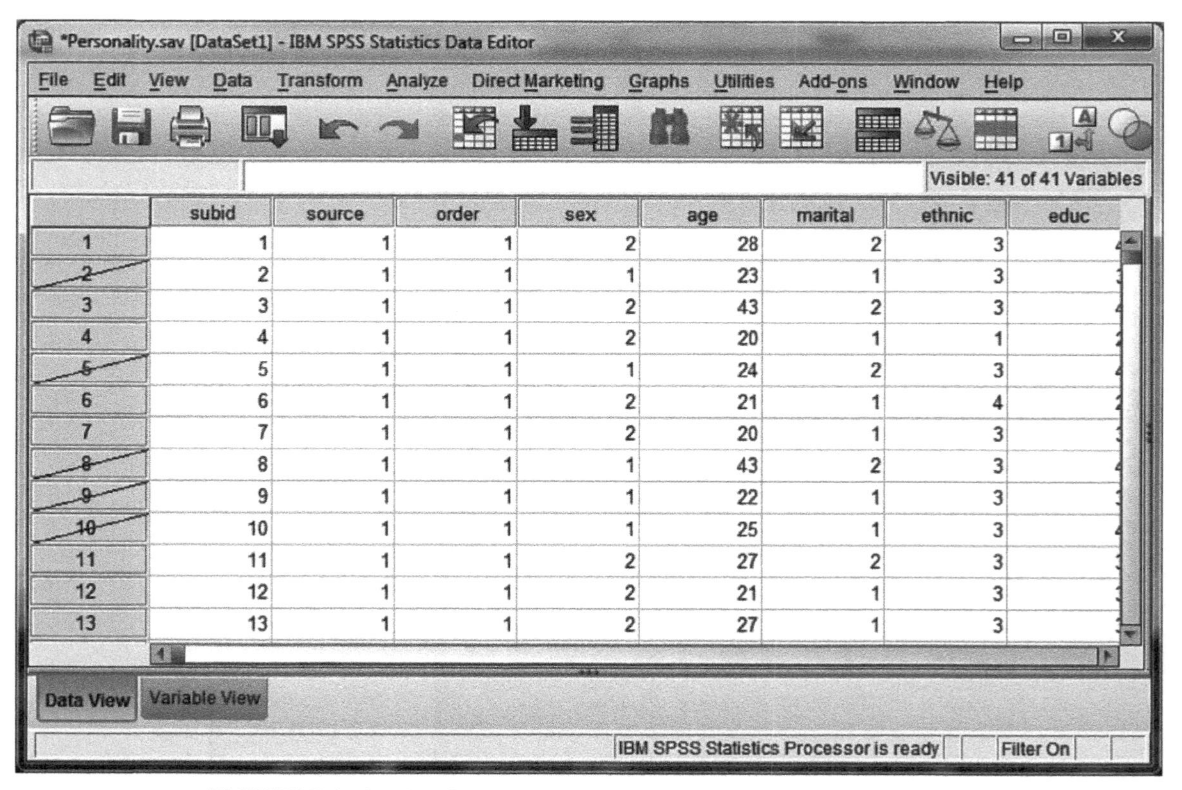

	subid	source	order	sex	age	marital	ethnic
1	1	1	1	2	28	2	3
2	2	1	1	1	23	1	3
3	3	1	1	2	43	2	3
4	4	1	1	2	20	1	1
5	5	1	1	1	24	2	3
6	6	1	1	2	21	1	4
7	7	1	1	2	20	1	3
8	8	1	1	1	43	2	3
9	9	1	1	1	22	1	3
10	10	1	1	1	25	1	3
11	11	1	1	2	27	2	3
12	12	1	1	2	21	1	3
13	13	1	1	2	27	1	3

FIGURE 7.6 The data file with the female cases selected.

- Double-click **sex** to move it from the **Variable List** to the panel at the top of the screen.
- Click the equal sign on the keypad.
- Click the number **2** on the keypad (this is the code for females).
- Click the **Continue** pushbutton at the bottom of the screen to return to the **Select Cases** dialog window.
- Click **OK** to perform the selection.

The result of the selection is shown in Figure 7.6. Note that cases 2, 6, 8, 9, and 10 (those with **sex** code **1**, indicating they are males) have a "cross-out" mark in their first column. This is a visual cue that these cases will be excluded from any subsequent data analysis. Note also in the lower right corner of the data file the presence of the expression **Filter On** to indicate that the data file is being filtered on some basis.

The expression **Filter On** may be understood because the selection of a subset of cases is mediated by a new variable named **filter_$** that IBM SPSS has generated. The new variable is placed as the last variable of the data file, but we have moved it next to **sex** (in the **Variable View**, we highlighted the row for **filter_$** and dragged it up to **sex**) to make it easier to see how the two variables interface.

The **sex** and **filter_$** variables are shown in Figure 7.7. The **filter_$** variable is in perfect synchrony with **sex**: where **sex** shows a code of **2**, the **filter_$** variable shows a

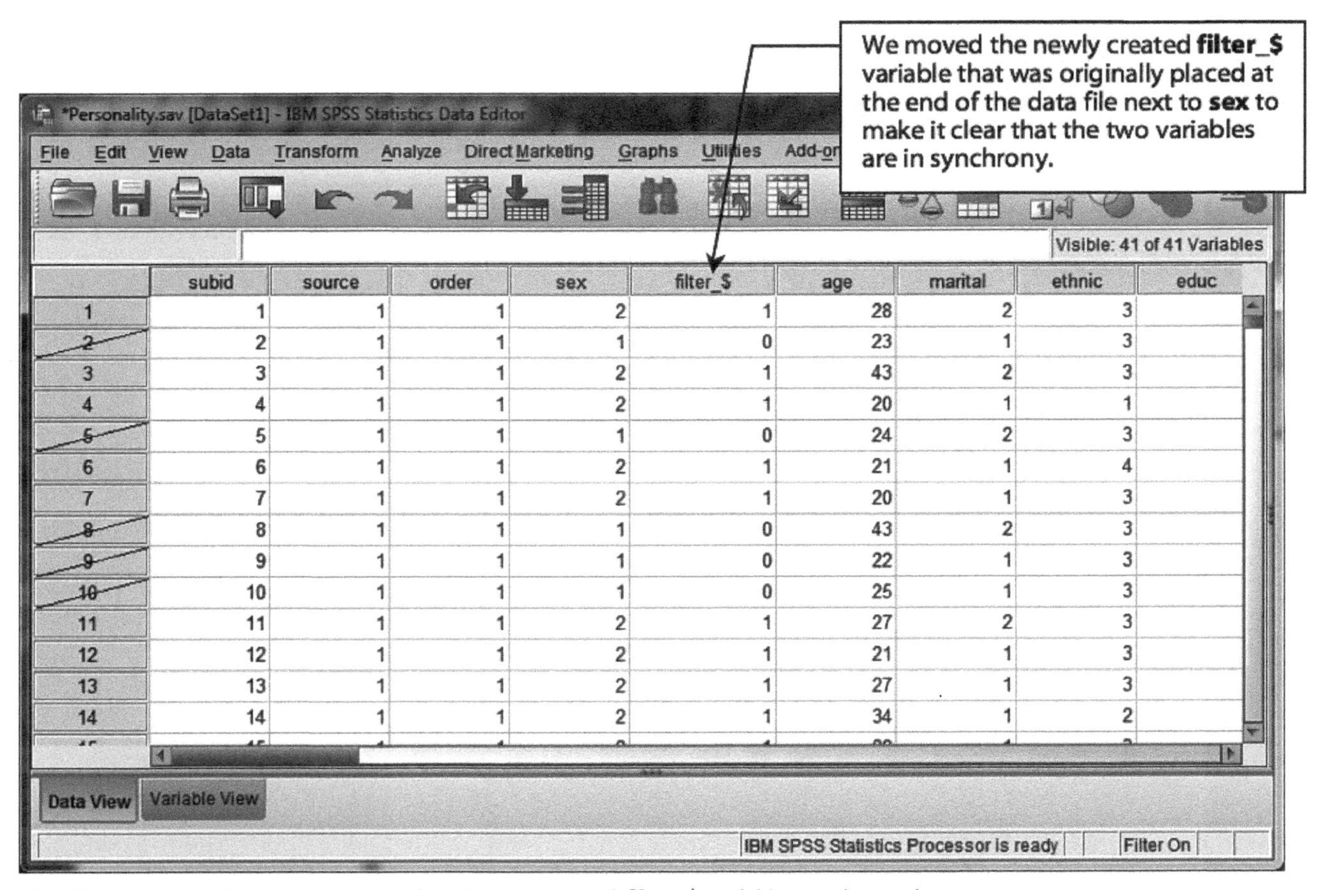

	subid	source	order	sex	filter_$	age	marital	ethnic	educ
1	1	1	1	2	1	28	2	3	
2	2	1	1	1	0	23	1	3	
3	3	1	1	2	1	43	2	3	
4	4	1	1	2	1	20	1	1	
5	5	1	1	1	0	24	2	3	
6	6	1	1	2	1	21	1	4	
7	7	1	1	2	1	20	1	3	
8	8	1	1	1	0	43	2	3	
9	9	1	1	1	0	22	1	3	
10	10	1	1	1	0	25	1	3	
11	11	1	1	2	1	27	2	3	
12	12	1	1	2	1	21	1	3	
13	13	1	1	2	1	27	1	3	
14	14	1	1	2	1	34	1	2	

FIGURE 7.7 Data file screenshot showing that the **sex** and **filter_$** variables are in synchrony; **sex** code **2** is associated with x variable code **1** (select) and **sex** code **1** is associated with x variable code **0** (do not select).

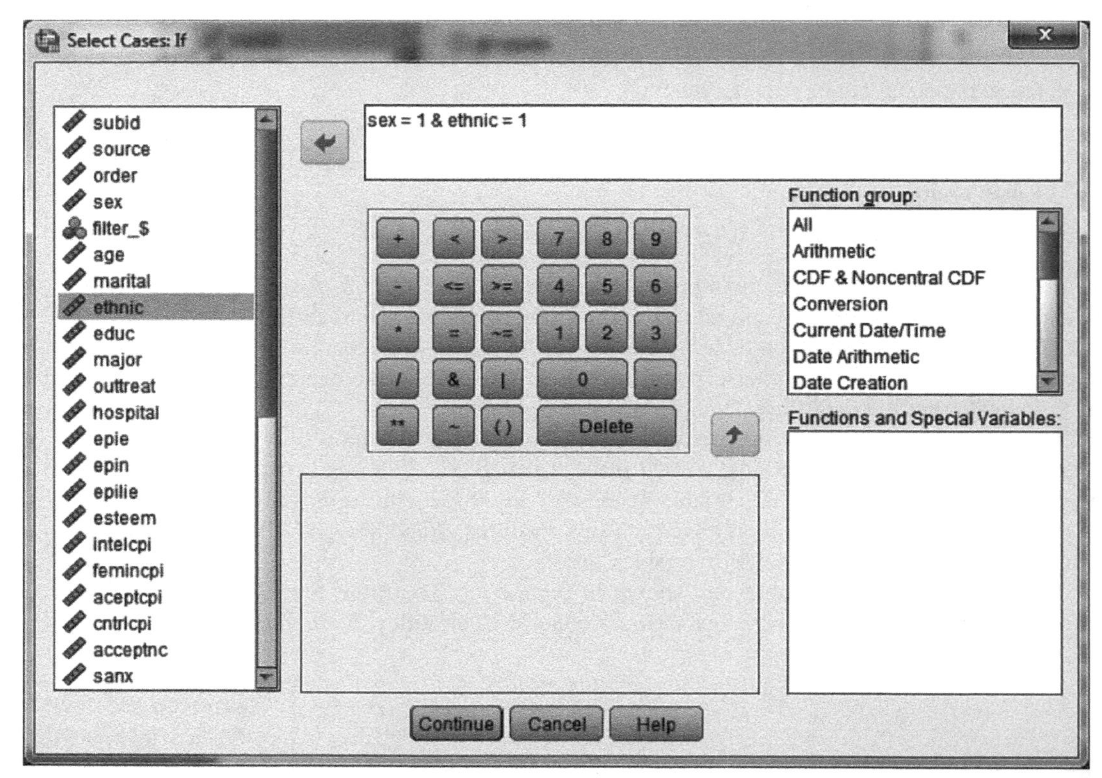

FIGURE 7.8 An example of a more complex selection of cases where we are selecting for Asian-American females.

code of **1** (representing that a case is selected), and where **sex** shows a code of **1**, the **filter_$** variable shows a code of **0** (representing that a case is not selected).

It is important that when finished with analyses on the selected cases, researchers should deactivate the selection. To include all cases in the analysis, select **Data → Select Cases**, check the first option for **All cases**, and click **OK**.

It is possible to configure more complicated selection conditions. For example, the researchers may want to analyze data from females who are Asian-American (where Asian-Americans are coded as **1** under the variable of **ethnic**). This additional specification can easily be accomplished by simply adding **& ethnic=1** in the **Select Cases: If** window as shown in Figure 7.8.

CHAPTER 8

Splitting Data Files

8.1 OVERVIEW

Selecting cases in a single category of a variable (e.g., selecting females) to be more extensively analyzed (see Section 7.3) is only one way to focus on a subset of the sample. In other contexts, each of the categories (e.g., both males and females) may be worthy of separate, individual examination. Using the **Select Cases** routine, we could select for one category of Variable A, carry out our analyses, turn that selection off, select for a second category of Variable A, carry out the same analyses, turn that selection off, select for a third category of Variable A, and so on. If that is the goal, it is easier to accomplish separate analyses using the **Split File** routine in IBM SPSS®.

8.2 THE GENERAL SPLITTING PROCESS

Splitting a data file in IBM SPSS has a very specific meaning. We identify a variable whose values will represent separate groups. For example, in our **Personality** data file, the variable for **sex** is coded **1**, **2**, and **9** for male, female, and missing, respectively. If we split this variable, we would create three subordinate data files, that is, the main data file would be first sorted and then split (virtually rather than physically) into three smaller data files. IBM SPSS would do this by sorting the file based on the splitting variable so that all cases with the same **sex** code would be vertically adjacent to each other in the main data file, and a filter variable, as we have seen in Section 7.3, would be created to represent each code. We note that it is possible to create a split based on the combination of two or more variables, but our goal here is to show the basics of the procedure based on a single variable.

Once the split is in effect, we could perform all of the statistical analyses that were on our research agenda. Any statistical analysis we invoked would automatically be performed at the same time for all of the subordinate data sets. Thus, with respect to the **sex** variable, we would obtain three analyses for each statistical procedure we used. Once we had completed all of our work with the **Split File**, we should recombine the subordinate files (turn off the split) and, if we wished, resort the file to match what it was originally (assuming that was the preferred ordering of the cases).

Performing Data Analysis Using IBM SPSS®, First Edition.
Lawrence S. Meyers, Glenn C. Gamst, and A. J. Guarino.

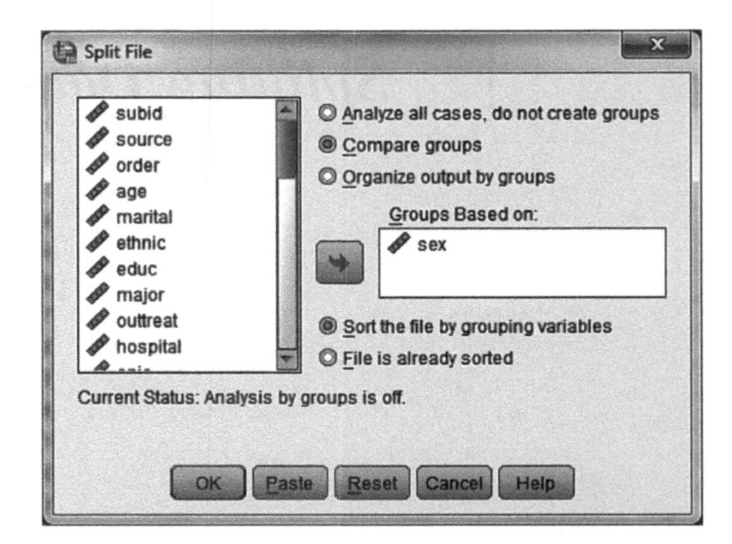

FIGURE 8.1
The **Split File** screen.

8.3 THE PROCEDURE TO SPLIT THE DATA FILE

We open our **Personality** data file and from the main menu select **Data ➔ Split File**. As shown in Figure 8.1, we have moved **sex** into the **Groups Based on** panel and clicked the buttons for **Compare groups** (this will consolidate the output) and **Sort the file by grouping variables** (the file, as may be seen in Figure 7.1, is currently sorted by **subid**). Click **OK**. The file is now split as shown in a confirmatory message in an output file.

8.4 THE DATA FILE AFTER THE SPLIT

The file is now sorted based on **sex**, with all cases whose code is **1** displayed first followed by those with the code **2** and those with the code **9**. The message **Split by sex** will appear in the lower right corner.

IBM SPSS has generated a new variable named **filter_$** as the way it identifies the subordinate data files (we saw this in Section 7.3 when we selected cases). As we have seen, the new variable is placed as the last variable of the data file, but we have moved it next to **sex** in Figure 8.2 to make it easier to see that the two variables are in synchrony. In IBM SPSS 20, the "**filter_$**" variable no longer appears; however, the lower right corner "**Split by**" message will still be seen.

Figure 8.2a shows the data file as it transitions from **sex** code **1** to **2** (from line 142 to 143), and Figure 8.2b shows the data file as it transitions from **sex** code **2** to **9** (from line 421 to 422). The **filter_$** variable is in perfect synchrony with **sex**: where **sex** transitions from **1** to **2**, **filter_$** transitions from **0** to **1**, and where **sex** transitions from **2** to **9**, **filter_$** transitions from **1** to empty cells (to represent missing values). Note that the **Split File** based on the **sex** variable is indicated in the lower right corner of the data file to remind users that it is in effect; in IBM SPSS 20, the lower right corner "**Split by**" message will still be seen.

8.5 STATISTICAL ANALYSES UNDER SPLIT FILE

Assume that we performed a **Bivariate Correlations** procedure on two variables in the **Personality** data file (openness as measured by the NEO Five-Factor Inventory named **neoopen** and self-esteem as measured by the Coopersmith Self-Esteem Inventory named

*Personality.sav [DataSet1] - IBM SPSS Statistics Data Editor

File Edit View Data Transform Analyze Direct Marketing Graphs Utilities Add-ons Window Help

Visible: 41 of 41 Variables

	subid	source	order	sex	filter_$	age	marital	ethnic
136	622	2	2	1	0	43	4	9
137	625	2	1	1	0	41	2	7
138	631	2	2	1	0	30	1	3
139	632	2	1	1	0	75	5	3
140	640	2	9	1	0	21	2	3
141	643	2	1	1	0	42	1	3
142	648	2	1	1	0	26	2	3
143	1	1	1	2	1	28	2	3
144	3	1	1	2	1	43	2	3
145	4	1	1	2	1	20	1	1
146	6	1	1	2	1	21	1	4
147	7	1	1	2	1	20	1	3
148	11	1	1	2	1	27	2	3

Data View Variable View

IBM SPSS Statistics Processor is ready Split by sex

(a)

*Personality.sav [DataSet1] - IBM SPSS Statistics Data Editor

File Edit View Data Transform Analyze Direct Marketing Graphs Utilities Add-ons Window Help

Visible: 41 of 41 Variables

	subid	source	order	sex	filter_$	age	marital	ethnic
414	654	2	1	2	1	54	2	3
415	655	2	1	2	1	53	4	3
416	656	2	1	2	1	43	4	4
417	658	2	1	2	1	25	1	4
418	659	2	1	2	1	99	4	3
419	660	2	1	2	1	43	2	3
420	662	2	1	2	1	38	5	4
421	664	2	1	2	1	33	1	2
422	36	1	1	9	.	99	9	9
423	620	2	1	9	.	99	9	9
424	635	2	2	9	.	99	9	9
425	636	2	1	9	.	99	9	9
426								

Data View Variable View

IBM SPSS Statistics Processor is ready Split by sex

(b)

FIGURE 8.2 Data file screenshots showing that the **sex** and **filter_$** variables are in synchrony for the transition from (a) **sex** code **1** to **2** and from (b) **sex** code **2** to **9**.

Correlations

sex			neoopen openness: neo	esteem self-esteem: coopersmith
1 male	neoopen openness: neo	Pearson Correlation	1	.281**
		Sig. (2-tailed)		.001
		N	142	142
	esteem self-esteem: coopersmith	Pearson Correlation	.281**	1
		Sig. (2-tailed)	.001	
		N	142	142
2 female	neoopen openness: neo	Pearson Correlation	1	.193**
		Sig. (2-tailed)		.001
		N	278	278
	esteem self-esteem: coopersmith	Pearson Correlation	.193**	1
		Sig. (2-tailed)	.001	
		N	278	279
9 missing	neoopen openness: neo	Pearson Correlation	.[a]	.[a]
		Sig. (2-tailed)		.
		N	0	0
	esteem self-esteem: coopersmith	Pearson Correlation	.[a]	.[a]
		Sig. (2-tailed)	.	
		N	0	1

**. Correlation is significant at the 0.01 level (2-tailed).
a. Cannot be computed because at least one of the variables is constant.

FIGURE 8.3 Correlations performed separately for males and females under **Split File**.

esteem) under the current **Split File** by **sex**. Figure 8.3 shows the results. For our current purposes, note that only separate results are obtained for the males (**sex** code **1**) and the females (**sex** code **2**). As there was only one case with a missing **sex** code, IBM SPSS was unable to compute a correlation for that group (shown in footnote **a** in the table). The potential usefulness of separate analyses is seen here, in that the Pearson *r* between openness and self-esteem was somewhat different for males and females (.28 and .19, respectively).

8.6 RESETTING THE DATA FILE

Once we have completed all of our analyses under **Split File**, we should reset the data file. Select **Data ➔ Split File**. Highlight the **sex** variable in the **Groups Based on** panel and then click the arrow to its left (now pointing back toward the **Variable List**, as shown in Figure 8.4). Then select the top radio (little round) button associated with **Analyze all cases, do not create groups** (see Figure 8.5) and click **OK**.

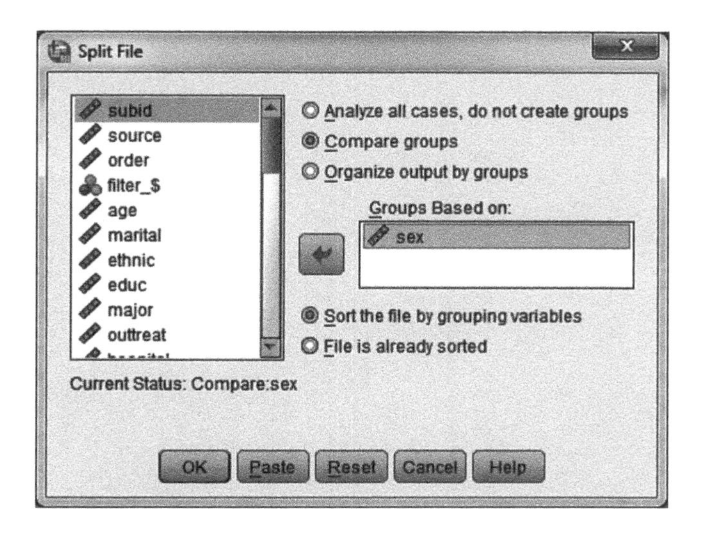

FIGURE 8.4

As the first step to resetting the data file, we highlight **sex** and click the arrow pointing toward the **Variable List** to return **sex** to that list.

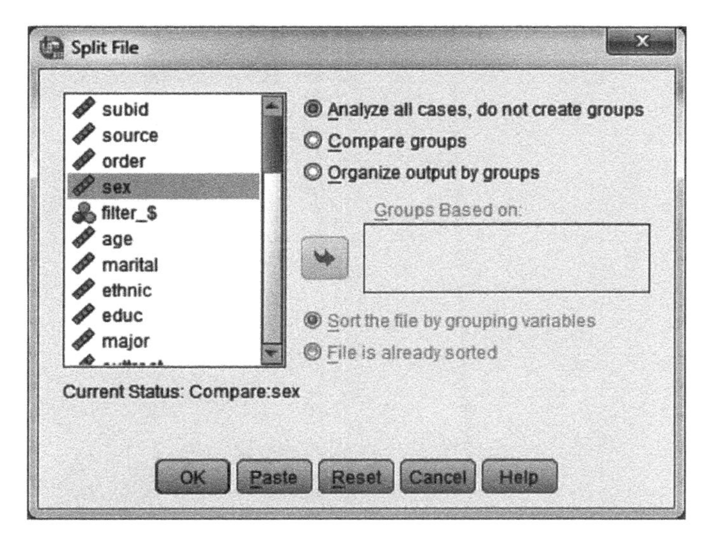

FIGURE 8.5

As the last step to resetting the data file, we select **Analyze all cases, do not create groups** and click **OK**.

Finally, we should delete the **filter_$** variable (if we are in IBM SPSS 19) and resort the data file. To accomplish the deletion, set the data file to **Variable View**, highlight the row corresponding to the **filter_$** variable, and select from the main menu **Edit ➔ Clear** to remove the variable. Then **Sort Cases** based on **subid** and save the data file in that form.

CHAPTER 9

Merging Data from Separate Files

9.1 OVERVIEW

It is not uncommon to have different sets of data each of which comprises a portion of the data for a particular study. Two circumstances when this happens are as follows:

- Different members of the research team have collected and/or entered data into different IBM SPSS® data files on all of the variables but for different cases. In such a situation, we want to combine all of the cases together into a single data file. IBM SPSS labels this as **Add Cases**.
- Information on different variables for the full set of cases is recorded or housed in different IBM SPSS data files. In such a situation, we want to combine all of the variables together into a single data file. IBM SPSS labels this as **Add Variables**.

9.2 ADDING CASES

We use the very simple (for illustration purposes) data file **Self Regard**, a screenshot of which is shown in Figure 9.1. It contains an identification variable named **subid**, two demographic variables (**sex** and **age**), and one personality variable (**regard**). The five cases in the file are sorted on **subid**.

Five additional cases, **subid # 6** through **subid # 10**, are contained in the file **Self Regard adds** currently residing in the flash drive. This additional data file, shown in Figure 9.2, contains the same variables as our **Self Regard** data file and is also sorted on the **subid** variable.

Our goal is to bring the data from **Self Regard adds** into **Self Regard**. Although we could copy and paste from one data file to another, very large data files are probably best combined using the IBM SPSS **Merge Files** procedure that we demonstrate here.

We open our **Self Regard** data file and from the main menu select **Data ➔ Merge Files ➔ Add Cases**. This opens the **Add Cases** dialog window (see Figure 9.3). We have selected **An external SPSS Statistics data file** because **Self Regard adds** is not currently open. We have navigated to the file on the flash drive and selected it.

Performing Data Analysis Using IBM SPSS®, First Edition.
Lawrence S. Meyers, Glenn C. Gamst, and A. J. Guarino.

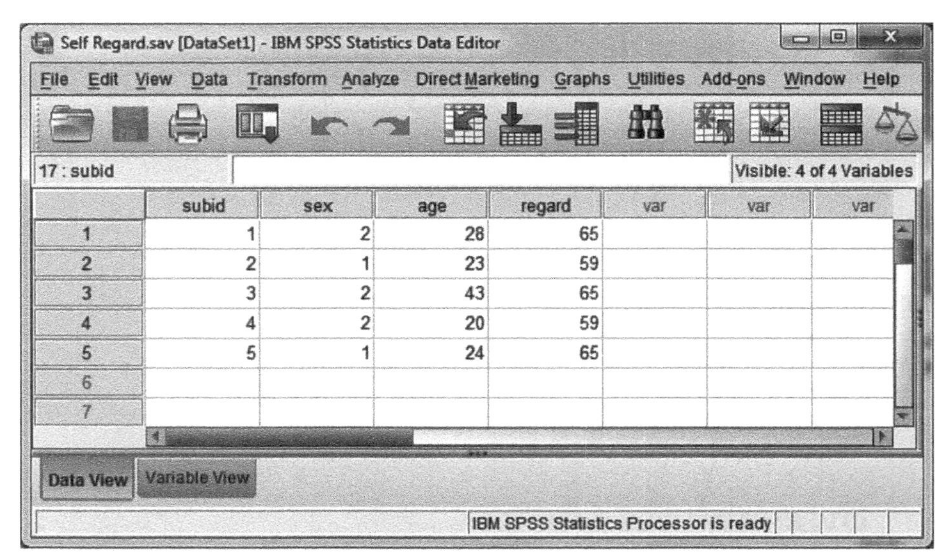

FIGURE 9.1 Our primary **Self Regard** data file.

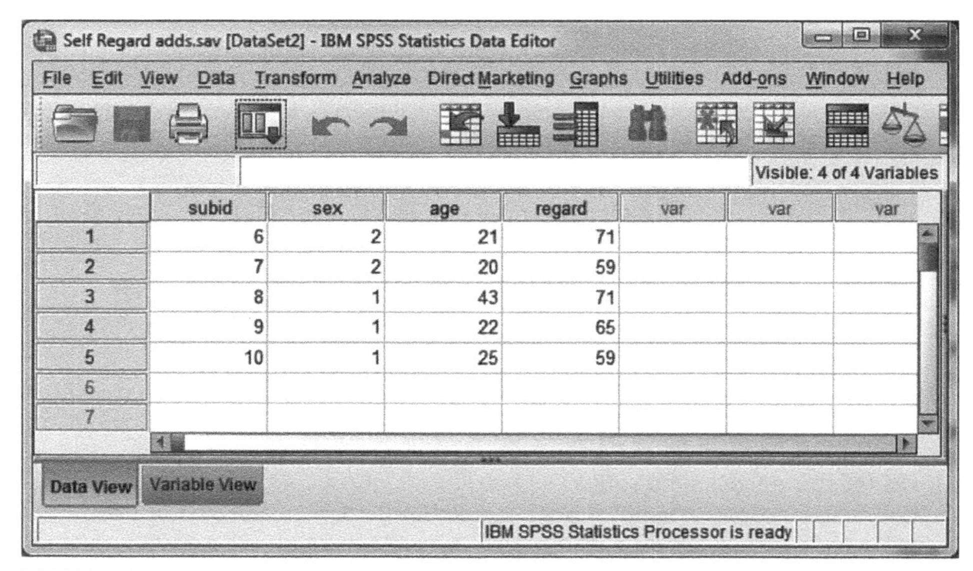

FIGURE 9.2 The **Self Regard adds** data file; the data here are to be added to the **Self Regard** data file.

Clicking **Continue** opens the main **Add Cases** dialog window as presented in Figure 9.4. The panel for **Unpaired Variables** contains variables that do not match across the two data files (e.g., names do not match, variables with the same name are defined as numeric in one field and as string in the other, sets of string variables have different defined lengths). As all of the variables match, this panel is empty.

The panel for **Variables in New Active Dataset** contains those variables that match across the two data files and will therefore be combined. We can remove any variables here that we wish; in our example, we retain all of the variables.

Clicking **OK** accomplishes the merge, as shown in Figure 9.5. We save this file under a different name (**Self Regard merged**) to preserve the original file by selecting **File → Save As** (see Figure 9.6).

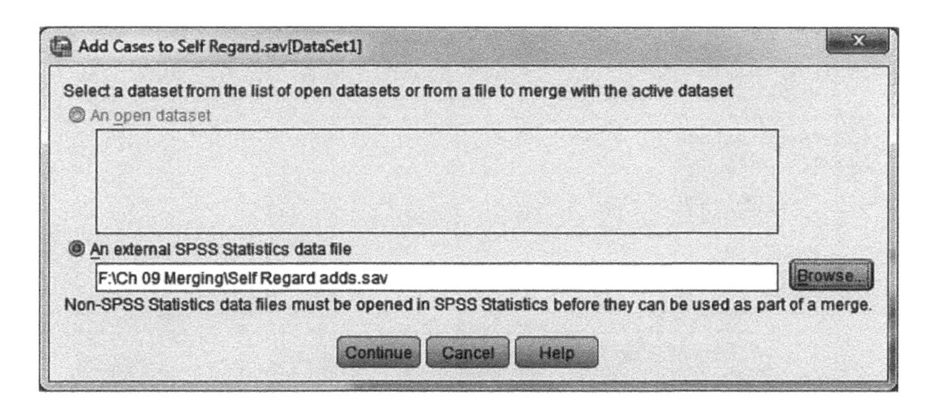

FIGURE 9.3

The initial **Add Cases** dialog screen.

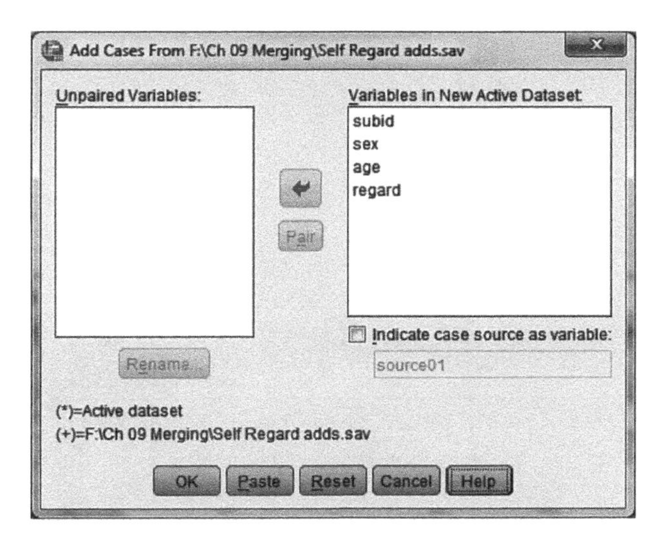

FIGURE 9.4

The main **Add Cases** dialog screen.

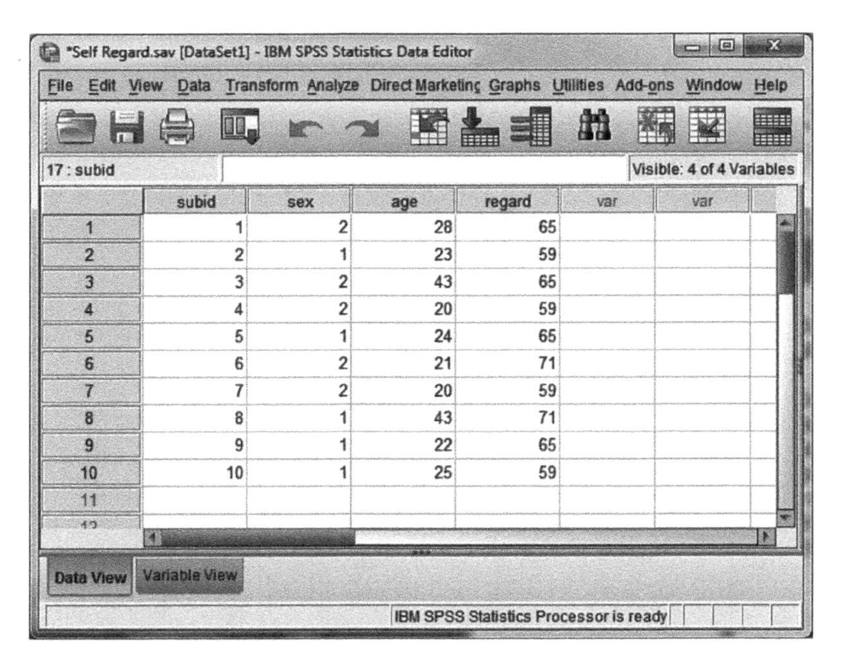

FIGURE 9.5

The additional cases have been added to **Self Regard**.

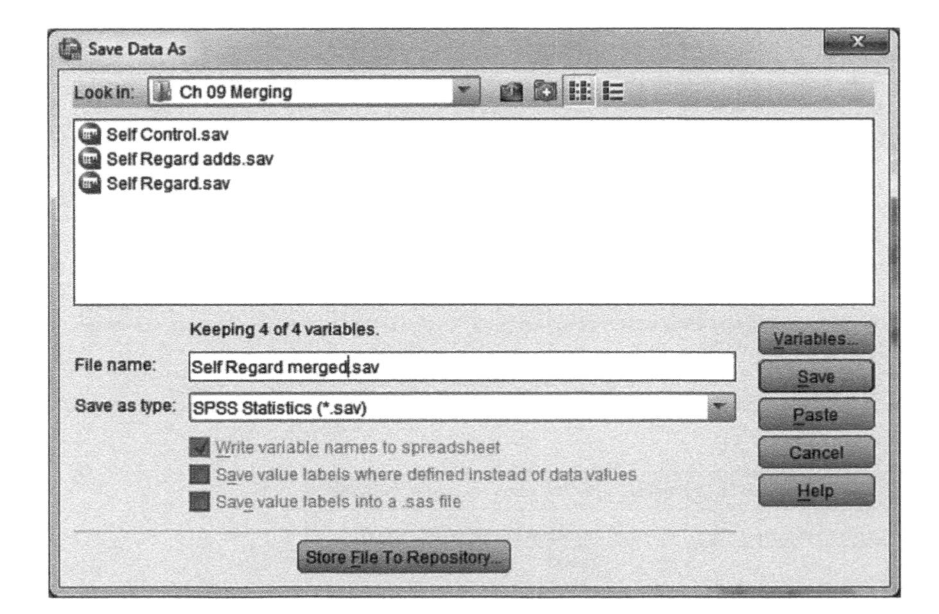

FIGURE 9.6

We save the merged data file under a new name.

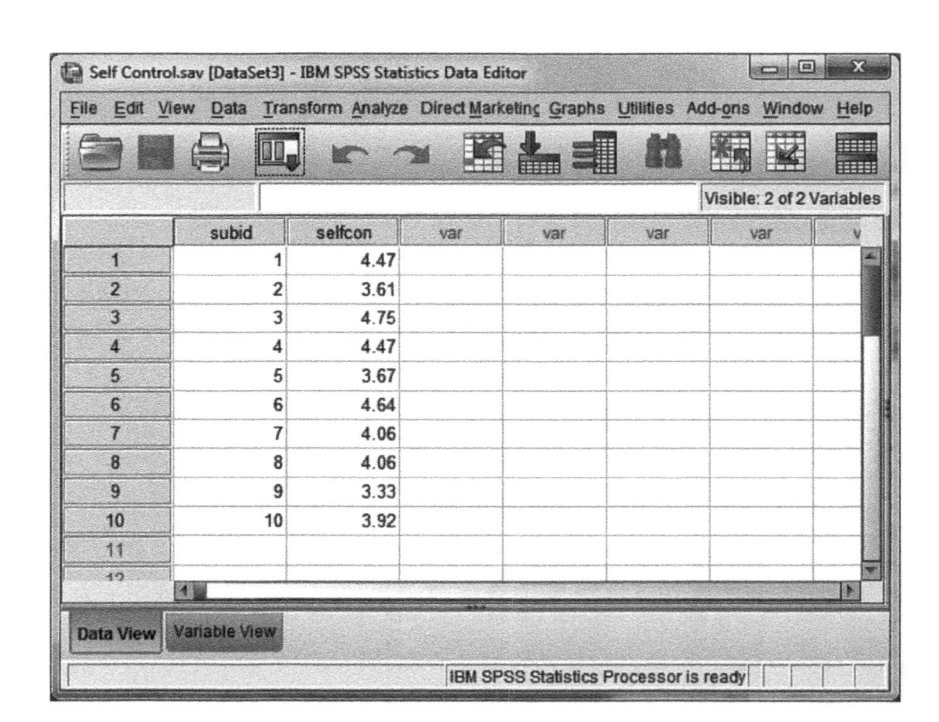

FIGURE 9.7

The **Self Control** data file; the data here are to be added to the **Self Regard merged** data file.

9.3 ADDING VARIABLES

To illustrate how to combine variables, we will merge the variable **selfcon** contained in the data file **Self Control** into the **Self Regard merged** data file we just built in Section 9.2. As shown in Figure 9.7, **Self Control** contains **subid** and **selfcon** with the same cases as those in the **Self Regard merged** data file ordered on the **subid** variable; this matches the ordering of the **Self Regard merged** data file (IBM SPSS requires that the variables must be sorted on the same basis in both files).

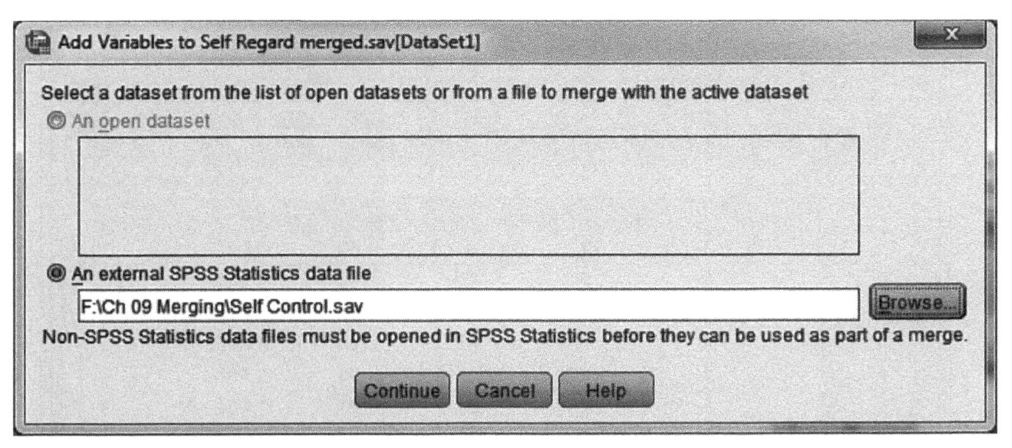

FIGURE 9.8 The initial **Add Variables** dialog screen.

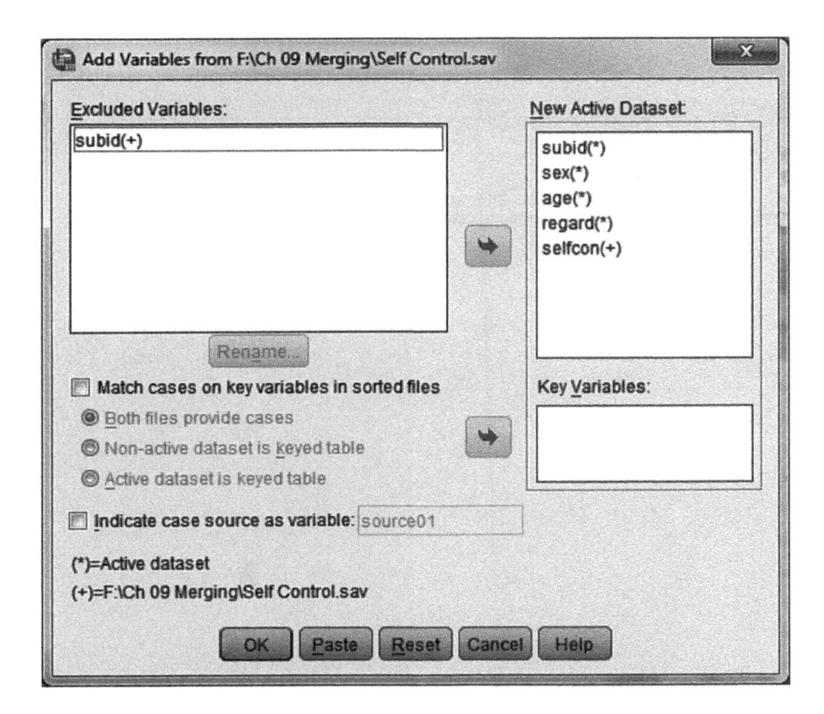

FIGURE 9.9 The main **Add Variables** dialog screen.

We open our **Self Regard merged** data file and from the main menu select **Data ➔ Merge Files ➔ Add Variables**. This opens the **Add Variables** dialog window (see Figure 9.8). We have selected **An external SPSS Statistics data file** because **Self Control** is not currently open, navigated to the file on the flash drive, and selected it.

Clicking **Continue** opens the main **Add Variables** dialog window as presented in Figure 9.9. The panel for **Excluded Variables** contains the variable(s) in **Self Control** that duplicate what we already have in **Self Regard merged** and therefore will not be included in the merge. The variables in the **New Active Dataset** list the variables from both files. Those marked with an asterisk (*) are already in **Self Regard merged**; the variable **selfcon** is marked with a plus sign (+) to indicate that it will be added to the set. The **Key Variables** panel is properly blank here, but if some cases were missing in one of the data files, we could identify the variable whose value can be used to match the cases so that the new variable(s) can be merged.

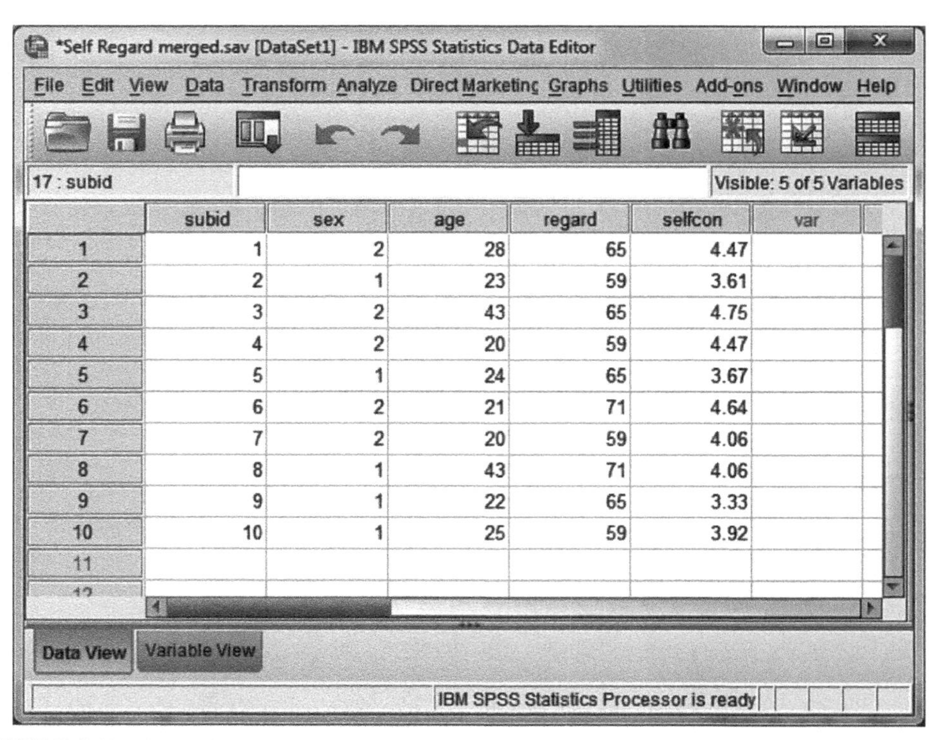

FIGURE 9.10 The additional variable **selfcon** has been added to **Self Regard merged**.

Clicking **OK** accomplishes the merge, as shown in Figure 9.10. Once again, this new data file should be saved.

PART 4

DESCRIPTIVE STATISTICS PROCEDURES

CHAPTER 10

Frequencies

10.1 OVERVIEW

Once the data are verified as correctly entered, one of the first steps researchers perform as part of their data analysis is generating descriptive statistics on the variables in the study. The **Frequencies** procedure in IBM SPSS® is one of the procedures available for this purpose.

In generating such statistics, it is important to distinguish between variables assessed on a nominal or categorical scale of measurement from those assessed on a quantitative (summative response, interval, or ratio) scale of measurement (Meyers, Gamst, & Guarino, 2013). For categorical variables, our only option is to determine the frequencies of cases classified into each category (e.g., the number of cases in each ethnicity category). Other descriptive statistics, such as the mean and standard deviation of such a variable with more than two categories, are not interpretable values and so should not be requested.

For quantitative variables, we often do have an interest in the number of cases represented by each value of the variable, but our interest usually diminishes with greater numbers of possible values. For example, we would be more interested in the number of cases choosing 1, 2, 3, 4, and 5 on a 5-point response scale (e.g., to determine that all scale points are being selected with reasonable frequency) than in the number of cases whose score on a measure of extraversion was 31, 32, 33, and so on all the way to 70. But we would always want to obtain other descriptive statistics providing us with information about the central tendency, variability, and shape of the distribution.

10.2 NUMERICAL EXAMPLE

The data we use for our example are extracted from a study of personality variables on 425 university students. Variables include demographics as well as personality measures. Data are contained in the data file named **Personality**.

Performing Data Analysis Using IBM SPSS®, First Edition.
Lawrence S. Meyers, Glenn C. Gamst, and A. J. Guarino.

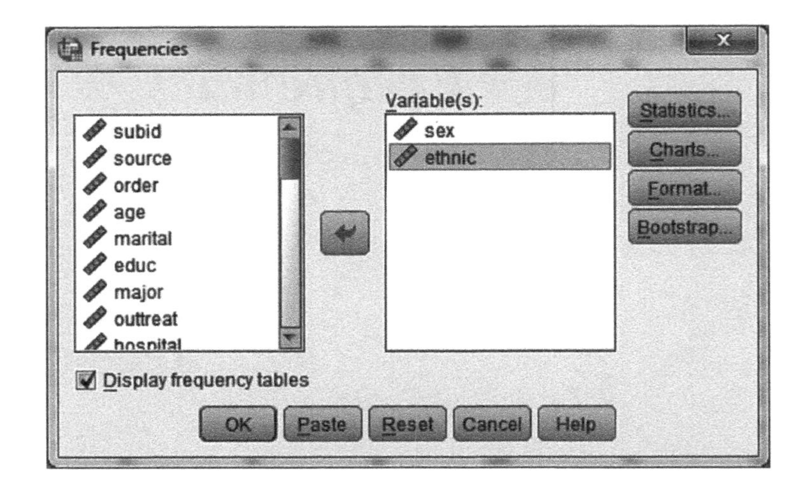

FIGURE 10.1
The main dialog window of **Frequencies**.

10.3 ANALYSIS SETUP: CATEGORICAL VARIABLES

We open the data file named **Personality** and from the main menu select **Analyze ➔ Descriptive Statistics ➔ Frequencies**. The **Frequencies** procedure is appropriate to use for both categorical and quantitative variables. To illustrate an analysis of categorical variables, we have moved **sex** and **ethnic** into the **Variable(s)** panel as shown in Figure 10.1. We have also checked the checkbox **Display frequency tables** below the **Variable List** panel. As the choices available in the **Statistics** screen apply to quantitative variables, we just click **OK** to perform the analysis.

10.4 ANALYSIS OUTPUT: CATEGORICAL VARIABLES

The frequency tables for the **sex** and **ethnic** variables are shown in Figure 10.2. Following is a brief description of the information contained in each column of the tables:

- The first column (not labeled) lists the values (categories) associated with each variable together with their labels (the ones that were typed in the **Variable View** of the data file when the data were entered). **Valid** values appear first in sequential order followed by **Missing** values (in this data file, the code **9** is used to indicate a missing value).
- *Frequency* provides a count of cases for each value. For example, there are **31** participants in the sample of Asian descent.
- *Percent* represents the percent of cases in each category *with respect to the total set of cases*. For example, the **31** participants of Asian descent represent **7.3**% of the total sample of **425**.
- *Valid Percent* represents the percent of cases in each category *with respect to the valid number of cases*. For example, the **31** participants of Asian descent represent **7.4**% of the **421** cases who have valid values on the **ethnic** variable.
- *Cumulative Percent* just continually adds in cases listed in order and computes the growing percent *with respect to the valid number of cases*. Because the coding of most categorical variables is arbitrary (as is true here where the groups are listed in alphabetic order), this computation is ordinarily not particularly useful for such nominal variables.

sex

		Frequency	Percent	Valid Percent	Cumulative Percent
Valid	1 male	142	33.4	33.7	33.7
	2 female	279	65.6	66.3	100.0
	Total	421	99.1	100.0	
Missing	9 missing	4	.9		
Total		425	100.0		

ethnic

		Frequency	Percent	Valid Percent	Cumulative Percent
Valid	1 asian	31	7.3	7.4	7.4
	2 black	11	2.6	2.6	10.1
	3 white	298	70.1	71.5	81.5
	4 hispanic	36	8.5	8.6	90.2
	5 native american	6	1.4	1.4	91.6
	6 pacific islander	6	1.4	1.4	93.0
	7 other	29	6.8	7.0	100.0
	Total	417	98.1	100.0	
Missing	9 missing	8	1.9		
Total		425	100.0		

FIGURE 10.2 Frequency tables for two categorical variables.

10.5 ANALYSIS SETUP: QUANTITATIVE VARIABLES

We open the data file named **Personality** and from the main menu select **Analyze ➔ Descriptive Statistics ➔ Frequencies**. To illustrate an analysis of quantitative variables, we have moved **neoopen** (a measure of openness to new experiences from the NEO Five-Factor Inventory; Costa & McCrae, 1992) and **neoneuro** (a measure of neuroticism from the NEO Five-Factor Inventory) into the **Variable(s)** panel as shown in Figure 10.3. We have also checked the checkbox **Display frequency tables** below the **Variable List** panel to show such output.

Selecting the **Statistics** pushbutton opens the **Statistics** screen shown in Figure 10.4. We select the following statistics to illustrate what is available in this procedure:

- Under **Percentile Values**, we select **Quartiles**.
- Under **Central Tendency**, we select **Mean** and **Median**.
- Under **Dispersion**, we select **Std. deviation**, **Minimum** (the lowest valid value in the distribution), **Maximum** (the highest valid value in the distribution), and **S.E. mean** (the standard error of the mean).
- Under **Distribution**, we select **Skewness** and **Kurtosis**.

We select **Continue** to return to the main dialog window, and click the **Charts** pushbutton. In the **Charts** window shown in Figure 10.5, we select **Histograms** and opt not to

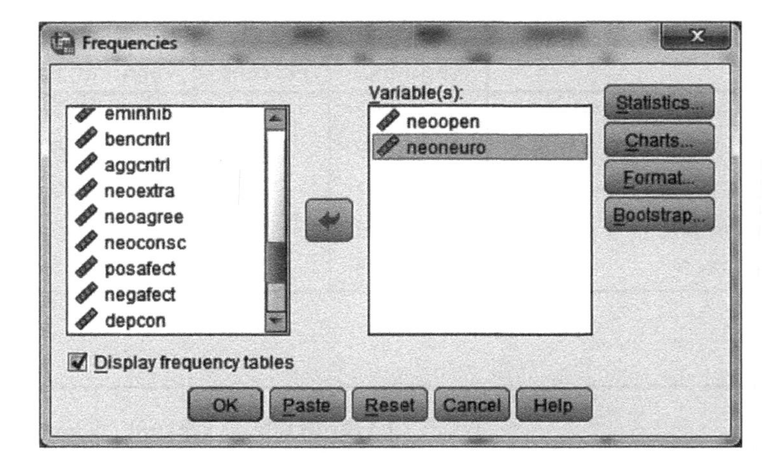

FIGURE 10.3
The main dialog window of **Frequencies**.

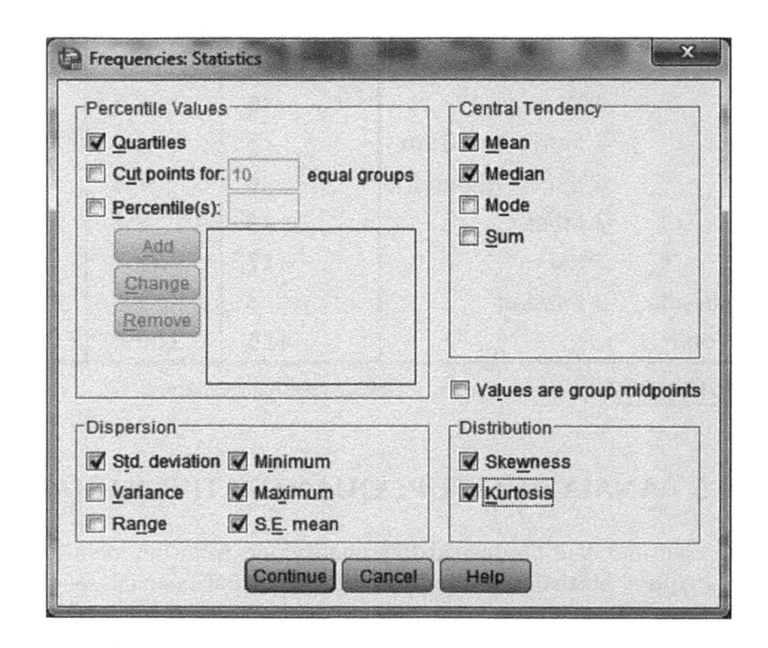

FIGURE 10.4
The **Statistics** screen of **Frequencies**.

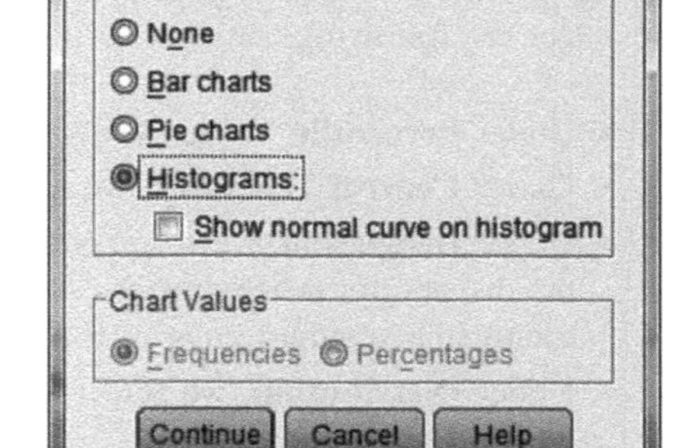

FIGURE 10.5
The **Charts** screen of **Frequencies**.

superimpose a normal curve on it (we do not check **Show normal curve on histogram**). We select **Continue** to once again return to the main dialog window and click **OK** to perform the analysis.

10.6 ANALYSIS OUTPUT: QUANTITATIVE VARIABLES

The descriptive statistics for **neoopen** and **neoneuro** is shown in Figure 10.6. The scores for these two variables are standardized as linear T scores (mean of 50 and a standard deviation of 10) based on the national norms. Given the means in the table, our sample appears to be average in neuroticism and half a standard deviation more open than the norm.

The standard error of the mean (**Std. Error of Mean**) is computed by dividing the standard deviation of the scores by the square root of N, where N is the valid sample size. Conceptually, the standard error of the mean represents the standard deviation of the sample means that we would obtain if we were to draw an infinite number of random samples (based on a sample size N) from the population. We presume that this distribution of (hypothetical) sample means is normally distributed. Given that we cannot actually draw an infinite number of samples and determine the mean of each so that we can compute their standard deviation, we use the above formula to estimate the value of this statistic.

One of main uses of the standard error of the mean is to serve as a basis for computing a confidence interval around the mean. To compute the 95% confidence interval, for example, we multiply the obtained value of the standard error of the mean (**.53233** for **neoopen** and **.54396** for **neoneuro**) by 1.96 (a z value of ± 1.96 subsumes 95% of the area under a normal distribution) and subtract and add those values from and to the

Statistics

		neoopen openness: neo	neoneuro neuroticism: neo
N	Valid	420	420
	Missing	5	5
Mean		55.4607	50.5394
Std. Error of Mean		.53233	.54396
Median		55.1448	49.2904
Std. Deviation		10.90952	11.14793
Skewness		-.157	.184
Std. Error of Skewness		.119	.119
Kurtosis		-.472	-.263
Std. Error of Kurtosis		.238	.238
Minimum		27.51	23.01
Maximum		79.05	84.77
Percentiles	25	48.1271	42.7201
	50	55.1448	49.2904
	75	63.5911	58.4888

FIGURE 10.6
The descriptive statistics output.

mean to identify the lower and upper values of the confidence interval, respectively. In the present instance, using **neoopen** to illustrate this, our computations are as follows:

- 95% Band = 0.53233 * 1.96 = 1.04337
- Lower confidence value = mean − 95% band = 55.4607 − 1.04337 = 54.42
- Upper confidence value = mean + 95% band = 55.4607 + 1.04337 = 56.50

Thus, if hypothetically we were to repeatedly draw 420 cases randomly from the population an infinite number of times, then 95% of the time the sample mean for openness is expected to be between 54.42 and 56.50. Another way to express this is to assert with 95% confidence that the population mean is in the range 54.42–56.50.

The **Minimum** and **Maximum** values in Figure 10.6 are decimal values because the scores are in standardized form based on the national norms (rounded to two decimal places). Thus, for **neoopen**, at least one case obtained a standard score of **27.51** and at least one case obtained a standard score of **79.05**. Values corresponding to the 25th, 50th, and 75th percentiles are shown in Figure 10.6 at the bottom of the table.

Skewness describes the degree of asymmetry exhibited by the distribution of scores. The normal curve (being symmetric) has a skewness value of 0. Positive skewness describes the situation where the bulk of scores is toward the relatively lower values of

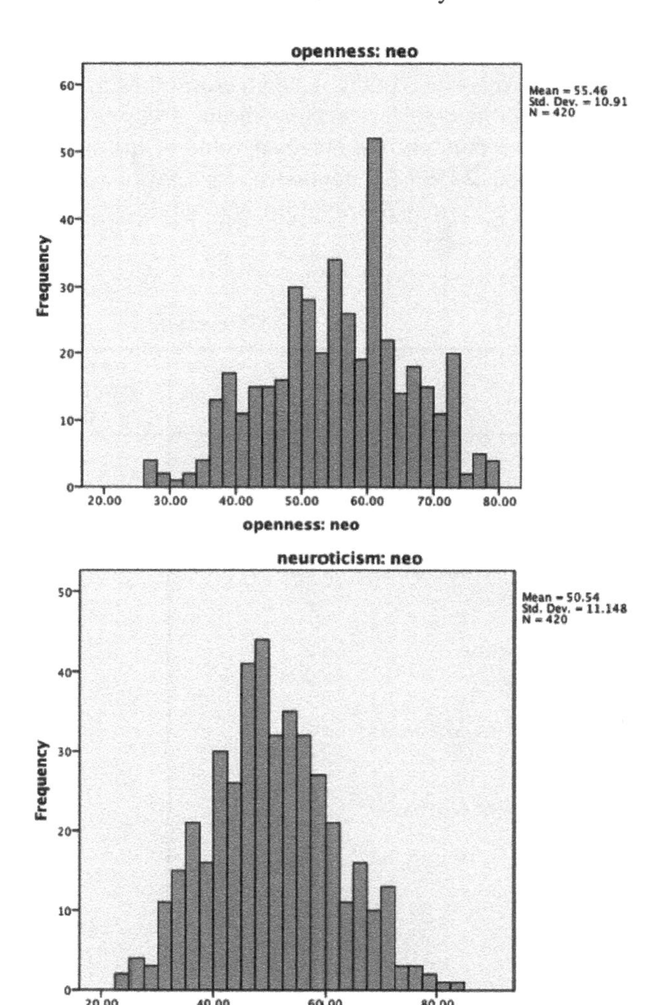

FIGURE 10.7

Histograms for the two quantitative variables.

the variable (the "tail" of the distribution points toward the positive end of the X-axis); negative skewness describes the situation where the bulk of scores is toward the relatively higher values of the variable (the "tail" of the distribution points toward the negative end of the X-axis).

Kurtosis describes the degree to which the distribution is compressed or flattened with respect to the normal curve. The normal curve has a kurtosis value of 0, known as *mesokurtosis*. Positive kurtosis, known as *leptokurtosis*, indicates that, compared to the normal curve, the distribution is more compressed toward the center; negative kurtosis, known as *platykurtosis*, indicates that, compared to the normal curve, the distribution is relatively flattened.

The results shown in Figure 10.6 indicate that skewness and kurtosis are well within ± 1.00 and thus suggest, respectively, that the distributions are (a) relatively symmetric and (b) neither very compressed nor spread out. This can be most clearly seen in the histograms presented in Figure 10.7. The distribution of openness exhibits a bit of leptokurtosis (negative kurtosis) as its value of −**.472** approaches the ± 0.5 threshold of what some researchers treat as suggestive of a bit of compression, but it really does not give the appearance of anything untoward; however, with a standard error of **.238**, its 95% band ($1.96 * 0.238 = 0.467$) places the lower confidence band at almost -1.00 (-0.938), a value that borders on what many would regard as at least mild compression of the distribution. In comparison, neuroticism with its kurtosis value of −**.263** is a bit more balanced.

The first and last parts of the frequency table for **neoopen** are shown in Figure 10.8, as the full table is quite lengthy. Each value in the data file is represented in the table together with the frequency with which that value occurred. For example, the value **30.95** occurred once but the value **37.82** occurred eight times (i.e., eight cases had that value). Totally, there were **420** valid values and **5** missing values.

neoopen openness: neo

		Frequency	Percent	Valid Percent	Cumulative Percent
Valid	27.51	2	.5	.5	.5
	27.89	2	.5	.5	1.0
	29.59	2	.5	.5	1.4
	30.95	1	.2	.2	1.7
	32.66	1	.2	.2	1.9
	33.00	1	.2	.2	2.1
	34.38	3	.7	.7	2.9
	34.70	1	.2	.2	3.1
	36.10	2	.5	.5	3.6
	36.41	3	.7	.7	4.3
	37.82	8	1.9	1.9	6.2
	38.11	4	.9	1.0	7.1
	73.90	3	.7	.7	97.4
	75.62	2	.5	.5	97.9
	77.29	4	.9	1.0	98.8
	77.34	1	.2	.2	99.0
	78.99	3	.7	.7	99.8
	79.05	1	.2	.2	100.0
	Total	420	98.8	100.0	
Missing	System	5	1.2		
Total		425	100.0		

FIGURE 10.8

Frequency tables for the two quantitative variables.

CHAPTER 11

Descriptives

11.1 OVERVIEW

In addition to the **Frequencies** procedure described in Chapter 10, IBM SPSS® also has the **Descriptives** procedure, which is applicable to only quantitative variables. It produces most of the same statistics as the **Frequencies** procedure but neither contains percentiles and nor provides graphic output. Its main virtues are that it allows us to transform raw scores to standardized (z) scores (see Chapter 13) and provides the output in a form that is convenient for researchers to compare variables.

11.2 NUMERICAL EXAMPLE

The data we use for our example are extracted from a study of personality variables on 425 university students. Variables include demographics as well as personality measures. Data are contained in the data file named **Personality**.

11.3 ANALYSIS SETUP

We open the data file named **Personality** and from the main menu select **Analyze ➔ Descriptive Statistics ➔ Descriptives**. We will generate descriptive statistics on the same variables used in Chapter 10 to allow readers to compare **Descriptives** with **Frequencies**. Thus, we move **neoopen** and **neoneuro** into the **Variable(s)** panel as shown in Figure 11.1. We have not checked **Save standardized values of variables** below the **Variable List** panel because we will do this as a part of Chapter 13.

Selecting the **Options** pushbutton opens the **Options** screen shown in Figure 11.2. We select the following statistics to illustrate what is available in this procedure:

- In the first (unlabeled) row in the window, we select **Mean**.
- Under **Dispersion**, we select **Std. deviation**, **Minimum** (the lowest valid value in the distribution), **Maximum** (the highest valid value in the distribution), and **S.E. mean** (the standard error of the mean).

Performing Data Analysis Using IBM SPSS®, First Edition.
Lawrence S. Meyers, Glenn C. Gamst, and A. J. Guarino.

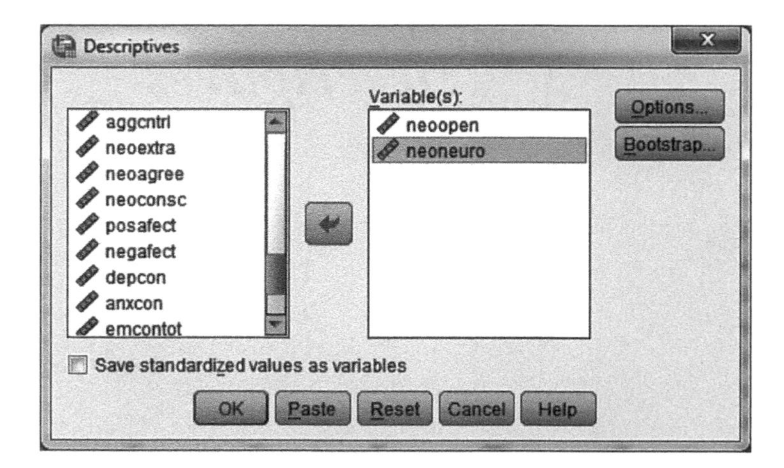

FIGURE 11.1
The main dialog window of **Descriptives**.

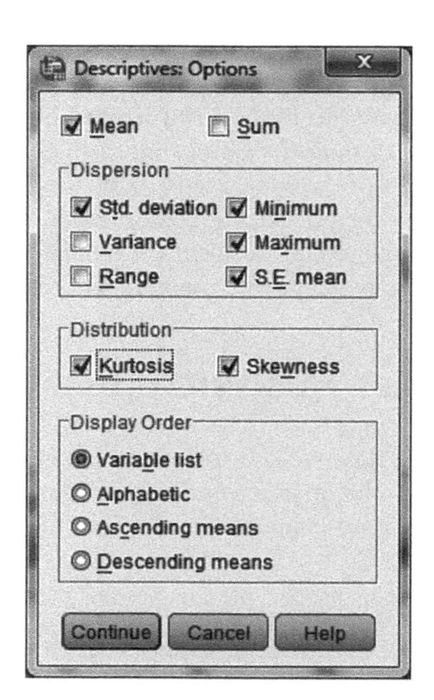

FIGURE 11.2
The **Options** screen of **Descriptives**.

Descriptive Statistics

	N	Minimum	Maximum	Mean		Std. Deviation	Skewness		Kurtosis	
	Statistic	Statistic	Statistic	Statistic	Std. Error	Statistic	Statistic	Std. Error	Statistic	Std. Error
neoopen openness: neo	420	27.51	79.05	55.4607	.53233	10.90952	-.157	.119	-.472	.238
neoneuro neuroticism: neo	420	23.01	84.77	50.5394	.54396	11.14793	.184	.119	-.263	.238
Valid N (listwise)	420									

FIGURE 11.3 The descriptive statistics output.

- Under **Distribution**, we select **Skewness** and **Kurtosis**.
- Under **Display Order**, we select **Variable list** (the variables could also be ordered alphabetically or by the value of their means).

We select **Continue** to return to the main dialog window and click **OK** to perform the analysis.

11.4 ANALYSIS OUTPUT

The results of the analysis are presented in Figure 11.3, which duplicate the values we obtained from the **Frequencies** procedure (see Figure 10.6).

CHAPTER 12

Explore

12.1 OVERVIEW

Explore is still another procedure in addition to the **Frequencies** and **Descriptives** procedures that is designed to generate descriptive statistics. **Explore** is also applicable to only quantitative variables but is able to produce more types of output than either of the other two procedures (except for generating standardized scores).

12.2 NUMERICAL EXAMPLE

The data we use for our example are extracted from a study of personality variables on 425 university students. Variables include demographics as well as personality measures. Data are present in the data file named **Personality**.

12.3 ANALYSIS SETUP

We open the data file named **Personality** and from the main menu select **Analyze Descriptive Statistics Explore**. We will generate descriptive statistics on the same variables used in Chapters 10 and 11 to allow readers to compare **Explore** with **Descriptives** and **Frequencies**. Thus, we move **neoopen** and **neoneuro** into the **Dependent List** panel of the main dialog window as shown in Figure 12.1.

We leave **Factor List** empty for our illustration, but this is a way to obtain separate descriptive statistics for each category of a nominal variable. For example, if we placed **sex** in that panel with **neoopen** and **neoneuro** in the **Dependent List** panel, we would obtain separate analyses for males and females on those two variables; this would yield the same result as executing a **Split File** by sex and running the **Explore** procedure on **neoopen** and **neoneuro**.

Selecting the **Statistics** pushbutton opens the **Statistics** screen shown in Figure 12.2. We select **Descriptives** (this displays a set of default descriptive statistics) and retain 95 in the panel for **Confidence Interval for Mean** (this calculates the lower and upper 95% confidence boundaries of the mean). We also select **Percentiles** to obtain a variety of

Performing Data Analysis Using IBM SPSS®, First Edition.
Lawrence S. Meyers, Glenn C. Gamst, and A. J. Guarino.

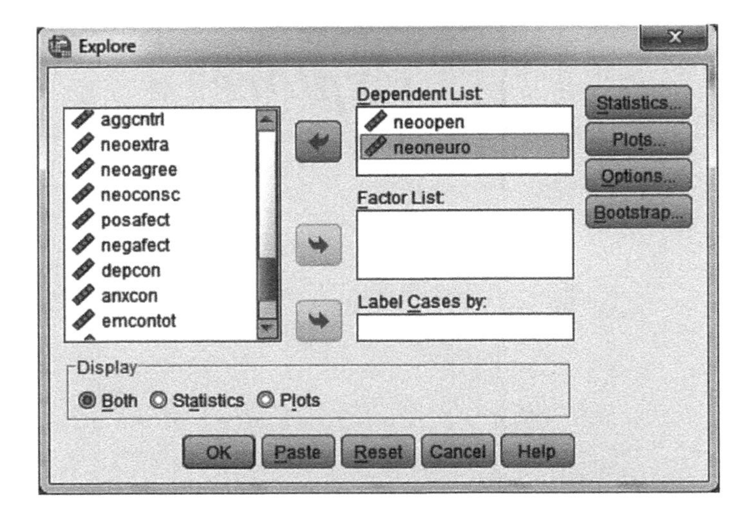

FIGURE 12.1
The main dialog window of **Explore**.

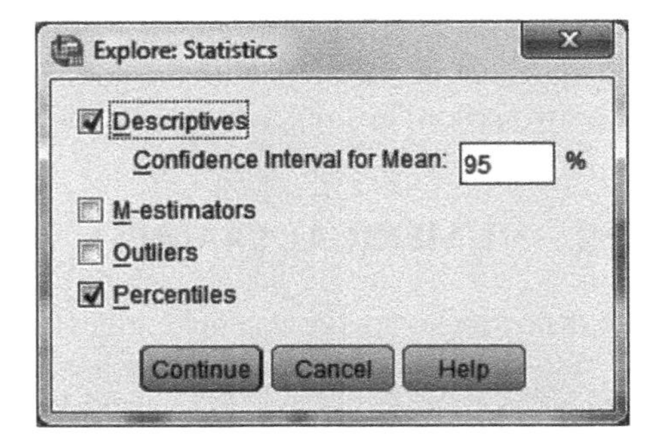

FIGURE 12.2
The **Statistics** screen of **Explore**.

percentile values in addition to the quartiles. We select neither **M-estimators** (these are a set of alternative ways to weigh cases that we do not address in this book) and nor **Outliers** (we address this in Chapter 18).

We select **Continue** to return to the main dialog window and select the **Plots** pushbutton to reach the screen shown in Figure 12.3. In the **Boxplots** area, we select **Factor levels together**. As we have not specified any factors in the main dialog window, this is a moot point; if we had a factor, then we could group the boxplots by either factors or the dependent variables. We also select **Histogram** in the **Descriptive** area so that we can compare this plot to that produced by the **Frequencies** procedure.

We select **Continue** to return to the main dialog window and select the **Options** pushbutton to reach the screen shown in Figure 12.4. We choose **Exclude cases pairwise**. This choice will allow the computation of the descriptive statistics for each variable to be based on different numbers of cases, that is, the number of valid values might differ from variable to variable. In multivariate analyses (where multiple variables are analyzed simultaneously), all cases must have valid values on all of the variables included in the analysis and thus listwise selection is in effect in those analyses (any case with a missing value on any of the variables in the analysis is removed from the analysis). We can use the same set of cases as would be in the multivariate analysis by selecting **Exclude cases listwise**. We select **Continue** to return to the main dialog window and click **OK** to perform the analysis.

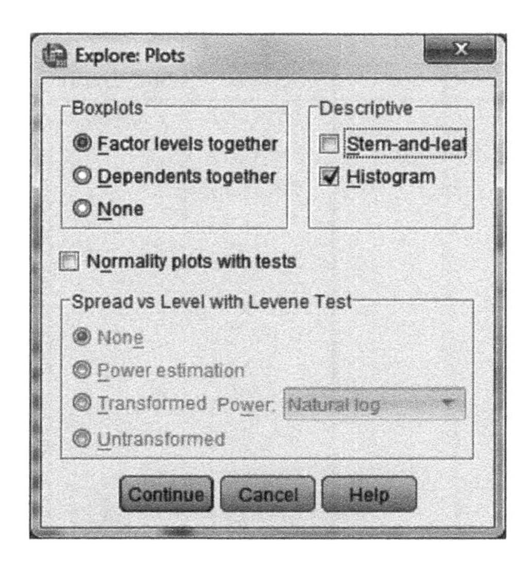

FIGURE 12.3
The **Plots** screen of **Explore**.

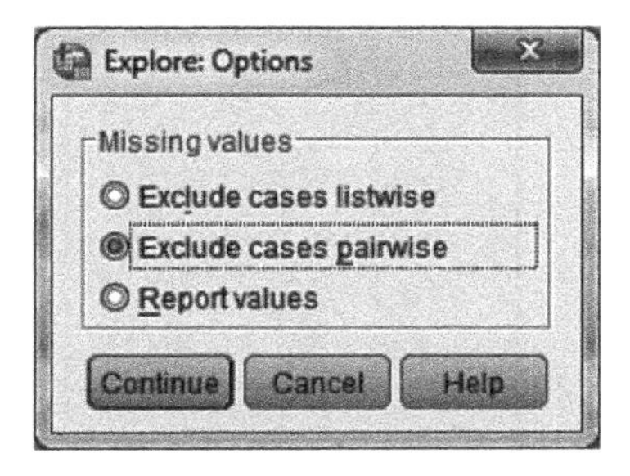

FIGURE 12.4
The **Options** screen of **Explore**.

12.4 ANALYSIS OUTPUT

The descriptive statistics are shown in Figure 12.5. The statistics duplicate the values we obtained from the **Frequencies** and **Descriptives** procedures, but we obtain additional statistics, which include the following.

- *95% Confidence Interval for the Mean Lower and Upper Bounds.* These values are computed by multiplying the standard error of the mean (not part of the output) by 1.96 and subtracting and adding those respective values to obtain the upper and lower boundaries, respectively. These values match (within rounding error) those we calculated in Chapter 10.
- *5% Trimmed Mean.* This is the mean of the variable when the top and bottom 5% of the scores have been excluded from the computation. The trimmed mean removes relatively extreme scores from both ends of the distribution. It is a very rough way to determine if there are outliers at one of the ends of the distribution based on the following reasoning:
  - If the relatively extreme scores were symmetrically present in both the lower and upper ends of the distribution, the trimmed mean should be very similar in value to the full mean.
  - If the relatively extreme scores were disproportional in the lower or higher end of the distribution, this trimmed mean should differ sufficiently from the full mean to be noticeable.
- *Interquartile Range (IQR).* The absolute difference between the scores corresponding to the first and third quartiles (the 25th and 75th percentiles).

Descriptives

			Statistic	Std. Error
neoopen openness: neo	Mean		55.4607	.53233
	95% Confidence Interval for Mean	Lower Bound	54.4143	
		Upper Bound	56.5071	
	5% Trimmed Mean		55.5686	
	Median		55.1448	
	Variance		119.018	
	Std. Deviation		10.90952	
	Minimum		27.51	
	Maximum		79.05	
	Range		51.55	
	Interquartile Range		15.46	
	Skewness		-.157	.119
	Kurtosis		-.472	.238
neoneuro neuroticism: neo	Mean		50.5394	.54396
	95% Confidence Interval for Mean	Lower Bound	49.4701	
		Upper Bound	51.6086	
	5% Trimmed Mean		50.4154	
	Median		49.2904	
	Variance		124.276	
	Std. Deviation		11.14793	
	Minimum		23.01	
	Maximum		84.77	
	Range		61.76	
	Interquartile Range		15.77	
	Skewness		.184	.119
	Kurtosis		-.263	.238

FIGURE 12.5 The descriptive statistics output.

Percentiles

		Percentiles						
		5	10	25	50	75	90	95
Weighted Average (Definition 1)	neoopen openness: neo	37.8179	39.8126	48.1271	55.1448	63.5911	70.2949	72.1821
	neoneuro neuroticism: neo	33.1099	37.1314	42.7201	49.2904	58.4888	66.3732	70.3154
Tukey's Hinges	neoopen openness: neo			48.1271	55.1448	63.5911		
	neoneuro neuroticism: neo			42.7201	49.2904	58.4888		

FIGURE 12.6 Percentile output.

The **Percentiles** output is shown in Figure 12.6. The first main row gives results for the 5th, 10th, 25th, 50th, 75th, 90th, and 95th percentiles for each variable. The second major row identifies the quartiles. These are called **Tukey's Hinges** after the eminent statistician John Tukey who introduced boxplots (Tukey, 1977); the so-called hinges in Figure 12.6 are the first and third quartiles that are shown in the boxplot (Meyers et al., 2013).

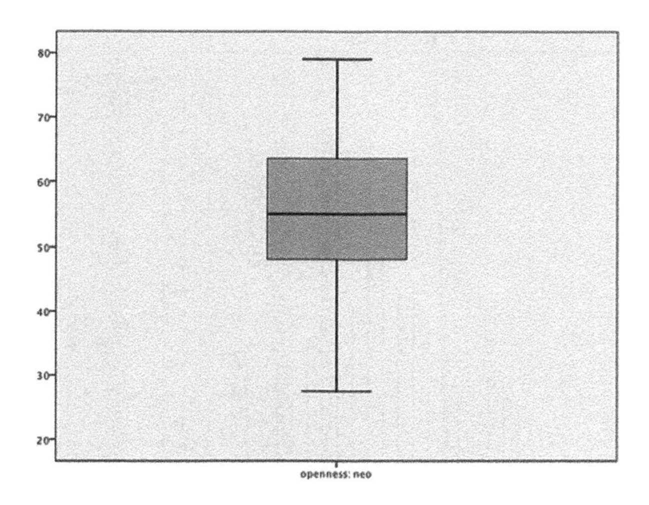

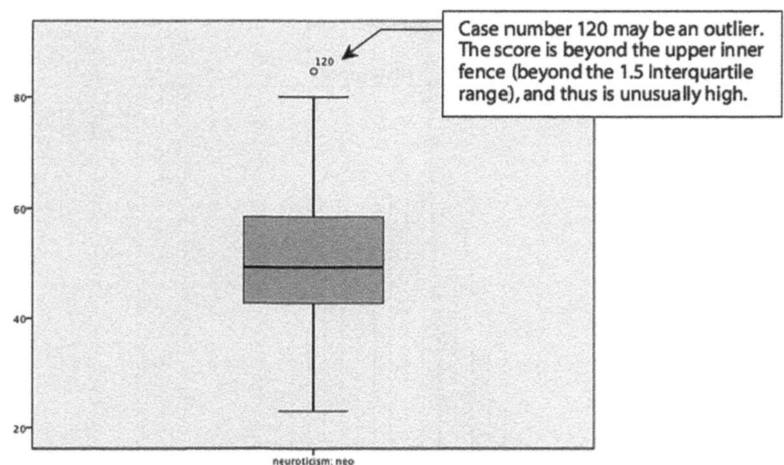

FIGURE 12.7
Boxplots for the two variables.

Tukey's hinges may be seen in the boxplots presented in Figure 12.7. The vertical axis in the IBM SPSS® plot depicts percentiles. Because percentiles indicate the proportion of scores that are above the given score, higher scores and thus higher percentiles are toward the top (a very high score would be in the 99th percentile and at the top of the plot).

The rectangle in the approximate center of each boxplot represents the middle portion of the distribution. The median (50th percentile) is marked by the heavy horizontal line and the lower and upper boundaries of the rectangle are approximately the first and third quartiles (25th and 75th percentiles, respectively, although Tukey conceived of them as the halfway points between the median and each end of the distribution); these three percentiles are computed by IBM SPSS as Tukey's hinges.

The vertical lines extending from the rectangle or "box" are the whiskers (hence the name *box and whiskers plot* sometimes given to the display), and their ends are the lower and upper inner fences, respectively. These fences are drawn at a distance of ± 1.5 IQR, which is roughly about ± 1.6 standard deviation units from the mean. Data points exceeding ± 1.5 IQR but less than ± 3.0 IQR are designated by circles in the IBM SPSS output; data points exceeding ± 3.0 IQR are designated by the letter E. In Figure 12.7, the neuroticism data point for Case 120 lies beyond the upper inner fence and is drawn as a circle; we thus know that Case 120 scored quite high on this measure and may qualify as an outlier.

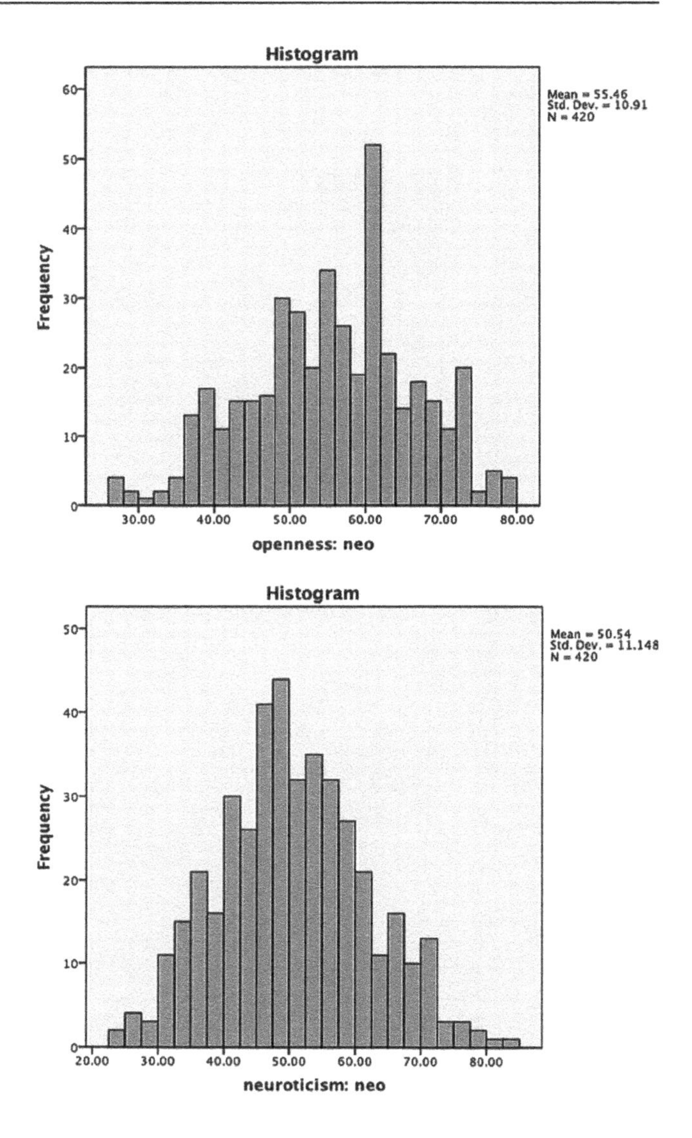

FIGURE 12.8

Histograms for the two variables.

The histograms are shown in Figure 12.8. These are identical to those produced by the **Frequencies** procedure presented in Chapter 10.

PART 5

SIMPLE DATA TRANSFORMATIONS

CHAPTER 13

Standardizing Variables to z Scores

13.1 OVERVIEW

In Chapter 11, we indicated that the **Descriptives** procedure can be used to transform the values of a variable to z scores. A z score is the basic standardized score and is a direct index of how many standard deviation units a given raw score is from the mean of the distribution (the distance from the mean in standard deviation units). It can be hand-calculated by subtracting the mean from the raw score and dividing that difference by the standard deviation: $z = (X - M)/\text{SD}$. Positive z scores indicate that the score is greater than the mean; negative z scores indicate that the score is less than the mean. Thus, a z score of 1.25 informs us that the raw score is greater than the mean by one and a quarter standard deviations, and a z score of -0.75 informs us that the raw score is less than the mean by three-quarters of a standard deviation.

We can compare these two cases in the sample in terms of how they performed on the variable: assuming higher values represent more of the construct being measured, we could say that the case whose z score is 1.25 appears to rate much higher on the construct than the case whose z score is -0.75. We can also use z scores to compare performance across measures within a single sample. For example, we can convert two variables to their z score equivalents to determine immediately how consistent a given case fared across the two variables.

13.2 NUMERICAL EXAMPLE

The data we use for our example are extracted from a study of personality variables on 425 university students. Data are present in the data file named **Personality**.

13.3 ANALYSIS SETUP

We open the data file named **Personality** and from the main menu select **Analyze → Descriptive Statistics → Descriptives**. To illustrate how to obtain z scores, we move **neoneuro** into the **Variable(s)** panel as shown in Figure 13.1. We have checked **Save standardized values of variables** below the **Variable List** panel but do not specifically

Performing Data Analysis Using IBM SPSS®, First Edition.
Lawrence S. Meyers, Glenn C. Gamst, and A. J. Guarino.

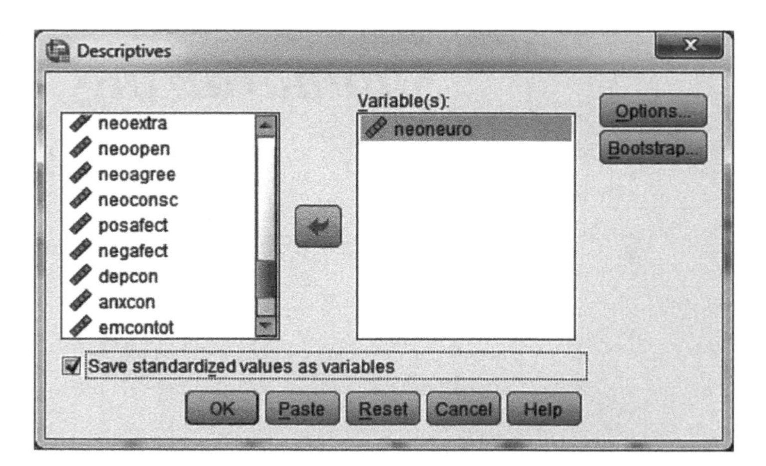

FIGURE 13.1
The main dialog window of **Descriptives**.

Descriptive Statistics

	N	Minimum	Maximum	Mean	Std. Deviation
neoneuro neuroticism: neo	420	23.01	84.77	50.5394	11.14793
Valid N (listwise)	420				

FIGURE 13.2 The output from the standardization operation.

ask for any descriptive statistics (the default statistics will be obtained). We then click **OK** to perform the analysis.

13.4 ANALYSIS OUTPUT

There are two results of performing the transformation. One set of results is shown in Figure 13.2. It is an output table providing some very limited descriptive information. The **Minimum**, **Maximum**, **Mean**, and **Std. Deviation** are summaries of the variable **neoneuro**; this is the default set of statistics preset in the **Options** dialog screen. On the basis of this mean and standard deviation, the z scores were calculated; that is, the scores are standardized with respect to their standing in the sample (they are not standardized with respect to any outside normative population).

The other result of performing the transformation is the creation of a new variable in the data file as shown in Figure 13.3. IBM SPSS® has generated the name of the new variable by placing an uppercase **Z** in front of the name of the original variable—it may not be elaborate but it is unambiguous. Hence, the new variable is named **Zneoneuro**, but we can easily change it in the **Variable View** screen if we wish. Thus, Case 1 has a standard score on neuroticism of **1.08192**; Case 2, **−1.80396**; and so on. This variable is now available for further analyses and saving the data file will preserve it.

13.5 DESCRIPTIVE STATISTICS ON ZNEONEURO

We illustrate the effect of a z score transformation by obtaining through the **Descriptives** procedure a few descriptive statistics on both the original **neoneuro** and the standardized **Zneoneuro** variables. The results are shown in Figure 13.4. We can see certain of the score equivalents from these results:

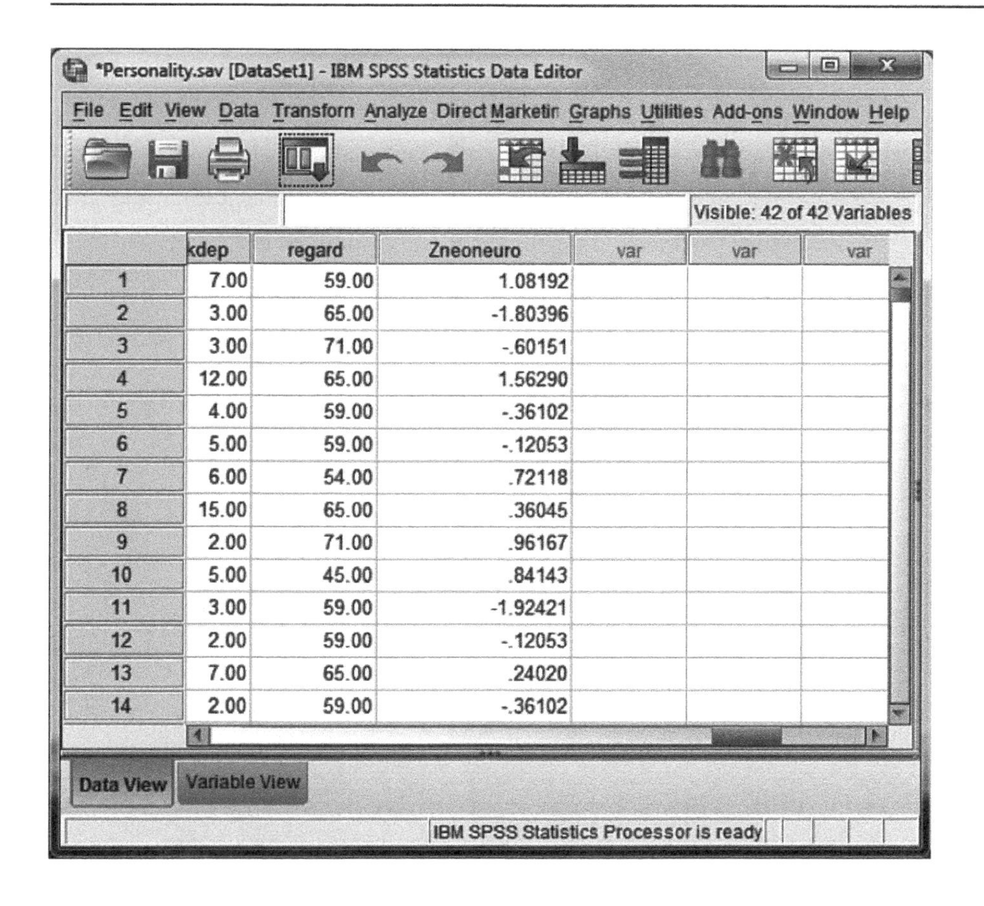

FIGURE 13.3
A portion of the data file showing the newly created standardized variable.

Descriptive Statistics

	N	Minimum	Maximum	Mean	Std. Deviation	Skewness		Kurtosis	
	Statistic	Statistic	Statistic	Statistic	Statistic	Statistic	Std. Error	Statistic	Std. Error
neoneuro neuroticism: neo	420	23.01	84.77	50.5394	11.14793	.184	.119	-.263	.238
Zneoneuro Zscore: neuroticism: neo	420	-2.46953	3.07059	.0000000	1.00000000	.184	.119	-.263	.238
Valid N (listwise)	420								

FIGURE 13.4 Descriptive statistics on both **neoneuro** and **Zneoneuro**.

- The raw score mean of **50.5394** has a z score of zero.
- A raw score of **23.01** has a z score of **−2.46953**.
- A raw score of **84.77** has a z score of **3.07059**.

One other feature worthy of mention is that a z score transformation does not alter the shape of the distribution. Thus, the skewness and kurtosis of the original variable are preserved in its z score transformation.

13.6 OTHER STANDARD SCORES

Although the **Descriptives** procedure can be used to perform the z score transformation, other standardized score systems are also commonly used. Examples of common standardized scales are the linear T scores (with a standardized mean of 50 and a standardized standard deviation of 10) and SAT scores (with a standardized mean of 500 and a standardized standard deviation of 100). To transform our z scores to one of these other standardized systems, we require a relatively simple computation that can be performed in IBM SPSS, the details of which are shown in Chapter 16.

CHAPTER 14

Recoding Variables

14.1 OVERVIEW

To recode a variable is to modify the values of a variable according to a rule specified by the user as part of the recoding operation. Both nominal and quantitative variables can be subjected to recoding to achieve various purposes.

Categorical variables are commonly recoded. In the process of data collection, we often ask for very specific information, such as the ethnicity of the individual or the region of the state or country in which a respondent resides. Because the responses are so specific, the number of cases in certain categories may be too low to analyze as a stand-alone group. We may also find that some categories may be, if not exactly the same, sufficiently similar to raise the possibility that they should not be treated as different categories. We often deal with these issues by combining cases from two or more categories into a larger group; that is, we recode this variable to create more expansive subsets of cases.

It is also not uncommon to recode quantitative variables that are measured on a summative response scale (e.g., a 5-point or 7-point scale). In certain analyses, it may be important that each of the response categories is represented by some minimum number of choices. Under those circumstances, response categories with very low endorsements (usually the lowest or highest categories) may be recoded into an adjacent category. This reduces the total number of categories but allows each remaining one to be associated with an adequate sample size to support some analyses.

Quantitative variables approximating continuous measurement may occasionally be recoded to eliminate outliers. In those instances, researchers will determine a minimum or maximum value of the variable that is permitted and recode values beyond that boundary to the minimum or maximum value allowed. For example, a response of 107 to a question concerning the number of doctor visits for different medical problems over the last 12 months might be considered out of range in a given data set. Researchers could either define this value as missing (the most likely action they would take) or, if they had reason to believe that the patient was responding in good faith by indicating that many visits, recode this value to some arbitrary maximum that they had validly established.

Performing Data Analysis Using IBM SPSS®, First Edition.
Lawrence S. Meyers, Glenn C. Gamst, and A. J. Guarino.

IBM SPSS® allows us to recode a variable into the same variable or recode a variable into a different variable. Here is what this means together with our recommendations:

- Recoding into the same variable instructs IBM SPSS to overwrite the original variable with the recoded values. Our recommendation to most users, but especially to those who are not an expert in using IBM SPSS, is to *never use this choice*. We most strongly recommend *never* overwriting an original variable; rather we urge users to create another variable containing the recoded values, thus preserving the original values in case they are ever needed for some future purpose.
- Recoding into a different variable instructs IBM SPSS to create a new variable representing the recoded values under a name designated by the user. This choice preserves the original variable and is our recommended strategy.

14.2 NUMERICAL EXAMPLE

The fictional data we use for our example are in the data file **Doctor Visits**. The data set represents a study of the number of medical visits patients have made to their family physician during the previous 12 months for different medical issues. The two variables of interest are as follows:

- The number of visits reported by patients is recorded under the variable **doctor_visits**. Because they were patients, the minimum number of visits is necessarily 1, as patients were selected from the records once they had at least one visit.
- Patients from a wide range of states across the United States were sampled. Each state was assigned an arbitrary code, and the value labels supplied in the **Variable View** show these data. This variable is named **state**.

14.3 ANALYSIS STRATEGY

In this fictional study, the goal of the researchers was to determine if there were differences in the number of visits by different places of residence. The most direct way to address this is with a one-way between-subjects analysis of variance (ANOVA), which we cover in Chapter 47. However, because the number of patients from each separate state is too low to support the ANOVA, we need to combine patients from various states together to form larger groups. For our present purposes, we adopt the following sequential strategy:

1. We obtain a summary of the number of patients in each state using the **Frequencies** procedure.
2. We recode the variable **state** into a different variable named **region** representing geographic regions. We use the specification for ranges of values in this analysis and show recoding for specific values in Section 14.7.
3. We obtain a count of the number of patients in each newly created region using **Frequencies**.
4. To generate descriptive statistics for each region, we could perform the analysis in the **Explore** procedure placing **doctor_visits** in the **Dependent List** and **region** in the **Factor List**. But because (a) the statistics produced by **Explore** are arrayed vertically and (b) the output for the regions would be listed vertically as well, the results would take up several pages and comparing the regions would be visually somewhat inconvenient. We therefore opt to split the file by **region**.

5. We obtain descriptive statistics on **doctor_visits** for each region using **Descriptives**. With **Split File** in effect, the output will be arrayed in a more condensed manner, with the output for each **region** occupying a row.

14.4 FREQUENCIES ANALYSIS

We open the data file named **Doctor Visits** and from the main menu select **Analyze ➔ Descriptive Statistics ➔ Frequencies**. Based on our discussion in Chapter 10, we have moved **state** into the **Variable(s)** panel, have checked **Display frequency table** below the **Variable List** panel, and we click **OK** to perform the analysis.

Figure 14.1 presents the frequency table for the **state** variable. Most states have patient counts in the range 5–10. The states have been coded somewhat systematically already, helping us to visualize the regions into which we can code them. The regions do not need to have equal numbers of cases as long as the count of cases is sufficient to support the ANOVA (we will not perform that analysis here). The regions that seem to capture the geographic grouping of the states we have represented in the data file are New England, Southeast, Southwest, Midwest, West, and Far West.

state

		Frequency	Percent	Valid Percent	Cumulative Percent
Valid	1 Maine	7	3.9	3.9	3.9
	2 Vermont	6	3.3	3.3	7.2
	3 New Hamshire	7	3.9	3.9	11.1
	4 Rhode Island	7	3.9	3.9	15.0
	5 Virginia	6	3.3	3.3	18.3
	6 North Carolina	7	3.9	3.9	22.2
	7 Georgia	8	4.4	4.4	26.7
	8 Florida	9	5.0	5.0	31.7
	9 Oklahoma	8	4.4	4.4	36.1
	10 Texas	9	5.0	5.0	41.1
	11 New Mexico	9	5.0	5.0	46.1
	12 Ohio	9	5.0	5.0	51.1
	13 Wisconsin	5	2.8	2.8	53.9
	14 Minnesota	9	5.0	5.0	58.9
	15 South Dakota	6	3.3	3.3	62.2
	16 Montana	7	3.9	3.9	66.1
	17 Wyoming	7	3.9	3.9	70.0
	18 Colorado	10	5.6	5.6	75.6
	19 Utah	5	2.8	2.8	78.3
	20 Idaho	5	2.8	2.8	81.1
	21 Washington	8	4.4	4.4	85.6
	22 Oregon	9	5.0	5.0	90.6
	23 California	7	3.9	3.9	94.4
	24 Nevada	10	5.6	5.6	100.0
	Total	180	100.0	100.0	

FIGURE 14.1 Frequency table for **state**.

14.5 RECODING AN ORIGINAL VARIABLE USING RANGES

From the main menu, selecting **Transform → Recode into Different Variables** brings us to the main **Recode** dialog window as shown in Figure 14.2. We have moved **state** into the **Numeric Variable → Output Variable** panel. The variable **state** is followed by a "?" that needs to be replaced by the name of the newly recoded variable. We name the recoded variable **region** in the **name** panel under **Output Variable**, and when we click the **Change** pushbutton our name will replace the "?" this is shown in Figure 14.3.

Selecting the **Old and New Values** pushbutton below the **Numeric Variable → Output Variable** panel brings us to the window where we specify our recoding (see Figure 14.4). To recode Maine, Vermont, New Hampshire, and Rhode Island (codes **1** through **4**), we select **Range LOWEST through value** because our lowest code is **1**.

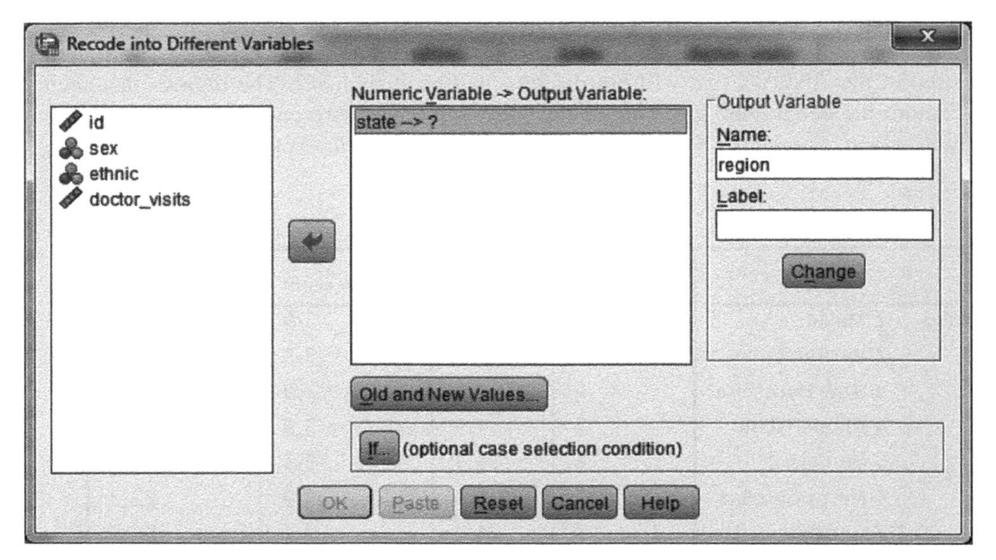

FIGURE 14.2 The main dialog window for **Recode into Different Variables**.

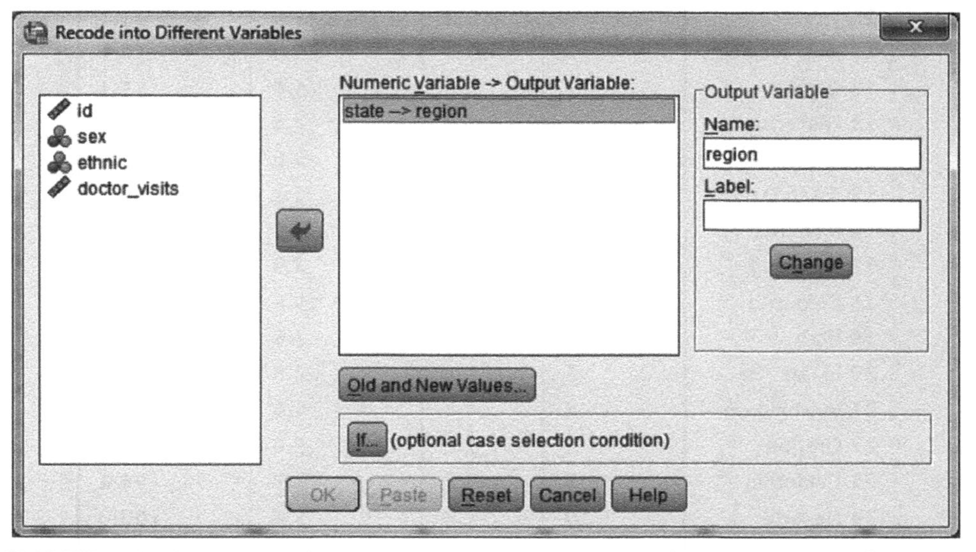

FIGURE 14.3 The name of the recoded variable is now in the **Numeric Variable → Output Variable** panel, indicating that **state** will be recoded into **region**.

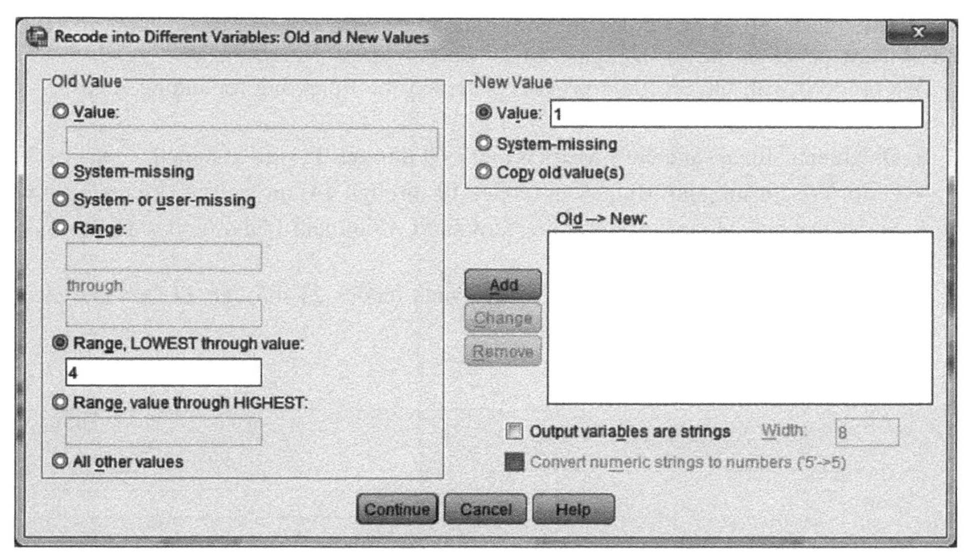

FIGURE 14.4 The first recoding is specified: the lowest code (code number 1) through code 4 for **state** will be recoded into code number 1 for **region**.

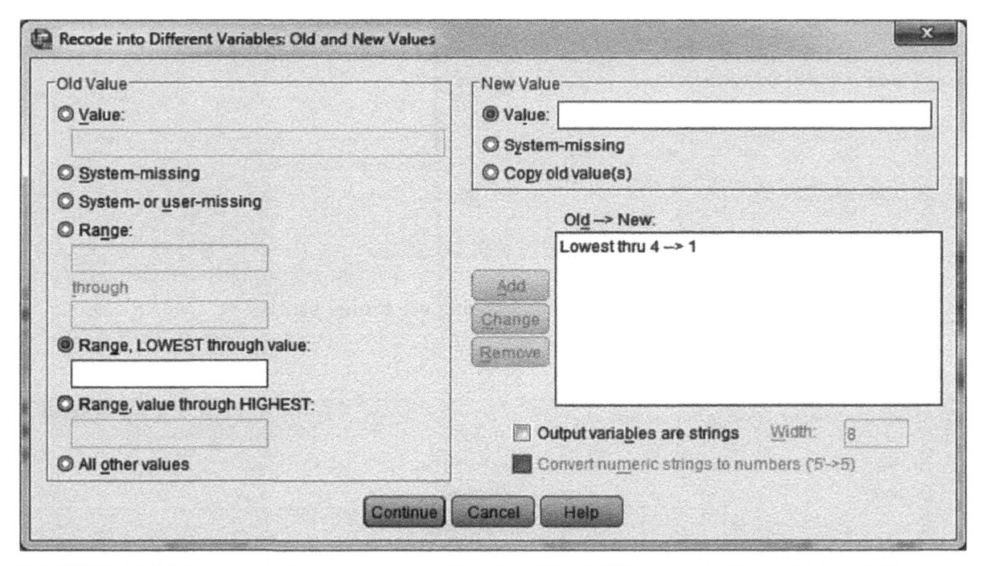

FIGURE 14.5 The recode is now registered in the **Old ➔ New** panel after clicking **Add**.

IBM SPSS will start with **1** and group the codes together into a new code. We type the number **4** in the panel as shown in Figure 14.4. In the **New Value** panel in the upper right portion of the window, we select **Value** and type the number **1**. Thus, the codes of **1**, **2**, **3**, and **4** of **state** will become the code **1** for **region**. Clicking **Add** brings this recode into the **Old ➔ New** panel as shown in Figure 14.5. There is no opportunity to add a label here, so we must wait until the recode is completed and then add the labels in the **Variable View**.

Our next set of states to combine is Virginia, North Carolina, Georgia, and Florida (codes **5** through **8**). We select **Range**, which has two panels, the upper panel for the lowest code of our set and the lower panel for the highest code of our set. We place the values of **5** and **8** in the respective panels, select **Value** and type the number **2** in the

New Value panel (all shown in Figure 14.6), and click **Add** to bring this recode into the **Old ➔ New** panel as shown in Figure 14.7.

We proceed with the recoding process grouping the following remaining states:

- Oklahoma, Texas, and New Mexico (codes **9** through **11**) are assigned a code of **3**.
- Ohio, Wisconsin, and Minnesota (codes **12** through **14**) are assigned a code of **4**.
- South Dakota, Montana, Wyoming, Colorado, Utah, and Idaho (codes **15** through **20**) are assigned a code of **5**.
- Washington, Oregon, California, and Nevada (codes **21** through **24**) are assigned a code of **6**.

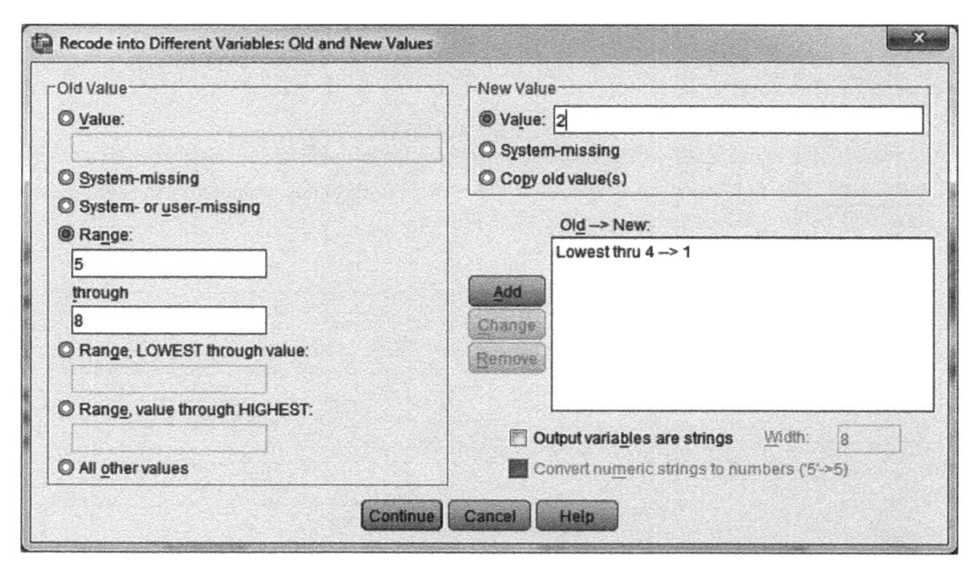

FIGURE 14.6 Recoding the second set of states using the **Range** panels.

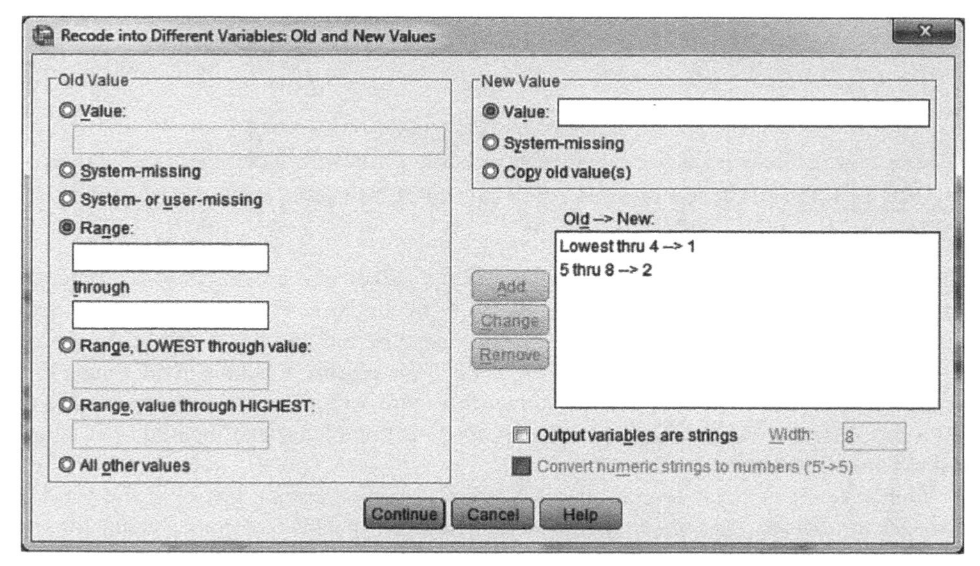

FIGURE 14.7 Recoding for the second set of states has now been specified.

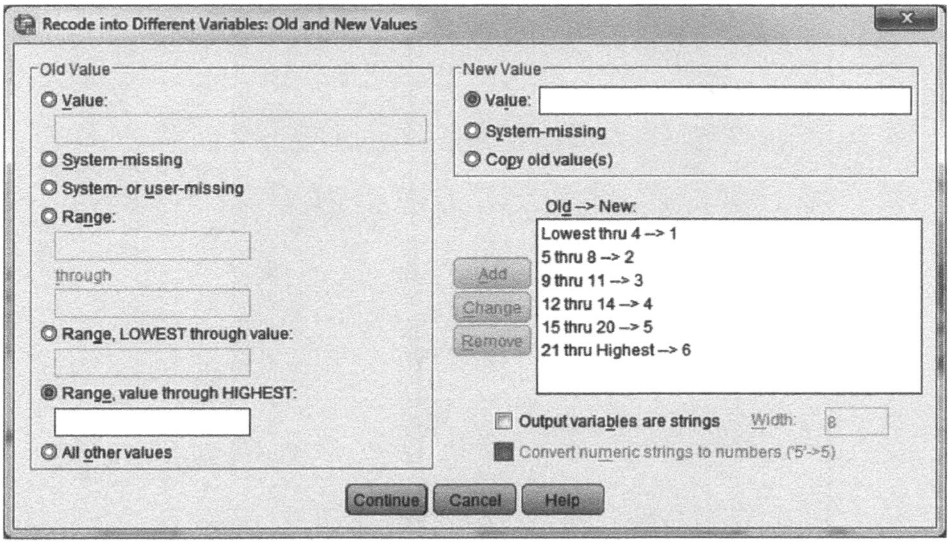

FIGURE 14.8 Recoding for all of the states has now been specified.

For this last recode, we can use **Range value through HIGHEST** by entering **21** in the panel. IBM SPSS will then incorporate into that recode the highest value code for the variable **state**. The resulting full set of recodes is shown in Figure 14.8. We click **Continue** to return to the main dialog window and click **OK** to perform the recode.

14.6 THE RESULTS OF THE RECODING

The case-by-case results of the recoding can be seen in the screenshot of the data file shown in Figure 14.9. A new variable named **region** has been added to the end of the data file. It shows as two decimal places because of the default in IBM SPSS. We switch to **Variable View**, set the decimal values of **region** to zero by clicking in the cell and toggling down (see Chapter 2), type in the labels of our regions (see Figure 14.10), and click **OK** to register those changes. We then save the data file.

We have performed a **Frequencies** analysis on **region**, having selected **Display frequency tables**, and the results are shown in Figure 14.11. The regions are relatively evenly represented in our sample, with the Midwest providing the lowest number of doctor visits and the West providing the highest.

To get a sense of any regional difference in visiting their physicians, we have performed a **Split File by region** as described in Chapter 8. We have then performed a **Descriptives** analysis on **doctor_visits** (placing that variable in the **Variable(s)** panel and asking for a range of descriptive statistics). The results of that analysis are shown in Figure 14.12. Generally, it appears from visual inspection that (in this fictional data set) patients in the Southeast visit their physicians almost twice as frequently as those in the Southwest and that, generally, patients in the Southeast, Northeast, and West tend to visit their physicians more frequently than those in the Southwest, Far West, and Midwest.

14.7 RECODING AN ORIGINAL VARIABLE USING INDIVIDUAL VALUES

To illustrate recoding using individual values, we will recode the variable named **ethnic** containing our code for ethnic categories. Obtaining a frequency table in the **Frequencies**

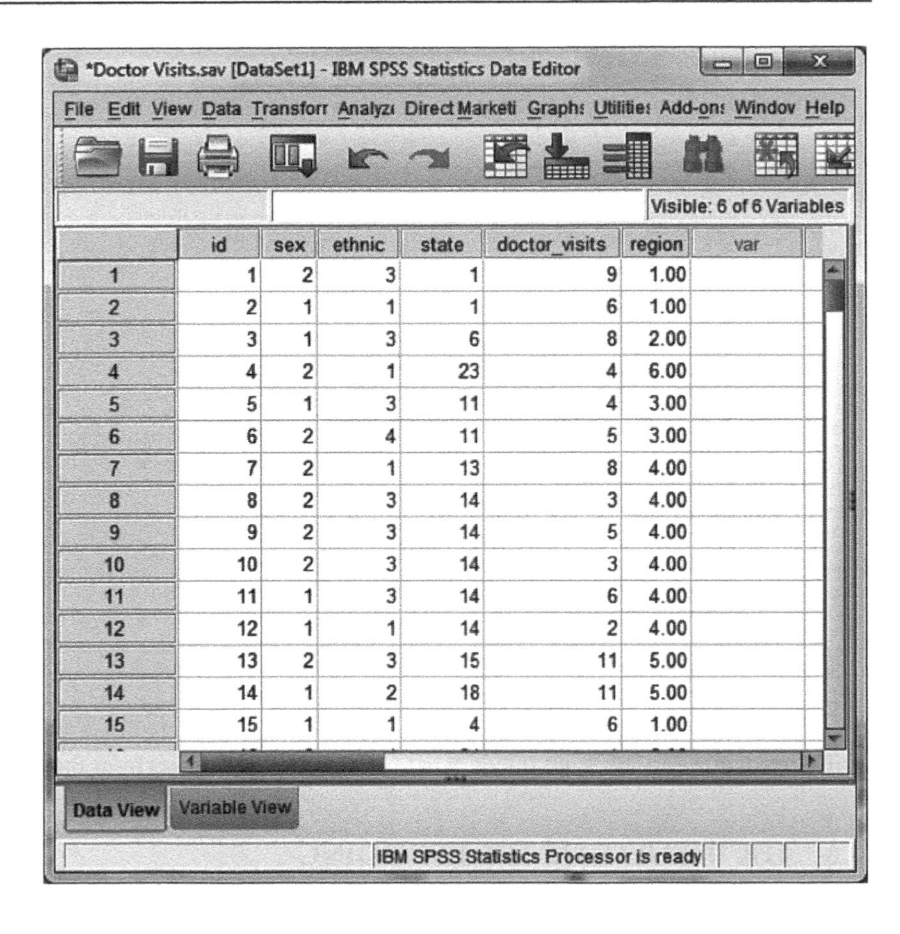

	id	sex	ethnic	state	doctor_visits	region	var
1	1	2	3	1	9	1.00	
2	2	1	1	1	6	1.00	
3	3	1	3	6	8	2.00	
4	4	2	1	23	4	6.00	
5	5	1	3	11	4	3.00	
6	6	2	4	11	5	3.00	
7	7	2	1	13	8	4.00	
8	8	2	3	14	3	4.00	
9	9	2	3	14	5	4.00	
10	10	2	3	14	3	4.00	
11	11	1	3	14	6	4.00	
12	12	1	1	14	2	4.00	
13	13	2	3	15	11	5.00	
14	14	1	2	18	11	5.00	
15	15	1	1	4	6	1.00	

FIGURE 14.9

The data file with the newly recoded variable added.

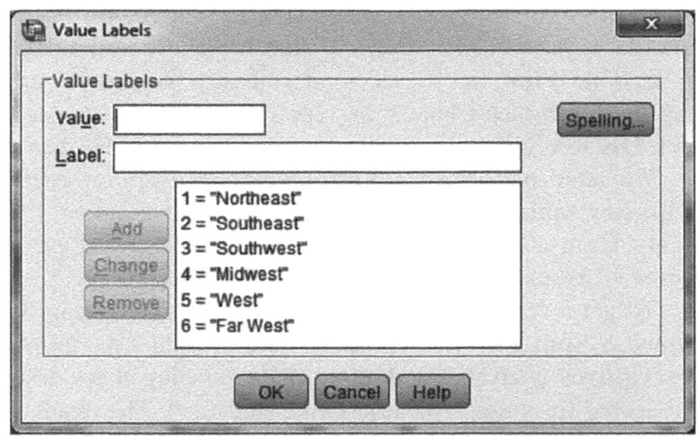

FIGURE 14.10

The value labels for the newly created **region** variable.

procedure yields the results shown in Figure 14.13. The group codes with their labels are displayed in the first column. Assume that the researchers decide to collapse these categories in the following manner to facilitate performing additional analyses on this variable: **Asian American** and **Pacific Islander** will be combined as will **Mexican American** and **Latino/a American**. Such recoding will change the numerical codes for all categories.

To recode the ethnic variable, we select **Transform ➔ Recode into Different Variables**. This brings us to the main recoding dialog window as shown in Figure 14.14. We have moved **ethnic** into the **Numeric Variable ➔ Output Variable** panel, named the

region

		Frequency	Percent	Valid Percent	Cumulative Percent
Valid	1 Northeast	27	15.0	15.0	15.0
	2 Southeast	30	16.7	16.7	31.7
	3 Southwest	26	14.4	14.4	46.1
	4 Midwest	23	12.8	12.8	58.9
	5 West	40	22.2	22.2	81.1
	6 Far West	34	18.9	18.9	100.0
	Total	180	100.0	100.0	

FIGURE 14.11 Frequency counts for the categories of **region**.

Descriptive Statistics

		N	Minimum	Maximum	Mean		Std. Deviation	Skewness		Kurtosis	
region		Statistic	Statistic	Statistic	Statistic	Std. Error	Statistic	Statistic	Std. Error	Statistic	Std. Error
1 Northeast	doctor_visits	27	5	11	8.74	.332	1.723	-.394	.448	-.689	.872
	Valid N (listwise)	27									
2 Southeast	doctor_visits	30	5	13	9.50	.324	1.776	-.633	.427	.429	.833
	Valid N (listwise)	30									
3 Southwest	doctor_visits	26	1	9	4.65	.490	2.497	-.123	.456	-1.202	.887
	Valid N (listwise)	26									
4 Midwest	doctor_visits	23	1	11	5.78	.544	2.610	-.008	.481	-.635	.935
	Valid N (listwise)	23									
5 West	doctor_visits	40	4	11	8.70	.348	2.198	-.771	.374	-.475	.733
	Valid N (listwise)	40									
6 Far West	doctor_visits	34	1	11	5.21	.454	2.649	.076	.403	-.835	.788
	Valid N (listwise)	34									

FIGURE 14.12 Descriptive statistics on the six regions after splitting the data file by **region** and performing a **Descriptives** analysis.

ethnic

		Frequency	Percent	Valid Percent	Cumulative Percent
Valid	1 Asian American	35	19.4	19.4	19.4
	2 Pacific Islander	5	2.8	2.8	22.2
	3 European American	77	42.8	42.8	65.0
	4 Mexican American	22	12.2	12.2	77.2
	5 Latino American	16	8.9	8.9	86.1
	6 African American	25	13.9	13.9	100.0
	Total	180	100.0	100.0	

FIGURE 14.13 A frequency table for **ethnic**.

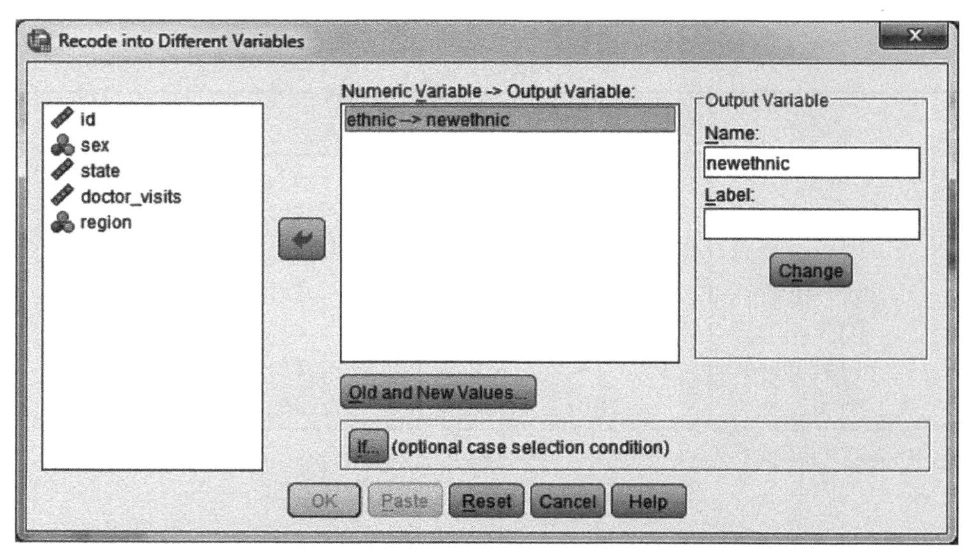

FIGURE 14.14 The variable **newethnic** has now been named.

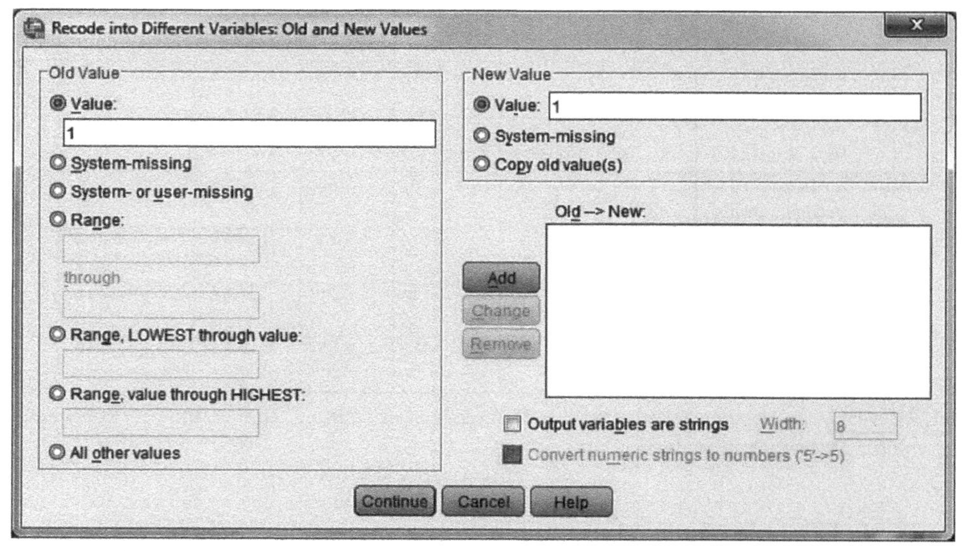

FIGURE 14.15 Recoding the value of 1 to 1.

recoded variable **newethnic** in the **name** panel under **Output Variable**, and clicked the **Change** pushbutton; the results of this are shown in Figure 14.14.

Selecting the **Old and New Values** brings us to the recoding specification screen (see Figure 14.15). In the **Value** panel in the **Old Value** area, we type **1**, and in the **Value** panel in the **New Value** area, we type **1** (see Figure 14.15); to finish we click **Add**. To combine **Pacific Islander** with **Asian Americans**, in the **Value** panel in the **Old Value** area, we type **2**, whereas in the **Value** panel in the **New Value** area, we type **1** (see Figure 14.16); to finish we click **Add**.

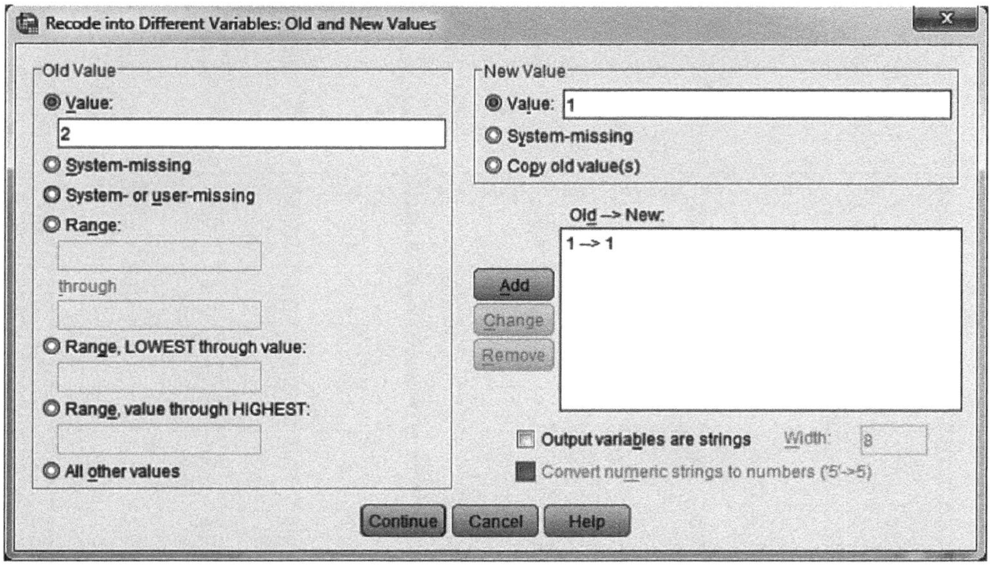

FIGURE 14.16 Recoding the value of 2 to 1.

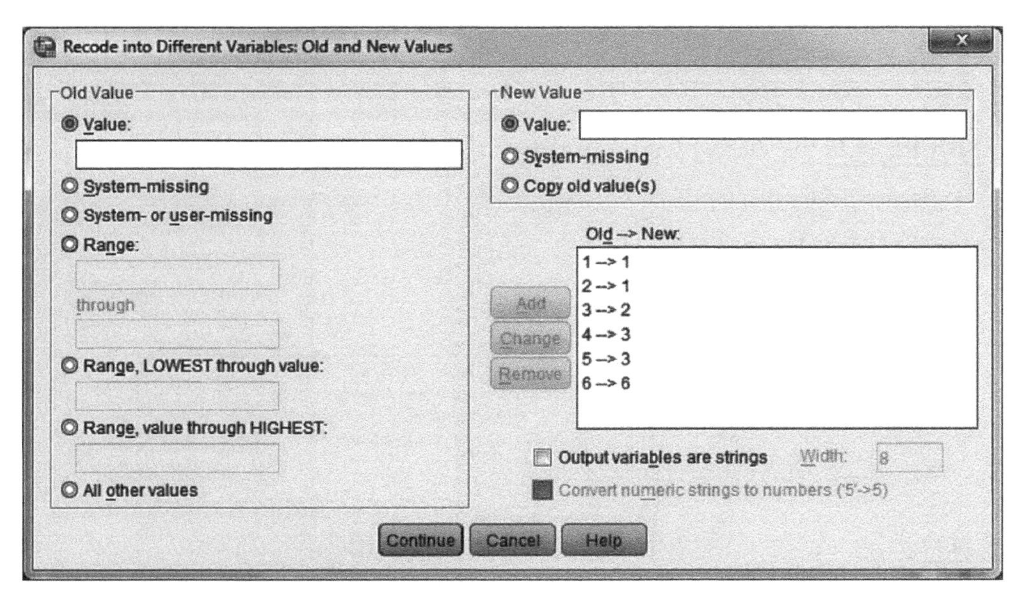

FIGURE 14.17 The recoding of **ethnic** to **newethnic** is ready to be executed.

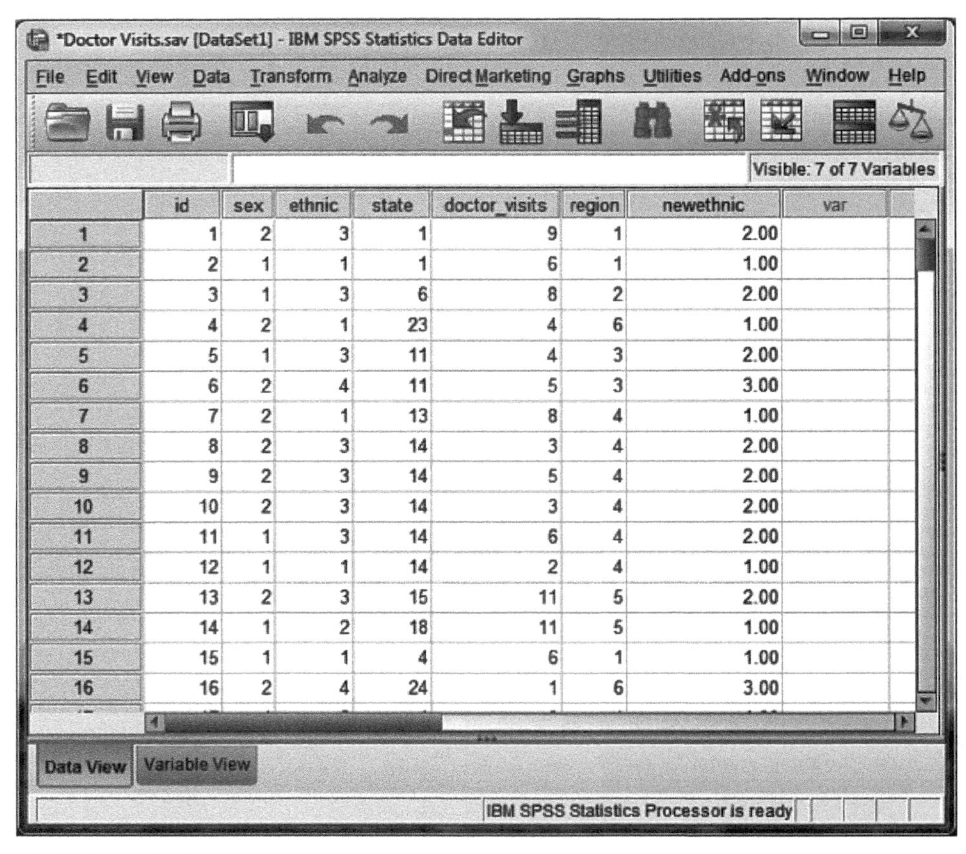

FIGURE 14.18 The recoded variable **newethnic** appears at the end of the data file.

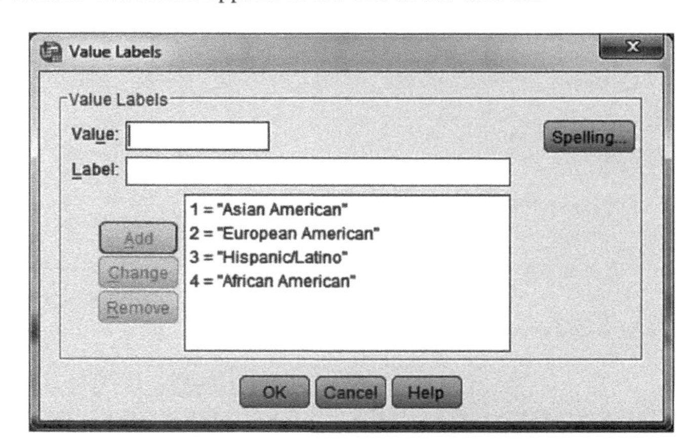

FIGURE 14.19
The value labels for **newethnic**.

newethnic

		Frequency	Percent	Valid Percent	Cumulative Percent
Valid	1 Asian American	40	22.2	22.2	22.2
	2 European American	77	42.8	42.8	65.0
	3 Hispanic/Latino	38	21.1	21.1	86.1
	4 African American	25	13.9	13.9	100.0
	Total	180	100.0	100.0	

FIGURE 14.20 A frequency table for **ethnic**.

We continue in this manner as follows:

- We recode **3** into **2**.
- We recode **4** into **3**.
- We recode **5** into **3**.
- We recode **6** into **6**.

The results of this typing are shown in Figure 14.17. We click **Continue** to return to the main dialog window and click **OK** to perform the recode. The recoded variable **newethnic** appears at the end of the data file (see Figure 14.18). We shift to **Variable View** to change the decimal places and supply value labels to **newethnic** (see Figure 14.19). A frequency table of **newethnic** from the **Frequencies** procedure (presented in Figure 14.20) shows the number of cases in each of the recoded categories with the new value labels.

CHAPTER 15

Visual Binning

15.1 OVERVIEW

Visual binning is an alternative recoding procedure that allows us to recode a quantitative variable into a discrete number of categories (bins or groups of cases). It is "visual" because IBM SPSS® displays a histogram (somewhat vertically compressed) and will mark the boundaries of our to-be-generated bins with vertical lines superimposed on the histogram to render it as a more visual process. Reducing a quantitative variable down to a handful of bins or ordered categories disregards the rich information contained in the data, and it is certainly not recommended as common practice, but there may be isolated instances where it may be worthwhile to perform exploratory analyses on a variable that has been reduced to four or five global ordered categories.

The **Visual Binning** procedure allows us to establish bins based on preset binning choices available in the dialog window. In one preset binning choice, we can establish categories based on standard deviation units. This choice would divide the distribution into four categories whose boundaries would be −1 SD, 0 SD, and +1 SD (these correspond to percentiles of 15.9, 50, and 84.1, respectively); such a strategy would result in groups of unequal size with the two middle groups having more cases than the two end groups. In another preset binning choice, we can create bins on the basis of percentiles to keep the group sizes approximately equal. For example, we can create five bins spaced 20 percentile points apart, we could create four bins spaced 25 percentile points (one quartile) apart, and so on.

15.2 NUMERICAL EXAMPLE

The data file we use for our example is **Personality**. We will subject the openness measure (**neoopen**) to the **Visual Binning** procedure, dividing the distribution into four groups based on quartiles. To set the stage, we have obtained the mean, standard deviation, minimum and maximum scores, and the quartile demarcations of **neoopen**, as well as its histogram, from the **Frequencies** procedure. This output is presented in Figure 15.1, where it can be seen that **neoopen** scores range from **27.51** to **79.05**, with the 25th, 50th, and 75th quartiles corresponding to values of **48.1271**, **55.1448**, and **63.5911**, respectively.

Performing Data Analysis Using IBM SPSS®, First Edition.
Lawrence S. Meyers, Glenn C. Gamst, and A. J. Guarino.

Statistics

neoopen openness: neo

N	Valid	420
	Missing	5
Mean		55.4607
Std. Deviation		10.90952
Minimum		27.51
Maximum		79.05
Percentiles	25	48.1271
	50	55.1448
	75	63.5911

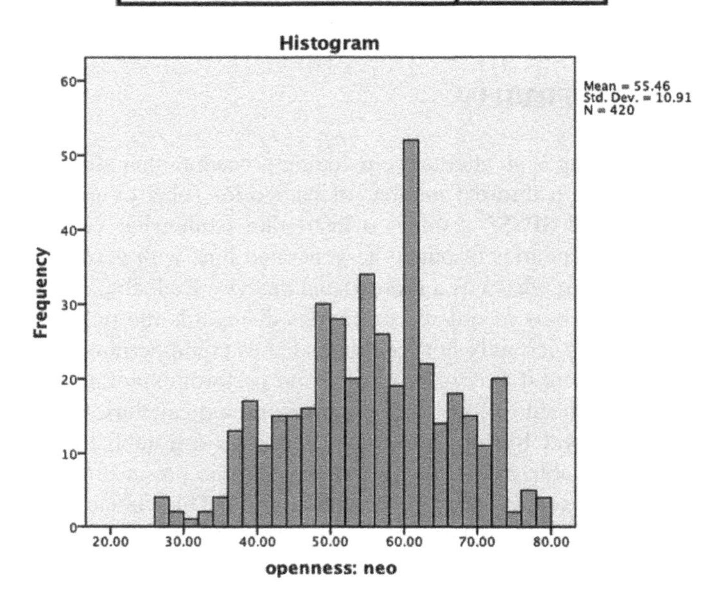

FIGURE 15.1

Some descriptive statistics for **neoopen** from the **Frequencies** procedure.

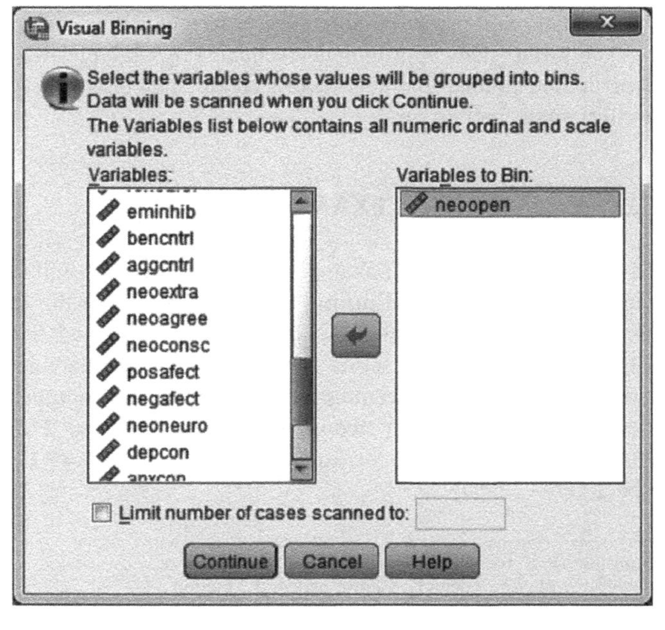

FIGURE 15.2

The **Visual Binning** initial dialog window.

15.3 ANALYSIS SETUP

We open the data file named **Personality** and from the main menu select **Transform ➔ Visual Binning**. This opens the initial dialog window shown in Figure 15.2. We move **neoopen** into the **Variables to Bin** panel and click **Continue** to reach the main dialog window.

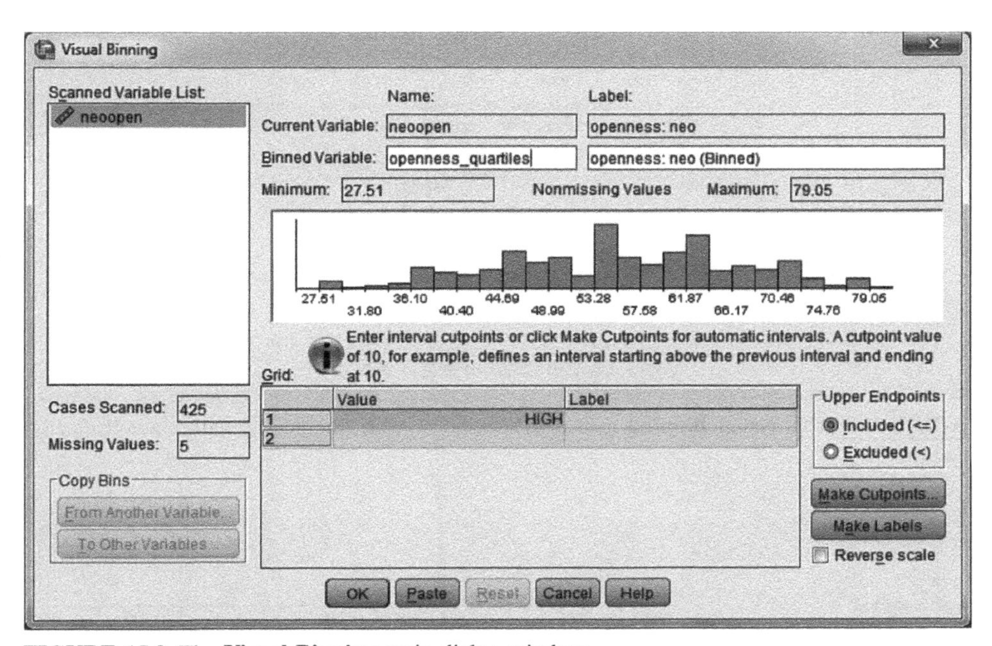

FIGURE 15.3 The **Visual Binning** main dialog window.

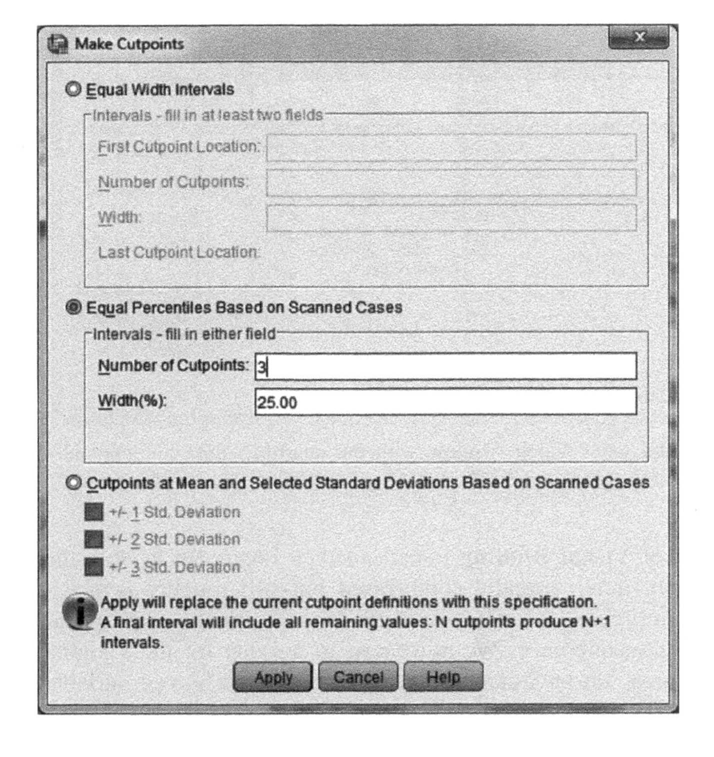

FIGURE 15.4

The **Make Cutpoints** dialog window of **Visual Binning** specifying quartile bins.

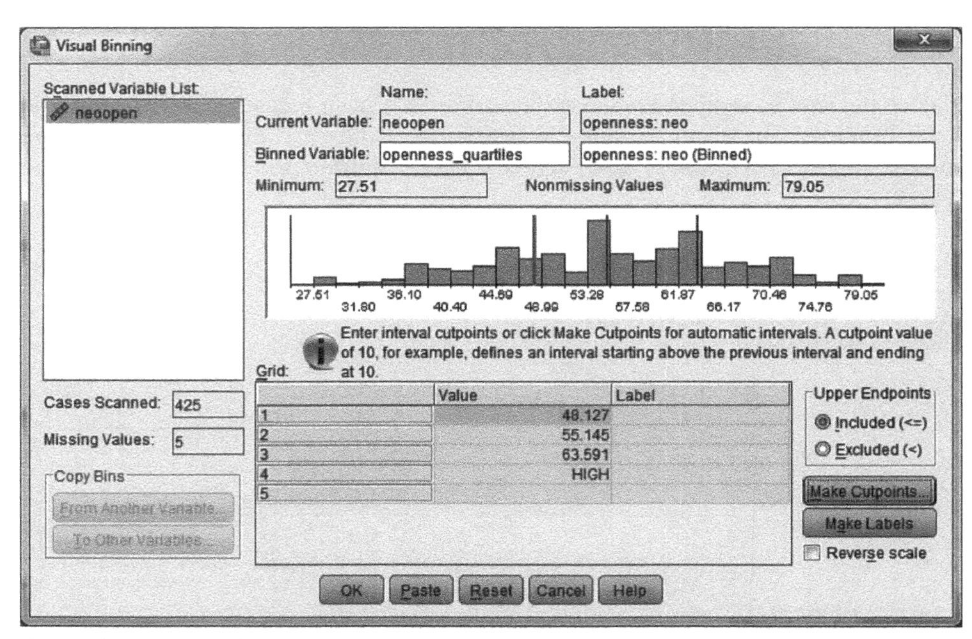

FIGURE 15.5 The **Visual Binning** main dialog window with the quartile values automatically filled in.

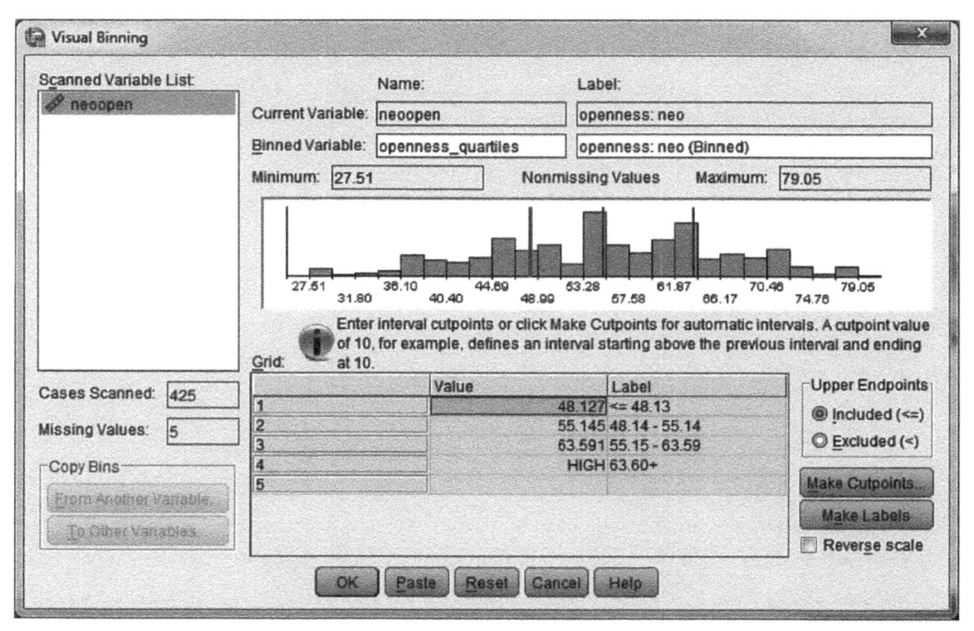

FIGURE 15.6 The **Visual Binning** main dialog window with the quartile values automatically filled in together with labels for the groups.

The main dialog window of **Visual Binning** is presented in Figure 15.3. Note the representation of the histogram; it is vertically compressed but still roughly approximates the full histogram seen in Figure 15.1. The minimum and maximum scores appear just above the histogram for easy reference. We must type in a name for the variable we will create; in the **Name** area, we have named our to-be-generated binned variable **openness_quartiles**.

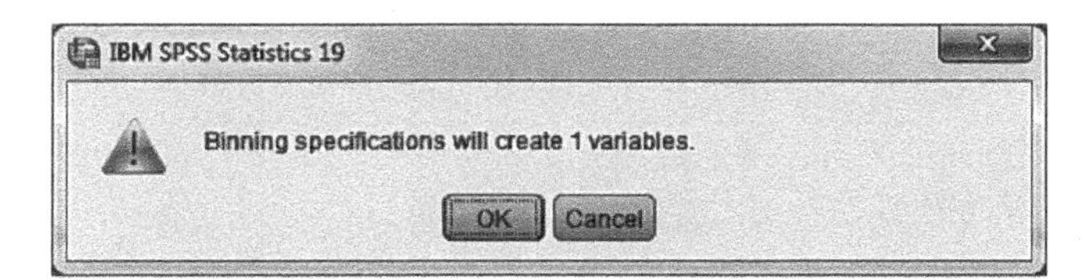

FIGURE 15.7
Verification screen for creating the new binned variable.

*Personality.sav [DataSet2] - IBM SPSS Statistics Data Editor

File Edit View Data Transform Analyze Direct Marketing Graphs Utilities Add-ons Window Help

Visible: 42 of 42 Variables

		anxcon	emcontot	angercon	selfcon	beckdep	regard	openness_quartiles	var
1	00	2.86	3.10	3.43	3.61	7.00	59.00	3	
2	00	2.71	2.52	1.86	3.67	3.00	65.00	4	
3	43	2.14	1.71	1.57	4.06	3.00	71.00	2	
4	86	2.57	2.71	2.71	3.33	12.00	65.00	4	
5	29	2.00	1.95	1.57	3.92	4.00	59.00	1	
6	57	2.14	1.86	1.86	3.83	5.00	59.00	2	
7	57	2.71	2.33	1.71	3.17	6.00	54.00	1	
8	57	3.14	3.43	3.57	4.00	15.00	65.00	1	
9	00	2.00	2.14	2.43	3.89	2.00	71.00	2	
10	57	2.71	2.76	3.00	3.61	5.00	45.00	2	
11	71	3.00	2.10	1.57	4.81	3.00	59.00	3	
12	29	2.00	2.19	2.29	4.06	2.00	59.00	2	
13	86	4.00	3.67	3.14	4.17	7.00	65.00	2	
14	71	2.29	1.81	1.43	4.25	2.00	59.00	4	
15	29	2.57	2.57	2.86	3.72	8.00	54.00	1	
16	43	3.43	2.67	2.14	3.53	2.00	65.00	2	
17	29	2.00	2.24	2.43	3.75	3.00	65.00	3	
18	00	1.86	2.67	3.14	3.08	11.00	33.00	3	
19	57	2.86	2.24	1.29	3.09	18.00	65.00	4	

FIGURE 15.8 Data file with **openness_quartiles** placed at the end.

We could type in our boundaries in the rows of the **Grid** under the column labeled as **Value**, but there is an easier way to accomplish the quartile categorization. We select the **Make Cutpoints** pushbutton to open the **Make Cutpoints** dialog window (see Figure 15.4). We select **Equal Percentiles Based on Scanned Cases**. In the **Number of Cutpoints** panel, we type **3** (this will result in four bins); IBM SPSS automatically fills in the **Width(%)** panel with **25.00** (which is the width of a quartile). The two-place decimal value is used because 100 may not be evenly divisible by the number of boundaries we specify. Click **Apply** to return to the main dialog window.

Upon returning to the main dialog window, we find that the percentile values have been automatically filled in the **Grid** under **Value** and that these quartile boundaries are represented in the histogram by colored vertical lines (see Figure 15.5). Clicking the **Make Labels** pushbutton creates unimaginative but very descriptive labels for our quartile groups as shown in Figure 15.6; knowing the exact percentile boundaries removes any ambiguity concerning how each bin is defined.

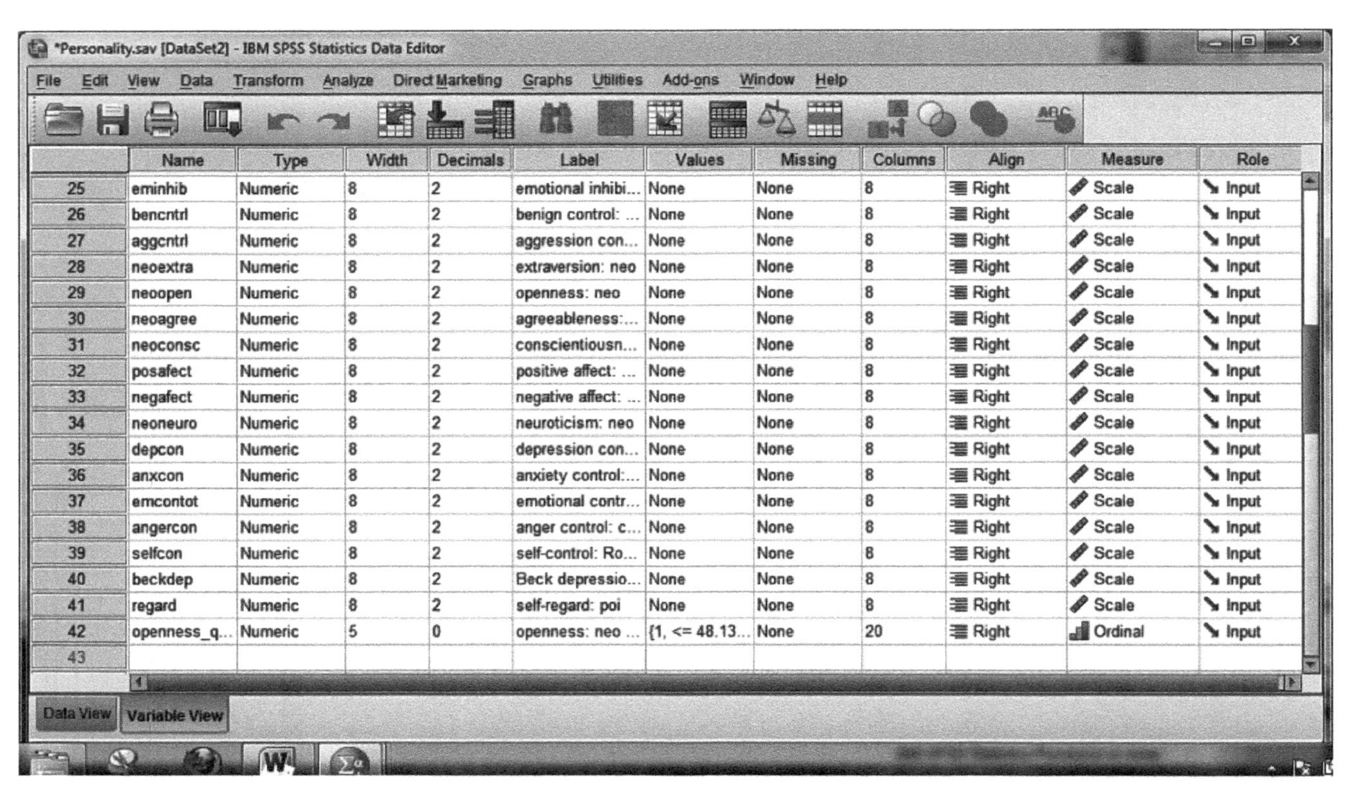

	Name	Type	Width	Decimals	Label	Values	Missing	Columns	Align	Measure	Role
25	eminhib	Numeric	8	2	emotional inhibi...	None	None	8	Right	Scale	Input
26	bencntrl	Numeric	8	2	benign control: ...	None	None	8	Right	Scale	Input
27	aggcntrl	Numeric	8	2	aggression con...	None	None	8	Right	Scale	Input
28	neoextra	Numeric	8	2	extraversion: neo	None	None	8	Right	Scale	Input
29	neoopen	Numeric	8	2	openness: neo	None	None	8	Right	Scale	Input
30	neoagree	Numeric	8	2	agreeableness:...	None	None	8	Right	Scale	Input
31	neoconsc	Numeric	8	2	conscientiousn...	None	None	8	Right	Scale	Input
32	posafect	Numeric	8	2	positive affect: ...	None	None	8	Right	Scale	Input
33	negafect	Numeric	8	2	negative affect: ...	None	None	8	Right	Scale	Input
34	neoneuro	Numeric	8	2	neuroticism: neo	None	None	8	Right	Scale	Input
35	depcon	Numeric	8	2	depression con...	None	None	8	Right	Scale	Input
36	anxcon	Numeric	8	2	anxiety control:...	None	None	8	Right	Scale	Input
37	emcontot	Numeric	8	2	emotional contr...	None	None	8	Right	Scale	Input
38	angercon	Numeric	8	2	anger control: c...	None	None	8	Right	Scale	Input
39	selfcon	Numeric	8	2	self-control: Ro...	None	None	8	Right	Scale	Input
40	beckdep	Numeric	8	2	Beck depressio...	None	None	8	Right	Scale	Input
41	regard	Numeric	8	2	self-regard: poi	None	None	8	Right	Scale	Input
42	openness_q...	Numeric	5	0	openness: neo ...	{1, <= 48.13...	None	20	Right	Ordinal	Input
43											

FIGURE 15.9 The **Variable View** showing the newly created binned variable.

Click **OK** to create the new variable. We will be presented with a verification screen (see Figure 15.7). Clicking **OK** executes the procedure. The new variable (**openness_quartiles**) has been placed at the end of the data file as seen in Figure 15.8. The values of **1** through **4** represent the group codes; **1** indicates the cases that are at or below the first quartile, **2** indicates the cases that are between the first quartile and the median (the second quartile), and so on. The labels appear in the **Variable View** (see Figure 15.9), and IBM SPSS has identified the new variable as an **Ordinal Measure**. It is probably best to save the data file now that this new variable has been added to it.

CHAPTER 16

Computing New Variables

16.1 OVERVIEW

When we compute a new variable, we identify one or more of the quantitative variables in our data file and specify a computation we wish to apply. When working with a single variable, for example, we could transform it to its natural log, compute its squared value, or place it in an algebraic expression. When working with a set of variables, we could compute, for example, a total score or a mean.

Whether working with single variables or sets of variables, the result of the computation is a new variable that is placed at the end of the data file once it has been computed. Each case will have a value on the new variable, provided a valid computation was possible; if the computation was not performed for a particular case for whatever reason (e.g., a variable had a missing value), then a **system missing** marker (a blank cell) will be seen in the data file for that case.

16.2 COMPUTING AN ALGEBRAIC EXPRESSION

In Chapter 13, we computed z scores for the variable **neoneuro** in the data file **Personality** and concluded the chapter by indicating that other standardized scores can be computed from z scores. We illustrate that computation here by obtaining the z scores through the **Descriptives** procedure and then generating linear T scores based on **Zneoneuro**. To transform a z score to a linear T score, we multiply the z score by 10 and add 50 to that result. When it has been computed, the linear T score will have a mean of 50 and a standard deviation of 10, and its distribution will mirror the z score distribution on which it was based.

We have begun the procedure by making the data file named **Personality** the active data file and generating the variable **Zneoneuro** as described in Chapter 13. To compute linear T scores, from the main menu select **Transform → Compute Variable**. This opens the **Compute Variable** dialog window shown in Figure 16.1.

The **Target Variable** panel is where we provide a name for our newly computed variable. We name the variable **Tneoneuro**. Selecting the **Type & Label** pushbutton under the panel brings us to the dialog screen shown in Figure 16.2. Although its name

Performing Data Analysis Using IBM SPSS®, First Edition.
Lawrence S. Meyers, Glenn C. Gamst, and A. J. Guarino.

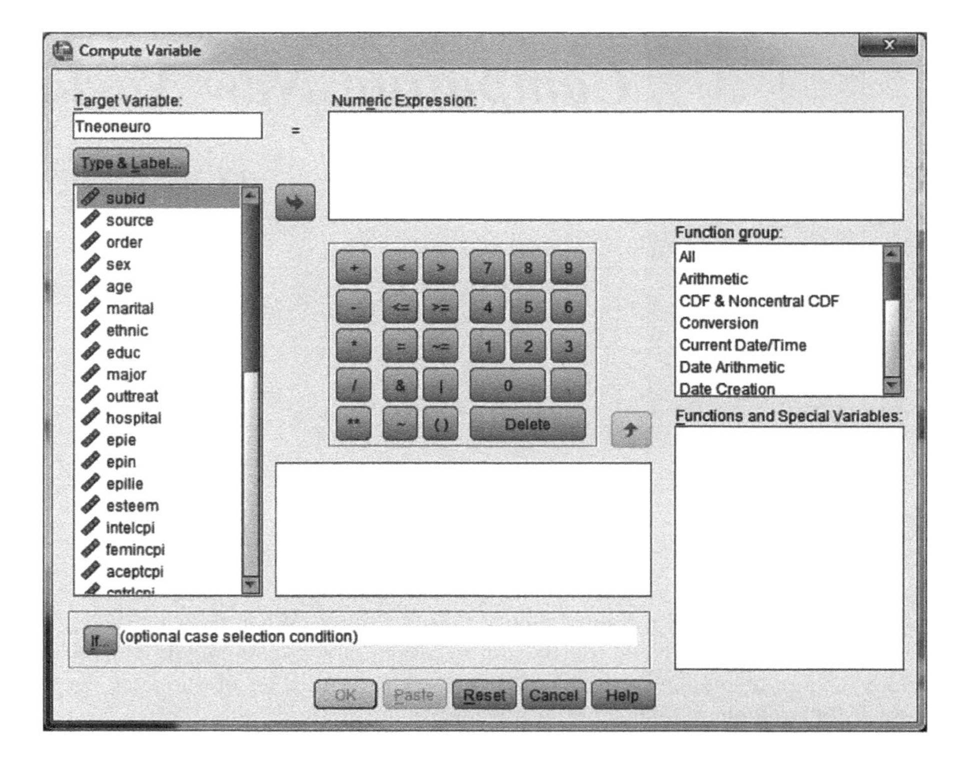

FIGURE 16.1

The **Compute Variable** dialog window with the name of our new variable provided.

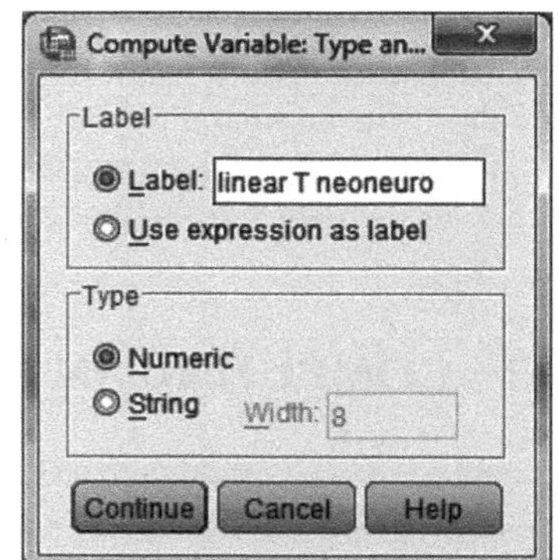

FIGURE 16.2

The **Type and Label** screen of the **Compute Variable** dialog window.

may be self-explanatory, we illustrate how to interact with this dialog screen by providing the label **linear T neoneuro** and retaining the default **Numeric** in the **Type** area of the window. Clicking **Continue** returns us to the **Compute Variable** dialog window.

We are now ready to specify our algebraic expression to generate the linear T values. The sequence needed to specify the computation is as follows:

1. Click the double parentheses on the symbol keypad. This will place a set of parentheses in the **Numeric Expression** panel.
2. Place the cursor inside the set of parentheses.
3. Double-click **Zneoneuro** from the **Variable List** panel. This will place the variable just before the cursor in the **Numeric Expression** panel.
4. We need to multiply the *z* score by 10. Click the single asterisk (this represents multiplication) on the symbol keypad. This will place an asterisk in the **Numeric Expression** panel immediately following **Zneoneuro**.

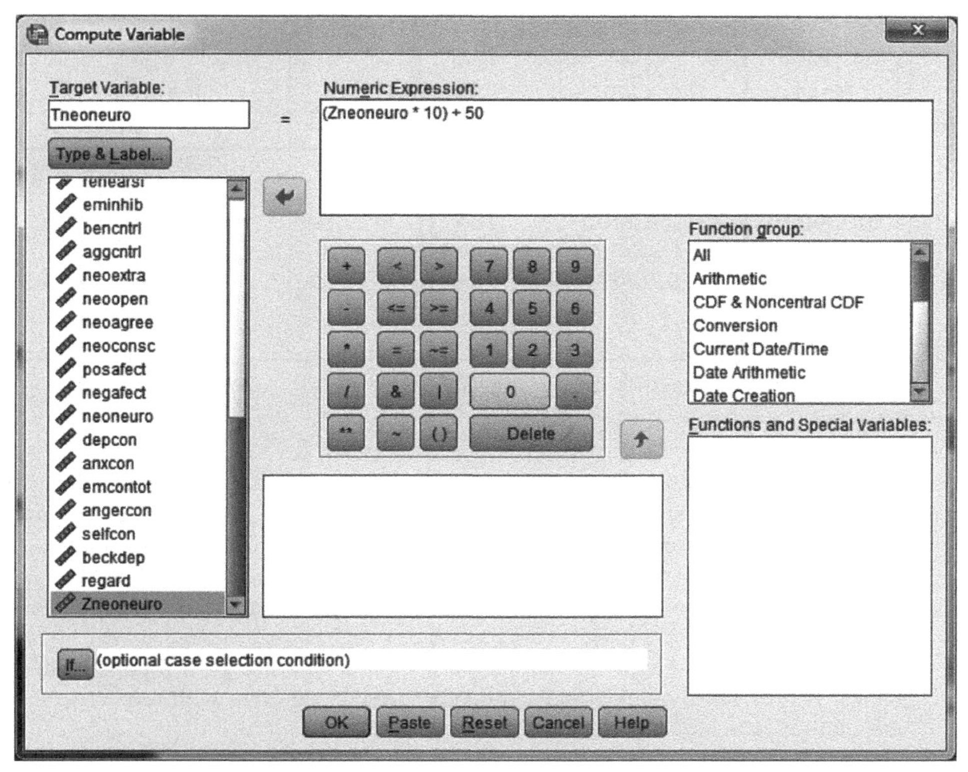

FIGURE 16.3 The algebraic expression to compute the linear T score is complete.

5. Click **1** and then **0** on the numeric keypad to place the number **10** after the asterisk in the **Numeric Expression** panel.
6. Move the cursor to the right of the closing parenthesis.
7. Click the plus sign on the symbol keypad.
8. Click **5** and then **0** on the numeric keypad to place the number **50** after the plus sign in the **Numeric Expression** panel.

The finished expression is shown in Figure 16.3. We click **OK** to perform the computation.

16.3 THE OUTCOME OF COMPUTING THE LINEAR T SCORES

To illustrate the outcome of our computation, we have analyzed **neoneuro**, **Zneoneuro**, and **Tneoneuro** in the **Descriptives** procedure. The limited set of descriptive statistics we generated is shown in Figure 16.4. As can be seen, the linear T score variable has a mean of **50** and a standard deviation of **10**. Neither the *z* score nor the linear T score transformation has changed the shape of the distribution, as the skewness and kurtosis values have remained constant throughout.

16.4 COMPUTING THE MEAN OF A SET OF VARIABLES

It is frequently the case that in the process of our data analysis, we need to combine a set of variables together. The simplest way to do this is to compute a mean or total

Descriptive Statistics

	N	Mean	Std. Deviation	Skewness		Kurtosis	
	Statistic	Statistic	Statistic	Statistic	Std. Error	Statistic	Std. Error
neoneuro neuroticism: neo	420	50.5394	11.14793	.184	.119	-.263	.238
Zneoneuro Zscore: neuroticism: neo	420	.0000000	1.00000000	.184	.119	-.263	.238
Tneoneuro linear T neoneuro	420	50.0000	10.00000	.184	.119	-.263	.238
Valid N (listwise)	420						

FIGURE 16.4 Descriptive statistics for the original variable and its z score and linear T score transformations.

score of the set assuming that they are all assessed on the same metric, for example, a 5-point scale or sales dollars per week (if they are assessed on different metrics, then they should first be converted to z scores and then combined).

Total scores and mean values are equivalent summaries of a set of variables. To obtain a total, we sum all of the scores. But there can be disadvantages to total scores:

- If the total score is based on summative response scales (e.g., a 5-point scale), then it is difficult to quickly and intuitively interpret the total with respect to the original rating scale.
- The total score must be based on the same number of variables for it to be comparable across cases. Thus, every case must have the same number of valid scores in order to participate in the analysis.

To obtain a mean value of a set of variables, we sum all of the scores on the variables and then divide by the number of scores. Mean scores can have certain advantages:

- If the mean score is based on summative response scales (e.g., a 5-point scale), then it is easy to quickly and intuitively interpret the mean with respect to the original rating scale.
- If there are a sufficient number of scores that are to be combined, then there are occasions when having a missing value on a very small percentage of the variables may still yield an interpretable and valid mean. IBM SPSS® has a way to accomplish this, which we will demonstrate.

16.5 NUMERICAL EXAMPLE OF COMPUTING THE MEAN OF A SET OF VARIABLES

The Aspirations Index (Kasser & Ryan, 1993, 1996) is an inventory that assesses the value placed on achieving certain extrinsic and intrinsic goals. We will compute the mean for one of the intrinsic subscales to illustrate the computation process. The 30-item version of the inventory represented in the data file named **Aspirations Index** was part of a larger project of one of our graduate students at the time (Leanne Williamson, now at the Ohio State University). A total of 310 university students completed the inventory. Respondents used a 9-point response scale to represent the degree to which the content was applicable to them. Items in the data file are named **aspire01** through **aspire30**.

The subscale we will generate is called Affiliation, and the items associated with the subscale are **aspire02**, **aspire08**, **aspire14**, **aspire20**, and **aspire26**. Item content includes having committed and enduring relationships, sharing life with a loved one, having good friends, and being loved by others and being loved in return. All of the items are positively worded so that no recoding before combining them is required.

16.6 THE COMPUTATION PROCESS

We open the data file **Aspirations Index** and from the main menu select **Transform → Compute Variable**. This opens the **Compute Variable** dialog window shown in Figure 16.5. IBM SPSS has a wide range of already structured functions to compute a variety of things. We have scrolled down the **Function group** panel to select **Statistical**. This action displays in the **Functions and Special Variables** panel the set of statistical functions available. We selected **Mean** and double-clicked it (we could have clicked the upward pointing arrow adjacent to the panel). These steps resulted in the generic **Mean** function being displayed in the **Numeric Expression** panel as shown in Figure 16.5.

The question marks in the **Mean** function inside the parentheses are to be replaced by variables whose mean is to be computed. The function is displayed in generic form as **Mean(?,?)** but can be customized in two ways. First, only two question marks (placeholders for variables) are shown in the generic display but more than two variables can be included.

The second aspect of customization concerns the stand-alone word **Mean**. As it is in the **Numeric Expression** panel, IBM SPSS requires that at least one of the variables in the set has a valid value. If this basic condition is met by a case, then a mean will be computed for that case. To use our Affiliation scale as an example, there are five items comprising the subscale; however, if a case has a valid value on only one of the variables (e.g., that case has missing values on four of the five items), IBM SPSS will

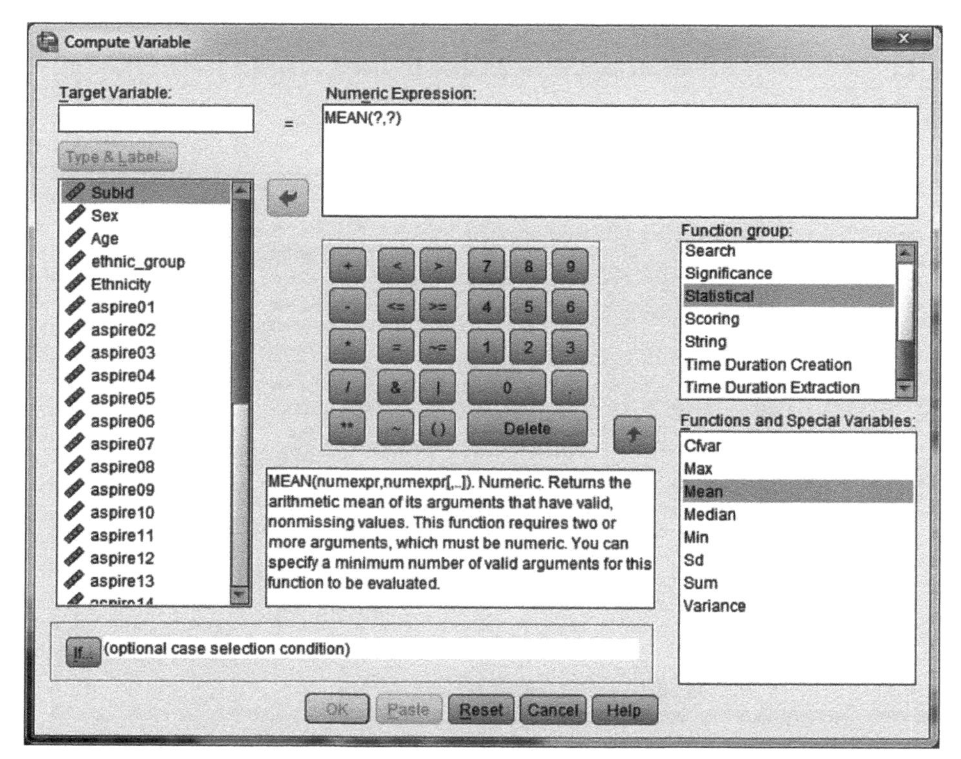

FIGURE 16.5 The generic form of the **Mean** function has been invoked.

still compute a mean for that case (the computed mean would be equal to the value on the item with a valid score). In such an extreme situation, it would be difficult to believe that the computed mean is representative of the set of five items, and using it could potentially compromise the internal and external validity of the whole data analysis.

This problem can be avoided in one of two ways. First, the researchers would have examined the data set and presumably discovered and probably removed cases with large amounts of missing data. But even then, an additional precaution is available in IBM SPSS: we can add the **.n** specification to the **Mean**. In such a specification, the **n** is the minimum number of valid values we require in order to permit a mean to be computed; if there are fewer than **n** valid values, then IBM SPSS would return a **system missing** value (a blank) rather than a mean. For example, if we were to specify **Mean.4** in computing the Affiliation subscale, there must be at least **4** valid (nonmissing) values in order to proceed with the computation (some researchers might use **Mean.5** here, as there are only five items and four of five valid values might be considered too liberal a criterion). We will use the specification **Mean.4** in our computation.

To create the specification in the **Compute Variable** window, we have named our subscale **affiliation**; as the variable name is sufficient and the default is a numeric variable, we have not accessed the **Type & Label** window. We have then moved our cursor into the **Numeric Expression** panel. This panel can be edited, and we have typed **.4** next to the word **Mean**, deleted the question marks, and double-clicked our variables associated with the subscale into the parentheses, being careful to separate them by (the required) commas. We have separated the variables with spaces following the commas to make it easier for humans to read the expression, but those spaces are optional for the software. The finished expression is shown in Figure 16.6. We click **OK** to perform the computation.

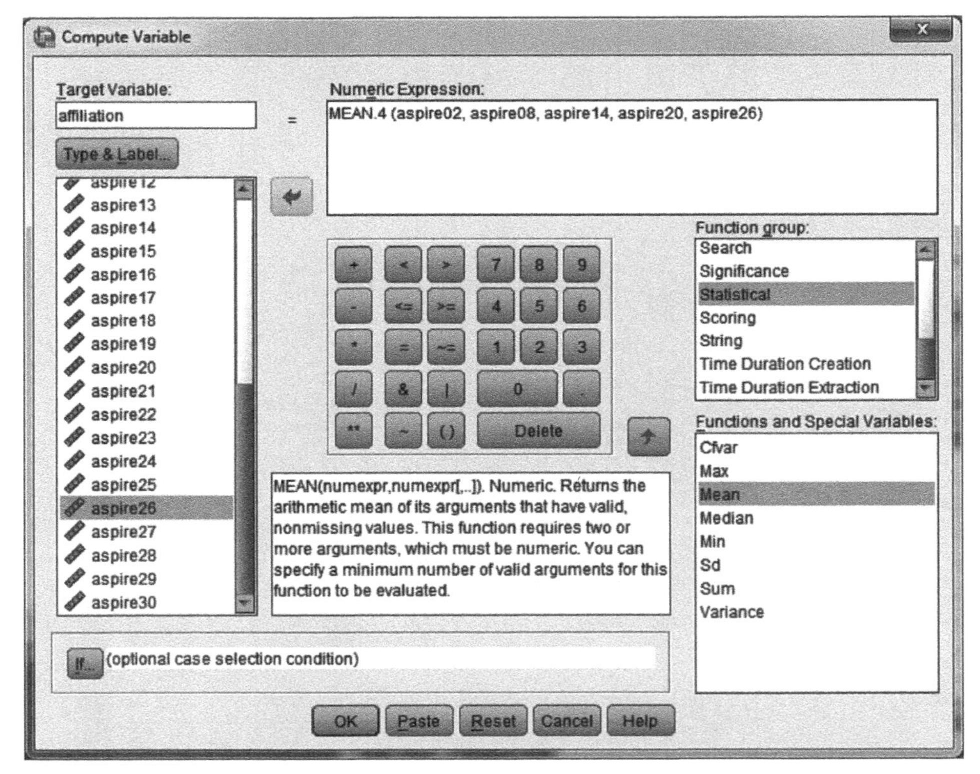

FIGURE 16.6 The **Compute Variable** dialog window with the name of our new variable, and the **Mean** function specified to require at least four valid values for a mean to be computed.

Descriptive Statistics

	N	Minimum	Maximum	Mean	Std. Deviation
aspire02 I will feel that there are people who really love me, and whom I love.	310	2	9	8.34	1.173
aspire08 I will have good friends that I can count on.	310	3	9	8.21	1.160
aspire14 I will share my life with someone I love.	310	1	9	8.46	1.186
aspire20 I will have committed, intimate relationships.	310	1	9	8.15	1.344
aspire26 I will have deep, enduring relationships.	309	3	9	8.23	1.196
affiliation	310	4.40	9.00	8.2787	.89890
Valid N (listwise)	309				

FIGURE 16.7 Descriptive statistics for the individual items and the subscale that they combine to form.

16.7 THE OUTCOME OF COMPUTING THE AFFILIATION SUBSCALE

To illustrate the outcome of our computation, we have analyzed the five individual items comprising the subscale and the **affiliation** subscale itself in the **Descriptives** procedure. The limited set of descriptive statistics we generated is shown in Figure 16.7. As can be seen, the subscale mean is the average of the individual item means and is based on the full set of 310 cases, even though there were only 309 valid cases for item **aspire26**; this is because we specified a minimum of four valid values for the subscale to be computed, a condition that was met for every case. Note also that the standard deviation of the subscale mean is lower than the standard deviations of the raw scores for the individual items (because the subscale mean is an average of the item means).

CHAPTER 17

Transforming Dates to Age

17.1 OVERVIEW

It is not uncommon in behavioral, social, and medical research studies to capture dates as one of the data elements. Perhaps the most familiar instance of this, and the one that we treat in this chapter, is birth date, but other instances can include the date of sale of a vehicle or an appliance, the date of the first medical treatment of a patient, and the date of the first day of employment with a particular organization.

Raw dates cannot be used as variables in most of the statistical procedures we use, but they can be transformed to a numeric variable representing the amount of elapsed time between that date and some other date. This time variable resulting from such a transformation represents ratio-level measurement that can appropriately be used in virtually all statistical analyses. For example, the time elapsed (e.g., number of years) between the date of birth and the current date is one's age.

17.2 THE IBM SPSS® SYSTEM CLOCK

IBM SPSS tracks the current date and time with the variable **$TIME**. The **$** symbol is used to represent a system variable, that is, a variable already defined by IBM SPSS and reserved for the given purpose; it becomes available when the software is active to keep track of information that may be required or may be potentially needed by the system. In its **Help** screen under **System Variables**, IBM SPSS informs us that the **$TIME** variable "...represents the number of seconds from midnight, October 14, 1582, to the date and time when the transformation command is executed..." It thus serves as a reference point for determining the current date and time. The date October 14, 1582, is far from arbitrary, by the way. When Pope Gregory XIII in 1582 ordered replacing the Julian calendar (created by Julius Caesar) with the improved one that we now call the Gregorian calendar, he removed the days 5 through 14 of October in that year to help with the transition from the old (and by then inaccurate calendar) to the new calendar; thus, October 14, 1582, does not exist for those of us who use the Gregorian calendar.

Performing Data Analysis Using IBM SPSS®, First Edition.
Lawrence S. Meyers, Glenn C. Gamst, and A. J. Guarino.

17.3 DATE FORMATS

The data file containing our date variable is named **Employee data**, and a screenshot of it is shown in Figure 17.1. Dates are associated with the **bdate** variable in the data file and denote the date of birth of each case. The second case, for example, has the birth date **05/23/1958**, corresponding to May 23, 1958. In the **Variable View** screen, **bdate** is specified under **Type** as **Date** (see Section 2.4), as shown in Figure 17.2, with its proper **American Date 10 Character Width** format.

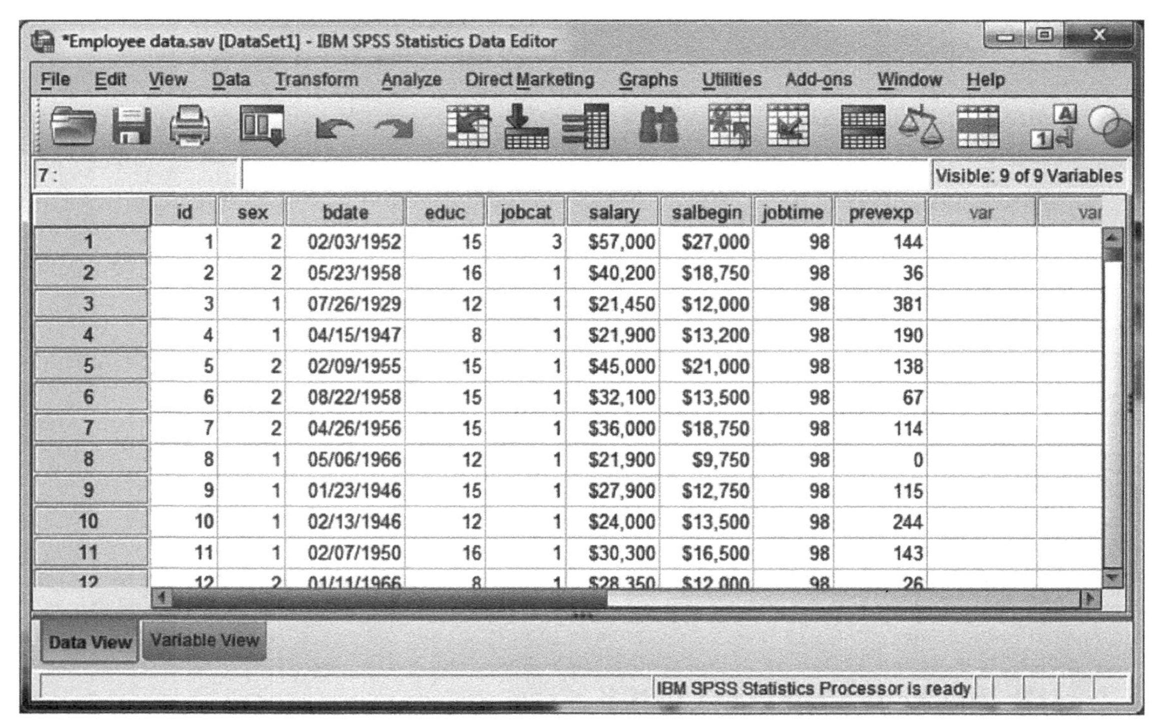

	id	sex	bdate	educ	jobcat	salary	salbegin	jobtime	prevexp	var	var
1	1	2	02/03/1952	15	3	$57,000	$27,000	98	144		
2	2	2	05/23/1958	16	1	$40,200	$18,750	98	36		
3	3	1	07/26/1929	12	1	$21,450	$12,000	98	381		
4	4	1	04/15/1947	8	1	$21,900	$13,200	98	190		
5	5	2	02/09/1955	15	1	$45,000	$21,000	98	138		
6	6	2	08/22/1958	15	1	$32,100	$13,500	98	67		
7	7	2	04/26/1956	15	1	$36,000	$18,750	98	114		
8	8	1	05/06/1966	12	1	$21,900	$9,750	98	0		
9	9	1	01/23/1946	15	1	$27,900	$12,750	98	115		
10	10	1	02/13/1946	12	1	$24,000	$13,500	98	244		
11	11	1	02/07/1950	16	1	$30,300	$16,500	98	143		
12	12	2	01/11/1966	8	1	$28,350	$12,000	98	26		

FIGURE 17.1 Data file.

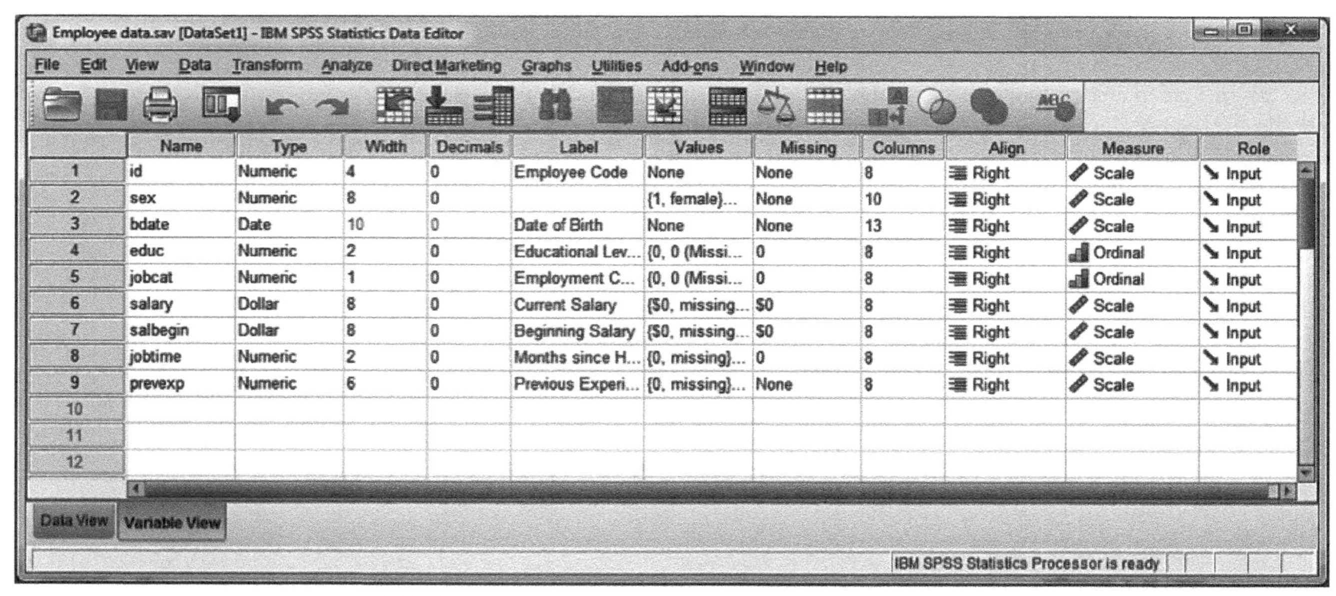

	Name	Type	Width	Decimals	Label	Values	Missing	Columns	Align	Measure	Role
1	id	Numeric	4	0	Employee Code	None	None	8	Right	Scale	Input
2	sex	Numeric	8	0		{1, female}...	None	10	Right	Scale	Input
3	bdate	Date	10	0	Date of Birth	None	None	13	Right	Scale	Input
4	educ	Numeric	2	0	Educational Lev...	{0, 0 (Missi...	0	8	Right	Ordinal	Input
5	jobcat	Numeric	1	0	Employment C...	{0, 0 (Missi...	0	8	Right	Ordinal	Input
6	salary	Dollar	8	0	Current Salary	{$0, missing...	$0	8	Right	Scale	Input
7	salbegin	Dollar	8	0	Beginning Salary	{$0, missing...	$0	8	Right	Scale	Input
8	jobtime	Numeric	2	0	Months since H...	{0, missing}...	0	8	Right	Scale	Input
9	prevexp	Numeric	6	0	Previous Experi...	{0, missing}...	None	8	Right	Scale	Input
10											
11											
12											

FIGURE 17.2 Data file showing **Variable View**.

Dates (as well as time) can be represented in a variety of formats in IBM SPSS data files. The two important points to bear in mind are as follows:

- The variable must be identified in the data file as a **Date** variable. This can be seen in Figure 17.2 for **bdate** in the second column under **Type**.
- The variable must be in one of the formats recognizable by IBM SPSS as representing a date.

The format for **bdate** in this data file is **American Date 10 Character Width** (mm/dd/yyyy) as can be seen in Figure 17.1. This format represents the fact that a total of 10 characters (this includes the slashes) are used with two digits denoting the month followed by two digits denoting the day followed by four digits denoting the year, with each field separated by a forward slash. Examples of other date formats are as follows:

- **American Date 8 Character Width** takes the form mm/dd/yy; an example is 05/23/58.
- **International Date 9 Character Width** takes the form dd-mmm-yy; an example is 23-MAY-58.
- **European Date 10 Character Width** takes the form dd.mm.yyyy; an example is 23.05.1958.

17.4 THE DATE AND TIME WIZARD

Dates are transformed to time by invoking the **Date and Time Wizard**. Our plan is to transform **bdate** into a variable representing age in years. We open **Employee data** and from the main menu select **Transform → Date and Time Wizard**. This opens the **Date and Time Wizard** window shown in Figure 17.3.

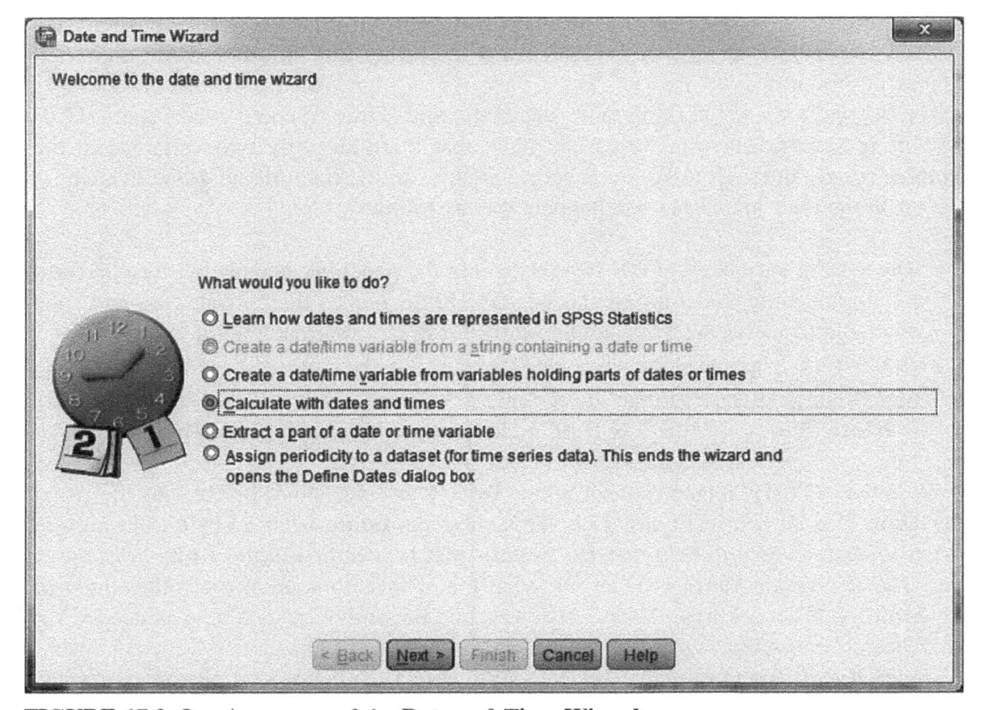

FIGURE 17.3 Opening screen of the **Date and Time Wizard**.

There are six choices presented to us in the initial **Date and Time Wizard** window, indicating the various operations that IBM SPSS can perform. Briefly, these choices are as follows:

- **Learn how dates and times are represented in SPSS Statistics** provides a brief review of the **Date and Time Wizard** operations.
- **Create a date/time variable from a string containing a date or time** allows us to convert a **String Variable Type** (e.g., a variable composed of words) to the **Date Variable Type** similar to the format in **bdate** in our example date file.
- **Create a date/time variable from variables holding parts of dates or times** is used to produce a date/time variable from a set of existing variables. For example, the data file may contain three separate variables of month, day, and year. This operation will combine the three variables into a single date/time variable.
- **Calculate with dates and times** provides two possible operations. The first operation adds a constant to a date (e.g., a month to an age) or calculates the duration of some process by subtracting a start time variable from an end time variable. The second operation will calculate a number of time units between two dates; for example, a birth date is subtracted from the present year to produce a chronological age. The latter operation is illustrated in this chapter.
- **Extract a part of a date or time variable** allows the researcher to select a single element from a date/time variable; for example, we can convert a variable containing the month, day, and year to a new variable containing only the month or the day or the year.
- **Assign periodicity to a dataset (for time series data)** creates a set of sequential dates that can be used with time series data.

For our purposes in this chapter where we wish to compute an age from a birth date, we select **Calculate with dates and times** and click the **Next** pushbutton (it may be a **Continue** pushbutton on a Mac). This brings us to the first of the three steps as shown in Figure 17.4. We select **Calculate the number of time units between two dates (e.g., calculate an age in years from a birthdate and another date)** and click the **Next** pushbutton.

The second step in working with the **Date and Time Wizard** (see Figure 17.5) allows us to specify our transformation. Note that there are only two variables in the **Variables** panel, although there are several variables in the data file, and one of them is not even in the data file. What has happened is as follows:

- The system variable **$TIME** is brought into the panel because we wish to perform a date and time transformation, and IBM SPSS makes the current date and time available to us because it is common to use it as a reference point.
- IBM SPSS displays only those variables in the data file that are in one of the many date and time formats it recognizes; the only such variable in the data file is **bdate** (it was specified as **Date** under **Type** in the **Variable View**).

We move **$TIME** into the panel under **Date 1** and we move **bdate** into the panel under **Date 2** as shown in Figure 17.6. This subtracts **bdate** from **$TIME** (the present time). This difference will be measured in units that is specified in the **Units** drop-down menu. The default of **Years** is what we wish for in this illustration, but other units of time (**Months**, **Weeks**, **Days**, **Hours**, **Minutes**, and **Seconds**) are available as needed for other research purposes.

Under **Result Treatment** we selected **Truncate to integer** as a person is n years old for the entire year until the day of their $n + 1$ birthday. Thus, fractional parts are

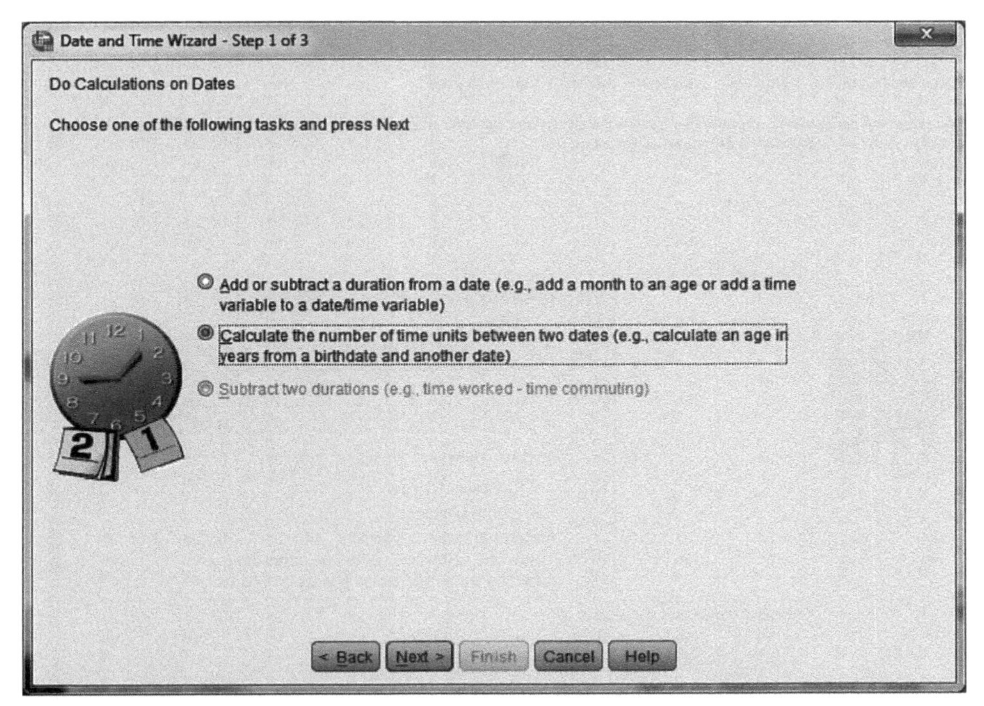

FIGURE 17.4 The first step in working with the **Date and Time Wizard**.

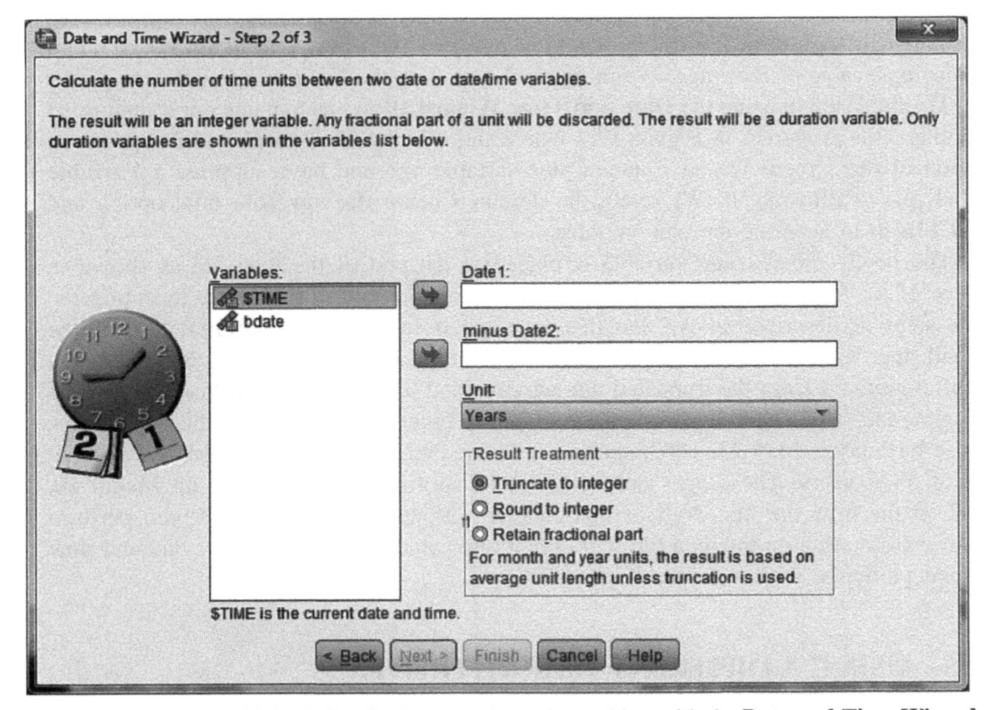

FIGURE 17.5 The initial window in the second step in working with the **Date and Time Wizard**.

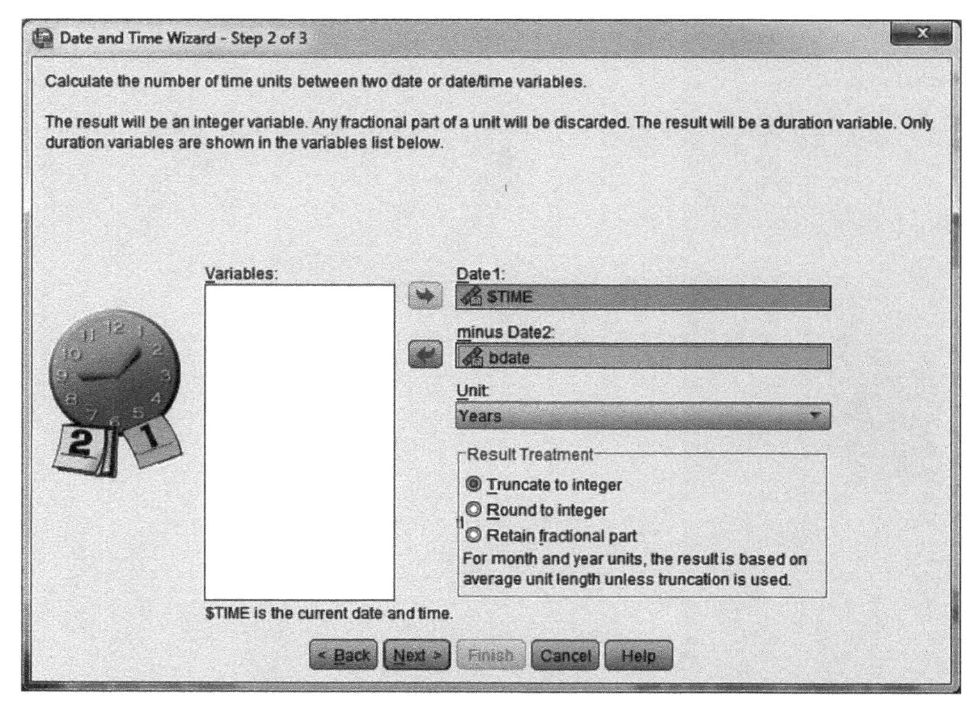

FIGURE 17.6 The specified window in the second step in working with the **Date and Time Wizard**.

ignored in reporting a truncated age. However, we could have chosen to round (cases over half-way to their next birthday would be assigned their next year of age) or to retain the decimal value of their age. When finished, we click the **Next** pushbutton.

The final step in using the **Date and Time Wizard** allows us to name our transformed variable. This is shown in Figure 17.7 where the calculation is displayed in upper left portion of the screen. We have named our variable **age** and have supplied a **Variable Label** just to illustrate it. We retain the default **Create the variable now** option and click **Finish** to generate the new variable.

The newly created **age** variable is placed at the end of the data file as shown in Figure 17.8. We have also performed a second transformation of **bdate**, repeating the same steps as described above, but this time specifying **Retain fractional part** in the second step. We named the variable that is based on this second transformation **exact_age** to differentiate it from the truncated **age** variable and to illustrate what this specification will generate. This **exact_age** variable is shown in Figure 17.9. Note that the second case whose birthday is May 23, 1958, shows a truncated age of 53 years and has a fractional age of 53.80 years. These ages are based on transformations performed on March 10, 2012, at the time the first draft of this chapter was written. Any readers who perform this transformation on the data file will obtain **age** values dependent on the day and time of their particular work session.

17.5 USING A DIFFERENT TIME REFERENT

Because **$TIME** sets the date and time reference to when the transformation is performed, the resulting computation using the current date and time can become (no pun intended) outdated. One way to avoid the problem is to do the following:

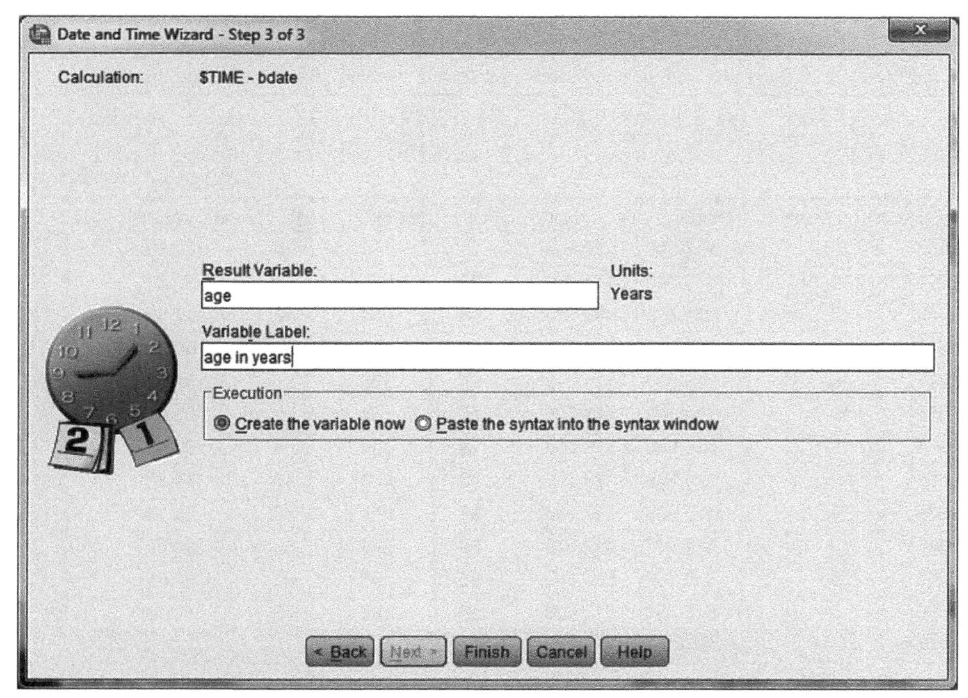

FIGURE 17.7 The final step in working with the **Date and Time Wizard**.

*Employee data.sav [DataSet1] - IBM SPSS Statistics Data Editor

File Edit View Data Transform Analyze Direct Marketing Graphs Utilities Add-ons Window Help

68 : Visible: 10 of 10 Variables

	id	sex	bdate	educ	jobcat	salary	salbegin	jobtime	prevexp	age	var
1	1	2	02/03/1952	15	3	$57,000	$27,000	98	144	60	
2	2	2	05/23/1958	16	1	$40,200	$18,750	98	36	53	
3	3	1	07/26/1929	12	1	$21,450	$12,000	98	381	82	
4	4	1	04/15/1947	8	1	$21,900	$13,200	98	190	65	
5	5	2	02/09/1955	15	1	$45,000	$21,000	98	138	57	
6	6	2	08/22/1958	15	1	$32,100	$13,500	98	67	53	
7	7	2	04/26/1956	15	1	$36,000	$18,750	98	114	56	
8	8	1	05/06/1966	12	1	$21,900	$9,750	98	0	46	
9	9	1	01/23/1946	15	1	$27,900	$12,750	98	115	66	
10	10	1	02/13/1946	12	1	$24,000	$13,500	98	244	66	
11	11	1	02/07/1950	16	1	$30,300	$16,500	98	143	62	
12	12	2	01/11/1966	8	1	$28,350	$12,000	98	26	46	

Data View Variable View

IBM SPSS Statistics Processor is ready

FIGURE 17.8 The newly created **age** variable is available at the end of the data file.

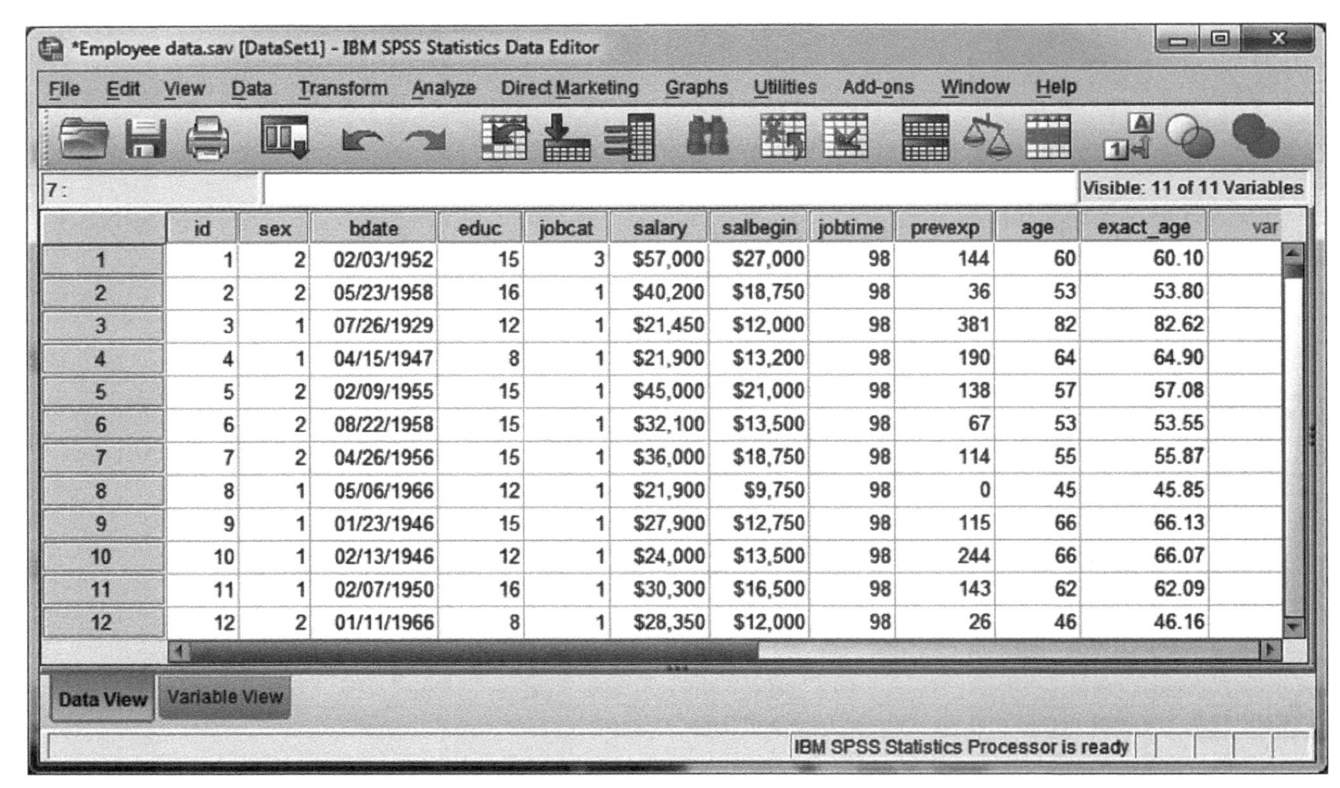

	id	sex	bdate	educ	jobcat	salary	salbegin	jobtime	prevexp	age	exact_age	var
1	1	2	02/03/1952	15	3	$57,000	$27,000	98	144	60	60.10	
2	2	2	05/23/1958	16	1	$40,200	$18,750	98	36	53	53.80	
3	3	1	07/26/1929	12	1	$21,450	$12,000	98	381	82	82.62	
4	4	1	04/15/1947	8	1	$21,900	$13,200	98	190	64	64.90	
5	5	2	02/09/1955	15	1	$45,000	$21,000	98	138	57	57.08	
6	6	2	08/22/1958	15	1	$32,100	$13,500	98	67	53	53.55	
7	7	2	04/26/1956	15	1	$36,000	$18,750	98	114	55	55.87	
8	8	1	05/06/1966	12	1	$21,900	$9,750	98	0	45	45.85	
9	9	1	01/23/1946	15	1	$27,900	$12,750	98	115	66	66.13	
10	10	1	02/13/1946	12	1	$24,000	$13,500	98	244	66	66.07	
11	11	1	02/07/1950	16	1	$30,300	$16,500	98	143	62	62.09	
12	12	2	01/11/1966	8	1	$28,350	$12,000	98	26	46	46.16	

FIGURE 17.9 The additional variable **exact_age** at the end of the data file.

- Select a fixed reference date.
- Create a new date variable (being sure to establish it as a **Date** variable under **Type** in the **Variable View** window).
- Assign the same reference date to all cases.

The variable would then appear in the **Variables** panel in the second step of the **Date and Time Wizard** (see Figure 17.5) and it can be moved into the **Date 1** panel (see Figure 17.6) to represent the variable from which the variable date (e.g., birth date) would be subtracted. This solution, of course, is not perfect either, as it computes an age with respect to an arbitrary reference date rather than the current date.

Descriptive Statistics

	N	Minimum	Maximum	Mean	Std. Deviation	Skewness		Kurtosis	
	Statistic	Statistic	Statistic	Statistic	Statistic	Statistic	Std. Error	Statistic	Std. Error
age age in years	473	41	83	54.91	11.794	.856	.112	-.579	.224
Valid N (listwise)	473								

FIGURE 17.10 Descriptive statistics for **age**.

17.6 USING AGE IN A STATISTICAL ANALYSIS

To illustrate that the **age** variable is available for statistical analysis, we have generated a small set of descriptive statistics using the **Descriptives** procedure. As can be seen in Figure 17.10, ages of the cases in this sample ranged from 41 to 83 with a mean of 54.91 and a standard deviation of 11.79. The distribution tended to be somewhat positively skewed and might be viewed as slightly platykurtic (a little flattened compared to the normal distribution).

PART 6

EVALUATING SCORE DISTRIBUTION ASSUMPTIONS

CHAPTER 18

Detecting Univariate Outliers

18.1 OVERVIEW

Chapters 19 and 20 deal with detecting *outliers*—extreme or unusual values—that may be found in the distribution of scores for a variable. These outliers are referred to as *univariate outliers* (covered in this chapter) when they reside on a single variable, and as *multivariate outliers* (covered in Chapter 19) when they are a function of multiple variables.

The detection of outliers among variables in a data set is essential because the results of many statistical procedures (e.g., multiple regression, analysis of variance) will become distorted when extreme scores are included in the analysis. Outliers can occur because of data entry errors, unusual circumstances (e.g., respondent fatigue or illness), and (more frustratingly) for no known cause.

There are a number of approaches to univariate outlier detection, and we will illustrate a popular technique that uses *boxplots* (also known as *box and whiskers plots*) to identify outliers. Figure 18.1 displays the generic structure of a box and whisker plot. Scores are represented on the *Y*-axis; in the plot produced by IBM SPSS®, percentiles rather than (as proxies for) raw scores are placed on the *Y*-axis.

Boxplots display a square or rectangle that represents the distribution of scores between the first quartile or 25th percentile and the third quartile or 75th percentile (these points are also known as *Tukey's hinges*). The distance between the first and third quartile is defined as the Interquartile Range (IQR). A horizontal line or bar inside the box represents the distribution's midpoint or median (50th percentile).

Beyond the first and third quartiles are vertical lines (called *whiskers*) that extend 1.5 IQR below and above and end at the lower and upper inner fences. Data points beyond the fences are considered outliers. IBM SPSS marks values that fall between ± 1.5 IQR and ± 3.0 IQR with open circles; these may be considered outliers in some contexts. Scores that are beyond ± 3.0 IQR are marked with asterisks and are ordinarily considered to be *extreme scores*.

18.2 NUMERICAL EXAMPLE

The present example is focused on Global Assessment of Functioning (**GAF**) scores for mental health clients. **GAF** scale values can range from 1 (*severe impairment*) to 100

Performing Data Analysis Using IBM SPSS®, First Edition.
Lawrence S. Meyers, Glenn C. Gamst, and A. J. Guarino.

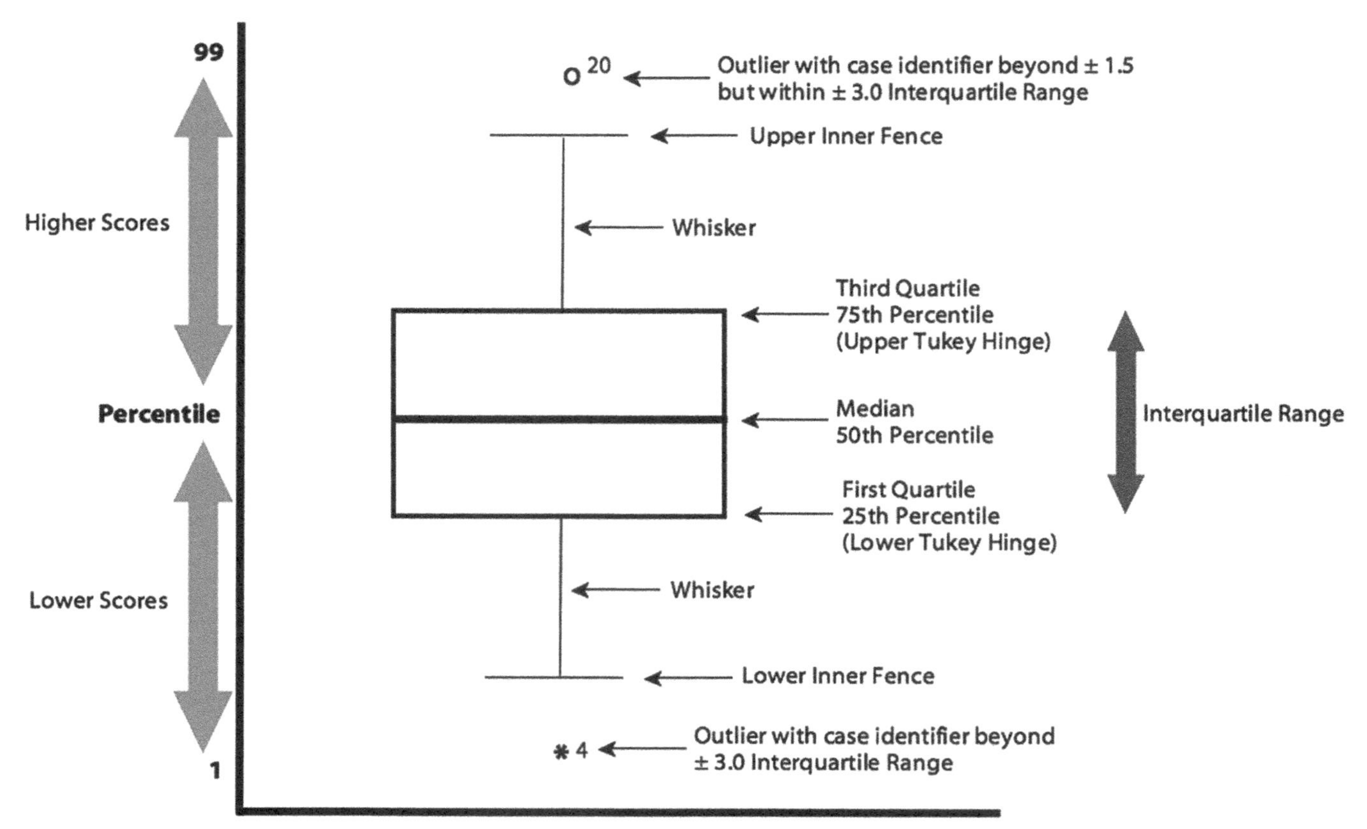

FIGURE 18.1 Generic structure of an IBM SPSS box and whiskers plot. Source: Modified from Meyers et al., 2013.

(*good general functioning*). Also included is the **sex** of the clients with females coded as **1** and males coded as **2**. We will look for univariate outliers on the quantitative (metric) variable **GAF** and again separately for each sex. The data can be found in the file named **GAF and sex**.

18.3 ANALYSIS SETUP: SAMPLE AS A WHOLE

Open the file named **GAF and sex** and, from the main menu, select **Analyze ➔ Descriptive Statistics ➔ Explore**, which produces the **Explore** main dialog window seen in Figure 18.2. From the variable list we have highlighted and moved the **GAF** variable to the **Dependent List** panel.

Selecting the **Statistics** pushbutton produces the **Statistics** dialog window (see Figure 18.3), and we have checked **Descriptives** and **Outliers**. Click **Continue** to return to the main dialog window and select the **Plots** pushbutton. This action produces the **Plots** dialog window shown in Figure 18.4. We have removed the default checkmark from **Stem-and-leaf** but have retained the **Boxplot** of **Factor levels together**. Click **Continue** to return to the main dialog window and click **OK** to perform the procedure.

18.4 ANALYSIS OUTPUT: SAMPLE AS A WHOLE

We refer readers to Chapter 12 for a discussion of the descriptive statistics produced by **Explore**, and will focus on the rest of the output here. The top table in Figure 18.5

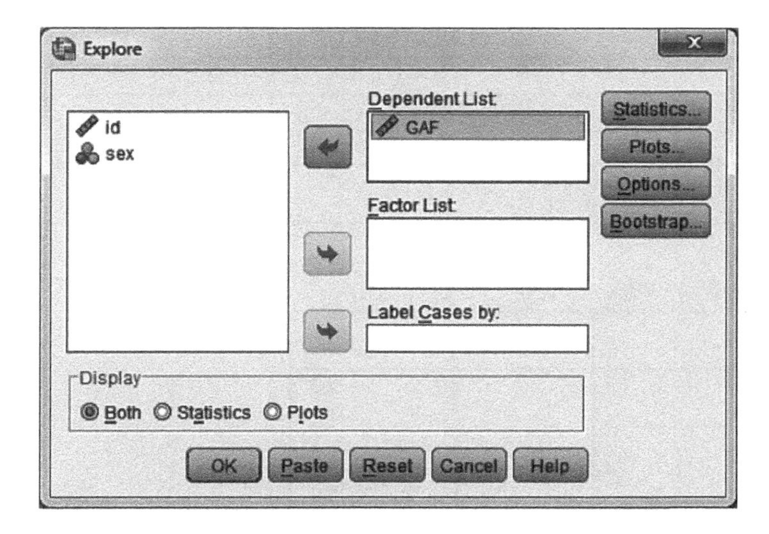

FIGURE 18.2
The main dialog window of **Explore**.

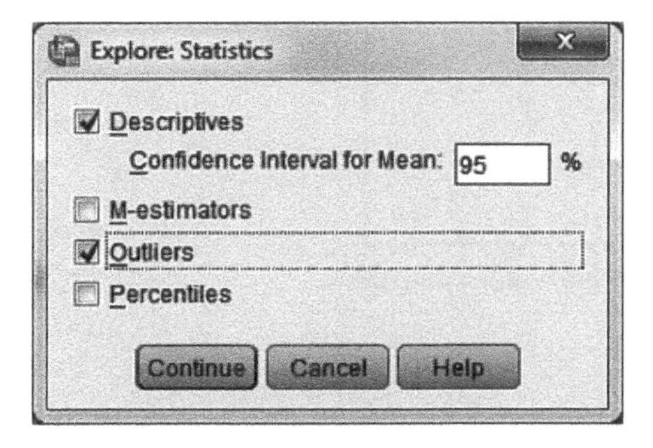

FIGURE 18.3
The **Statistics** dialog window of **Explore**.

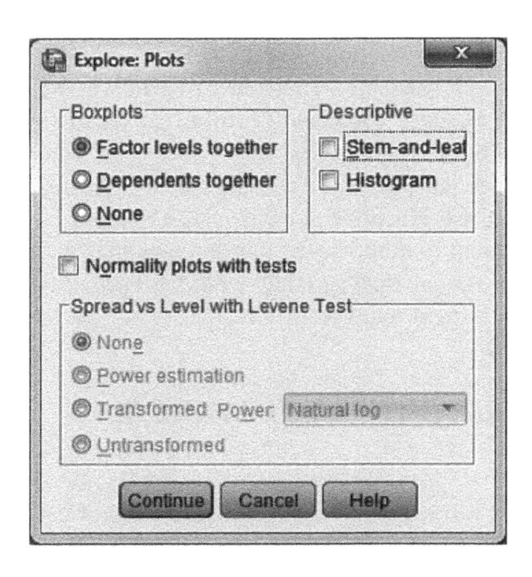

FIGURE 18.4
The **Plots** dialog window of **Explore**.

presents the **Case Processing Summary**, shown here to report a count of cases ($N =$ **25** with no missing cases), as we are not showing the descriptive statistics output.

The bottom table in Figure 18.5 presents the **Extreme Values** output, providing the five highest and five lowest values on the **GAF** variable along with the case number corresponding to each value. These values are not necessarily outliers; rather, they are simply the highest and lowest values in the distribution. We see from the output, for

Case Processing Summary

	Cases					
	Valid		Missing		Total	
	N	Percent	N	Percent	N	Percent
GAF Global Assessment of Functioning	25	100.0%	0	.0%	25	100.0%

Extreme Values

			Case Number	Value
GAF Global Assessment of Functioning	Highest	1	20	70
		2	6	60
		3	5	55
		4	15	55
		5	23	55
	Lowest	1	4	20
		2	24	40
		3	13	40
		4	17	42
		5	2	44

FIGURE 18.5 **Case Processing Summary** and **Extreme Values** output.

example, that **Case Number 20** had the highest **GAF** value of **70** and that **Case Number 6** had the second highest **GAF** value of **60**. Furthermore, **Case Number 4** had the lowest **GAF** value of **20** and **Case Numbers 24** and **13** had the second lowest **GAF** value of **40**.

Figure 18.6 displays the boxplot for the **GAF** variable. There are two univariate outliers for this distribution of **GAF** scores. The score for **Case Number 20** (with a **GAF** score of **70**) is marked with a circle. IBM SPSS identified it as worthy of note because it is more than 1.5 IQR beyond the third quartile; however, the circle indicates that it also lies inside 3.0 IQR. The score for **Case Number 4** (with a **GAF** score of **20**) is marked with an asterisk and indicates it is an extreme score that is more than 3.0 IQR beyond the first quartile. Both of these cases on this variable would be possible candidates for elimination from a subsequent statistical analysis.

18.5 ANALYSIS SETUP: CONSIDERING THE CATEGORICAL VARIABLE OF SEX

It is often useful to examine univariate outliers with respect to the levels of a categorical variable on which data have been collected. We illustrate this using the **sex** variable. From the main menu select **Analyze ➔ Descriptive Statistics ➔ Explore**, which produces the IBM SPSS Explore main dialog window, shown in Figure 18.7. We configure the analysis exactly as just described, but with the following exception: from the variable list we have moved the **GAF** variable to the **Dependent List** panel and we have moved the **sex** variable to the **Factor List** panel. By including a **Factor** in our analysis, each portion of the output will be provided separately for each level of the **Factor** (separately for females and males in our example).

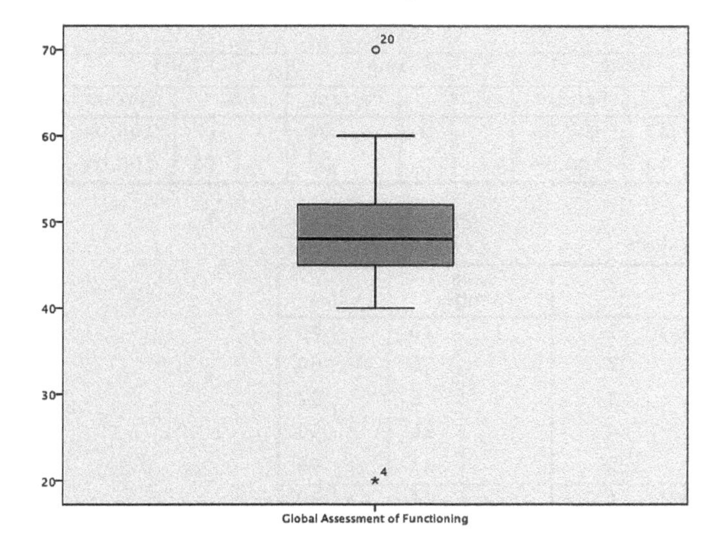

FIGURE 18.6
The box and whiskers plot.

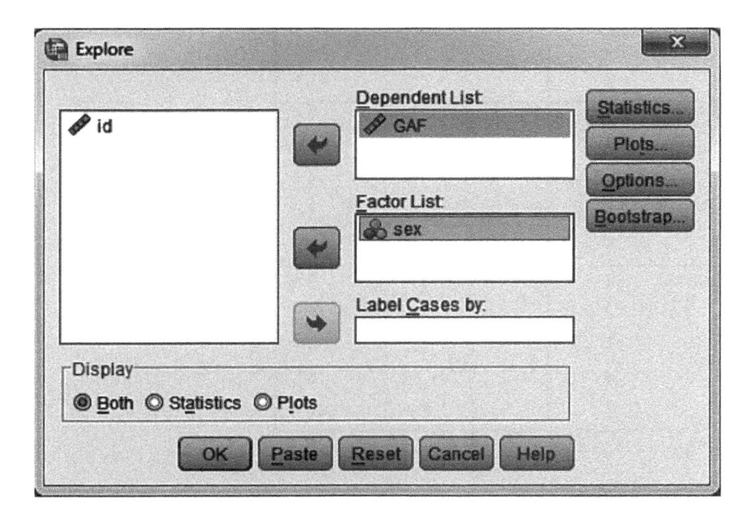

FIGURE 18.7
The main dialog window of **Explore**.

18.6 ANALYSIS OUTPUT: CONSIDERING THE CATEGORICAL VARIABLE OF SEX

The top table in Figure 18.8 presents the **Case Processing Summary** separately for females and males. We note that there are 11 females and 14 males with no missing values. The bottom table in Figure 18.8 presents the **Extreme Values** output, providing the five highest and five lowest values on the **GAF** variable separately for females and males, along with the case number corresponding to each value. We see from the output that the highest (a **GAF** score of **70**) and lowest (a **GAF** score of **20**) values in the output for the sample as a whole were associated with females. IBM SPSS also informs us by way of footnotes that some of the fifth-ranked values in the table have a frequency greater than 1 in the distribution.

Figure 18.9 presents the boxplots for **GAF** separately by the respondent **sex**. From these boxplots, we see the visual depiction of what we learned from examining the **Extreme Values** output, namely, both univariate outliers previously identified are found

Case Processing Summary

	sex	Cases					
		Valid		Missing		Total	
		N	Percent	N	Percent	N	Percent
GAF Global Assessment of Functioning	1 female	11	100.0%	0	.0%	11	100.0%
	2 male	14	100.0%	0	.0%	14	100.0%

Extreme Values

	sex			Case Number	Value
GAF Global Assessment of Functioning	1 female	Highest	1	20	70
			2	6	60
			3	8	52
			4	16	50
			5	11	48
		Lowest	1	4	20
			2	24	40
			3	13	40
			4	2	44
			5	25	45[a]
	2 male	Highest	1	5	55
			2	15	55
			3	23	55
			4	9	54
			5	1	50[b]
		Lowest	1	17	42
			2	14	45
			3	12	45
			4	3	45
			5	18	48[c]

a. Only a partial list of cases with the value 45 are shown in the table of lower extremes.
b. Only a partial list of cases with the value 50 are shown in the table of upper extremes.
c. Only a partial list of cases with the value 48 are shown in the table of lower extremes.

FIGURE 18.8 **Case Processing Summary** and **Extreme Values** output for each level of **sex**.

GAF Global Assessment of Functioning

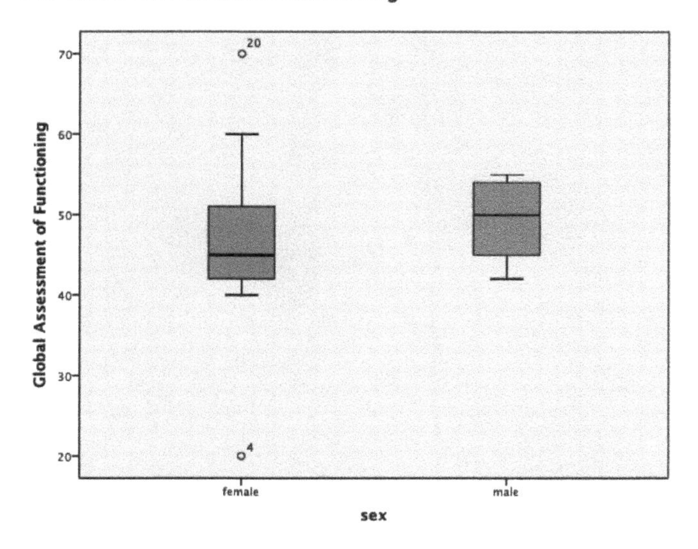

FIGURE 18.9
The box and whiskers plot for each level of **sex**.

among the female respondents. A circle now represents the **GAF** score for **Case Number 4**. This is because the IQR that is its frame of reference for the **GAF** score of **20** is based only on the female clients, and such a score falls inside the range 3.0 IQR for the female group.

This output raises an important issue. Researchers may decide to remove cases with outliers from subsequent analyses based, at least in part, on the boxplot results. If so, we suggest that they do so on the results for the sample as a whole. It is our view that researchers consider the results based on examining the separate levels of the categorical variable as value-added information. Measures of variability based on only a portion of the scores are less stable, and it is ill advised to make decisions about removing cases taking only a subset of the data into account.

CHAPTER 19

Detecting Multivariate Outliers

19.1 OVERVIEW

Multivariate outliers are cases with an extreme combination of values on two or more variables. Detecting this sort of outlier is in some sense more "subtle" than detecting univariate outliers because the combination of variables generating the multivariate outlier is not as readily apparent in the data set as recognizing a univariate outlier in the listing of values for a single variable. Adding to the less obvious nature of identifying multivariate outliers is that a case may not be a univariate outlier on any of the variables tied to the multivariate outlier. For example, we might have an age variable and a variable representing the number of academic peer-reviewed journal articles the individuals have published over their lifetime. Age might range from 13 to 75 years, and the number of publications might range from 0 to 210. The distribution of these individual variables (e.g., their minimum and maximum values) may not be particularly noteworthy or unusual, but a 14 year old with 147 publications would likely be a multivariate outlier either because that case was a child prodigy or (more likely) because one of those values represented a data entry (or data recording) error.

19.2 THE MAHALANOBIS DISTANCE

One common procedure used to identify multivariate outliers is to compute the *Mahalanobis distance* associated with each case. The Mahalanobis distance is the multivariate distance between two points and can be most easily imagined by considering a scatterplot of two variables. In a traditional scatterplot, the *X*- and *Y*-axes intersect at 90 degrees and cases are depicted by data points corresponding to their X and Y coordinates. It is possible to locate the X,Y coordinate of the center of that scatterplot. This central position is known as a *centroid* and can be conceived of as a multivariate mean or average of the multivariate distribution of coordinates. The distance between the centroid and any of the other data points (e.g., using a ruler) is called the Euclidian distance.

When we apply the Mahalanobis distance strategy to measure distance, the process takes into account the correlation between the variables. In the simplified situation of two variables, if the two variables were correlated then the axes would not intersect at

Performing Data Analysis Using IBM SPSS®, First Edition.
Lawrence S. Meyers, Glenn C. Gamst, and A. J. Guarino.

90 degrees and the scatterplot would therefore take on a shape somewhat different from that of our traditional scatterplot. The distance between the centroid and any of the other data points is called the Mahalanobis distance. It is scaled analogously to a z score but in a multivariate manner (Raykov & Marcoulides, 2008).

One very convenient property of the Mahalanobis distances is that their squared values can be approximately described by a chi-square distribution with degrees of freedom equal to the number of variables on which the multivariate combination is based (Varmuza & Filzmoser, 2009). The **Linear Regression** procedure can be used to compute the squared Mahalanobis distance for each case based on a specified set of variables. We then evaluate these values for statistical significance against the appropriate chi-square distribution using a stringent alpha level of .001 (Kline, 2011) because of the large number of evaluations (one for each case) that are involved. Cases whose squared Mahalanobis distance exceeds the critical chi-square value can be considered multivariate outliers and can become possible candidates for exclusion from subsequent data analyses.

19.3 NUMERICAL EXAMPLE

The present example examines mental health clients' **GAF** scores, and their gender (**sex**) and **age**. We generate the Mahalanobis distance values using the **Linear Regression** procedure. We then determine if there are any multivariate outliers based on this combination of the three variables by viewing the distance values and attempting to diagnose which subset of variables might be primarily generating the multivariate outliers. The data can be found in the file named **GAF sex age**.

19.4 ANALYSIS SETUP: LINEAR REGRESSION

Open the IBM SPSS® save file named **GAF sex age** and from the main menu select **Analyze → Regression → Linear**, which produces the main **Linear Regression** dialog window shown in Figure 19.1. We have moved **sex**, **GAF**, and **age** to the **Independent(s)** panel. The results of the Mahalanobis distance calculations are based only on the three variables in the **Independent(s)** panel. Because the **Linear Regression** procedure requires a **Dependent** variable in order to perform the full regression analysis (even though our interest is exclusively in obtaining the squared Mahalanobis distances), we have arbitrarily designated **id** as the dependent variable and moved it to the **Dependent** panel; the dependent variable in this full regression analysis (**id**) is irrelevant, and the main multiple regression results are meaningless and should be ignored.

Clicking the **Save** pushbutton produces the **Save** dialog window, where we have activated the **Mahalanobis** checkbox in the **Distances** panel (see Figure 19.2). This specification will create a new variable in the data file that will be filled in with the squared Mahalanobis distance for each case. Click **Continue** to return to the main dialog window and click **OK** to perform the analysis.

19.5 ANALYSIS OUTPUT: LINEAR REGRESSION

Our focus is exclusively on the new variable that now appears at the end of the data file (see Figure 19.3) and we will ignore the meaningless regression results. The new variable is called **MAH_1** (named generically by IBM SPSS as the first set of saved Mahalanobis distances in our analysis session). It represents the squared Mahalanobis

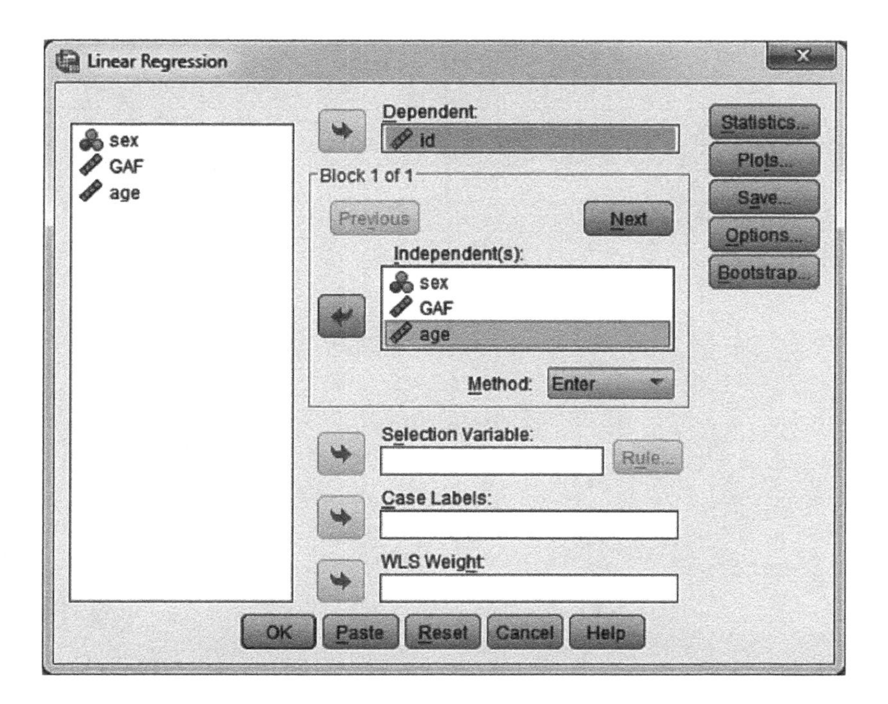

FIGURE 19.1
The main **Linear Regression** dialog window.

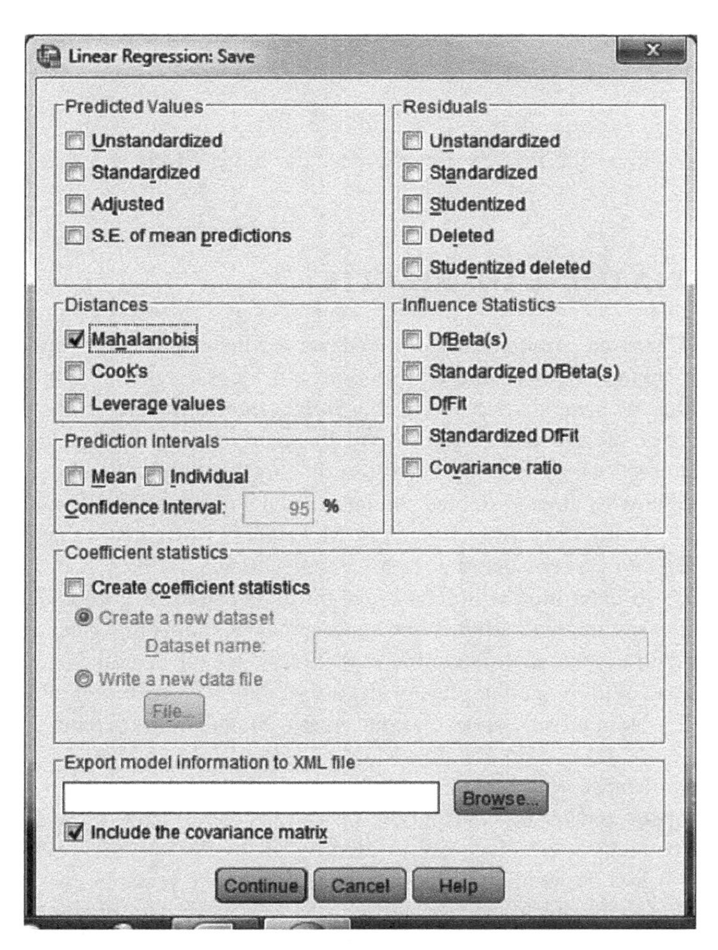

FIGURE 19.2
The **Save** window of **Linear Regression**.

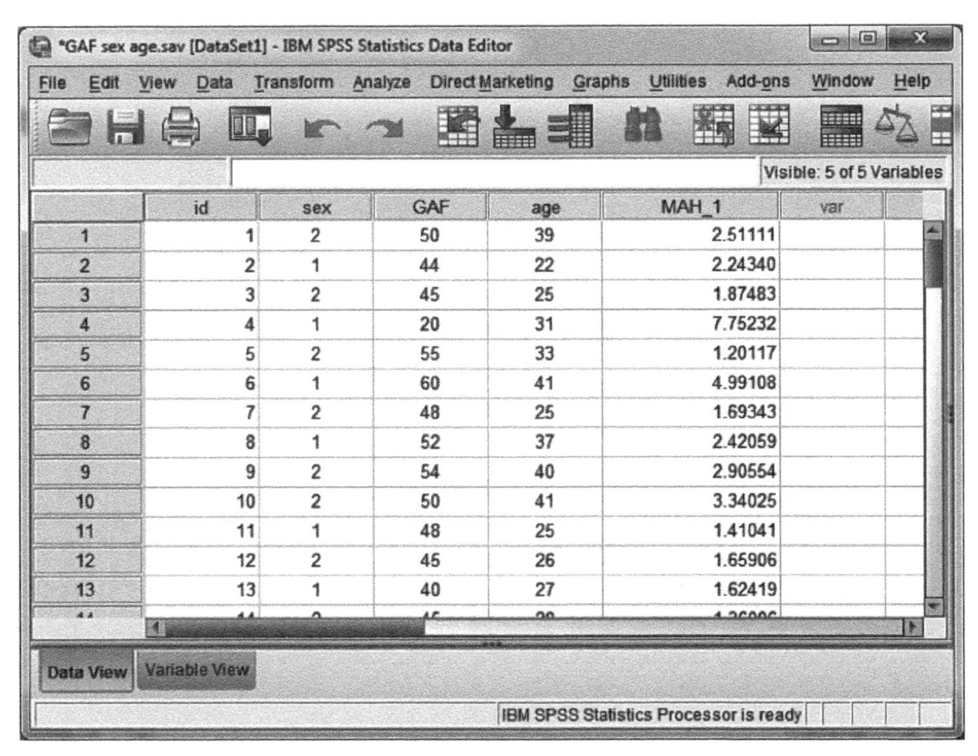

FIGURE 19.3 The data file with the saved **MAH_1** variable.

distance for each case. For example, the first case in the data file has a **MAH_1** value of **2.51111**.

19.6 STRATEGIES TO EXAMINE THE RESULTS

To determine which cases, if any, are multivariate outliers, we evaluate these squared Mahalanobis distance values against a chi-square distribution with degrees of freedom equal to the number of variables we have clicked over to the **Independent(s)** panel in the analysis setup (three variables in the present example). We therefore consult the *Table of Critical Values* for chi-square shown in Table A.1 and use the stringent alpha level of $p = .001$. In the present example with three variables, we have three degrees of freedom. Consulting the table for **df** = **3**, any case with a squared Mahalanobis distance value equal to or greater than **16.266** can be considered a multivariate outlier.

There are alternative ways to determine which cases might be considered as multivariate outliers based on the chi-square evaluation of the squared Mahalanobis distances. The IBM SPSS **Explore** procedure is one alternative but it provides only the five highest (and lowest) values of a variable. With large data files or in case of data files with several outliers, we may not be able to obtain all values beyond our critical Mahalanobis distance value in the **Explore** output.

The squared Mahalanobis distance values cannot fall below zero; thus, we are interested in examining only the upper end of the distribution. Given this asymmetric nature of our inquiry, there is a convenient alternative way to identify all of the outliers. We can **Sort Cases** (see Chapter 7) in a descending order by the new **MAH_1** variable; this will place the highest squared Mahalanobis distance values at the top of the data file and allow us to see them simply by viewing the data file in **Data View** (see Chapter 2).

19.7 EXAMINING THE DATA

From the main menu select **Data ➔ Sort Cases**, move **MAH_1** into the **Sort by** panel, select **Descending** to place the highest value at the top of the spreadsheet (shown in Figure 19.4), and click **OK**. The completed sorting can be seen in the **Data View** mode of the data file in Figure 19.5, where it can be seen that only one case has a **MAH_1** value exceeding our critical value of **16.266**. That one case is **ID # 20**, whose **MAH_1** value is **17.99652**.

Looking across the row for **ID # 20** provides us with a sense of why that case might be an outlier—that client was female (**sex** code of **1**), **21** years old (probably one of the youngest cases in the data set), and had a **GAF** score of **90** (suggesting someone who

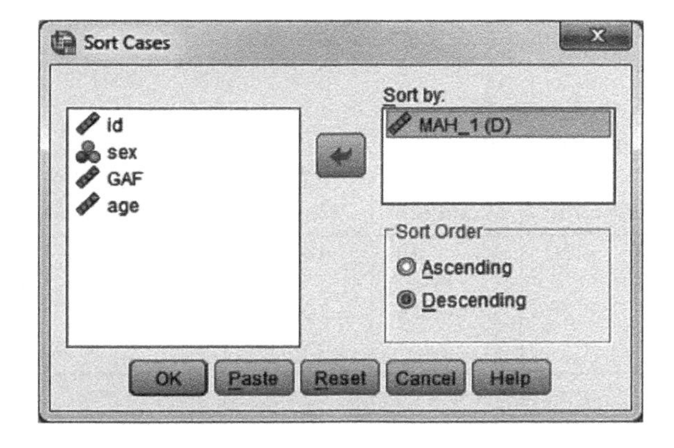

FIGURE 19.4
The main **Sort Cases** dialog window.

*GAF sex age.sav [DataSet1] - IBM SPSS Statistics Data Editor

File Edit View Data Transform Analyze Direct Marketing Graphs Utilities Add-ons Window Help

Visible: 5 of 5 Variables

	id	sex	GAF	age	MAH_1	var
1	20	1	90	21	17.99652	
2	26	2	82	40	10.42064	
3	30	2	79	27	8.11465	
4	4	1	20	31	7.75232	
5	38	1	56	45	6.78726	
6	39	2	73	38	5.92816	
7	46	2	46	44	5.21453	
8	6	1	60	41	4.99108	
9	23	2	55	44	4.91814	
10	34	1	63	23	3.94985	
11	19	2	50	20	3.44982	
12	57	2	38	23	3.41922	
13	58	1	40	38	3.39834	

Data View Variable View

IBM SPSS Statistics Processor is ready

FIGURE 19.5 The data file after sorting the saved **MAH_1** variable.

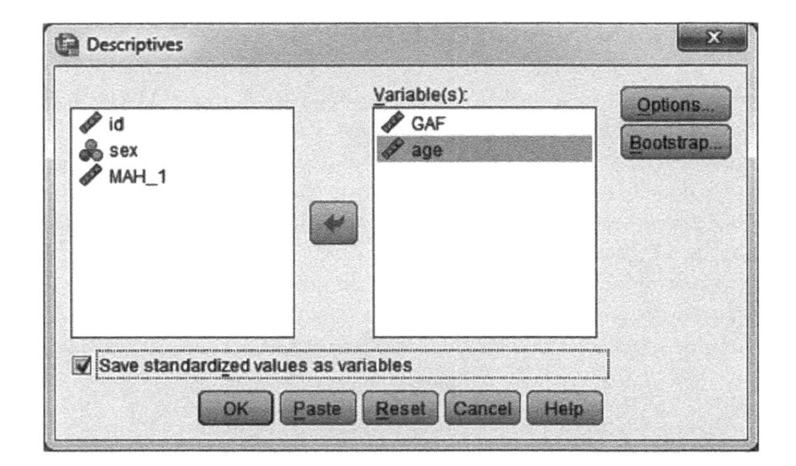

FIGURE 19.6
The main **Descriptives** dialog window.

Descriptive Statistics

	N	Minimum	Maximum	Mean	Std. Deviation
GAF Global Assessment of Functioning	63	20	90	49.43	11.052
age Respondent Age	63	20	45	29.95	6.786
Valid N (listwise)	63				

FIGURE 19.7 The default descriptive statistics output.

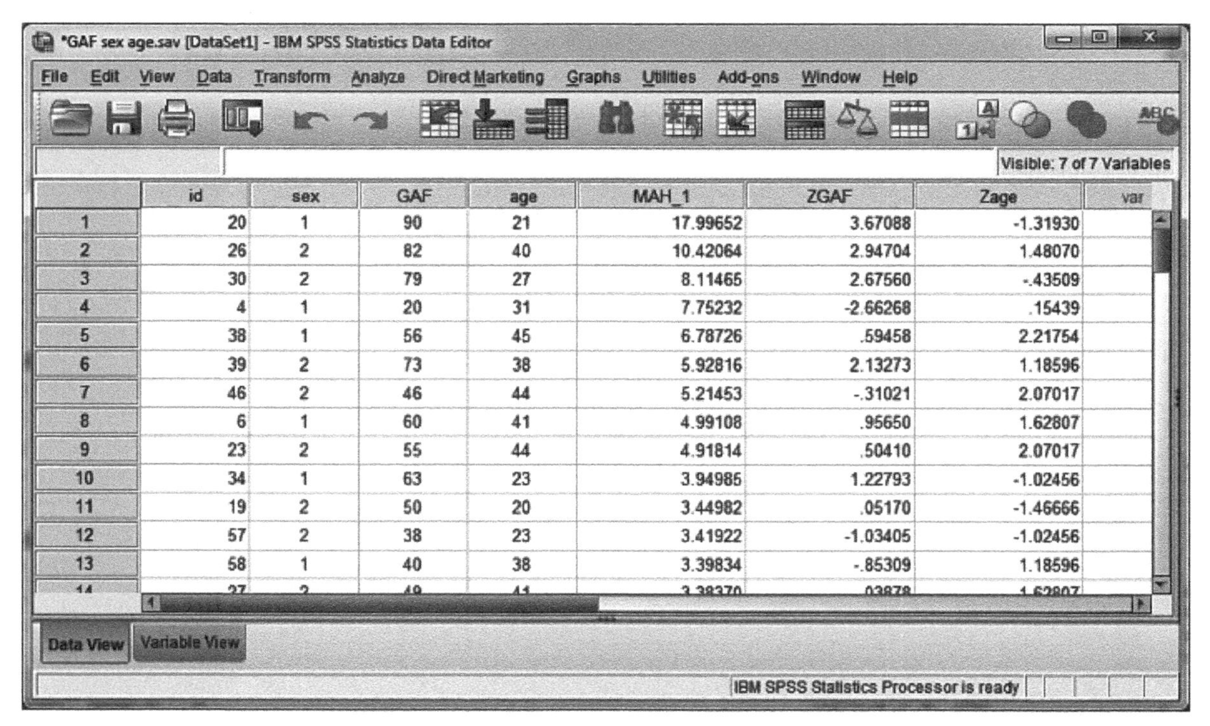

	id	sex	GAF	age	MAH_1	ZGAF	Zage	var
1	20	1	90	21	17.99652	3.67088	-1.31930	
2	26	2	82	40	10.42064	2.94704	1.48070	
3	30	2	79	27	8.11465	2.67560	-.43509	
4	4	1	20	31	7.75232	-2.66268	.15439	
5	38	1	56	45	6.78726	.59458	2.21754	
6	39	2	73	38	5.92816	2.13273	1.18596	
7	46	2	46	44	5.21453	-.31021	2.07017	
8	6	1	60	41	4.99108	.95650	1.62807	
9	23	2	55	44	4.91814	.50410	2.07017	
10	34	1	63	23	3.94985	1.22793	-1.02456	
11	19	2	50	20	3.44982	.05170	-1.46666	
12	57	2	38	23	3.41922	-1.03405	-1.02456	
13	58	1	40	38	3.39834	-.85309	1.18596	

FIGURE 19.8 The data file still sorted on the saved **MAH_1** variable but showing the standardized **GAF** and **age** variables.

is very high functioning). To see how far these two quantitative scores are from their respective univariate means, we can standardize the **age** and **GAF** scores as described in Chapter 13. Briefly, we select **Descriptive Statistics → Descriptives**, move **GAF** and **age** into the **Variable(s)** panel, check the box for **Save standardized values as variables** (shown in Figure 19.6), and click **OK**.

The default descriptive statistics for the two variables are shown in Figure 19.7. **GAF** has a mean of **49.43** with a standard deviation of **11.052**, whereas **age** has a mean of **29.95** with a standard deviation of **6.786**. Thus, the **GAF** score of **ID # 20** of **90** is more than three standard deviation units higher than the mean, and the **age** of **ID # 20** is more than one standard deviation unit lower than the mean.

These extreme scores can be seen very clearly and more precisely in Figure 19.8 where **ID # 20** is still in the first row in our sorted data file. We can now easily see that her **GAF** score of **90** is **3.67088** standard deviation units higher than the mean, and her **age** ($z = -$**1.31930**) is **1.31930** standard deviation units below the mean. This combination (and not her sex, as there is almost an equal number of females and males) of a very high **GAF** score and a somewhat, but not outrageously, young age produced a multivariate outlier. Depending on the research context, this case might or might not be excluded from subsequent data analysis.

CHAPTER 20

Assessing Distribution Shape: Normality, Skewness, and Kurtosis

20.1 OVERVIEW

A normal distribution is a frequency distribution that represents the relative number of occurrences at each value of a variable. The shape of the distribution resembles a bell, and so gets its nickname as the bell-shaped curve. We have drawn the normal distribution in Figure 20.1. Many statistical procedures covered in this chapter assume that the errors associated with the scores on the dependent variable are normally distributed (see Gamst et al. (2008)); when this assumption is met, the dispersion of the scores themselves also tends to be normally distributed. Most of the quantitative (metric) variables that we use in our research (e.g., age, weight, blood pressure, monthly income) approximate this ideal.

Normal distributions have a variety of properties, including the following:

- The normal distribution is horizontally symmetric around the mean of distribution (each side is a mirror image of the other side).
- The middle value represents the mean, median, and mode of the distribution.
- The standard deviation is the distance between the mean and the inflection point (change of the direction of slope) on the side of the curve (see Figure 20.1).
- Standard deviation is an interval-level scale of measurement. Knowing the distance of one standard deviation allows us to fill in the rest of the X-axis. The count of standard deviation units in terms of distance from the mean is a z score scale.
- The mean has a z score of zero.
- Between ± 1.00 standard deviation units, there is approximately 68.26% of the area (or scores); this range corresponds to percentile scores of approximately 16–84.
- Between ± 1.96 standard deviation units, there is approximately 95% of the area; this range corresponds to percentile scores of approximately 2.5–97.5.
- Between ± 3.00 standard deviation units, there is approximately 99% of the area; this range corresponds to percentile scores of approximately 0.1–99.9.
- As the distance from the mean increases, the curve approaches but never reaches the X-axis.

Performing Data Analysis Using IBM SPSS®, First Edition.
Lawrence S. Meyers, Glenn C. Gamst, and A. J. Guarino.

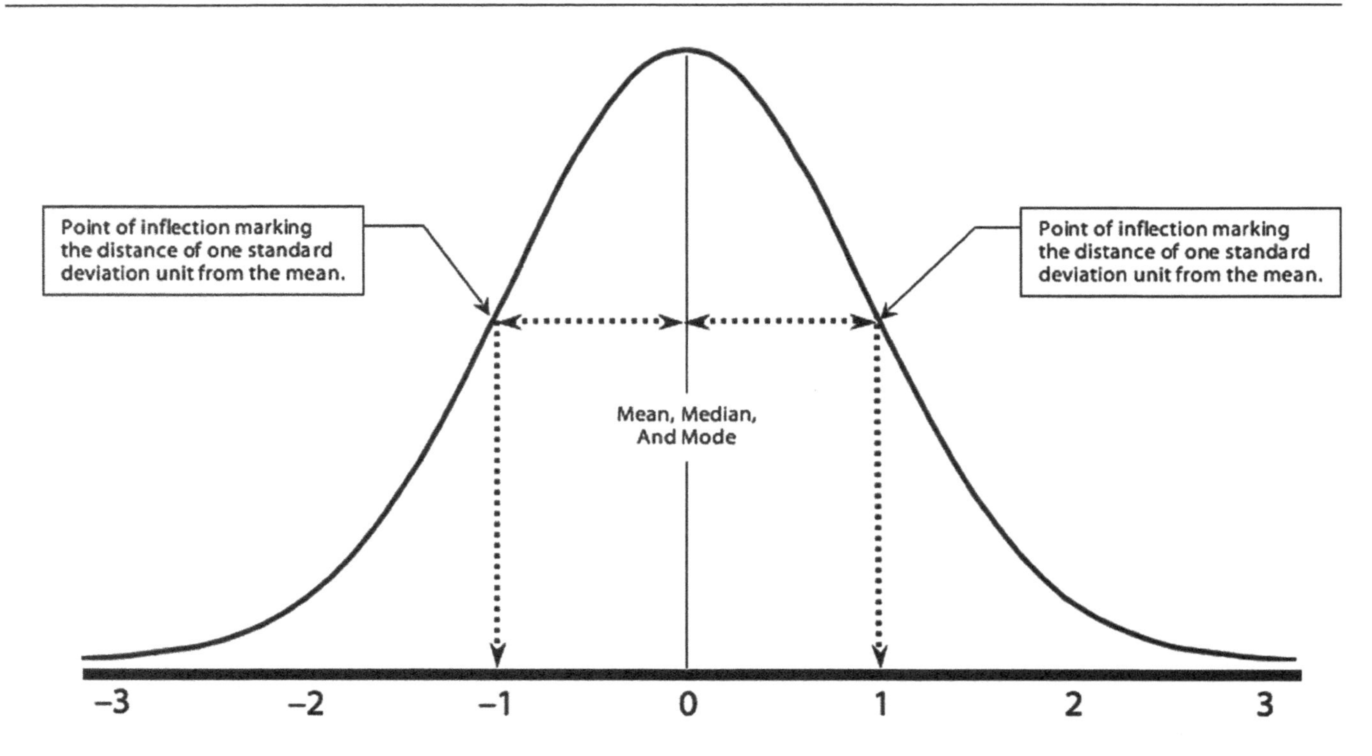

FIGURE 20.1 The normal curve.

Normality can be assessed in three ways.

- We can use the Shapiro–Wilk and/or the Kolmogorov–Smirnov procedures to test the null hypothesis that our distribution is normal in shape.
- We can assess the skewness and kurtosis of our distribution.
- We can assess normality graphically by visually examining histograms and normal probability plots.

IBM SPSS® provides two statistical tests for a distribution's normality: the *Shapiro–Wilk* and the *Kolmogorov–Smirnov* tests. The Shapiro–Wilk test evaluates the null hypothesis that a sample distribution comes from a normally distributed population. Thus, a statistically significant result ($p < .05$) is indicative of a normality violation. The Kolmogorov–Smirnov test standardizes a sample distribution and compares it with a reference standard normal distribution. A statistically significant outcome ($p < .05$) indicates a possible normality violation. However, these tests are quite powerful and with large sample sizes will return a statistically significant result even when the departure from normality is not practically important.

When distributions of scores deviate from normality, it is useful to determine their *skewness* and *kurtosis*. Skewness describes the degree of a distribution's symmetry. A normal curve has a skewness of zero. Skewness values close to zero (less than ± 1.00) are considered indicative of a relatively symmetric distribution (Meyers et al., 2013). Values beyond this range may indicate either *positive skewness*, where the scores are packed around the lower end of the distribution, or *negative skewness*, where the scores are clustered around a distribution's high end.

Kurtosis refers to whether the distribution of scores is compressed (peaked) or flat relative to the normal curve. A normal curve has a kurtosis of zero. Kurtosis values less than ±1.00 are typically considered to be in the range of the normal curve; values outside that range may be considered either *leptokurtic*, they are relatively more peaked distributions (positive kurtosis), or *platykurtic*, they are relatively flatter distributions (negative kurtosis).

Histograms are a graphical means of displaying the frequency distribution of a quantitative or metric variable's distribution of scores. Values of the quantitative variable are represented in the *X*-axis and the frequency of occurrence is represented in the *Y*-axis. IBM SPSS can also superimpose a normal curve on the histogram for a quick visual inspection. Typically, categorical (nonmetric) variables are profiled with bar charts rather than histograms.

In a normal probability plot, the observed values of a variable are rank-ordered and plotted against an expected normal distribution of values. A normal distribution produces values that fall on a straight diagonal line, and the plotted data values are compared to this diagonal. If the data values follow the diagonal line, then normality is assumed (Meyers et al., 2013; Stevens, 2009).

20.2 NUMERICAL EXAMPLE

The present example examines mental health clients' **GAF** scores and their **age**. We will examine the distribution of scores for each variable and assess for any normality violations of these two quantitative (metric) variables. The data can be found in the file named **GAF and age**.

20.3 ANALYSIS STRATEGY

We make use of two of the three descriptive statistics procedures in IBM SPSS, **Frequencies** and **Explore**. There is much the same output available in both. We use the **Frequencies** procedure to generate the skewness and kurtosis information and to obtain a histogram with the normal curve superimposed on it; we use the **Explore** procedure to generate tests of normality and to produce the normal probability plot.

20.4 ANALYSIS SETUP: FREQUENCIES

We open the data file named **GAF and age** and from the main menu select **Analyze → Descriptive Statistics → Frequencies**, which produces the **Frequencies** dialog window shown in Figure 20.2. We have moved **GAF** and **age** to the **Variable(s)** panel and disengaged the **Display frequency tables** checkbox.

Selecting the **Statistics** pushbutton produces the **Statistics** dialog window shown in Figure 20.3, where we have activated only the **Skewness** and **Kurtosis** checkboxes. Clicking **Continue** returns us to the main dialog window.

Selecting the **Charts** pushbutton produces the **Charts** dialog window (see Figure 20.4). We have activated the **Histograms** and **Show normal curve on histogram** options. Click **Continue** to return to the main dialog window and click **OK** to perform the analysis.

20.5 ANALYSIS OUTPUT: FREQUENCIES

The skewness and kurtosis output is shown in Figure 20.5. For the **GAF** variable, both skewness (**.247**) and kurtosis (**−.331**) values are well within the ±1.00 criterion; thus the

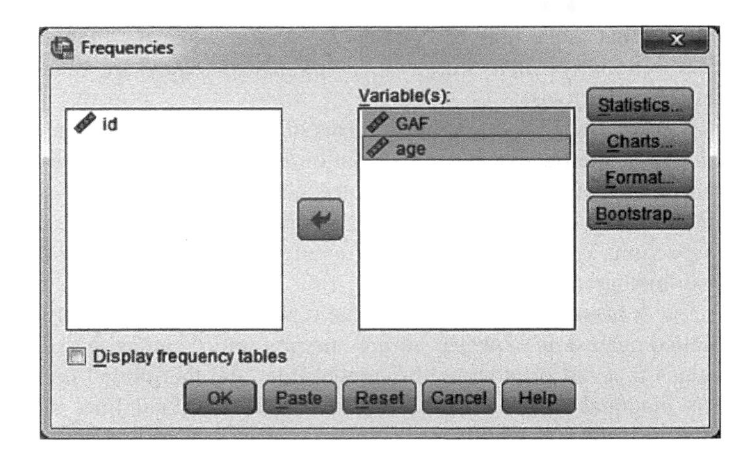

FIGURE 20.2
The main dialog window of **Frequencies**.

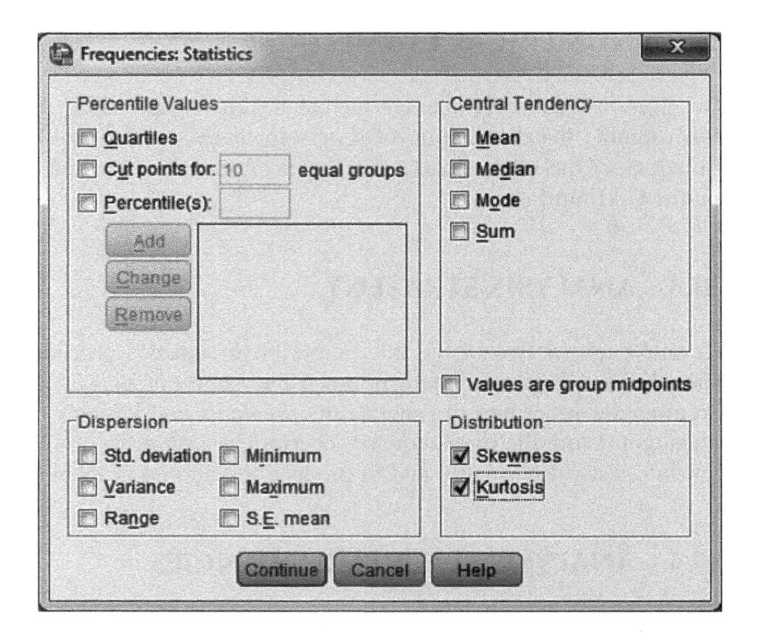

FIGURE 20.3
The **Statistics** dialog window of **Frequencies**.

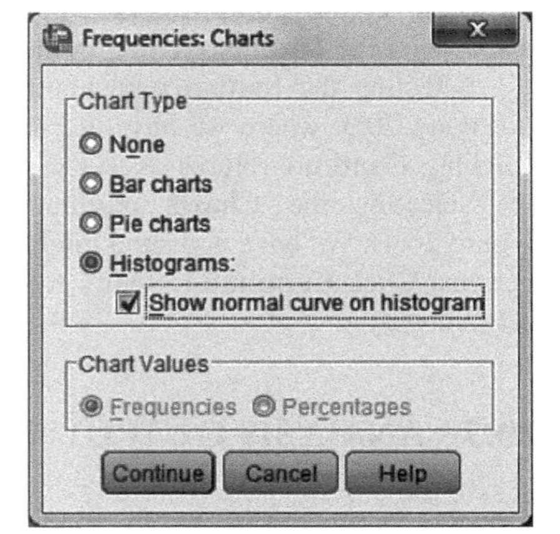

FIGURE 20.4
The **Charts** dialog window of **Frequencies**.

Statistics

		GAF Global Assessment of Functioning	age
N	Valid	38	38
	Missing	0	0
Skewness		.247	1.075
Std. Error of Skewness		.383	.383
Kurtosis		-.331	.379
Std. Error of Kurtosis		.750	.750

FIGURE 20.5

The skewness and kurtosis output.

distribution is relatively symmetric and appears to be neither particularly compressed nor flattened with respect to the normal curve. The kurtosis value for **age** (**.379**) is also within the ±1.00 criterion, but **age** does appear to be positively skewed with a value of **1.075**.

In examining the values of skewness and kurtosis, it is important to take into account their standard errors. We use the standard error, in general, to establish a confidence interval around a statistic. For example, by multiplying the standard error by 1.96 and subtracting that value from and adding that value to the value of the statistic, we can identify the lower and upper limits of the 95% confidence interval. We can then use that 95% confidence interval to evaluate the obtained skewness and kurtosis values.

With respect to **GAF**, the standard error of skewness is **.383**, and 1.96 * **.383** is .751. Subtracting and adding that value to **.247** yields a 95% confidence interval of −.504 to .998. It would appear that a skewness value of zero falls within the confidence interval and thus suggests that the obtained skewness value is not statistically different from zero. The standard error of kurtosis for **GAF** is **.750**, and 1.96 * **.750** is 1.470. Subtracting and adding that value to **−.331** yields a 95% confidence interval of −1.801 to .419. It would appear that a kurtosis value of zero also falls within its confidence interval and thus suggests that the obtained kurtosis value is not statistically different from zero.

With respect to **age**, the standard error of skewness is **.383**, and 1.96 * **.383** is .751. Subtracting and adding that value to **1.075** yields a 95% confidence interval of .324–1.826. It would appear that a skewness value of zero does not fall within the confidence interval, thus suggesting that the distribution of **age** is not symmetric. The standard error of kurtosis for **age** is **.750**, and 1.96 * **.750** is 1.470. Subtracting and adding that value to **.379** yields a 95% confidence interval of −1.091 to.1.849. It would appear that a kurtosis value of zero does fall within the confidence interval and thus suggests that the obtained kurtosis value is not statistically different from zero.

Figure 20.6 presents frequency histograms for the **GAF** and **age** variables. The horizontal axis of each graph is drawn in 10-point increments of **GAF** scores and **age** values, respectively. The vertical axis represents frequency of occurrence. The normal curve superimposed on each histogram provides a convenient visual reference. It appears from the histograms that **GAF** scores closely approximate a normal distribution, but, as we noted earlier, the **age** variable is noticeably positively skewed.

20.6 ANALYSIS SET UP: EXPLORE

The next set of analyses utilizes the **Explore** procedure to produce the Kolmogorov–Smirnov and Shapiro–Wilk normality tests and normal (Q-Q) or probability plots. From the main menu selecting **Analyze ➔ Descriptive Statistics ➔ Explore** produces the **Explore** dialog window shown in Figure 20.7, where we have moved **GAF** and **age** over to the **Dependent List** panel.

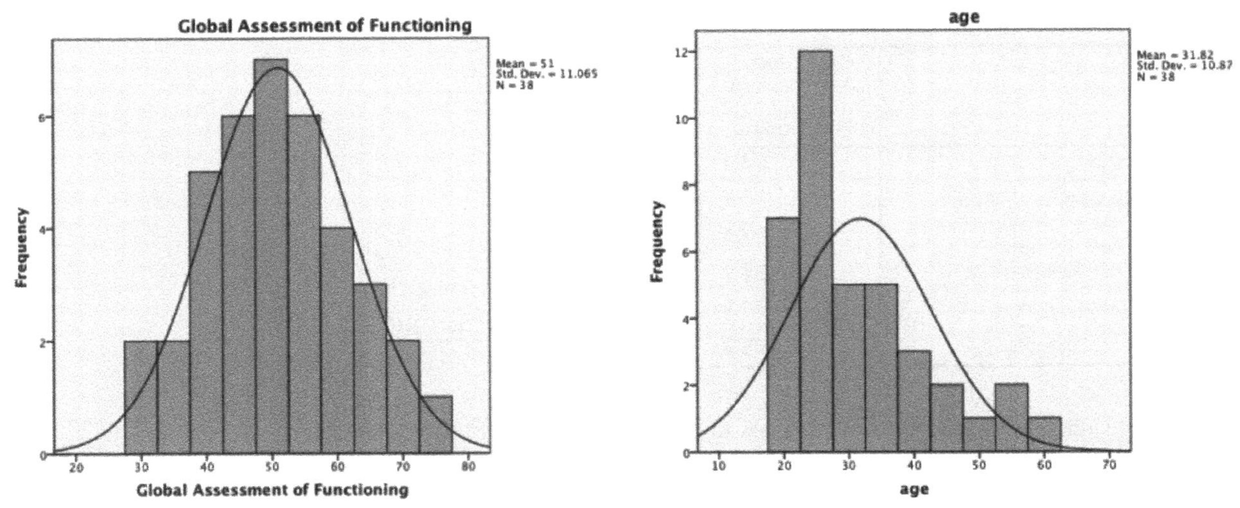

FIGURE 20.6 Histograms for **GAF** and **age**.

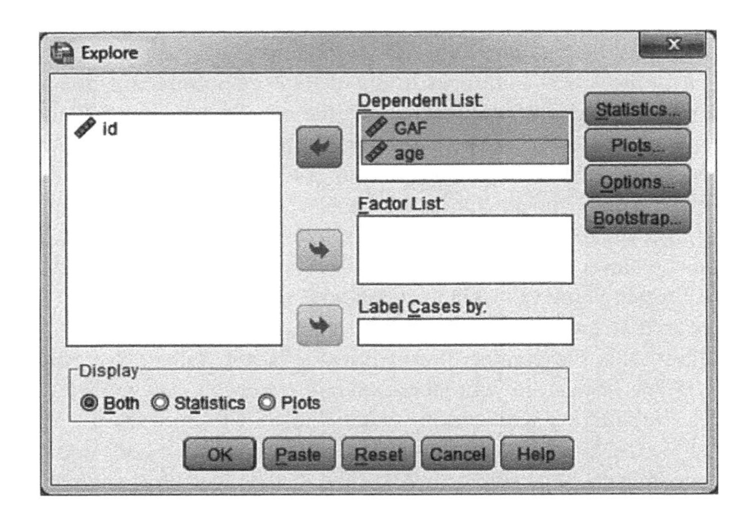

FIGURE 20.7
The main dialog window of **Explore**.

Clicking the **Plots** pushbutton produces the **Plots** dialog window shown in Figure 20.8. We have activated **None** in the **Boxplots** panel (as we have no need for boxplots in this example) and have not selected either plot in the **Descriptive** panel. However, we did activate the **Normality plots with tests** option. Click **Continue** to return to the main dialog window and click **OK** to perform the analysis.

20.7 ANALYSIS OUTPUT: EXPLORE

Figure 20.9 displays the **Kolmogorov–Smirnov** and the **Shapiro–Wilk** normality tests. Neither test indicated that the **GAF** distribution differed from normality (p = **.200** and **.793**, respectively). Conversely, for the **age** variable, both tests were statistically significant (p = **.007** and **.001**, respectively), indicating a normality violation for that variable. This confirms our suspicions when we evaluated skewness and kurtosis with respect to their standard errors.

Figure 20.10 displays the normal probability (or Q-Q) plots for the **GAF** and **age** variables (IBM SPSS also produces a de-trended normal Q-Q plot that removes the linear

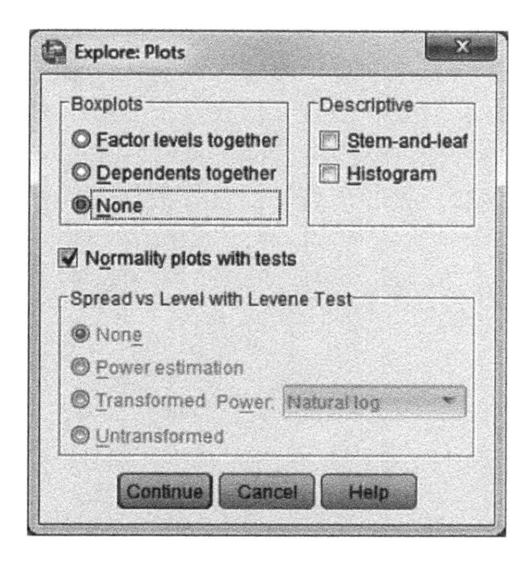

FIGURE 20.8
The **Plots** dialog window of **Frequencies**.

Tests of Normality

	Kolmogorov–Smirnov[a]			Shapiro–Wilk		
	Statistic	df	Sig.	Statistic	df	Sig.
GAF Global Assessment of Functioning	.089	38	.200*	.982	38	.793
age	.171	38	.007	.880	38	.001

a. Lilliefors Significance Correction
*. This is a lower bound of the true significance.

FIGURE 20.9 Outcomes of the **Kolmogorov–Smirnov** and the **Shapiro–Wilk** normality tests.

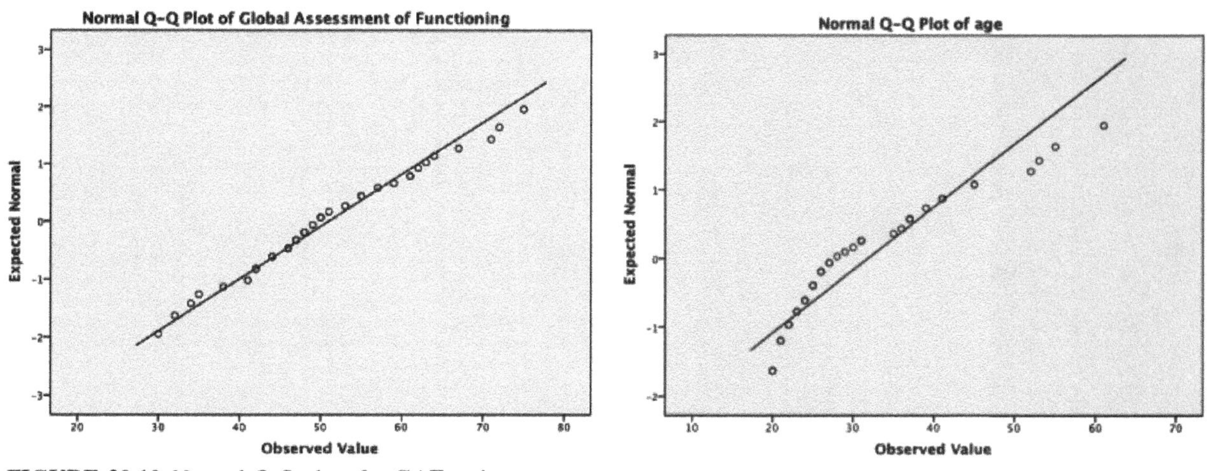

FIGURE 20.10 Normal Q-Q plots for **GAF** and **age**.

trend in the data). This is a normal probability plot or Q-Q plot (the Q stands for *quantiles*, which are points taken at regular intervals in a cumulative distribution). The diagonal line angled from the lower left to the upper right represents an idealized normal distribution. An observed distribution of scores (represented by circles) is superimposed on the diagonal line. A completely normal distribution of scores would have its circles lined up exactly along the diagonal line. As we can see in Figure 20.10, the **GAF** distribution of scores lines up fairly close to the diagonal line, whereas the **age** distribution deviates somewhat from the diagonal line, confirming the previous normality test results violation.

CHAPTER 21

Transforming Data to Remedy Statistical Assumption Violations

21.1 OVERVIEW

A *data transformation* is a mathematical procedure that can be used to modify or adjust variables that violate the statistical assumptions of normality, linearity, and homoscedasticity, or have unusual patterns of univariate or multivariate outliers (Meyers, Gamst, & Guarino, 2009, 2013; Osborne, 2008; Quinn & Keough, 2002). These transformations of the raw data often help reduce distortions (i.e., skewness, kurtosis) found in the shape or dispersion of a distribution of scores, but should be employed judiciously due to their mixed benefits. On the one hand, they can help reduce a distribution's skewness (this also brings outliers closer to the bulk of the scores) and kurtosis. On the other hand, the resulting transformed data values may change the scale of measurement of the construct and the values may be difficult to interpret (see Gamst et al. (2008) and Meyers et al. (2013)).

There are a variety of transformations that are commonly used by researchers to bring the distribution closer to the underlying assumptions of their statistical analyses. If we were to consider transforming the variable X, then some of these transformations are as follows:

- Square root of X
- Base 10 logarithm of X
- Inverse of X (1 divided by X)
- Reflect X (multiply X by −1)
- Square X
- Cube X

Transforming the scores of a variable necessitates performing some mathematical operation on those scores to create a new variable; that is, after the transformation, each case would have a score on the original variable and another score on a newly computed variable. These newly computed scores would be related to the original ones in that they would be the square root, logarithm, or whatever other transformation was performed on that original score.

Performing Data Analysis Using IBM SPSS®, First Edition.
Lawrence S. Meyers, Glenn C. Gamst, and A. J. Guarino.

21.2 NUMERICAL EXAMPLE

The present example is from a hypothetical health care study. One interest motivating this study is to examine the frequency of a sample of 173 clients using the available services of a particular health care provider. One of the variables in that study was the number of visits per year made by the members to their family physician. We have isolated this variable named **doc_visits** in the data file named **health care use**. Our interest is in adjusting its distribution in preparation for further data analysis.

21.3 ANALYSIS STRATEGY

Using transformations to correct skewness often succeeds in correcting kurtosis as well. The variable **doc_visits** exhibits substantial positive skew (a preponderance of lower scores) and positive kurtosis (it is leptokurtic, i.e., the distribution is compressed relative to the normal curve). For our first analysis, we will generate the descriptive statistics for this distribution to serve as our baseline when we evaluate the transformations we use in an effort to make the distribution more normal.

There are three types of transformations that are ordinarily applied to reduce the magnitude of positive skew (Osborne, 2008; Meyers et al., 2009). These work by "drawing in" the larger magnitude scores toward the bulk of the scores that are in the lower magnitude area. It should be noted that different transformations (squaring and cubing the values) targeted to correct negatively skewed distributions work by "pushing out" the scores of somewhat greater magnitude.

We apply in turn the three transformations to ameliorate positive skew by using the **Compute Variable** procedure to build new variables that will represent the transformed values. We then perform descriptive statistics on the set of transformed variables to see how our efforts to normalize the distribution fared. The three transformations we use are as follows:

- *Square Root Transformation.* The effect of taking a square root of an original value is obtaining a smaller value, and the size of this effect is proportional to the magnitude of the original value. For example, the square root of 4 is 2, a difference of two scale units, but the square root of 100 is 10, a difference of 90 scale units. Thus, larger scores are drawn closer to the lower scores.
- *Log Transformation.* Logarithms use a base that is raised to a power to represent a positive number (the log cannot be calculated either for a negative number or for 0). Among the common bases that are used is base 10, although other bases (base 2, base e) are often used as well. Consider this example to see how it works. The logarithm of 100 is the value that the base (we are using base 10 here) must be raised to obtain 100; that is, 10 to what power equals 100? Here, the answer is 2, and so the logarithm of 100 in base 10 is 2. This sort of transformation also draws larger scores inward toward the lower scores.
- *Reflected Inverse Transformation.* As noted earlier, the inverse of X is 1/X. This makes large values small and small values large, thus "flipping" the scaling of the variable. Because it is an effective transformation in many circumstances, researchers like using it for severely skewed distributions but dislike the reversal of the scaling metric. To deal with this reversal, we add one extra step when using this transformation: we first multiply the variable by −1 to "reflect" it. By reflecting the variable, we are "flipping" or reversing the scoring metric in advance so that when we take the inverse and get the flip or reverse, we will be back to the original way in which the variable was scored. We know it sounds a

little convoluted but the end result with all the flipping is that the variable lands right-side up.

21.4 ANALYSIS SETUP: FREQUENCIES OF THE ORIGINAL doc_visits VARIABLE

Open the data file named **health care use** and from the main menu select **Analyze → Descriptive Statistics → Frequencies**. This produces the main **Frequencies** dialog window (screenshots for this general setup can be found in Chapters 10 and 20), where we have moved **doc_visits** to the **Variable(s)** panel and have also disengaged the **Display frequency tables** checkbox. In the **Statistics** dialog window (not shown), we have activated **Mean**, **Std. deviation**, **Skewness**, and **Kurtosis**. We have also requested in the **Charts** dialog window a **Histogram** but have not requested that the normal curve be shown on it.

21.5 ANALYSIS OUTPUT: FREQUENCIES OF THE ORIGINAL doc_visits VARIABLE

The output is shown in Figure 21.1. The mean of **4.63** visits each year is close to the national average of the United States (about 3.6 visits per year), but both the skewness (**1.620**) and kurtosis (**2.408**) are substantial. The skewness can be seen in the histogram where most of the health care clients have a relatively few visits per year but there are some who have quite a few. We can also see the kurtosis in the histogram in that the distribution is relatively compressed in the region of less frequent visits. In general, the distributional assumptions for most of our statistical procedures are not met here.

21.6 ANALYSIS SETUP: SQUARE ROOT TRANSFORMATION

From the main menu select **Transform → Compute Variable**, which produces the **Compute Variable** dialog window (readers can review Chapter 16 for a refresher). We have supplied the name **square_root_visits** for the new variables in the **Target Variable** panel.

IBM SPSS® has a built-in square root function housed in the **Function group** of **Arithmetic**. To access it, we highlight **Arithmetic** in the **Function group** panel and scroll down to **Sqrt** in the **Functions and Special Variables** panel. This is shown in Figure 21.2.

Double-clicking **Sqrt** (or clicking the up arrow to the left of the panel) moves the function into the **Numeric Expression** panel at the top of the dialog window. It appears as **SQRT(?)**, with the "?" highlighted, and double-clicking **doc_visits** replaces the highlighted "?" with our **doc_visits** variable as shown in Figure 21.3. Clicking **OK** performs the computation and places the new variable (**square_root_visits**) at the end of the data file (see Figure 21.4).

21.7 ANALYSIS SETUP: LOG BASE 10 TRANSFORMATION

We engage in an analogous procedure to perform the log base 10 transformation. We select **Transform → Compute Variable**, name the to-be-transformed variable **log10_visits**, scroll down to **Lg10** in the **Arithmetic** function set of the **Function group**, double-click **Lg10** into the **Numeric Expression** panel, and double-click **doc_visits** to place it in the expression (see Figure 21.5). Clicking **OK** performs the computation.

Statistics

doc_visits number of physician visits per year

N	Valid	173
	Missing	0
Mean		4.63
Std. Deviation		3.542
Skewness		1.620
Std. Error of Skewness		.185
Kurtosis		2.408
Std. Error of Kurtosis		.367

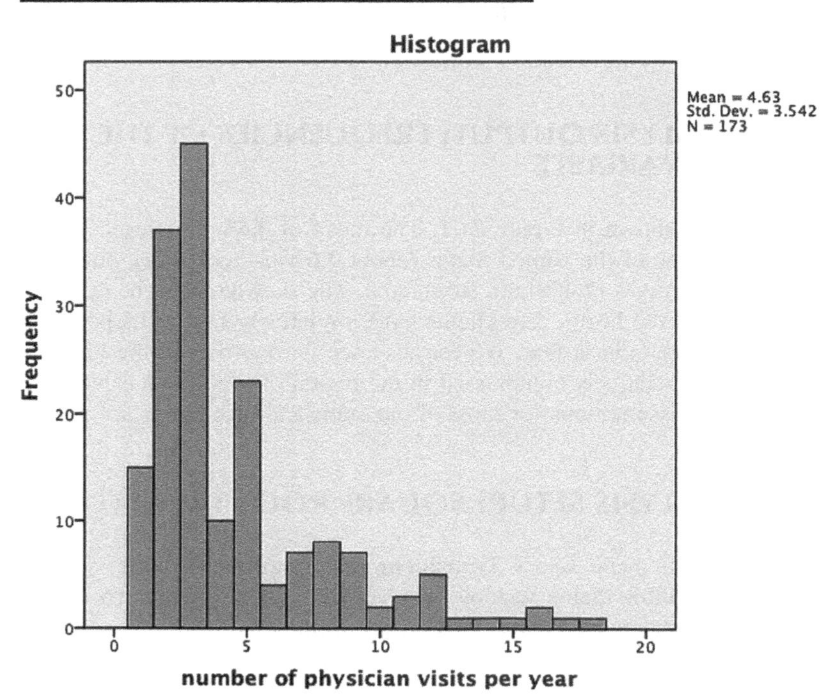

FIGURE 21.1
Descriptive statistics and histogram for **doc_visits**.

21.8 ANALYSIS SETUP: REFLECTED INVERSE TRANSFORMATION

The reflected inverse transformation is accomplished in two successive computations:

- We first multiply **doc_visits** by −1 to create a reflected variable.
- We then perform an inverse transformation on the reflected variable.

There is no established function to reflect a variable, so to reflect **doc_visits**, we select **Transform → Compute Variable**, name the to-be-transformed variable **reflected_visits**, double-click **doc_visits** into the **Numeric Expression** panel, and type * **(−1)** next to it to multiply the variable by −1 as shown in Figure 21.6. Clicking **OK** performs the computation and places the new variable (**reflected_visits**) at the end of the data file.

We now take the inverse of the reflected variable to create the reflected inverse transformation. In the **Compute Variable** window, name the new variable **reflected_inverse_visits**, type **1/** in the **Numeric Expression** panel, highlight **reflected_visits**, and

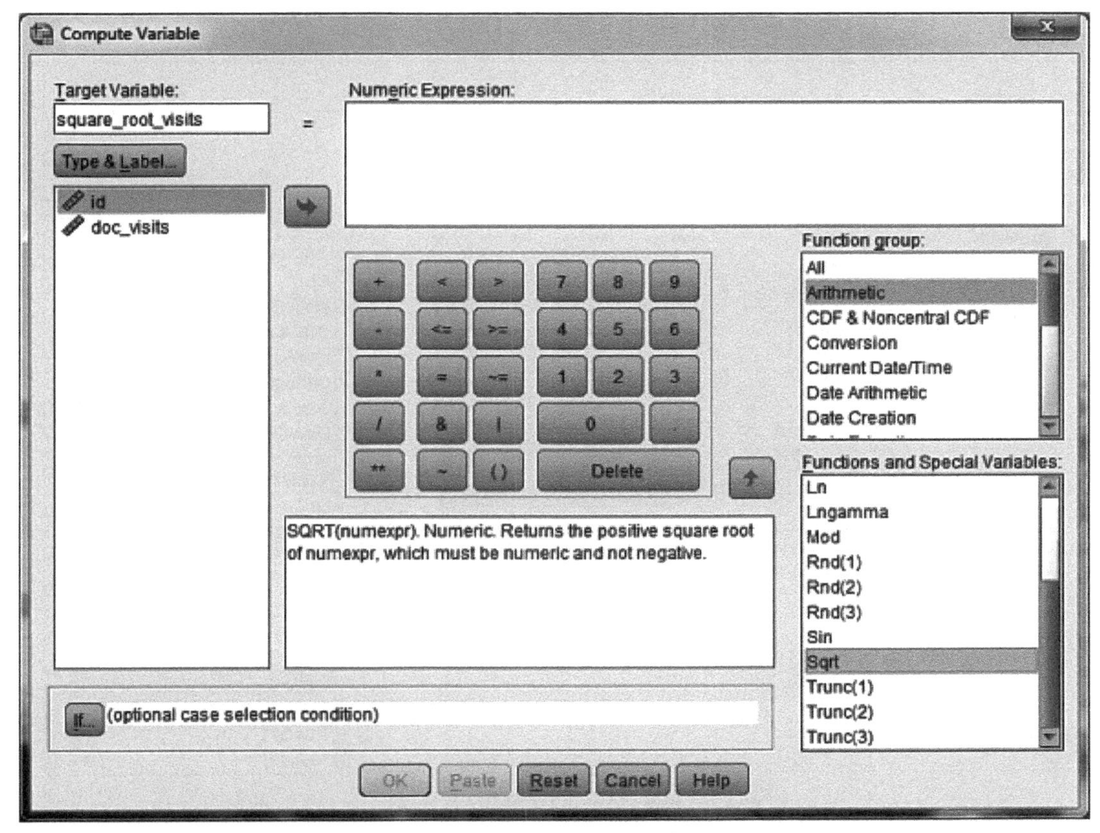

FIGURE 21.2 Accessing the square root (**Sqrt**) function in **Compute Variables**.

then click the right-facing arrow to the left of the **Numeric Expression** panel to move **reflected_visits** into it, as shown in Figure 21.7. Clicking **OK** performs the computation and places the new variable (**reflected_inverse_visits**) at the end of the data file.

A screenshot of a portion of the data file is presented in Figure 21.8. For each case, we see the original **doc_visits** variable together with the three transformed variables (as well as the intermediate **reflected_visits** variable). As we indicated earlier, one of the disadvantages of using transformations is the potential difficulty of directly interpreting them in terms of the original variable. For example, the client with **id** code **175** registered **10** doctor visits, a value that is quite understandable. Our transformed values for that case are less directly understandable, in that the **10** doctor visits show transformed values of **3.16**, **1.00**, and −**.10** for **square_root_visits**, **log10_visits**, and **reflected_inverse_visits**, respectively.

21.9 ANALYSIS SETUP: FREQUENCIES OF ALL OF THE TRANSFORMED VARIABLES

We can now check our work to see if our transformations have reduced the original skewness or kurtosis. From the main menu select **Analyze → Descriptive Statistics → Frequencies**. This produces the main **Frequencies** dialog window (screenshots for this general setup can be found in Chapters 10 and 20), where we have moved **doc_visits**, **square_root_visits**, **log10_visits**, and **reflected_inverse_visits** to the **Variable(s)** panel.

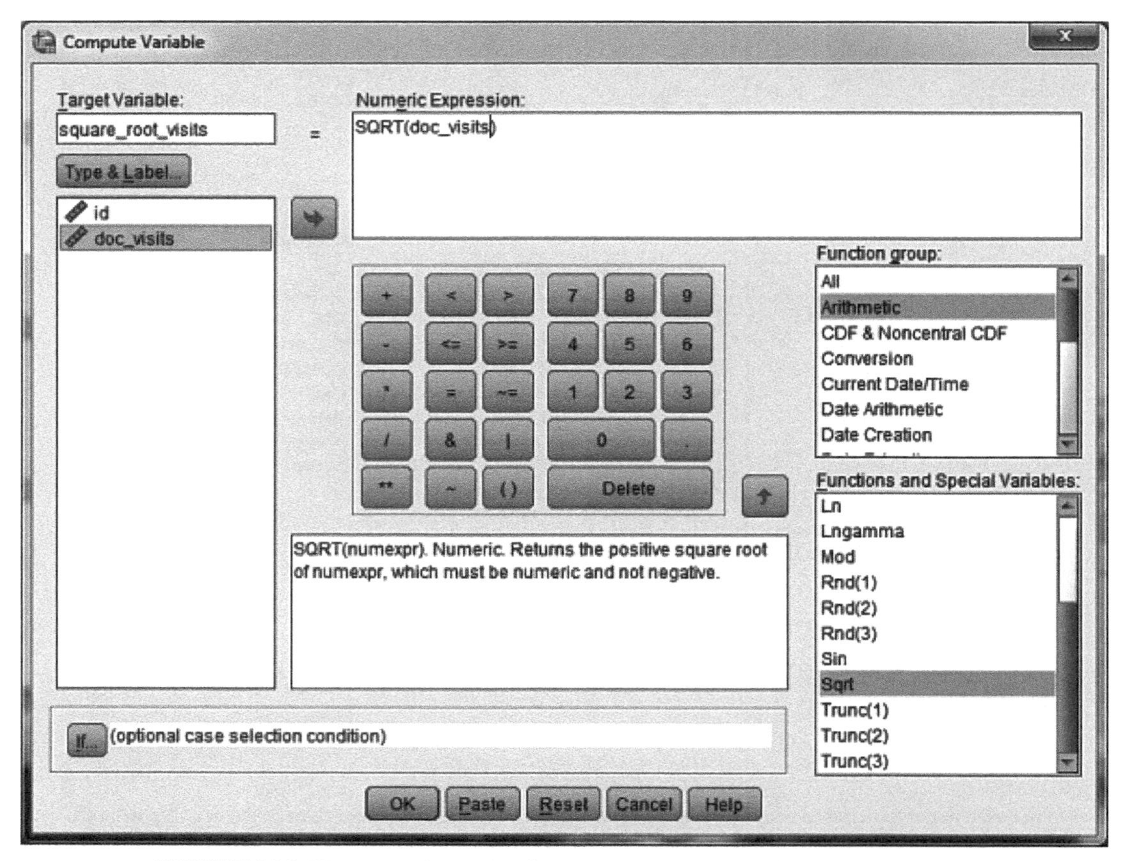

FIGURE 21.3 We are ready to take the square root of **doc_visits**.

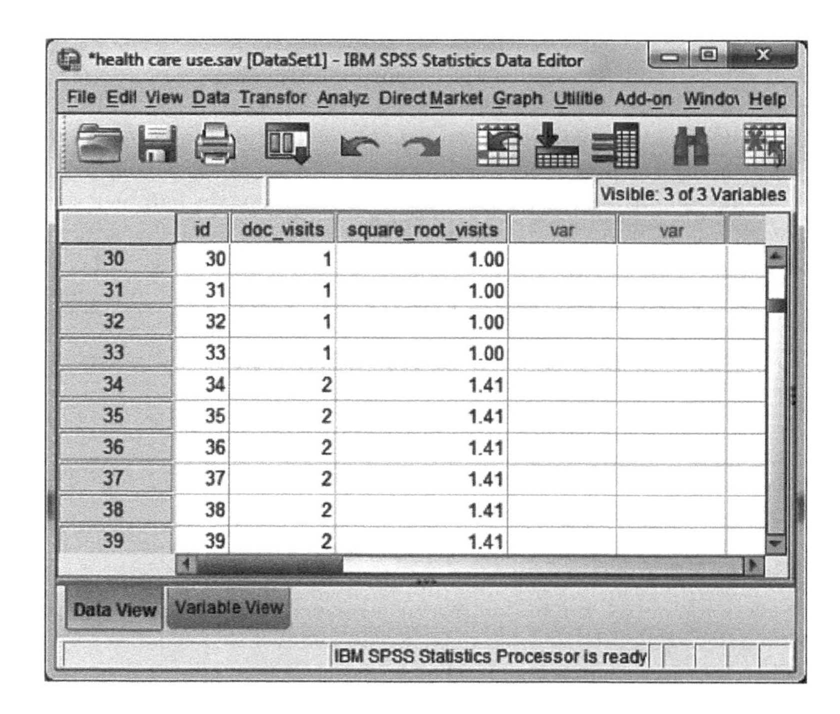

	id	doc_visits	square_root_visits	var	var
30	30	1	1.00		
31	31	1	1.00		
32	32	1	1.00		
33	33	1	1.00		
34	34	2	1.41		
35	35	2	1.41		
36	36	2	1.41		
37	37	2	1.41		
38	38	2	1.41		
39	39	2	1.41		

FIGURE 21.4
A portion of the data file with the inclusion of **square_root_visits**.

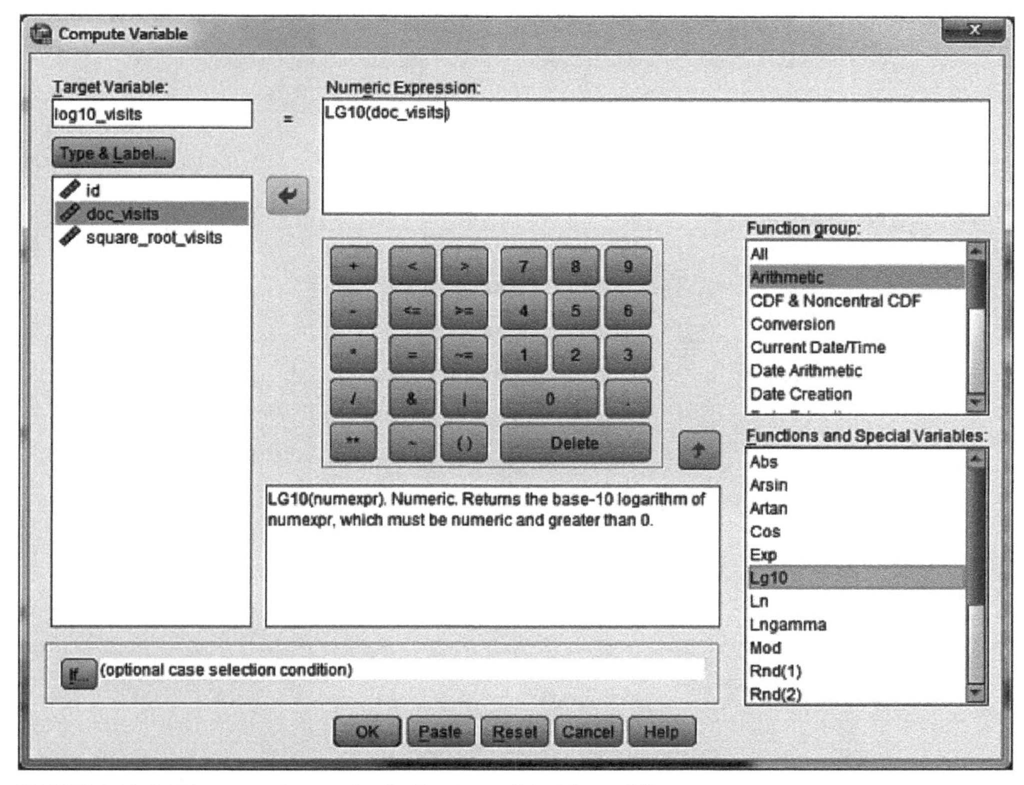

FIGURE 21.5 We are ready to take the log base 10 of **doc_visits**.

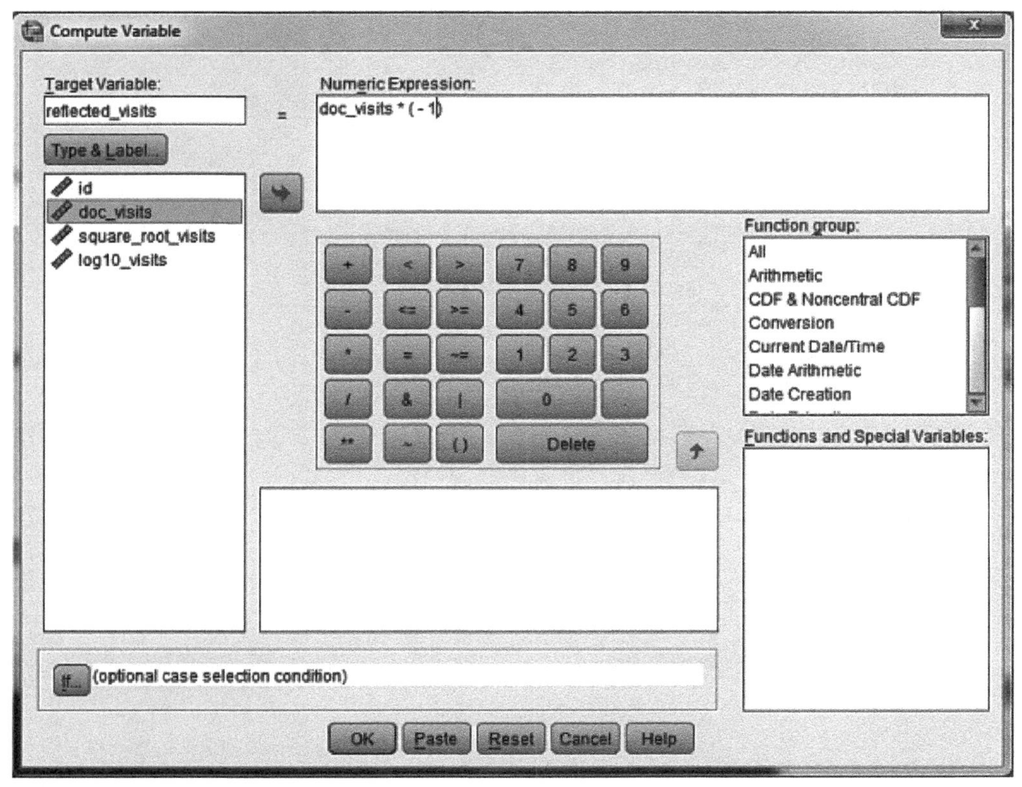

FIGURE 21.6 Reflecting **doc_visits**.

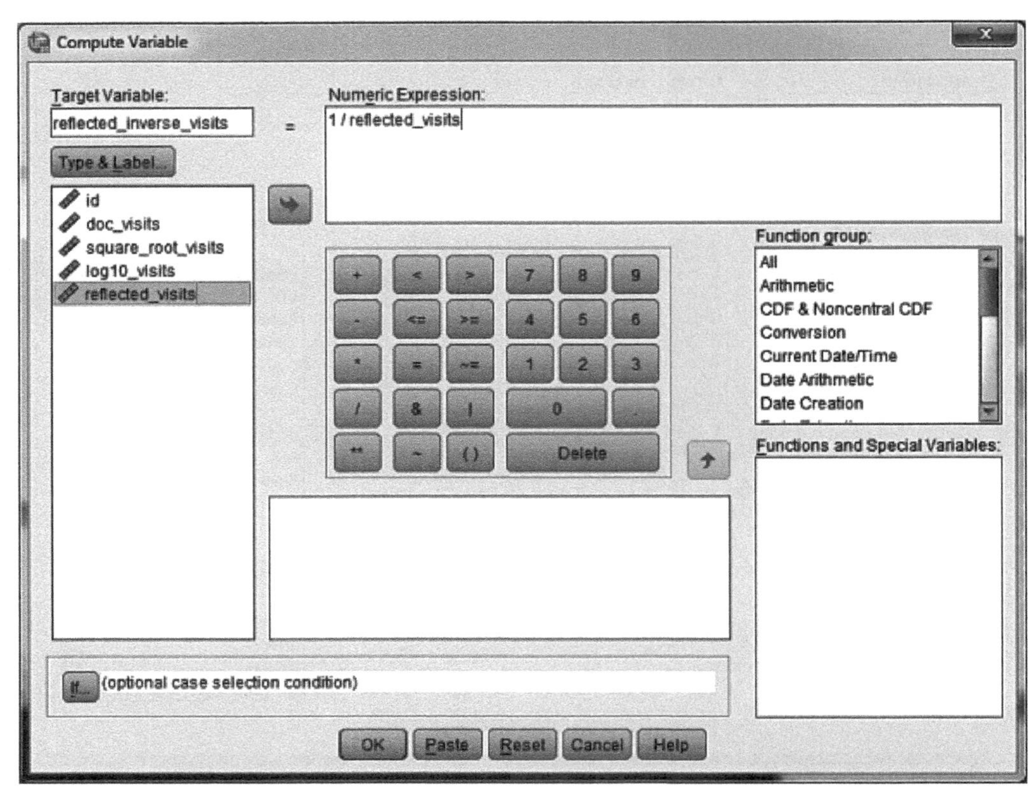

FIGURE 21.7 Inversing the reflected variable to produce a reflected inverse transformation.

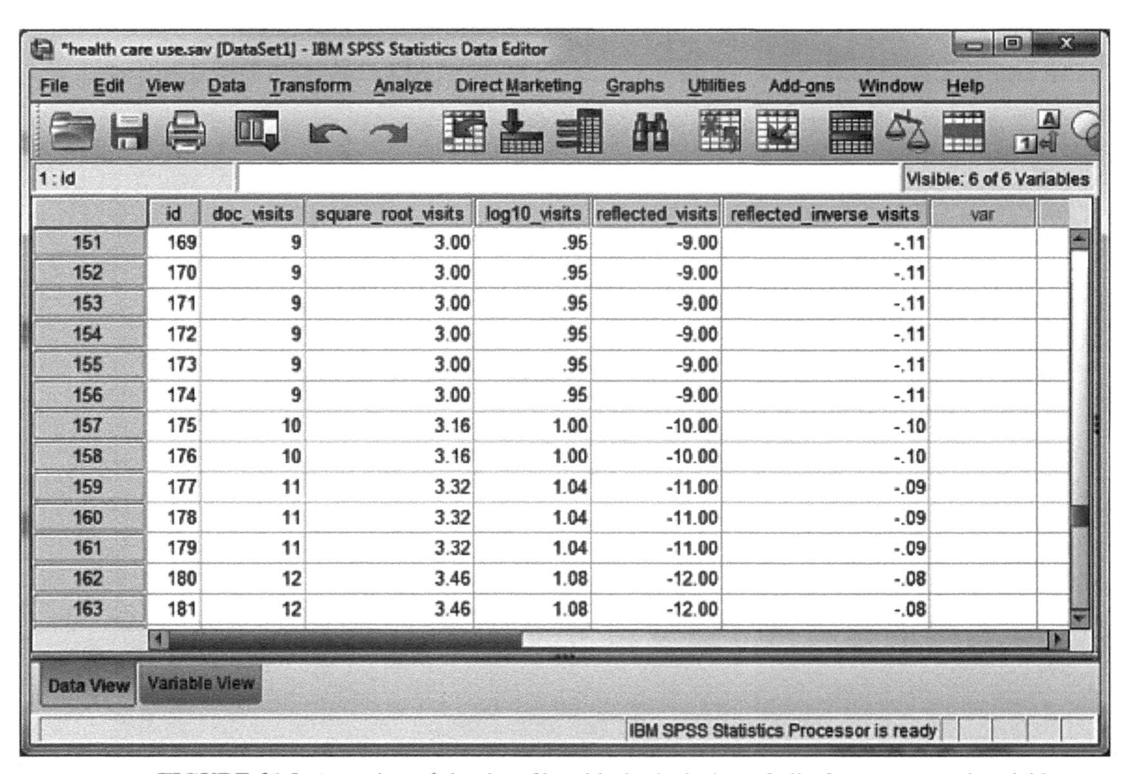
*health care use.sav [DataSet1] - IBM SPSS Statistics Data Editor

File Edit View Data Transform Analyze Direct Marketing Graphs Utilities Add-ons Window Help

1 : id — Visible: 6 of 6 Variables

	id	doc_visits	square_root_visits	log10_visits	reflected_visits	reflected_inverse_visits	var
151	169	9	3.00	.95	-9.00	-.11	
152	170	9	3.00	.95	-9.00	-.11	
153	171	9	3.00	.95	-9.00	-.11	
154	172	9	3.00	.95	-9.00	-.11	
155	173	9	3.00	.95	-9.00	-.11	
156	174	9	3.00	.95	-9.00	-.11	
157	175	10	3.16	1.00	-10.00	-.10	
158	176	10	3.16	1.00	-10.00	-.10	
159	177	11	3.32	1.04	-11.00	-.09	
160	178	11	3.32	1.04	-11.00	-.09	
161	179	11	3.32	1.04	-11.00	-.09	
162	180	12	3.46	1.08	-12.00	-.08	
163	181	12	3.46	1.08	-12.00	-.08	

Data View | Variable View

IBM SPSS Statistics Processor is ready

FIGURE 21.8 A portion of the data file with the inclusion of all of our computed variables.

Statistics

		doc_visits number of physician visits per year	square_root_visits	log10_visits	reflected_inverse_visits
N	Valid	173	173	173	173
	Missing	0	0	0	0
Mean		4.63	2.0221	.5577	-.3491
Std. Deviation		3.542	.73776	.30469	.24460
Skewness		1.620	.924	.168	-1.435
Std. Error of Skewness		.185	.185	.185	.185
Kurtosis		2.408	.348	-.447	1.784
Std. Error of Kurtosis		.367	.367	.367	.367

FIGURE 21.9 Descriptive statistics for the original and transformed variables.

In the **Statistics** dialog window (not shown), we have requested **Mean**, **Std. deviation**, **Skewness**, and **Kurtosis**. We have also requested in the **Charts** dialog window a **Histogram**.

21.10 ANALYSIS OUTPUT

Figure 21.9 displays the **Statistics** results from our **Frequencies** analysis. The first column describes our original **doc_visits** variable, and the issue is how much improvement is obtained with each transformation. The square root transformation certainly improved the shape of the distribution; skewness was lowered from **1.620** to **.924**, and kurtosis was lowered from **2.408** to **.348**. Thus, the skewness is close to the ±1.00 guideline but the value for kurtosis is relatively close to that of the normal distribution.

The log base 10 transformation appeared to be a bit stronger than the square root transformation; skewness was lowered from **1.620** to **.168**, and kurtosis was lowered from **2.408** to **−.447**. Taking the log of the variable reduced skewness down to a negligible amount but actually overshot the zero mark for kurtosis, driving it negative or mildly platykurtic (a little flattened). Nonetheless, these values are also acceptable.

The reflected inverse transformation just was not effective in this context. It drove the skewness all the way to **−1.435** but only lowered the kurtosis to a level that was still leptokurtic. Given a choice of these three transformations, we would be inclined to go with the log transformation for this particular data set.

PART 7

BIVARIATE CORRELATION

CHAPTER 22

Pearson Correlation

22.1 OVERVIEW

In its most general sense, a correlation indexes the extent to which the variables in the analysis are related. There are several correlation coefficients applicable to the research conducted in the behavioral, social, and medical sciences, but the most widely used is the Pearson Product Moment Correlation, usually referred to as the Pearson correlation or just the Pearson r. It assumes that the variables represent approximately interval measurement and that they are approximately normally distributed; outliers can seriously distort the value of the correlation, and so should be appropriately handled before data analysis.

The Pearson r was developed by Karl Pearson (1896) based on the initial development of the idea by Sir Francis Galton (1886, 1888). It assesses the degree to which two variables are linearly related. To the extent that the relationship between the two variables is not linear (e.g., a U-shaped function), the Pearson r will substantially underestimate how strongly the two variables are associated (in case of a symmetric U-shaped function, the Pearson r will be zero).

To say that two variables are related means that they covary. One way to think of covariation is that variation in one variable is synchronous with that of the other variable. For example, cases with higher values on one variable might tend to have lower values on the other variable. A related way to think of covariation is that values of one variable are predictable by some margin better than chance from the knowledge of the corresponding values on the other. The Pearson r^2 indexes the strength of the relationship, that is, the amount of variance shared between the two variables.

22.2 NUMERICAL EXAMPLE

The data we use for our example are extracted from a study of personality variables on 425 university students. Data are present in the data file named **Personality**. We focus on two of the personality variables in this chapter: **beckdep** as a measure of depression and **regard** as a measure of self-regard.

Performing Data Analysis Using IBM SPSS®, First Edition.
Lawrence S. Meyers, Glenn C. Gamst, and A. J. Guarino.

22.3 ANALYSIS SETUP: CHECKING FOR LINEARITY

Because the Pearson *r* assesses the degree to which a pair of variables is linearly related, it is useful to make a rough determination of the degree to which the relation is linear before calculating the Pearson *r*. This can be accomplished by examining the scatterplot of the two variables.

To obtain a scatterplot, we open the data file named **Personality** and from the main menu select **Graphs → Legacy Dialogs → Scatter/Dot**, which produces the window shown in Figure 22.1. We select **Simple Scatter** and click **Define** to open the **Simple Scatter** dialog screen presented in Figure 22.2. We will arbitrarily place **beckdep** in the **Y Axis** panel and **regard** in the **X Axis** panel. Clicking **OK** produces the scatterplot.

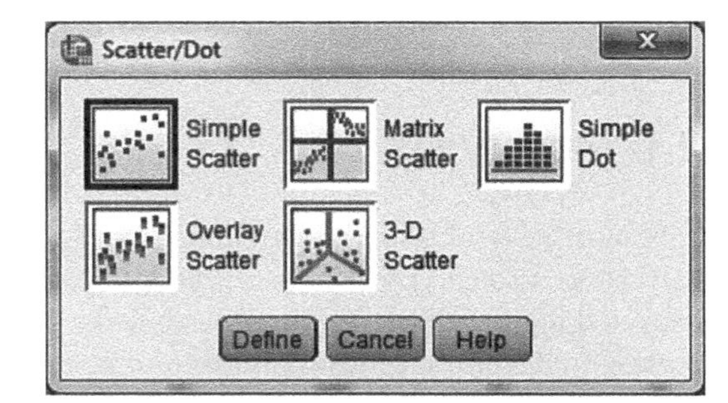

FIGURE 22.1
Scatter/Dot selection screen.

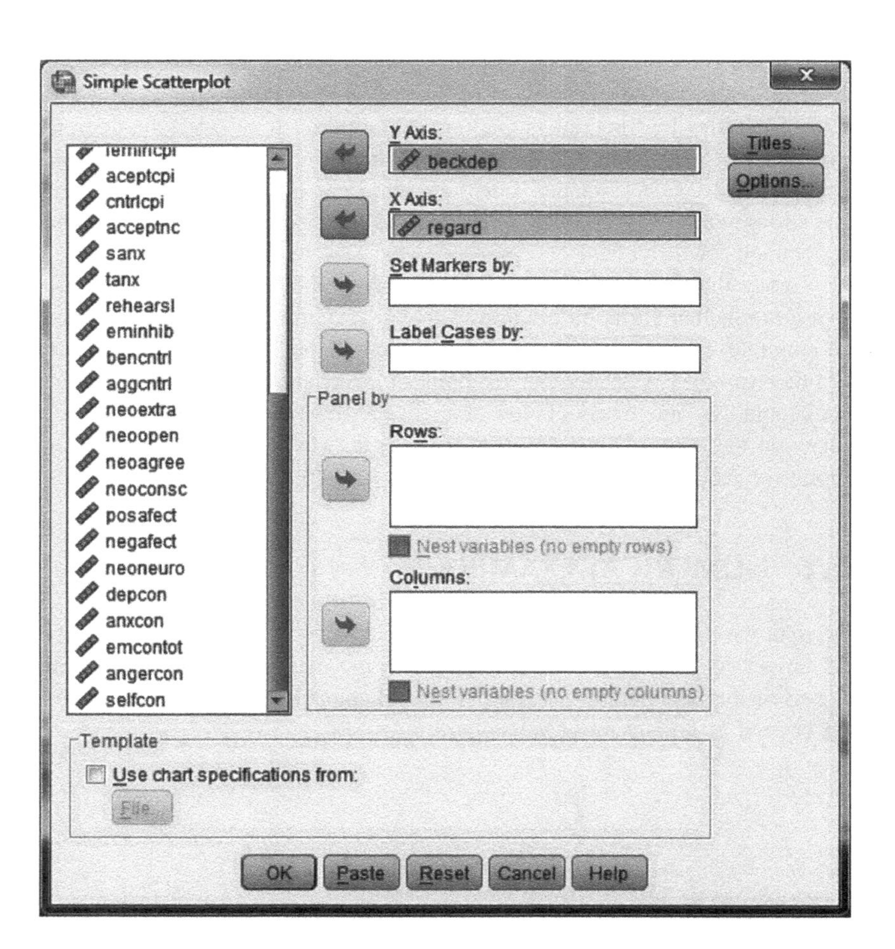

FIGURE 22.2
The **Simple Scatter** dialog screen.

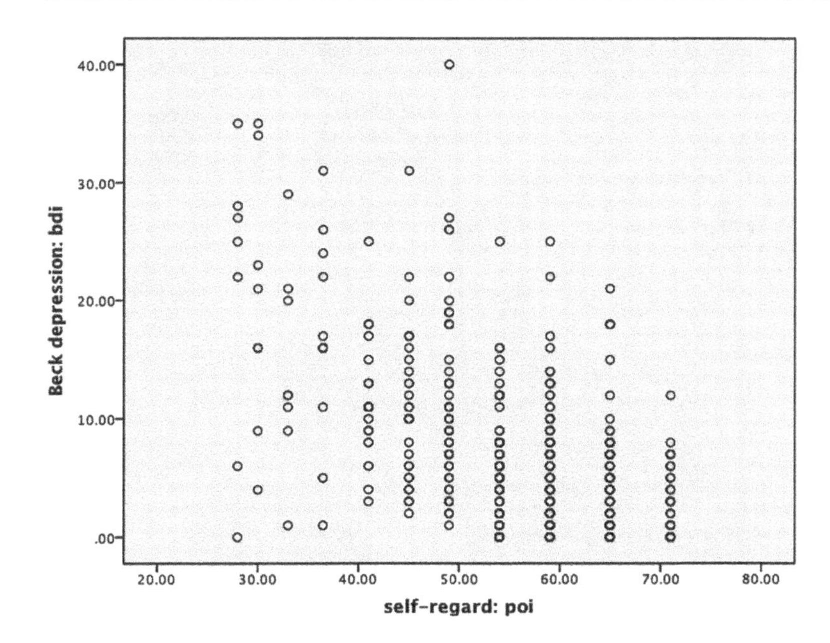

FIGURE 22.3
Scatterplot of **beckdep** and **regard**.

22.4 ANALYSIS OUTPUT: CHECKING FOR LINEARITY

The scatterplot of **beckdep** and **regard** is presented in Figure 22.3. Visual inspection suggests that the two variables are related linearly. It also appears that the line of best fit is angled from the upper left to the lower right, indicating a negative (inverse) relationship.

22.5 ANALYSIS SETUP: CORRELATING A SINGLE PAIR OF VARIABLES

With the appearance of linearity, we can confidently generate the Pearson correlation between **beckdep** and **regard**. From the main menu select **Analyze → Correlate → Bivariate**. This opens the main **Bivariate Correlations** window shown in Figure 22.4. We move **beckdep** and **regard** into the **Variables** panel. We retain the default specifications of asking only for **Pearson** in the **Correlation Coefficients** area, requesting **Two-tailed** under **Test of Significance**, and ask that IBM SPSS® **Flag significant correlations** (it will use asterisks to represent statistical significance).

Selecting the **Options** pushbutton allows us to obtain the **Means and standard deviations** of the variables in the analysis (see Figure 22.5). We retain the **Missing Values** specification of **Exclude cases pairwise**. Click **Continue** to return to the main dialog window and click **OK** to perform the analysis.

22.6 ANALYSIS OUTPUT: CORRELATING A SINGLE PAIR OF VARIABLES

The output of the analysis is presented in Figure 22.6. Descriptive statistics are displayed in the top table and the correlation in the bottom table. Correlation results are presented in a "square" matrix, with a value of 1.00 on the diagonal. The three entries inside the cell(s) showing the intersection of **beckdep** and **regard** provide the following information in the following order:

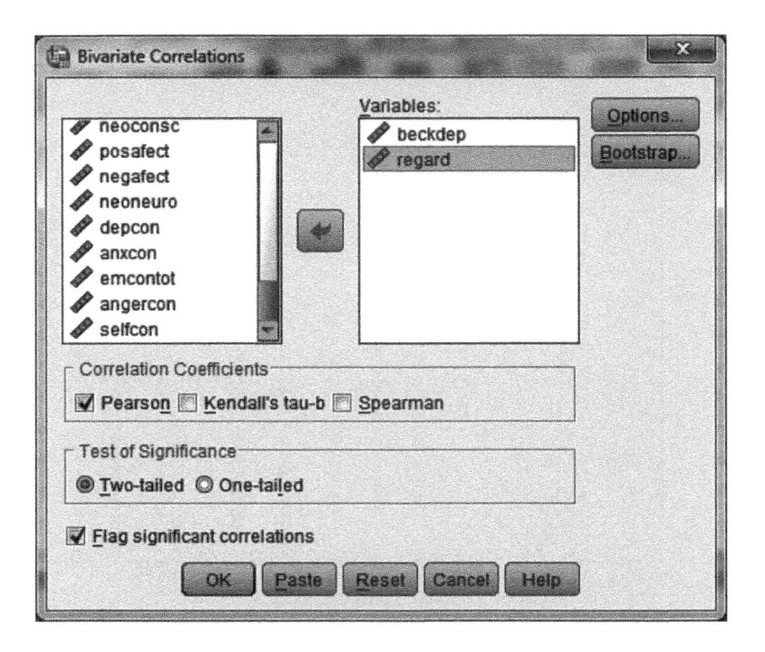

FIGURE 22.4
The main **Bivariate Correlations** dialog window.

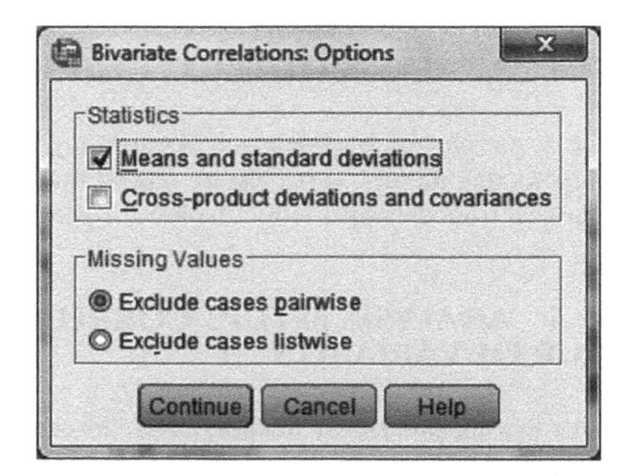

FIGURE 22.5
The **Options** window of **Bivariate Correlations**.

1. The Pearson correlation between the two variables is −**.539**.
2. The probability of that correlation value with an N of **421** (degrees of freedom = $N - 2$ or 419) occurring by chance, given the null hypothesis is true, is less than .001 (shown as **.000** in the output but reported as $p < .001$). We can see the exact probability by double-clicking **.000** in the output table (see Section 5.4).
3. There were **421** cases with valid values on both variables.

22.7 CORRELATING SEVERAL PAIRS OF VARIABLES

To show the correlation output for several pairs of variables, we have added **neoneuro** (neuroticism), **posafect** (positive affect), and **neoconsc** (consientiousness) to the set of variables but otherwise specified the analysis as described. We show the correlation matrix in Figure 22.7.

The diagonal of the matrix can be easily seen angled from upper left to lower right. With a square matrix, values above and below the diagonal are redundant. All of the

Descriptive Statistics

	Mean	Std. Deviation	N
beckdep Beck depression: bdi	7.4703	6.98858	421
regard self-regard: poi	55.2227	10.54974	422

Correlations

		beckdep Beck depression: bdi	regard self-regard: poi
beckdep Beck depression: bdi	Pearson Correlation	1	-.539**
	Sig. (2-tailed)		.000
	N	421	421
regard self-regard: poi	Pearson Correlation	-.539**	1
	Sig. (2-tailed)	.000	
	N	421	422

**. Correlation is significant at the 0.01 level (2-tailed).

FIGURE 22.6 Descriptive statistics and the Pearson correlation output.

Correlations

		beckdep Beck depression: bdi	regard self-regard: poi	neoneuro neuroticism: neo	posafect positive affect: mpq	neoconsc conscientiousness: neo
beckdep Beck depression: bdi	Pearson Correlation	1	-.539**	.608**	-.432**	-.235**
	Sig. (2-tailed)		.000	.000	.000	.000
	N	421	421	419	421	419
regard self-regard: poi	Pearson Correlation	-.539**	1	-.582**	.411**	.246**
	Sig. (2-tailed)	.000		.000	.000	.000
	N	421	422	420	422	420
neoneuro neuroticism: neo	Pearson Correlation	.608**	-.582**	1	-.441**	-.379**
	Sig. (2-tailed)	.000	.000		.000	.000
	N	419	420	420	420	420
posafect positive affect: mpq	Pearson Correlation	-.432**	.411**	-.441**	1	.230**
	Sig. (2-tailed)	.000	.000	.000		.000
	N	421	422	420	422	420
neoconsc conscientiousness: neo	Pearson Correlation	-.235**	.246**	-.379**	.230**	1
	Sig. (2-tailed)	.000	.000	.000	.000	
	N	419	420	420	420	420

**. Correlation is significant at the 0.01 level (2-tailed).

FIGURE 22.7 Descriptive statistics and the Pearson correlation output for a set of five variables.

correlations are statistically significant given an alpha level of .05. These correlations are based on sample sizes ranging from 419 to 422 due to the **Exclude cases pairwise** specification in the **Options** dialog window. Under pairwise exclusion, cases missing a value on one of the two variables involved in one correlation calculation (and who are thus excluded from that calculation) are eligible to be included in another calculation if they have valid values on those two variables.

CHAPTER 23

Spearman Rho and Kendall Tau-b Rank-Order Correlations

23.1 OVERVIEW

When the assumptions underlying the Pearson correlation cannot be met, there are nonparametric correlation procedures that can be implemented as an alternative. Generally, nonparametric procedures do not compare a sample parameter (e.g., the sample mean) to a population parameter (e.g., the population mean) and make fewer assumptions about the distribution (e.g., they do not require a normal distribution). Furthermore, nonparametric procedures are applicable to variables assessed on categorical (e.g., frequency) or ordinal scales of measurement (Marascuilo & McSweeney, 1977). Two of the most commonly used alternatives to the Pearson correlation are built into the **Bivariate Correlation** procedure in IBM SPSS®: the Spearman rho and the Kendall tau-b correlations. Both are applicable to variables measured on ordinal scales and assume only monotonicity of the relationship.

Ordinal measurement conveys only *greater than* or *less than* information and is represented by ranked data. In applying the Spearman rho or the Kendall tau-b correlations, the values of the X and Y variables are first independently rank-ordered (although the Kendall tau-b can be applied to raw data as well). The ranks are then substituted for the raw score values for each pair of scores, and the computation for each is then performed. Neither correlation thus requires that the data approximate interval measurement, and neither correlation requires that the variables be approximately normally distributed. By tapping into only ordinal information, the influence of outliers that can seriously and adversely affect the Pearson r is circumvented by the Spearman rho and the Kendall tau-b.

A monotonic relationship between two variables exists when higher values of one variable are consistently associated with higher values of the other or when higher values of one variable are consistently associated lower values of the other. A linear relationship, as assessed by the Pearson r, is a special case of a monotonic relationship, but some other nonlinear relationships (e.g., a tilted J shaped function) are also monotonic and can therefore be assessed by the Spearman rho and the Kendall tau-b correlations. The key to monotonicity is that the association does not show a reversal someplace along the continuum (e.g., higher values of X are associated with higher values on Y up to some point and then still higher values of X are associated with lower values of Y).

Performing Data Analysis Using IBM SPSS®, First Edition.
Lawrence S. Meyers, Glenn C. Gamst, and A. J. Guarino.

23.2 THE SPEARMAN RHO CORRELATION

The Spearman rho correlation was introduced by Sir Charles Spearman (1904b) primarily but not exclusively as a way to minimize the adverse influence of outliers in computing a Pearson *r*. If we were to transform two variables to their rank-order equivalents and apply the computational procedure for the Pearson correlation, we would obtain the value of the Spearman rho. Thus, many researchers think of the Spearman rho as a close relative of the Pearson *r*.

In calculating the Spearman rho, we determine the difference between the ranks of each data pair. These differences are then squared and summed and that value is entered into the computational formula. Because large differences in ranks are magnified when squared, the presence of large rank differences will result in relatively lower values of rho. Values of rho closer to ± 1.00 indicate that the rankings on X and Y are relatively similar to each other; values of rho closer to .00 indicate the rankings of X and Y are relatively independent of each other.

23.3 THE KENDALL TAU-b CORRELATION

Maurice G. Kendall (1938, 1948) introduced the tau correlation as an alternative approach to rank-order correlation. There are three variations of the statistic, tau-a, tau-b, and tau-c, but in the IBM SPSS **Bivariate Correlation** procedure only tau-b (applicable for tied ranks as well as rankings with no ties for two variables) is available.

As was true for the Spearman rho, the values of the two variables are first independently rank-ordered. Then the cases are listed based on their rank order on X, starting with Rank 1.

To compute the Kendall tau-b, the number of concordant and discordant pairs below each case is counted based on the Y variable. A concordant pair is one for which the ranking on Y is lower than the given case. It is labeled as concordant because cases lower in the listing are ranked lower on X (i.e., the basis of the listing), and if the Y value is also ranked lower, then the relationship is concordant (consistent or consonant). A discordant pair is one for which the ranking on Y is higher than the given case. It is labeled as discordant because cases lower in the listing are ranked lower on X (that is the basis of the listing), and if the Y value is ranked higher, then the relationship is discordant (inconsistent or dissonant).

Values of tau-b closer to $+1.00$ indicate that the rankings of X and Y are ordered in a very similar manner (i.e., they are mostly concordant); values of tau-b closer to -1.00 indicate that the rankings of X and Y are ordered inversely (i.e., they are mostly discordant); values of tau-b closer to .00 indicate the rankings of X and Y are relatively independent of each other. There are some suggestions (e.g., Howell, 2010; Wilcox, 2012) that Kendall's correlation is a better population estimate than Spearman's, and thus may be somewhat preferred, but most textbooks discussing both usually do not indicate a preference between them.

23.4 NUMERICAL EXAMPLE WITHOUT TIES

We use the data file named **Spearman Kendall No Ties** containing two generic variables named **X** and **Y**. These variables have each independently been rank-ordered and have been listed based on the **X** ranking. For example, the first case is ranked first on **X** but second on **Y**. A screenshot of the data file is presented in Figure 23.1.

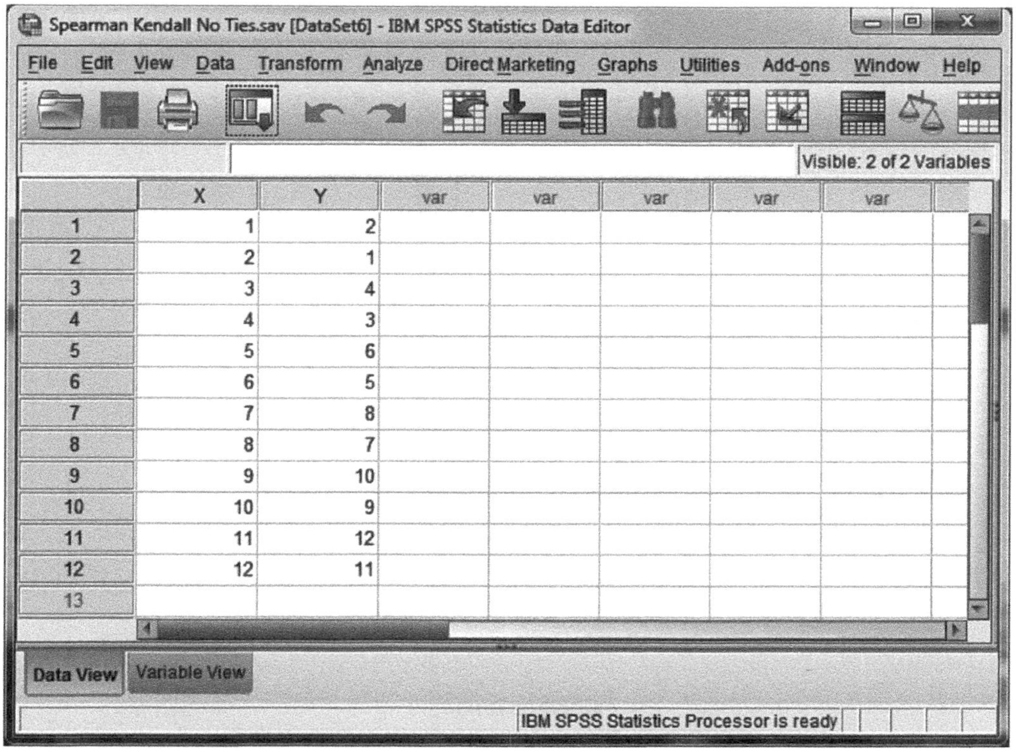

Spearman Kendall No Ties.sav [DataSet6] - IBM SPSS Statistics Data Editor

File Edit View Data Transform Analyze Direct Marketing Graphs Utilities Add-ons Window Help

Visible: 2 of 2 Variables

	X	Y	var	var	var	var	var
1	1	2					
2	2	1					
3	3	4					
4	4	3					
5	5	6					
6	6	5					
7	7	8					
8	8	7					
9	9	10					
10	10	9					
11	11	12					
12	12	11					
13							

Data View Variable View

IBM SPSS Statistics Processor is ready

FIGURE 23.1 The Kendall No Ties data file.

23.5 ANALYSIS SETUP

The setup for the analysis is similar to how we obtained the Pearson correlation (which we will also request here for comparison purposes). From the main menu, we select **Analyze ➔ Correlate ➔ Bivariate**. This opens the main **Bivariate Correlations** window as shown in Figure 23.2. We move **X** and **Y** into the **Variables** panel requesting **Pearson**, **Kendall's tau-b**, and **Spearman** in the **Correlation Coefficients** area. We retain the default specifications requesting **Two-tailed** under **Test of Significance** and asking that IBM SPSS **Flag significant correlations**. Click **OK** to perform the analysis.

23.6 ANALYSIS OUTPUT

The output of the analysis is shown in Figure 23.3, with the parametric (Pearson r) and two nonparametric correlations provided in separate output tables. As indicated earlier, the Pearson and Spearman correlations return the same value on the ranked data. Kendall's tau-b is noticeably lower than the Spearman rho; this is a very typical result as long as the rank differences are not extreme. In both instances, however, it appears that the variations in the ranks of X and Y are very much in synchrony (concordance) with each other. All correlations are statistically significant under an alpha level of .05; the Pearson r of **.958** and the Spearman rho of **.958** are at $p < .001$ and the Kendall tau-b of **.818** is also at $p < .001$.

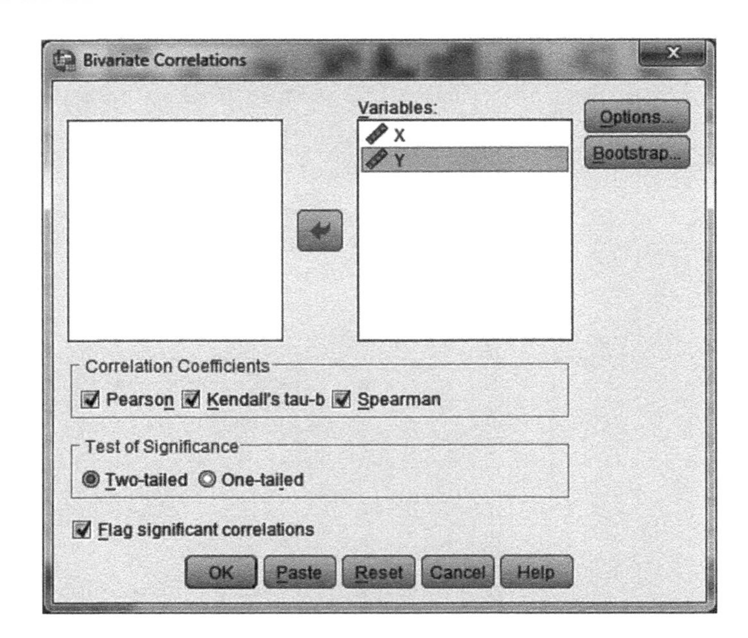

FIGURE 23.2

The main **Bivariate Correlations** dialog window.

Correlations

		X	Y
X	Pearson Correlation	1	.958**
	Sig. (2-tailed)		.000
	N	12	12
Y	Pearson Correlation	.958**	1
	Sig. (2-tailed)	.000	
	N	12	12

**. Correlation is significant at the 0.01 level (2-tailed).

Correlations

			X	Y
Kendall's tau_b	X	Correlation Coefficient	1.000	.818**
		Sig. (2-tailed)	.	.000
		N	12	12
	Y	Correlation Coefficient	.818**	1.000
		Sig. (2-tailed)	.000	.
		N	12	12
Spearman's rho	X	Correlation Coefficient	1.000	.958**
		Sig. (2-tailed)	.	.000
		N	12	12
	Y	Correlation Coefficient	.958**	1.000
		Sig. (2-tailed)	.000	.
		N	12	12

**. Correlation is significant at the 0.01 level (2-tailed).

FIGURE 23.3

The Pearson, Kendall tau-b, and Spearman correlations based on data with no ties.

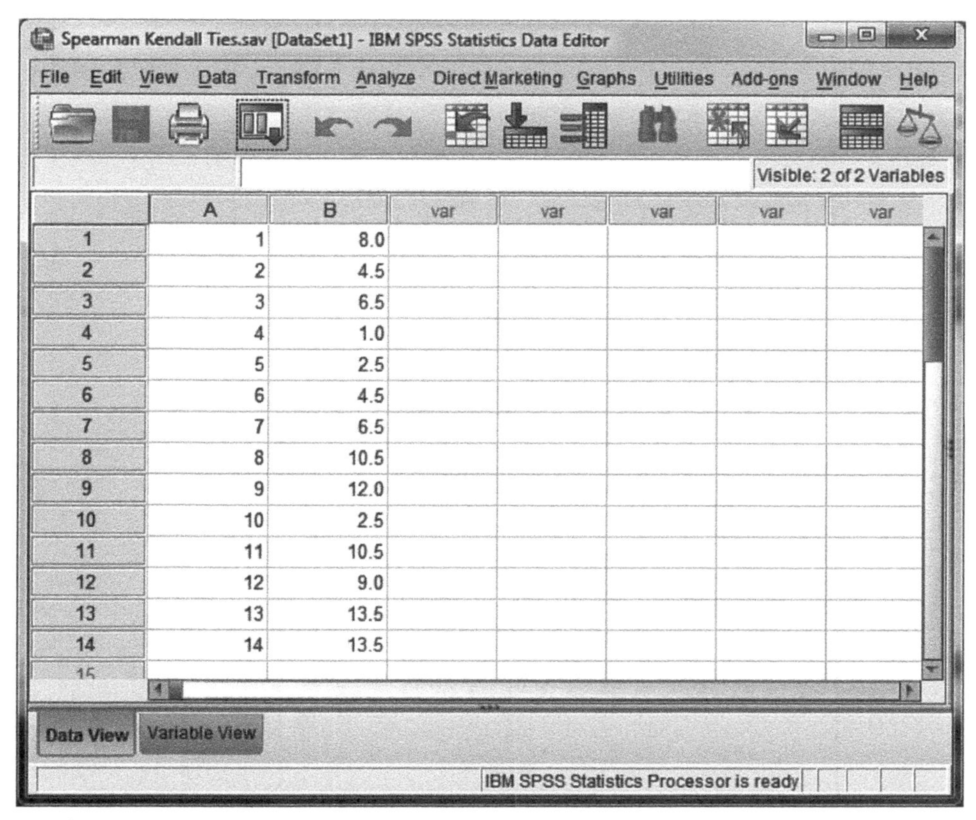

	A	B	var	var	var	var	var
1	1	8.0					
2	2	4.5					
3	3	6.5					
4	4	1.0					
5	5	2.5					
6	6	4.5					
7	7	6.5					
8	8	10.5					
9	9	12.0					
10	10	2.5					
11	11	10.5					
12	12	9.0					
13	13	13.5					
14	14	13.5					

FIGURE 23.4 The Kendall Ties data file.

23.7 NUMERICAL EXAMPLE WITH TIES

To demonstrate the differences among the three correlations when there are ties in the rankings, we use the data file named **Spearman Kendall Ties** containing two generic variables named **A** and **B**. These variables have each independently been rank-ordered and listed based on the **A** ranking. Thus, Case number 1 is ranked first on **A** but eighth on **B**. A screenshot of the data file is presented in Figure 23.4.

Ties in ranks are handled by assigning each of the tied ranks an average of the ranks that would have been used in the absence of a tie. For example, Case number 2 and Case number 6 were tied on the **B** measure and were each given the average rank of **4.5**. As can be seen in the data file, 10 of the 14 cases were involved in tied scores on the **B** variable.

23.8 ANALYSIS SETUP AND OUTPUT

The setup for the analysis is identical to our description in Section 23.5, and the output of the analysis is shown in Figure 23.5. Once again the Pearson r and the Spearman rho are equal in value, as the algorithm for calculating the Spearman used by IBM SPSS accounts for the presence of ties. For these same data, the Kendall tau-b, also designed to accommodate ties, returns a noticeably lower correlation value. All correlations are statistically significant under an alpha level of .05; the Pearson r of **.632** and the Spearman rho of **.632** are at p = **.015** and the Kendall tau-b of **.475** is at p = **.021**. The choice

Correlations

		A	B
A	Pearson Correlation	1	.632*
	Sig. (2-tailed)		.015
	N	14	14
B	Pearson Correlation	.632*	1
	Sig. (2-tailed)	.015	
	N	14	14

*. Correlation is significant at the 0.05 level (2-tailed).

Correlations

			A	B
Kendall's tau_b	A	Correlation Coefficient	1.000	.475*
		Sig. (2-tailed)	.	.021
		N	14	14
	B	Correlation Coefficient	.475*	1.000
		Sig. (2-tailed)	.021	.
		N	14	14
Spearman's rho	A	Correlation Coefficient	1.000	.632*
		Sig. (2-tailed)	.	.015
		N	14	14
	B	Correlation Coefficient	.632*	1.000
		Sig. (2-tailed)	.015	.
		N	14	14

*. Correlation is significant at the 0.05 level (2-tailed).

FIGURE 23.5

The Pearson, Kendall tau-b, and Spearman correlations based on data with ties.

between reporting the Spearman rho or the Kendall tau-b falls to the preferences of the researchers, as each is reported in the journal literature. Our slight preference is for the Kendall tau-b.

PART 8

REGRESSING (PREDICTING) QUANTITATIVE VARIABLES

CHAPTER 24

Simple Linear Regression

24.1 OVERVIEW

Pearson correlation and simple linear regression are two ways to express the same fundamental idea. Pearson correlation generally addresses the degree to which two variables are associated or related, whereas simple linear regression addresses the prediction of one variable based on the other (a specific way to characterize a relationship). Galton (1886) first discussed the idea of regression (regression toward the level of "mediocrity" or what we now think of as regression toward the mean) and 2 years later (Galton, 1888) broached the idea of covariation or what we now call correlation.

Unlike bivariate correlation where the two variables have the same status in the analysis, the two variables play different roles in simple linear regression. The value of one of the variables is being predicted; this variable is the *dependent*, *criterion*, or *outcome* variable and is generically denoted as the Y variable. The other variable is used to predict the dependent variable; this variable is the *independent* or *predictor* variable and is generically denoted as the X variable.

Simple linear regression derives its name from the following reasons:

- It is *regression* because we are predicting the value of Y from one quantitative X variable. Prediction is represented by a regression equation or *model* specifying the predicted value of Y as a function of X.
- It is *simple* because we use a single predictor or independent variable (X) to predict the value of outcome, criterion, or dependent variable (Y).
- It is *linear* because we are generating a linear regression model, that is, the equation or model that results from the regression analysis represents a straight line.

The regression model is generated by a *least squares* algorithm; that is, the linear prediction function describes where the sum of the squared deviations around the function is minimal and thus produces the best fit line relating X and Y. This is why the procedure is sensitive to outliers; distances from the center of the distribution are squared and large distances (which define outliers) that are squared are "disproportionally large" and thus draw the regression function toward them.

The model generated by the regression procedure is provided in both raw score and standardized score form. The raw score (unstandardized) model (Predicted $Y = \mathbf{a} + \mathbf{b}X$)

Performing Data Analysis Using IBM SPSS®, First Edition.
Lawrence S. Meyers, Glenn C. Gamst, and A. J. Guarino.

includes a value for the constant (**a**), which is the *Y* intercept of the function (the predicted value of *Y* when $X = 0$), and a regression coefficient (**b**), which is the weight given to *X* to maximize the predictability of *Y*. The regression coefficient is the slope of the linear function and represents the amount of change expected in *Y* for each unit change in *X*.

The standardized model (Predicted Y_z = **beta** X_z) applies to *X* and *Y* when each has been transformed to *z* scores. It does not include a value for the constant, as the standardized model passes through the origin (the *Y* intercept is zero). The standardized regression coefficient is labeled as *beta*. This model is automatically produced by IBM SPSS® when the raw score model is being generated. In simple linear regression, the beta coefficient is the Pearson correlation coefficient.

24.2 NUMERICAL EXAMPLE

We use the data file named **Exercise** containing fictional data from a study predicting the degree to which employees of a large firm are committed to following a physical exercise regimen. In addition, the researchers also assessed the degree to which the employees dieted as well as some personal variables (their need for social affiliation, their self-acceptance, their self-esteem, and the esteem they felt in their own body). The names of the variables are self-descriptive, and a screenshot of a portion of the data file is presented in Figure 24.1. For the present example, we use **selfesteem** to predict **exercise_commitment**.

Exercise.sav [DataSet3] - IBM SPSS Statistics Data Editor

File Edit View Data Transform Analyze Direct Marketing Graphs Utilities Add-ons Window Help

Visible: 7 of 7 Variables

	CaseID	exercise_commitment	diet_intensity	social_affiliation_needs	selfacceptance	selfesteem	bodyesteem
1	1	3.43	2.88	99	59	33	94
2	2	2.29	3.25	104	55	42	90
3	3	1.71	2.88	92	68	42	100
4	4	3.29	2.63	87	61	34	99
5	5	3.71	3.25	112	71	43	122
6	6	3.29	3.38	101	59	40	86
7	7	3.71	4.63	111	49	37	112
8	8	1.43	1.38	100	62	44	75
9	9	2.43	2.00	109	58	44	102
10	10	3.86	4.25	97	72	42	83
11	11	5.00	5.00	101	68	33	126
12	12	2.43	2.75	109	70	40	96
13	13	4.00	4.00	94	67	28	119
14	14	2.86	3.50	81	48	35	88

Data View Variable View

IBM SPSS Statistics Processor is ready

FIGURE 24.1 The data file.

24.3 ANALYSIS SETUP

From the main menu, we select **Analyze → Regression → Linear**. This opens the main **Linear Regression** window as shown in Figure 24.2. We move **exercise_commitment** into the **Dependent** panel and **selfesteem** into the **Independent(s)** panel. We retain **Enter** in the **Method** drop-down menu (this deals with how the variables are to be entered into the model as it is built—the issue is not relevant with just a single predictor).

Selecting the **Statistics** pushbutton opens the **Statistics** screen (see Figure 24.3). Most of these are more relevant to multiple linear regression, but it provides an opportunity here to address them in a simpler manner than we do in our Chapter 26. We check **Estimates** (to obtain the regression coefficients), **Model fit** (to obtain the R^2 and adjusted R^2), **Descriptives** (to obtain the descriptive statistics), and **Part and partial correlations** (to obtain the zero-order, partial, and semipartial correlations).

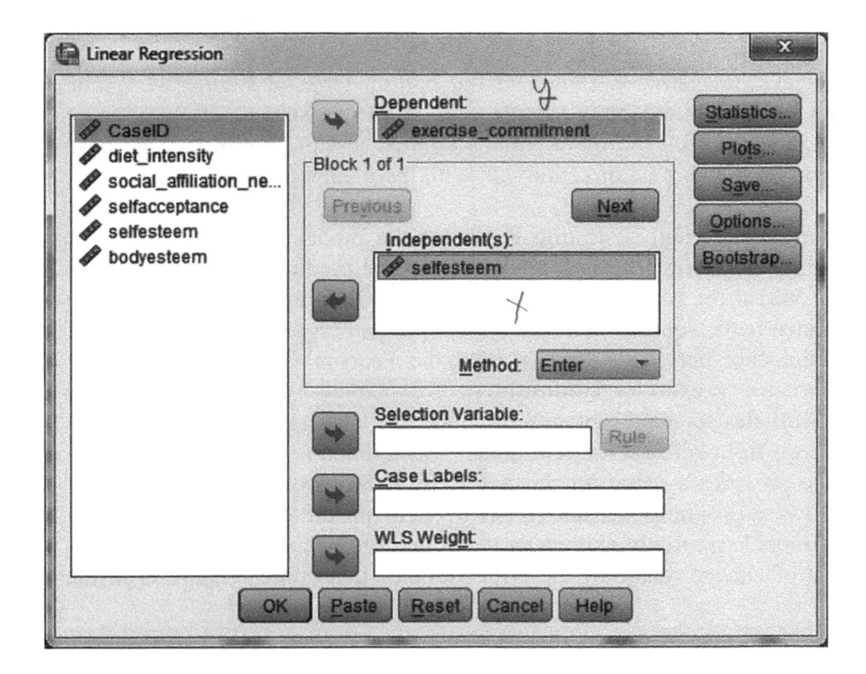

FIGURE 24.2

The main **Linear Regression** dialog window.

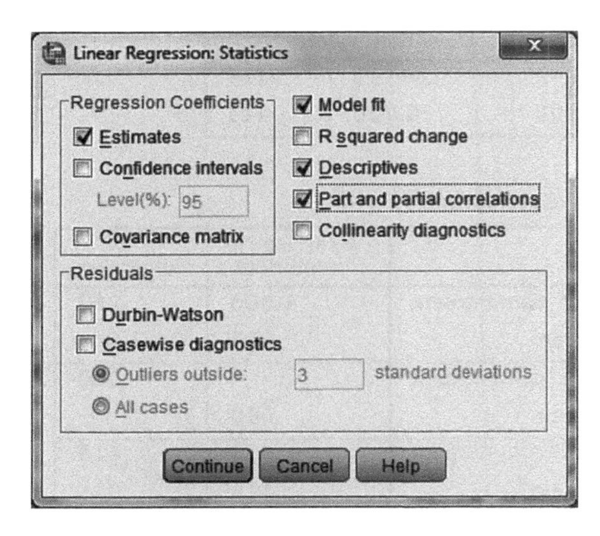

FIGURE 24.3

The **Statistics** screen of **Linear Regression**.

Click **Continue** to return to the main dialog window and click **OK** to perform the analysis.

24.4 ANALYSIS OUTPUT

The descriptive statistics (means and standard deviations) for the two variables are shown in the top table of Figure 24.4. As was seen in the screenshot of the data file and is apparent from the descriptive statistics, **selfesteem** and **exercise_commitment** are assessed on quite different scales, even though both are quantitatively measured. The variable of **exercise_commitment** was computed by averaging across survey items that were associated with a 5-point summative response scale, with higher values indicating greater commitment; the mean of **3.3742** suggests somewhat more than neutral commitment to exercise. The variable of **selfesteem** varied between 20 and 50, with higher values indicating greater levels of self-esteem; the mean of **39.08** suggests a relatively high level of self-esteem.

The square correlation matrix is shown in the bottom table of Figure 24.4. The Pearson correlation of **.262** based on a sample size of **415** is statistically significant at $p < .001$. IBM SPSS has used the more lenient **1-tailed** test of significance because prediction is going to be in only one direction, but with the *p* value showing at **.000**, that is not an issue here.

Figure 24.5 shows the result of testing the fit of the model. In the top table, **R** represents the multiple correlation; because there is only one predictor variable, **R** takes on the same value (**.262**) as the Pearson *r* (the Pearson *r* is the limiting case of the multiple correlation). **R Square** is the squared multiple correlation and represents the strength of the predictive relationship; here, its value is **.069** (the Pearson r^2) and suggests that about 7% of the variance of **exercise_commitment** is explained by **selfesteem**. We did not request in the **Statistics** screen to obtain the change in R^2 because with only a single predictor we would go from zero (before the model was applied to the data) to **.069**.

IBM SPSS also provides a value for the **Adjusted R Square**. This is supplied by most statistical software programs because R^2 is a bit of an inflated estimate of how much variance the model can really explain in that some of the prediction it subsumes is taking advantage of chance (some of the error variance that coincidentally is in the

Descriptive Statistics

	Mean	Std. Deviation	N
exercise_commitment	3.3742	.87191	415
selfesteem	39.08	6.706	415

Correlations

		exercise_commitment	selfesteem
Pearson Correlation	exercise_commitment	1.000	.262
	selfesteem	.262	1.000
Sig. (1-tailed)	exercise_commitment	.	.000
	selfesteem	.000	.
N	exercise_commitment	415	415
	selfesteem	415	415

FIGURE 24.4

Descriptive statistics and the correlation matrix.

Model Summary

Model	R	R Square	Adjusted R Square	Std. Error of the Estimate
1	.262[a]	.069	.066	.84248

a. Predictors: (Constant), selfesteem

ANOVA[b]

Model		Sum of Squares	df	Mean Square	F	Sig.
1	Regression	21.593	1	21.593	30.422	.000[a]
	Residual	293.139	413	.710		
	Total	314.732	414			

a. Predictors: (Constant), selfesteem
b. Dependent Variable: exercise_commitment

FIGURE 24.5 Testing the fit of the model.

direction of prediction). For this reason, an adjustment (based on sample size and the number of predictor variables in the model) is made to R^2 that provides a more realistic (a lower) estimate (sometimes referred to as *shrinkage*) of how well the model is performing (Meyers et al., 2013) and should always be reported in conjunction with R^2. In the present example, because of the high ratio of cases to predictors, R^2 was expected to shrink by a minimal amount (from **.069** to **.066**).

The bottom table in Figure 24.5 provides a test of the statistical significance of the regression model using a one-way between-subjects ANOVA. The effect evaluated in the ANOVA is the regression model (labeled **Regression** in the summary table), and it has one degree of freedom because that is how many predictors are in the model. The total degrees of freedom are equal to $N - 1$ or **414**, leaving **413** degrees of freedom for the error term. The model accounts for a significant amount of dependent variable variance, $F(1, 413) = \mathbf{30.42}$, $p < .001$. If the **Regression** effect was not statistically significant, we would have concluded that we could predict no better than chance and thus we would stop examining the results of the analysis at that point.

The eta square value in this ANOVA (a frequently used index of strength of effect in the ANOVA family of designs) is equal to the sum of squares associated with the **Regression** effect (this is the regression model) divided by the sum of squares associated with the **Total** variance, which yields **21.593/314.732** or **.069**; this is the same value as R^2 because the ANOVA and linear regression are just different expressions of the general linear model.

Figure 24.6 presents the **Coefficients** of the variable in the regression model as well as some other information. There are three coefficients provided under the **Correlations** heading in the table. These are briefly noted as follows:

- **Zero-order** Correlation. This is the Pearson r between the predictor and the dependent variable. It is labeled as *zero-order* to represent the fact that there are no covariates taken into account. In a multiple regression analysis, the set of predictor variables in the model serve as covariates in evaluating the effect of any given predictor.
- **Partial** Correlation. This is the correlation between the predictor and the variance not explained by the other predictors in the model (the residual variance). With no other predictors in this model, the partial correlation is equal to the Pearson r.
- **Part** Correlation. This is a short name for the *semipartial correlation*. Squaring this value to obtain the *squared semipartial correlation* informs us of how much

Coefficients[a]

Model		Unstandardized Coefficients		Standardized Coefficients	t	Sig.	Correlations		
		B	Std. Error	Beta			Zero-order	Partial	Part
1	(Constant)	2.043	.245		8.345	.000			
	selfesteem	.034	.006	.262	5.516	.000	.262	.262	.262

a. Dependent Variable: exercise_commitment

FIGURE 24.6 The regression coefficients together with the correlations.

variance of the dependent variable is explained exclusively by the predictor given that there are (usually) other predictors in the model; that is, the squared semipartial correlation is said to assess the amount of dependent variable variance uniquely explained by the given predictor (taking into account the other predictors in the model). Again, with no other predictors in this model, the semipartial correlation is equal to the Pearson r.

The Y intercept (labeled as the **constant** in the table) and the unstandardized regression coefficient for **selfesteem** are shown under **Unstandardized Coefficients** together with the standard error for each estimate. The unstandardized (raw score) regression coefficient for **selfesteem** is **.034** (this is the slope of the unstandardized model) and is statistically significant ($p < .001$) as assessed by a t test. Because it is positive, we can say that an increase of one unit on the **selfesteem** measure is expected to be associated with an increase in **exercise_commitment** of 0.034 units. This may seem like a very small increase in **exercise_commitment** until we recall (see Figure 24.4) that **selfesteem** scores average about 39 and **exercise_commitment** scores average about 3.4.

The beta (standardized) regression coefficient for **selfesteem** is **.262** (the slope of the standardized model); this is also the value of the Pearson r between **selfesteem** and **exercise_commitment**. On the basis of the value of this coefficient, we can say that an increase of 1.00 z score (one standard deviation unit) on the **selfesteem** measure is expected to be associated with an increase in **exercise_commitment** of **.262** z score (standard deviation) units. The unstandardized and standardized coefficients represent alternative but comparable ways to express the slope of the regression function.

The **Constant** is the Y intercept of the unstandardized regression equation and represents the predicted value of Y for an X score of zero. Here, a **selfesteem** score of zero (were such a score possible) would be associated with an **exercise_commitment** score of **2.043**.

We can summarize the two models by writing out the regression equations. The unstandardized simple regression model is as follows:

$$\textbf{exercise\_commitment} = \textbf{2.043} + (\textbf{.034})(\textbf{selfesteem})$$

The standardized simple regression model is as follows:

$$\textbf{exercise\_commitment}_z = (\textbf{.262})(\textbf{selfesteem}_z)$$

24.5 THE *Y* INTERCEPT ISSUE

It is worthwhile noting that with a possible range on the **selfesteem** variable of 20–50, a **selfesteem** score of zero is an out-of-range value. Because of this, we would not interpret the Y intercept in this model in any empirical (meaningful) way. In other research studies, it is possible that a score of zero on the predictor variable is a valid value and so an empirical or meaningful interpretation of the Y intercept would then be possible.

The lack of empirical meaning associated with the Y intercept does not often matter to researchers because their focus is on the R^2 and adjusted R^2 values, as well as on the regression coefficients informing them of the slope of the regression function. But there are situations in which this lack of empirical meaning is, if not exactly problematic, then certainly not an efficient use of all the information potentially contained in the regression model. One very useful way to deal with this issue is to *center* the predictor, and we address this in Chapter 25.

CHAPTER 25

Centering the Predictor Variable in Simple Linear Regression

25.1 OVERVIEW

As we indicated in Chapter 24, the Y intercept (the constant) in the unstandardized regression model is the predicted value of the dependent (criterion) variable when the score on the independent (predictor) variable is zero. However, it is not unusual for zero to be an out-of-range value on the predictor variable. Examples include predictor variables assessed on summative response scales (e.g., 5-point scales using response values of 1 through 5), test scores measured on standardized scales (e.g., intelligence, achievement tests), and various biological medical/health measures (e.g., heart rate, body mass index). Even when a zero score on the independent (predictor) variable is a valid value, it is often the case that it is an unusual or an unlikely score, given the measurement scale or the distribution of the predictor variable. When a zero value on the predictor is either not valid or is an unusual score, the value of the Y intercept has no or very little empirical utility.

One approach to providing meaning or utility to the Y intercept without altering the results of a simple regression analysis is to *center* the predictor variable. To center a variable is to create from the original values of the variable a new variable with a mean of zero. We accomplish this by transforming the raw score to a deviation score using the **Compute** procedure to subtract a reference score, such as the mean of the distribution, from the score recorded for each case (i.e., Deviation Score = Obtained Score − Mean). Therefore, any case whose score is equal to the mean has a deviation score of zero.

A *z* score is an elaborate form of a centered score; it is centered on the mean, and any *z* score represents the distance between a score and the mean of the distribution. It is more elaborate than a simple deviation score because, by dividing the distance of the score from the mean by the standard deviation, the distance between each data point and the mean is expressed in standard deviation units.

Centering is a strategy that is commonly applied when dealing with interaction effects in regression (Cohen, Cohen, West, & Aiken, 2003) and when performing an analysis involving multilevel modeling (Meyers et al., 2013). In the context of simple linear regression, there are two straightforward options for centering:

- *Mean centering* (more precisely called *grand mean centering* in the context of multilevel modeling) transforms the raw scores of a variable into deviations (distances) from the overall mean of that variable. In such centering, the mean is the

Performing Data Analysis Using IBM SPSS®, First Edition.
Lawrence S. Meyers, Glenn C. Gamst, and A. J. Guarino.

center of the transformed distribution and becomes the new zero value. This is the strategy we illustrate in this chapter.

- *Reference-score centering* requires us to select a particular score as the reference value that presumably has a special meaning to the researchers. An example of this might be the passing score on a written examination taken by applicants to become licensed as an electrician; such an examination score is unlikely to be the mean of the test score distribution but may actually be of greater use to a state licensing agency as the reference score in a regression study predicting on-the-job performance based on licensing examination performance. Such centering transforms the raw scores into deviations (distances) from that reference value. In such centering, the reference score is the center of the transformed distribution and becomes the new zero value.

25.2 NUMERICAL EXAMPLE

We use the data file named **BMI and pulse rate** containing fictional data from a medical study. Body mass index can be used to judge whether someone is overweight. Adolphe Quetelet originally developed this index in 1832 as the Quetelet Index, which was named the body mass index in 1972 by Ancel Keys (Eknoyan, 2008). Quetelet, in his quest to describe the average person, wished to determine if the normal curve could be applied to characteristics of human beings (Stigler, 1986).

Body mass index is the ratio of weight in kilograms to squared height in meters or, equivalently, weight in pounds multiplied by 703 to squared height in inches. Body mass index categories are as follows: underweight is less than 18.5, normal is between 18.5 and 24.9, overweight is between 25 and 29.9, obese is between 30 and 39.9, and morbidly obese is 40 or greater. Higher levels of body mass index pose a potentially greater health risk. One standard gauge of health status is heart rate, and body mass index may be related to (i.e., predict) that health factor. In the data file, we have recorded the heart rate under the variable of **pulse_rate** and body mass index under the variable of **BMI** for 40 medical patients.

25.3 ANALYSIS STRATEGY

Our analysis strategy is as follows:

- We perform a descriptive statistics procedure to determine the mean of **BMI** so that we can perform the centering operation.
- We center the **BMI** scores using a **Compute** procedure to build (and save) to the data file a new centered variable (subtracting the **BMI** mean from the **BMI** score for each case).
- We use the original **BMI** variable to predict **pulse_rate** in a simple linear regression analysis so that we can compare these results to those of the centered analysis.
- We use the centered **BMI** variable to predict **pulse_rate** in a simple linear regression analysis to show what the difference is in a centered procedure.

25.4 OBTAINING DESCRIPTIVE STATISTICS ON THE PREDICTOR VARIABLE

We open the data file named **BMI and pulse rate**, and from the main menu select **Analyze ➔ Descriptive Statistics ➔ Descriptives**. We move **BMI** into the **Variable(s)**

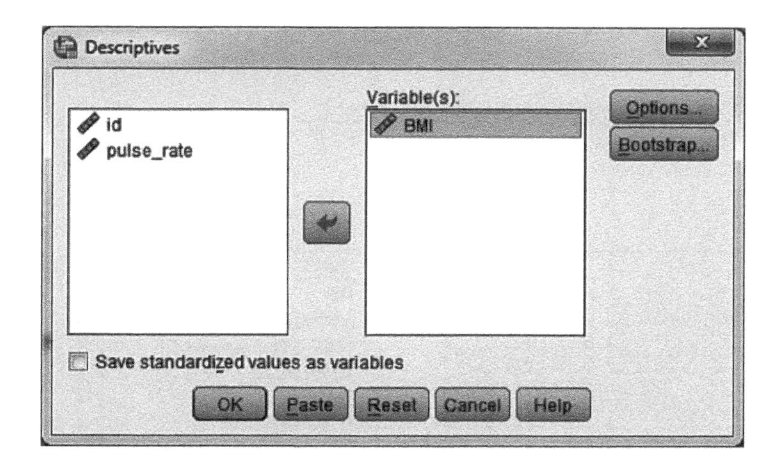

FIGURE 25.1

The main dialog window of **Descriptives**.

Descriptive Statistics

	N	Minimum	Maximum	Mean	Std. Deviation
BMI body mass index	40	18.00	34.10	25.3100	4.83347
Valid N (listwise)	40				

FIGURE 25.2

Descriptive statistics for **BMI**.

panel as shown in Figure 25.1. As the default statistics produced include the mean, we click **OK** to perform the analysis.

The results of the analysis are shown in Figure 25.2. Our interest here is in the mean, which is **25.31**. In our centering computation, we will subtract this value from each **BMI** score.

25.5 COMPUTING THE CENTERED PREDICTOR VARIABLE

As described in Chapter 16, from the main menu select **Transform ➔ Compute Variable**. This opens the **Compute Variable** dialog window shown in Figure 25.3. We have named the new variable **BMI_centered** in the **Target Variable** panel. We have then moved the variable **BMI** into the **Numeric Expression** panel, clicked the minus sign (second row, leftmost key), and then typed **25.31** in the **Numeric Expression** panel. This expression will subtract the value of **25.31** from each **BMI** score. Click **OK** to perform the procedure.

A screenshot of a portion of the data file with the newly computed **BMI_centered** variable in place is shown in Figure 25.4. Patient number 3, for example, has a **BMI_centered** value of **−6.31**. This value was obtained by subtracting the **BMI** mean of **25.31** from patient 3's **BMI** score of **19.00** (**19.00** − **25.31** = **−6.31**). The negative value informs us that the score is below the mean, and the value of **6.31** informs us that the score is **6.31** body mass units from the mean.

To illustrate the effects of a centering operation on the distribution of a variable, we have analyzed both **BMI** and **BMI_centered** in the **Descriptives** procedure (where we have requested more than the default statistics). The results of this analysis are shown in Figure 25.5. The mean has shifted from **25.31** for **BMI** to **.0000** for **BMI_centered** and the accompanying minimum and maximum values have also changed (as the values for **BMI_centered** are now deviations from the mean). More to the point, the fundamental variability attributes of the distribution (standard deviation, skewness, and kurtosis) are unchanged from the original variable as a result of this centering transformation; thus, the centered distribution is simply shifted intact horizontally from the original location around the original mean to a new location centered on zero but is otherwise identical.

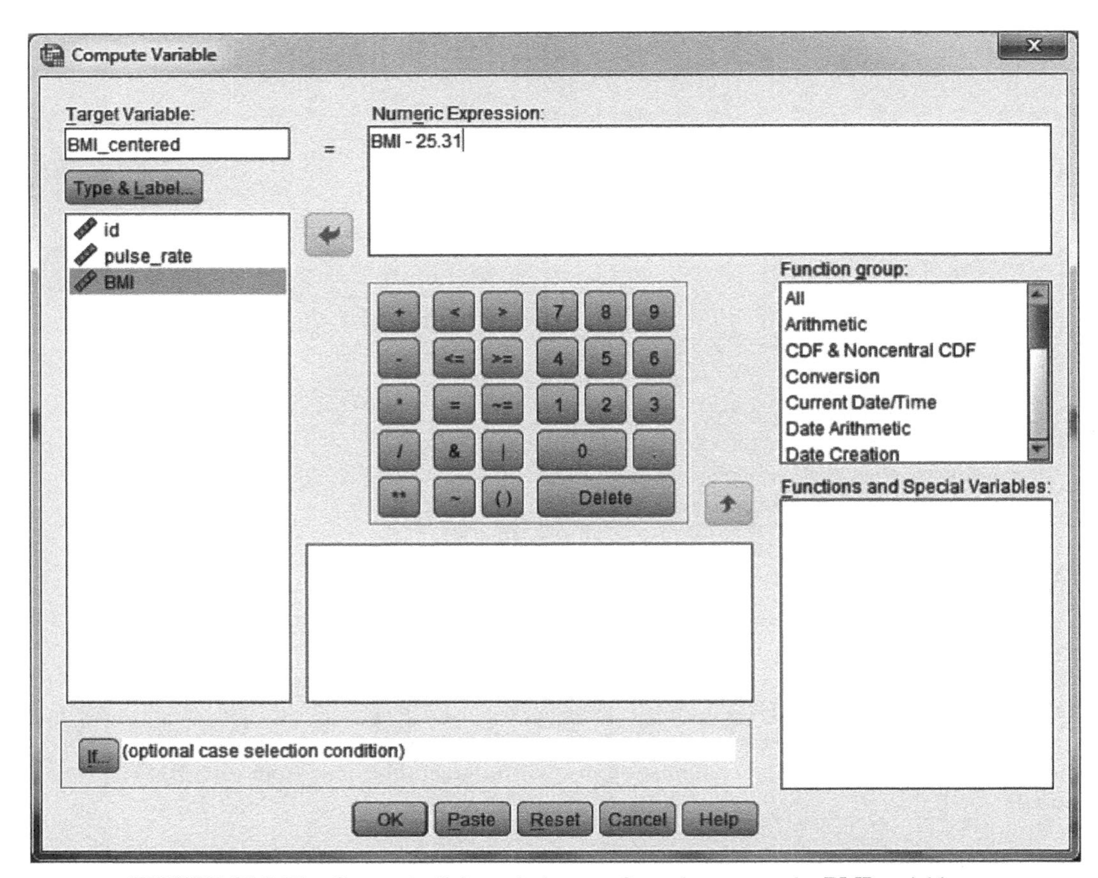

FIGURE 25.3 The **Compute** dialog window configured to center the **BMI** variable.

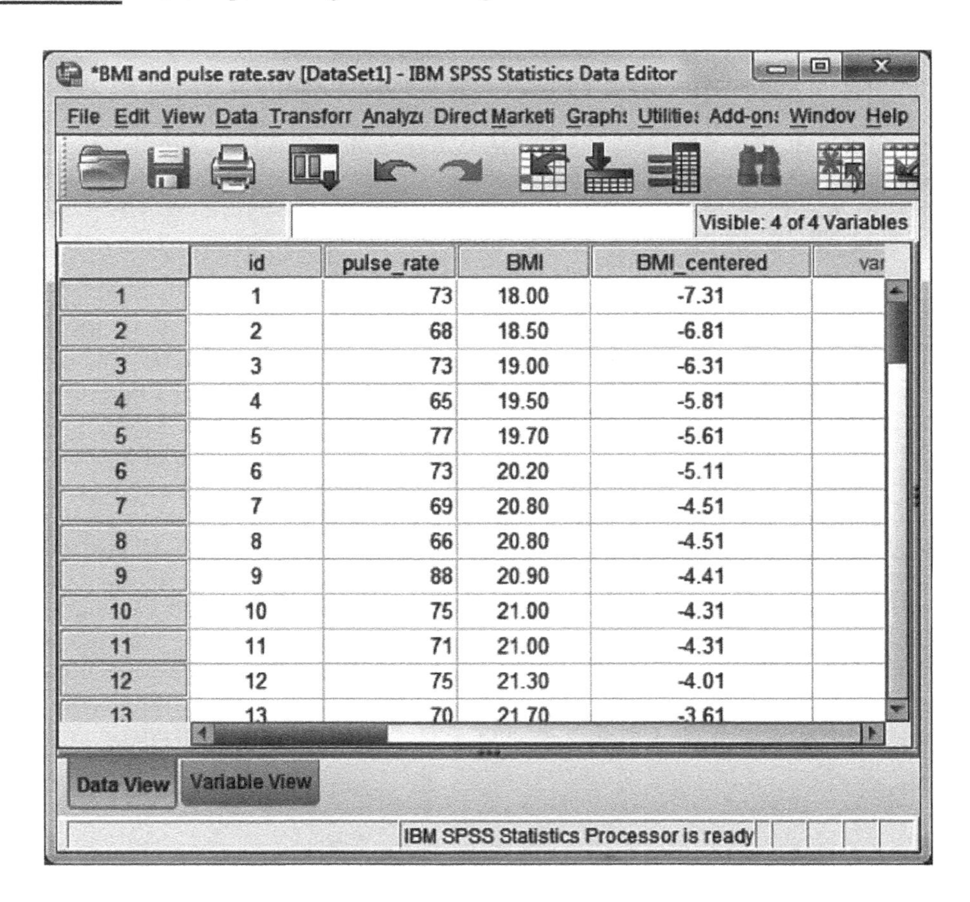

FIGURE 25.4
A portion of the data file with the newly computed **BMI_centered** variable in place.

Descriptive Statistics

	N	Minimum	Maximum	Mean		Std. Deviation	Skewness		Kurtosis	
	Statistic	Statistic	Statistic	Statistic	Std. Error	Statistic	Statistic	Std. Error	Statistic	Std. Error
BMI body mass index	40	18.00	34.10	25.3100	.76424	4.83347	.319	.374	-1.227	.733
BMI_centered	40	-7.31	8.79	.0000	.76424	4.83347	.319	.374	-1.227	.733
Valid N (listwise)	40									

FIGURE 25.5 Descriptive statistics for **BMI** and **BMI_centered**.

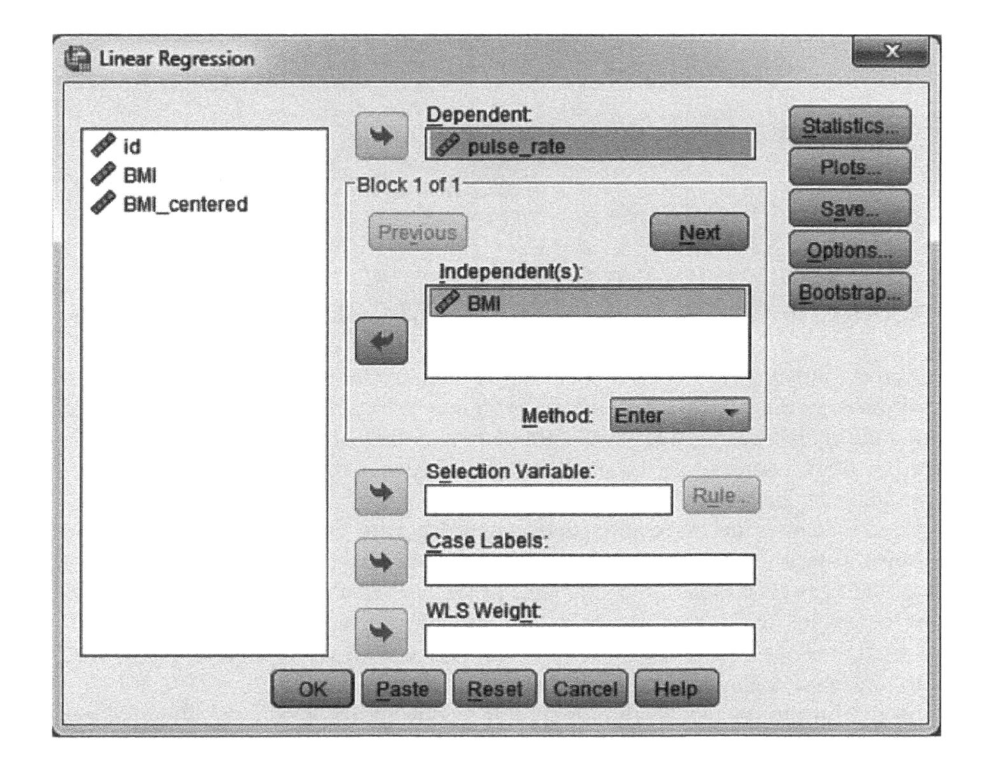

FIGURE 25.6
The main dialog window of **Linear Regression**.

25.6 ANALYSIS SETUP: SIMPLE LINEAR REGRESSION USING BMI AS THE PREDICTOR

From the main menu, we select **Analyze ➔ Regression ➔ Linear**. This opens the main **Linear Regression** window as shown in Figure 25.6. We move **pulse_rate** into the **Dependent** panel and **BMI** into the **Independent(s)** panel.

In the **Statistics** window (see Figure 25.7) we check **Estimates**, **Model fit**, **R squared change**, **Descriptives**, and **Part and partial correlations**. Click **Continue** to return to the main dialog window and click **OK** to perform the analysis.

25.7 ANALYSIS SETUP: SIMPLE LINEAR REGRESSION USING BMI_CENTERED AS THE PREDICTOR

We set up the analysis exactly as described earlier but with one exception. In this analysis, we use **BMI_centered** as the independent variable.

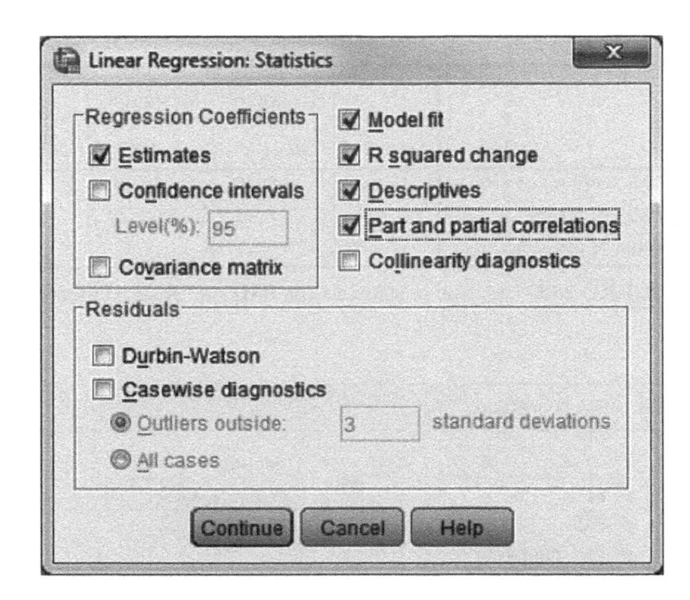

FIGURE 25.7

The **Statistics** dialog window of **Linear Regression**.

25.8 ANALYSIS OUTPUT FROM BOTH REGRESSION ANALYSES

The descriptive statistics (means and standard deviations) for the dependent and independent variables in the regression analysis are shown in Figure 25.8, with the top table representing the analysis using **BMI** as the predictor and the bottom table representing the analysis using **BMI_centered** as the predictor. This matches the output of our previous descriptive statistics analysis.

Figure 25.9 shows the correlation between **pulse_rate** and **BMI** (top table) and between **pulse_rate** and **BMI_centered** (bottom table). Note that centering did not affect the value of the Pearson correlation coefficient, which was **.866** in both analyses. Pearson correlation represents the degree to which two variables covary, and because the variability of **BMI** did not change when we transformed it to **BMI_centered**, **pulse_rate** covaries to the same degree with both of them.

The **Model Summary** tables are presented in Figure 25.10. Again, the effectiveness of model is identical using the original (top table) and centered (bottom table) predictor variables. The R^2 is **.749** and the adjusted R^2 is **.743**. It therefore appears that body mass index, whether centered (**BMI_centered**) or not (**BMI**), accounted for approximately three-quarters of the variance of **pulse_rate**.

Descriptive Statistics

	Mean	Std. Deviation	N
pulse_rate	82.40	11.388	40
BMI body mass index	25.3100	4.83347	40

Descriptive Statistics

	Mean	Std. Deviation	N
pulse_rate	82.40	11.388	40
BMI_centered	.0000	4.83347	40

FIGURE 25.8

Descriptive statistics.

Correlations

		pulse_rate	BMI body mass index
Pearson Correlation	pulse_rate	1.000	.866
	BMI body mass index	.866	1.000
Sig. (1-tailed)	pulse_rate	.	.000
	BMI body mass index	.000	.
N	pulse_rate	40	40
	BMI body mass index	40	40

Correlations

		pulse_rate	BMI_centered
Pearson Correlation	pulse_rate	1.000	.866
	BMI_centered	.866	1.000
Sig. (1-tailed)	pulse_rate	.	.000
	BMI_centered	.000	.
N	pulse_rate	40	40
	BMI_centered	40	40

FIGURE 25.9 The Pearson correlation of the variables.

Model Summary

					Change Statistics				
Model	R	R Square	Adjusted R Square	Std. Error of the Estimate	R Square Change	F Change	df1	df2	Sig. F Change
1	.866[a]	.749	.743	5.776	.749	113.597	1	38	.000

a. Predictors: (Constant), BMI body mass index

Model Summary

					Change Statistics				
Model	R	R Square	Adjusted R Square	Std. Error of the Estimate	R Square Change	F Change	df1	df2	Sig. F Change
1	.866[a]	.749	.743	5.776	.749	113.597	1	38	.000

a. Predictors: (Constant), BMI_centered

FIGURE 25.10 **Model Summary** table.

The results of ANOVA, testing the statistical significance of the regression model are shown in Figure 25.11. Still again, the original **BMI** variable and its centered counterpart (**BMI_centered**) produced identical results. The regression model was associated with an F ratio of **113.597** and, with **1** and **38** degrees of freedom, was statistically significant ($p < .001$). The eta square value was **3789.836/5057.600** or .749, the same value shown earlier for R^2.

The detailed results of the analyses containing the regression coefficients are presented in Figure 25.12. The unstandardized simple regression model with **BMI** as the predictor variable is as follows:

$$\textbf{pulse\_rate} = 30.781 + (2.039)(\textbf{BMI})$$

ANOVA[b]

Model		Sum of Squares	df	Mean Square	F	Sig.
1	Regression	3789.836	1	3789.836	113.597	.000[a]
	Residual	1267.764	38	33.362		
	Total	5057.600	39			

a. Predictors: (Constant), BMI body mass index
b. Dependent Variable: pulse_rate

ANOVA[b]

Model		Sum of Squares	df	Mean Square	F	Sig.
1	Regression	3789.836	1	3789.836	113.597	.000[a]
	Residual	1267.764	38	33.362		
	Total	5057.600	39			

a. Predictors: (Constant), BMI_centered
b. Dependent Variable: pulse_rate

FIGURE 25.11 ANOVA testing of the statistical significance of the regression model.

Coefficients[a]

Model		Unstandardized Coefficients		Standardized Coefficients	t	Sig.	Correlations			Collinearity Statistics	
		B	Std. Error	Beta			Zero-order	Partial	Part	Tolerance	VIF
1	(Constant)	30.781	4.929		6.245	.000					
	BMI body mass index	2.039	.191	.866	10.658	.000	.866	.866	.866	1.000	1.000

a. Dependent Variable: pulse_rate

Coefficients[a]

Model		Unstandardized Coefficients		Standardized Coefficients	t	Sig.	Correlations			Collinearity Statistics	
		B	Std. Error	Beta			Zero-order	Partial	Part	Tolerance	VIF
1	(Constant)	82.400	.913		90.226	.000					
	BMI_centered	2.039	.191	.866	10.658	.000	.866	.866	.866	1.000	1.000

a. Dependent Variable: pulse_rate

FIGURE 25.12 Regression coefficients table.

The standardized simple regression model is as follows:

$$\mathbf{pulse\_rate}_z = (.866)(\mathbf{BMI}_z)$$

The unstandardized simple regression model with **BMI_centered** as the predictor variable is as follows:

$$\mathbf{pulse\_rate} = 82.400 + (2.039)(\mathbf{BMI})$$

The standardized simple regression model is as follows:

$$\mathbf{pulse\_rate}_z = (.866)(\mathbf{BMI}_z)$$

The zero-order, partial, and semipartial correlations are the same in both analyses because they all represent (in simple linear regression with a single predictor) the Pearson r. The beta (standardized) regression coefficient is also equal to the Pearson r and thus takes on the same value in both analyses, and the unstandardized regression coefficient is the same as well in both analyses, that is, **2.039**. Thus, the slope of the regression function is the same regardless of whether **BMI** or **BMI_centered** is used as the predictor. This makes sense because the general nature of the relationship between heart rate and body mass index is not affected by the centering operation. We may then say (based on the unstandardized model) that every unit gain in body mass index is associated on average with approximately 2 more heartbeats per minute.

The only value in these analyses that has been changed by the centering operation is the Y intercept (the constant in the raw score regression model). Using the original variable of **BMI**, a body mass index of zero corresponds to an expected heart rate of **30.781**, values that have no real-world meaning (it is likely that someone with a body mass index of zero, should that impossibility ever occur, would not likely have a pulse at all). However, the Y intercept for the centered model is **82.40**, not coincidentally the mean of the **pulse_rate** variable. Translating that value into an empirical context simply requires us to recall that **BMI_centered** is a mean centered variable; thus, a value of zero on this predictor represents the mean of **BMI** (which we know from Figure 25.5 is **25.31**). We thus immediately learn from this outcome that patients with a body mass index of **25.31** are expected to have on average a heart rate of **82.40**. It is this immediate interpretation of the meaning of the Y intercept that is gained from the centering operation (with no change in any of the other regression results) that makes centering so useful in situations that warrant its use.

CHAPTER 26

Multiple Linear Regression

26.1 OVERVIEW

Multiple linear regression is an extension of simple linear regression in that there are two or more predictors that are included in the model. The raw score model thus contains multiple **b**X terms, one for each predictor, and the standardized model contains multiple **beta** X_z terms. The linear function is fit by using the least-squares algorithm, and the weights associated with the predictors are those that maximize the prediction of Y.

The most common way of performing a multiple regression analysis is using the *standard* (sometimes called the simultaneous) method in which all variables are entered into the model in a single (albeit complicated) step. Each predictor is evaluated with all other variables presumed to be in the model; thus, the other predictors act as covariates with respect to the predictor that is being evaluated. The weights are known as *partial regression coefficients* because they are computed with respect to the other predictors in the model, and so even adding or subtracting a single variable from the set of predictors can potentially change the value of the partial regression coefficients by a substantial margin.

In some contexts, researchers may have reason to simplify a multiple regression model by selecting only the "best" predictors, that is, only those predictors that are significantly predictive of the criterion variable when controlling for all the other predictors. For example, certain predictors may be very resource intensive to use, may be uncomfortably intrusive, or in combination with the other predictors make less than optimal theoretical sense. The idea of using a reduced predictor set is to perform virtually the same amount of predictive work explaining the variance of the dependent variable as the full set of predictor variables, but the outcome must have pragmatic or theoretical utility for researchers to justify using the resulting model.

The **Linear Regression** procedure includes a set of step methods (forward, backward, and stepwise) each used to generate a single parsimonious model. These methods retain only those predictors of the set that are statistically significant rather than including the entire set in the model. Briefly, the step methods are as follows:

- The *forward* method adds variables to the model one variable (step) at a time. To be entered, a predictor must account for a statistically significant amount of dependent variable variance not already explained by other variables (the residual variance) that are in the model. Once in the model, it remains in, even if its predictive power is diminished in the presence of predictors entered later. The

Performing Data Analysis Using IBM SPSS®, First Edition.
Lawrence S. Meyers, Glenn C. Gamst, and A. J. Guarino.

process ends when no remaining predictors can explain a significant amount of the residual variance.

- The *backward* method enters all of the variables at first (just as in the standard method), but then removes variables one (step) at a time. When all nonsignificant predictors are removed from the model, the process stops.
- The *stepwise* method adds predictor variables to the model one step at a time. To be entered, a predictor must account for a statistically significant amount of dependent variable variance not already explained by other variables (the residual variance) that are in the model. Upon entering the third and all subsequent predictors, the stepwise method removes any variable that no longer explains a significant amount of dependent variable variance in the presence of the other predictors. Removed variables can find their way back into the model in the later steps. The process ends when no remaining predictors can explain a significant amount of the residual variance. Of the three step methods, this is the most frequently used.

Despite their appropriate use in certain limited situations, these step methods have lost favor over the last quarter century as we increasingly became aware of their weaknesses, including applying the incorrect degrees of freedom and thus increasing the chances of committing more Type I errors than the nominal alpha level would suggest, capitalizing on sampling error and thus resulting in poor generalizability, and often failing to identify the "best" set of predictors producing the largest explained variance (Cohen, Cohen, West, & Aiken, 2003; Meyers et al., 2013; Tabachnick & Fidell, 2007).

An alternative to the traditional step methods was introduced in IBM SPSS® version 19, the **Automatic Linear Modeling** procedure within the **Regression** module. This procedure performs an *all-possible-subsets* analysis. The all-possible-subsets method is considered to be more sophisticated and credible than the step methods (e.g., Huberty, 1989; Thompson, 1995, 2006). While the step methods construct a single model by adding or subtracting one predictor at a time, the all-possible-subsets method, as the name implies, calculates all the possible combinations of the predictors and presents competing models for researchers to consider.

Even with a small set of predictors, there are many possible subsets of predictors. The number of possible combinations (i.e., all-possible-subsets) is $(2^p) - 1$, where p is the number of predictors (we subtract 1 to eliminate the zero variable subset). For example, if there are five predictors, the number of all possible subsets is equal to $(2^5) - 1$ or 31. However, the output will be limited to the top 10 models, and, if the solution is relatively straightforward, the best model will often match the one produced by the stepwise method. As was true for the step procedures, researchers using this approach need to exercise judgment in selecting the most appropriate model for their needs based on both theoretical and pragmatic concerns.

26.2 NUMERICAL EXAMPLE

We use the data file named **Exercise** that was used in Chapter 24 in discussing simple linear regression. For the present example, we will use all five of the variables (except **CaseID**) to predict **exercise_commitment**.

26.3 ANALYSIS STRATEGY

We perform three analyses each using a different strategy of building the regression model: the standard method, the stepwise method, and the all-possible-subsets method. The first two will be performed in the **Linear Regression** procedure and the third will be performed in the **Automatic Linear Modeling** procedure.

26.4 ANALYSIS SETUP: STANDARD METHOD

From the main menu we select **Analyze → Regression → Linear**. This opens the main **Linear Regression** window as shown in Figure 26.1. We move **exercise_commitment** into the **Dependent** panel and all of the other variables into the **Independent(s)** panel. We retain **Enter** in the **Method** drop-down menu, as this will enter all of the variables on a single step as called for by the standard method.

Selecting the **Statistics** pushbutton opens the **Statistics** screen (see Figure 26.2). We check **Estimates** (to obtain the regression coefficients), **Model fit** (to obtain the R^2 and adjusted R^2), **R squared change** (to show this output for illustration purposes), **Descriptives** (to obtain the descriptive statistics), and **Part and partial correlations** (to

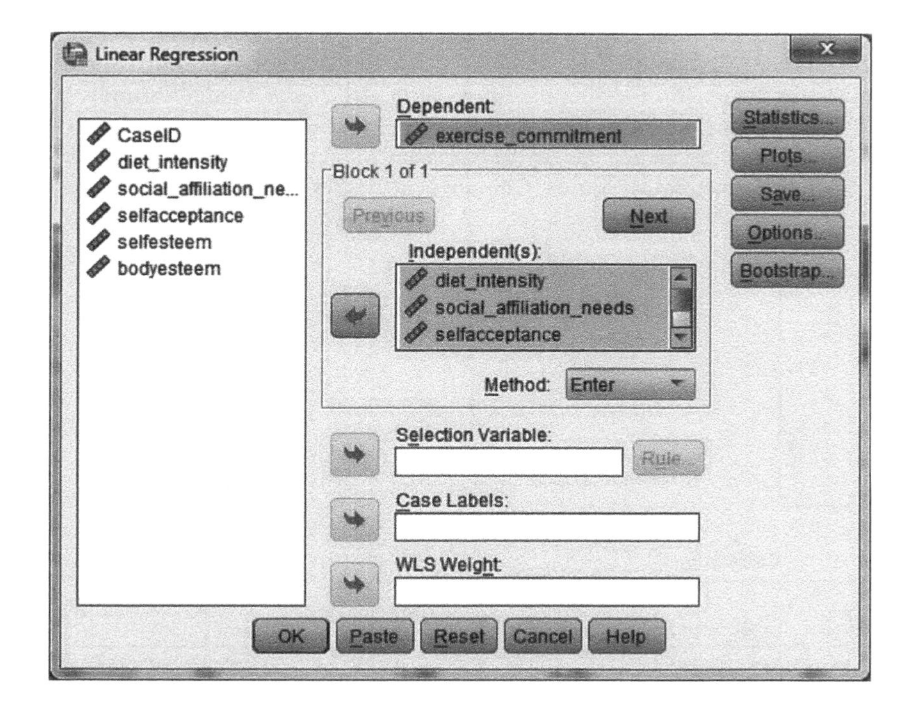

FIGURE 26.1
The main **Linear Regression** dialog window.

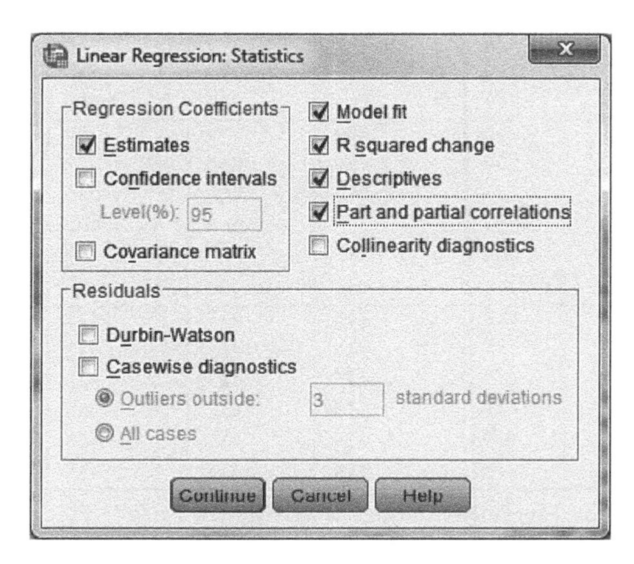

FIGURE 26.2
The **Statistics** screen of **Linear Regression**.

obtain the zero-order, partial, and semipartial correlations). Click **Continue** to return to the main dialog window and click **OK** to perform the analysis.

26.5 ANALYSIS OUTPUT: STANDARD METHOD

The descriptive statistics (means and standard deviations) for the six variables (five predictors and one dependent variable) and the square correlation matrix are shown in Figure 26.3. The six variables are assessed on quite different scales, even though all are quantitatively measured. The Pearson correlations are shown in the first major row of the **Correlations** table, and their corresponding probability levels are shown in the second major row. **Diet_Intensity** and **bodyesteem** are more highly correlated with the dependent variable of **exercise_commitment** than the other variables (**.464** and **.409**, respectively). Within the set of predictors, **selfesteem** and **bodyesteem** are more highly correlated than the other pairs.

Figure 26.4 shows the result of testing the fit of the model. In the **Model Summary** table, we see that the multiple correlation (**R**) is **.591**, with a corresponding value of **R Square** of **.350**, suggesting that 35% of the variance of **exercise_commitment** is explained by the set of predictors. **R Square Change** is also **.350**, as we went from zero

Descriptive Statistics

	Mean	Std. Deviation	N
exercise_commitment	3.3742	.87191	415
diet_intensity	3.1943	.75827	415
social_affiliation_needs	100.62	12.198	415
selfacceptance	61.24	7.118	415
selfesteem	39.08	6.706	415
bodyesteem	108.10	19.908	415

Correlations

		exercise_ commitment	diet_intensity	social_ affiliation_ needs	selfacceptanc e	selfesteem	bodyesteem
Pearson Correlation	exercise_commitment	1.000	.464	.177	.180	.262	.409
	diet_intensity	.464	1.000	.215	.243	.073	.104
	social_affiliation_needs	.177	.215	1.000	.306	.340	.245
	selfacceptance	.180	.243	.306	1.000	.352	.205
	selfesteem	.262	.073	.340	.352	1.000	.529
	bodyesteem	.409	.104	.245	.205	.529	1.000
Sig. (1-tailed)	exercise_commitment	.	.000	.000	.000	.000	.000
	diet_intensity	.000	.	.000	.000	.068	.017
	social_affiliation_needs	.000	.000	.	.000	.000	.000
	selfacceptance	.000	.000	.000	.	.000	.000
	selfesteem	.000	.068	.000	.000	.	.000
	bodyesteem	.000	.017	.000	.000	.000	.
N	exercise_commitment	415	415	415	415	415	415
	diet_intensity	415	415	415	415	415	415
	social_affiliation_needs	415	415	415	415	415	415
	selfacceptance	415	415	415	415	415	415
	selfesteem	415	415	415	415	415	415
	bodyesteem	415	415	415	415	415	415

FIGURE 26.3 Descriptive statistics and the correlation matrix.

35% 2x+a

Model Summary

Model	R	R Square	Adjusted R Square	Std. Error of the Estimate	Change Statistics				
					R Square Change	F Change	df1	df2	Sig. F Change
1	.591[a]	.350	.342	.70744	.350	43.974	5	409	.000

a. Predictors: (Constant), bodyesteem, diet_intensity, selfacceptance, social_affiliation_needs, selfesteem

ANOVA[b]

Model		Sum of Squares	df	Mean Square	F	Sig.
1	Regression	110.038	5	22.008	43.974	.000[a]
	Residual	204.693	409	.500		
	Total	314.732	414			

a. Predictors: (Constant), bodyesteem, diet_intensity, selfacceptance, social_affiliation_needs, selfesteem
b. Dependent Variable: exercise_commitment

FIGURE 26.4 Testing the fit of the model.

before the model was generated to the full value because all predictors were entered into the model in the first step. The **Adjusted R Square** value is .**342**, and represents some R^2 shrinkage as a result of including five predictors in the model.

The **ANOVA** table in Figure 26.4 provides a test of the statistical significance of the regression model using a one-way between-subjects ANOVA. The regression model has five degrees of freedom because that is the number of predictors in the model. The total degrees of freedom are equal to $N - 1$ or **414**, leaving **409** degrees of freedom for the error term. The model accounts for a significant amount of dependent variable variance, $F(1, 409) = \mathbf{43.97}$, $p < .001$. The eta square value is equal to **Regression** variance divided by **Total** variance, which yields **110.038/314.732** or .350. Note that this is the same value as the R^2 because the ANOVA and linear regression are just different expressions of the general linear model.

Figure 26.5 presents the **Coefficients** table for the variables in regression model. The three columns on the far right of the table show **Correlations**:

- The **Zero-order** correlations are the Pearson r values of the predictors with the criterion variable. For example, the Pearson r between **diet_intensity** and **exercise_commitment** is .**464**. They are being labeled by IBM SPSS as *zero-order partial correlations* (with the term *partial correlation* omitted) because no covariates are being used in evaluating the strength of those relationships (which thus reduces to the limiting case of the Pearson r).

Coefficients[a]

Model		Unstandardized Coefficients		Standardized Coefficients	t	Sig.	Correlations		
		B	Std. Error	Beta			Zero-order	Partial	Part
1	(Constant)	.076	.378		.202	.840			
	diet_intensity	.495	.048	.431	10.319	.000	.464	.454	.411
	social_affiliation_needs	-.001	.003	-.017	-.376	.707	.177	-.019	-.015
	selfacceptance	-.001	.005	-.010	-.226	.821	.180	-.011	-.009
	selfesteem	.008	.007	.061	1.211	.226	.262	.060	.048
	bodyesteem	.015	.002	.338	7.162	.000	.409	.334	.286

a. Dependent Variable: exercise_commitment

FIGURE 26.5 The regression coefficients together with the correlations.

- The **Partial** correlations are the correlations between each predictor and the residual variance of the dependent variable when statistically controlling for all of the other predictors. For example, **diet_intensity** is correlated **.454** with that portion of **exercise_commitment** not explained by the other predictors (these other predictors are being used as covariates in assessing that relationship). These partial correlations are *fourth-order partial correlations* in that the other four predictors act as covariates when the statistic for each predictor is being computed.
- The **Part** correlations are the *semipartial correlations* representing the *unique* association between each predictor and the dependent variable. We typically square these values for interpretation purposes. For example, the squared semipartial correlation between **diet_intensity** and **exercise_commitment** is $.411^2$ or approximately .17. We can therefore say that controlling for all of the other weighted predictors in the regression model, **diet_intensity** uniquely accounts for approximately 17% of the variance of **exercise_commitment** (note that some additional variance of **exercise_commitment** associated with **diet_intensity** is also associated with some other predictor(s) and thus does not count toward the value of the squared semipartial correlation of **diet_intensity**).

The columns on the left of the **Coefficients** table in Figure 26.5 present information concerning the regression coefficients. It can be recalled from Chapter 24 that the raw score (unstandardized) regression coefficient for **selfesteem** is **.034** and is statistically significant as the single predictor of **exercise_commitment**. In the companionship of the other four predictors, however, **selfesteem**'s partial regression coefficient is **.008**, a weight not significantly different from zero ($p = .226$). This is a good illustration of the relativistic nature of the values of regression coefficients. The weights of the predictors in the model are determined to be those that, when joined in combination with the other predictors, maximally predict the dependent variable; these weights often do not reflect the predictive capability of each variable in isolation.

Only two of the predictor variables are statistically significant in this model, **diet_intensity** and **bodyesteem**, the two that, coincidentally, had the highest correlations with **exercise_commitment**; this can also be seen in the relatively higher values of the unstandardized and standardized (beta) coefficients associated with these variables. We can thus say the following:

- When controlling for the other predictors in the model, an increase of one unit of **diet_intensity** is expected to be associated with a **.495** unit gain in **exercise_commitment**.
- When controlling for the other predictors in the model, an increase of one unit of **bodyesteem** is expected to be associated with a **.015** unit gain in **exercise_commitment**.

We can also glean a sense of the dynamics underlying the weights associated with these variables, two aspects of which are as follows:

- We noted that **selfesteem** and **bodyesteem** were relatively highly correlated, yet **bodyesteem** was a significant predictor in the model and **selfesteem** was not. It is likely that much of the prediction work done by **selfesteem** was redundant, with the predictive work done by **bodyesteem**, and only the latter was thus given credit for that work.
- The semipartial correlations indicate that **diet_intensity** and **bodyesteem** were able to contribute much more unique explanatory power than the other variables. Based on the squared values, **diet_intensity** and **bodyesteem** uniquely explained

about 17% (.**411**^{2} = .1689) and 8% (.**286**^{2} = .0818), respectively, of the variance of **exercise_commitment**.

One statistic that is not provided in the IBM SPSS output is the structure coefficient for each predictor. This statistic is the correlation between the individual predictor and the predictor variate (the weighted linear composite of the predictors) and can be used, when the set of predictor variables warrant it, to help interpret the underlying latent construct represented by the predictor model (e.g., a weighted linear composite of personality variables is likely to represent some meaningful underlying dimension). Thompson has particularly emphasized the importance of this statistic in regression (Courville & Thompson, 2001; Thompson & Borrello, 1985), but it is commonly used to interpret other latent constructs in, for example, principal components/factor analysis and discriminant function analysis.

The structure coefficients in regression can be computed by dividing the Pearson *r* values for each predictor by the multiple correlation (*R*), which is **.591** in our example. Thus, the structure coefficients for **diet_intensity**, **social_affiliation_needs**, **selfacceptance**, **selfesteem**, and **bodyesteem** are .785, .299, .305, .443, and .692, respectively. In this situation, the two statistically significant predictors also had relatively substantial structure coefficients (and thus drive the interpretation). The latent construct underlying the model thus appears to be valuing a trim and attractive body, a desire that appears to drive a commitment to exercise.

26.6 ANALYSIS SETUP: STEPWISE METHOD

We will briefly present the setup and output for the stepwise method, highlighting only the differences between this and the standard method. In the main **Linear Regression** window, shown in Figure 26.6, we move **exercise_commitment** into the **Dependent** panel and all of the other variables into the **Independent(s)** panel and select **Stepwise** from the **Method** drop-down menu.

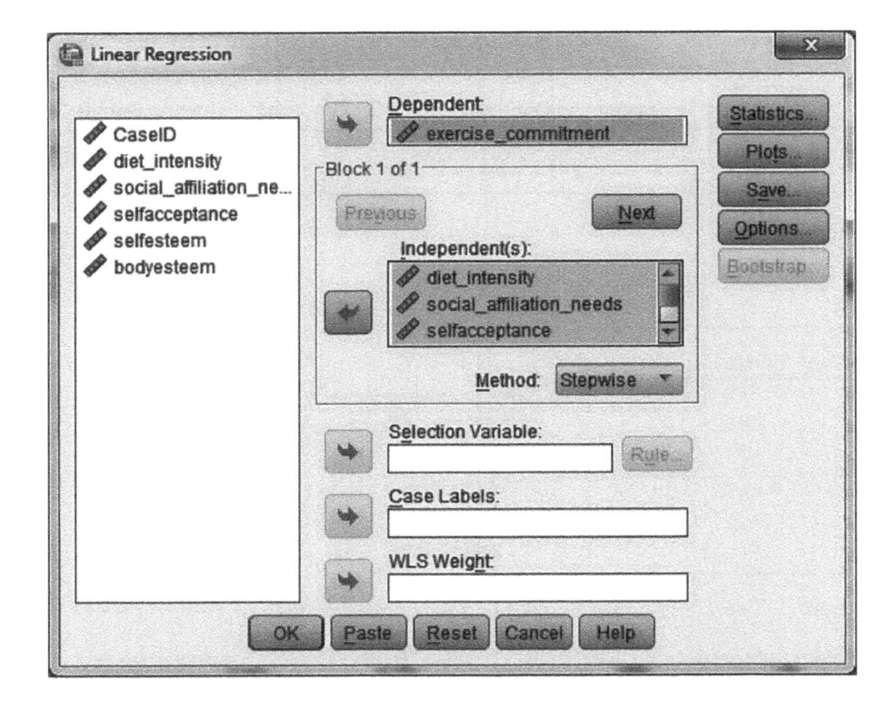

FIGURE 26.6
The main **Linear Regression** dialog window.

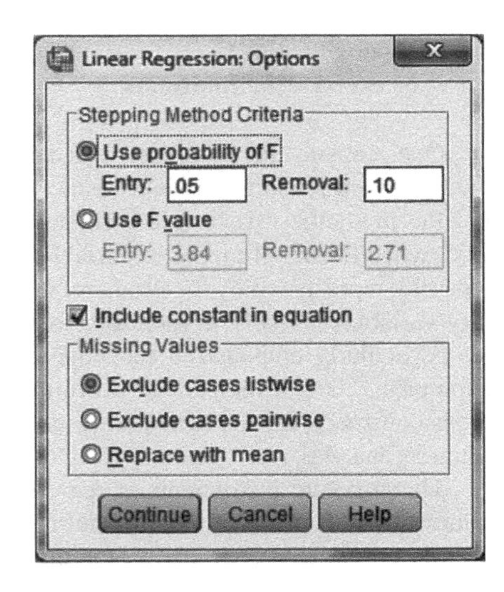

FIGURE 26.7
The **Options** screen of **Linear Regression**.

We configure the **Statistics** screen as we did for the standard regression analysis, but we open the **Options** screen (see Figure 26.7) just to show the **Stepping Method Criteria**. We retain the default **Entry** and **Removal** criteria of **.05** and **.10**, respectively (having this sort of a difference between them prevents the analysis from getting caught in an infinite loop of entry and removals). Click **Continue** to return to the main dialog window and click **OK** to perform the analysis.

26.7 ANALYSIS OUTPUT: STEPWISE METHOD

The **Model Summary** table is shown in Figure 26.8. There are two models, with the second built on the first. The variables contained in each model are shown in the footnotes

Model Summary

Model	R	R Square	Adjusted R Square	Std. Error of the Estimate	Change Statistics				
					R Square Change	F Change	df1	df2	Sig. F Change
1	.464[a]	.216	.214	.77311	.216	113.574	1	413	.000
2	.589[b]	.347	.344	.70614	.132	83.048	1	412	.000

a. Predictors: (Constant), diet_intensity
b. Predictors: (Constant), diet_intensity, bodyesteem

ANOVA[c]

Model		Sum of Squares	df	Mean Square	F	Sig.
1	Regression	67.883	1	67.883	113.574	.000[a]
	Residual	246.849	413	.598		
	Total	314.732	414			
2	Regression	109.293	2	54.647	109.592	.000[b]
	Residual	205.438	412	.499		
	Total	314.732	414			

a. Predictors: (Constant), diet_intensity
b. Predictors: (Constant), diet_intensity, bodyesteem
c. Dependent Variable: exercise_commitment

FIGURE 26.8 Testing the fit of the model.

to the **Model Summary** table. The first model contained only **diet_intensity** as the best predictor of **exercise_commitment**; the second model added **bodyesteem**. Because there were no additional models, we learn that no additional predictors could significantly improve the R^2 based on those two predictors.

The first model explained close to 22% of the variance ($R^2 = .216$). The second model incremented the R^2 by **.132** (**R Square Change**) to give a final R^2 of **.347** and an adjusted R^2 of **.344**, thus explaining about 34% of the variance with just two predictors. Note that this is about the same percentage of variance explained by the full model (because it contained three nonsignificant predictors) and may provide a sense of why the stepwise method had such appeal. However, if the full set of variables had been chosen with care based on existent theories and on the research literature, then removing (not entering) variables from regression models strictly based on statistical decisions made by the software should be done only for very carefully considered reasons (e.g., the need to predict in a limited resource context with as few variables as possible).

The ANOVA results are shown in the bottom table in Figure 26.8. The degrees of freedom for **Regression** is a count of the predictors in the model; thus, the first model has one degree of freedom because there is only one predictor variable in it and the second model has two degrees of freedom because there are two predictor variables in it. Both models are statistically significant, again not a surprise given that this is the stepwise method (and only statistically significant predictors are added to the model).

Figure 26.9 presents the **Coefficients** of the variables in regression model. The partial regression coefficients for **selfesteem** and **bodyesteem** are very close in value to what we saw in the standard regression output, and each uniquely accounts for a

Coefficients[a]

Model		Unstandardized Coefficients		Standardized Coefficients	t	Sig.	Correlations		
		B	Std. Error	Beta			Zero-order	Partial	Part
1	(Constant)	1.668	.164		10.142	.000			
	diet_intensity	.534	.050	.464	10.657	.000	.464	.464	.464
2	(Constant)	.081	.230		.354	.724			
	diet_intensity	.490	.046	.426	10.653	.000	.464	.465	.424
	bodyesteem	.016	.002	.365	9.113	.000	.409	.410	.363

a. Dependent Variable: exercise_commitment

Excluded Variables[c]

Model		Beta In	t	Sig.	Partial Correlation	Collinearity Statistics
						Tolerance
1	social_affiliation_needs	.081[a]	1.810	.071	.089	.954
	selfacceptance	.072[a]	1.600	.110	.079	.941
	selfesteem	.229[a]	5.423	.000	.258	.995
	bodyesteem	.365[a]	9.113	.000	.410	.989
2	social_affiliation_needs	-.005[b]	-.120	.905	-.006	.904
	selfacceptance	.002[b]	.051	.960	.002	.908
	selfesteem	.053[b]	1.125	.261	.055	.720

a. Predictors in the Model: (Constant), diet_intensity
b. Predictors in the Model: (Constant), diet_intensity, bodyesteem
c. Dependent Variable: exercise_commitment

FIGURE 26.9 The regression coefficients together with the correlations.

reasonable percentage of the variance of **exercise_commitment**. Variables excluded from the analysis are shown in the bottom table.

26.8 ANALYSIS SETUP: AUTOMATIC LINEAR MODELING

From the main menu, we select **Analyze → Regression → Automatic Linear Modeling**. This opens the main **Automatic Linear Modeling** window as shown in Figure 26.10, with the default of **Use predefined roles** activated. We change this to the **Use custom field assignments** button (see Figure 26.11). This shifts all the variables to the **Fields** panel located on the left side of the **Automatic Linear Modeling** window. We then move **exercise_commitment** into the **Target** panel and move all of the other variables (except **CaseID**) to the **Predictor (Inputs)** panel as shown in Figure 26.12.

Selecting the **Build Options** tab changes the left panel to **Select an item** with **Objectives** highlighted as shown in Figure 26.13. We retain the default of **Create a standard model** (an explanation of the selected objective appears in the **Description** panel below the model options).

Selecting the **Basics** choice in the **Select an item** panel opens the **Basics** window with **Automatically prepare data** already checked as the default (see Figure 26.14).

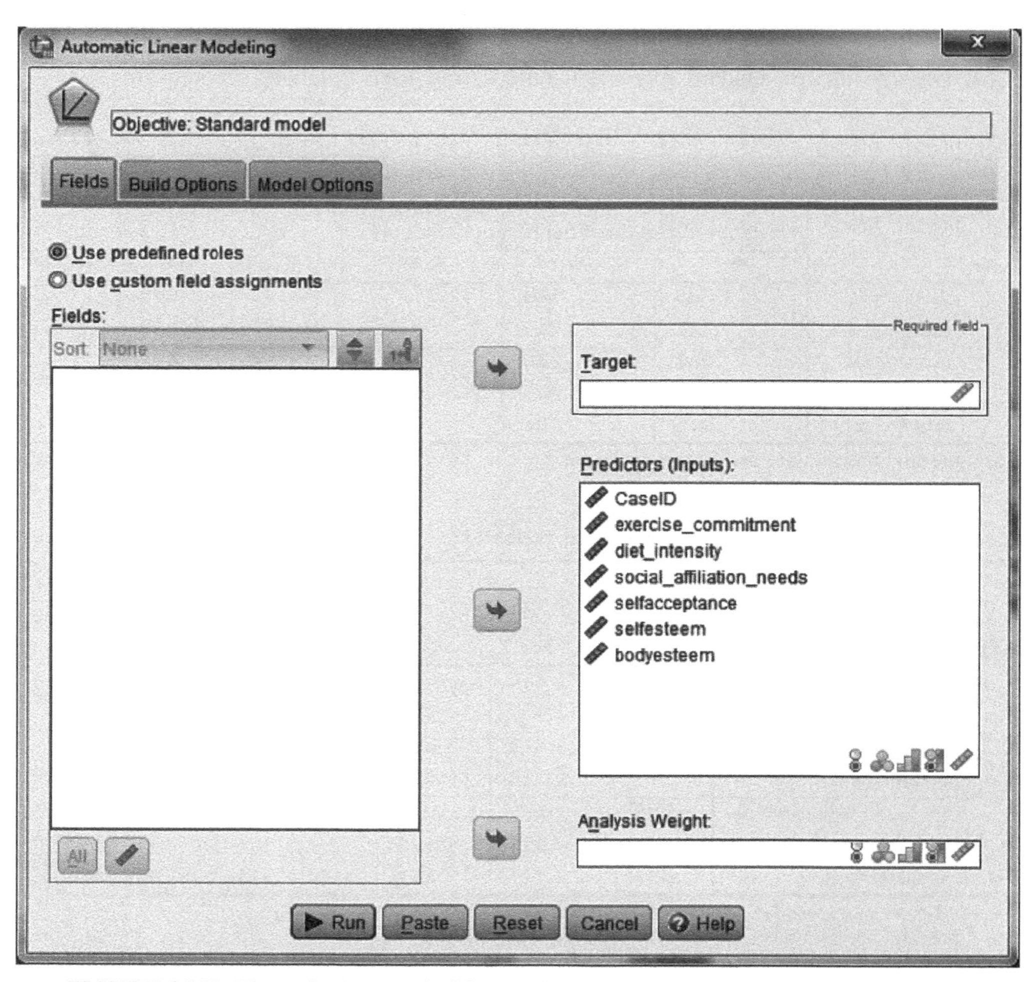

FIGURE 26.10 The main **Automatic Linear Modeling** window.

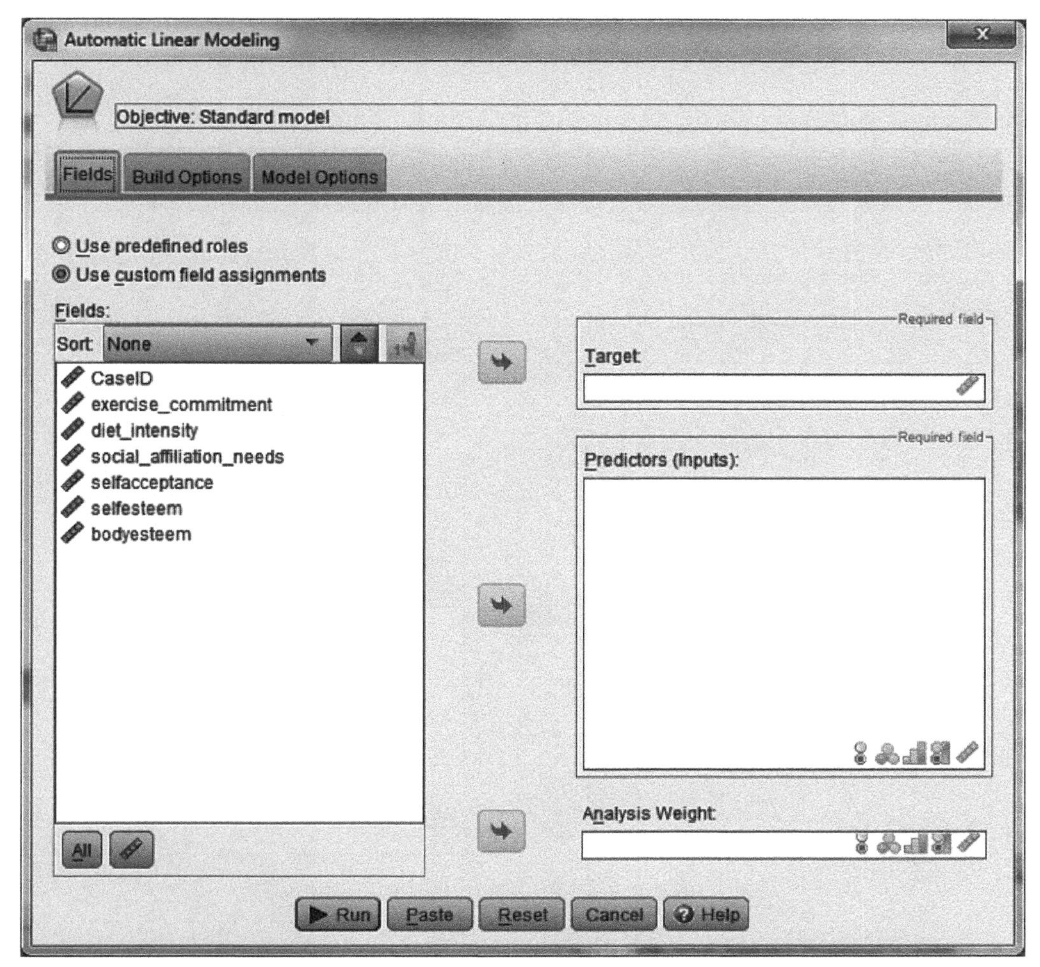

FIGURE 26.11 The main **Automatic Linear Modeling** window with **Use custom field assignments** selected.

We retain that default, as we have no missing values in our data set and our quantitative variables have lots of values (see later discussion). The operations that will be performed, as needed, are as follows:

- **Date and Time handling**. Variables identified as a **Date** variable will be transformed into continuous (**scale**) variables.
- **Adjustment of measurement level**. In the **Variable View**, there are three **Measure** designations—**Scale** for continuous variables, **Ordinal** for ranked data, and **Nominal** for categorical variables. The adjustments made here are (a) if a predictor **Scale** variable has less than five distinct values, it will be changed to an **Ordinal Measure** and (b) **Ordinal** predictors with more than 10 distinct values are changed to **Scale Measures**.
- **Outlier handling**. Continuous (**Scale**) predictors with values beyond a designated cutoff value (e.g., three standard deviations from the mean) are converted to the cutoff value. Researchers should be sure that they want to handle outliers in this manner, rather than using an alternative method (e.g., treat them as missing), before accepting **Automatically prepare data**. For our data set, we accept the cutoff replacement procedure.

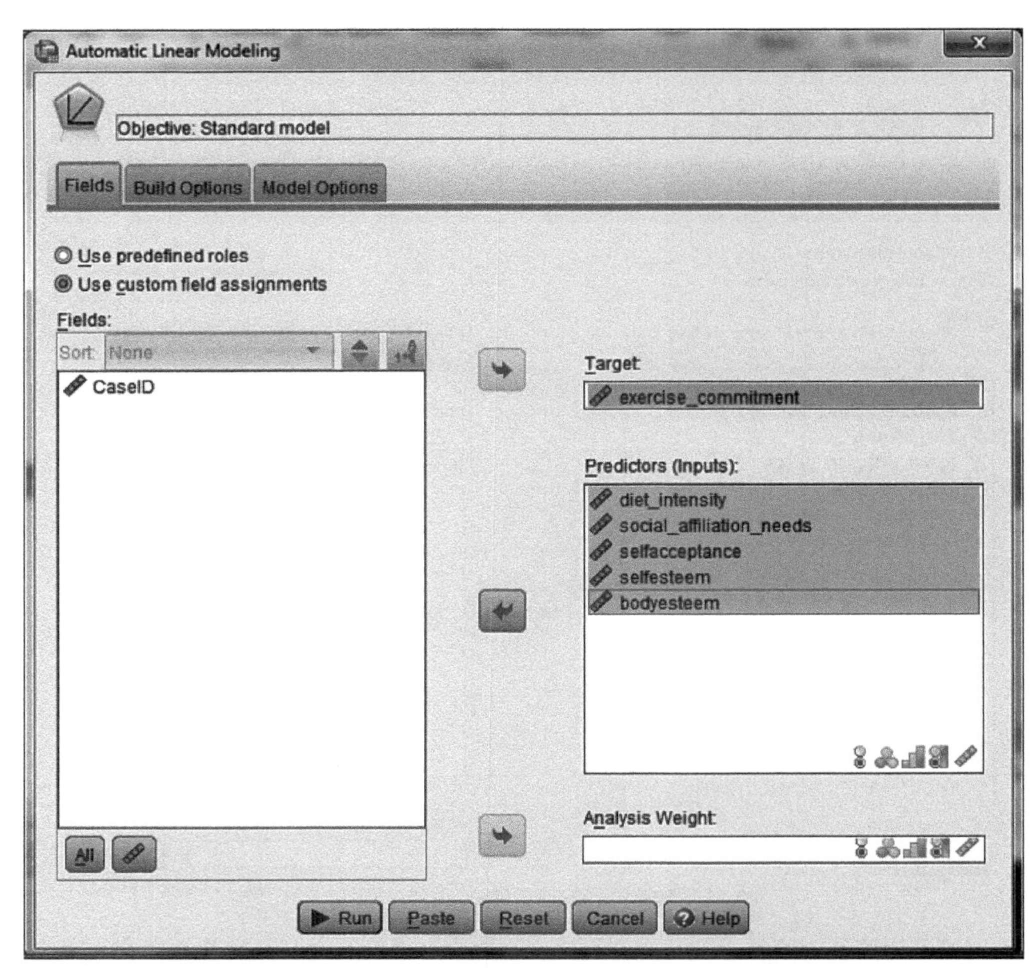

FIGURE 26.12 The main **Automatic Linear Modeling** window configured for our analysis.

- **Missing value handling**. For predictor variables that are **Nominal**, **Ordinal**, and **Scale**, missing entries are replaced with the variable's mode, median, and mean, respectively. However, not only can data be missing in different patterns (not all of which are completely at random), there are a variety of strategies for dealing with missing data, with one of the least preferred methods being mean substitution (e.g., Enders, 2010; Graham, 2009; McKnight, McKnight, Sidani, & Figuerdo, 2007; Meyers et al., 2013). Inserting the mean in place of missing scale values reduces variability; this in turn inappropriately shrinks the standard error; lower standard errors in turn increase the frequency of committing Type I errors. If there are missing data in the data set, researchers need to determine if they do or do not wish to accept **Automatically prepare data** with its mean substitution strategy. As an alternative approach, researchers can first perform one of the more preferred strategies for handling missing data (assuming it is justified based on the missing values patterns) and then, with no missing data issues, accept the **Automatically prepare data** option, as mean substitution would not be invoked.
- **Supervised merging**. Categorical variables (**Nominal**) that have equivalent associations on the **Target** (dependent) variable are merged to produce a more parsimonious model.

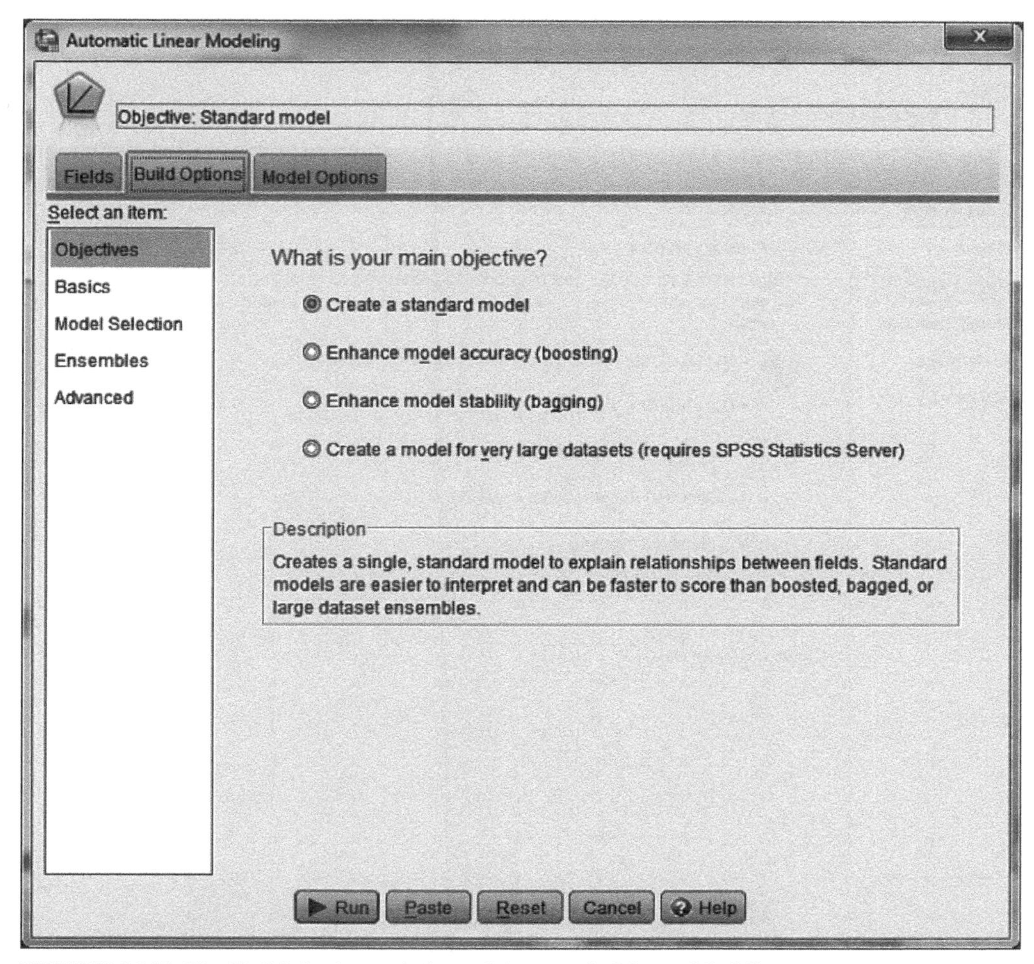

FIGURE 26.13 The **Build Options** window of **Automatic Linear Modeling**.

- **Confidence level**. Computes the confidence intervals of the coefficients in the models. The default is a 95% confidence interval, and we retain that.

Highlight the **Model Selection** choice in the **Select an item** panel. As shown in Figure 26.15, select from the drop-down menu **Best subsets** to replace the default of **Forward stepwise** in the **Model selection method** panel. In the **Best Subsets Selection** area, at the bottom of the screen, in the **Criteria for entry/removal** panel, we accept the default of **Information Criterion (AICC)**; this choice identifies the subsets of the predictors producing the "best" models and is useful in comparing competing models. There are two other choices on the drop-down menu for **Criteria for entry/removal**. Briefly, the **Adjusted R-squared** criterion identifies the "best" models based on the largest **Adjusted R-squared**. The **Overfit Prevention Criterion (ASE)** is based on the fit (average squared error) of the overfit prevention set (a random subsample of ~30% of the original dataset).

We would move to the **Ensembles** choice only if we had selected one of the other options in the **Objectives** window and we have no need to deal with the **Advanced** choice. The **Model Options** tab allows us to save the predicted values to the data set and export the model as a *.zip* file. As we do not intend to make use of either of these options, we are finished configuring the analysis and click the **Run** pushbutton to perform the analysis.

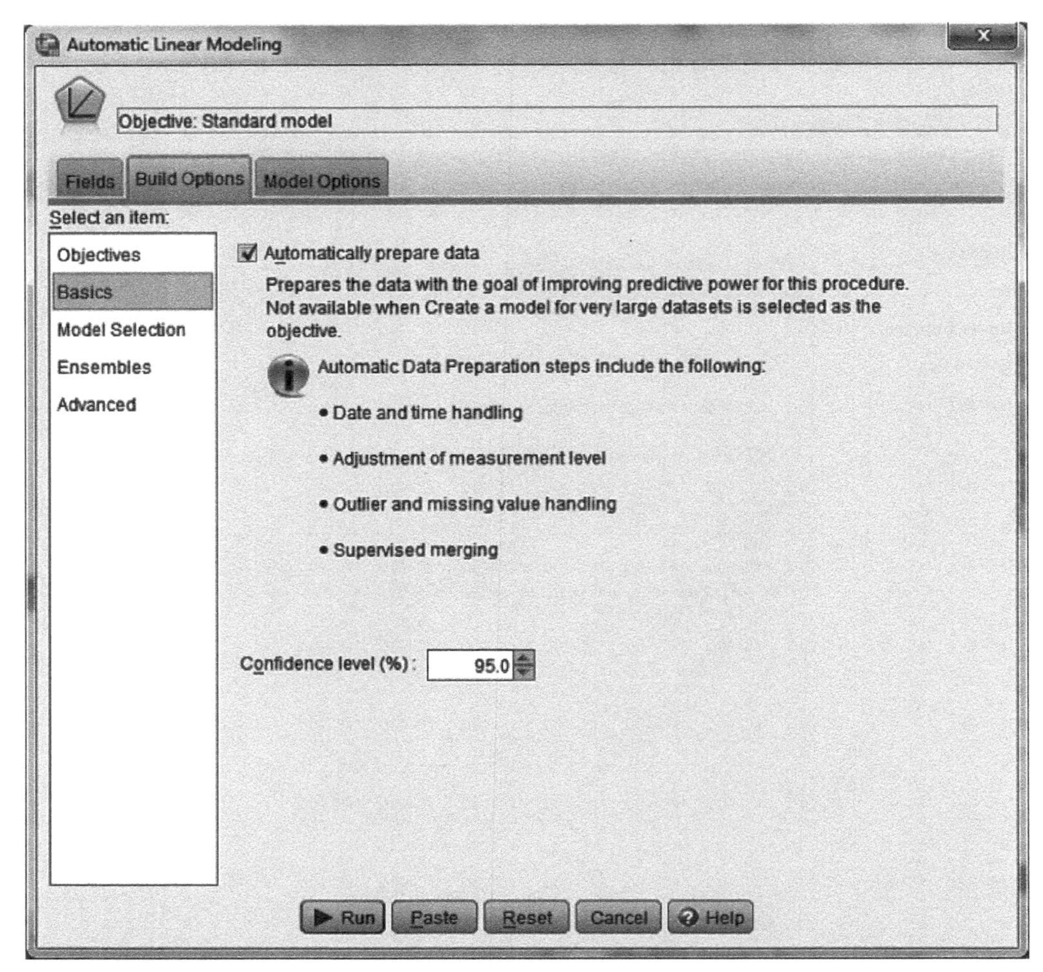

FIGURE 26.14 The **Basics** window of **Automatic Linear Modeling**.

26.9 ANALYSIS OUTPUT: AUTOMATIC LINEAR MODELING

The output opens in the **Model Viewer** window by displaying the **Model Summary** and the **Accuracy** bar graph as shown in Figure 26.16. This format is different from most of the other outputs we cover in this book but is more visually oriented. The left vertical panel with the thumbnails is our access to different portions of the output. There is no thumbnail for the **Model Summary** output, as it is literally an overview of the analysis setup.

The first thumbnail showing a horizontal bar graph is highlighted as a default. This is the **Accuracy** bar graph and it is already displayed below the **Model Summary** in the opening window. It is a visual depiction of the amount of variance explained by the best fitting model. Double-clicking the bar graph activates it, and moving the cursor across the bar reveals this information: **Adjusted R square = .344** (see Figure 26.17). This number also appears to the left of the bar graph in the opening window but is unlabeled. Also not displayed in the opening window is the model (recall that there are 31 possible models) to which the adjusted R^2 value applies.

Activating the **Accuracy** bar graph also activates the vertical list of thumbnails in the leftmost panel. A scrollbar has been added to the panel (this can be seen in Figure 26.17),

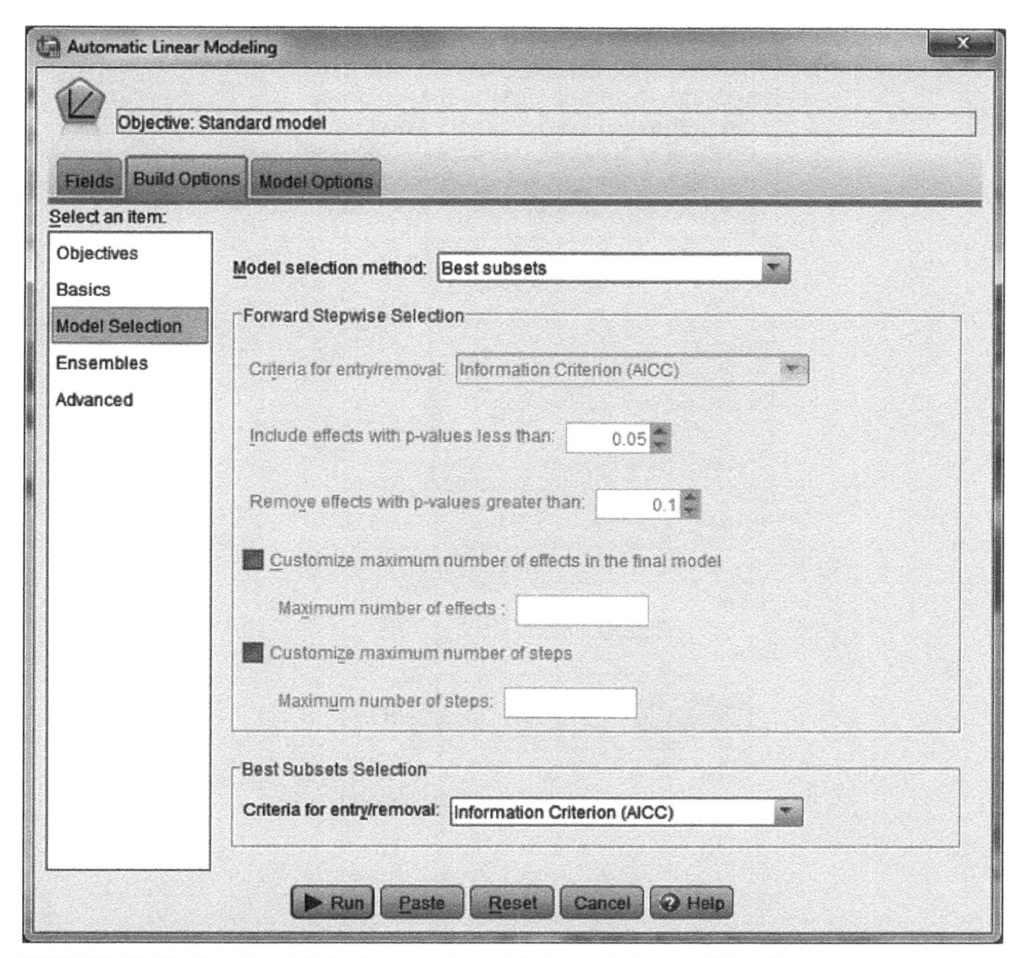

FIGURE 26.15 The **Model Selection** window of **Automatic Linear Modeling**.

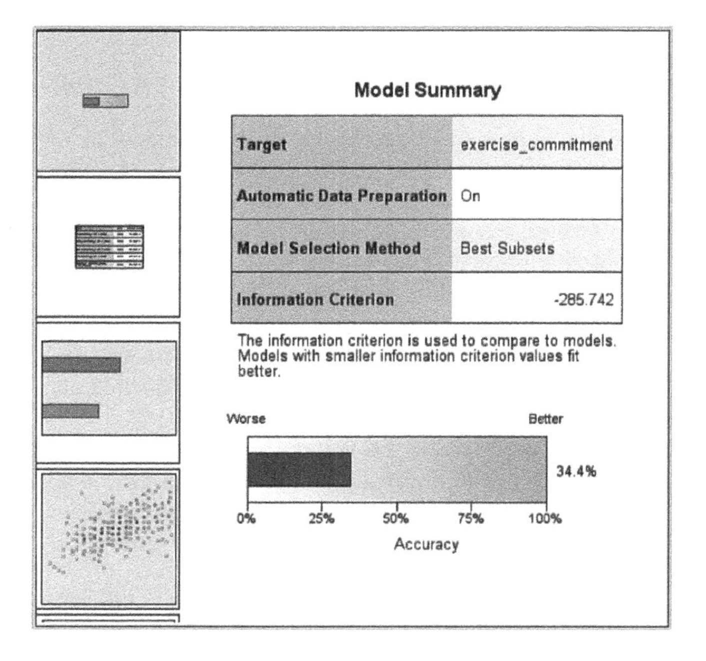

FIGURE 26.16
The **Model Summary** output of **Automatic Linear Modeling**.

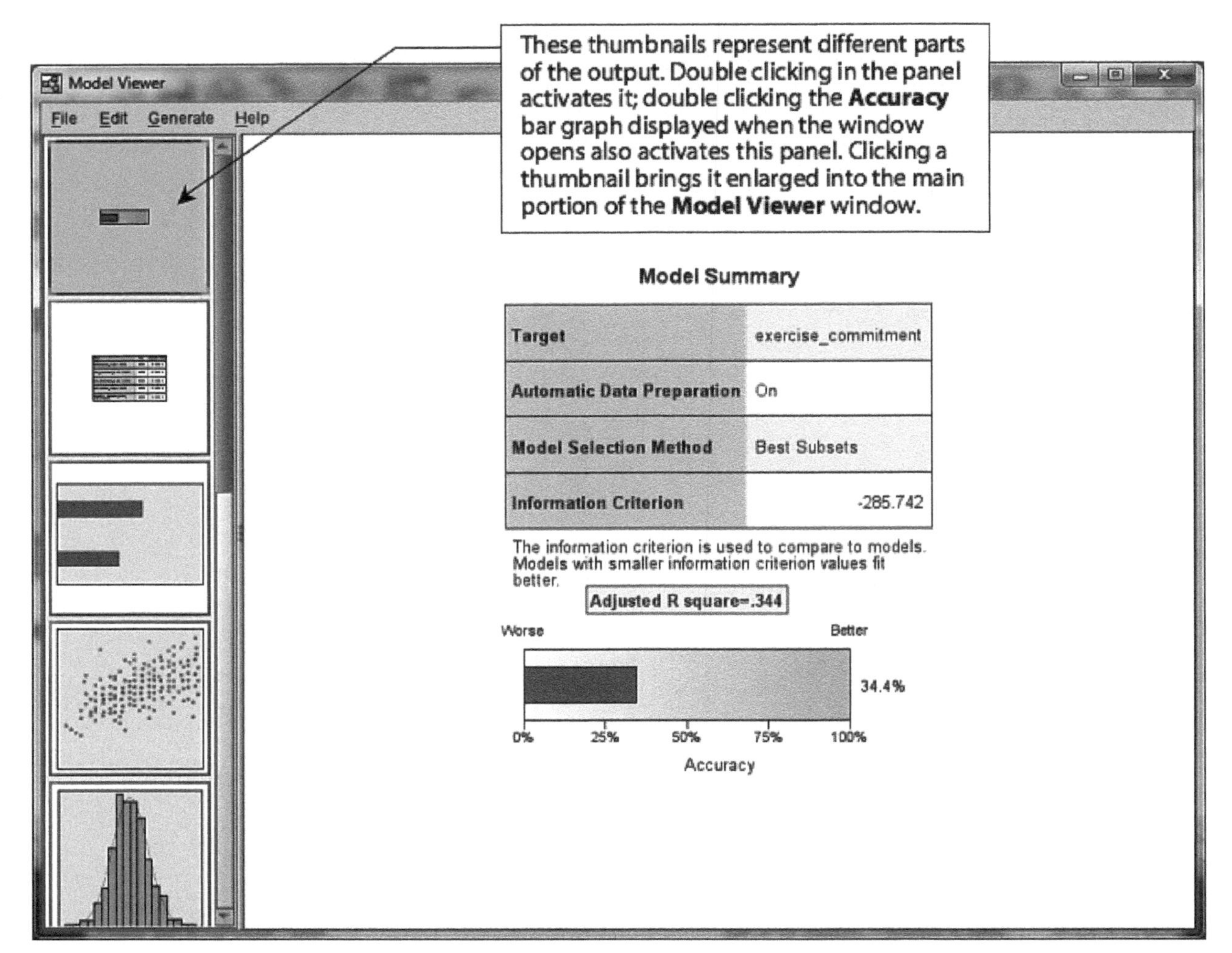

FIGURE 26.17 The **Accuracy** output of **Automatic Linear Modeling**.

as there are more thumbnails than that can be viewed at once in the window, and placing the cursor over a thumbnail provides a label for the output it represents.

The second thumbnail from the top, shown in Figure 26.18, reports the actions of the **Automatic Data Preparation** function. In our example, **Trim outliers** was the only transformation needed for all the **Predictors**. Scores greater than the absolute value of a z score of 3.0 were replaced with 3.0 because we indicated in the analysis setup that three standard deviations distance from the mean was the designated cutoff value.

We recommend to next select the next-to-last thumbnail (it resembles a little grid). This is the **Model Building Summary** and gives us the needed overview of the results. We show this output in Figure 26.19. Although there were 31 possible models to evaluate, IBM SPSS thankfully shows us only the top 10 models (a set that is typically much larger than we need to see).

All of the available predictor variables are listed in the first column; they are named as being **transformed** in that the evaluations of fit are made on the standardized model to place all variables on the same (z score) metric. Each column under **Model** is one of the top 10 best fitting models ordered by its value on the **Information Criterion**. Lower (more negative) values indicate a better fit.

The top-rated model (**Model 1**) has an **Information Criterion** value of **−285.742**. The variables included as predictors in that model are checked (**diet_intensity** and

Automatic Data Preparation

Target: exercise_commitment

Field	Role	Actions Taken
(bodyesteem_transformed)	Predictor	Trim outliers
(diet_intensity_transformed)	Predictor	Trim outliers
(selfacceptance_transformed)	Predictor	Trim outliers
(selfesteem_transformed)	Predictor	Trim outliers
(social_affiliation_needs_transformed)	Predictor	Trim outliers

If the original field name is X, then the transformed field is displayed as (X_transformed). The original field is excluded from the analysis and the transformed field is included instead.

FIGURE 26.18 The **Automatic Data Preparation** output of **Automatic Linear Modeling**.

Model Building Summary

Target: exercise_commitment

		Model									
		1	2	3	4	5	6	7	8	9	10
Information Criterion		-285.742	-284.982	-283.727	-283.716	-283.147	-282.989	-281.701	-281.115	-238.156	-236.163
Effect	diet_intensity_transformed	✓	✓	✓	✓	✓	✓	✓	✓	✓	✓
	bodyesteem_transformed	✓	✓	✓	✓	✓	✓	✓	✓		
	selfesteem_transformed		✓			✓	✓		✓	✓	✓
	social_affiliation_needs_transformed			✓		✓		✓	✓		
	selfacceptance_transformed				✓		✓	✓	✓		✓

The model building method is Best Subsets using the Information Criterion.
A checkmark means the effect is in the model.

FIGURE 26.19 The **Model Building Summary** output of **Automatic Linear Modeling**.

bodyesteem); this is the same model generated by the stepwise procedure. It is marked by a rectangle to inform us that the numerical results in the other parts of the output are keyed to this model. Thus, the adjusted R^2 value seen in the **Accuracy** bar graph applies to this model. Other models, with different combinations of predictors, are shown in the **Model Building Summary**.

The **Effects** icon (a circle with two lines—because there are only two predictors in the model—pointing to clock positions 7:00 and 11:00) and its window are shown in Figure 26.20. By selecting **Table** from the drop-down **Style** menu at the lower left portion of the window to get out of **Diagram** mode, we see the more familiar ANOVA summary table shown in Figure 26.21. These are the same values as those produced by the final stepwise model.

The **Coefficients** icon (a circle with three lines pointing to clock positions 7:00, 9:00, and 11:00) is just below the **Effects** icon. Its **Table** mode is shown in Figure 26.22. The unstandardized partial regression coefficients are presented for each predictor together with a test of the statistical significance of each; these coefficients are also the same as those obtained in the stepwise solution. Both predictors contribute statistically significant amounts of prediction to the model.

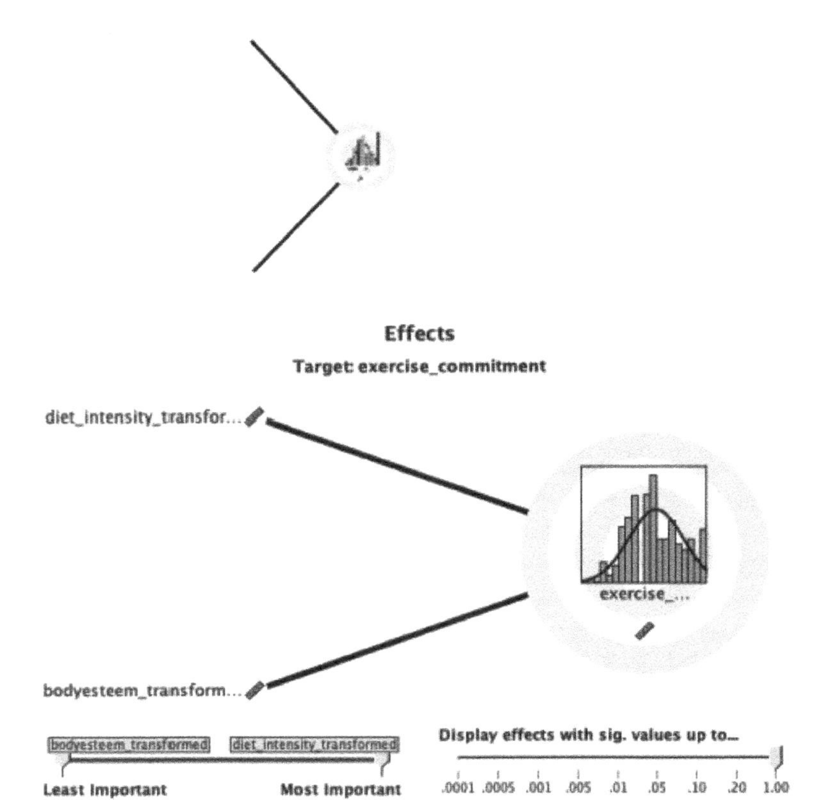

FIGURE 26.20 The **Effects** output of **Automatic Linear Modeling** in **Diagram** mode.

Effects

Target: exercise_commitment

Source	Sum of Squares	df	Mean Square	F	Sig.
	109.293	2	54.647	109.592	.000
Residual	205.438	412	0.499		
Corrected Total	314.732	414			

bodyesteem_transformed diet_intensity_transformed

Least Important Most Important

Display effects with sig. values up to...

.0001 .0005 .001 .005 .01 .05 .10 .20 1.00

FIGURE 26.21 The **Effects** output of **Automatic Linear Modeling** in **Table** mode.

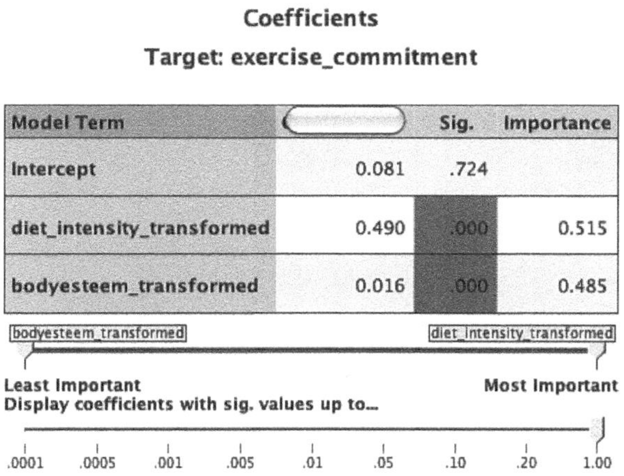

Coefficients

Target: exercise_commitment

Model Term		Sig.	Importance
Intercept	0.081	.724	
diet_intensity_transformed	0.490	.000	0.515
bodyesteem_transformed	0.016	.000	0.485

bodyesteem_transformed diet_intensity_transformed

Least Important Most Important

Display coefficients with sig. values up to...

.0001 .0005 .001 .005 .01 .05 .10 .20 1.00

FIGURE 26.22

The **Coefficients** output of **Automatic Linear Modeling** in **Table** mode.

Also reported in the **Coefficients** table is a value called **Importance**. As described in their historical overview, Johnson and LeBreton, (2004) suggest that researchers have been engaged for more than three-quarters of a century in the search for ways to characterize the relative importance of each predictor in a regression model. From the older and simpler standbys, such as the values of the regression coefficients, magnitudes of the squared semipartial correlations, and change in R^2, alternative indexes have emerged, such as usefulness (Darlington, 1968), dominance analysis (Budescu, 1996), and relative weight analysis (Johnson, 2000; Tonidandel & LeBreton, 2011).

The strategy adopted by IBM SPSS uses the sum of squares for the residual (the error variance in the ANOVA table) as a signal regarding importance. To the extent that a predictor is important in the model, leaving it out of the model should produce a substantial increase in the residual sum of squares; to the extent that a predictor is not important in the model, leaving it out of the model should produce a minor increase in the residual sum of squares. Briefly, in implementing this strategy, IBM SPSS computes a series of models excluding each predictor in each successive model, records the sum of squares associated with the residuals for each model, adds $1/p$ to each residual where p is the total number of predictors, and then determines the ratio of each subtotal to the grand total (these ratios will sum to 1.00).

The implementation described previously results in normalized predictor importance values and appears to be what is reported in the **Automatic Linear Modeling** procedure as the relative **Importance** of each predictor. Because the residual sums of squares can be generated from the squared semipartial correlations, this method is related to Darlington's (1968) usefulness index (we are grateful to one of our colleagues, Professor Greg Hurtz of the California State University, Sacramento, for his instruction on this relative importance topic).

The last column of the **Coefficients** table reports the results of this relative importance analysis: the relative importance of **diet_intensity** is **.515** and the relative importance of **bodyesteem** is **.485**. Thus, in the top-rated model, **diet_intensity** was somewhat more important a predictor than **bodyesteem** in predicting **exercise_commitment**.

Having a description of the model that was determined to best fit the data, we can view the analysis of the residuals. Residuals in this context are the differences between the predicted and observed values of Y, and they are typically in a standardized form to overcome differences in the metric of the different predictors. One of the assumptions underlying regression is that the residuals are normally distributed. The **Residuals** icon is a histogram, and clicking it gives us a histogram—a frequency distribution of standardized residuals (see Figure 26.23). The normal curve is also superimposed on the histogram so that we can visually determine how closely the histogram approximates it. In our

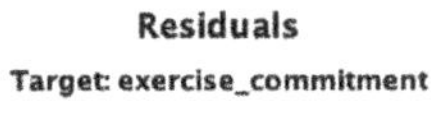

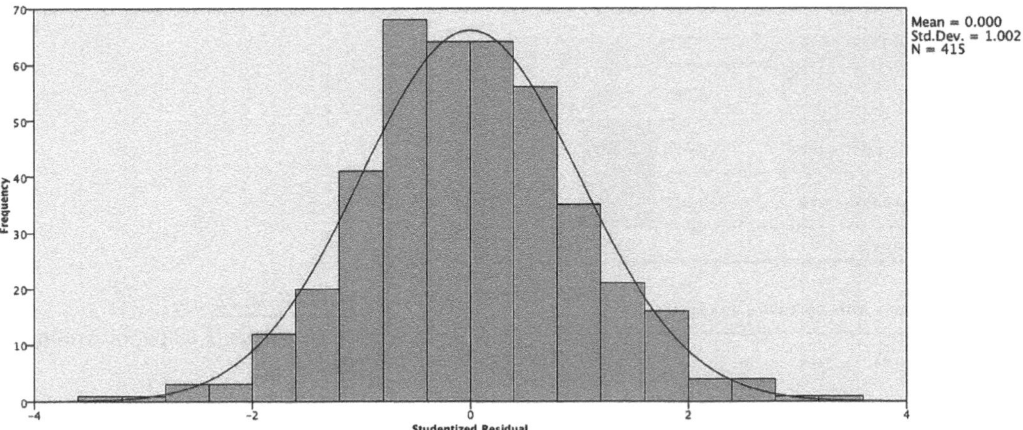

The histogram of Studentized residuals compares the distribution of the residuals to a normal distribution.The smooth line represents the normal distribution.The closer the frequencies of the residuals are to this line, the closer the distribution of the residuals is to the normal distribution.

FIGURE 26.23 The **Residuals** output of **Automatic Linear Modeling** in **Diagram** mode.

Residuals

Target: exercise_commitment

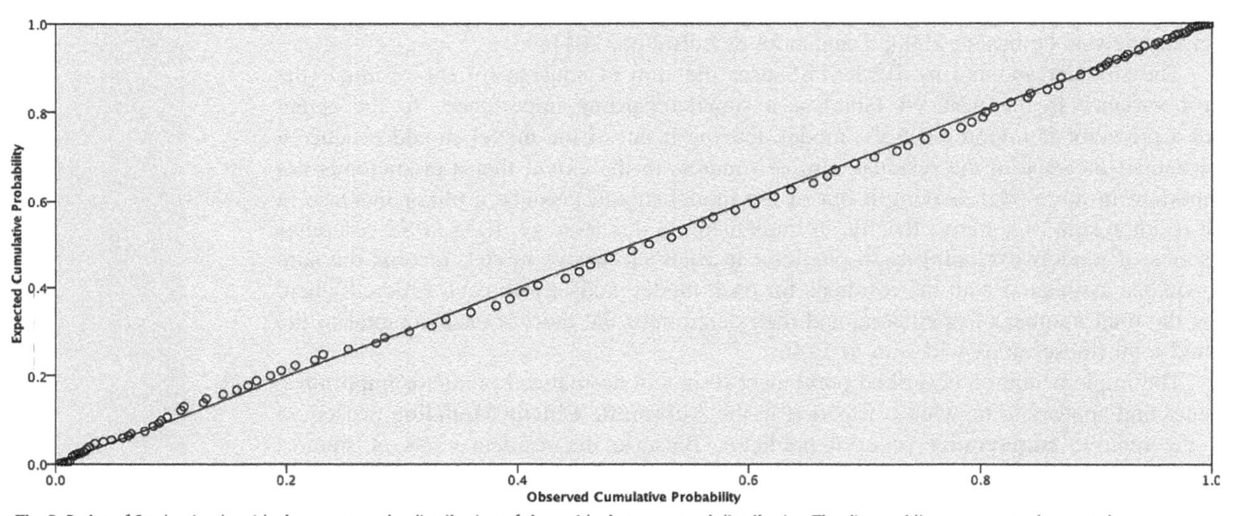

The P-P plot of Studentized residuals compares the distribution of the residuals to a normal distribution.The diagonal line represents the normal distribution.The closer the observed cumulative probabilities of the residuals are to this line, the closer the distribution of the residuals is to the normal distribution.

FIGURE 26.24 The **Residuals** output of **Automatic Linear Modeling** in **P-P Plot** mode.

example, the residuals are clearly normally distributed indicating no assumption violation.

Select the **P-P Plot** from the drop-down **Style** menu at the lower left portion of the **Residuals** window to get out of **Histogram** mode and to obtain a **P-P Plot**. **P** stands for "probability" in the name of the type of plot, and we are presented with the view of the plot of the **Expected Cumulative Probability** on the *Y*-axis against the **Observed Cumulative Probability** on the *X*-axis, shown in Figure 26.24, as another way to view the information contained in the histogram. The diagonal line in the plot is a reference line indicating how the data points should fall if the residuals were normally distributed; the example plot, which conveys the same information as the histogram, confirms that the residuals are normally distributed.

CHAPTER 27

Hierarchical Linear Regression

27.1 OVERVIEW

Hierarchical linear regression is an extension of standard multiple linear regression, with a conceptual element resembling the step procedures. The key factor in hierarchical regression is that, in contrast to the step regression procedures where the researchers leave all decisions about entry to the software, the researchers play an active role in structuring the analysis within the hierarchical strategy. In return for such an investment, researchers are able to statistically control for the effect of predictors when it makes theoretical, empirical, or common sense to do so.

In hierarchical regression, predictors are entered in order in subsets or blocks. A subset can be composed of just a single variable or several variables. Each block of variables can be controlled by any method available (simultaneously as defined by the standard method, stepwise, and so on). Because some variables are given "primacy" over others in order of entry, the procedure is referred to as hierarchical linear regression.

The primary advantage of using such a blocking or hierarchical strategy is that variables entered in earlier blocks serve as covariates for those entered later. For example, we might wish to ask respondents to participate in a survey about their commitment to an exercise schedule. But because it is a "socially correct" behavior, those respondents sensitive to social expectations might be inclined to tell us that they are somewhat more committed to such a schedule than they perhaps might truly be. It may also be the case that those respondents highly committed to dieting to lose weight might also be inclined to use exercise in such an effort.

One way to deal with the potential confounding of social desirability in this design and to take account of the extent to which respondents were also dieting is to use measures of social desirability and dieting as covariates in the analysis. The dynamics for doing this in the context of hierarchical regression might be as follows:

- In the first block, enter the measure of social desirability. This variable will thus account for a certain amount of variance of the dependent variable of commitment to exercise.
- In the second block, enter a measure of dieting intensity. With social desirability already in the model, this newly entered variable must target the residual variance of commitment to exercise (whatever social desirability did not explain). Thus,

Performing Data Analysis Using IBM SPSS®, First Edition.
Lawrence S. Meyers, Glenn C. Gamst, and A. J. Guarino.

social desirability acts as a covariate in that it statistically accounts for whatever variance of the dependent variable it can; the measure of dieting intensity is forced to do predictive work not already under the auspices of the verbal ability variable. At the same time, the dieting intensity variable acts as a covariate in the evaluation of social desirability and thus could affect its regression weight in the model.

- In the third block, enter other personality variables that might explain the variance of measure of commitment to exercise not explained by either social desirability or dieting intensity. Each variable is evaluated, with the others acting as covariates.

27.2 NUMERICAL EXAMPLE AND ANALYSIS STRATEGY

We use the data file named **Exercise** that was used in Chapter 24, and we will continue to predict **exercise_commitment**. However, we will enter the variables in the following three ordered blocks, with each block using the simultaneous method of entry (assume that these decisions were based on solid theoretical and empirical reasons):

1. In Block 1, we enter **social_affiliation_needs** to control for social desirability effects.
2. In Block 2, we enter **diet_intensity** to control for the effects of diet on exercise. It will be evaluated, with the effects of social desirability having been statistically controlled. At the same time, **social_affiliation_needs** will be evaluated (actually, reevaluated) controlling for the effects of **diet_intensity**.
3. In Block 3, we enter the remaining esteem/acceptance variables. Each of these variables will be evaluated, with the effects of all of the other variables having been statistically controlled.

27.3 ANALYSIS SETUP

From the main menu, we select **Analyze ➔ Regression ➔ Linear**. This opens the main **Linear Regression** window as shown in Figure 27.1. We move **exercise_commitment** into the **Dependent** panel and **social_affiliation_needs** into the **Independent(s)** panel. Note that at the very top of the **Independent(s)** panel just below the **Dependent** panel we see **Block 1 of 1**. IBM SPSS® will keep track of our blocks in this way and allow us to toggle back and forth with the **Next** and **Previous** pushbuttons if we need to make any changes before performing the analysis. We retain **Enter** in the **Method** drop-down menu. This completes our specification for the first block. Click **Next** to configure the second block.

We are now in **Block 2 of 2**. As seen in Figure 27.2, the dependent variable remains in its panel but the **Independent(s)** panel has been emptied in preparation for our specification. We move **diet_intensity** into the **Independent(s)** panel, retain the **Enter** method, and click **Next** to configure the third block.

We are now in **Block 3 of 3**. As shown in Figure 27.3, the dependent variable still remains in its panel, but the **Independent(s)** panel has once again been emptied in preparation for our specification. We now move **selfacceptance**, **selfesteem**, and **bodyesteem** into the **Independent(s)** panel to complete our set of predictors. We retain the **Enter** method as we wish to enter all three variables simultaneously into the model. We ask for the same output in the **Statistics** window as in Chapter 24, and click **OK** in the main dialog window to perform the analysis.

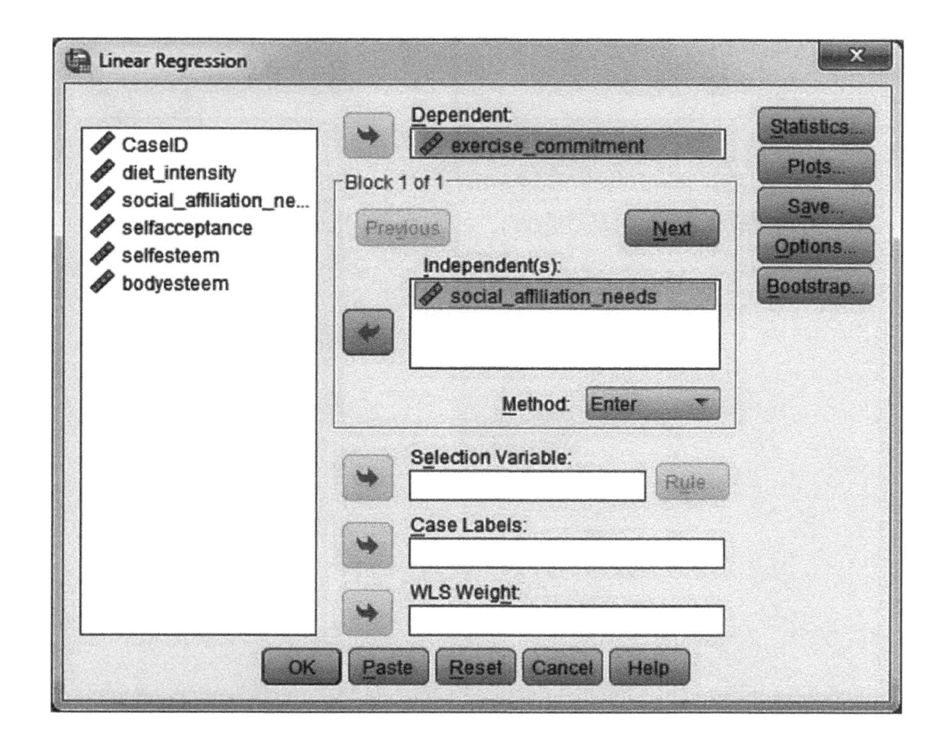

FIGURE 27.1
The first block of the hierarchical regression analysis is configured.

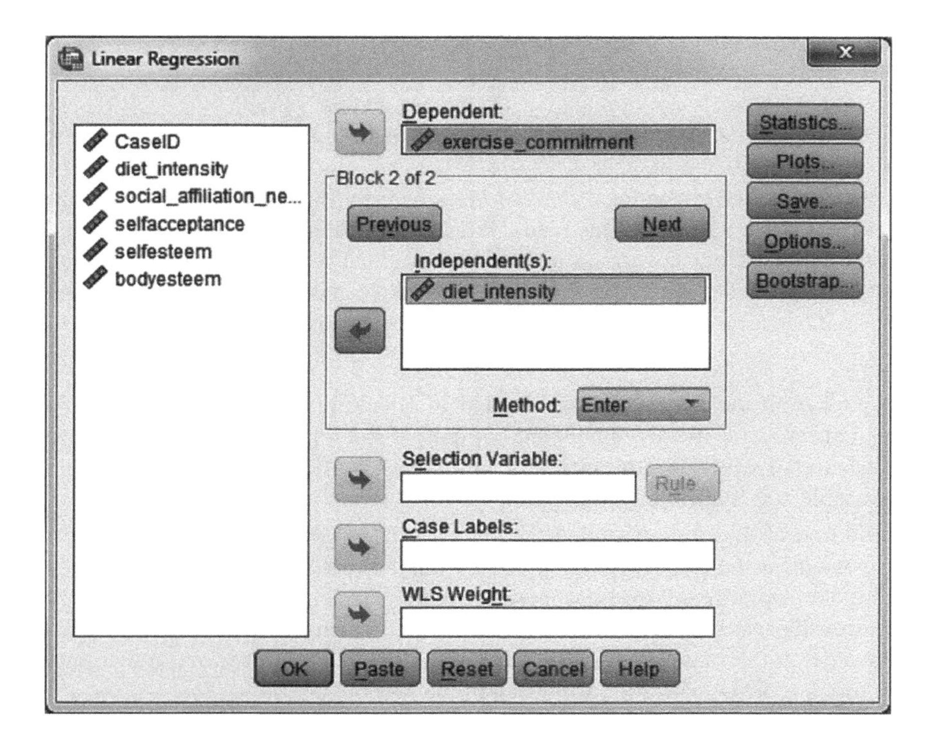

FIGURE 27.2
The second block of the hierarchical regression analysis is configured.

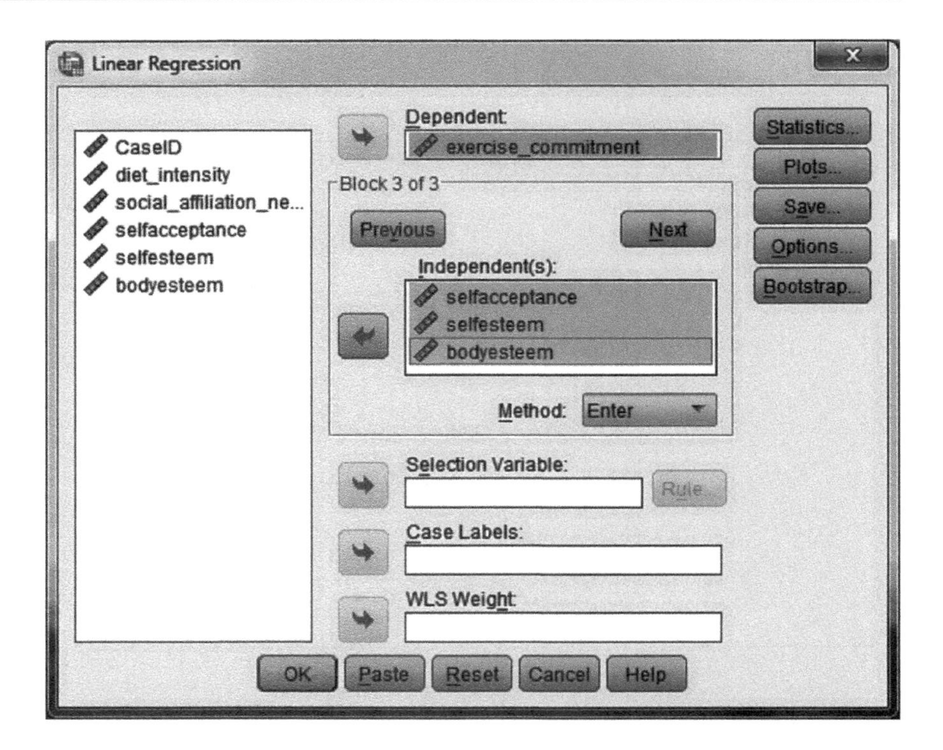

FIGURE 27.3
The third block of the hierarchical regression analysis is configured.

27.4 ANALYSIS OUTPUT

The relevant output from the hierarchical regression analysis is presented in Figure 27.4. There are three models in our analysis because IBM SPSS treated each block as a separate model. Note that the models are cumulative, as we are adding variables in each block; the final model matches exactly the results from the standard multiple regression analysis and is interpreted in the same manner.

What we gain with the hierarchical analysis is an opportunity to observe the dynamics of the interplay of these variables. Our focus, in addition to interpreting the last model, is on the contribution of our covariates and the value-added we obtain when including more variables to the model; we therefore key on **R Square** and **R Square Change** (and the tests of significance) associated with the three blocks. We note the following:

- It appears from the **Model Summary** that in Block 1, **social_affiliation_needs** accounted for a statistically significant but quite a small proportion (about 3%) of the variance of **exercise_commitment**. From the **Coefficients** table, we see that this variable was statistically significant ($p < .001$).
- The addition of **diet_intensity** into the model in Block 2 made a relatively major improvement in the prediction ability of the model, accounting for an additional 19% of the variance of **exercise_commitment**. The prediction contribution of **diet_intensity** was statistically significant ($p < .001$) controlling for **social_affiliation_needs**; at the same time, **social_affiliation_needs** became at best a marginal predictor ($p =$ **.071**), with **diet_intensity** in the model.
- Finally, the esteem/acceptance variables added in Block 3 contributed a statistically significant amount of prediction over and above the effects of **social_affiliation_needs** and **diet_intensity**, adding another almost 13% in explaining

Model Summary

Model	R	R Square	Adjusted R Square	Std. Error of the Estimate	Change Statistics				
					R Square Change	F Change	df1	df2	Sig. F Change
1	.177[a]	.031	.029	.85925	.031	13.289	1	413	.000
2	.471[b]	.222	.218	.77099	.191	100.969	1	412	.000
3	.591[c]	.350	.342	.70744	.128	26.781	3	409	.000

a. Predictors: (Constant), social_affiliation_needs
b. Predictors: (Constant), social_affiliation_needs, diet_intensity
c. Predictors: (Constant), social_affiliation_needs, diet_intensity, bodyesteem, selfacceptance, selfesteem

ANOVA[d]

Model		Sum of Squares	df	Mean Square	F	Sig.
1	Regression	9.811	1	9.811	13.289	.000[a]
	Residual	304.920	413	.738		
	Total	314.732	414			
2	Regression	69.830	2	34.915	58.737	.000[b]
	Residual	244.902	412	.594		
	Total	314.732	414			
3	Regression	110.038	5	22.008	43.974	.000[c]
	Residual	204.693	409	.500		
	Total	314.732	414			

a. Predictors: (Constant), social_affiliation_needs
b. Predictors: (Constant), social_affiliation_needs, diet_intensity
c. Predictors: (Constant), social_affiliation_needs, diet_intensity, bodyesteem, selfacceptance, selfesteem
d. Dependent Variable: exercise_commitment

Coefficients[a]

Model		Unstandardized Coefficients		Standardized Coefficients	t	Sig.	Correlations		
		B	Std. Error	Beta			Zero-order	Partial	Part
1	(Constant)	2.104	.351		5.997	.000			
	social_affiliation_needs	.013	.003	.177	3.645	.000	.177	.177	.177
2	(Constant)	1.153	.329		3.506	.001			
	social_affiliation_needs	.006	.003	.081	1.810	.071	.177	.089	.079
	diet_intensity	.514	.051	.447	10.048	.000	.464	.444	.437
3	(Constant)	.076	.378		.202	.840			
	social_affiliation_needs	-.001	.003	-.017	-.376	.707	.177	-.019	-.015
	diet_intensity	.495	.048	.431	10.319	.000	.464	.454	.411
	selfacceptance	-.001	.005	-.010	-.226	.821	.180	-.011	-.009
	selfesteem	.008	.007	.061	1.211	.226	.262	.060	.048
	bodyesteem	.015	.002	.338	7.162	.000	.409	.334	.286

a. Dependent Variable: exercise_commitment

FIGURE 27.4 Output from the hierarchical regression analysis.

the variance of **exercise_commitment**. The effect of **diet_intensity** remained statistically significant ($p < .001$) but the effect of **social_affiliation_needs** in the full model became trivial ($p = .707$).

CHAPTER 28

Polynomial Regression

28.1 OVERVIEW

Most of the statistical procedures we cover in this book assume that the relationship between variables can be described by a completely linear function; that is, a straight line (e.g., $Y = \mathbf{a} + \mathbf{b}X$) is the best fitting model to represent the relationship between two (or more) variables. There are situations, however, where the relationship between variables may be somewhat more complex; we address in this chapter a situation in which the variables are related in a polynomial manner.

Polynomial expressions involve variables that are raised to some positive whole-number power. Our ordinary raw score measures represent variables that are raised to the power of 1 (technically, a first-order polynomial); any number raised to the first power has a value equal to that number (e.g., $9^1 = 9$), and so we conventionally omit the exponent when writing numbers. If we raise a variable to the power of 2 (X^2), then the squared value represents a quadratic expression (a second-order polynomial), and if we raise a variable to the power of 3 (X^3), then we would be talking about a cubic function (a third-order polynomial); it is quite unusual in the behavioral, medical, and biological research fields to deal with a fourth-order (quartic functions), fifth-order (quintic functions), or higher order polynomials.

Polynomial variables can be entered in a regression equation. For example, we could envision some variable Y predicted to be a function of the square of X (e.g., $Y = \mathbf{a} + \mathbf{b}X + \mathbf{b}X^2$). Such a model is an *intrinsically linear model* (Meyers et al., 2013) because we add the terms together as required in a linear function but one (or more) of the terms that we include in addition is not itself linear (e.g., it is quadratic). If we were to plot such a function, the result would not be a straight line. For example, quadratic functions have one "bend" or curvature and have the possibility of one inflection or turning point in them; an idealized depiction of a quadratic function would be a U-shaped or inverted U-shaped plot. A cubic function has an additional curvature and the possibility of having up to two inflection points; an idealized depiction of a cubic function would be an N-shaped or inverted N-shaped plot.

Polynomial functions sometimes very dramatically apply to the lives of human beings. One well-known example of a polynomial function applied to daily life is the relationship between the quality of performance and the level of stress, anxiety, or general arousal that we experience as we perform some task. The general principle is that increased arousal up to some point facilitates task performance but hinders performance

Performing Data Analysis Using IBM SPSS®, First Edition.
Lawrence S. Meyers, Glenn C. Gamst, and A. J. Guarino.

beyond that point (i.e., performance is an inverted U-shaped function of arousal). This is generally known as the Yerkes–Dodson law, so named because Yerkes and Dodson (1908) first introduced it over a century ago when the two researchers attempted to teach mice to select the lighter chamber and avoid the darker one on penalty of electric shock. Over the more than hundred years since its introduction, this inverted U-shaped function of performance and arousal or stress has become part of our common knowledge.

Polynomial regression analysis uses as predictors quadratic and (potentially) cubic or higher polynomials in addition to the linear variable on which the polynomial predictors are based. It is ordinarily performed as a hierarchical analysis, with the linear term entered in the first block, the quadratic term added to the model in the second block, the cubic term added to the model in the third block, and so on. The polynomial function (model) that best fits the data, given the set of predictors, is based on the least-squares algorithm. We evaluate the amount of R^2 change (and its statistical significance) associated with each additional polynomial element to determine the viability of each model and to select the model with the best combination of parsimony and utility.

28.2 NUMERICAL EXAMPLE

The data for our fictional study of the Yerkes–Dodson law is present in the file named **performance and stress**. College students were given a set of 20 puzzles that needed to be solved. The number of puzzles correctly solved is provided under the variable of **performance**. Test-taking conditions were such that the students experienced different levels of stress; the amount of stress they were experiencing is recorded under the variable of **stress_level**.

28.3 ANALYSIS STRATEGY

Our goal is to predict **performance** from **stress_level** and, given what we know of the Yerkes–Dodson law, we would expect an inverted U-shaped function to relate these two variables (assuming a sufficiently wide range of the level of stress experienced by the participants and that the puzzles were of a moderate difficulty level). Thus, we anticipate the need to include at least a quadratic component in our regression model. Our analysis strategy follows this sequence:

- We obtain a scatterplot to visually evaluate the relationship between the two variables, expecting to see an inverted U-shaped array.
- We **Compute** any polynomial predictor variables (e.g., squared **stress_level**, cubic **stress_level**) judged to be needed based on our visual examination of the scatterplot.
- We use the **Linear Regression** procedure to perform a hierarchical regression analysis using the linear variable (original **stress_level**) and the polynomial variables in a sequence of models.

28.4 OBTAINING THE SCATTERPLOT

We open the data file named **performance and stress** and from the main menu select **Graphs ➔ Legacy Dialogs ➔ Scatter/Dot**. As described in Chapter 22, this opens the **Scatter/Dot** window shown in Figure 28.1. Select **Simple Scatter** and click **Define** to open the **Simple Scatter** dialog screen shown in Figure 28.2. We place **performance**

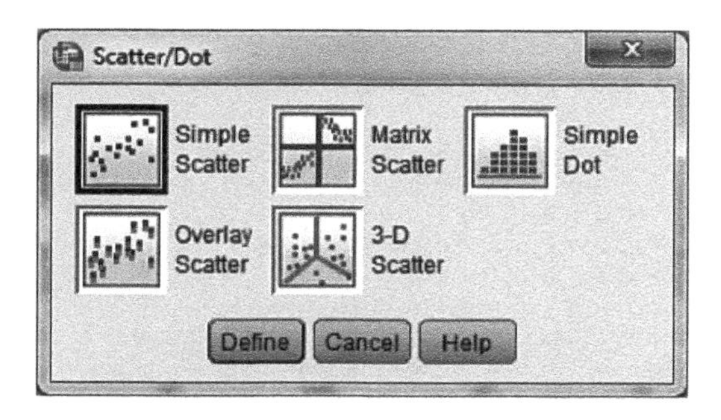

FIGURE 28.1
Scatter/Dot selection screen.

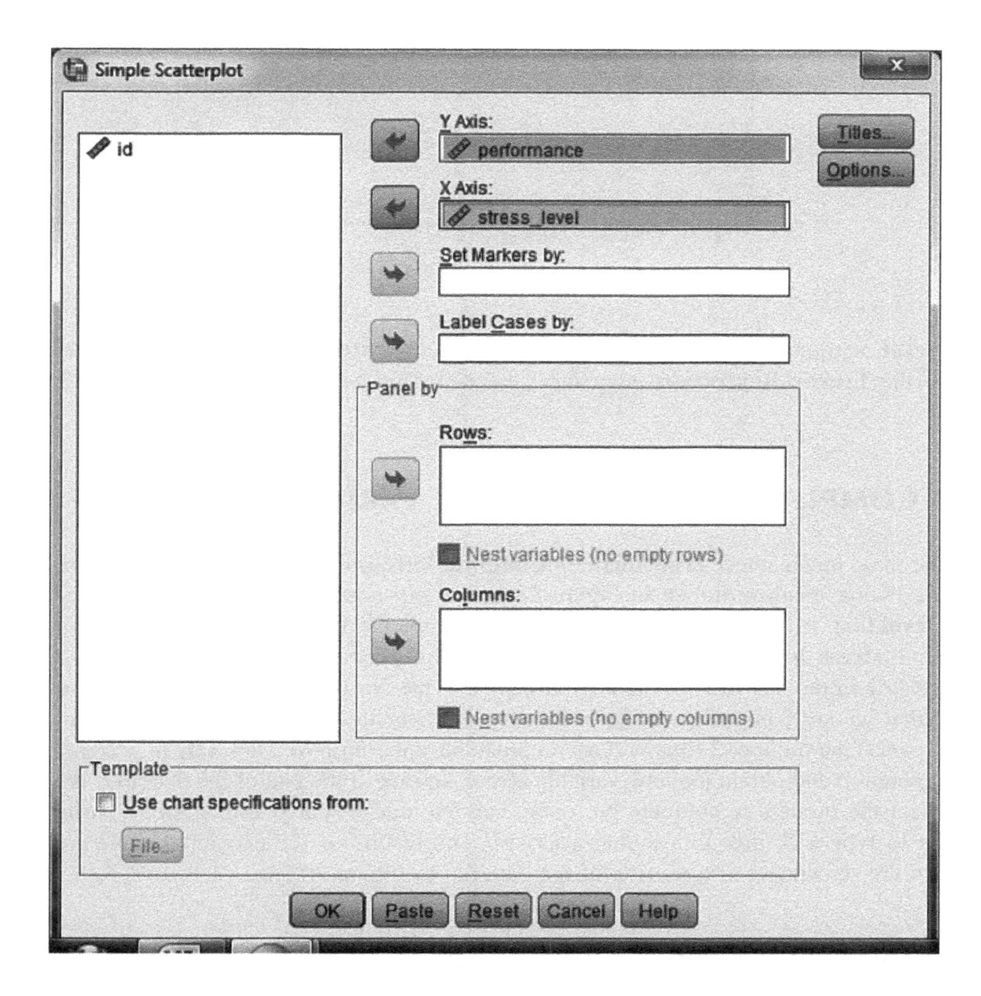

FIGURE 28.2
The **Simple Scatter** dialog screen.

in the **Y Axis** panel and **stress_level** in the **X Axis** panel. Clicking **OK** produces the scatterplot.

The scatterplot of **performance** and **stress_level** is presented in Figure 28.3. Visual inspection suggests that the two variables are related in a mostly quadratic manner. A straight line's (a linear function) fit to the data points (coordinates) would be roughly parallel to the *X*-axis; thus, there appears to be no viable linear component to the relationship. Because the scatterplot is not perfectly symmetric (it "sticks out" a little toward the right), suggesting a somewhat different rate of curvature in that part of the data array, there is the possibility that there might be a weak cubic component to the relationship. We will therefore use three predictors in our hierarchical analysis: the linear variable of

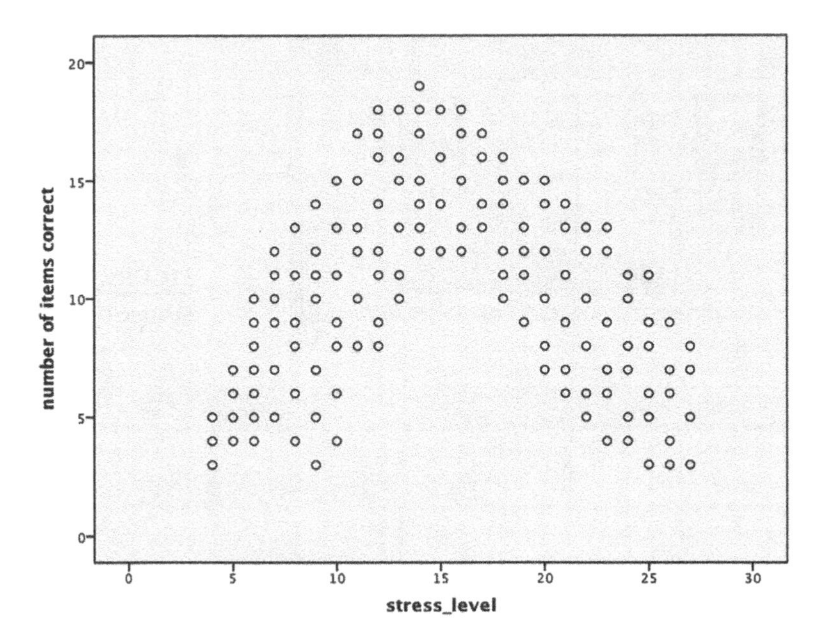

FIGURE 28.3

Scatterplot of test performance as a function of level of stress.

stress_level, a squared **stress_level** variable, and a cubic **stress_level** variable. To use them in the **Linear Regression** procedure, we will need to compute these latter two variables.

28.5 COMPUTING THE POLYNOMIAL VARIABLES

From the main menu select **Transform → Compute Variable**. This opens the **Compute Variable** dialog window shown in Figure 28.4. We will compute the squared value of **stress_level** first. In the **Target Variable** panel, we name the variable **stress_square**. We now move **stress_level** into the **Numeric Expression** panel, click the double-asterisk key (first key in the last row on the key panel below the **Numeric Expression** panel) to specify that we are computing an exponent or power function, and type (or select from the key panel) the number **2** (this will accomplish the squaring). We click **OK** to perform the computation and obtain the new variable **stress_square** at the end of the data file. We then repeat the process to compute our cubic variable that we have named **stress_cube** (be sure to type a **3** after the double asterisk). The results of these computations are shown in the screenshot of a portion of the data file in Figure 28.5.

28.6 ANALYSIS SETUP: LINEAR REGRESSION

From the main menu, we select **Analyze → Regression → Linear** to open the main **Linear Regression** window. We will perform a hierarchical regression analysis as described in Chapter 27. Briefly, we move **performance** into the **Dependent** panel and **stress_level** into the **Independent(s)** panel as **Block 1 of 1**. Click **Next** and move **stress_square** into the **Independent(s)** panel for the second block. Click **Next** and move **stress_cube** into the **Independent(s)** panel for the third block. The setup for the third block is shown in Figure 28.6. In the **Statistics** window, we ask for **Regression Coefficients Estimates**, **Model fit**, **R squared change**, **Descriptives**, and **Part and partial correlations** (see Figure 28.7). Click **Continue** to return to the main dialog window and click **OK** to perform the analysis.

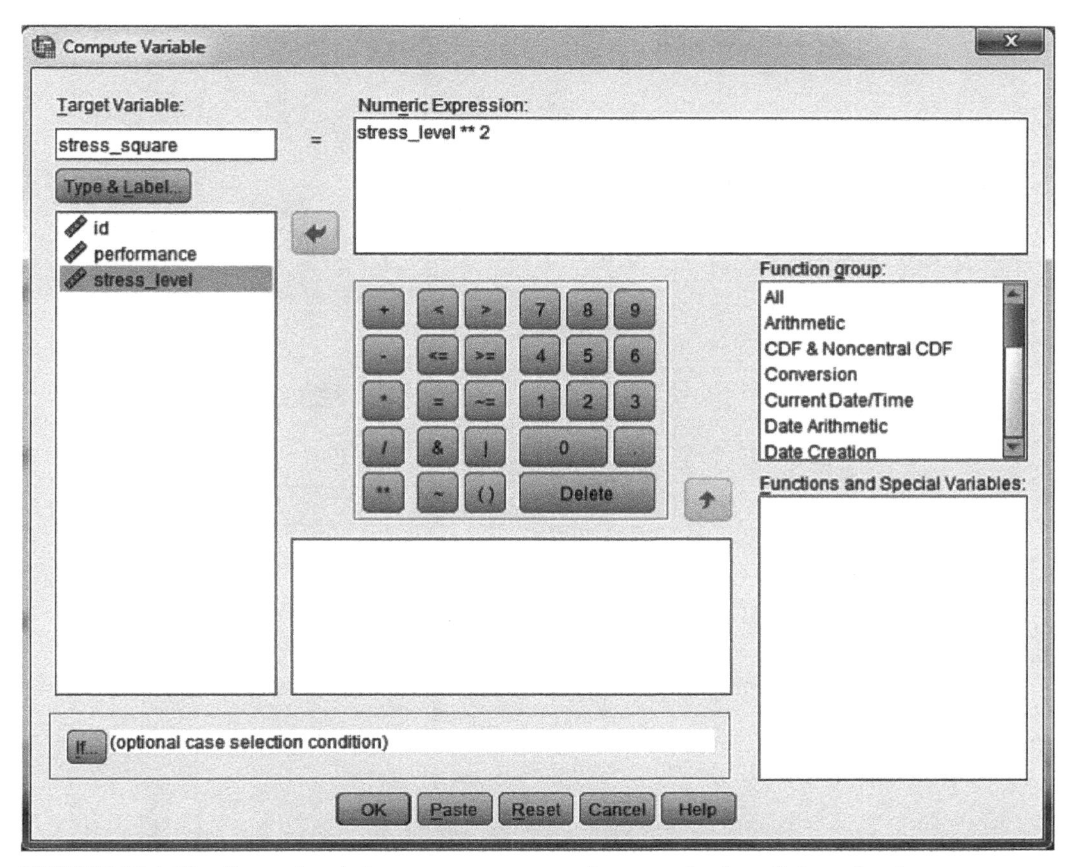

FIGURE 28.4 The **Compute** window setup to compute the squared value of **stress_level**.

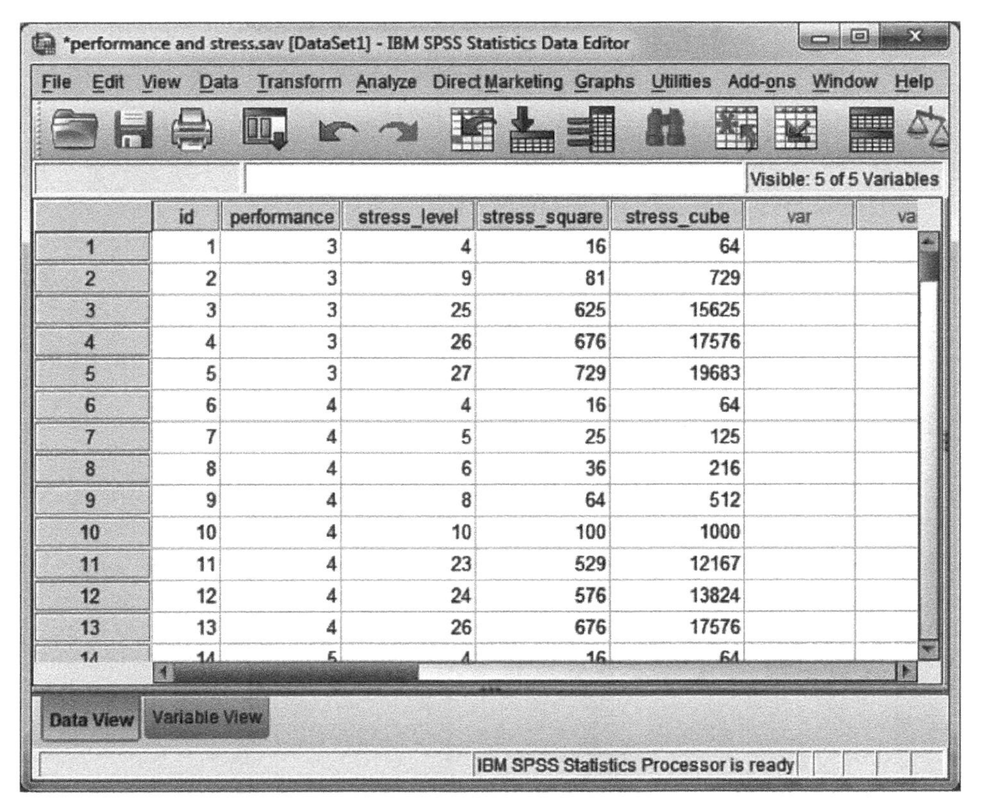

	id	performance	stress_level	stress_square	stress_cube	var
1	1	3	4	16	64	
2	2	3	9	81	729	
3	3	3	25	625	15625	
4	4	3	26	676	17576	
5	5	3	27	729	19683	
6	6	4	4	16	64	
7	7	4	5	25	125	
8	8	4	6	36	216	
9	9	4	8	64	512	
10	10	4	10	100	1000	
11	11	4	23	529	12167	
12	12	4	24	576	13824	
13	13	4	26	676	17576	
14	14	5	4	16	64	

FIGURE 28.5 A portion of the data file showing the newly computed polynomial variables.

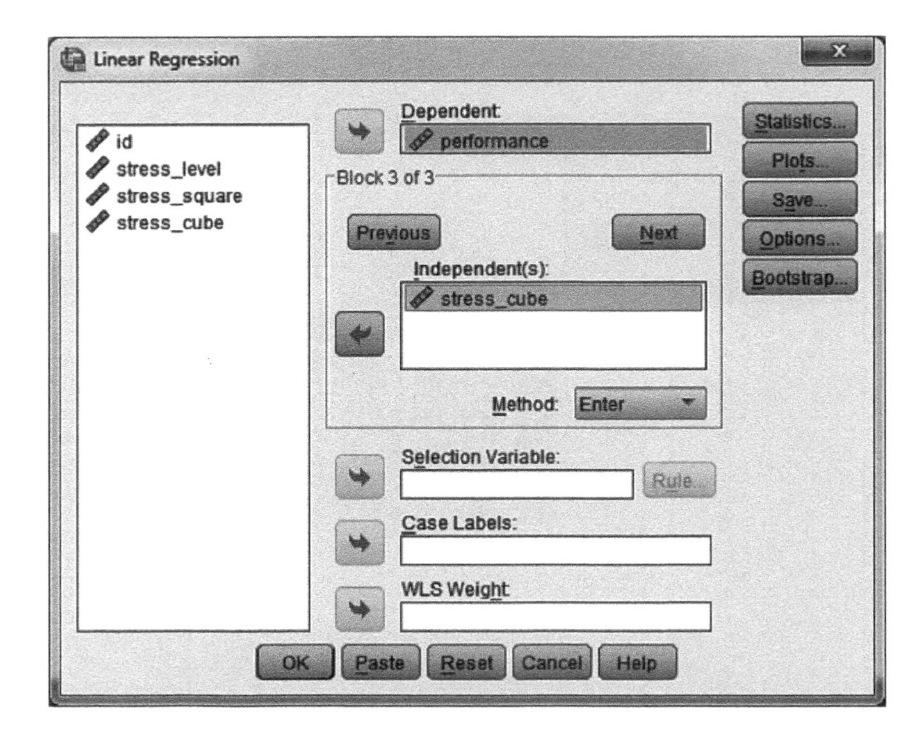

FIGURE 28.6
The main dialog window of **Linear Regression** in the third block of the hierarchical analysis.

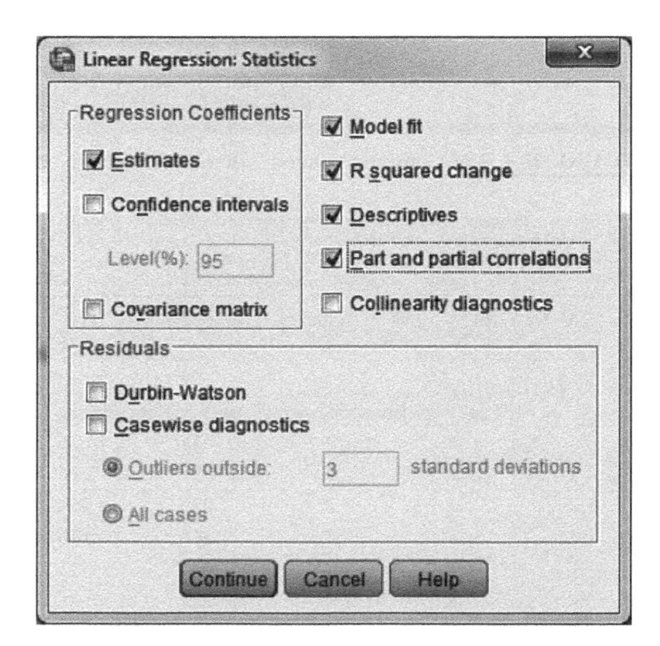

FIGURE 28.7
The **Statistics** dialog window of **Linear Regression**.

28.7 ANALYSIS OUTPUT: LINEAR REGRESSION

The **Model Summary** output from the hierarchical regression analysis is presented in Figure 28.8. There are three models in our analysis, in that each block is treated as a separate model, although the models are cumulative as we are adding variables in each block. The first model contains only the linear **stress_level** variable and is not statistically significant (p = **.625**); this is consistent with our expectations, given the scatterplot. The R^2 of **.002** tells a vivid story.

Model Summary

Model	R	R Square	Adjusted R Square	Std. Error of the Estimate	Change Statistics				
					R Square Change	F Change	df1	df2	Sig. F Change
1	.042[a]	.002	-.006	4.182	.002	.241	1	134	.625
2	.712[b]	.508	.500	2.948	.506	136.589	1	133	.000
3	.716[c]	.512	.501	2.945	.005	1.279	1	132	.260

a. Predictors: (Constant), stress_level
b. Predictors: (Constant), stress_level, stress_square
c. Predictors: (Constant), stress_level, stress_square, stressl_cube

FIGURE 28.8 The **Model Summary** output of the **Linear Regression** analysis.

ANOVA[d]

Model		Sum of Squares	df	Mean Square	F	Sig.
1	Regression	4.207	1	4.207	.241	.625[a]
	Residual	2343.175	134	17.486		
	Total	2347.382	135			
2	Regression	1191.394	2	595.697	68.537	.000[b]
	Residual	1155.988	133	8.692		
	Total	2347.382	135			
3	Regression	1202.485	3	400.828	46.213	.000[c]
	Residual	1144.898	132	8.673		
	Total	2347.382	135			

a. Predictors: (Constant), stress_level
b. Predictors: (Constant), stress_level, stress_square
c. Predictors: (Constant), stress_level, stress_square, stressl_cube
d. Dependent Variable: performance number of items correct

FIGURE 28.9 **ANOVA** table for the **Linear Regression** analysis.

When the quadratic term (**stress_square**) was added to the model, the R^2 changed by **.506** to reach a value of **.508**. This change in R^2 was statistically significant ($p < .001$). The third model added the cubic term (**stress_cube**) to the model, and the R^2 change of **.005** to boost the R^2 to **.512** was not statistically significant (p = **.260**). Thus, the relationship between performance and stress appears to be best described by a quadratic function.

Figure 28.9 presents the ANOVA summary table for each model. Note that the model with the cubic term is still statistically significant, even though the cubic term added no significant enhancement to the amount of variance that was explained—it is statistically significant because the quadratic variable, which is doing all of the predictive work, is contained in the third model.

The **Coefficients** output is shown in Figure 28.10. Two points are worth noting here:

- The regression coefficient for the linear term is statistically significant in the second model, even though there is no linear relationship between the two variables. This is because the best fitting curve is one that weights both the linear and quadratic terms, with coefficients whose values are significantly different from zero so that the quadratic function can optimally fit the data.
- The beta coefficients in the second and third models are well in excess of 1.00. This is because the linear, quadratic, and cubic variables are, not surprisingly,

Coefficients[a]

Model		Unstandardized Coefficients		Standardized Coefficients	t	Sig.	Correlations		
		B	Std. Error	Beta			Zero-order	Partial	Part
1	(Constant)	10.626	.900		11.802	.000			
	stress_level	-.026	.054	-.042	-.490	.625	-.042	-.042	-.042
2	(Constant)	-4.489	1.441		-3.116	.002			
	stress_level	2.336	.206	3.760	11.360	.000	-.042	.702	.691
	stress_square	-.075	.006	-3.868	-11.687	.000	-.172	-.712	-.711
3	(Constant)	-7.509	3.034		-2.475	.015			
	stress_level	3.104	.709	4.994	4.378	.000	-.042	.356	.266
	stress_square	-.131	.049	-6.706	-2.649	.009	-.172	-.225	-.161
	stressl_cube	.001	.001	1.642	1.131	.260	-.263	.098	.069

a. Dependent Variable: performance number of items correct

FIGURE 28.10 **Coefficients** table for the **Linear Regression** analysis.

very highly correlated with each other (e.g., **stress_level** and **stress_square** are correlated .983); standardized regression coefficients can easily exceed 1.00 when the predictors are relatively strongly correlated. This multicollinearity should likely cause researchers to abandon analyses containing ordinary variables (they would exclude one or more of the predictors from the analysis and then perform the analysis again), but we accept this (disconcerting) feature in a polynomial analysis by not interpreting the regression coefficients as we usually do but instead we focus on the overall nature of the relationship (e.g., the obtained relationship here was a quadratic one) and the R^2 information.

CHAPTER 29

Multilevel Modeling

29.1 OVERVIEW

When we use a multiple regression procedure to predict the value of a quantitative dependent variable, one of the presumptions underlying the treatment of the data is that the scores of the cases on the dependent variable are independent of each other (technically, their errors are independent); this is one of the assumptions of the general linear model. However, there are many circumstances when that independence assumption is violated. For example, we may focus on the self-assessment of quality of life by patients in different hospitals as in the example we use for this chapter. In this context, the self-reports of patients within a hospital may be more correlated than those from patients selected at random from the sample. Other illustrations of this (based on some examples from Snijders & Bosker, 2012) are as follows:

- The performance on some achievement test of school children within a school may be more correlated than those from children enrolled in different schools.
- The political attitudes of residents within a neighborhood may be more correlated than those from residents of different neighborhoods.
- The productiveness of employees working within a particular office may be more correlated than those from different offices.

In these examples, the patients, school children, neighborhood residents, and employees are *hierarchically structured* (also called *nested* or *clustered*) within groups, where each group in the sample is a hospital, school, neighborhood, or office, respectively.

The important point here is that the data supplied by the cases (e.g., personal assessment of their quality of life) may be partly a function of the particular group (hospital) with which they are associated. To the extent that this is true, the scores of the cases on the dependent variable are not independent; that is, the scores of patients within the same hospital may be more related than scores selected from patients from different hospitals. This is what we label as a *nesting* or *clustering* effect, and to the extent that this effect is observed, the research design needs to take this into effect in order to validly evaluate the results of a study. One way to take this clustering effect into account in the statistical analysis is by using a *multilevel modeling* procedure.

Performing Data Analysis Using IBM SPSS®, First Edition.
Lawrence S. Meyers, Glenn C. Gamst, and A. J. Guarino.

Several different fields have labeled the general technique we use here by many different labels, partly depending on the way it is used (Norman & Streiner, 2008; Vogt, 2007). Among these labels are hierarchical linear modeling, mixed modeling, nested modeling, growth curve analysis, and contextual modeling. We follow the lead of Vogt (2007) in using the term *multilevel modeling*, and we use the **Linear Mixed Models** procedure in IBM SPSS® to perform the analysis. More complete recent coverage of this complex topic can be found in Bickel (2007), Goldstein (2011), Hox (2010), Hox and Roberts (2010), Raudenbush and Bryk (2002), and Stevens (2009).

Multilevel modeling takes into account the clustering or nesting effect of the responses of individuals being more related on the dependent variable within groups than between groups. We therefore use multilevel modeling to analyze the data at multiple levels (i.e., from both the individual level and group level). If we consider the individual cases as the most basic (microscopic) level of analysis (called Level 1 in the context of multilevel modeling), then group membership is a higher order (more macroscopic) level of analysis (called Level 2 in the context of multilevel modeling). It is possible that these (Level 2) groups are themselves nested within an even higher level of organization, although our example will be restricted to only a two-level hierarchy. Multilevel modeling is a generalization of the multiple regression procedure allowing for the assessment of individual observations while controlling for group effects as well as assessing group effects while controlling for individual observations.

To estimate these variance components of the model, the IBM SPSS **Linear Mixed Models** procedure uses a *maximum likelihood* algorithm; the default variation, which we use in our analyses, is a *restricted maximum likelihood* method that takes into account degrees of freedom associated with the fixed effects in making these estimates. As one illustration of a research context where the data might exhibit a hierarchical structure, and the one we use for our numerical example, the assessment of quality of life by patients in a hospital may be partially explained from their level of optimism concerning their future (a Level 1 variable). Well over 100 hospitals were involved in this fictional study, and it is possible that, for a variety of reasons, patients in one hospital may have reported a higher quality of life than those in another hospital. Thus, not only the scores of the individual patients may be predictive of the degree of quality of life that is reported but the particular hospital with which the patients are associated may also be a viable predictor (e.g., patients in Hospitals A and B report on average a higher quality of life than those in Hospitals C and D). It may therefore be necessary to incorporate hospital as a Level 2 variable in the prediction model.

In this example, the quality-of-life assessments of patients within a given hospital may on average be more alike (correlated) than those of patients selected at random from the entire sample, that is, the Level 2 variable may be related to (predictive of) quality-of-life judgments. This correlation is labeled as *clustering* (not to be confused with the clustering procedures described in Chapters 59 and 60). To the extent that clustering is obtained, the assumption of independence (that the patients in the study are independent of each other with respect to the dependent variable) is violated.

Multilevel modeling allows us to accomplish two goals with respect to this clustering effect. The first goal it allows us to accomplish is that it provides a convenient way to determine the level of clustering in the data set. We do this by generating an *intraclass correlation* (ICC); see Meyers et al. (2013) for a more complete discussion of this coefficient and see Chapter 37 for another application of this statistic. Using a two-level hierarchy for our illustration, the ICC estimates the amount of the dependent variable variance that can be explained by the Level 2 variable. An ICC of zero would indicate that we observe no Level 2 effect on the dependent variable, with the implication that we do not need to conduct the multilevel modeling procedure (i.e., ordinary regression analysis would be sufficient, as the assumption of independence would have been met).

However, as the value of the ICC increases, the more necessary multilevel modeling becomes.

The ICC's threshold to conduct a multilevel modeling is debated. Lee (2000) has suggested that an ICC greater than .10 requires multilevel methods. Others have reported that even very small levels of correlation can justify the use of multilevel modeling (Reise & Duan, 2003), partly because the adverse consequences of violating the assumption of independence are substantial (Meyers et al., 2013), including biased estimates of the parameters of the model, alpha level inflation, and a greater risk of drawing an improper conclusion.

The second goal multilevel modeling allows us to accomplish is to permit us to use this clustering effect as a covariate in our analysis. By doing so, we can evaluate the contributions of the other predictors in the data set when we statistically control for the Level 2 effect.

Multilevel modeling is substantially more complex than multiple regression analysis. Because there are at least two levels in the hierarchical structure, researchers can use different combinations of predictors to address different issues; that is, researchers can configure alternative models to address different research issues. As a consequence of this flexibility, each model addresses a somewhat different issue, and a common analysis strategy used in multilevel modeling is to increase the complexity of the model in a series of analyses.

We face two issues when performing a multilevel analysis: treating predictors as either fixed or random effects or both and if/how to center the predictor variables. We deal with these two issues briefly in turn.

Generally, treating an effect as a *fixed effect* focuses the results of the analysis on the particular values represented in the data file, whereas treating an effect as a *random effect* allows the results of the analysis to be generalized to a wider range of values than that is represented in the data file. In multilevel modeling, fixed effects focus on the differences in the intercepts (means) and slopes, whereas random effects focus on variance that is explained by the variables. Furthermore, in IBM SPSS, only categorical variables can be defined as a **Factor**, and only **Factors** can be assigned as fixed effects; quantitative or dichotomously coded categorical variables can both qualify as **Covariates**, and **Covariates** can be assigned as either fixed or random effects.

In multilevel modeling, it is traditional to center the quantitative predictor variables. In our numerical example, the only quantitative predictor variable in the data set is **optimism**. As we discussed in Chapter 25, to center a variable involves subtracting a mean from each score.

The two types of centering commonly used for Level 1 variables are *grand mean centering* and *group mean centering*. For grand mean centering, we subtract the grand (total sample) mean from the score for each case; thus, the centered mean for the sample becomes zero and the scores for the cases represent their distances from the grand mean. For group mean centering, we subtract each individual group mean from the score on the variable for those cases within that particular group; thus, each group has a centered mean of zero and the scores for the cases indicate their distances from their respective group mean. Deciding on the type of centering to impose on a Level 1 variable is a complex decision that has been discussed elsewhere (see Enders & Tofighi (2007) and Meyers et al. (2013)). In our example, we center our Level 1 variable (patient optimism score) on the grand mean.

In centering Level 2 (and higher level) quantitative predictors, we focus on the group means of the quantitative predictors rather than the scores of the individual cases; that is, the group means are treated as the "scores" from which a mean is subtracted. Each case within a particular group is first assigned the value of the mean of the group for a specified quantitative predictor variable. We then center these group mean "scores" by subtracting the grand mean from each. Differences among these centered "scores" thus reflect group differences. By using the **optimism** variable in our example to illustrate this process, this Level 2 centering operation is accomplished as follows:

- We first determine the mean of **optimism** for each hospital (group).
- In the data file, we create a new variable and assign the group mean to each patient associated with the respective hospital, thus creating a Level 2 optimism variable (e.g., all patients associated with Hospital A are assigned the mean optimism score for Hospital A).
- We then center the Level 2 optimism variable by subtracting the grand mean of optimism from each Level 2 optimism score. Thus, each patient associated with a given hospital will have the same value for the centered Level 2 optimism variable. To the extent that we observe differences among these scores, we would infer that hospitals differ in the degree of optimism reported by their patients.

29.2 NUMERICAL EXAMPLE

This fictional example provides the results of 4500 patients from 142 different hospitals, and a screenshot of a portion of the data file named **patient optimism** is shown in Figure 29.1. The outcome (dependent) variable is the score for these patients on the *Patient-Reported Outcome Measures* (PROMs), an inventory that can range from 0 to 100, with higher scores indicating greater perceived health and quality of life; the variable is named **prom_dv** to make it clear that it is the dependent variable. The 142 hospitals

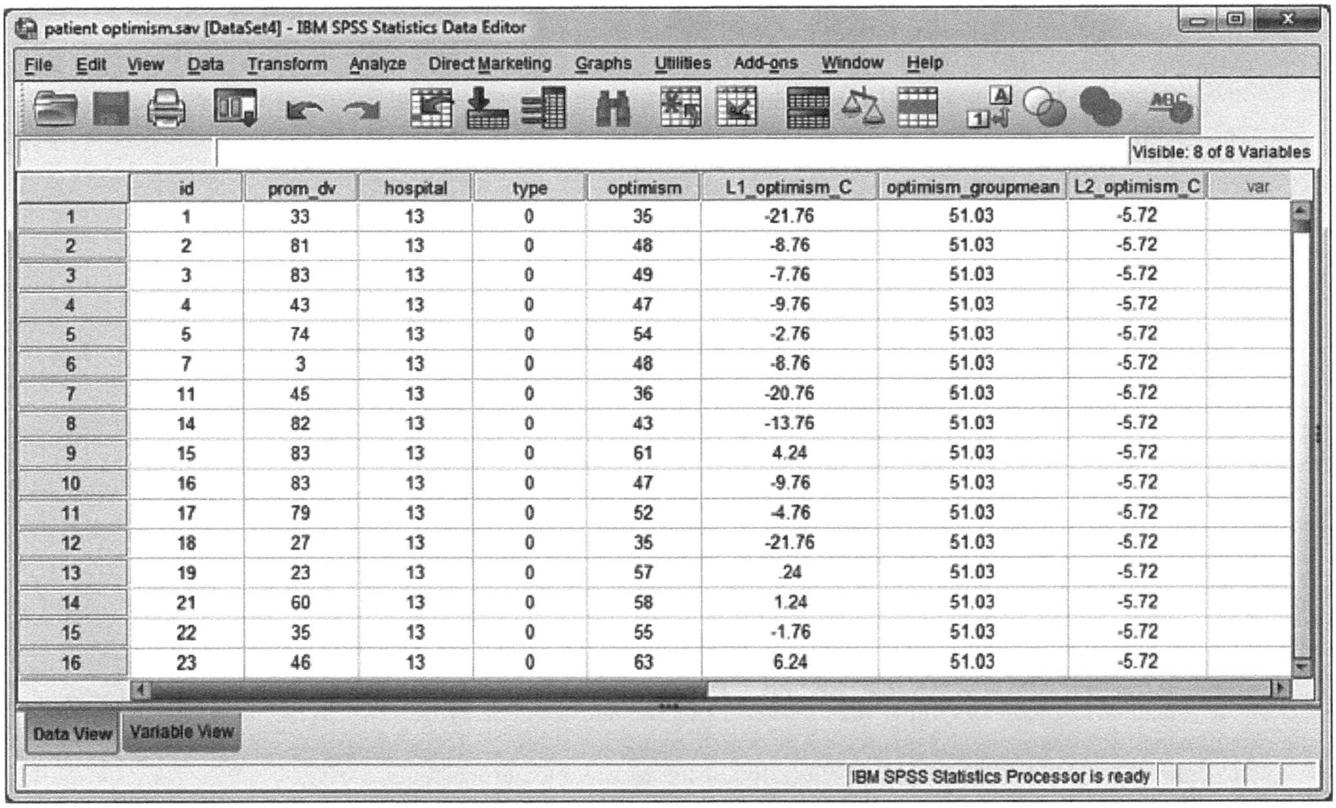

patient optimism.sav [DataSet4] - IBM SPSS Statistics Data Editor

File Edit View Data Transform Analyze Direct Marketing Graphs Utilities Add-ons Window Help

Visible: 8 of 8 Variables

	id	prom_dv	hospital	type	optimism	L1_optimism_C	optimism_groupmean	L2_optimism_C	var
1	1	33	13	0	35	-21.76	51.03	-5.72	
2	2	81	13	0	48	-8.76	51.03	-5.72	
3	3	83	13	0	49	-7.76	51.03	-5.72	
4	4	43	13	0	47	-9.76	51.03	-5.72	
5	5	74	13	0	54	-2.76	51.03	-5.72	
6	7	3	13	0	48	-8.76	51.03	-5.72	
7	11	45	13	0	36	-20.76	51.03	-5.72	
8	14	82	13	0	43	-13.76	51.03	-5.72	
9	15	83	13	0	61	4.24	51.03	-5.72	
10	16	83	13	0	47	-9.76	51.03	-5.72	
11	17	79	13	0	52	-4.76	51.03	-5.72	
12	18	27	13	0	35	-21.76	51.03	-5.72	
13	19	23	13	0	57	.24	51.03	-5.72	
14	21	60	13	0	58	1.24	51.03	-5.72	
15	22	35	13	0	55	-1.76	51.03	-5.72	
16	23	46	13	0	63	6.24	51.03	-5.72	

Data View Variable View

IBM SPSS Statistics Processor is ready

FIGURE 29.1 A portion of the data file.

represent the hierarchical structure of the data; these identification codes are present in the categorical variable of **hospital**. We have also dichotomously categorized these hospitals under the variable **type** into either **non-teaching** (coded as **0** in the data file) or **teaching** (coded as **1** in the data file) hospitals.

The one quantitative predictor variable in the data file is **optimism**, with higher scores indicating more optimism. This variable has been grand-mean-centered at Level 1 under the variable name **L1_optimism_C**. Scores on this variable reflect the difference between the individual patient optimism score and the total sample (grand) mean of optimism (**L1_optimism_C** = **optimism** − grand mean). The mean of **L1_optimism_C**, that is, the mean of the sample for the grand-mean-centered variable, is zero.

29.3 ANALYSIS STRATEGY

We start by showing how to build the Level 2 centered optimism variable. This entails the following three steps:

- We need to assign each patient a score on the Level 2 variable equal to the mean optimism score of the group. This is accomplished using the **Aggregate** function.
- We then calculate the grand mean (this had to be done for the Level 1 centering operation but we generate it again to show it to readers).
- We subtract the grand mean from the aggregated scores for each patient generated in the first step.

We then perform the set of multilevel modeling analyses. It is common practice in performing a multilevel modeling analysis to examine a set of models, starting with a simpler model and generally building to more complex models (Hox, 2010). We configure the following set of five models:

- *The Unconditional Model.* The first step in any multilevel modeling analysis is to assess the amount of variance of the dependent variable that is associated with the Level 2 variable(s). In our example, the ICC describes the amount of **prom_dv** score variance that is explained by differences among the 142 hospitals in the study. This unconditional model serves as a baseline that we use to evaluate subsequent models.
- *A Mixed Level 1 Model.* The first predictive model includes the **Fixed** and **Random Effects** of (individual) Level 1 centered optimism predicting **prom_dv** scores.
- *A Mixed Level 2 Model.* This model incorporates the **Fixed** and **Random Effects** of the type of hospital together with the (individual) Level 1 optimism to predict **prom_dv** scores.
- *A Hierarchical Model.* This model adds the optimism level of the hospital (the Level 2 centered variable) to the mixed Level 2 model.
- *An Interaction Model.* This model adds the interaction effect of the type of hospital by the Level 2 optimism level of the hospital to the hierarchical model.

29.4 AGGREGATING THE OPTIMISM VARIABLE

The most direct way to assign the mean value for a given hospital to all of the patients associated with that hospital is to *aggregate* the **optimism** variable by **hospital**. To aggregate a variable is to assign to each member of a group a particular value computed by a designated function (e.g., a mean).

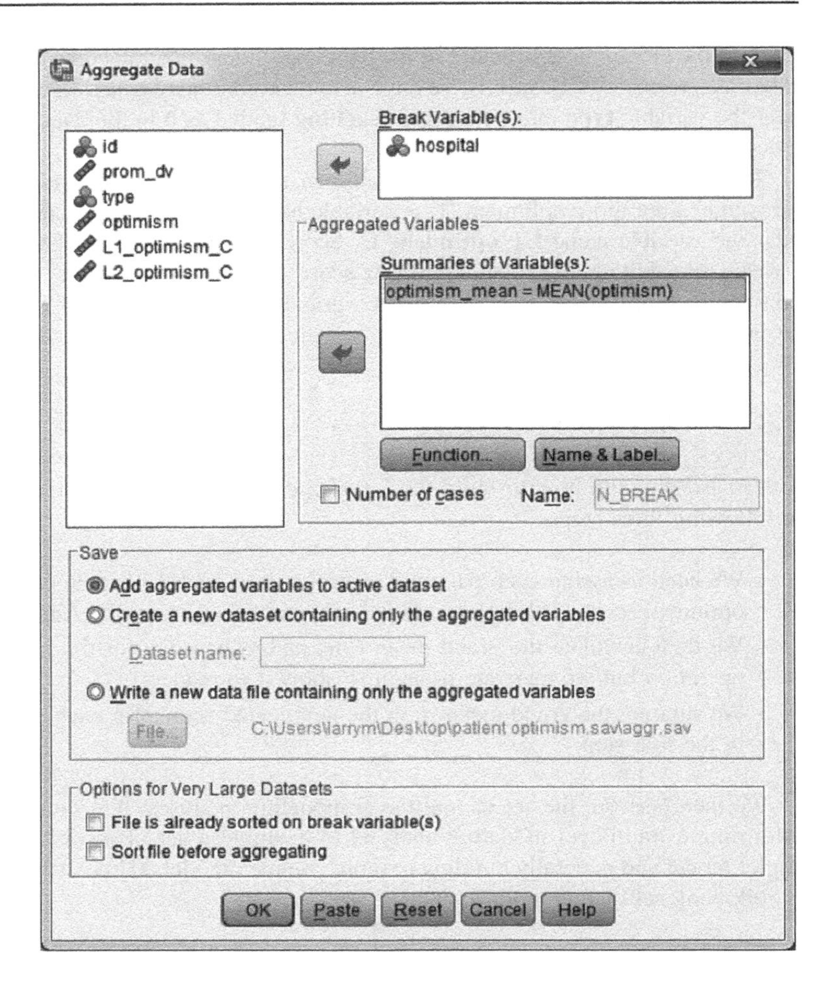

FIGURE 29.2
The main **Aggregate** window.

We open **patient optimism** and from the main menu, we select **Analyze ➔ Data ➔ Aggregate** to open the **Aggregate** window shown in Figure 29.2. We move **hospital** into the **Break Variable(s)** panel. The **Brake Variable** is the basis for defining the groups; here, each **hospital** will be a group.

We move **optimism** into the **Summaries of Variable(s)** panel. This is the variable whose values we wish to aggregate. When we move the variable into the panel, IBM SPSS automatically does two things:

- It specifies the mean as the default function. This is represented by the expression **MEAN(optimism)** following the equal sign. The variable in parentheses is the variable to be aggregated (and the one we moved into the panel).
- It assigns a generic name to the to-be-aggregated variable. Here, that name is **optimism_mean**.

We select the **Function** pushbutton. This action opens the **Aggregate Function** window (see Figure 29.3); we present this just to show readers what it looks like (the default function—the mean—is what we wish to compute). There are several ways by which we can aggregate our variable, including calculating the median, the sum, and the standard deviation of the scores within each level of our **Brake Variable**. Here, we want to calculate the mean **optimism** value for each **hospital**, and so we accept the default function and select **Continue**.

Selecting the **Name & Label** pushbutton opens the **Name and Label** window shown in Figure 29.4. We create the name **optimism_groupmean** to reflect the idea that it is

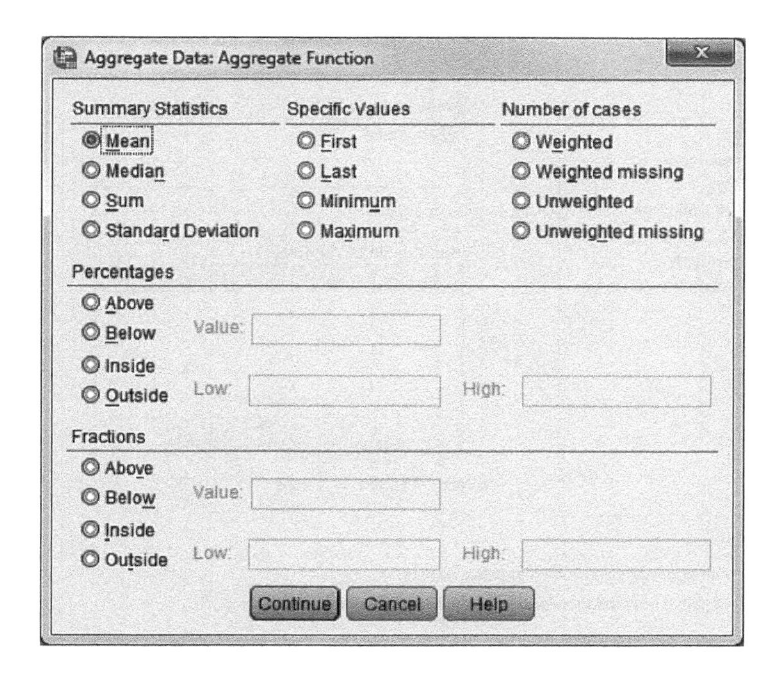

FIGURE 29.3

The **Aggregate Function** window with **Mean** selected.

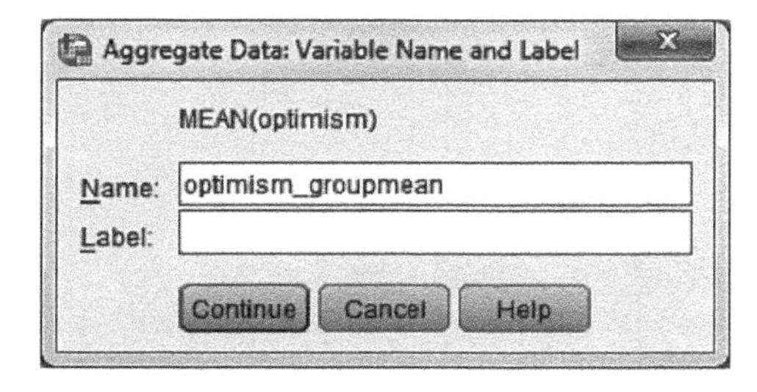

FIGURE 29.4

The **Aggregate Name and Label** window.

a Level 2 variable and click **Continue**. Our configured **Aggregate** window is shown in Figure 29.5, where we click **OK**.

29.5 CENTERING THE LEVEL 2 optimism_groupmean VARIABLE

To center the Level 2 **optimism_groupmean** variable, we must subtract the grand mean of optimism from each score. We therefore need to determine the value of the grand mean, and we do so by analyzing the original **optimism** variable and **optimism_groupmean** in the **Descriptives** procedure. The results, shown in Figure 29.6 for both the **optimism_groupmean** and the **optimism** variables, yielded a grand mean for **optimism_groupmean** of **56.7567** and a grand mean for **optimism** of 56.76. The differences in the **Minimum**, **Maximum**, and **Std. Deviation** between **optimism_groupmean** and **optimism** are obtained because the former variable represents the group means and the latter variable represents the individual scores.

We then **Compute** our centered Level 2 variable (to be named **L2_optimism_C**) by subtracting the grand mean (**56.7567**) from the **optimism_groupmean** variable as shown in Figure 29.7 (because the **optimism_groupmean** variable represents only an intermediate step, we will not retain this variable in the variable list for the multilevel modeling analyses in order to simplify the visual display).

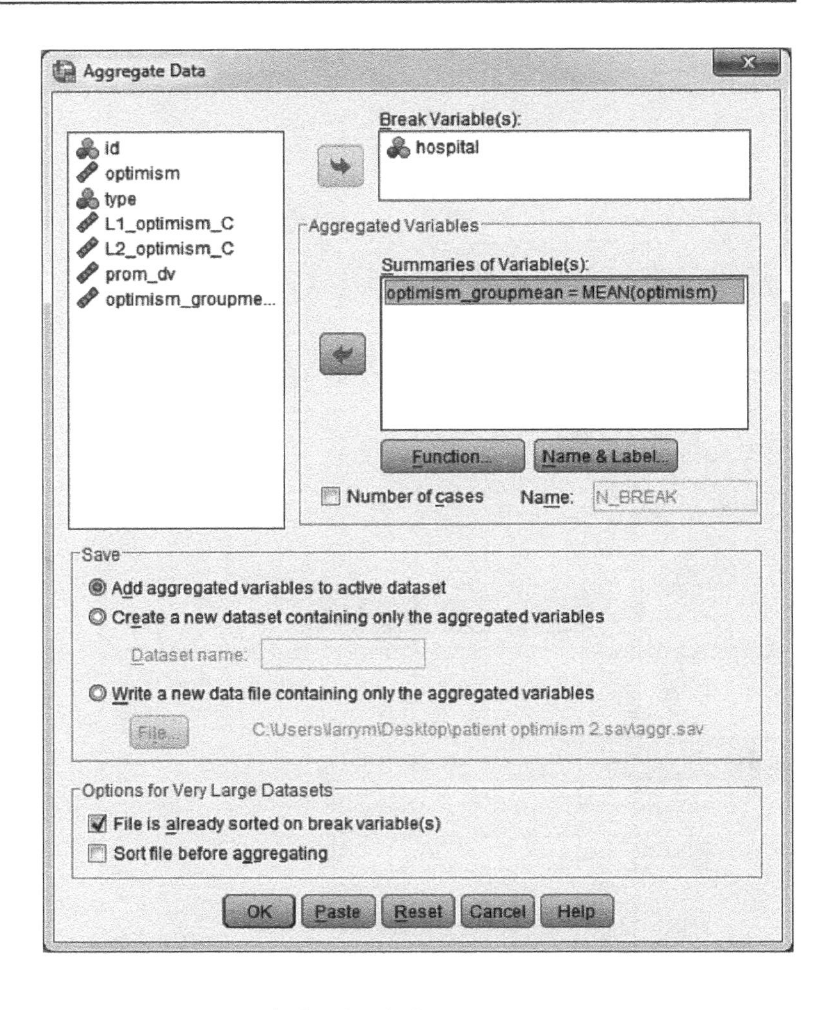

FIGURE 29.5

The main **Aggregate** window with the to-be-aggregated variable named.

Descriptive Statistics

	N	Minimum	Maximum	Mean	Std. Deviation
optimism_groupmean	4500	42.53	68.60	56.7567	5.35599
optimism	4500	21	95	56.76	11.165
Valid N (listwise)	4500				

FIGURE 29.6 The **Descriptive Statistics** summary table.

29.6 ANALYSIS SETUP: UNCONDITIONAL MODEL

We are now ready to start evaluating our multilevel models, the first of which is the unconditional model. With the **patient optimism** data file open, we select from the main menu **Analyze ➔ Mixed Models ➔ Linear**. This action displays the initial **Specify Subjects and Repeated** window shown in Figure 29.8. In this window, we designate the clustering variable. We move **hospital** to the **Subjects** panel to designate that our patients are organized (clustered/nested) by being associated with different hospitals. This specifies that any differences among the hospitals will be statistically controlled in the multilevel analysis (i.e., it will serve as a covariate). Because we do not have a longitudinal variable, we leave the **Repeated** panel empty. Select the **Continue** pushbutton to reach the main **Linear Mixed Models** window.

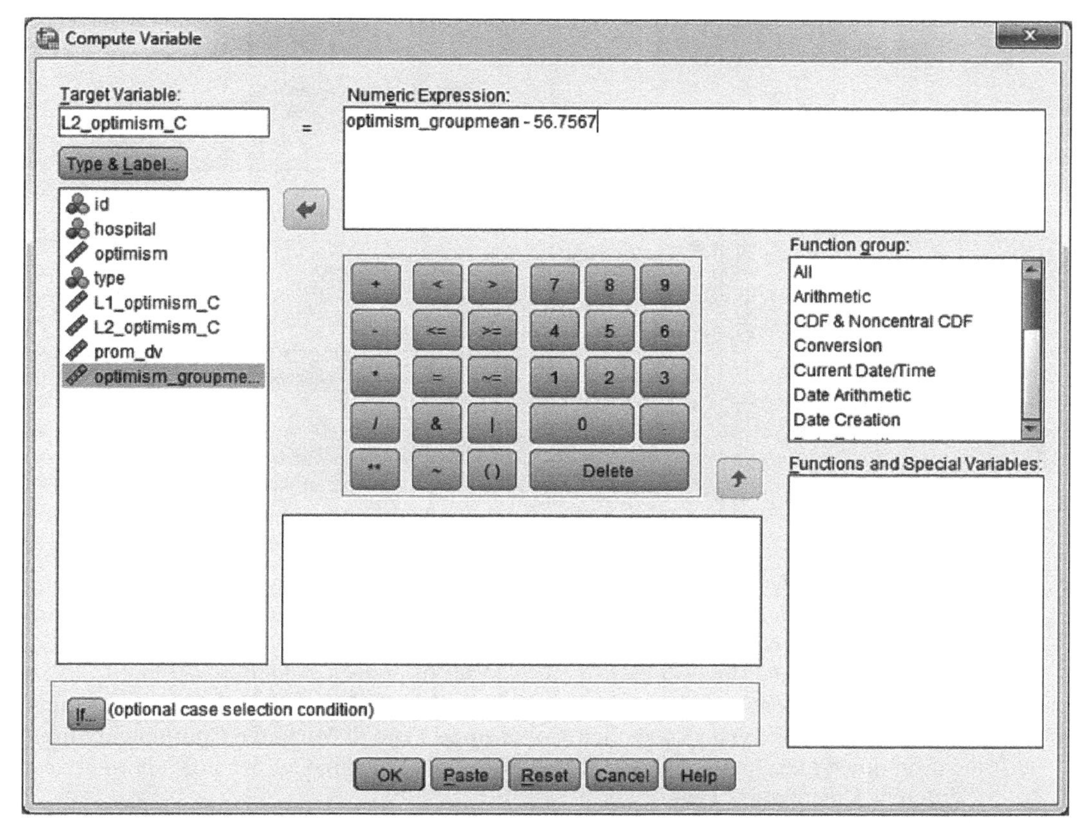

FIGURE 29.7 Centering the Level 2 optimism variable by subtracting the grand mean from the aggregated **optimism_groupmean** variable.

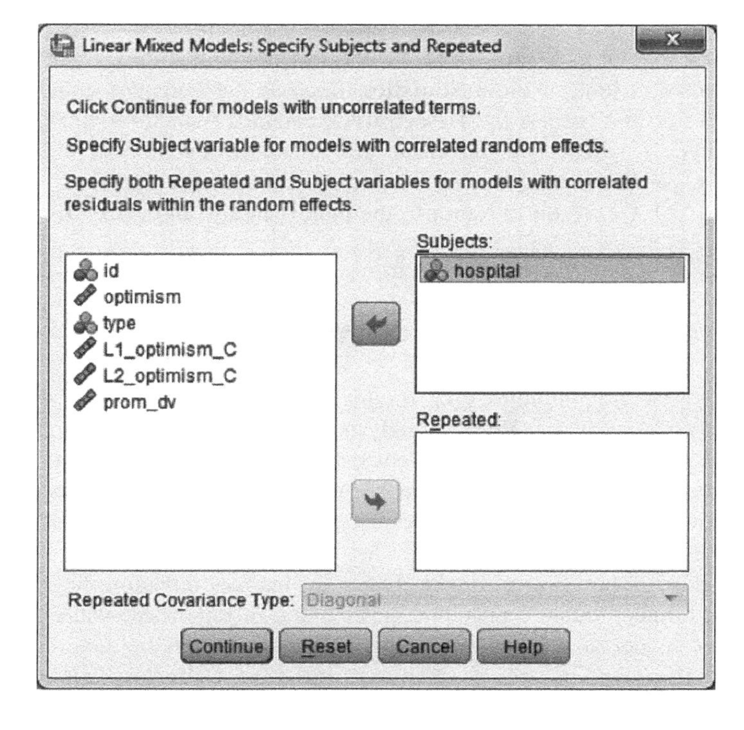

FIGURE 29.8

The initial **Specify Subjects and Repeated** window.

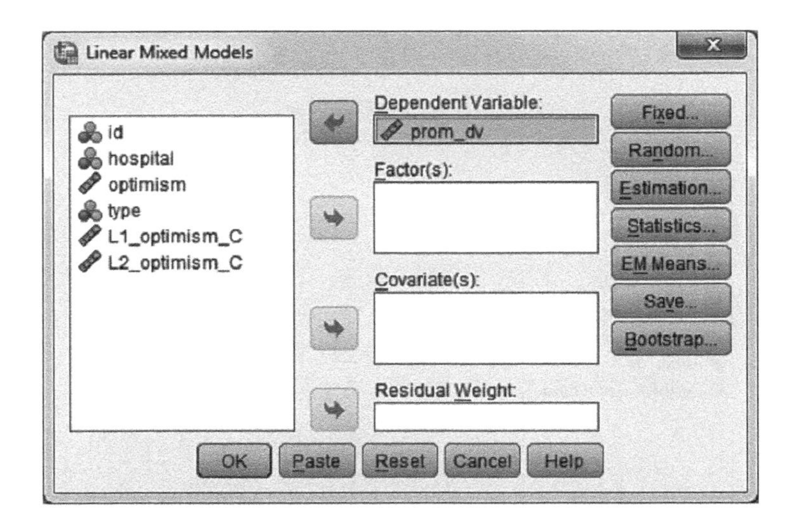

FIGURE 29.9
The main **Linear Mixed Models** window set up for the unconditional model.

The main **Linear Mixed Models** window is shown in Figure 29.9. We place **prom_dv** in the **Dependent Variable** panel but do not specify any predictors (they would be placed in the **Factor(s)** or **Covariate(s)** panel). The only effect in the model will be the **Subjects** variable of **hospital**.

We presume that the hospitals in our data set have been randomly selected from all hospitals in the country; given this presumption, we will define **hospital** as representing a random effect. We retain the default **Covariance Type** of **Variance Components** from the drop-down menu; once we start to add predictors to the model, we may opt to invoke a different **Covariance Type**.

To designate **hospital** as a random effect, we select the **Random** pushbutton; this selection opens the screen shown in Figure 29.10. In the **Random Effects** area toward the top of the window, we check **Include Intercept**. In the **Subject Groupings** area in the bottom part of the window, we move **hospital** into the **Combinations** panel. Click **Continue** to return to the main window and select the **Statistics** pushbutton.

The **Statistics** window is shown in Figure 29.11. Check **Parameter estimates** and **Tests for covariance parameters** under **Model Statistics**. Because the output is quite lengthy, for our purposes, we do not request the **Descriptive Statistics** (which provides for each hospital the number of patients and the mean and standard deviation of the dependent variable) or the **Case Processing Summary** (which provides the number of patients for each hospital). Click **Continue** to return to the main window and click **OK** to perform the analysis.

29.7 ANALYSIS OUTPUT: UNCONDITIONAL MODEL

The **Model Dimension** table is shown in Figure 29.12. It reports the number of parameters that were assessed in the analysis; in the present analysis, there were **3** parameters (indicated in the bottom **Total** row). The **Fixed Effect** estimated was the **Intercept** (which represents the grand mean), **Random Effect** (degree to which the intercepts vary among the different hospitals in the **Subject Variable** named **hospital**), and **Residual** (the unexplained variance of the dependent variable).

Figure 29.13 presents the **Information Criteria**. These are indexes reflecting how well the model fits the data. Smaller values represent a better fitting model. These values are useful in comparing one model to another. Norusis (2012) recommends the use of the **Akaike Information Criterion (AIC)** and the **Schwarz Bayesian Criterion (BIC)** to compare models, with lower scores indicating a more accurate model.

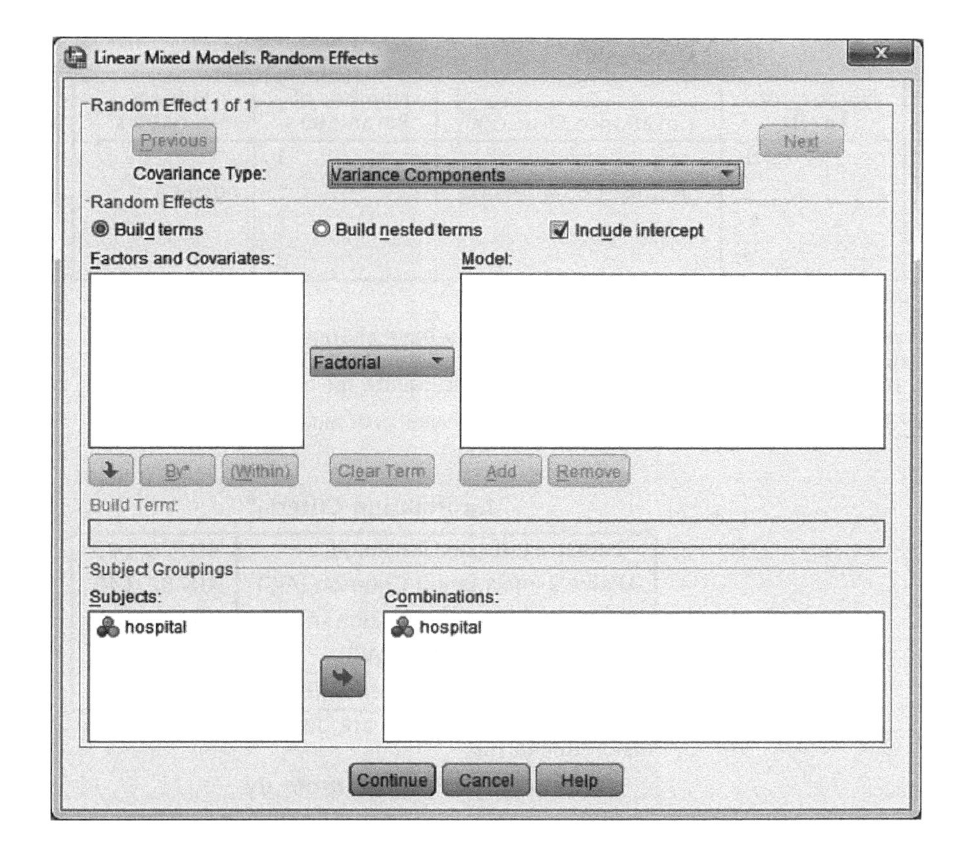

FIGURE 29.10
The **Random Effects** window set up for the unconditional model.

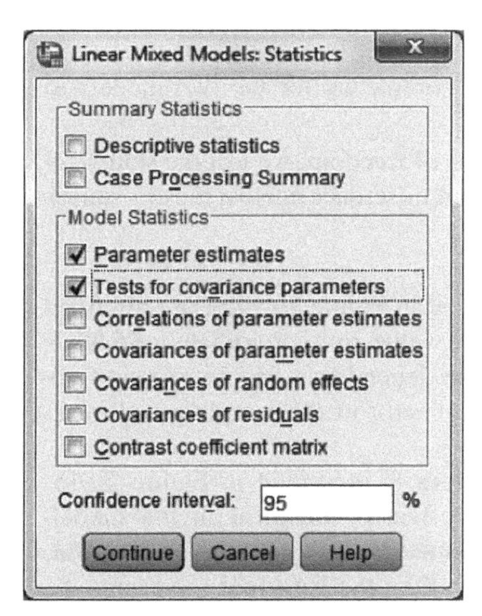

FIGURE 29.11
The **Statistics** window.

We can also use the **−2 Restricted Log Likelihood** value to determine if there is a statistically significantly improvement between two competing nested models (i.e., one model is a subset of the other). This strategy will be applied as we build increasingly more complex models in our series of analyses by adding another predictor in each step; thus, the earlier model will be nested within the later model. To test whether the later model has significantly improved the fit, a *chi-square difference test* is performed in the following manner:

Model Dimension[a]

		Number of Levels	Covariance Structure	Number of Parameters	Subject Variables
Fixed Effects	Intercept	1		1	
Random Effects	Intercept[b]	1	Variance Components	1	hospital
Residual				1	
Total		2		3	

a. Dependent Variable: prom_dv.

b. As of version 11.5, the syntax rules for the RANDOM subcommand have changed. Your command syntax may yield results that differ from those produced by prior versions. If you are using version 11 syntax, please consult the current syntax reference guide for more information.

FIGURE 29.12 The **Model Dimension** output for the unconditional model.

Information Criteria[a]

-2 Restricted Log Likelihood	40778.149
Akaike's Information Criterion (AIC)	40782.149
Hurvich and Tsai's Criterion (AICC)	40782.152
Bozdogan's Criterion (CAIC)	40796.972
Schwarz's Bayesian Criterion (BIC)	40794.972

The information criteria are displayed in smaller-is-better forms.

a. Dependent Variable: prom_dv.

FIGURE 29.13 The **Information Criteria** for the unconditional model.

- We subtract the values of −2 **Restricted Log Likelihood** for the two models to obtain an absolute difference score.
- We subtract the values for the total number of parameters for the two models to obtain the absolute difference.
- Using the difference in parameters as our degrees of freedom, we test the statistical significance of the −2 **Restricted Log Likelihood** difference against the chi-square distribution available in Table A.1.

The table of **Estimates of Fixed Effects** is presented in Figure 29.14. The only fixed effect in the unconditional model is the intercept. The value of approximately **56.06** is the estimated population grand mean parameter for the dependent variable of **prom_dv** based on the model, a value that is within the margin of error of the sample grand mean of 56.46.

The table of **Estimates of Covariance Parameters** is presented in Figure 29.15. These are variance estimates based on the random effect of **hospital** in the model. The intercept represents the amount of variance explained by the presence of **hospital**, which is **86.579037** and is statistically significant ($p < .001$); the **Residual** represents the

Estimates of Fixed Effects[a]

Parameter	Estimate	Std. Error	df	t	Sig.	95% Confidence Interval Lower Bound	95% Confidence Interval Upper Bound
Intercept	56.061397	.855013	138.203	65.568	.000	54.370798	57.751997

a. Dependent Variable: prom_dv.

FIGURE 29.14 The **Estimates of Fixed Effects** for the unconditional model.

Estimates of Covariance Parameters[a]

Parameter		Estimate	Std. Error	Wald Z	Sig.	95% Confidence Interval	
						Lower Bound	Upper Bound
Residual		476.298593	10.205299	46.672	.000	456.710746	496.726542
Intercept [subject = hospital]	Variance	86.579037	12.435913	6.962	.000	65.335484	114.729841

a. Dependent Variable: prom_dv.

FIGURE 29.15 The **Estimates of Covariance Parameters** for the unconditional model.

remaining unexplained variance of **476.298593**. The total variance is **562.87762**, the sum of those two values.

The ICC represents the percentage of variance explained by hospital and is computed as **86.579037/562.87762**. The resulting covariance value is **15.38156**, indicating that **hospital** accounted for approximately 15.38% of the variance of **prom_dv**. This value for the ICC informs us that scores on **prom_dv** may be said to cluster within hospitals (i.e., patients from the same hospital report more similar scores) to a sufficient extent for us to judge that the observations violate the assumption of independence. Because there is a sufficiently strong Level 2 clustering effect, we are therefore well advised to continue using **hospital** as a covariate in our series of analyses on **prom_dv** scores.

These results also inform us that there is still quite a bit of work remaining for us to do. With only a little more than 15% of total variance explained by **hospital**, a little less than 85% of the variance of **prom_dv** remains unexplained, and some of that variance may be associated with the other variables in our data file. Thus, a series of additional multilevel analyses appear warranted.

29.8 ANALYSIS SETUP: MIXED LEVEL 1 MODEL

The first predictive model we examine is the mixed Level 1 model, where we will use as a single predictor the (individual) Level 1 centered optimism variable. With the **patient optimism** data file open, we select from the main menu **Analyze → Mixed Models → Linear**. This action displays the initial **Specify Subjects and Repeated** window already described in Section 29.6, which we configure in the same way to control for the effect of **hospital** as our clustering variable. Select the **Continue** pushbutton to reach the main **Linear Mixed Models** window.

The main **Linear Mixed Models** window is shown in Figure 29.16. We place **prom_dv** in the **Dependent Variable** panel and place **L1_optimism_C** in the **Covariate(s)** panel.

Selecting the **Fixed** pushbutton opens the **Fixed Effects** window presented in Figure 29.17. We move **L1_optimism_C** from the **Factors and Covariates** panel to the **Model** panel by highlighting it and clicking the **Add** pushbutton. This will allow us to evaluate the average intercept and slope of the **L1_optimism_C** scores predicting **prom_dv**. The **Include intercept** box is already checked, and we retain this specification. Click **Continue** to return to the main window and select the **Random** pushbutton.

The **Random Effects** screen is shown in Figure 29.18. The top portion of the screen concerns the **Random Effects**. In the drop-down menu, for **Covariance Type**, select **Unstructured**. Covariance here deals with the relationship between the intercept and slope parameters of **L1_optimism_C**, and we select **Unstructured** because the nature of the relationship is unknown. We then check the **Include intercept** box. Finally, highlight **L1_optimism_C** and select the **Add** pushbutton; this will move **L1_optimism_C** into the **Model** window. In the bottom portion of the screen is the **Subject Groupings**, where

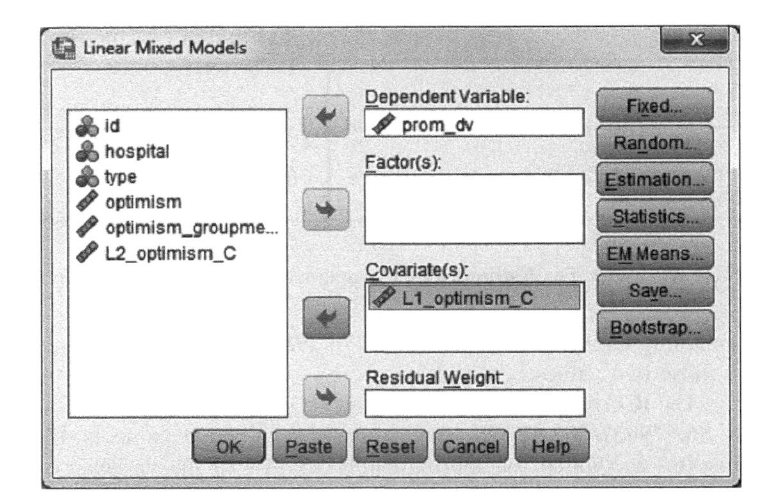

FIGURE 29.16
The main **Linear Mixed Models** window set up for the mixed Level 1 model.

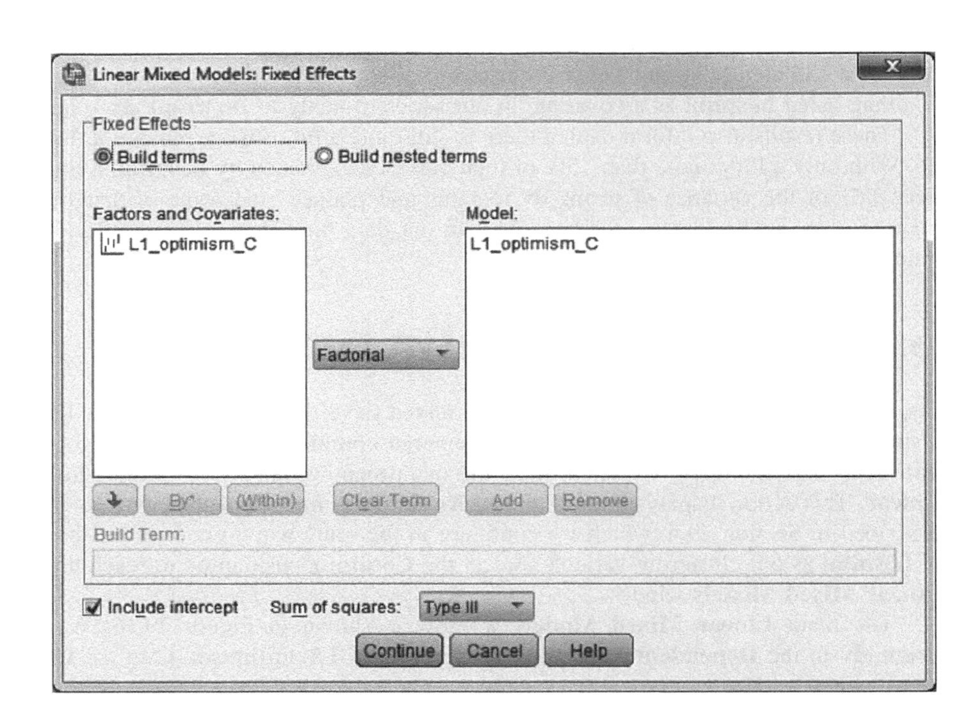

FIGURE 29.17
The **Fixed Effects** window set up for the mixed Level 1 model.

we have moved **hospital** from the **Subjects** panel to the **Combinations** panel. Click **Continue** to return to the main window.

We configure the **Statistics** window as described in Section 29.6 (requesting **Parameter estimates** and **Tests for covariance parameters** under **Model Statistics**). Click **OK** to perform the analysis.

29.9 ANALYSIS OUTPUT: MIXED LEVEL 1 MODEL

The **Model Dimension** table is shown in Figure 29.19. There were **6** parameters in this model. Also documented in the table is the fact that we used an **Unstructured Covariance Structure**.

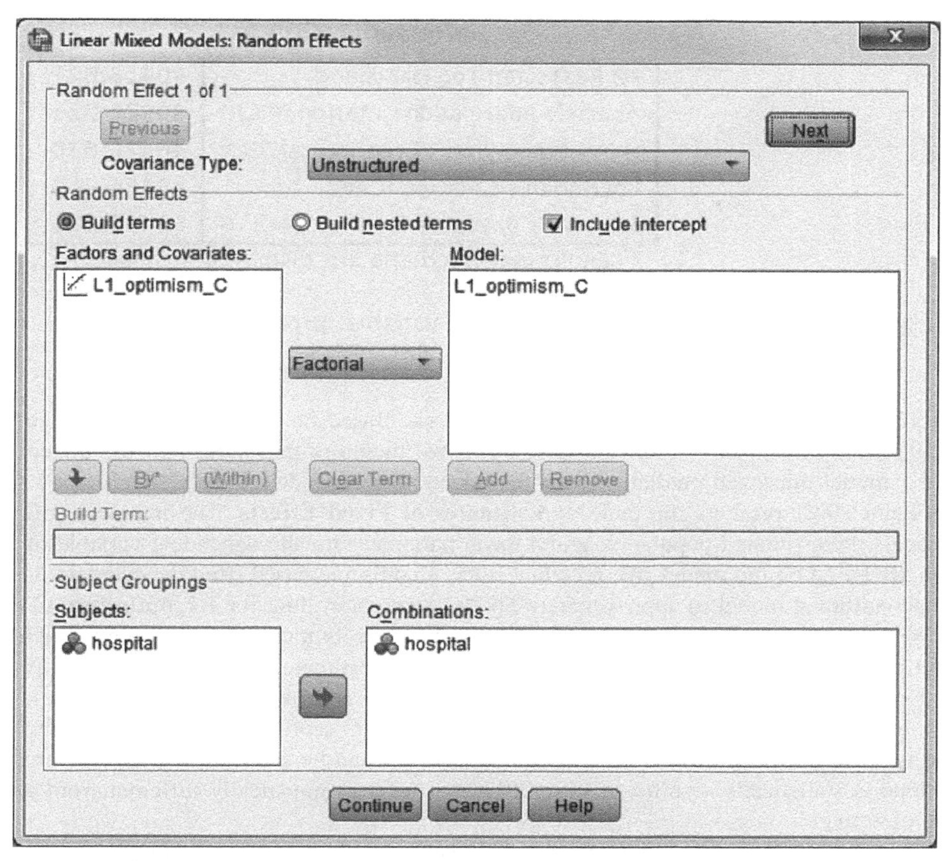

FIGURE 29.18 The **Random Effects** window set up for the mixed Level 1 model.

Model Dimension[a]

		Number of Levels	Covariance Structure	Number of Parameters	Subject Variables
Fixed Effects	Intercept	1		1	
	L1_optimism_C	1		1	
Random Effects	Intercept + L1_optimism_C[b]	2	Unstructured	3	hospital
Residual				1	
Total		4		6	

a. Dependent Variable: prom_dv.

b. As of version 11.5, the syntax rules for the RANDOM subcommand have changed. Your command syntax may yield results that differ from those produced by prior versions. If you are using version 11 syntax, please consult the current syntax reference guide for more information.

FIGURE 29.19 The **Model Dimension** output for the mixed Level 1 model.

Figure 29.20 presents the **Information Criteria**. Both the **AIC** and the **BIC** in the **Information Criteria** table report smaller values than the previous analyses, indicating a better model fit.

We can also perform a chi-square difference test. The **−2 Restricted Log Likelihood** values for the unconditional model and our mixed Level 1 model are **40778.149** and **40418.621**, respectively, with a difference of 359.528. The degrees of freedom for these two models are **3** and **6**, respectively, with a difference of 3. We therefore evaluate the

Information Criteria[a]

-2 Restricted Log Likelihood	40418.621
Akaike's Information Criterion (AIC)	40426.621
Hurvich and Tsai's Criterion (AICC)	40426.630
Bozdogan's Criterion (CAIC)	40456.266
Schwarz's Bayesian Criterion (BIC)	40452.266

The information criteria are displayed in smaller-is-better forms.

a. Dependent Variable: prom_dv.

FIGURE 29.20
The **Information Criteria** for the mixed Level 1 model.

chi-square difference with three degrees of freedom. Based on Table A.1, a difference of 359.528 is statistically significant ($p < .001$). We therefore may infer that the mixed Level 1 model improved prediction over the unconditional model.

Figure 29.21 presents the table of **Estimates of Fixed Effects**. The intercept still represents the estimated population grand mean parameter for the dependent variable of **prom_dv** based on the model. We note that it has slightly increased from the estimate of the unconditional model to approximately **56.29** when controlling for **L1_optimism_C**.

We now have an additional entry in the table (the output for the unconditional model reported only an intercept value). The parameter estimate for **L1_optimism_C** is the slope of the function, and we interpret the slope of approximately **.60** as follows: every increase of one scale value in the **L1_optimism_C** score increases (because the slope is positive) the **prom_dv** score by about **.60**. As can be seen in the **Sig**. column, the slope is statistically significant ($p < .001$); that is, it is statistically different from a slope of zero.

The table of **Estimates of Covariance Parameters** is presented in Figure 29.22. These are variance estimates based on the random effects of **hospital** in the model. Incorporating the patients' individual **optimism** scores into the model reduced the **Residual**

Estimates of Fixed Effects[a]

Parameter	Estimate	Std. Error	df	t	Sig.	95% Confidence Interval	
						Lower Bound	Upper Bound
Intercept	56.286180	.670919	127.854	83.894	.000	54.958638	57.613722
L1_optimism_C	.601095	.035523	130.164	16.921	.000	.530817	.671372

a. Dependent Variable: prom_dv.

FIGURE 29.21 The **Estimates of Fixed Effects** for the mixed Level 1 model.

Estimates of Covariance Parameters[a]

Parameter		Estimate	Std. Error	Wald Z	Sig.	95% Confidence Interval	
						Lower Bound	Upper Bound
Residual		442.438587	9.620978	45.987	.000	423.978007	461.702966
Intercept + L1_optimism_C [subject = hospital]	UN (1,1)	47.406351	7.893186	6.006	.000	34.206683	65.699505
	UN (2,1)	-.450082	.298203	-1.509	.131	-1.034549	.134386
	UN (2,2)	.033586	.018446	1.821	.069	.011446	.098552

a. Dependent Variable: prom_dv.

FIGURE 29.22 The **Estimates of Covariance Parameters** for the mixed Level 1 model.

from about **476.58** to about **442.44** (it became about 7.16% smaller), signifying that **L1_optimism_C** explains more of the within-hospital variance than we could in the unconditional model.

The next three rows are labeled as **UN (1,1)**, **UN (2,1)**, and **UN (2,2)**, in which the **UN** stands for the **Unstructured Covariance Type** that we specified in the **Random Effects** dialog window. The numbers in parentheses after **UN** refer to coordinates in a particular output table that we requested. The three specific parameters of the variance in the order that we suggest examining these results are as follows:

- **UN (1,1)** represents the variance estimate (with a value of approximately **47.41**) of the **prom_dv** intercepts, with **L1_optimism_C** in the model. In this example, the **prom_dv** intercepts vary significantly ($p < .001$) among the hospitals.
- **UN (2,2)** represents the variance estimate (with a value of approximately **0.03**) of the **prom_dv** slopes, with **L1_optimism_C** in the model. This parameter failed to achieve statistical significance ($p =$ **.069**), indicating that the hospitals have roughly equivalent slopes.
- **UN (2,1)** represents the covariance estimate between the **prom_dv** slopes and intercepts (with a value of approximately **−0.45**), with **L1_optimism_C** in the model. This output assesses whether higher values of the intercept were associated with steeper or flatter slopes. This parameter also failed to achieve statistical significance ($p =$ **.131**), indicating that there was no relationship between the values of the intercepts and the values of the slopes. Because the random effect provides no significant relationship to the **prom_dv** score variance, we will simplify the remaining models by not specifying the random effects of **L1_optimism_C**.

29.10 ANALYSIS SETUP: MIXED LEVEL 2 MODEL

With the exception of removing the random effects of **L1_optimism_C**, we are ready to further raise the complexity level of our multilevel model. We do this by now taking into consideration the effects of the type of hospital in the sample (recall that we have both nonteaching and teaching hospitals represented). With the **patient optimism** data file open, we select from the main menu **Analyze ➔ Mixed Models ➔ Linear**. This action displays the initial **Specify Subjects and Repeated** window where we define **hospital** as our clustering variable. Select the **Continue** pushbutton to reach the main **Linear Mixed Models** window.

The main **Linear Mixed Models** window is shown in Figure 29.23. We place **prom_dv** in the **Dependent Variable** panel and **L1_optimism_C** and **type** in the **Covariate(s)** panel.

Selecting the **Fixed** pushbutton opens the **Fixed Effects** window (see Figure 29.24). We move **L1_optimism_C** and **type** from the **Factors and Covariates** panel to the **Model** panel and make sure that the **Include intercept** box is checked.

In the top portion of the **Random Effects** screen shown in Figure 29.25, we select **Unstructured** as our **Covariance Type** and move **type** into the **Model** panel (and do not specify **L1_optimism_C** as a random effect as indicated at the end of Section 29.9). In the bottom portion of the screen, we have moved **hospital** from the **Subjects** panel to the **Combinations** panel. Click **Continue** to return to the main window.

We configure the **Statistics** window as described in Section 29.6 (requesting **Parameter estimates** and **Tests for covariance parameters** under **Model Statistics**). Click **OK** to perform the analysis.

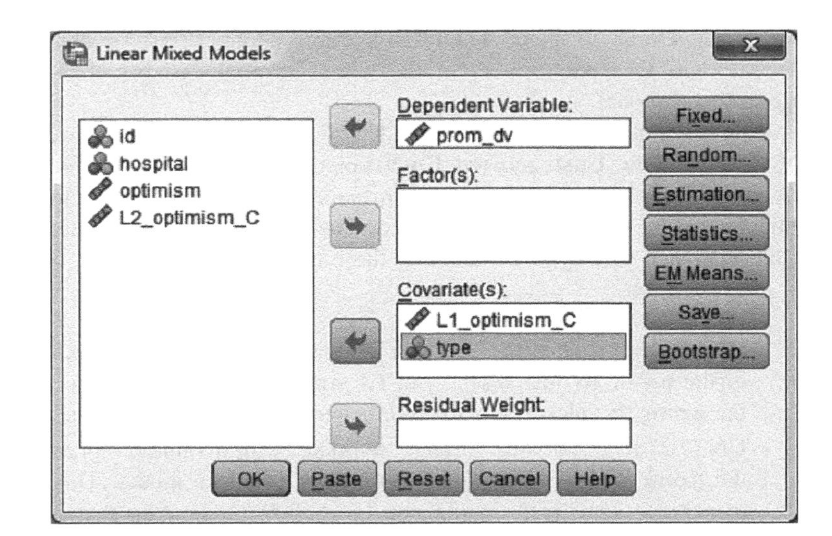

FIGURE 29.23
The main **Linear Mixed Models** window setup for the mixed Level 2 model.

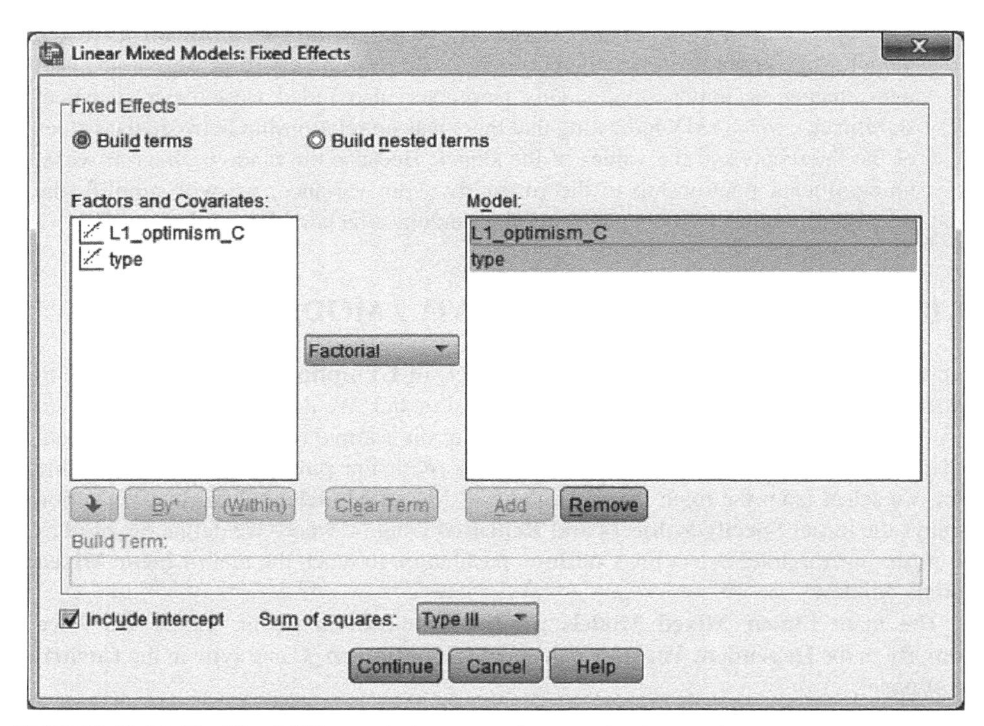

FIGURE 29.24 The **Fixed Effects** window setup for the mixed Level 2 model.

29.11 ANALYSIS OUTPUT: MIXED LEVEL 2 MODEL

The **Model Dimension** table is shown in Figure 29.26. There were **7** parameters in this mixed Level 2 model. Also documented in the table is our use of an **Unstructured Covariance Structure**.

Figure 29.27 presents the **Information Criteria**, including both the **AIC** and the **BIC**. These values are smaller than those in the mixed Level 1 model, suggesting a better model fit.

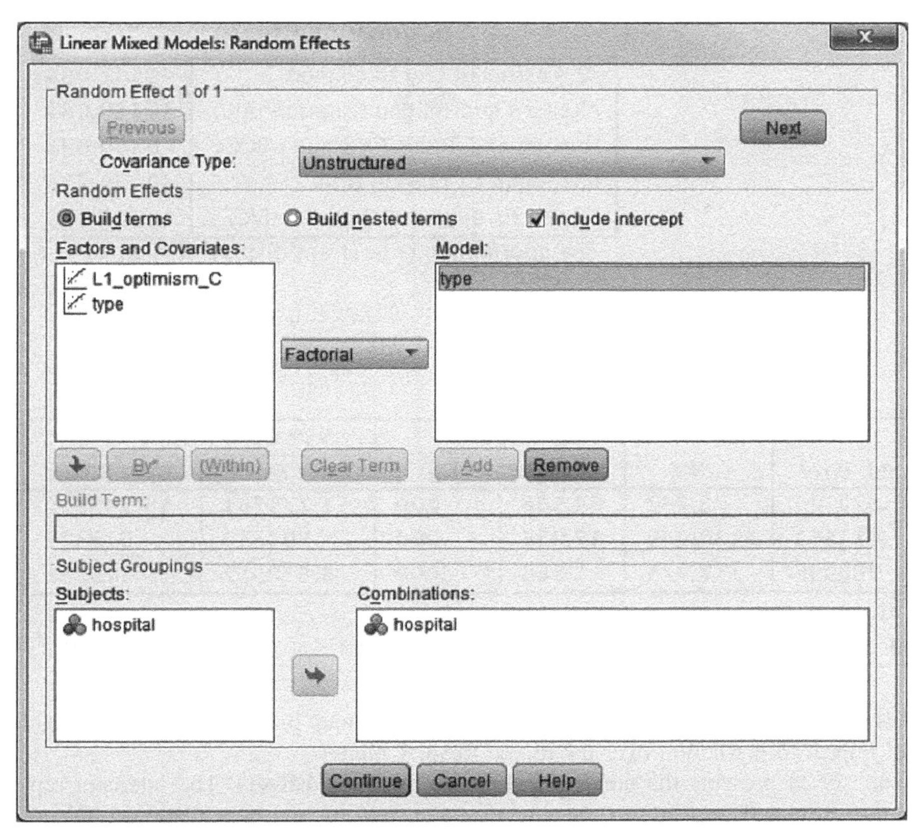

FIGURE 29.25 The **Fixed Effects** window setup for the mixed Level 2 model.

Model Dimension[a]

		Number of Levels	Covariance Structure	Number of Parameters	Subject Variables
Fixed Effects	Intercept	1		1	
	L1_optimism_C	1		1	
	type	1		1	
Random Effects	Intercept + type[b]	2	Unstructured	3	hospital
Residual				1	
Total		5		7	

a. Dependent Variable: prom_dv.

b. As of version 11.5, the syntax rules for the RANDOM subcommand have changed. Your command syntax may yield results that differ from those produced by prior versions. If you are using version 11 syntax, please consult the current syntax reference guide for more information.

FIGURE 29.26 The **Model Dimension** output for the mixed Level 2 model.

We can also perform a chi-square difference test. The **−2 Restricted Log Likelihood** values for the mixed Level 1 model and our mixed Level 2 model are **40418.621** and **40342.084**, respectively, with a difference of 76.537. The degrees of freedom for these two models are **6** and **7**, respectively, with a difference of 1. We therefore evaluate the chi-square difference with one degree of freedom. Based on Table A.1, the difference of

Information Criteria[a]

-2 Restricted Log Likelihood	40342.084
Akaike's Information Criterion (AIC)	40350.084
Hurvich and Tsai's Criterion (AICC)	40350.093
Bozdogan's Criterion (CAIC)	40379.729
Schwarz's Bayesian Criterion (BIC)	40375.729

The information criteria are displayed in smaller-is-better forms.

a. Dependent Variable: prom_dv.

FIGURE 29.27

The **Information Criteria** for the mixed Level 2 model.

Estimates of Fixed Effects[a]

Parameter	Estimate	Std. Error	df	t	Sig.	95% Confidence Interval	
						Lower Bound	Upper Bound
Intercept	52.907960	.820175	78.747	64.508	.000	51.275361	54.540559
L1_optimism_C	.566469	.031552	4180.049	17.953	.000	.504610	.628329
type	7.018122	1.260996	117.775	5.566	.000	4.520957	9.515288

a. Dependent Variable: prom_dv.

FIGURE 29.28 The **Estimates of Fixed Effects** for the mixed Level 2 model.

76.537 is statistically significant ($p < .001$). We therefore may infer that the mixed Level 2 model improved prediction over the mixed Level 1 model.

Figure 29.28 presents the table of **Estimates of Fixed Effects**. The intercept represents the estimated population grand mean parameter for the dependent variable of **prom_dv** based on the model now containing both **L1_optimism_C** and **type**. We note that it has increased from the estimate of the mixed Level 1 model to approximately **52.91**.

The parameter slope estimate for **L1_optimism_C** is approximately **.57** and indicates that every increase of one scale value in the **L1_optimism_C** score increases the **prom_dv** score by about **.57** when controlling for **type**. As can be seen in the **Sig.** column, the slope is statistically significant ($p < .001$).

Hospital **type** is associated with a parameter estimate of about 7.02. This is a dichotomously coded variable with values of **0** for nonteaching hospitals and **1** for teaching hospitals. Because the parameter is positive and because a value of **1** is greater than the value of **0**, we interpret the slope to indicate that patients in teaching hospitals report **prom_dv** scores of about 7.02 (on average) that is higher than the **L1_optimism_C** scores of patients in nonteaching hospitals.

The table of **Estimates of Covariance Parameters** is presented in Figure 29.29. These are variance estimates based on the random effects of **hospital** in the model. Incorporating the patients' individual **optimism** scores into this model reduced the **Residual** very slightly from about **442.44** in the mixed Level 1 model to about **437.91** (it became about 1% smaller), signifying that the model containing both **L1_optimism_C** and **type** explains just a little more of the within-hospital variance than we could in the mixed Level 1 model.

The **Unstructured** covariance components indicate the following:

- **UN (1,1)** represents the variance estimate (with a value of approximately **39.53**) of the **prom_dv** intercepts, with **L1_optimism_C** and **type** in the model. The **prom_dv** intercepts vary significantly ($p < .001$) among the hospitals.
- **UN (2,2)** represents the variance estimate (with a value of approximately **88.54**) of the **prom_dv** slopes, with **L1_optimism_C** and **type** in the model. This parameter

Estimates of Covariance Parameters[a]

Parameter		Estimate	Std. Error	Wald Z	Sig.	95% Confidence Interval	
						Lower Bound	Upper Bound
Residual		437.905790	9.408154	46.545	.000	419.848989	456.739176
Intercept + type [subject = hospital]	UN (1,1)	39.530591	9.131638	4.329	.000	25.136465	62.167357
	UN (2,1)	-40.533368	19.116722	-2.120	.034	-78.001454	-3.065281
	UN (2,2)	88.535053	36.545735	2.423	.015	39.423394	198.827518

a. Dependent Variable: prom_dv.

FIGURE 29.29 The **Estimates of Covariance Parameters** for the mixed Level 2 model.

was statistically significant ($p = .015$), indicating that the slopes vary significantly among the hospitals.

- **UN (2,1)** represents the covariance estimate between the **prom_dv** slopes and intercepts (with a value of approximately **−40.53**), with **L1_optimism_C** and **type** in the model. This indicates that there is a significant ($p = .034$) covariance (correlation) between the slope and the intercept. Because the valence of this parameter is negative, we interpret it as indicating that higher values of **type** (recall that nonteaching and teaching hospitals were coded as **0** and **1**, respectively) are associated with smaller variances; thus, teaching hospitals have less steep slopes than nonteaching hospitals.

29.12 ANALYSIS SETUP: HIERARCHICAL MODEL

We now add **L2_optimism_C** to the model to evaluate the average optimism level of the hospital based on the possibility that this Level 2 variable, a characteristic of the hospital, can enhance our prediction of **prom_dv**.

We open the **patient optimism** data file and select from the main menu **Analyze → Mixed Models → Linear**. We define **hospital** as our clustering variable in the initial **Specify Subjects and Repeated** window and select the **Continue** pushbutton to reach the main **Linear Mixed Models** window.

The main **Linear Mixed Models** window is shown in Figure 29.30. We place **prom_dv** in the **Dependent Variable** panel and **L1_optimism_C**, **type**, and **L2_optimism_C** in the **Covariate(s)** panel.

Selecting the **Fixed** pushbutton opens the **Fixed Effects** window shown in Figure 29.31. We move **L1_optimism_C**, **type**, and **L2_optimism_C** from the **Factors and Covariates** panel to the **Model** panel and make sure that the **Include intercept** box is checked. We use the same specifications for **Random Effects** and **Statistics** as we used in the mixed Level 2 model analysis.

29.13 ANALYSIS OUTPUT: HIERARCHICAL MODEL

The **Model Dimension** table is shown in Figure 29.32. There were **8** parameters in this mixed Level 2 model.

Figure 29.33 presents the **Information Criteria**, including both the **AIC** and the **BIC**. These values are smaller than in the mixed Level 2 model, suggesting a better model fit.

We can also perform a chi-square difference test. The **−2 Restricted Log Likelihood** values for the mixed Level 2 model and this hierarchical model are **40342.084** and **40308.216**, respectively, with a difference of 33.868. The degrees of freedom for these

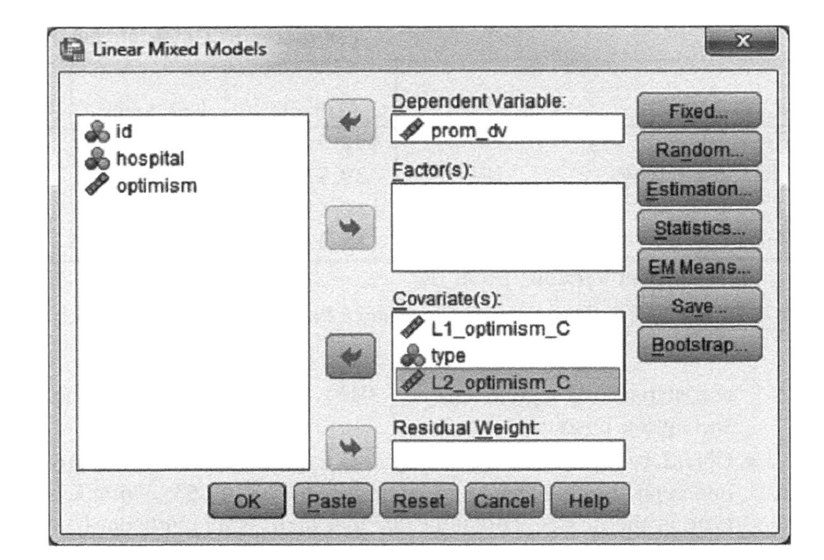

FIGURE 29.30
The main **Linear Mixed Models** window setup for the hierarchical model.

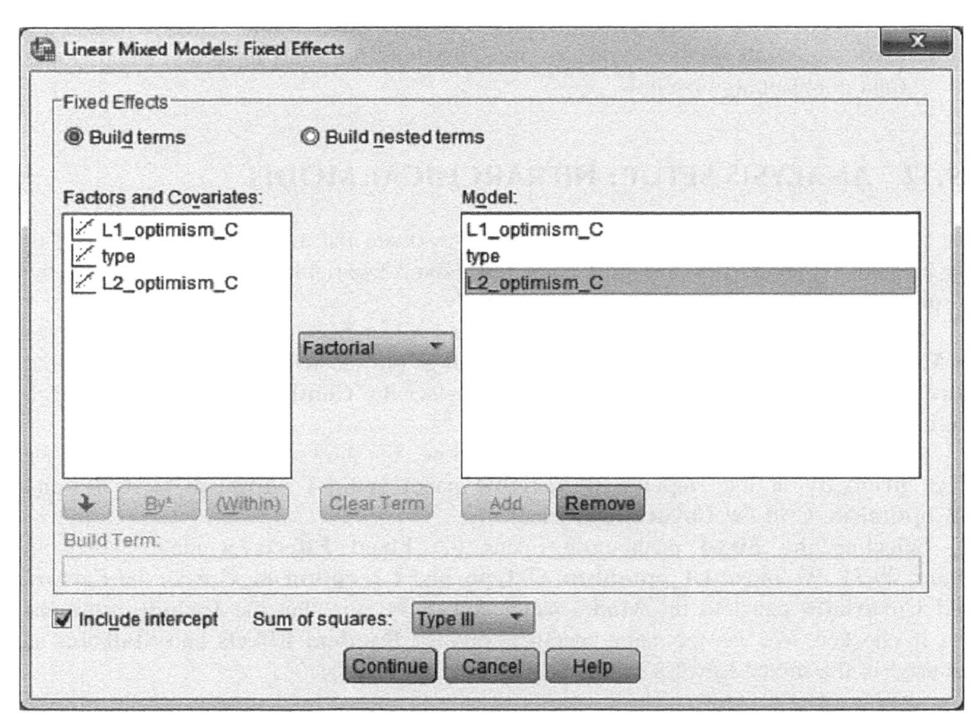

FIGURE 29.31 The **Fixed Effects** window setup for the hierarchical model.

two models are **7** and **8**, respectively, with a difference of 1. We therefore evaluate the chi-square difference with one degree of freedom. Based on Table A.1, the difference of 33.868 is statistically significant ($p < .001$). We therefore may infer that the hierarchical model improved prediction over the mixed Level 2 model.

Figure 29.34 presents the table of **Estimates of Fixed Effects**. The intercept represents the estimated population grand mean parameter for the dependent variable of **prom_dv** based on the model now containing **L1_optimism_C**, **type**, and **L2_optimism_C**. Its value is now approximately **54.12**.

Model Dimension[a]

		Number of Levels	Covariance Structure	Number of Parameters	Subject Variables
Fixed Effects	Intercept	1		1	
	L1_optimism_C	1		1	
	type	1		1	
	L2_optimism_C	1		1	
Random Effects	Intercept + type[b]	2	Unstructured	3	hospital
Residual				1	
Total		6		8	

a. Dependent Variable: prom_dv.

b. As of version 11.5, the syntax rules for the RANDOM subcommand have changed. Your command syntax may yield results that differ from those produced by prior versions. If you are using version 11 syntax, please consult the current syntax reference guide for more information.

FIGURE 29.32 The **Model Dimension** output for the hierarchical model.

Information Criteria[a]

-2 Restricted Log Likelihood	40308.216
Akaike's Information Criterion (AIC)	40316.216
Hurvich and Tsai's Criterion (AICC)	40316.225
Bozdogan's Criterion (CAIC)	40345.860
Schwarz's Bayesian Criterion (BIC)	40341.860

The information criteria are displayed in smaller-is-better forms.

a. Dependent Variable: prom_dv.

FIGURE 29.33
The **Information Criteria** for the hierarchical model.

Estimates of Fixed Effects[a]

Parameter	Estimate	Std. Error	df	t	Sig.	95% Confidence Interval Lower Bound	95% Confidence Interval Upper Bound
Intercept	54.122072	.734037	81.442	73.732	.000	52.661689	55.582455
L1_optimism_C	.522389	.032565	4386.097	16.042	.000	.458546	.586232
type	4.545301	1.211160	125.283	3.753	.000	2.148318	6.942284
L2_optimism_C	.722658	.112798	171.193	6.407	.000	.500003	.945313

a. Dependent Variable: prom_dv.

FIGURE 29.34 The **Estimates of Fixed Effects** for the hierarchical model.

The parameter slope estimate for **L1_optimism_C** is approximately **.52** and is statistically significant ($p < .001$). It indicates that every increase of one scale value in the **L1_optimism_C** score increases the **prom_dv** score by about **.52** when controlling for **type** and **L2_optimism_C**.

Hospital **type** is associated with a parameter estimate of about 4.55 ($p < .001$). We interpret the slope to indicate that patients in teaching hospitals report **prom_dv** scores of about 4.55 (on average) that is higher than those of patients in nonteaching hospitals, controlling for **L1_optimism_C** and **L2_optimism_C**.

Estimates of Covariance Parameters[a]

Parameter		Estimate	Std. Error	Wald Z	Sig.	95% Confidence Interval	
						Lower Bound	Upper Bound
Residual		437.762926	9.400126	46.570	.000	419.721334	456.580029
Intercept + type [subject = hospital]	UN (1,1)	25.533087	6.920918	3.689	.000	15.009953	43.433749
	UN (2,1)	-34.765359	13.194063	-2.635	.008	-60.625248	-8.905470
	UN (2,2)	82.744132	25.275530	3.274	.001	45.469897	150.574158

a. Dependent Variable: prom_dv.

FIGURE 29.35 The **Estimates of Covariance Parameters** for the hierarchical model.

The parameter slope estimate for **L2_optimism_C** is approximately **.72** and is statistically significant ($p < .001$). It indicates that every increase of one scale value in the **L2_optimism_C** score increases the **prom_dv** score by about **.72** when controlling for **type** and **L1_optimism_C**.

The table of **Estimates of Covariance Parameters** is presented in Figure 29.35. Adding the **Hospital Level optimism** scores into the model reduced the **Residual** only slightly from about **437.91** to about **437.76**, suggesting little variance is being explained between hospitals. The values in **UN (1,1)**, **UN (2,2)**, and **UN (2,1)** have all been reduced in value, reflecting the inclusion of **L2_optimism_C**. However, the results among the covariances remain the same as in the previous model.

29.14 ANALYSIS SETUP: INTERACTION MODEL

The last model in our series builds on the hierarchical model with one additional predictor. It is possible that the degree of optimism characterizing a given hospital (represented by **L2_optimism_C**) may be *moderated* by hospital **type**; that is, the slope of the function predicting **prom_dv** from **L2_optimism_C** may be different for nonteaching and teaching hospitals. This would be represented by a statistically significant *interaction* between **type** and **L2_optimism_C**.

We open the **patient optimism** data file and select from the main menu **Analyze ➔ Mixed Models ➔ Linear**. We define **hospital** as our clustering variable in the initial **Specify Subjects and Repeated** window and select the **Continue** pushbutton to reach the main **Linear Mixed Models** window.

The main **Linear Mixed Models** window is shown in Figure 29.36. We place **prom_dv** in the **Dependent Variable** panel and **L1_optimism_C**, **type**, and **L2_optimism_C** in the **Covariate(s)** panel.

Selecting the **Fixed** pushbutton opens the **Fixed Effects** window is shown in Figure 29.37. We move **L1_optimism_C**, **type**, and **L2_optimism_C** from the **Factors and Covariates** panel to the **Model** panel and make sure that the **Include intercept** box is checked.

To include the **type*L2_optimism_C** interaction in the model, we take the following steps:

- From the drop-down menu between the **Factors and Covariates** panel and the **Model** panel, we select **Interaction**. This is shown in Figure 29.37.
- While holding down the **Control** or **Shift** key, we highlight **type** and **L2_optimism_C** (see Figure 29.38).
- We select the **Add** pushbutton to create the **type*L2_optimism_C** interaction variable that will appear in the **Model** panel (see Figure 29.39). This will determine the average intercept and slope of this interaction effect while controlling for **L1_optimism_C**, **type**, and **L2_optimism_C** on **prom_dv** scores.

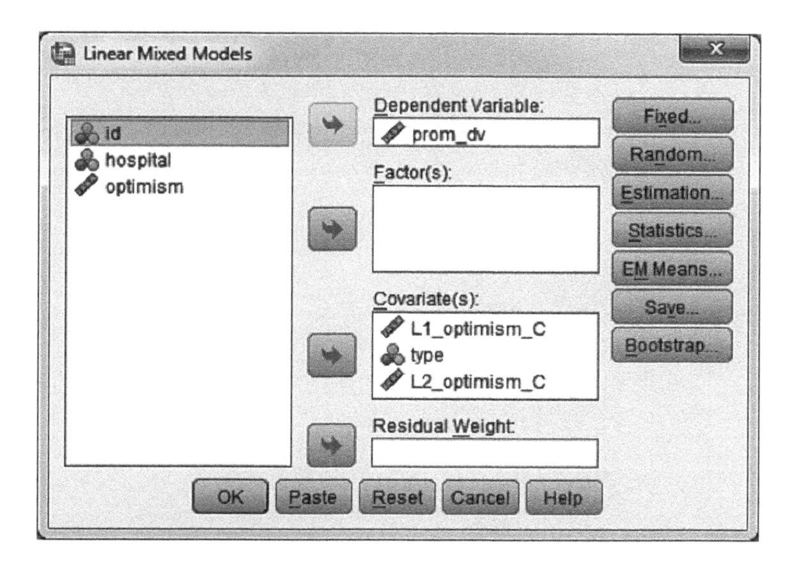

FIGURE 29.36 The main **Linear Mixed Models** window setup for the interaction model.

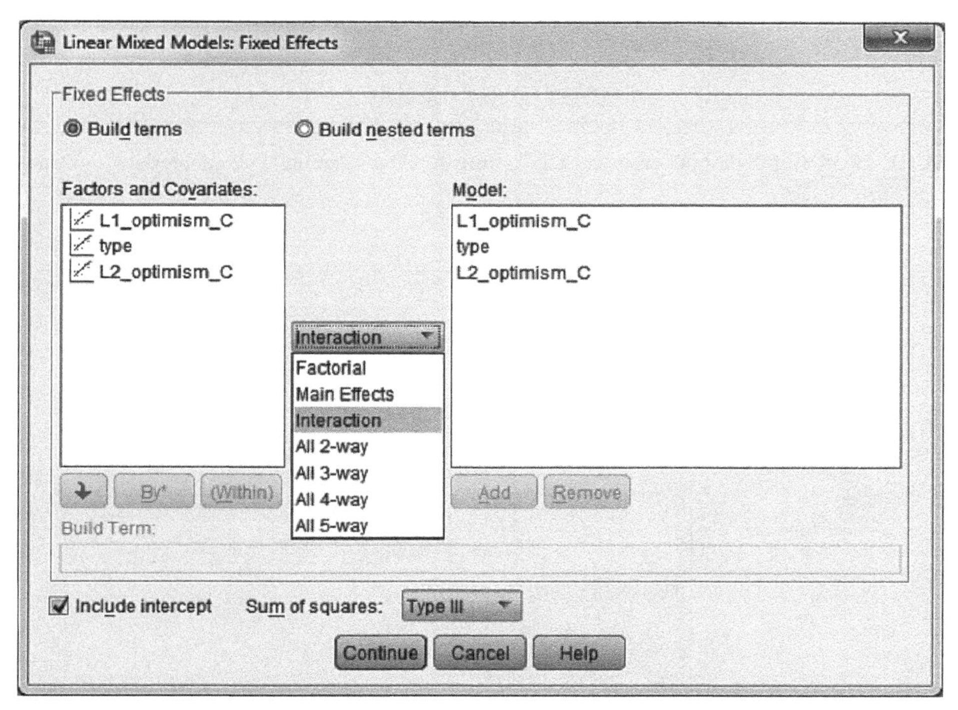

FIGURE 29.37 Select **Interaction** in the **Fixed Effects**.

We use the same specifications for **Random Effects** and **Statistics** as we used in the mixed Level 2 model analysis.

29.15 ANALYSIS OUTPUT: INTERACTION MODEL

The **Model Dimension** table is shown in Figure 29.40. There were **9** parameters in this mixed Level 2 model.

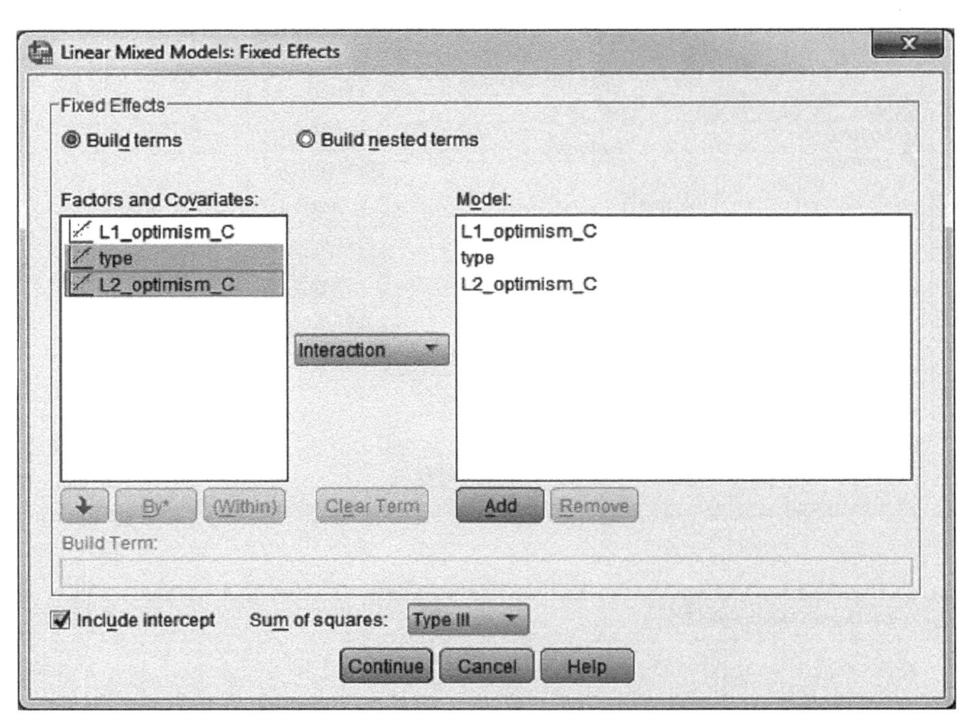

FIGURE 29.38 Highlight both **type** and **L2_optimism_C** by selecting while depressing the **Control** or **Shift** key.

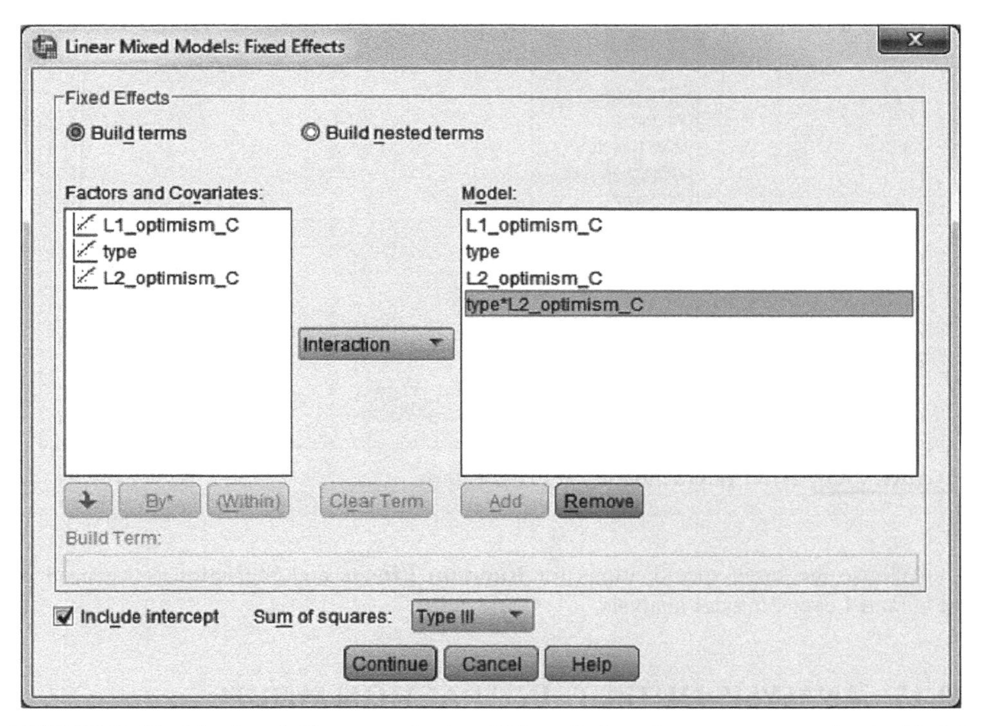

FIGURE 29.39 Selecting **Add** creates the **type*L2_optimism_C** term to appear in the **Model** panel.

Model Dimension[a]

		Number of Levels	Covariance Structure	Number of Parameters	Subject Variables
Fixed Effects	Intercept	1		1	
	L1_optimism_C	1		1	
	type	1		1	
	L2_optimism_C	1		1	
	type * L2_optimism_C	1		1	
Random Effects	Intercept + type[b]	2	Unstructured	3	hospital
Residual				1	
Total		7		9	

a. Dependent Variable: prom_dv.

b. As of version 11.5, the syntax rules for the RANDOM subcommand have changed. Your command syntax may yield results that differ from those produced by prior versions. If you are using version 11 syntax, please consult the current syntax reference guide for more information.

FIGURE 29.40 The **Model Dimension** output for the interaction model.

Information Criteria[a]

-2 Restricted Log Likelihood	40308.429
Akaike's Information Criterion (AIC)	40316.429
Hurvich and Tsai's Criterion (AICC)	40316.438
Bozdogan's Criterion (CAIC)	40346.072
Schwarz's Bayesian Criterion (BIC)	40342.072

The information criteria are displayed in smaller-is-better forms.

a. Dependent Variable: prom_dv.

FIGURE 29.41
The **Information Criteria** for the interaction model.

Estimates of Fixed Effects[a]

						95% Confidence Interval	
Parameter	Estimate	Std. Error	df	t	Sig.	Lower Bound	Upper Bound
Intercept	54.286888	.751795	77.164	72.210	.000	52.789924	55.783852
L1_optimism_C	.522458	.032566	4385.911	16.043	.000	.458613	.586303
type	4.583932	1.210262	122.898	3.788	.000	2.188272	6.979591
L2_optimism_C	.821607	.150584	88.817	5.456	.000	.522392	1.120822
type * L2_optimism_C	-.217574	.219609	138.222	-.991	.324	-.651802	.216654

a. Dependent Variable: prom_dv.

FIGURE 29.42 The **Estimates of Fixed Effects** for the interaction model.

Estimates of Covariance Parameters[a]

Parameter		Estimate	Std. Error	Wald Z	Sig.	95% Confidence Interval	
						Lower Bound	Upper Bound
Residual		437.804027	9.401891	46.566	.000	419.759083	456.624701
Intercept + type [subject = hospital]	UN (1,1)	25.442105	6.925166	3.674	.000	14.923196	43.375476
	UN (2,1)	-34.711825	13.167860	-2.636	.008	-60.520357	-8.903294
	UN (2,2)	82.562031	25.160162	3.281	.001	45.434218	150.029852

a. Dependent Variable: prom_dv.

FIGURE 29.43 The **Estimates of Covariance Parameters** for the interaction model.

Figure 29.41 presents the **Information Criteria**, including both the **AIC** and the **BIC**. These values increased very slightly over the hierarchical model, suggesting no improvement in model fit.

We can also perform a chi-square difference test. The **−2 Restricted Log Likelihood** values for the mixed hierarchical model and the interaction model are **40308.216** and **40308.429**, respectively, with a difference of 0.213. The degrees of freedom for these two models are **8** and **9**, respectively, with a difference of 1. We therefore evaluate the chi-square difference with one degree of freedom, and a chi-square value of less than 1.00 is not statistically significant. We therefore may infer that the interaction model offers no improved prediction over the hierarchical model.

Figure 29.42 presents the table of **Estimates of Fixed Effects**. The results are virtually the same as those for the hierarchical model, in that the **type*L2_optimism_C** interaction failed to achieve statistical significance ($p = .324$).

The table of **Estimates of Covariance Parameters** is presented in Figure 29.43. Adding the **type*L2_optimism_C** interaction into the model barely affected the values of the parameters shown in the table from those associated with the hierarchical model.

PART 9

REGRESSING (PREDICTING) CATEGORICAL VARIABLES

CHAPTER 30

Binary Logistic Regression

30.1 OVERVIEW

In contrast to ordinary least-squares linear regression where the dependent or outcome variable is quantitative, we use logistic regression when we wish to predict the value of a categorical dependent variable (e.g., success for failure in a training program or medical regime). When only two outcome categories are assessed, the procedure is called *binary logistic regression*; when three or more outcome categories are involved, the procedure is called *multinomial logistic regression.* Predictor variables can be any combination of quantitative and binary variables. Details regarding the nature of the logistic regression model are described in other sources (e.g., Menard, 2010; Meyers et al., 2013).

The two possible outcomes in binary logistic regression represent two different groups of cases, and the results of the analysis are framed in terms of the likelihood of a case being in (coded in the data file as) one of the groups as opposed to the other. As such, it is important to carefully formulate the group (categorical) coding schema for the outcome variable in building the data file. The two groups (outcomes) and category codes are as follows:

- The *response* or *target* group represents the desired or expected outcome (e.g., success). It is this category to which prediction is directed. This category should be given a code of **1** for the outcome variable.
- The *reference* or *control* group represents the alternative outcome (e.g., failure). This category should be given a code of **0** for the outcome variable.

It is not uncommon for binary variables to be used as predictors in logistic regression analysis (e.g., sex, family history of a certain medical condition), and the results of the analysis are framed in terms of one of the categories being more or less likely to achieve the target outcome. The analysis describes the likelihood of achieving the target outcome for the *focus* category with respect to the other category, which is the *reference* category. For example, if the predictor variable is **sex** and we wished to describe the results in terms of females being more (or less) likely to achieve the target outcome, then **female** would be the focus category and **male** would be the reference category. If we wish to describe the results in terms of males being more (or less) likely to achieve the target outcome, then **male** would be the focus category and **female** would be the reference

Performing Data Analysis Using IBM SPSS®, First Edition.
Lawrence S. Meyers, Glenn C. Gamst, and A. J. Guarino.

category. The statistical outcome of one would be the inverse of the other. IBM SPSS® provides us with a dialog window to select which category (the lower or higher code) is to be designated as the reference category.

At a simple level, the result of interest to most researchers is the *odds ratio* associated with each predictor. Assume that we have a target outcome (e.g., success). Odds represent the ratio of number of successes to number of failures for any particular group of cases. To illustrate the way in which we use odds ratios in logistic regression, let us assume these two situations:

- The odds of females being successful (as opposed to failing) are 10 to 1.
- The odds of males being successful (as opposed to failing) are 5 to 1.

Now, if we focus on the females and compute the odds ratio, we place the odds associated with females in the numerator and those associated with males in the denominator. For our example, this yields a ratio of 10 : 5, with the odds ratio for females being 2.00. We interpret this odds ratio to indicate that the odds for females being successful are twice the odds of males being successful. If **female** is the focus category and **male** is the reference category, the odds ratio for **sex** shown in the logistic regression results would be 2.00.

By the same reasoning as just described, the odds ratio for males represents the odds associated with males divided by the odds associated with females. Here, the ratio is 5 : 10 or 0.50. We interpret this odds ratio to indicate that the odds for males being successful are half of those for females. If **male** is the focus category and **female** is the reference category, the odds ratio for **sex** shown in the logistic regression results would be 0.50.

It is important to note that in many contexts, having the odds of 5 to 1 associated with success in a program, as is the case with the males in our example, might be considered to represent a wonderful opportunity for success. But this favorable situation is overshadowed by the 10 to 1 odds of success associated with the females. Our point is that the odds ratio does not convey information about the absolute level of the odds but only how the odds associated with one group compare with those of another.

30.2 NUMERICAL EXAMPLE

The data for our fictional study represent a set of at-risk junior high school children in a special educational early intervention program. We use student sex as well as the degree of encouragement students received from their families as predictors of program success. Under the variable of **graduated**, success (the target category) is defined as the children having graduated and entering a high school (coded as **1** for **graduated**) and failure (the reference category) is defined as the children not having graduated (coded as **0** for **not graduated**). In this example, **sex** of child is coded as **0** (**male**) and **1** (**female**), and we will designate **female** (having the "higher" or "last" **sex** code) as the focus group and **male** as the reference group (having the "lower" or "first" **sex** code) in the analysis setup. Also measured on a scale ranging from 0 through 45 is **family_encouragement**, with higher values representing more encouragement. The data file is named **graduation**, and a screenshot of a portion of the data file is shown in Figure 30.1.

30.3 ANALYSIS SETUP

From the main menu, we select **Analyze ➔ Regression ➔ Binary Logistic** to open the main **Logistic Regression** window shown in Figure 30.2. We move **graduated** into

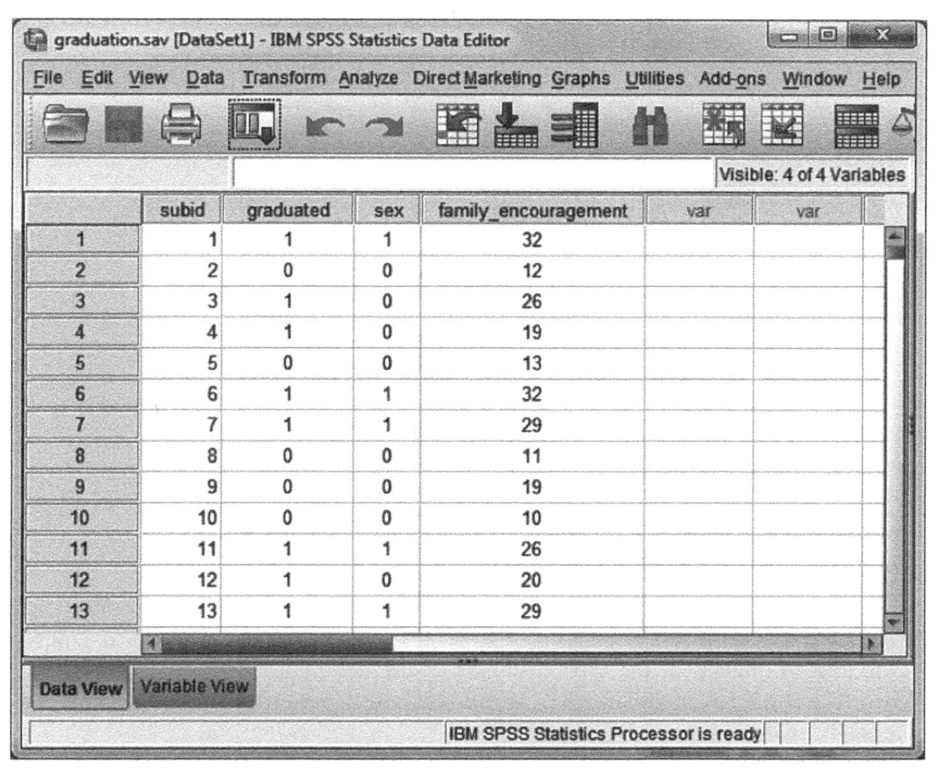

	subid	graduated	sex	family_encouragement	var	var
1	1	1	1	32		
2	2	0	0	12		
3	3	1	0	26		
4	4	1	0	19		
5	5	0	0	13		
6	6	1	1	32		
7	7	1	1	29		
8	8	0	0	11		
9	9	0	0	19		
10	10	0	0	10		
11	11	1	1	26		
12	12	1	0	20		
13	13	1	1	29		

FIGURE 30.1 A portion of the data file.

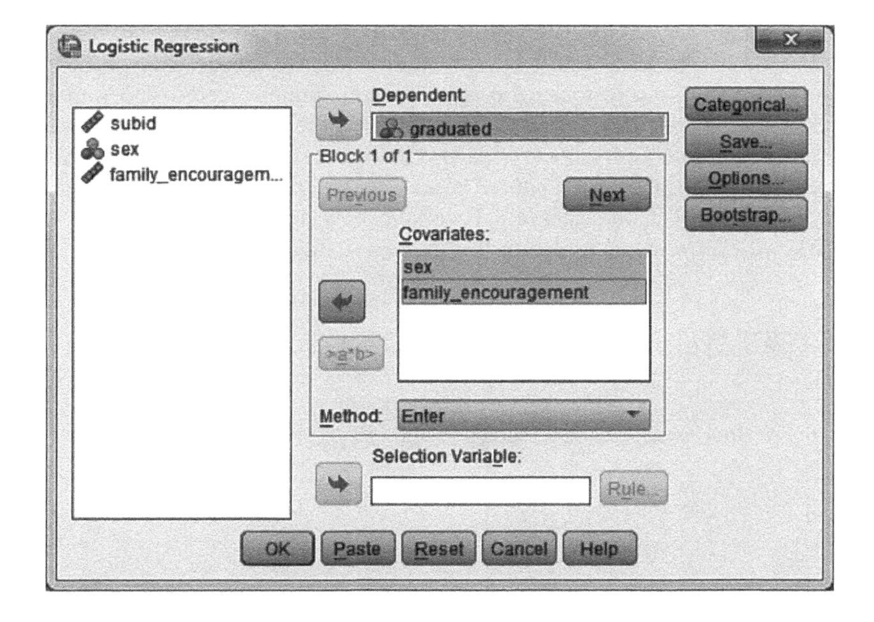

FIGURE 30.2 The main **Logistic Regression** dialog window.

the **Dependent** panel and **sex** and **family_encouragement** into the **Covariates** panel to register them as predictor variables. We retain the default **Method** of **Enter** to perform a standard logistic regression analysis where both variables are entered together in a single block (various step procedures are available in the **Method** drop-down menu).

When variables are moved into the **Covariates** panel, the **Categorical** pushbutton becomes available. Selecting the **Categorical** pushbutton opens the **Define Categorical**

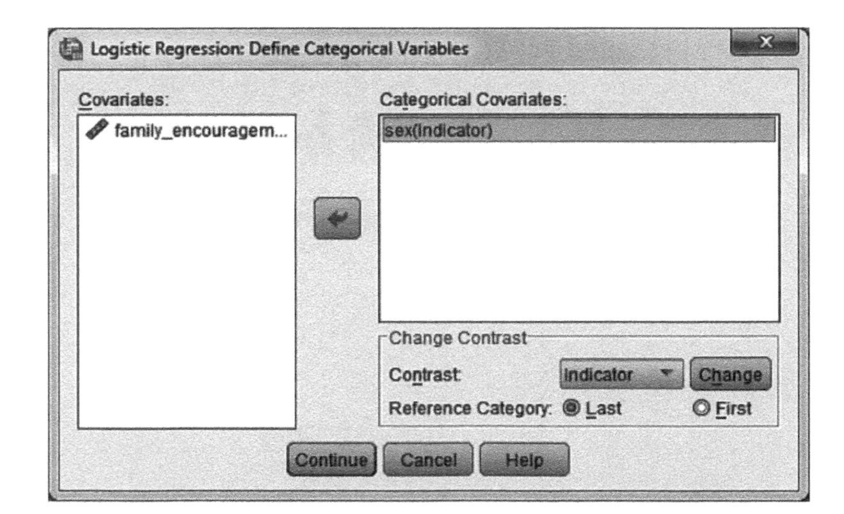

FIGURE 30.3

The **Define Categorical Variables** dialog window of **Logistic Regression**.

Variables window that allows us to specify our focus and reference groups for any categorical predictor variable. This window is shown in Figure 30.3. We move **sex** into the **Categorical Covariates** panel, and the expression **(indicator)** appears after the name of the variable. **Indicator** is the type of contrast that we wish to use; it contrasts the presence of a category with its absence, and this is the first step in specifying how we want the analysis to treat **sex**.

The second step in specifying how we want the analysis to treat **sex** is to identify the focus and reference categories. IBM SPSS calls for the specification of the **Reference Category** (by implication, the other category of a binary variable will be the focus category). Males and females are arbitrarily coded as **0** and **1**, respectively. In our analysis, we wish to have **female** as the focus category and **male** as the reference category; in our coding schema, **female** is the **last** category (the highest numeric code) and **male** is the **first** category (the lowest numeric code). We therefore select **first** to make the code of **0** (**male**) the **Reference Category**, and click **Change** to register that with IBM SPSS. Upon clicking **Change**, the specification is added to the name of the variable as shown in Figure 30.4, so that our categorical variable now appears in the panel as **sex(Indicator(first))**. Click **Continue** to return to the main dialog window.

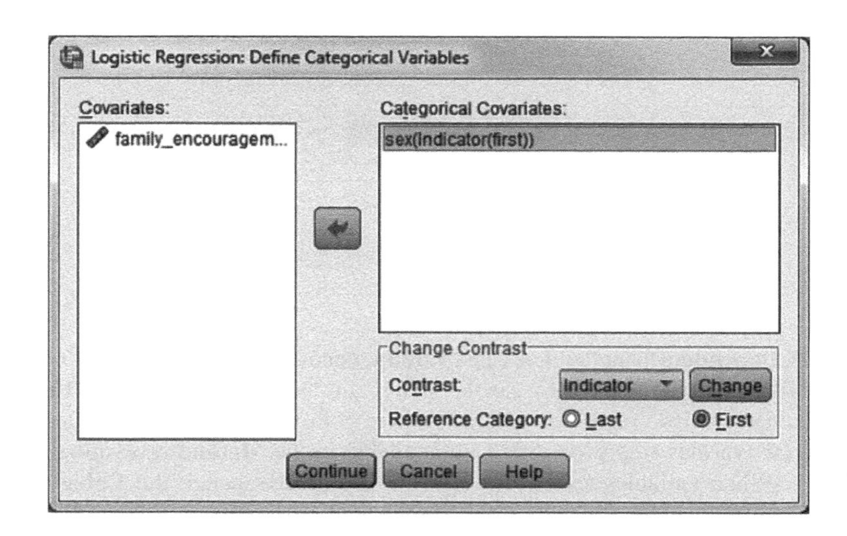

FIGURE 30.4

The **Define Categorical Variables** dialog window of **Logistic Regression** after specifying the **Reference Category** as **First**.

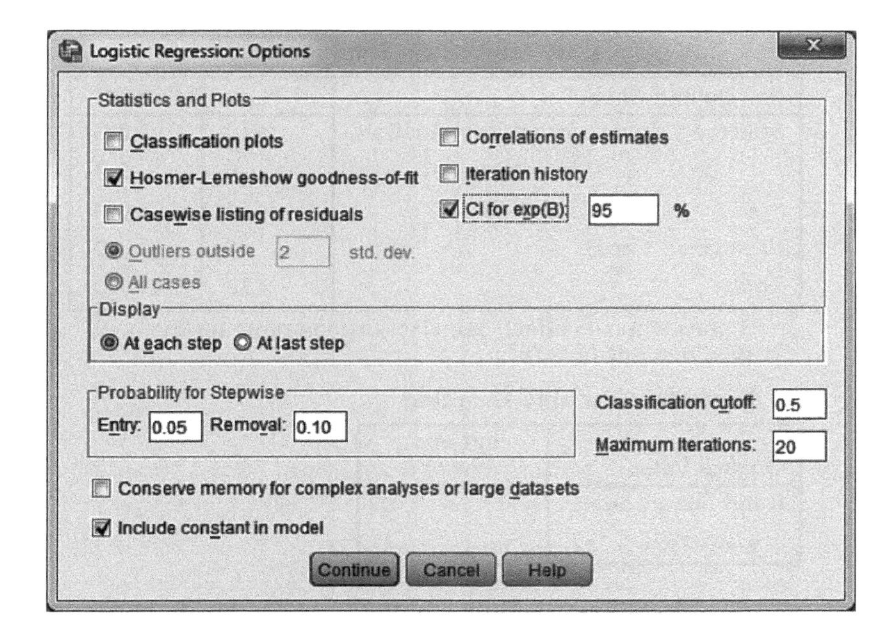

FIGURE 30.5
The **Options** dialog window of **Logistic Regression**.

In the **Options** window, we ask for the **Hosmer-Lemeshow goodness-of-fit** statistics as well as the 95% confidence interval for the odds ratio (**CI for exp(B)**). The **Classification cutoff** is the probability we need to achieve in order to classify a case as belonging to the target group. Probabilities can range between 0 and 1, and the default in IBM SPSS is **.5**. This means that if a case has a **.5** or greater probability of being in the target group (based on the final logistic model), it is classified (predicted to be) in the target group; if a case has less than a **.5** probability of being in the target group, it is classified (predicted to be) in the reference group. For this analysis, we will keep the default **Classification cutoff** at **.5**, but we will deal explicitly with this specification in Chapter 31. Click **Continue** to return to the main dialog window and click **OK** to perform the analysis (Figure 30.5).

30.4 ANALYSIS OUTPUT

Figure 30.6 shows what can be called administrative output. The **Case Processing Summary** table displays information on the number of cases in the analysis. The **Dependent Variable Encoding** shows the "internal recoding" of our binary outcome variable. IBM SPSS does this recoding because not all researchers may assign codes of **0** and **1** to the outcomes. The software thus assigns the lower code (whatever was used by the researchers) a value of **0** and the higher code a value of **1**; IBM SPSS ignored the fact that we had already anticipated the need to code as **0** and **1**, and so it performed its recoding anyway.

We also obtain a count of males and females in the **Categorical Variables Codings** table together with their codes. The codes under **Parameter coding** reflect our specification of the reference group; the category shown as **.000** (**male**) is treated as the reference category in the analysis.

Figure 30.7 presents the results for the intercept-only model, computed with only the constant in the equation but with none of the predictor variables; it is called the **Step 0** model (the model as of **Block 0**), as there are no predictor variables in the equation yet. The **Classification Table** simply provides counts of the number of cases in each binary

Case Processing Summary

Unweighted Cases[a]		N	Percent
Selected Cases	Included in Analysis	410	98.6
	Missing Cases	6	1.4
	Total	416	100.0
Unselected Cases		0	.0
Total		416	100.0

a. If weight is in effect, see classification table for the total number of cases.

Dependent Variable Encoding

Original Value	Internal Value
0 did not graduate	0
1 graduated	1

Categorical Variables Codings

		Frequency	Parameter coding (1)
sex	0 male	193	.000
	1 female	217	1.000

FIGURE 30.6
Administrative output.

Block 0: Beginning Block

Classification Table[a,b]

Observed			Predicted: graduated 0 did not graduate	Predicted: graduated 1 graduated	Predicted: Percentage Correct
Step 0	graduated	0 did not graduate	0	138	.0
		1 graduated	0	272	100.0
	Overall Percentage				66.3

a. Constant is included in the model.
b. The cut value is .500

Variables in the Equation

		B	S.E.	Wald	df	Sig.	Exp(B)
Step 0	Constant	.679	.105	42.153	1	.000	1.971

Variables not in the Equation

			Score	df	Sig.
Step 0	Variables	sex(1)	85.028	1	.000
		family_encouragement	92.627	1	.000
	Overall Statistics		139.639	2	.000

FIGURE 30.7 The **Beginning Block** output.

outcome. It is a prediction table, with the **Observed** cases in the rows and the **Predicted** group membership represented by the columns.

With only the intercept in the model, our prediction is based exclusively on the frequencies in that table: **138** cases did not graduate and **272** cases did graduate. Thus, if we had no additional information, our best single guess is that a program participant would have graduated (more graduated than did not), and our classification (predictions) would be correct **66.3**% of the time.

The middle table in Figure 30.7 shows the **Variables in the Equation**; of course, in the intercept-only model, we have no predictor variables, so the only factor in the model is the intercept (shown as **Constant**). The odds ratio, shown as **Exp(B)**, has a value of **1.971**. This is because **272** is **1.971** times (approximately twice) as large as **138** (two-thirds of the sample graduated and one-third did not). The odds ratio informs us that a random program participant is **1.971** times (almost twice) as likely to have graduated as not. The bottom table labeled as **Variables not in the Equation** reminds us that these predictor variables have yet to be entered into the model.

Figure 30.8 shows the tables evaluating the model, with the predictors included. This is **Block 1** or **Step 1** and is our only step because we entered both of our predictors together. The **Omnibus Tests of Model Coefficients** contains the model chi-square, a statistical test of the null hypothesis that all the predictor coefficients are zero. It is equivalent to the overall F test in linear regression. The **Model** chi-square value (in the last row) is **157.368**, and with **2** degrees of freedom (there are two predictors in the model), we have a statistically significant amount of prediction ($p < .001$).

The **Model Summary** table provides three indexes of how well the logistic regression model fits the data. With all the variables in the model, the goodness-of-fit **−2 Log likelihood** statistic is **366.403**. We do not usually interpret this statistic directly but use it to compare different logistic models. The **Cox and Snell** pseudo $\boldsymbol{R^2}$ is **.351** and the **Nagelkerke** pseudo **R Square**, which is always the higher of the two, is **.442**. On the basis of the **Nagelkerke** pseudo **R Square**, we would thus conclude that about 44% of the variance associated with graduation is explained by our predictor variables.

The **Hosmer and Lemeshow Test** provides a formal test assessing whether the predicted probabilities match the observed probabilities. We hope to obtain a nonsignificant p value for this test because the goal of the research is to derive predictors that will accurately predict the actual probabilities. In this example, the goodness-of-fit statistic is **11.204**; it is tested as a chi-square value and is associated with a p value of **.190**, indicating an acceptable match between predicted and observed probabilities.

The **Contingency Table for Hosmer and Lemeshow Test**, shown in the bottom table in Figure 30.8, demonstrates more details of the **Hosmer and Lemeshow** test. This output has divided the data into 10 groups based on the outcome variable. These groups are defined by increasing rates of graduation (called "steps" in the table). For example, the first group (**Step 1**) represents those students least likely to graduate. The observed frequencies were that **33** cases did not graduate and **9** cases did graduate. The observed and the expected frequencies (based on the prediction model) match reasonably well for all of the steps and is a desirable result.

The **Classification Table** is presented in Figure 30.9. The overall predictive accuracy is **80.0**% now that we have our predictors in the model, although the model predicted better for graduation (**87.1**%) than for not having graduated (**65.9**%). Recall from **Block 0** that without considering any of our predictors, the likelihood or probability of a correct prediction was **66.3**%; thus, our predictors certainly contributed to successful prediction.

The fine-tuned results of the analysis are presented in the **Variables in the Equation** table shown in Figure 30.10. The table presents for each predictor the raw score partial regression coefficient (written as an uppercase **B** by IBM SPSS) and its standard error (**S.E.**). These coefficients indicate the amount of change expected in the log odds when there is a one-unit change in the predictor variable, with all the other variables in the

Block 1: Method = Enter

Omnibus Tests of Model Coefficients

		Chi-square	df	Sig.
Step 1	Step	157.368	2	.000
	Block	157.368	2	.000
	Model	157.368	2	.000

Model Summary

Step	-2 Log likelihood	Cox & Snell R Square	Nagelkerke R Square
1	366.403[a]	.319	.442

a. Estimation terminated at iteration number 5 because parameter estimates changed by less than .001.

Hosmer and Lemeshow Test

Step	Chi-square	df	Sig.
1	11.204	8	.190

Contingency Table for Hosmer and Lemeshow Test

		graduated = 0 did not graduate		graduated = 1 graduated		
		Observed	Expected	Observed	Expected	Total
Step 1	1	33	36.122	9	5.878	42
	2	21	21.011	8	7.989	29
	3	22	22.559	14	13.441	36
	4	23	15.522	8	15.478	31
	5	9	12.109	23	19.891	32
	6	13	11.967	30	31.033	43
	7	7	7.613	31	30.387	38
	8	4	4.148	27	26.852	31
	9	3	2.556	29	29.444	32
	10	3	4.393	93	91.607	96

FIGURE 30.8 The omnibus results and the model fit information.

model held constant (see Meyers et al., 2013). A coefficient close to 0 suggests that there is no change in the outcome variable associated with the predictor variable.

The **Sig**. column represents the *p* value for testing whether a predictor is significantly associated with graduation controlling for the other predictor(s). The logistic coefficients can be used in a manner similar to linear regression coefficients to generate predicted values. In this example, the model is as follows:

$$\textbf{graduation} = -3.256 + 1.970\ (\textbf{sex}) + 0.143\ (\textbf{family\_encouragement})$$

Because **family_encouragement** is a quantitative variable that can take a range of values, the expression **0.143 (family_encouragement)** can also take on a wide range of values. But because **sex** is coded in a binary manner, there are only two values possible

Classification Table[a]

Observed			Predicted: graduated: 0 did not graduate	Predicted: graduated: 1 graduated	Predicted: Percentage Correct
Step 1	graduated	0 did not graduate	91	47	65.9
		1 graduated	35	237	87.1
	Overall Percentage				80.0

a. The cut value is .500

FIGURE 30.9 The **Classification Table**.

Variables in the Equation

		B	S.E.	Wald	df	Sig.	Exp(B)	95% C.I.for EXP(B) Lower	95% C.I.for EXP(B) Upper
Step 1[a]	sex(1)	1.970	.272	52.422	1	.000	7.170	4.207	12.221
	family_encouragement	.143	.019	54.939	1	.000	1.154	1.111	1.198
	Constant	-3.256	.445	53.468	1	.000	.039		

a. Variable(s) entered on step 1: sex, family_encouragement.

FIGURE 30.10 The **Variables in the Equation**.

for the expression **1.970** (**sex**): for females (coded as **1**), the value of the expression will be **1.970*1** or 1.970; for males (coded as **0**), the value of the expression will be **1.970*0** or 0.

The **Exp(B)** column provides the odds ratios associated with each predictor (adjusting for the other predictor), with the 95% confidence interval associated with each provided in the final two columns. The adjusted odds ratio for **sex** is **7.170**, with a 95% confidence interval of **4.207–12.221**. This odds ratio indicates that in this sample, the odds of females (because they were the focal group) graduating are **7.170** times the odds of males graduating, controlling for **family_encouragement**.

The adjusted odds ratio for **family_encouragement** is **1.154**, with a confidence interval of **1.111–1.198**. This is a quantitatively measured variable, and so we interpret this odds ratio of **1.154** to mean that an increase of 1 in the **family_encouragement** measure increases the odds of graduation over the odds for not graduating by **1.154** times, controlling for **sex**.

We can apply the odds ratio to any two scores on the **family_encouragement** variable. For example, we can say when controlling for **sex** that the odds of graduation for a student with a **family_encouragement** score of 24 are **1.154** times greater than the graduation odds for a student whose **family_encouragement** score is 23, as well as 2.308 (computed as 2***1.154**) times greater than the graduation odds for a student whose **family_encouragement** score is 22 and 3.462 (computed as 3***1.154**) times greater than the graduation odds for a student whose **family_encouragement** score is 21, and so on.

CHAPTER 31

ROC Analysis

31.1 OVERVIEW

A receiver operator characteristic (ROC, pronounced "R-O-C" rather than "rock") analysis has been used for half a century in the context of signal detection and decision theory (see Meyers et al. (2013) for a brief history). Generally, it allows us to evaluate a range of decision rules that we may use to predict with which one of two possible outcomes or groups of particular cases are associated. It is particularly well suited to partner with binary logistic regression, and we treat the topic here as an extension of that topic.

ROC analysis requires that we have an underlying quantitative outcome measure, and in binary logistic regression it is the predicted probability of a case being in the target group. The decision to classify cases into the target and reference groups is made on the basis of this predicted probability. In the analysis described in Chapter 30, we retained the default **Classification cutoff** of **.5**. Cases associated with a **.5** or greater probability of being in the target group were classified (predicted to be) in the target group, whereas cases associated with less than **.5** probability of being in the target group were classified (predicted to be) in the reference group. However, a **Classification cutoff** of **.5** may not necessarily be the most appropriate decision cutoff because different circumstances might cause us to favor one type of decision strategy over another. Given a particular circumstance, we can use an ROC analysis to determine if a more appropriate cutoff should be established.

In an ROC analysis, the classification results (see the **Classification Table** in Figure 31.1) are characterized in decision-making language. The 2 × 2 table shows the rows as the observed outcomes and the columns as the predicted outcomes. The first row and column represent the reference (the *negative*) outcome (**did not graduate**), and the second row and column represent the target (the *positive*) outcome (**graduated**). Reading left to right for the first and then the second row, the cells represent the following:

- *True Negatives* (Observed, **did not graduate**; Predicted, **did not graduate**). These are the cases who have been correctly predicted as having the negative (**did not graduate**) outcome. The proportion of all reference group members that is represented by these cases is known as *specificity*.
- *False Positives* (Observed, **did not graduate**; Predicted, **graduated**). These are the cases with the negative (**did not graduate**) outcome (reference group members)

Performing Data Analysis Using IBM SPSS®, First Edition.
Lawrence S. Meyers, Glenn C. Gamst, and A. J. Guarino.

Classification Table[a]

Observed			Predicted: graduated 0 did not graduate	Predicted: graduated 1 graduated	Percentage Correct
Step 1	graduated	0 did not graduate	91	47	65.9
		1 graduated	35	237	87.1
	Overall Percentage				80.0

a. The cut value is .500

FIGURE 31.1 The **Classification Table** with the default **cut value** at **.5**.

who have incorrectly been predicted to have the positive outcome. The proportion of all reference group members that is represented by these cases is known as *(1 − specificity)*.

- *False Negatives* (Observed, **graduated**; Predicted, **did not graduate**). These are members of the target (**graduated**) group who have been predicted to have a negative (**did not graduate**) outcome.
- *True Positives* (Observed, **graduated**; Predicted, **graduated**). These are the cases who have been correctly predicted as having the positive outcome (**graduated**). The proportion of all target group members that is represented by these cases is known as *sensitivity*.

The success of correctly predicting group membership depends on the location of the decision criterion. But prediction will never be perfect (if it were, we would not be doing this analysis), and there are trade-offs involved as in any decision process. Generally, setting a higher probability as our decision criterion (**Classification cutoff**) will improve the rate of true negatives (predicting membership in the reference group) but at the expense of generating more false negatives (predicting **graduated** outcomes to be **did not graduate** outcomes). On the other hand, a lower decision criterion will give us a better true positive rate but will increase the number of false positives.

The decision criterion turns out to be a coordinate on an ROC curve (we will generate one in our analysis), a bow-shaped function with sensitivity (true positive rate) on the *Y*-axis and 1 − specificity (false-positive rate) on the *X*-axis. The amount of the bow of the curve (the area under the curve) is a gauge of how well the logistic model can differentiate the groups; more bow (larger areas under the curve) reflects greater differentiation.

IBM SPSS® provides in its output a set of coordinates of the ROC curve together with the probabilities associated with correctly predicting membership in the target group. The default coordinate used by IBM SPSS is the one that corresponds most closely to .5. The probabilities in the output are not continuous because they match to whole numbers of cases rather than fractions of cases (e.g., we correctly classify 237 cases as having **graduated** rather than 237.2 cases). The ROC output allows us to examine all available coordinates and select a predicted probability as the potentially more appropriate decision criterion.

31.2 NUMERICAL EXAMPLE

We use the example from Chapter 30 as our base, with the outcome variable of **graduated** and the predictors of **sex** and **family_encouragement**. The data file is named **graduation**.

31.3 ANALYSIS STRATEGY

We perform the analysis in three stages. First, we run the identical binary logistic regression analysis described in Chapter 30, except that we now save the predicted probabilities of target group membership to the data file; it is these predicted probabilities that the decision criterion (**Classification cutoff**) refers to. Second, we use these predicted probabilities as a variable in an ROC analysis. Third, we select a different decision criterion based on the results of the ROC analysis and perform another binary logistic regression analysis using the revised **Classification cutoff**.

31.4 BINARY LOGISTIC REGRESSION ANALYSIS: DEFAULT CLASSIFICATION CUTOFF

From the main menu, we select **Analyze ➔ Regression ➔ Binary Logistic** to open the main **Logistic Regression** window. We configure the analysis exactly as was done in Chapter 30 with the following addition. In the **Save** dialog window (see Figure 31.2), we check **Probabilities** in the **Predicted Values** area. Click **Continue** to return to the main dialog window and click **OK** to perform the analysis.

This analysis resulted in the predicted probability of target group membership being saved to the data file as can be seen in Figure 31.3 that shows a portion of the data file. IBM SPSS has named this saved variable as **PRE_1**; this generic name translates to "predicted values, first analysis."

Recall that the default **Classification cutoff** is **.5**, and the relationship of **PRE_1** to **.5** has driven predicted group membership of the cases. For example, **subid 1** with a predicted target (**graduated**) group membership of **.96425** (a value greater than or equal to **.5**) would be (correctly) classified in the target group, whereas **subid 2** with a predicted target group membership of **.17678** (a value less than **.5**) would be (correctly) classified in the reference (**did not graduate**) group.

The classification table from the analysis is shown in Figure 31.1. It is exactly the same as that obtained in the analysis described in Chapter 30.

31.5 ROC ANALYSIS: SETUP

From the main menu, we select **Analyze ➔ ROC Curve** to open the main **ROC Curve** dialog window as shown in Figure 31.4. We move **PRE_1** into the **Test Variable** panel

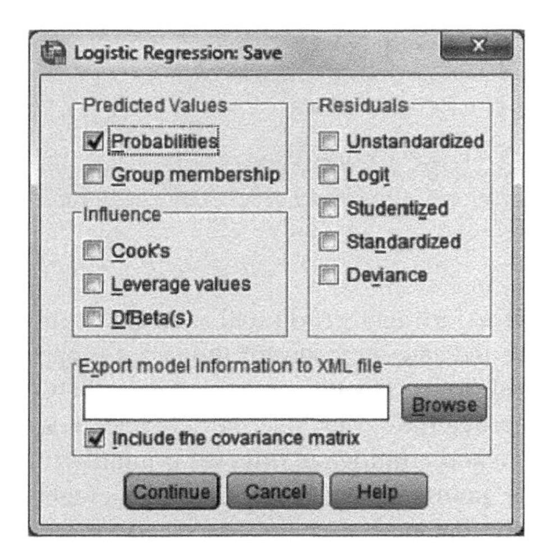

FIGURE 31.2

The **Save** dialog window of **Logistic Regression**.

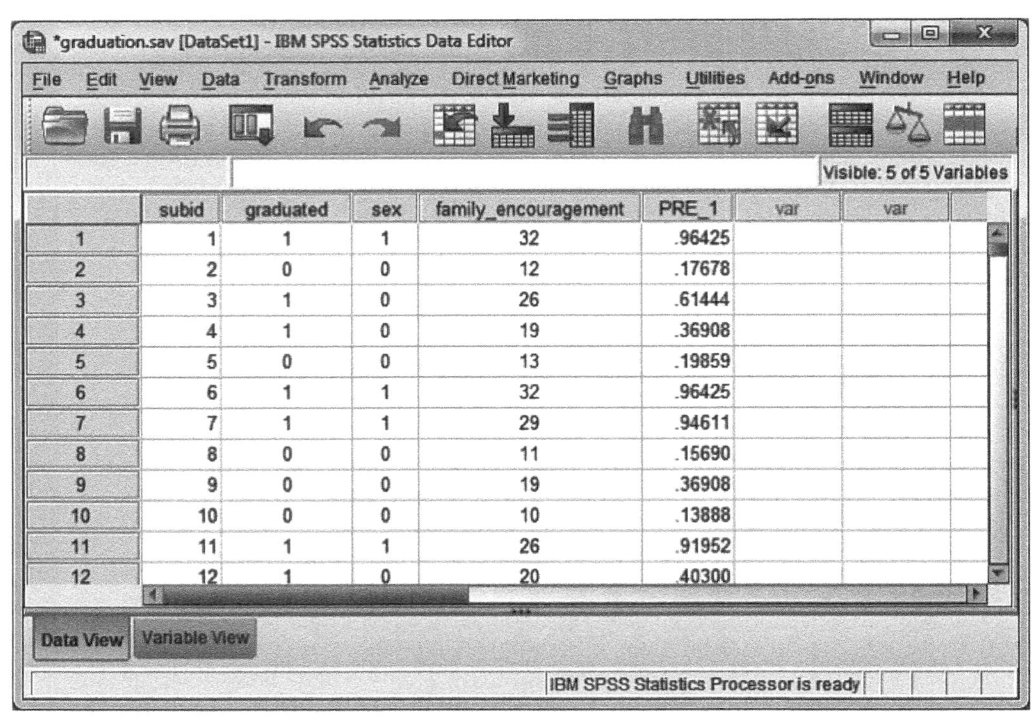

FIGURE 31.3 A portion of the data file showing the saved **PRE_1** variable.

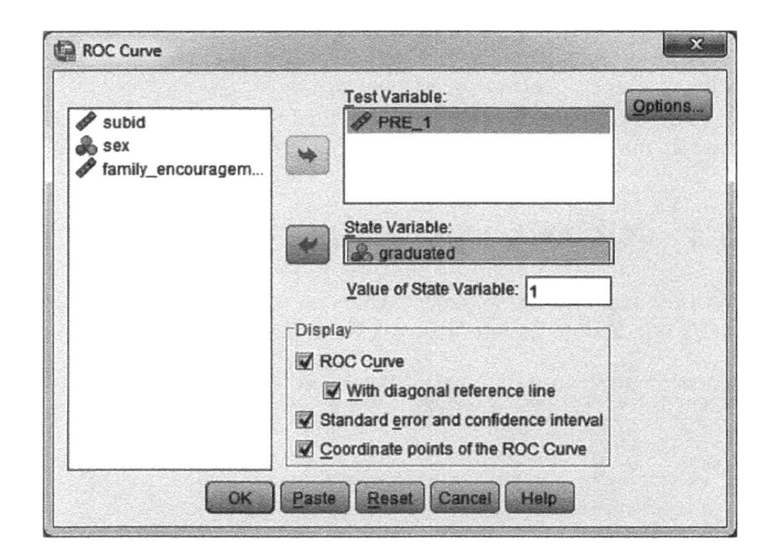

FIGURE 31.4
The main **ROC Curve** dialog window.

(these are the probabilities of target group membership) and **graduated** into the **State Variable** panel (this is the outcome variable). We indicate the code for the target group by entering **1** in the **Value of State Variable** panel. In the **Display** area, we select **ROC Curve** to display the curve, **With diagonal reference line** to include the reference line that is both traditional and useful, **Standard error and confidence interval** primarily to obtain the area under the curve, and **Coordinate points of the ROC Curve** to obtain our possible decision criteria. Click **OK** to perform the analysis.

31.6 ROC ANALYSIS: OUTPUT

The **ROC Curve** is presented in Figure 31.5. **Sensitivity** (true positive rate) is represented in the *Y*-axis, and **1 − Specificity** (false-positive rate) is represented in the *X*-axis. The curve is bow shaped, indicating that the logistic function can differentiate the outcome groups; the diagonal reference line visually depicts the situation where the logistic model has no differentiating capability. The curve does not show the individual coordinates that compose it (but those values are in a table provided in the output).

The table below the curve provides the value for the area under the curve. Larger areas indicate better differentiation by the logistic model. Meyers et al. (2013) provide interpretation guidelines of areas under the curve (.5's indicate no discrimination; .6's, poor discrimination; .7's, acceptable/good discrimination; .8's, very good discrimination; and .9's, excellent discrimination); using these guidelines, an area of **.851** with a confidence interval of **.813–.889** would be considered to represent very good discrimination.

The coordinates of the ROC curve are shown in Figure 31.6. This information interfaces with the **Classification Table** (see Figure 31.1) in the following way. We know from the **Classification Table** that **87.1%** of the cases in the target group were correctly classified. This cell in the **Classification Table** represents true positives and is called **Sensitivity** in the ROC analysis. We have marked out the row containing this value in the table in Figure 31.6 (**Sensitivity** = **.871**), and it corresponds to a predicted

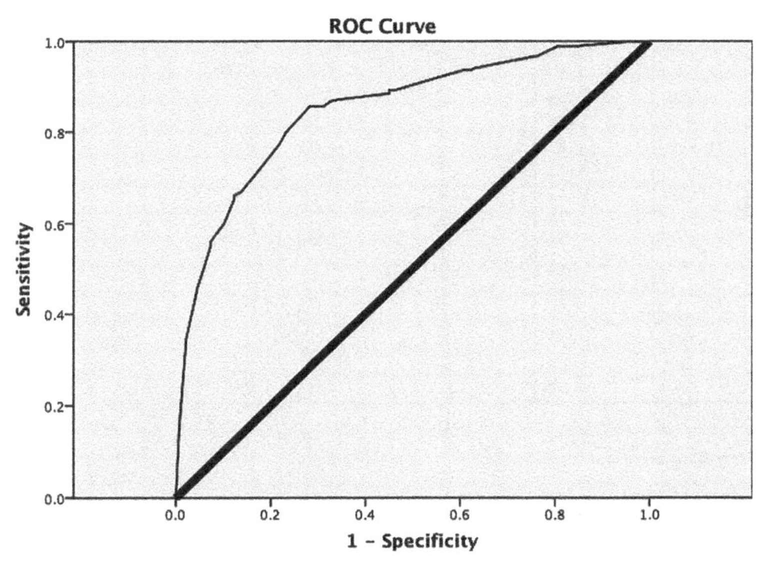

Area Under the Curve

Test Result Variable(s):PRE_1 Predicted probability

Area	Std. Error[a]	Asymptotic Sig.[b]	Asymptotic 95% Confidence Interval	
			Lower Bound	Upper Bound
.851	.019	.000	.813	.889

The test result variable(s): PRE_1 Predicted probability has at least one tie between the positive actual state group and the negative actual state group. Statistics may be biased.

a. Under the nonparametric assumption
b. Null hypothesis: true area = 0.5

FIGURE 31.5

The **ROC Curve** and the **Area Under the Curve**.

Coordinates of the Curve

Test Result Variable(s):PRE_1 Predicted probability

Positive if Greater Than or Equal To[a]	Sensitivity	1 - Specificity
.0000000	1.000	1.000
.0464985	1.000	.986
.0696374	1.000	.964
.1111302	.996	.920
.1478911	.989	.841
.1668427	.989	.833
.1876865	.989	.804
.2075295	.967	.761
.2322816	.967	.754
.2619223	.963	.739
.2868721	.941	.630
.3016161	.938	.623
.3208267	.938	.616
.3527496	.938	.609
.3819184	.893	.464
.3988760	.893	.449
.4204336	.890	.449
.4556147	.886	.449
.4912438	.871	.341
.5226910	.868	.326
.5405267	.857	.312
.5624066	.857	.297
.5931363	.857	.290
.6103469	.857	.283
.6271499	.798	.232
.6749427	.772	.217
.7209591	.706	.167
.7454681	.669	.138
.7716438	.662	.123
.7871285	.654	.123
.7987477	.599	.101
.8181266	.559	.072
.8472212	.548	.072
.8735717	.452	.043
.9004964	.449	.043
.9289251	.346	.022
.9422174	.342	.022
.9551810	.154	.007
1.0000000	.000	.000

The test result variable(s): PRE_1 Predicted probability has at least one tie between the positive actual state group and the negative actual state group.

a. The smallest cutoff value is the minimum observed test value minus 1, and the largest cutoff value is the maximum observed test value plus 1. All the other cutoff values are the averages of two consecutive ordered observed test values.

The default decision criterion of .5 did not fall at a place representing whole numbers of cases. IBM SPSS used the closest coordinate whose predicted probability was **.4912438**, a value pretty close to .5. We know this is the actual **cut value** because the **Sensitivity** (true positive rate) of **.871** matches the percentage of correctly classified target (**graduated**) group members in the **Classification Table** shown in Figure 31.3.

FIGURE 31.6

The table presenting the coordinates of the ROC curve.

probability value of **.4912438**, which is the actual decision criterion (the nominal cutoff value of .5 did not correspond to a whole number in the classification process).

The value of **.871** is one of the coordinates (the contribution of **Sensitivity**). The other coordinate is **.341** and represents the contribution of **1 – Specificity** (the false-positive rate). This can also be derived from the **Classification Table**. False positives are cases in the reference (**did not graduate**) group who were misclassified as target group (**graduated**) members. From Figure 31.1, we note that **47** of the **138** reference cases (47 + 91 = 138) were false positives, and **47/138** is **.341**, the value appearing as the **1 – Specificity** coordinate.

The coordinates table also allows us to evaluate alternative **Classification cutoffs**. As can be seen in the table, decision criteria near .2 and lower will result in a very high true positive (**Sensitivity**) rate, but the rate of false positives (**1 – Specificity**) would be extremely high as well; the reverse pattern may be seen with decision criteria near .94 or higher. Where researchers place the decision criterion really depends on what the value they place on the various outcomes. For example, when identifying true positives is a matter of life and death (perhaps in certain medical research studies) and if a high rate of false positives can be tolerated, then setting a very low decision criterion may be appropriate.

For our fictional data set, we are dealing with at-risk kids. Let us assume, even though our true positive rate is **.871**, that we want to achieve approximately a .90 hit rate. It appears from the coordinates table that a predicted probability of **.3819184** is associated with a **Sensitivity** (true positive) rate of **.893** and that a predicted probability of **.3527496** is associated with a **Sensitivity** (true positive) rate of **.938**; the corresponding rates of false positives (**1 – Specificity**) would then be **.464** and **.609**, respectively. Given the substantial jump in the false-positive rate, and given that **.893** is pretty close to our goal of .9, we select a **Classification cutoff** of **.3819184** for a follow-up logistic regression analysis.

31.7 BINARY LOGISTIC REGRESSION ANALYSIS: REVISED CLASSIFICATION CUTOFF

From the main menu, we select **Analyze ➔ Regression ➔ Binary Logistic** to open the main **Logistic Regression** window. We configure the analysis exactly as was done in Chapter 30, with the following exception. In the **Options** dialog window (see Figure 31.7), we now replace the default of **.5** with a **Classification cutoff** of **.3819184** (only a part of which can be seen in the panel). Click **Continue** to return to the main dialog window and click **OK** to perform the analysis.

The results of the analysis produced the identical logistic model to what was achieved in Chapter 30 and in Section 31.4, and we will not display those results (readers can

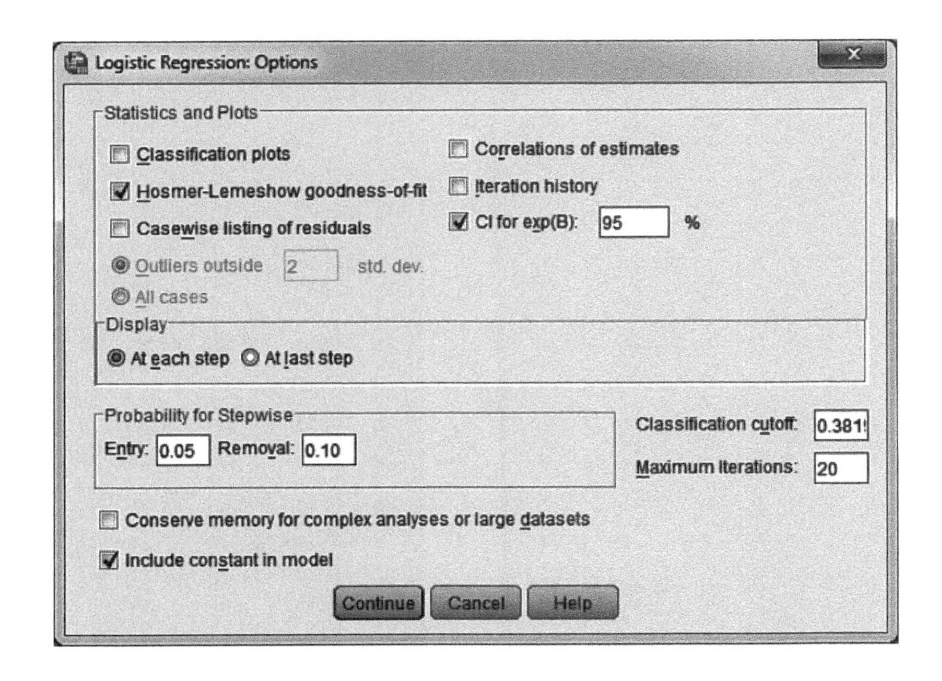

FIGURE 31.7

The **Options** dialog window of **Logistic Regression**.

Classification Table[a]

Observed			Predicted: graduated: 0 did not graduate	Predicted: graduated: 1 graduated	Predicted: Percentage Correct
Step 1	graduated	0 did not graduate	74	64	53.6
		1 graduated	29	243	89.3
	Overall Percentage				77.3

a. The cut value is .382

FIGURE 31.8 The **Classification Table** with the default **cut value** at **.382**.

verify this for themselves). The one set of results affected by revising the **Classification cutoff** is the classification portion of the analysis, and this is shown in Figure 31.8.

It can be seen in the table that our projected true positive rate of **.893** is precisely matched in our results. Furthermore, the false-positive rate of **.464** is also matched: **64** of **138** is 46.4%. Interestingly, our **Overall Percentage** of correct classification fell only slightly from **80%** previously to **77.3%** here. Thus, using this alternative decision threshold, we would be able to identify almost 90% of the students who will likely graduate based on the prediction model containing **sex** and the level of **family_encouragement** they receive.

CHAPTER 32

Multinominal Logistic Regression

32.1 OVERVIEW

Multinominal logistic regression extends binary logistic regression as covered in Chapter 30 to the situation where there are three or more outcome categories. Predictor variables can still be any combination of quantitative and binary variables.

Despite the presence of three or more outcome categories, logistic regression still calls for a binary prediction of group membership (we predict target group membership with respect to a reference group). This apparent contradiction is accommodated within the multinomial analysis by designating one of the groups (in the analysis setup) as the reference group. Each of the other groups serves as a target group and is compared to this reference group. Thus, with three outcome categories, two separate (binary logistic) sets of parameter estimates (the raw score coefficients and the odds ratios) are generated, one contrasting one of the outcomes to the reference category and another contrasting the other of the outcomes to the reference category. However, in the classification portion of the analysis, all outcome categories are considered together in that classification coefficients are generated for and applied to all groups, with the group achieving the highest score for each case determining the group to which that case is predicted to belong.

32.2 NUMERICAL EXAMPLE

Our example represents a portion of a data set provided by one of our graduate students, Kristine Christianson at the California State University, Sacramento; we have modified the data set somewhat for the purposes of our example. The data file is named **Kristine troubled students**.

Assume that we have identified three groups of students (under the outcome variable of **group**): students who are functioning at a level that is lower than optimal (**lower functioning** characterized by them experiencing relatively lower self-esteem and reporting relatively poorer interpersonal relationships), coded as **1** in the data file; students who appear to be **disorganized** in their personal and academic lives, coded as **2** in the data file; and students who appear to be functioning at a high level (**high functioning**), coded as **3** in the data file. We will designate the **high functioning** group as the reference group against which each of the other groups will be contrasted (serving as a target group).

Performing Data Analysis Using IBM SPSS®, First Edition.
Lawrence S. Meyers, Glenn C. Gamst, and A. J. Guarino.

Two quantitative variables and one categorical variable will be used as predictors. One quantitative predictor variable, parental criticism (**parent_criticism**), represents the extent to which the students believed that their parents are overly critical of them; scores range from 1 to 5, with higher scores reflecting more (perceived) parent criticism. The other quantitative predictor variable, a suppressive coping style (**suppressive_coping**), represents the tendency to deny problems and to avoid dealing with problems; scores range from 1 to 5, with higher scores reflecting more suppression.

The one categorical predictor variable is a characterization of the students in terms of the confidence they appear to exhibit; it is named **binary_confidence** in the data file. Students are classified as either exhibiting **high confidence** (coded as **0**) or **low confidence** (coded as **1**). This coding schema (intentionally chosen by us even if it seems as though it should be the reverse) automatically causes the **high confidence** group (the group with the lower code, i.e., **0**) to be the focus group and the **low confidence** group (the group with the higher code, i.e., **1**) to be the reference group.

32.3 ANALYSIS SETUP

From the main menu, we select **Analyze ➔ Regression ➔ Multinomial Logistic** to open the main **Multinomial Logistic Regression** window shown in Figure 32.1. We move **group** to the **Dependent** panel. IBM SPSS® has automatically designated the last category (ascending, i.e., from lower to higher values) in the coding schema as the **Reference Category**.

Although we do want this last category (**high functioning**, coded as **3** in the data file) to serve as our reference category, we select the **Reference Category** pushbutton to show that dialog screen; this is presented in Figure 32.2. If we had wished the lowest coded category to be the **Reference Category**, we would have selected **First Category**; if we had wished any other category to be the **Reference Category** (the middle category is the only other one in our example, but with more than three groups, there would be several to choose from), we would have selected **Custom** and typed in the group code. Click **Continue** to return to the main dialog window.

We now move **binary_confidence** into the **Factors** panel because it is a categorical variable, and we move **parent_criticism** and **suppressive_coping** into the **Covariate(s)**

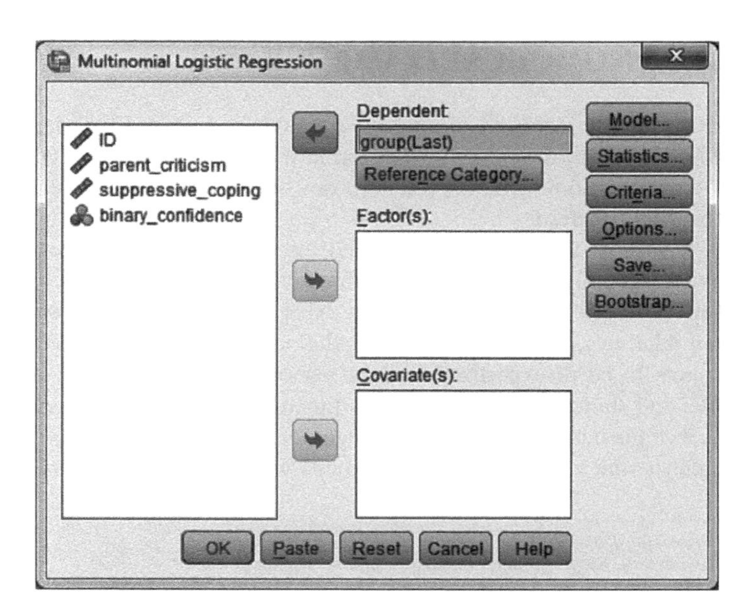

FIGURE 32.1

The main **Multinomial Logistic Regression** window.

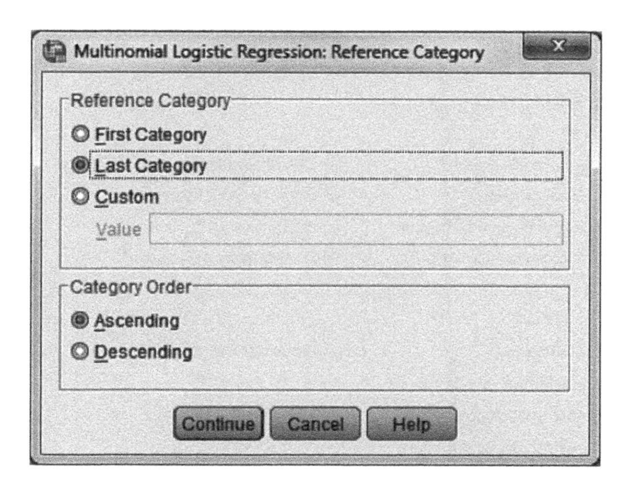

FIGURE 32.2

The **Reference Category** screen of **Multinomial Logistic Regression**.

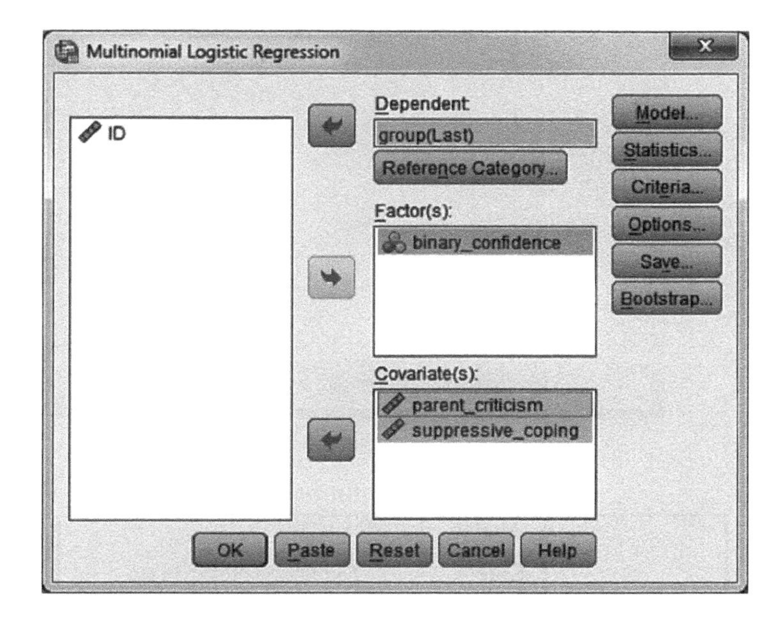

FIGURE 32.3

The main **Multinomial Logistic Regression** window now configured.

panel, as they are quantitative variables. These three variables are now specified as the predictors in our logistic regression analysis. This is shown in Figure 32.3.

Selecting the **Statistics** pushbutton gives us access to the **Statistics** window. As shown in Figure 32.4, we have checked **Case processing summary**, under **Model** we have marked the **Pseudo R-square**, **Step summary**, **Model fitting information**, and **Classification table**. Under **Parameters**, we have checked **Estimates** and **Likelihood ratio tests**. Click **Continue** to return to the main dialog window and click **OK** to perform the analysis.

32.4 ANALYSIS OUTPUT

The **Model Fitting Information** table in Figure 32.5 shows the model fit information. The **Final** model has a **−2 Log Likelihood** value of **605.91**, which is statistically significant ($p < .001$). This informs us that we can predict at a better than chance level using our set of predictors. The **Nagelkerke Pseudo R-Square** value, as seen in the bottom table

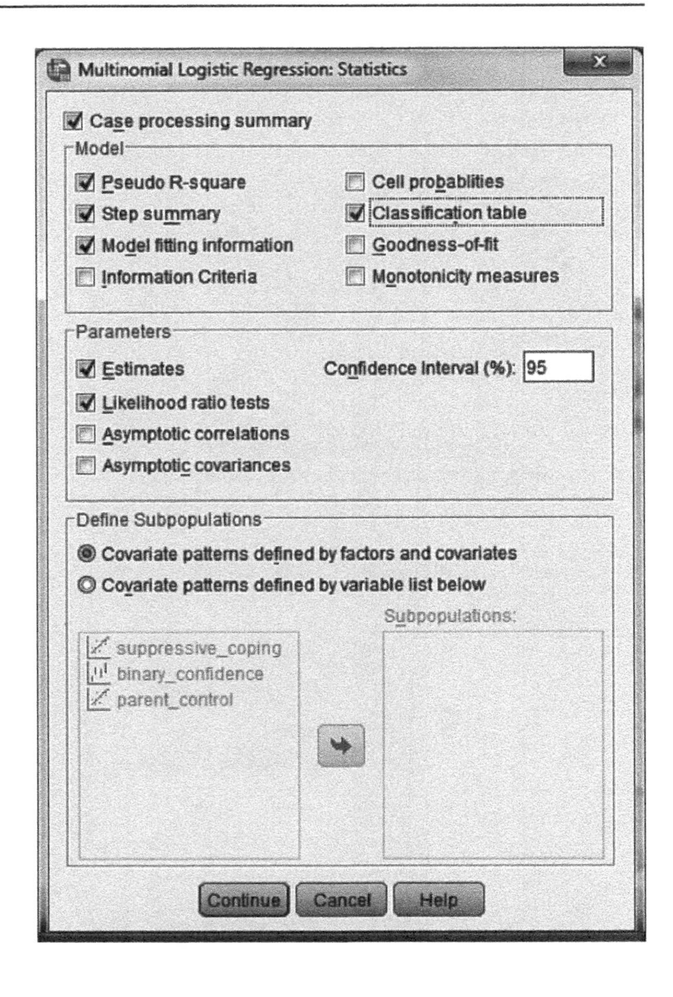

FIGURE 32.4
The **Statistics** window of **Multinomial Logistic Regression**.

Model Fitting Information

Model	Model Fitting Criteria	Likelihood Ratio Tests		
	-2 Log Likelihood	Chi-Square	df	Sig.
Intercept Only	722.41			
Final	605.91	116.50	6	.000

Pseudo R-Square

Cox and Snell	.245
Nagelkerke	.282
McFadden	.137

FIGURE 32.5
The **Model Fitting Information** table and **Pseudo R-Square** table.

in Figure 32.5, is **.282**. We interpret this to indicate that we are able to account for approximately 28% of the variance associated with student **group**.

Figure 32.6 shows the **Likelihood Ratio Tests**, presenting the consequences of removing one of the predictors from the final model. The **Likelihood Ratio Test** is what results when an effect is removed, and the reduced model is tested for statistical

Likelihood Ratio Tests

Effect	Model Fitting Criteria	Likelihood Ratio Tests		
	-2 Log Likelihood of Reduced Model	Chi-Square	df	Sig.
Intercept	605.91	.000	0	.
parent_criticism	615.30	9.388	2	.009
suppressive_coping	657.42	51.514	2	.000
binary_confidence	612.79	6.879	2	.032

The chi-square statistic is the difference in -2 log-likelihoods between the final model and a reduced model. The reduced model is formed by omitting an effect from the final model. The null hypothesis is that all parameters of that effect are 0.

a. This reduced model is equivalent to the final model because omitting the effect does not increase the degrees of freedom.

FIGURE 32.6

The **Likelihood Ratio Tests**.

significance with a chi-square procedure. Each of the last three rows considers removing the particular predictor named in the **Effect** column; hence, each is associated with **2** degrees of freedom because there are two predictors remaining in the model once the given effect is removed. For example, removing **parent_criticism** yields a **−2 Log Likelihood** value of **615.30** and with a corresponding chi-square value of **9.388**, results in a statistically significant predictive model based on the other two predictors (p = **.009**). As can be seen, the model remains statistically significant when each of the predictors is removed in turn.

The results of using the model to predict student **group** are shown in the **Parameter Estimates** table in Figure 32.7. Recall that our reference group was the **high functioning** group. Each of the major rows reports the results of a contrast between one of the other groups and the **high functioning** group.

The **B** column provides the raw score coefficients (adjusted for the presence of the other predictors in the model) associated with each of the predictors, and the standard

Parameter Estimates

group[a]		B	Std. Error	Wald	df	Sig.	Exp(B)	95% Confidence Interval for Exp(B)	
								Lower Bound	Upper Bound
1 lower functioning	Intercept	-4.424	.608	52.909	1	.000			
	parent_criticism	.434	.143	9.252	1	.002	1.544	1.167	2.042
	suppressive_coping	1.192	.201	35.263	1	.000	3.294	2.222	4.882
	[binary_confidence=.00]	-.672	.287	5.461	1	.019	.511	.291	.897
	[binary_confidence=1.00]	0[b]	.	.	0	.	.	.	.
2 disorganized	Intercept	-3.290	.555	35.090	1	.000			
	parent_criticism	.150	.140	1.146	1	.284	1.162	.883	1.529
	suppressive_coping	1.065	.193	30.400	1	.000	2.901	1.987	4.236
	[binary_confidence=.00]	-.554	.272	4.131	1	.042	.575	.337	.980
	[binary_confidence=1.00]	0[b]	.	.	0	.	.	.	.

a. The reference category is: 3 high functioning.
b. This parameter is set to zero because it is redundant.

FIGURE 32.7 The **Parameter Estimates** table.

error (**Std. Error**) of these statistics is shown next to the coefficients. These partial regression coefficients are tested for statistical significance by means of the **Wald** test, and the outcome of these tests is shown in the **Sig**. column. The odds ratio, which is the primary part of the output that we ordinarily interpret, is shown as **Exp(B)**.

The first major row labeled **lower functioning** contrasts the **lower functioning** group with the **high functioning** group. The raw score coefficients associated with **parent_criticism** and **suppressive_coping** are both positive and statistically significant. The direction of the effects are reasonable in that it makes sense that students who have lower levels of self-esteem and who have interpersonal difficulties would tend to more strongly believe that their parents were overly critical and would tend to deny problems and avoid dealing with problems more so than students who are functioning well (these coefficients and odds ratios for a given group are interpreted with respect to the reference group).

The odds ratios, adjusted for the other predictors in the model, yield an interpretation of the dynamics of these predictors. For example, **parent_criticism** is associated with an adjusted odds ratio of **1.544**; this informs us that an increase of 1 scale point in the **parent_criticism** measure increases the odds of a student being in the **lower functioning** group by **1.544** versus the odds of being in the **high functioning** group, controlling for the other predictors. By the same token, **suppressive_coping** is associated with an adjusted odds ratio of **3.294**; this informs us that an increase of 1 in the **suppressive_coping** measure increases the odds of a student being in the **lower functioning** group by **3.294** versus the odds of being in the **high functioning** group, controlling for the other predictors.

The results of our categorical **binary_confidence** predictor are shown in the bottom two rows of the **lower functioning** portion of the **Parameter Estimates** table. Two rows are used, as IBM SPSS explicitly acknowledges both of the binary codes, but only the row for **binary_confidence** = **.00** shows viable values. This reminds us that the code of **0** represents the focus group and that the code of **1** represents the reference group; values are associated with the focus group in the output.

We note that the partial raw regression coefficient is negative, and this is one way for researchers to reassure themselves that the coding schema is in accord with their intentions. Here, we would expect that more confident students (coded as **0** to designate them as the focus group) are less likely to be in the **lower functioning** group than in the **high functioning** group; thus, the negative coefficient is reasonable. The adjusted odds ratio of **.511** reinforces this assessment. We interpret the odds ratio as indicating that the odds for **high confidence** students (the focus group) to be in the **lower functioning** group are about half the odds for them to be in the **high functioning** group, controlling for the other predictors.

The second major row in Figure 32.7 labeled as **disorganized** contrasts the **disorganized** group with the **high functioning** group. As was true for the previous analysis, the raw score coefficients associated with **parent_criticism** and **suppressive_coping** are both positive and statistically significant. The adjusted odds ratio for **parent_criticism** is **1.162**, informing us that an increase of 1 in the **parent_criticism** measure increases the odds of a student being in the **disorganized** group by **1.162** versus the odds of being in the **high functioning** group, controlling for the other predictors.

The adjusted odds ratio for **suppressive_coping** is **2.901**. This informs us that an increase in 1 in the **suppressive_coping** measure increases the odds of a student being in the **disorganized** group by almost three times the odds of being in the **high functioning** group, controlling for the other predictors.

The results for our categorical **binary_confidence** predictor are shown in the bottom two rows of the **disorganized** portion of the **Parameter Estimates** table. We note that the partial raw regression coefficient is negative here as well, an outcome that makes sense in that we would probably expect that more confident students are less likely to

Classification

Observed	Predicted			
	1 lower functioning	2 disorganized	3 high functioning	Percent Correct
1 lower functioning	42	16	40	42.9%
2 disorganized	28	9	63	9.0%
3 high functioning	21	13	182	84.3%
Overall Percentage	22.0%	9.2%	68.8%	56.3%

FIGURE 32.8 The **Classification** table.

be in the **disorganized** group than in the **high functioning** group. The adjusted odds ratio of **.575** reinforces this assessment. The odds ratio indicates that the odds for **high confidence** students (the focus group) to be in the **disorganized** group are **.575** the odds for **lower confidence** students to be in the **high functioning** group, controlling for the other predictors.

The **Classification Table** is also shown in Figure 32.8. It displays how well the model classifies cases into the three categories of the outcome variable. Overall, the predictive accuracy is 56.3%, but that percentage differed considerably across the student **groups**. Those in the **high functioning** group were most accurately predicted (**84.3%**), those in the **lower functioning** group were next most accurately predicted (**42.9%**), and those in the **disorganized** group were most poorly predicted (**9.0%**).

PART 10

SURVIVAL ANALYSIS

CHAPTER 33

Survival Analysis: Life Tables

33.1 OVERVIEW

Chapters 33–35 examine a family of techniques variously labeled as *survival analysis*, *failure analysis*, or *event history analysis* that present the analysis of *Life Tables*, the *Kaplan–Meier* (or *product-limit*) method, and the *Cox Regression* method. Survival analysis (the generic term we will use throughout these chapters) examines the time interval between two events. These events can range from mundane (e.g., the starting and stopping of a newspaper subscription, purchasing and replacing a tire) to important events (e.g., onset of a disease to death, birth to first sexual experience, release from jail to arrest or conviction for another crime).

At least two factors complicate the examination of these time intervals. First, the event of primary interest (e.g., first sexual experience, being convicted of another crime) does not occur for all cases during the study period; for example, not all of the adolescents under study become sexually active during the study time frame. Second, the period of observation can vary from one case to another; for example, over a 3-year period, tire purchases can vary from 1 to 36 months.

Owing to these complicating factors, simply calculating a mean or median time between the occurrences of two events is not sufficient to glean the useful information from the study. Instead, two primary types of survival analyses have evolved to examine duration data. One approach is to use actuarial (Merrell, 1947) or product-limit (i.e., Kaplan–Meier) life tables that partition a period into smaller time intervals, such as weeks, months, or years (Kaplan & Meier, 1958; Kleinbaum, 1996; Lee & Wang, 2003; Norusis, 2012). Probabilities can then be estimated for the occurrence of an event at a given interval. The life table method is presented in this chapter, and the Kaplan–Meier method is described in Chapter 34. A second approach, the Cox regression or Cox proportional hazards model (Cox, 1972; Cox & Oakes, 1984; Singer & Willett, 1993, 2003; Tekle & Vermunt, 2012), attempts to predict survival time from covariates or independent variables; this method is presented in Chapter 35.

33.2 NUMERICAL EXAMPLE

The present example examines a hypothetical data set for evaluating how long taekwondo students continue to enroll (survive) in martial arts training as measured in months. Three

Performing Data Analysis Using IBM SPSS®, First Edition.
Lawrence S. Meyers, Glenn C. Gamst, and A. J. Guarino.

variables are provided in this data set. Students are each given an identification number under the variable **student**. The variable **months_in_course** represents the number of months a student was enrolled in taekwondo training classes (until they dropped out) during an 18-month block of time; thus, students will have been enrolled in the training class for different amounts of time. Whether the students represented a **drop out** (coded as **0**) or were **still attending** (coded as **1**) is provided by the variable **survival_status**; such a variable is often thought of as a *censorship* variable. Observations are censored when we have incomplete information about the situation, and this can happen under several different circumstances. One such circumstance is when a case does not exhibit or experience the event of interest. In the present example, the event of interest is dropping out of the class. Thus, a code of **1** representing continuing enrollment is thought of as indicating (perhaps anti-intuitively) a *censored event or observation* and a code of **0** representing withdrawal from the class is thought of as indicating an *uncensored event or observation*. The data can be found in the file named **taekwondo training**.

33.3 ANALYSIS SETUP

Open **taekwondo training** and from the main menu select **Analyze➔ Survival➔ Life Tables**; this produces the **Life Tables** main dialog window shown in Figure 33.1. We have moved **months_in_course** to the **Time** panel at the top right portion of the dialog window. It is also necessary to specify the period over which we wish to perform the analysis. For the **Display Time Intervals** panel, we have entered **0 through 18 by 1** to request that the first 18 months are displayed **one** month at a time.

We have moved **survival_status** into the **Status** panel. Upon doing so, the expression **(? ?)** appears next to the variable, as we need to supply to IBM SPSS® the code or codes representing the occurrence of the event of primary interest. Selecting the **Define Event** pushbutton produces the **Define Event for Status Variable** dialog window shown in Figure 33.2. In this relatively simple data file, the code of **0** for the variable **survival_status** identifies **drop out**, the event of primary interest. We therefore select **Single value** and type **0** in that panel. After clicking **Continue**, we see a **(0)** after the variable name **survival_status** in the **Status** panel (Figure 33.3).

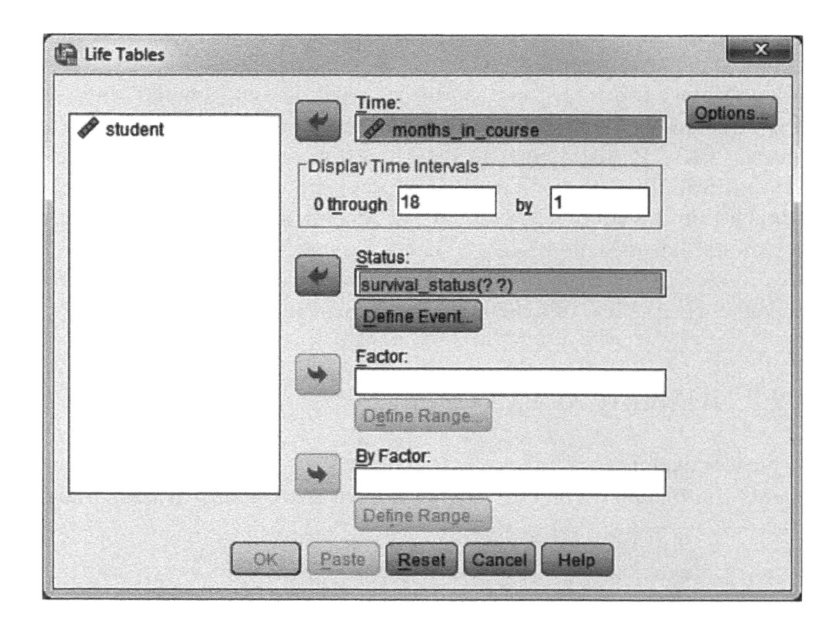

FIGURE 33.1
The main dialog window for **Life Tables**.

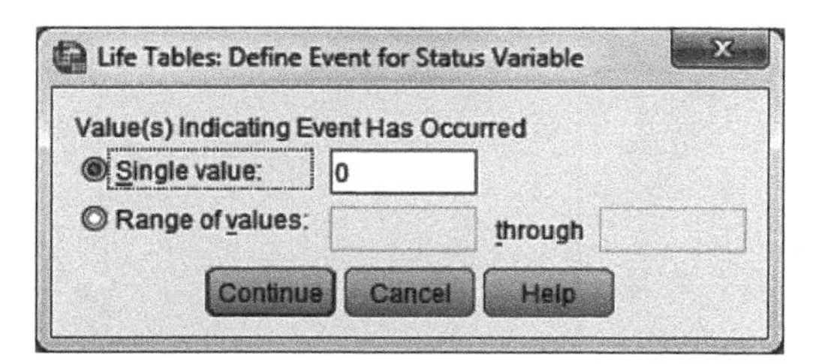

FIGURE 33.2
The **Define Event for Status Variable** window for **Life Tables**.

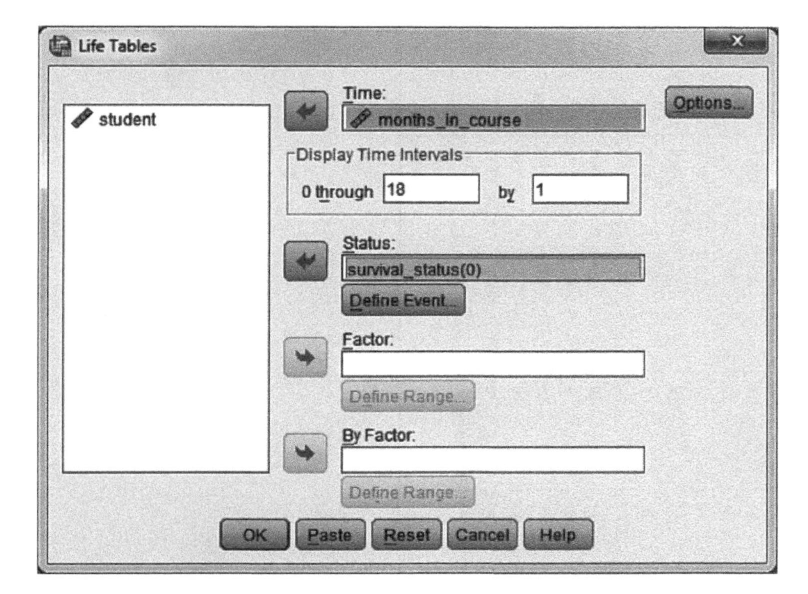

FIGURE 33.3
The main dialog window for **Life Tables** with **survival_status** now defined.

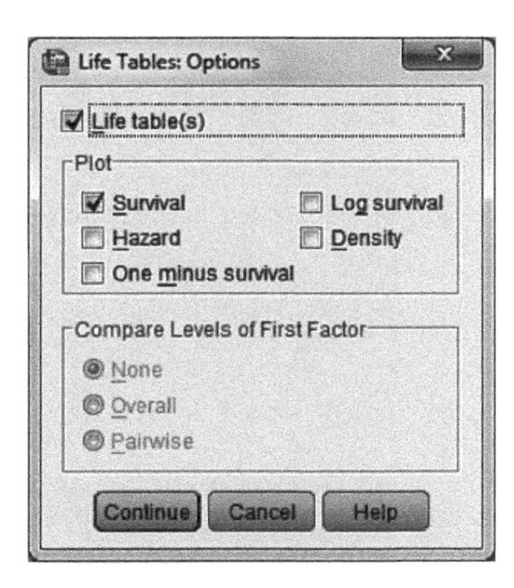

FIGURE 33.4
The **Options** window for **Life Tables**.

Selecting the **Options** pushbutton opens the **Options** dialog window shown in Figure 33.4. We check **Life table(s)** and **Survival** to obtain the cumulative survival function. We do not have different groups of cases, and so **Compare Levels of First Factor** options are not available. Select **Continue** to return to the main dialog window and click **OK** to perform the analysis.

Survival Variable: months_in_course

Life Table[a]

Interval Start Time	Number Entering Interval	Number Withdrawing during Interval	Number Exposed to Risk	Number of Terminal Events	Proportion Terminating	Proportion Surviving	Cumulative Proportion Surviving at End of Interval	Std. Error of Cumulative Proportion Surviving at End of Interval	Probability Density	Std. Error of Probability Density	Hazard Rate	Std. Error of Hazard Rate
0	20	0	20.000	0	.00	1.00	1.00	.00	.000	.000	.00	.00
1	20	0	20.000	0	.00	1.00	1.00	.00	.000	.000	.00	.00
2	20	1	19.500	1	.05	.95	.95	.05	.051	.050	.05	.05
3	18	0	18.000	0	.00	1.00	.95	.05	.000	.000	.00	.00
4	18	0	18.000	0	.00	1.00	.95	.05	.000	.000	.00	.00
5	18	0	18.000	1	.06	.94	.90	.07	.053	.051	.06	.06
6	17	1	16.500	0	.00	1.00	.90	.07	.000	.000	.00	.00
7	16	0	16.000	1	.06	.94	.84	.08	.056	.054	.06	.06
8	15	1	14.500	0	.00	1.00	.84	.08	.000	.000	.00	.00
9	14	1	13.500	0	.00	1.00	.84	.08	.000	.000	.00	.00
10	13	2	12.000	0	.00	1.00	.84	.08	.000	.000	.00	.00
11	11	2	10.000	0	.00	1.00	.84	.08	.000	.000	.00	.00
12	9	2	8.000	0	.00	1.00	.84	.08	.000	.000	.00	.00
13	7	1	6.500	0	.00	1.00	.84	.08	.000	.000	.00	.00
14	6	0	6.000	1	.17	.83	.70	.15	.140	.129	.18	.18
15	5	0	5.000	0	.00	1.00	.70	.15	.000	.000	.00	.00
16	5	1	4.500	0	.00	1.00	.70	.15	.000	.000	.00	.00
17	4	1	3.500	2	.57	.43	.30	.20	.400	.203	.80	.52
18	1	0	1.000	1	1.00	.00	.00	.00	.000	.000	.00	.00

a. The median survival time is 17.50

FIGURE 33.5 The **Life Table** output.

33.4 ANALYSIS OUTPUT

The **Life Table** is shown in Figure 33.5. This table consists of 13 columns of summary calculations. Each row is a 1-month period; the particular temporal period or interval is shown under **Interval Start Time**. The intervals always start at Time 0 and include the period up to 1 month. The second interval starts at 1 month but less than 2 months, and so on for 18 months. In our analysis setup, we instructed IBM SPSS to increment in 1-month intervals over the course of 18 months. With extended periods in their data, researchers could opt to increment over longer time intervals.

The **Number Entering Interval** lists the number of cases who continue or survive to the beginning of the current interval. We can see that **20** students (who comprised this example) continued for at least 2 months. At the 3-month interval, this number dropped to **18** indicating that two students did not survive till the end of the 3-month interval because they stopped coming to class or had only been with the taekwondo studio for less than 3 months.

The **Number Withdrawing during Interval** represents the number of cases who enter the interval minus one-half of those who withdrew during the interval. The **Number Exposed to Risk** estimates the *effective sample size* or how many cases were observed during the entire interval. For example, at the start of the second month interval, we have $(20 - (0.5 \times 1) = 19.5$. An assumption is made that when a case withdraws, it does so at the midpoint of the interval; hence, each case is weighted by one-half.

The **Number of Terminal Events** represents the number of cases for which the alternative or nonsurvival event (dropout) occurs within the interval. For example, at the start of the second month, we have one student dropping out of taekwondo training.

The **Proportion Terminating** represents the probability that nonsurvival (dropout) will occur during the interval for any case surviving to the beginning of that interval. For example, for the 16 cases entering the start of the seventh month of training, the probability that a student who has completed 7 months of training will quit at 8 months is .06, or 1/16 (the number of terminal events/number exposed to risk). The **Proportion Surviving** is calculated as $1 -$ proportion terminating. This value represents the probability that a person who enters an interval will leave the interval without the event (dropout) occurring. For example, for the 10-month interval we have $1 - .06 = .94$.

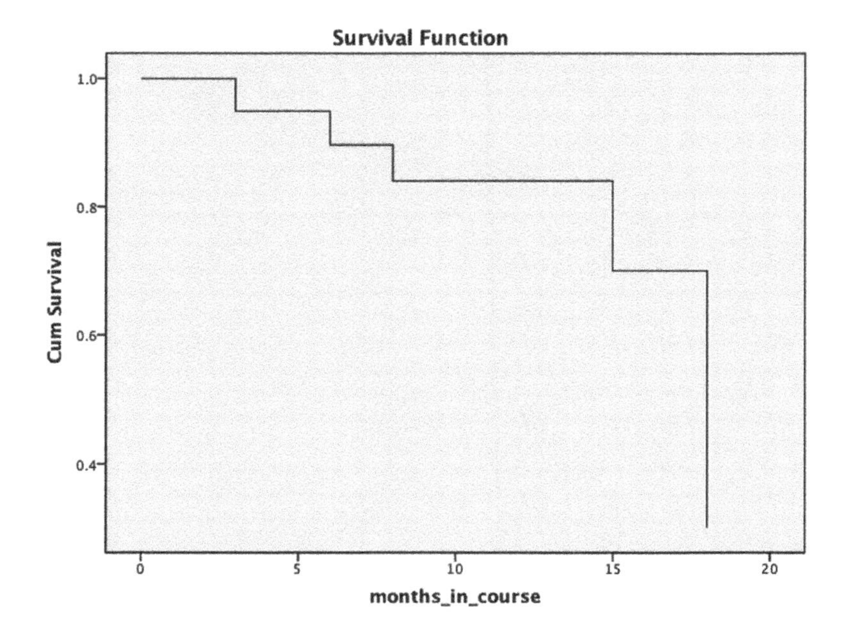

FIGURE 33.6
The **Survival Function** plot.

The **Cumulative Proportion Surviving at End of Interval** estimates the probability of surviving till the end of an interval. For example, the cumulative proportion of students "surviving" through the end of their twelfth month of martial arts instruction is .84 or 84%. This figure is often reported in survival analysis reports. The **Std. Error of Cumulative Proportion Surviving at End of Interval** provides a basis to estimate a confidence interval for this cumulative proportion.

The **Probability Density** function estimates the probability of a case experiencing nonsurvival (dropout) within the interval. The **Std. Error of Probability Density** estimates the variability of the density function. The **Hazard Rate** measures the instantaneous rate of change (dropout) at a specific time for cases surviving to the start of the interval. By multiplying the hazard rate by a small time increment, an approximation of nonsurvival (dropout) is provided for that specific time duration. The hazard rate is modeled in the Cox regression procedure (see Chapter 35). The **Std. Error of Hazard Rate** estimates the variability of the hazard rate.

IBM SPSS reports the **median survival time** as a footnote to its **Life Table**. In the present example, the median survival time is 17.50 months. This represents the point in time at which half the cases fail to survive (dropout).

The **Survival Function** or life table plot is shown in Figure 33.6. This plot is a visual depiction of the **Cumulative Proportion Surviving at End of Interval** column from the **Life Table** for each time interval. The cumulative proportion is represented on the *Y*-axis, and the time interval in months is represented on the *X*-axis.

CHAPTER 34

The Kaplan–Meier Survival Analysis

34.1 OVERVIEW

This second chapter in our survival analysis trilogy examines the Kaplan–Meier (or product-limit) method of producing life tables and survival functions. The Kaplan–Meier procedure (similar to the actuarial Life Tables procedure) provides proportions of cases surviving at various time intervals, survival functions, and tests of group differences. While the **Life Tables** procedure is often used with large sample sizes, the **Kaplan-Meier** procedure is typically the method of choice, particularly with smaller sample sizes.

34.2 NUMERICAL EXAMPLE

The present example examines an extension of the hypothetical taekwondo student enrollment data from Chapter 33. Five variables are provided in this data set contained in the data file named **taekwondo training K-M**. Students are each given an identification number under the variable **student**. The variable **months_in_course** represents the number of months a student was enrolled in taekwondo training classes (until they dropped out) during an 18-month period; thus, students will have been enrolled in the training class for different amounts of time. The variable **Survival_status** is an indicator of censorship, with students defined as **still attending** (coded as **1**, a censored event) or as a **drop out** (coded as **0**, an uncensored event). Student gender is coded as **1** for male and **2** for female under **sex_of_student**. Three different levels of instructors are also represented under **instructor_level**: black belt coded as **1**, master coded as **2**, and grand master coded as **3**.

34.3 ANALYSIS STRATEGY

In the previous **Life Tables** analysis (Chapter 33), we demonstrated how to estimate a cumulative survival function for one group of taekwondo students. In this chapter, we demonstrate how to compare survival functions for several groups (males and females) of taekwondo students. An additional set of analyses will examine survival functions by

Performing Data Analysis Using IBM SPSS®, First Edition.
Lawrence S. Meyers, Glenn C. Gamst, and A. J. Guarino.

student gender, which is nested or stratified by the student's type of instructor (black belt, master, and grand master).

34.4 ANALYSIS SETUP: COMPARING MALES AND FEMALES

Open the data file **taekwondo training K-M** and from the main menu select **Analyze ➔ Survival ➔ Kaplan-Meier**; this produces the **Kaplan-Meier** main dialog window shown in Figure 34.1.

We have moved **months_in_course** to the **Time** panel and **survival_status** to the **Status** panel. When a variable is entered into the **Status** panel, the **Define Event** push-button is activated. Clicking **Define Event** opens the **Define Event For Status Variable** window (see Figure 34.2). We enter **0** in the **Single value** panel to indicate that **drop out** is the event of interest.

Clicking **Continue** returns us to the main dialog window where we find that IBM SPSS® has placed a **(0)** after the variable **survival_status** in the **Status** panel (see Figure 34.3). If we wish to compare survival functions of two or more groups, we specify that variable as a **Factor**; it often but not always represents some causal treatment effect. For this example, we have moved the categorical variable **sex_of_student** into

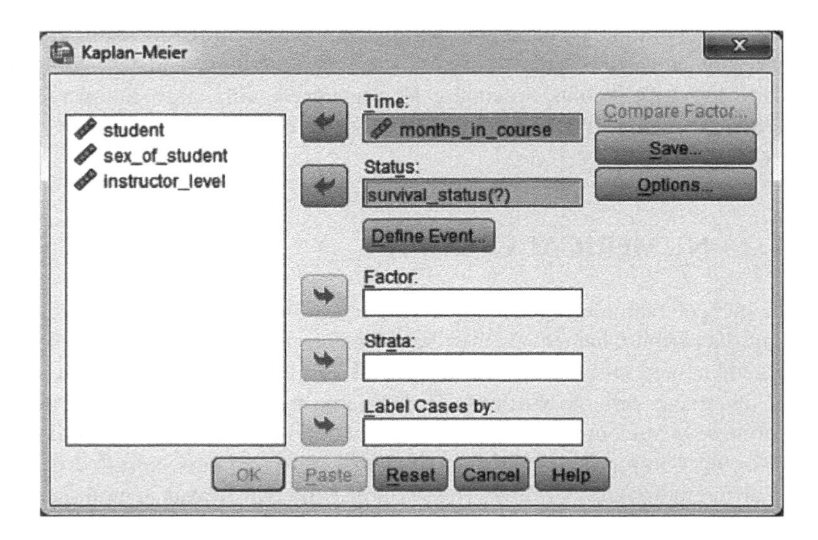

FIGURE 34.1
The main **Kaplan-Meier** dialog window ready to define **survival_status**.

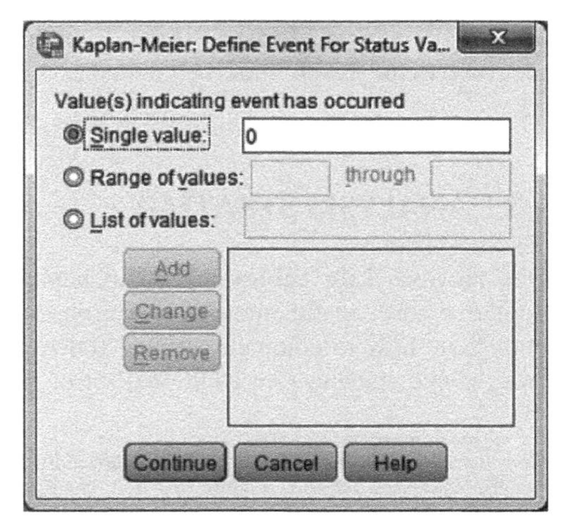

FIGURE 34.2
The **Define Event For Status Variable** window.

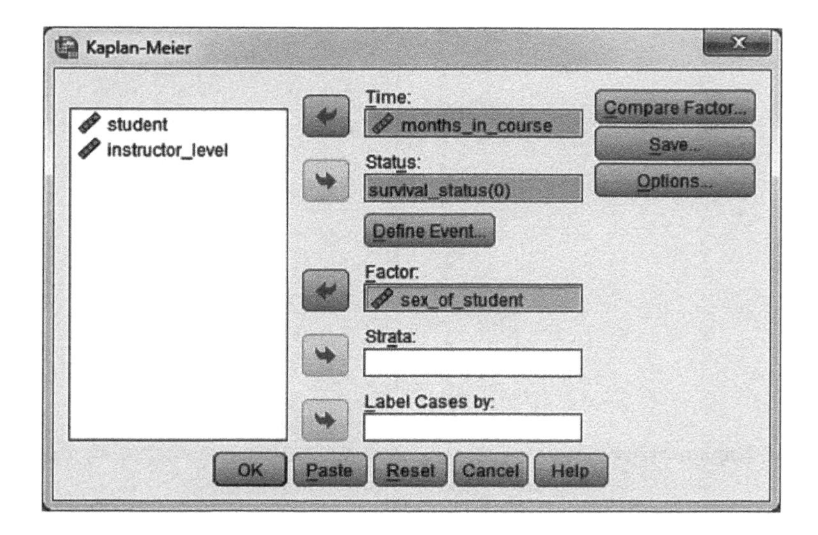

FIGURE 34.3

The main **Kaplan-Meier** dialog window.

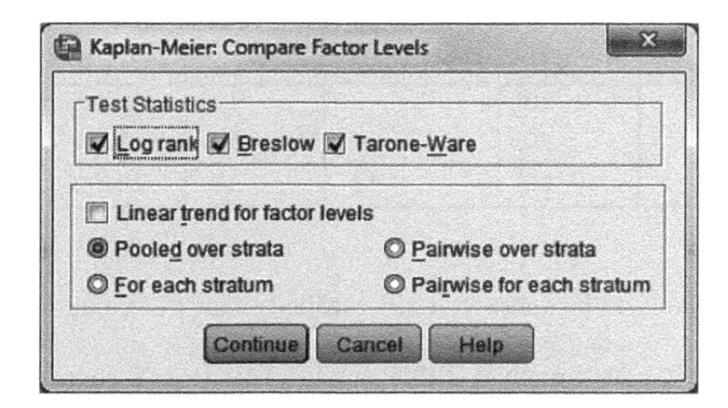

FIGURE 34.4

The **Compare Factor Levels** screen of **Kaplan-Meier**.

the **Factor** panel; this action has activated the **Compare Factor** pushbutton at the top right of the dialog window as shown in Figure 34.3. Clicking this pushbutton produces the **Compare Factor Levels** screen shown in Figure 34.4. We have activated the **Log rank**, **Breslow**, and **Tarone–Ware** options in the **Test Statistics** panel and also selected the **Pooled over strata** option in the lower panel. Clicking **Continue** returns us to the main dialog window.

From the main dialog window, clicking the **Options** pushbutton produces the **Options** dialog window as shown in Figure 34.5. We have retained the defaults in the **Statistics** panel: **Survival table(s)** and **Mean and median survival**, and we have activated **Survival** in the **Plots** panel. Click **Continue** to return to the main dialog window and click **OK** to perform the analysis.

34.5 ANALYSIS OUTPUT: COMPARING MALES AND FEMALES

Figure 34.6 provides the **Case Processing Summary** table that depicts the total number of cases for each **sex_of_student** group broken down by the observed number of events (**N of Events**) and the censored count. The event of interest is **drop out**, and there were six and seven of them for males and females, respectively; the censored event was **still attending**, and there were 24 males and 13 females. The last column provides the **Percent** of **Censored** observations showing that 80% of the males and 65% of the females remained in the course, and the number of events (or dropouts) for each group.

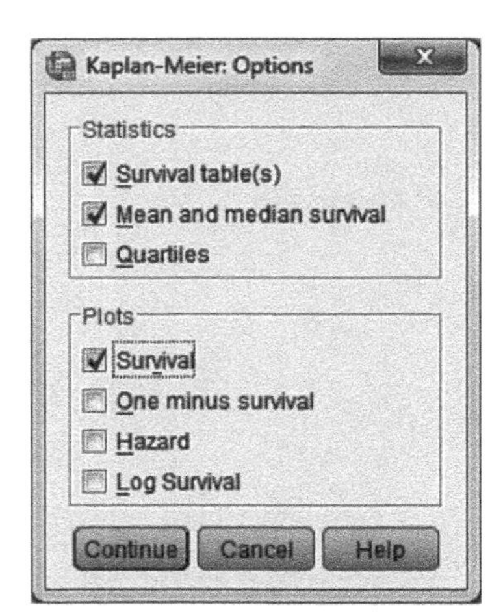

FIGURE 34.5
The **Options** window of **Kaplan-Meier**.

Case Processing Summary

sex_of_student	Total N	N of Events	Censored	
			N	Percent
1 male	30	6	24	80.0%
2 female	20	7	13	65.0%
Overall	50	13	37	74.0%

FIGURE 34.6
Case Processing Summary.

Female students at 35% were more likely to be censored (still enrolled) than were males at 20%.

Figure 34.7 presents the Kaplan–Meier **Survival Table** output. This table's seven columns of summary calculations are considerably more abbreviated than the 13-column **Life Table** we encountered in Chapter 33. The first column labeled **Student Gender** partitions the male and female cases in ascending order based on their individual survival time. Each case occupies its own row in the table. The second column labeled **Time** provides each case's survival time in months. The third column named **Status** indicates whether a case is censored (still attending class) or uncensored (having experienced the event of dropping out of the class).

Columns four and five provide an estimate and standard error of the **Cumulative Proportion Surviving at the Time**. This value provides the proportion of cases that have not reached the terminal event by the end of the interval. For example, approximately 96% of the males (1/30) remained in the course after the second month. The sixth column labeled **N of Cumulative Events** reports the number of cases that have experienced the terminal event (**drop out**) from time 1 until a particular time. The last column labeled **N of Remaining Cases** indicates the number of cases at that particular time who have not experienced the terminal event (**drop out**).

Figure 34.8 reports the mean and median survival time for male and female taekwondo students. We note that the median survival time for females is **17**.

Figure 34.9 presents the **Overall Comparisons** table that provides three overall (omnibus) tests of the equality of survival times across groups. The **Log Rank (Mantel–Cox)** tests the equality of survival functions by weighting all time points equally. The **Breslow (Generalized Wilcoxon)** tests the equality of survival functions by weighting all the points by the number of cases at risk at each point in time. Lastly, the **Tarone–Ware** tests the equality of survival functions by weighting all the points by the square root of the number of cases at risk at each point in time.

Survival Table

Student Gender		Time	Status	Cumulative Proportion Surviving at the Time		N of Cumulative Events	N of Remaining Cases
				Estimate	Std. Error		
Male	1	1.000	Still Attending	.	.	0	29
	2	1.000	Still Attending	.	.	0	28
	3	2.000	Drop-out	.964	.035	1	27
	4	2.000	Still Attending	.	.	1	26
	5	3.000	Drop-out	.927	.050	2	25
	6	3.000	Still Attending	.	.	2	24
	7	3.000	Still Attending	.	.	2	23
	8	4.000	Still Attending	.	.	2	22
	9	4.000	Still Attending	.	.	2	21
	10	4.000	Still Attending	.	.	2	20
	11	5.000	Drop-out	.881	.065	3	19
	12	6.000	Still Attending	.	.	3	18
	13	6.000	Still Attending	.	.	3	17
	14	6.000	Still Attending	.	.	3	16
	15	8.000	Still Attending	.	.	3	15
	16	9.000	Drop-out	.822	.083	4	14
	17	9.000	Still Attending	.	.	4	13
	18	9.000	Still Attending	.	.	4	12
	19	10.000	Still Attending	.	.	4	11
	20	10.000	Still Attending	.	.	4	10
	21	10.000	Still Attending	.	.	4	9
	22	12.000	Still Attending	.	.	4	8
	23	12.000	Still Attending	.	.	4	7
	24	12.000	Still Attending	.	.	4	6
	25	14.000	Drop-out	.685	.143	5	5
	26	15.000	Still Attending	.	.	5	4
	27	16.000	Drop-out	.514	.183	6	3
	28	16.000	Still Attending	.	.	6	2
	29	16.000	Still Attending	.	.	6	1
	30	18.000	Still Attending	.	.	6	0
Female	1	2.000	Drop-out	.950	.049	1	19
	2	4.000	Drop-out	.900	.067	2	18
	3	5.000	Still Attending	.	.	2	17
	4	6.000	Still Attending	.	.	2	16
	5	8.000	Still Attending	.	.	2	15
	6	8.000	Still Attending	.	.	2	14
	7	10.000	Still Attending	.	.	2	13
	8	10.000	Still Attending	.	.	2	12
	9	11.000	Still Attending	.	.	2	11
	10	11.000	Still Attending	.	.	2	10
	11	12.000	Still Attending	.	.	2	9
	12	12.000	Still Attending	.	.	2	8
	13	13.000	Still Attending	.	.	2	7
	14	14.000	Drop-out	.771	.132	3	6
	15	15.000	Drop-out	.643	.161	4	5
	16	17.000	Drop-out	.	.	5	4
	17	17.000	Drop-out	.386	.171	6	3
	18	17.000	Still Attending	.	.	6	2
	19	18.000	Drop-out	.193	.161	7	1
	20	18.000	Still Attending	.	.	7	0

FIGURE 34.7 The **Survival Table** for individual male and female students.

Means and Medians for Survival Time

sex_of_student	Mean[a]				Median			
			95% Confidence Interval				95% Confidence Interval	
	Estimate	Std. Error	Lower Bound	Upper Bound	Estimate	Std. Error	Lower Bound	Upper Bound
1 male	14.850	1.087	12.720	16.981	.	.	.	.
2 female	15.343	1.136	13.116	17.570	17.000	1.328	14.397	19.603
Overall	15.042	.778	13.517	16.566	17.000	.705	15.618	18.382

a. Estimation is limited to the largest survival time if it is censored.

FIGURE 34.8 **Means and Medians for Survival Time.**

Overall Comparisons

	Chi-Square	df	Sig.
Log Rank (Mantel-Cox)	.011	1	.917
Breslow (Generalized Wilcoxon)	.119	1	.731
Tarone-Ware	.098	1	.754

Test of equality of survival distributions for the different levels of sex_of_student.

FIGURE 34.9
Overall Comparisons based on the three tests of statistical significance of the equality of survival times across groups.

We recommend following the results of the log rank test, as it is associated with greater statistical power under most circumstances (see Norusis (2012) and Prentice & Marek (1979)), but in our example, the significance values of all three tests are all greater than .05. We therefore conclude that there is not a statistically significant difference between male and female survival time.

These findings are reinforced in the **Survival Functions** shown in Figure 34.10. This plot gives a visual representation of the life tables. The horizontal axis represents the number of months enrolled in a taekwondo class or time to event. The vertical axis shows the probability of survival or not dropping out of taekwondo instruction.

34.6 ANALYSIS SETUP: COMPARING MALES AND FEMALES WITH STRATIFICATION

In this example, we will perform the same analysis as before, except that we now include a stratification variable of **instructor_level**. This analysis will compare the two **sex_of_student** groups over strata or each of the three levels of **instructor_level**. Open the data file **taekwondo training K-M** and from the main menu select **Analyze ➔ Survival ➔ Kaplan-Meier**; this produces the **Kaplan-Meier** main dialog window shown in Figure 34.11. As we did for the previous analysis, we have moved **months_in_course** to the **Time** panel, **survival_status** to the **Status** panel (and set **Define Event For Status Variable** to **0**), and **sex_of_student** to the **Factor** panel. In the **Strata** box, we have moved **instructor_level** to enable us to examine the two genders within each of the instructor types.

As shown in Figure 34.12, in the **Compare Factor Levels** screen, we have activated the **Log rank**, **Breslow**, and **Tarone-Ware** options in the **Test Statistics** panel and selected the **Pooled over strata** option in the lower panel. We will activate (in separate runs) four of the factor level options: **Pooled over strata, For each stratum, Pairwise over strata**, and **Pairwise for each stratum** to demonstrate the output for each option, with this being the first run. Note that each option must be run in separate IBM SPSS Kaplan–Meier analyses.

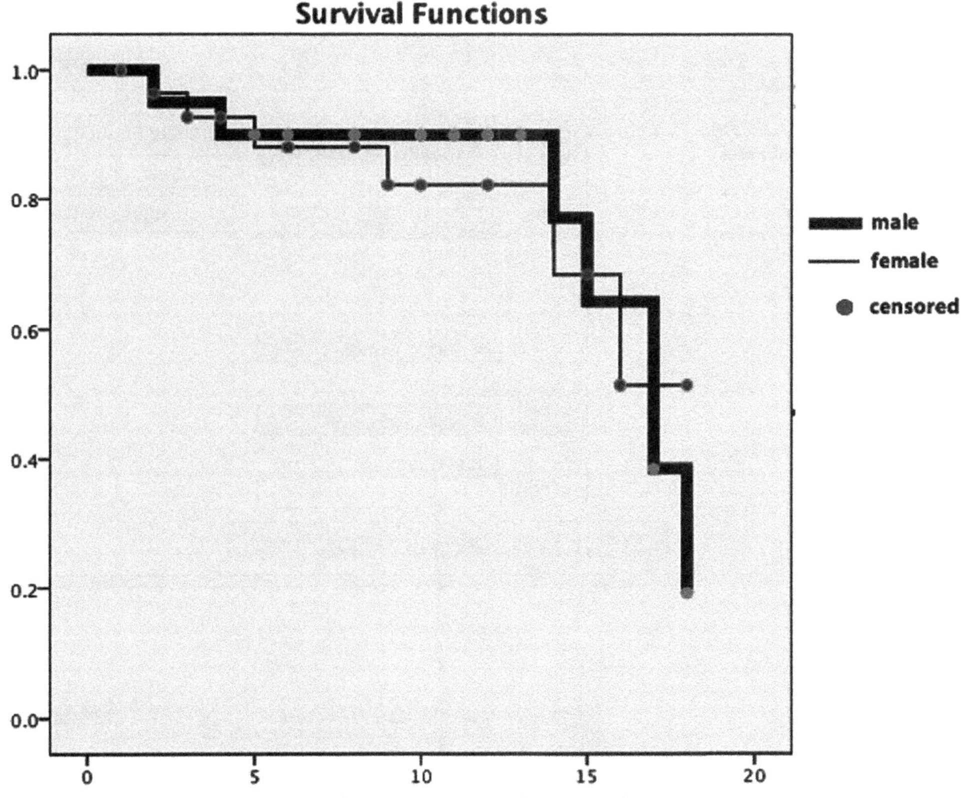

FIGURE 34.10 **Survival Functions** of male and female students.

In the **Options** dialog window, shown in Figure 34.13, we have activated only the **Survival** checkbox in the **Plots** panel (to simplify the output). Clicking **Continue** returns us to the main dialog window and clicking **OK** performs the analysis.

34.7 ANALYSIS OUTPUT: COMPARING MALES AND FEMALES WITH STRATIFICATION

Figure 34.14 displays the **Case Processing Summary** that partitions the 50 cases among the two levels of **Sex** and three levels of **Instructor**. The levels of the **Factor** variable (**sex_of_student**) are tabulated within each level of the **Strata** variable (**instructor_level**), and so subtotals (in the **Overall** rows) are provided only for each **instructor_level**.

Figure 34.15, Figure 34.16, Figure 34.17, and Figure 34.18 are **Test Statistics** output based on four separate IBM SPSS **Kaplan-Meier** runs where we activated (one at a time) one of the four **Compare Factor Levels** options, namely, **Pooled over strata, Pairwise over strata, For each stratum**, and **Pairwise for each stratum**, to give a sense of what the output looks like.

Figure 34.15 presents the **Overall Comparisons** when the **Compare Factor Levels** option is set at **Pooled over strata**. The three statistical tests (**Log Rank**, **Breslow**, **Tarone–Ware**) test for the equality of survival functions for the two levels of **sex_of_student**, but the results are different from our earlier analysis (see Figure 3.9) because they have been adjusted for the **Strata** variable **instructor_level**. None of the tests reached

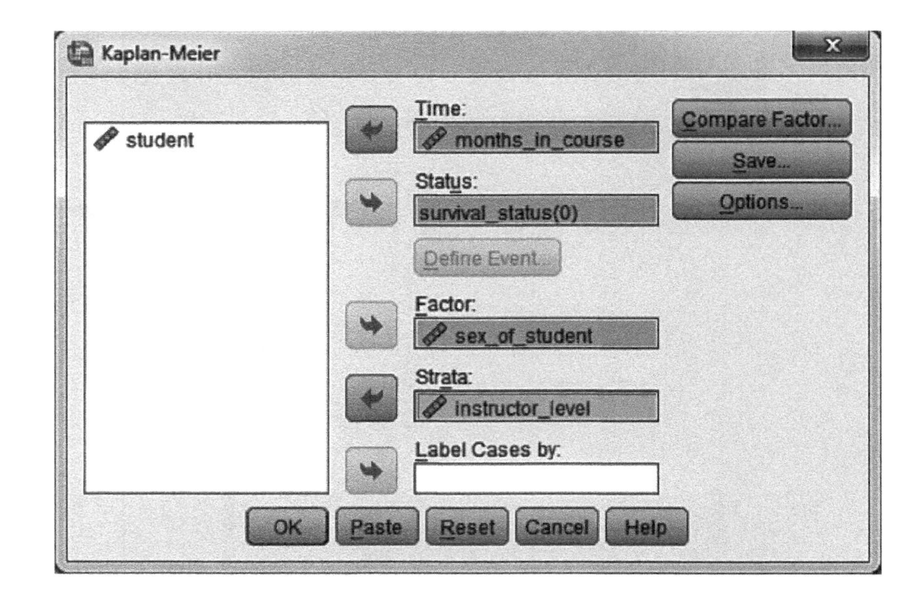

FIGURE 34.11

The main **Kaplan-Meier** dialog window set up with a **Strata** variable.

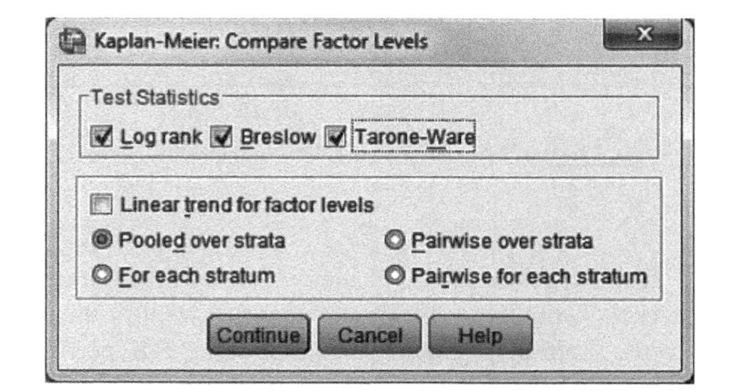

FIGURE 34.12

The **Compare Factor Levels** screen of **Kaplan-Meier**.

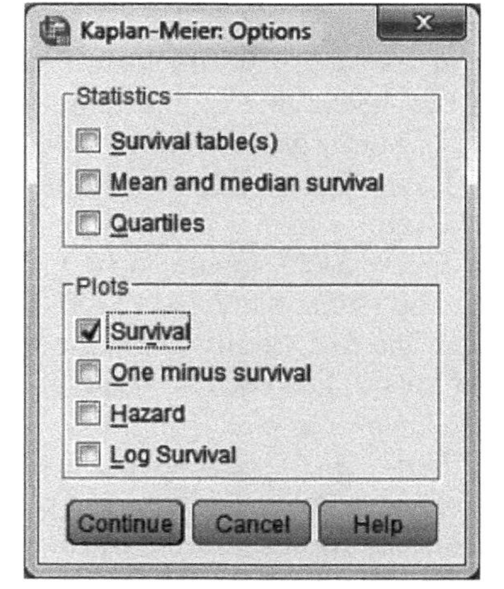

FIGURE 34.13

The **Options** window of **Kaplan-Meier**.

Case Processing Summary

instructor_level	sex_of_student	Total N	N of Events	Censored N	Censored Percent
1 blackbelt	1 male	14	3	11	78.6%
	2 female	5	2	3	60.0%
	Overall	19	5	14	73.7%
2 master	1 male	7	1	6	85.7%
	2 female	6	1	5	83.3%
	Overall	13	2	11	84.6%
3 grandmaster	1 male	9	2	7	77.8%
	2 female	9	4	5	55.6%
	Overall	18	6	12	66.7%
Overall	Overall	50	13	37	74.0%

FIGURE 34.14 **Case Processing Summary.**

Overall Comparisons[a]

	Chi-Square	df	Sig.
Log Rank (Mantel-Cox)	.046	1	.831
Breslow (Generalized Wilcoxon)	.027	1	.869
Tarone-Ware	.042	1	.838

Test of equality of survival distributions for the different levels of sex_of_student.

a. Adjusted for instructor_level.

FIGURE 34.15
Overall Comparisons based on the three tests of statistical significance of the equality of survival times across **sex_of_student** groups adjusted for **instructor_level**.

Pairwise Comparisons[a]

	sex_of_student	1 male Chi-Square	1 male Sig.	2 female Chi-Square	2 female Sig.
Log Rank (Mantel-Cox)	1 male			.046	.831
	2 female	.046	.831		
Breslow (Generalized Wilcoxon)	1 male			.027	.869
	2 female	.027	.869		
Tarone-Ware	1 male			.042	.838
	2 female	.042	.838		

a. Adjusted for instructor_level.

FIGURE 34.16 The adjusted **Pairwise Comparisons** of males and females based on the three tests of statistical significance of the equality of survival times collapsed across **instructor_level**.

Overall Comparisons

instructor_level		Chi-Square	df	Sig.
1 blackbelt	Log Rank (Mantel-Cox)	.104	1	.747
	Breslow (Generalized Wilcoxon)	.255	1	.614
	Tarone-Ware	.188	1	.665
2 master	Log Rank (Mantel-Cox)	.500	1	.480
	Breslow (Generalized Wilcoxon)	.500	1	.480
	Tarone-Ware	.500	1	.480
3 grandmaster	Log Rank (Mantel-Cox)	.089	1	.766
	Breslow (Generalized Wilcoxon)	.065	1	.798
	Tarone-Ware	.000	1	.994

Test of equality of survival distributions for the different levels of sex_of_student.

FIGURE 34.17 **Overall Comparisons** of **sex_of_student** based on three tests of statistical significance of the equality of survival times for each **instructor_level**.

Pairwise Comparisons

			1 male		2 female	
	instructor_level	sex_of_student	Chi-Square	Sig.	Chi-Square	Sig.
Log Rank (Mantel-Cox)	1 blackbelt	1 male			.104	.747
		2 female	.104	.747		
	2 master	1 male			.500	.480
		2 female	.500	.480		
	3 grandmaster	1 male			.089	.766
		2 female	.089	.766		
Breslow (Generalized Wilcoxon)	1 blackbelt	1 male			.255	.614
		2 female	.255	.614		
	2 master	1 male			.500	.480
		2 female	.500	.480		
	3 grandmaster	1 male			.065	.798
		2 female	.065	.798		
Tarone-Ware	1 blackbelt	1 male			.188	.665
		2 female	.188	.665		
	2 master	1 male			.500	.480
		2 female	.500	.480		
	3 grandmaster	1 male			.000	.994
		2 female	.000	.994		

FIGURE 34.18 **Pairwise Comparisons** of males and females based on the three tests of statistical significance of the equality of survival times for each **instructor_level**.

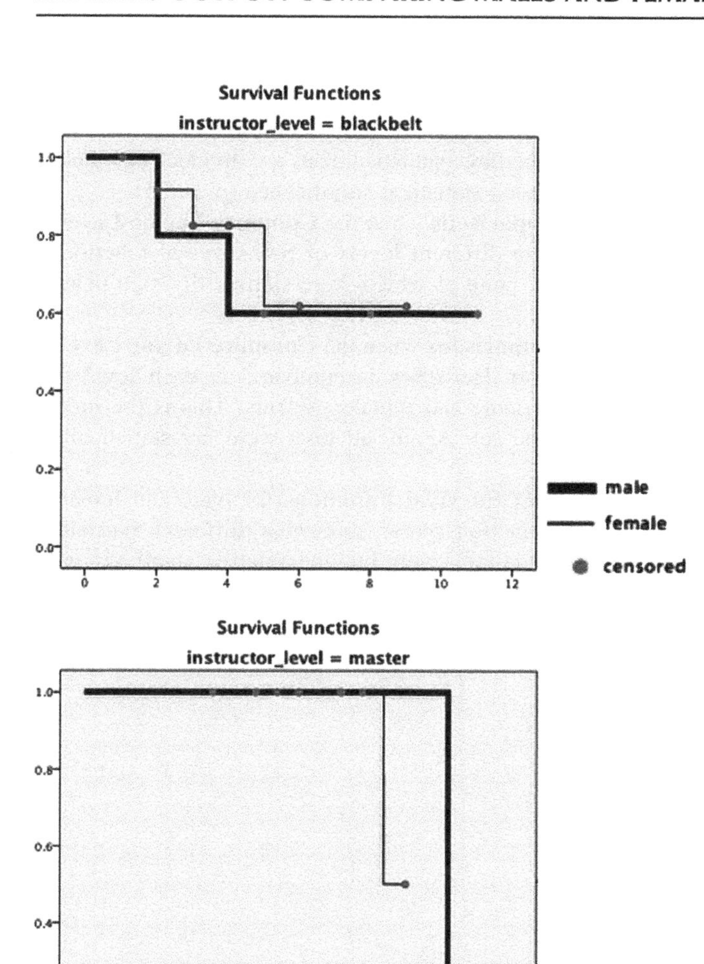

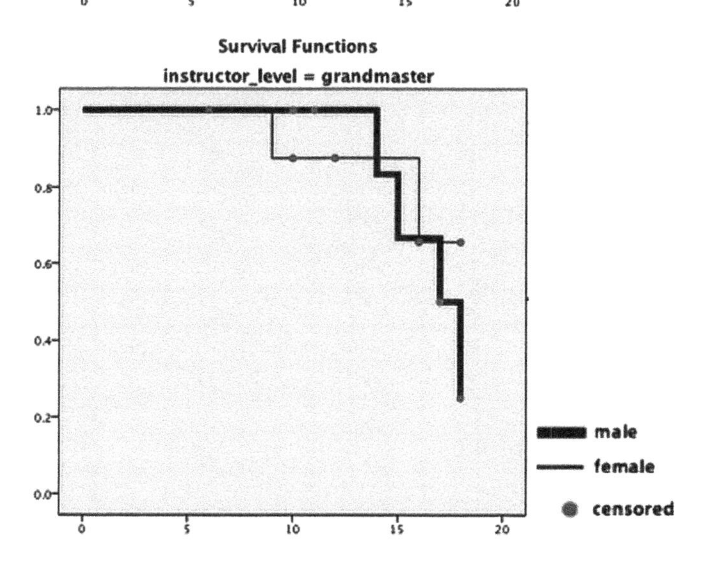

FIGURE 34.19

Survival Functions of male and female students for each **instructor_level**.

statistical significance ($p > .05$), indicating overall equivalence of survival functions for males and females.

Figure 34.16 presents the **Pairwise Comparisons** when the **Compare Factor Levels** option is set at **Pairwise over strata**. Here the three statistical tests are divided separately for males and females, none of which achieved statistical significance ($p > .05$).

Figure 34.17 provides the **Overall Comparisons** when the **Compare Factor Levels** option is set at **For each stratum**. Here the different levels of **Sex** survival functions are compared within each **instructor_level**, none of which were statistically significant ($p > .05$).

Figure 34.18 presents the **Pairwise Comparisons** when the **Compare Factor Levels** option is set at **Pairwise for each stratum**. Here **Sex** is compared at each level of **Instructor** for the three statistical tests for male and female students. This is the most comprehensive (microscopic) analysis of the set. Again, all tests were not statistically significant ($p > .05$).

Figure 34.19 presents the Kaplan–Meier **Survival Functions** for male and female students at each **instructor_level**. The functions look somewhat different partially because the students are at different proficiency levels but the relative small sample sizes in some of the cells make those frequencies a good deal less stable bases for projecting survival rates.

CHAPTER 35

Cox Regression

35.1 OVERVIEW

This final chapter in our survival analysis trilogy examines the Cox regression procedure (or Cox proportional hazards model), as it pertains to time interval data that contain censored observations. It was proposed by Cox (1972) to predict the occurrence of an event of interest at a particular time. This procedure is useful for modeling the time to a specified event (e.g., drop out of class). The Cox regression procedure assumes that the independent or predictor variables (called *covariates* by IBM SPSS®) are *time-constant* (i.e., they do not vary as a function of time). IBM SPSS also offers a separate procedure for Cox regression with a Time-Dependent Covariate that we do not cover.

The model developed by Cox (1972) can be formulated in terms of a *cumulative survival function* (i.e., the proportion of cases surviving to a particular time point) and the *hazard function or ratio* (i.e., the rate per unit time a case will experience the event given it survived to that point). Due in part to the pedagogical parsimony of the hazard function, the Cox model is typically expressed in hazard form as the Cox proportional hazards model (see Norusis (2012)).

The key ingredients for IBM SPSS **Cox Regression** are the following: a dichotomously coded **Status** variable that serves as the dependent measure; a **Time** variable (either continuous or categorical) that assesses the duration to the event defined by the **Status** variable; and **Covariates**, independent or predictor variables that may be either categorical or continuous. Interaction terms can also be used. Categorical covariates with three or more levels are automatically converted into a set of dummy variables, with one level being used as a reference category.

35.2 NUMERICAL EXAMPLE

The present example further extends our hypothetical taekwondo student enrollment data from Chapter 34. Six variables are provided in this data set contained in the data file named **taekwondo training CR**. Students are each given an identification number under the variable **student**. The variable **months_in_course** represents the number of months a student was enrolled in taekwondo training classes (until they dropped out) during an 18-month period; thus, students will have been enrolled in the training class for

Performing Data Analysis Using IBM SPSS®, First Edition.
Lawrence S. Meyers, Glenn C. Gamst, and A. J. Guarino.

different amounts of time. The variable **Survival_status** is an indicator of censorship, with students defined as **still attending** (coded as **1**, a censored event) or as a **drop out** (coded as **0**, an uncensored event). Student gender is coded as **1** for male and **2** for female under **sex_of_student**. Three different levels of instructors are also represented under **instructor_level**: black belt coded as **1**, master coded as **2**, and grand master coded as **3**. The final variable is **age** and refers to the age of the student in years.

35.3 ANALYSIS SETUP

Open the data file named **taekwondo training CR** and from the main menu select **Analyze ➔ Survival ➔ Cox Regression**; this produces the **Cox Regression** main dialog window as shown in Figure 35.1.

We have moved **months_in_course** to the **Time** panel and **survival_status** to the **Status** panel. When a variable is entered into the **Status** panel, the **Define Event** push-button is activated. Clicking **Define Event** opens the **Define Event for Status Variable** window (see Figure 35.2). We enter **0** in the **Single value** panel to indicate that **drop out** is the event of interest.

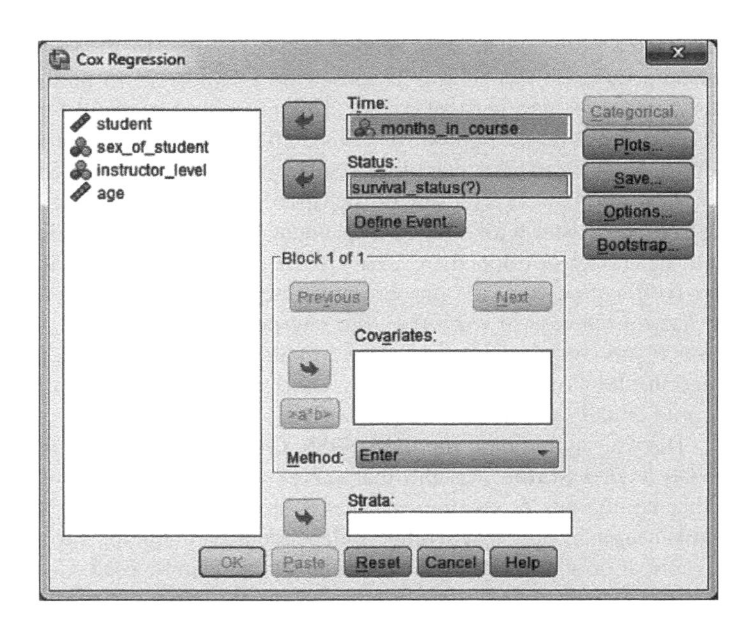

FIGURE 35.1

The main **Cox Regression** dialog window ready to define **survival_status**.

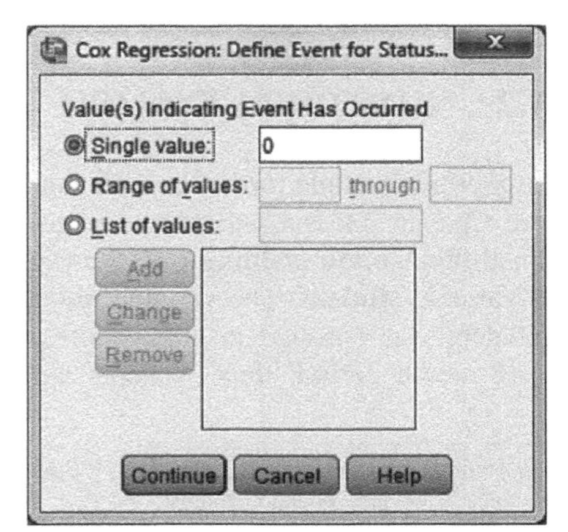

FIGURE 35.2

The **Define Event for Status Variable** window.

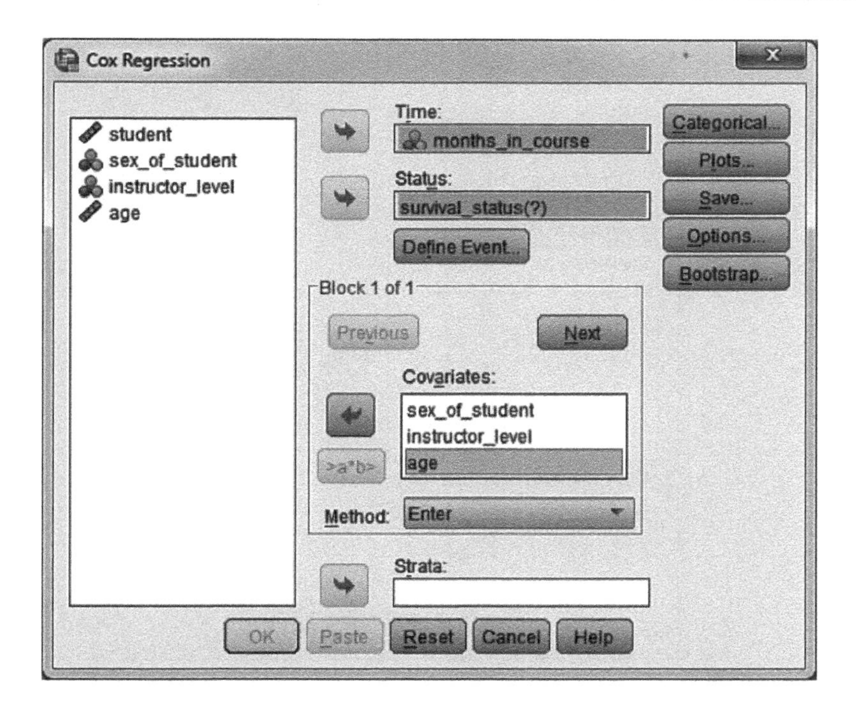

FIGURE 35.3

The main **Kaplan-Meier** dialog window with the **Covariates** specified.

Clicking **Continue** returns us to the main dialog window where we find that IBM SPSS has placed a **(0)** after the variable **survival_status** in the **Status** panel (see Figure 35.3). In the **Covariates** panel, we have clicked over our independent or predictor variables (**sex_of_student, instructor_level**, and **age**). We retain the default of **Enter** in the **Method** drop-down menu to produce a standard regression solution.

Selecting the **Categorical** pushbutton produces the **Define Categorical Covariates** dialog window (see Figure 35.4). We have moved **instructor_level** and **sex_of_student** to the **Categorical Covariates** panel because they are categorical variables. As can be seen at the bottom of the window, IBM SPSS uses as a default the last (highest numeric code) as the reference category in its dummy coding operation (relevant for variables with three or more categories). We leave that as the default (highlighting the variable activates the choice, thus allowing us to modify it). **Age** remains in the **Covariates** panel,

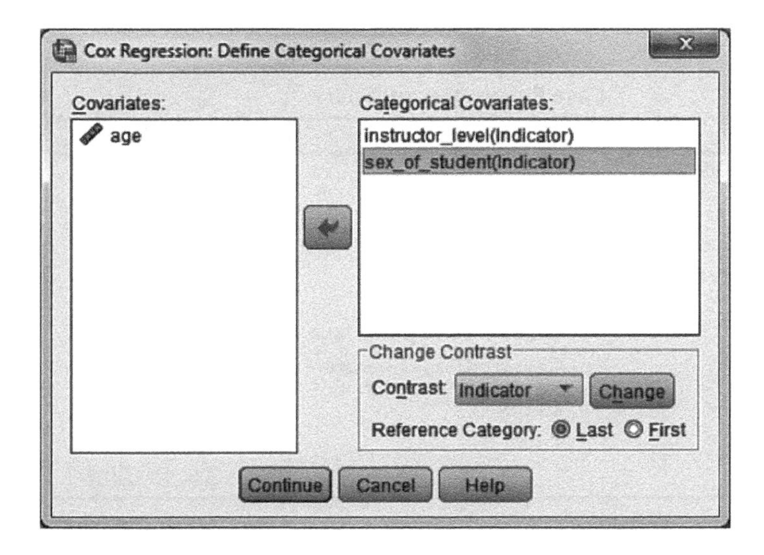

FIGURE 35.4

The **Define Categorical Covariates** window of **Cox Regression**.

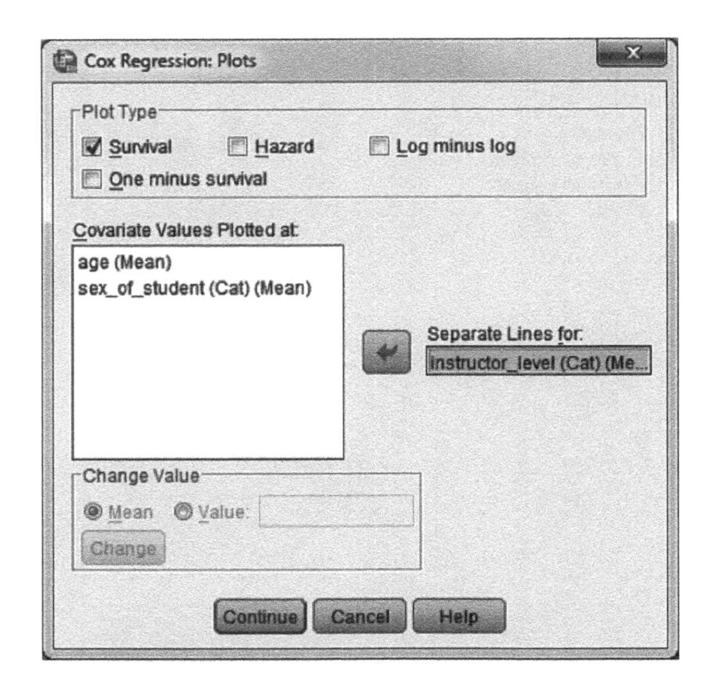

FIGURE 35.5
The **Plots** window of **Cox Regression**.

as it is a continuous variable. Clicking **Continue** brings us back to the **Cox Regression** main dialog window.

Clicking the **Plots** pushbutton produces the **Plots** dialog window shown in Figure 35.5, where we have activated **Survival** in the **Plot Type panel**. In the panel labeled **Covariate Values Plotted at**, we have moved **Instructor_level (Cat) (Mean)** to the **Separate Lines for** panel, which will produce separate survival functions for each level of **Instructor**. Clicking **Continue** returns us to the **Cox Regression** main dialog window and clicking **OK** performs the analysis.

35.4 ANALYSIS OUTPUT

Figure 35.6 provides the **Case Processing Summary** table that provides the number of cases that experienced the **drop out** event (N = **13**), the number of censored (**still attending**) cases (N = **35**), and the number of cases with missing values (N = **0**).

Case Processing Summary

		N	Percent
Cases available in analysis	Event[a]	13	26.0%
	Censored	35	70.0%
	Total	48	96.0%
Cases dropped	Cases with missing values	0	.0%
	Cases with negative time	0	.0%
	Censored cases before the earliest event in a stratum	2	4.0%
	Total	2	4.0%
Total		50	100.0%

a. Dependent Variable: months_in_course

FIGURE 35.6
Case Processing Summary.

Categorical Variable Codings[b,c]

		Frequency	(1)	(2)
sex_of_student[a]	1=male	30	1	
	2=female	20	0	
instructor_level[a]	1=blackbelt	19	1	0
	2=master	13	0	1
	3=grandmaster	18	0	0

a. Indicator Parameter Coding
b. Category variable: sex_of_student
c. Category variable: instructor_level

FIGURE 35.7
Categorical Variable Codings.

Figure 35.7 provides the **Categorical Variable Codings** table that depicts the dummy coding schema used by IBM SPSS for all of the categorical variables. Briefly (see Meyers et al. (2013) for a more complete description), dummy coding involves assigning 1's and 0's to the categories of a variable, with the number of code sets equal to one less than the number of categories. Dummy coding requires a reference category that is assigned the code 0; we have accepted the default of **Last** category (see Figure 35.4). Categories designated as the reference category have each of the other groups compared to it in the regression analysis.

For **sex_of_student** (whose coding of **1** and **2** can be seen in the **Categorical Variable Codings** table), **female** has the highest code value (with **2**) and thus becomes the reference category as shown in the column labeled **(1)** for the first (and only) set of dummy codes. **Males** will therefore be compared to **females** in the regression analysis.

Our **instructor_level** variable has three categories, also shown in the **Categorical Variable Codings** table. With the code **3**, **grandmaster** becomes the reference category; this has caused it to be assigned **0** in both coding sets, with **blackbelt** and **master** each assigned a dummy code **1** in one of the two dummy code sets. Because **grandmaster** is the reference category, **blackbelt** and **master** will each be compared to it in the analysis.

Figure 35.8 presents the initial **Block 0: Beginning Block** analysis. This presents the **Omnibus Tests of Model Coefficients**, which is realized with the **−2 Log Likelihood** statistic of **73.398**. The −2 Log Likelihood assesses **Block 0** where all covariates are yet to be entered into the analysis; hence, it plays a similar role as the total sum of squares does in the analysis of variance. All future blocks of covariates entered into the model are assessed on the basis of the *reduction* of this statistic.

Figure 35.9 provides the output for **Block 1: Method = Enter**. The **Method = Enter** heading reminds us that we have elected to enter our covariates (four in the present case) simultaneously. The **Omnibus Tests of Model Coefficients** table leads off with a **−2 Log Likelihood** value **63.018**. This likelihood ratio statistic is compared to the previous **(Block 0)** value, and its reduction is assessed with chi-square statistic with four degrees of freedom (due to the four covariates). In the portion of the table labeled as **Change From Previous Step**, the **Chi-square** value is **10.38**. It is calculated as the difference in **−2 Log Likelihood** from **Block 0** to **Block 1** (**73.398 − 63.018 = 10.38**). The chi-square value, which is statistically significant ($p = .034$), indicates that our four covariates significantly affect the probability of dropping out of taekwondo class (the hazard rate).

Block 0: Beginning Block

Omnibus Tests of Model Coefficients

-2 Log Likelihood
73.398

FIGURE 35.8
Block 0 Omnibus Test output.

Block 1: Method = Enter

Omnibus Tests of Model Coefficients[a]

-2 Log Likelihood	Overall (score)			Change From Previous Step			Change From Previ	
	Chi-square	df	Sig.	Chi-square	df	Sig.	Chi-square	df
63.018	11.766	4	.019	10.380	4	.034	10.380	

a. Beginning Block Number 1. Method = Enter

FIGURE 35.9 The **Omnibus** regression model with the covariates entered.

Variables in the Equation

	B	SE	Wald	df	Sig.	Exp(B)
sex_of_student	.019	.630	.001	1	.976	1.019
instructor_level			6.918	2	.031	
instructor_level(1)	3.147	1.222	6.638	1	.010	23.274
instructor_level(2)	.310	.863	.129	1	.719	1.364
age	.014	.020	.490	1	.484	1.014

FIGURE 35.10 The table of regression coefficients.

Figure 35.10 provides the **Variables in the Equation** table. This table presents the information for each covariate on separate rows. The column labeled **B** reports the unstandardized regression coefficients. Analogous to multiple regression, the unstandardized coefficients need be standardized so that we may compare the relative contributions of the covariates; this standardization is accomplished in survival analysis as a log transformation shown in the **Exp(B)** column.

The next four columns display the standard error of **B** (shown as **SE**), the outcome of the **Wald** test of statistical significance, the degrees of freedom (**df**), and the significance value of the coefficient (**Sig.**). The test used to evaluate the statistical significance of each covariate is the **Wald** statistic; it is distributed as a chi-square distribution and is computed as $(\mathbf{B/SE})^2$ (Katz, 2006).

The only statistically significant effect is the **instructor_level** covariate, a categorical variable with three levels. The **instructor_level** row indicates there is a significant overall effect ($p =$ **.031**), indicating that at least one of the **instructor_level** dummy coded variables is statistically significant (at least one significantly predicted the event of interest). Examining the two dummy **instructor_level** variables further informs us that only **instructor_level(1)** was statistically significant ($p =$ **.010**). Referring to Figure 35.7, we can see in the **Categorical Variable Codings** table that this dummy coded variable is represented by **blackbelt** compared to **grandmaster**.

The column labeled **Exp(B)** presents the results of the log transformations of the regression coefficients. It is known as the *hazard ratio*. Values at or near 1.00 indicate no significant change in the chances of observing the event of interest, with changes in the predictor variable. For a quantitative covariate as an example, the regression coefficient for **age** is .014. On the basis of the natural logarithm transformation, we compute **Exp(B)** as $2.72^{.014} =$ **1.014**. **Exp(B)** is the predicted change in the hazard ratio for every metric increase in the covariate while controlling for all the other covariates in the model. With the **Exp(B)** for age being **1.014**, we can say that the hazard ratio increases 1.014 for every increase in years; thus, for example, a 5-year increase equals a 5.07 hazard ratio while controlling for all the other covariates.

The **Exp(B)** value associated with **instructor_level(1)** is **23.274**. This was one part of a dummy coded variable contrasting **blackbelt** with the reference group **grandmaster**.

We interpret this effect as follows: students taught by **blackbelt** instructors were **23.274** (95% confidence interval: 2.12–255.10) times more likely to drop out than those taught by **grandmasters**.

There were no significant differences in **drop out** rate for **instructor_level(2)** (the **master** vs. **grandmaster** dummy variable); thus, the dropout rates associated with these two types of instructors were comparable. Note that if we computed the 95% confidence interval for the **instructor_level(2)** regression coefficient, it would subsume the value of 1.00, thus suggesting that **master** and **grandmaster** instructors are comparable in terms of their **drop out** rate.

Figure 35.11 provides the **Covariate Means and Pattern Values** table. This table displays the average value of each covariate or predictor variable. **instructor_level(1)** represents the first dummy variable for **instructor_level** and has a mean of **.354**. This variable was coded **1** for black belt instructor and **0** for all other instructors. The second dummy variable, **instructor_level(2)**, was coded **1** for master instructors and **0** for all other instructors and its mean was **.271**. The mean **age** was **20.896**, and the **sex_of_student** had a mean of **.583**. The columns labeled **Pattern 1, 2, 3** respect the means for **age** and **sex_of_student** and provide the coding scheme for the dummy variables **(1,0), (0,1) (0,0)**. These are used to produce the graph in Figure 35.13.

Figure 35.12 provides a single survival graph for all covariates together. The function is calculated on the basis of the mean of all covariates.

Covariate Means and Pattern Values

	Mean	Pattern		
		1	2	3
sex_of_student	.583	.583	.583	.583
instructor_level(1)	.354	1.000	.000	.000
instructor_level(2)	.271	.000	1.000	.000
age	20.896	20.896	20.896	20.896

FIGURE 35.11
Means of the variables.

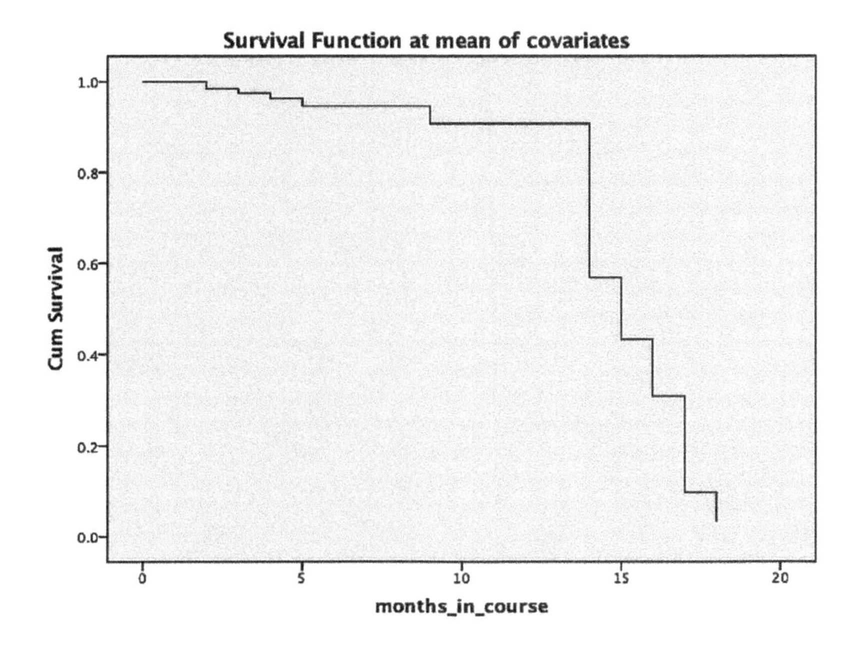

FIGURE 35.12
Survival Function at the mean of the predictors.

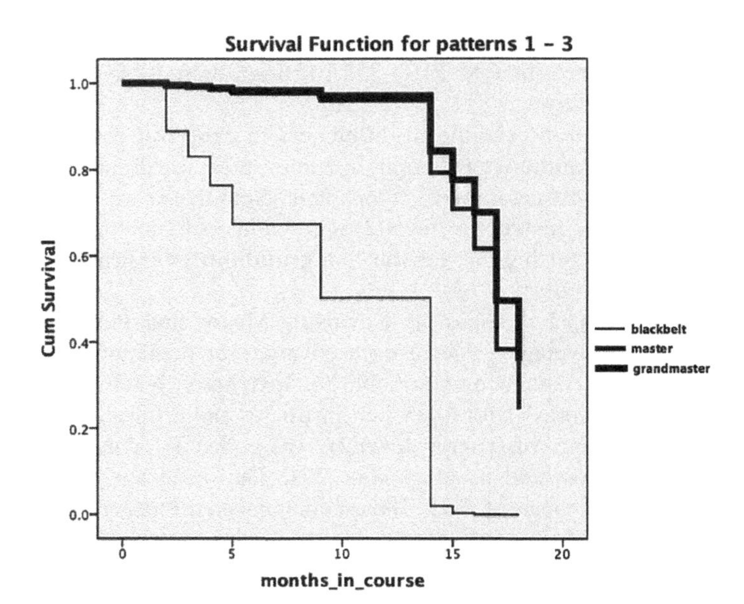

FIGURE 35.13
Survival functions by level of instructor.

Figure 35.13 depicts the survival functions for each **instructor_level**. Notice the higher rates of drop out associated with black-belt-level instruction as compared to master-level or grand-master-level instruction; this is why this variable was a statistically significant predictor in the regression analysis.

PART 11

RELIABILITY AS A GAUGE OF MEASUREMENT QUALITY

CHAPTER 36

Reliability Analysis: Internal Consistency

36.1 OVERVIEW

Over a century ago, Spearman (1904a, 1904b, 1907) suggested that the observed results of our measurement operations contain a mixture of both the true value of the construct and measurement error. Thus began an approach to the nature and quality of measurement known as the Classical Test Theory.

In the prototypical testing situation, we have developed a set of test items to measure some construct (e.g., an academic domain of knowledge, attitudes toward a health care organization, a personality characteristic) and have collected responses to these items from a sample of respondents. The issue facing us is to determine if these items (or which subset of these items) can be combined to form a scale that measures the construct with an acceptable degree of quality (i.e., with a minimally acceptable amount of measurement error).

One result of the quest to develop procedures that assess measurement quality (see Nunnally & Bernstein (1994) and Traub (1997)) has been the development of the *reliability coefficient*. A reliability coefficient theoretically represents the proportion of true score variance present in the total variance (true score plus error variance) of test scores (Lord & Novick, 1968; Nunnally & Bernstein, 1994). Reliability ranges between .00 and 1.00, with .00 indicating that none of the observed variance is due to true score variance (all of the observed variance is due to measurement error) and 1.00 indicating that observed scores are composed only of true score variance (there is no measurement error). All measurement operations ultimately serve the goal of achieving validity (allowing us to draw appropriate inferences from our measures), and a critical requirement to achieve validity is to measure our construct in a manner that is relatively free of measurement error, that is, to have a relatively reliable measurement procedure (Pedhazur & Schmelkin, 1991).

Because we cannot directly measure the true score (as it is a theoretical entity and is presumably embedded in error variance), we need to gauge reliability indirectly. One approach to this indirect approach has involved assessing the consistency of performance on at least two measurement occasions, an approach subsuming both *test–retest reliability* and *parallel forms reliability*. Test–retest involves cases tested at least twice over time, and parallel forms reliability involves cases completing two or more scales. Both evaluate consistency in part by examining the correlations of test scores between the testing occasions.

Performing Data Analysis Using IBM SPSS®, First Edition.
Lawrence S. Meyers, Glenn C. Gamst, and A. J. Guarino.

This chapter concerns those situations where we have just one measurement opportunity. We still need to assess consistency of performance, and our focus is thus on the responses to the items. In simplified terms, we expect that if each of the items is an indicator of the construct that is presumably being assessed by the scale, then respondents should be *internally consistent* in their responses to (endorsement of) the items because the items are at least partially measuring a common construct, that is, the responses to the items should be correlated. Moderation and common sense are the keys to assess item correlations and to combine them to form a scale assessing a particular construct.

- If the items are all correlated near 1.00, then we may conclude that we are asking the same question over and over, and this is a poor measurement technique.
- If the items are not correlated at all, then they have nothing in common, and because it is the construct underlying the scale that they are supposed to have in common, there is no core construct that is being measured. This situation is to be avoided.
- If some of the items are negatively (but modestly) correlated, then we must recode them so that all are oriented in the same direction of the construct before we combine items; that is, higher scores on the items should uniformly represent more knowledge, more positive attitudes, more of a characteristic, and so on. This recoding issue is often encountered because it is common practice to reverse-word several scale items to thwart acquiescence bias (Robinson, Shaver, & Wrightsman, 1991).

Developing an effective reliability index has been an ongoing process. In the early part of past century, it was common to use a *split-half* method where we divided the items into two sets (for the purposes of calculation), scored each set for each respondent, and correlated these subset scores; higher correlations were taken as indicating greater reliability. However, different splits often resulted in different correlations, leading to some ambiguity of interpretation. This procedure evolved into development by Kuder and Richardson (1937) of a general reliability coefficient known as the KR-20 coefficient because it was presented in the twentieth formula of their monograph. Its limiting feature was that it applied only to dichotomously scored items.

Finally, Cronbach (1951) developed a generalized version of KR-20 known as *coefficient alpha* applicable to most item-scoring systems, and this has become the most widely used reliability index; it yields the lower bound of the internal consistency of a scale. But care needs to be taken in working with this coefficient. Following are some of the pitfalls of interpreting coefficient alpha (Cortina, 1993):

- Although coefficient alpha is intended to be used to describe the internal consistency of homogeneous (unidimensional) scales, it is possible for multidimensional scales to produce acceptable values of coefficient alpha because there may be present subsets of strongly related items; thus, a relatively high coefficient alpha does not necessarily signify unidimensionality.
- Coefficient alpha is sensitive to the number of items on the scale. All else equal, longer scales will drive up (potentially inflate) the reliability of the scale. Modern psychometric researchers look to reduce an item set down to a focused subset of items by removing those that are "weaker" indicators of the construct and by countering item groups exhibiting local item dependence either by using a single item from that group or by forming within the scale minitests known as testlets (Steinberg & Thissen, 1996; Yen, 1993). To supplement coefficient alpha, we can also evaluate the *average inter-item correlation* (e.g., Clark & Watson, 1995), a statistic that is not sensitive to the size of the item set.

A reliability coefficient is a general index of the degree to which a scale is free of measurement error. As such, we can establish broad guidelines for quality and compare scales on the index. For example, a coefficient alpha of .70 would be about as low as we would want to see for a measure, and it would be desirable for scales to exceed .80 when considering both the number of items and sample size (Clark & Watson, 1995; Nunnally & Bernstein, 1994). However, if a measure is used to make a decision on an individual, very high reliabilities (i.e., in the range .90 or higher) are mandated (Gay, 2010).

The amount of estimated measurement error can be converted to scale score units by computing the standard error of measurement (multiply the square root of 1.00 − reliability coefficient by the standard deviation of the test scores) and building a (95%) confidence interval around it. As the standard error of measurement varies across the range of test scores (e.g., Feldt & Qualls, 1996; Lord, 1984), the conditional standard error of measurement for specific scale scores should be calculated when making decisions based on scale performance (i.e., whether a case has attained a criterion or threshold score, such as passing or failing or being grouped into a particular category).

36.2 NUMERICAL EXAMPLE

To illustrate how to perform a reliability analysis, we use the 12 items comprising the conscientiousness scale of the NEO Five-Factor Inventory (Costa & McCrae, 1992). The items are named by their item number in the inventory (e.g., **neo15**), and we assess characteristics such as being well organized, working hard to meet goals and responsibility, and meeting responsibilities on time. Items are evaluated on a 5-point *Strongly Disagree* to *Strongly Agree* response scale and are later converted to a 0–4 numerical score. Our data reflect any recoding that was needed to accommodate reverse-worded items so that the scoring of all the items is oriented toward higher levels of conscientiousness. Because the construct is well understood and because of the care and expertise that went into the development of the items, what we see is a scale that exhibits a high level of reliability in this data set. The data are provided in the data file named **conscientiousness items**.

36.3 ANALYSIS SETUP

Open **conscientiousness items** and from the main menu select **Analyze → Scale → Reliability Analysis**; this produces the **Reliability** main dialog window shown in Figure 36.1. We have moved all 12 **neo** variables (**neo5–neo60**) to the **Items** panel and typed **conscientiousness** in the **Scale label** panel as a label for the output. It is important to recognize that the **Reliability** procedure will not permanently compute a scale score and save it to the data file (for which we need to use a **Compute** procedure) but only combines the items for the duration of the analysis; it therefore is a projection of what the reliability of the scale would be on the basis of this set of items. We retain **Alpha** in the **Model** drop-down menu (split half and other computations are available as well in this menu).

Selecting the **Statistics** pushbutton opens the **Statistics** dialog window shown in Figure 36.2. In the **Descriptives** area, we check **Item** (this generates descriptive statistics for the items so that we can determine whether each has sufficient variance to merit inclusion on the scale) and **Scale if item deleted** (this generates several statistics we use to evaluate the item set). In the **Inter-item** area, we check **Correlations** (this generates a correlation matrix of the items), and in the **Summaries** area we check **Correlations**. Select **Continue** to return to the main dialog window and click **OK** to perform the analysis.

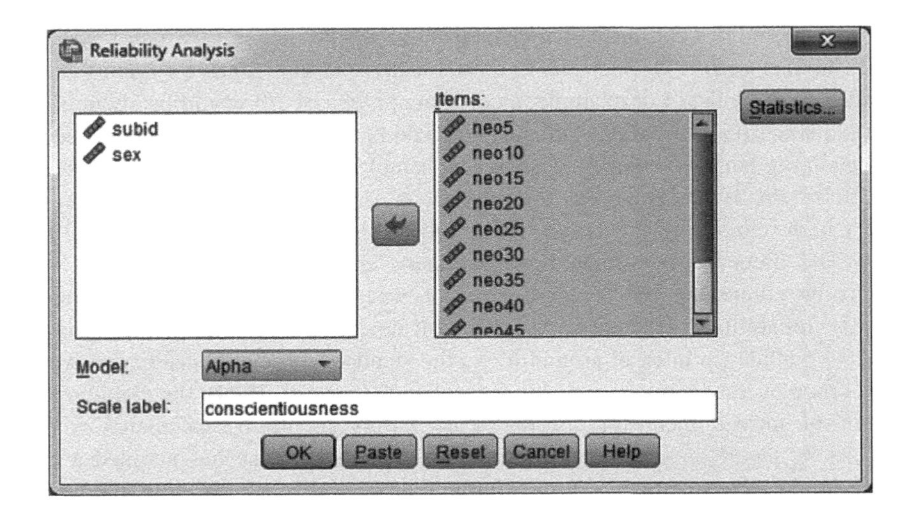

FIGURE 36.1
The main dialog window of **Reliability**.

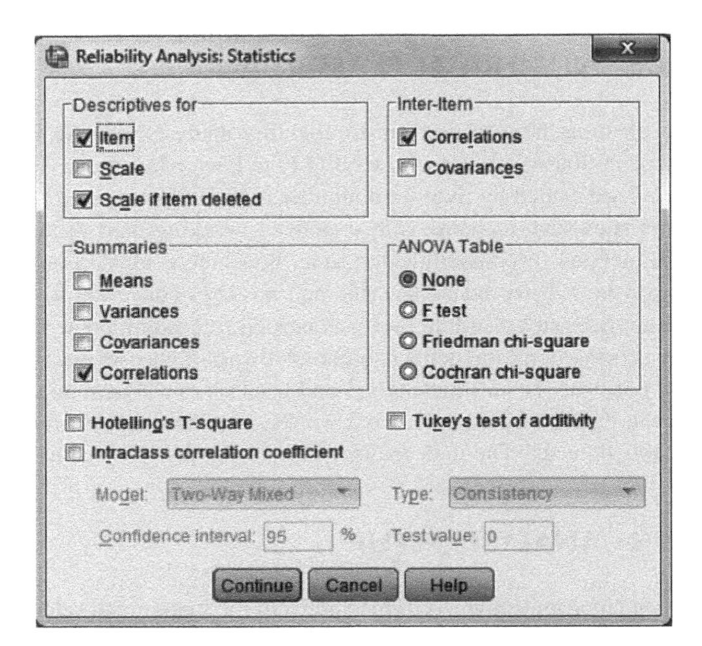

FIGURE 36.2
The **Statistics** dialog window of **Reliability**.

36.4 ANALYSIS OUTPUT

The top (**Case Processing**) table in Figure 36.3 shows us that our sample size for the analysis is 407. The bottom (**Item Statistics**) table presents the means and standard deviations of the items. Ideally, items rated on a 5-point scale should yield a standard deviation near 1.00, and for the most part, these items conform roughly to this expectation; thus, each item demonstrates sufficient variance to meaningfully correlate it with the other items.

The middle (**Reliability Statistics**) table gives us the short answer we were seeking at the outset of the analysis. Because these items are not in standardized form, we refer to the column labeled **Cronbach's Alpha** where we note that the coefficient has a value of **.806**. For a 12-item scale (shown under **N of Items**), this would be considered to reflect good reliability.

Figure 36.4 presents the correlation matrix of the items and, below it, a summary of the correlation statistics. We are looking primarily for two features in the correlation

Scale: conscientiousness

Case Processing Summary

		N	%
Cases	Valid	407	95.8
	Excluded[a]	18	4.2
	Total	425	100.0

a. Listwise deletion based on all variables in the procedure.

Reliability Statistics

Cronbach's Alpha	Cronbach's Alpha Based on Standardized Items	N of Items
.806	.817	12

Item Statistics

	Mean	Std. Deviation	N
neo5	2.61	1.032	407
neo10	2.38	1.131	407
neo15	2.22	.946	407
neo20	3.15	.603	407
neo25	2.55	.940	407
neo30	1.80	1.158	407
neo35	3.05	.769	407
neo40	3.14	.820	407
neo45	2.36	1.141	407
neo50	2.97	.730	407
neo55	2.55	1.033	407
neo60	2.84	.842	407

FIGURE 36.3

Case Processing Summary, coefficient alpha, and **Item Statistics**.

matrix: the correlations are neither too small to yield much common variance nor too large (in the .70's or greater) to indicate duplication of item and there are no negative correlations (as they all need to be oriented in the same direction). As we would expect from a well-respected scale with a coefficient alpha of approximately .81, the correlations in the matrix represent what we would hope to find.

In the table of correlation summary statistics, our focus is on the **Mean** of **.271**; this is the average correlation of the items on the scale. One of the criteria for a high-quality scale assessing a broadly defined construct recommended by Clark and Watson (1995) is that the average inter-item correlation be in the range .15–.20 (more narrowly defined constructs should correlate to a somewhat greater extent). The NEO Five-Factor conscientiousness scale appears to exceed this criterion, again suggesting a scale of high reliability.

A set of useful statistics is presented in the table of **Item-Total Statistics** shown in Figure 36.5. The table is used primarily for diagnostics to determine the contribution of each item (mostly by showing the consequences of deleting the item from the set). Of lesser interest are the first two columns (**Scale Mean if Item Deleted** and **Scale Variance if Item Deleted**); IBM SPSS® has totaled the item scores for each case and computed a scale mean. It then shows the scale mean if the item on the row is deleted from the set.

Inter-Item Correlation Matrix

	neo5	neo10	neo15	neo20	neo25	neo30	neo35	neo40	neo45	neo50	neo55	neo60
neo5	1.000	.243	.238	.212	.240	.174	.215	.079	.097	.173	.396	.111
neo10	.243	1.000	.138	.271	.383	.350	.376	.238	.258	.408	.410	.282
neo15	.238	.138	1.000	.215	.187	.154	.189	.155	.178	.185	.237	.182
neo20	.212	.271	.215	1.000	.288	.201	.346	.261	.168	.290	.244	.371
neo25	.240	.383	.187	.288	1.000	.312	.571	.244	.269	.431	.356	.372
neo30	.174	.350	.154	.201	.312	1.000	.298	.149	.249	.243	.352	.140
neo35	.215	.376	.189	.346	.571	.298	1.000	.306	.312	.402	.309	.399
neo40	.079	.238	.155	.261	.244	.149	.306	1.000	.480	.349	.256	.260
neo45	.097	.258	.178	.168	.269	.249	.312	.480	1.000	.345	.300	.149
neo50	.173	.408	.185	.290	.431	.243	.402	.349	.345	1.000	.346	.356
neo55	.396	.410	.237	.244	.356	.352	.309	.256	.300	.346	1.000	.167
neo60	.111	.282	.182	.371	.372	.140	.399	.260	.149	.356	.167	1.000

Summary Item Statistics

	Mean	Minimum	Maximum	Range	Maximum / Minimum	Variance	N of Items
Inter-Item Correlations	.271	.079	.571	.492	7.232	.010	12

FIGURE 36.4 Correlation matrix and a summary of the correlation statistics.

Item-Total Statistics

	Scale Mean if Item Deleted	Scale Variance if Item Deleted	Corrected Item-Total Correlation	Squared Multiple Correlation	Cronbach's Alpha if Item Deleted
neo5	29.00	35.791	.338	.204	.804
neo10	29.24	32.857	.532	.322	.785
neo15	29.40	36.560	.313	.123	.805
neo20	28.46	37.437	.438	.237	.796
neo25	29.06	33.885	.573	.419	.781
neo30	29.81	34.064	.416	.215	.798
neo35	28.57	35.103	.586	.426	.783
neo40	28.47	36.137	.429	.306	.795
neo45	29.25	33.976	.433	.313	.796
neo50	28.65	35.648	.557	.347	.786
neo55	29.07	33.424	.548	.354	.783
neo60	28.77	36.177	.410	.283	.796

FIGURE 36.5 The table of **Item-Total Statistics**.

The **Squared Multiple Correlation** is the result of a multiple regression procedure where the item on the row is being predicted by a weighted linear composite of the other variables. For example, the **Squared Multiple Correlation** associated with predicting **neo5** from the other items is **.204**; thus, approximately 20% of the variance of **neo5** is explained by a weighted linear composite of the other items. This statistic is useful to determine if there is sufficient multicollinearity in the data set to warrant exclusion of a variable. We ordinarily have the same goal here as when we examined the correlation matrix for correlations in the .70's; the difference is that the correlation matrix showed us bivariate (pairwise) correlations, whereas the multivariate correlation statistic is a way to discern multicollinearity.

Probably the column that attracts the primary interest of researchers is the **Corrected Item-Total Correlation**. The **Corrected Item-Total Correlation** is a Pearson correlation between the item score and the total of the remaining items in the set. It is "corrected" in that the total score does not include the item in question. All else equal, corrected item-total correlations can be interpreted as follows: .10's are acceptable, .20's are good, .30's are very good, and .40's or better are considered to be extremely good (assuming an average inter-item correlation in the .20's or .30's). We interpret such correlations to indicate that higher scores on the item are associated with higher scale scores; as such, the item is taken to be a good indicator of the construct. All of the obtained corrected item-total correlations shown in Figure 36.5 are at least very good.

Corrected item-total correlations in the range .00–.10 indicate that there is not much relationship between item performance and scale score. Unless its content was capturing a unique and important aspect of the construct, such items are candidates for deletion or rewriting.

Negative corrected item-total correlations are "red flags." They signify that the item and the scale are oriented in different directions and that the analysis should be stopped. The likelihood is that this is a reverse-worded item that was overlooked in the recoding process, and this is an easy fix (the reliability analysis should them be performed again with the properly coded item replacing the one that was not properly coded). If the problem is subtler, the item needs to be removed from the set.

Cronbach's Alpha if Item Deleted indicates the reliability coefficient consequences of removing an item. This information needs to be used carefully by the researchers. In a short inventory, removing an item can somewhat reduce the value of the alpha coefficient, as the statistic is sensitive to the number of items on the scale. With good quality items in a short scale, as shown in Figure 36.5, the alpha coefficient does not change dramatically, although it tends to get slightly lowered for most of the items. This statistic will provide a more dramatic result when the corrected item-total correlation is negative or even in the zero range.

CHAPTER 37

Reliability Analysis: Assessing Rater Consistency

37.1 OVERVIEW

There are numerous situations where we call on raters or judges to evaluate or classify the performance of individuals or other entities. For example, we may need to quantitatively evaluate the interview performance of prospective job applicants, the seriousness of a set of symptoms, if youthful offenders were or were not candidates to be placed in a certain treatment program, or if someone passed or failed an oral examination for licensure as a therapist. In all of these situations, we do not ordinarily rely on a single rater and run the risk of obtaining idiosyncratic evaluations, even for someone who is carefully trained in the rating system. Instead, we convene a panel of two or more raters and rely on the judgments of such raters to the extent that they exhibit an acceptable level of consistency in their assessments.

We focus on two procedures for assessing rater consistency. For ratings that are made on a quantitative scale of measurement (often 5- or 7-point summative response scales), we ordinarily use the *intraclass correlation* (ICC), a statistic that is available in the **Reliability** procedure of IBM SPSS® and that we also make use of in multilevel modeling (see Chapter 29). For a categorical rating system when we have only two raters, we use the *kappa coefficient*, a statistic that is available in the **Crosstabs** procedure of IBM SPSS; with more than two raters, an extension of kappa has been provided by Fleiss (1971) but this computation is not available in IBM SPSS. A couple of other variations of kappa to handle the prevalence of ratings for one of the categories and substantial differences between the raters have been reviewed by Di Eugenio and Glass (2004), but these are also not available in IBM SPSS.

The ICC

The ICC can be dated back to Pearson (1901a) in describing roommates rating their living quarters (Gonzalez & Griffin, 2004), but Shrout and Fleiss (1979) popularized it as a way to assess rater reliability. It can be calculated from a one-way within-subjects (Treatment × Subjects) ANOVA where the raters represent the within-subjects variable and the cases that are being evaluated represent the subjects (Guildford & Fruchter, 1978). There are no firm guidelines for interpreting the magnitude of the ICC in terms of rater consistency, but generally, it is desirable to achieve an ICC value of .70 or higher.

Performing Data Analysis Using IBM SPSS®, First Edition.
Lawrence S. Meyers, Glenn C. Gamst, and A. J. Guarino.

The IBM SPSS data file needs to be structured with the raters as the variables (columns or Treatment effect) and the cases evaluated as the rows. We could therefore compute Pearson correlations between pairs of raters, but this would not be an adequate way to evaluate rater consistency. Pearson correlations assess only covariation without concern about any magnitude differences between the raters; thus, the ratings of two raters may yield a substantial correlation but demonstrate large absolute differences in the magnitudes of their ratings. The ICC takes into account the absolute levels of agreement as well as the covariation of the ratings (IBM SPSS provides us with a choice on this matter).

It is possible to use different ANOVA models in the computation of the ICC. Here is a brief summary of these options:

- A *one-way random effects model* is used when different raters have evaluated the cases in the data file. This model is relatively infrequently used; it usually comes into play when there are large numbers of cases to evaluate and a single set of raters is unable to perform all of the evaluations.
- A *two-way random effects model* is used when all raters have evaluated every case. The decision to use this model hinges on the raters being considered a random sample of all possible judges. Most authors (e.g., Berkman & Reise, 2012; Howell, 2010) suggest that this model is applicable to most situations. Theoretically, by treating the raters as a random effect, we can generalize the ICC to a more general population of raters.
- A *two-way fixed effect model* is used when all raters have evaluated every case. The decision to use this model hinges on the raters being considered to be unique or that they are the only set of judges who are relevant. Theoretically, by treating the raters as a fixed effect, we cannot generalize the ICC to a more general population of raters but are simply describing the behavior of the raters in the study.

Deciding between the two-way random and two-way fixed effect is sometimes relatively straightforward and sometimes not so straightforward. For example, in basic research studies (e.g., rating videos of parent–child interactions) where the raters can be students or staff members, it is clear that they comprise a (maybe not quite random) sample and thus (sort of) satisfies the random effects model. In certain applied settings (e.g., there are only three staff analysts in the Exam Development unit whose job is to rate employment applications), the argument that the ICC needs to be descriptive of the particular set of raters seems persuasive and thus this situation appears to satisfy the fixed effect model. Between these two extremes lies a range of situations requiring careful consideration on the part the researchers in terms of which model to use. The random effects model will occasionally yield somewhat lower values of the ICC than the fixed effect model and thus may represent the more conservative approach.

IBM SPSS produces two versions of the ICC, and the difference between these has, in our personal experience, been the basis for some confusion. These two versions, and when they should be used, are as follows:

- The *single measures* ICC should be used when the ratings in the data file represent the scale value selected by the rater. This is a situation that will almost always be in effect, and thus, the single measures ICC should almost always be used.
- The *average measures* ICC should be used when the ratings in the data file represent average values of a set of raters. This is a situation that will rarely be in effect.

It has been suggested that if the average of the raters will be used to test hypotheses, then the average measures ICC could be used (e.g., Hallgren, 2012). However, assessing the stability of the mean rating is quite different from describing the extent to which the raters exhibit consistency. The value of the ICC for average measures will always be higher than what is obtained for the single measures computation. This is because the average measures ICC is computed by placing the single measures ICC into the Spearman–Brown prophecy formula (Shrout & Fleiss, 1979) that was originally designed by Brown (1910) and Spearman (1910) to project what the full-test correlation might be based on a split-half reliability estimate.

It is also possible to select between a *consistency* and an *absolute agreement* computation. The consistency computation does not take into account the differences between the raters and so provides a value of the ICC that is very close to the mean of the pairwise correlations between the raters. The absolute agreement option does take into account the differences between the raters. As one of the advantages of using an ICC is that it takes rater differences as well as rater covariance into account, we almost always choose the absolute agreement option.

Cohen's Kappa Coefficient

The kappa coefficient was proposed by Jacob Cohen (1960) to index the degree of agreement exhibited by two raters in their dichotomous categorization of the performance of the cases. Because we are dealing with assignment of one of the two available categories to the cases (e.g., pass or fail), the data gleaned from the ratings are in the form of frequencies. Although the frequency data are structured within a two-way chi-square contingency table, contrasting observed frequencies with those expected on the basis of the null hypothesis (see Chapter 63), the kappa coefficient calculation controls for the amount of agreement we would expect on the basis of chance. By controlling for chance agreement, kappa is a more appropriate way to assess rater consistency than simply computing the percentage of cases on which the two raters agreed in their categorization (which does not correct for chance).

The value of the kappa coefficient can vary from −1.00, indicating poorer than chance agreement, to 0.00, indicating exactly chance agreement, to 1.00, indicating perfect agreement (Fleiss & Cohen, 1973). The decimal value can be interpreted as the percent agreement between raters controlling for chance. Based partly on criteria suggested by Krippendorff (1980), at least in applied situations, values below about .70 might be considered not acceptable, those between .70 and the low to middle .80's might be treated as borderline acceptable, and those higher than the middle .80's might be considered acceptable to good depending on where they lie in that range.

37.2 NUMERICAL EXAMPLE: ICC

The fictional data for this example are provided in the data file named **job candidate 1st round ratings**. A total of 23 candidates in a city police force tested for a promotion to Lieutenant. As one of the components of the selection process, they were asked to study a complex scenario for 30 minutes and then present their resource deployment and contingencies plan to a panel of three subject-matter experts. The panel members were selected from a process that asked Police Captains from comparably sized cities in nearby states to take part in this testing and were provided with extensive training in this evaluation process before the testing began. After listening to each candidate's presentation and asking what questions were allowed by the testing process, each rater made a rating of the candidate's performance on a 7-point scale ranging from *Relatively Poor* to *Excellent*. The ratings are provided under the variables **rater1**, **rater2**, and **rater3**. It was anticipated that after the ratings were statistically analyzed, the panel members would be given the opportunity to discuss their ratings as a group and make any changes they deemed appropriate before the ratings were finalized.

37.3 ANALYSIS SETUP: ICC

Open **job candidate 1st round ratings** and from the main menu select **Analyze ➔ Scale ➔ Reliability Analysis**; this produces the **Reliability** main dialog window shown in Figure 37.1. We have moved **rater1**, **rater2**, and **rater3** to the **Items** panel.

Selecting the **Statistics** pushbutton opens the **Statistics** dialog window shown in Figure 37.2. In the **Inter-item** area, we check **Correlations** (this generates a correlation matrix of the items), and in the **ANOVA Table** area we check **F test** (this generates a one-way ANOVA so that we can determine if there are rater differences). We check the **Intraclass correlation coefficient** checkbox. From the **Model** drop-down menu, we select **Two-Way Random** (as these raters are a sample of those who were available

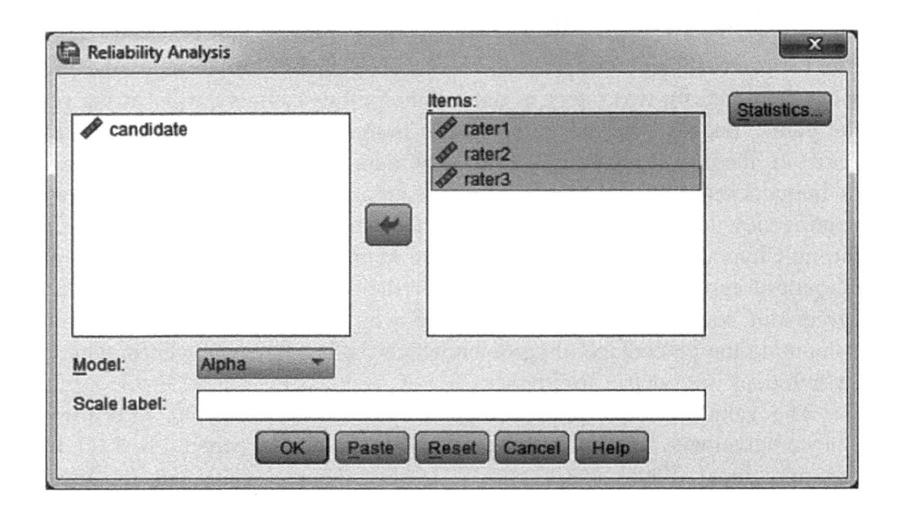

FIGURE 37.1
The main dialog window of **Reliability**.

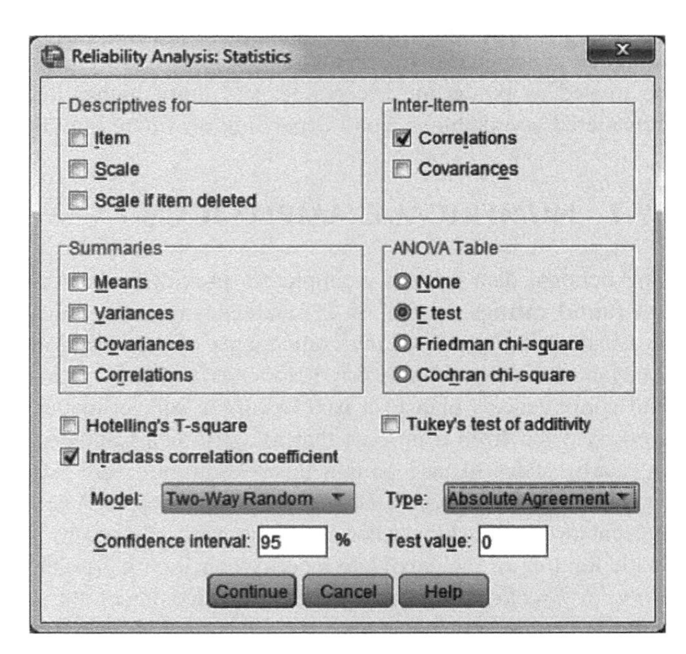

FIGURE 37.2
The **Statistics** dialog window of **Reliability**.

and thus comprise a random effect), and from the **Type** drop-down menu, we select **Absolute Agreement** (as we wish to have the ICC reflect rater differences as well as their covariance). Select **Continue** to return to the main dialog window and click **OK** to perform the analysis.

37.4 ANALYSIS OUTPUT: ICC

The top table in Figure 37.3 shows the correlation matrix for the raters. It is named **Inter-Item Correlation Matrix** because often researchers use the IBM SPSS **Reliability** procedure to determine the value of coefficient alpha (see Chapter 36), and in such an analysis, the items are the variables in the analysis. As can be seen from the matrix, there is a good deal of covariation permeating the ratings.

The **ANOVA** summary table is also shown in Figure 37.3. This is a one-way within-subjects design (see Chapter 51). The **Between People** effect represents the cases in the analysis (the job candidates in our example), and IBM SPSS (and most researchers) will not bother computing an F ratio for this effect (it ordinarily signifies individual differences).

The **Between Items** effect generally represents the within-subjects treatment effect and here addresses the differences among our raters. The F ratio of **41.008** informs us that we do have statistically significant rater differences ($p < .001$). Were we sufficiently interested in this, we would have performed the stand-alone ANOVA with multiple comparison tests (as described in Chapter 51); here, we just note that there are rater differences. Given that this is the first round of ratings and that a discussion of the ratings of these panel members is yet to take place with the opportunity (and likelihood) that some ratings will be changed, this may not present a difficulty for the selection process.

Figure 37.4 presents the two ICCs. The one in which we are interested is the **Single Measures** coefficient. Its value of **.710** is suggestive that there is some work the raters must do to get their ratings more in sync, as **.710** may be at the lower end of acceptability in some research contexts but is not likely to be judged as sufficient in the selection environment. Again, rater discussion should result in an improvement in the ICC.

Inter-Item Correlation Matrix

	rater1	rater2	rater3
rater1	1.000	.883	.853
rater2	.883	1.000	.878
rater3	.853	.878	1.000

ANOVA

		Sum of Squares	df	Mean Square	F	Sig
Between People		160.000	22	7.273		
Within People	Between Items	28.203	2	14.101	41.008	.000
	Residual	15.130	44	.344		
	Total	43.333	46	.942		
Total		203.333	68	2.990		

Grand Mean = 3.67

FIGURE 37.3 Rater correlation matrix and the ANOVA summary table.

Intraclass Correlation Coefficient

	Intraclass Correlation[a]	95% Confidence Interval		F Test with True Value 0			
		Lower Bound	Upper Bound	Value	df1	df2	Sig
Single Measures	.710[b]	.230	.890	21.149	22	44	.000
Average Measures	.880	.472	.960	21.149	22	44	.000

Two-way random effects model where both people effects and measures effects are random.

a. Type A intraclass correlation coefficients using an absolute agreement definition.
b. The estimator is the same, whether the interaction effect is present or not.

FIGURE 37.4 The **Single Measures** and **Average Measures** ICCs.

37.5 NUMERICAL EXAMPLE: KAPPA

The fictional data for this example are provided in the data file named **juvenile offender ratings**. A total of 56 juveniles (**ward** in the data file) who have been incarcerated for committing crimes were independently evaluated by two forensic psychologists for placement in a community work program that would have the offenders leave the detention facility during the daytime hours. The evaluation was performed to determine if the wards represented a sufficiently low risk (in terms of committing another crime or of escaping while they were off-site) to be placed in the work program. The psychologists each produced a binary rating for each youth under the variables **rater1** and **rater2** in the data file. The categories in the rating system were **remain on grounds** (coded as **1** in the data file) and **off-site work allowed** (coded as **2** in the data file).

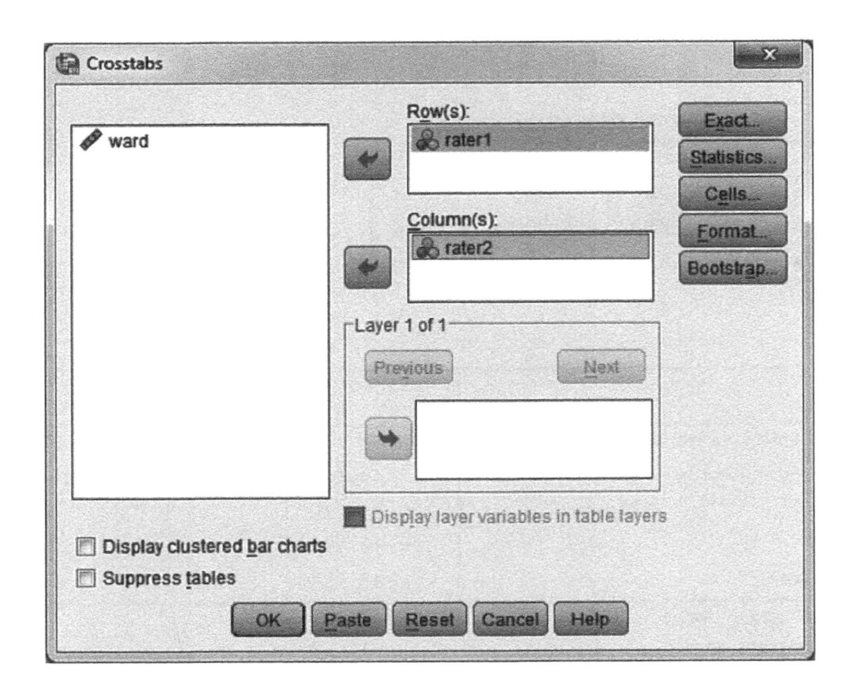

FIGURE 37.5
The main dialog window of **Crosstabs**.

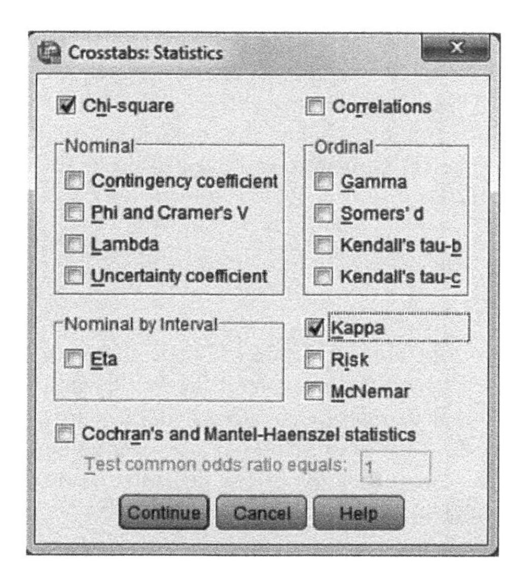

FIGURE 37.6
The **Statistics** dialog window of **Crosstabs**.

37.6 ANALYSIS SETUP: KAPPA

Open **juvenile offender ratings** and from the main menu select **Analyze ➔ Descriptives ➔ Crosstabs**; this produces the **Crosstabs** main dialog window shown in Figure 37.5. We have arbitrarily moved **rater1** to the **Row(s)** panel and **rater2** to the **Column(s)** panel.

A more extensive treatment of the **Crosstabs** procedure is discussed in Chapter 63. Nonetheless, we wish to obtain more than just the bare kappa statistic. Selecting the **Statistics** pushbutton opens the **Statistics** dialog window shown in Figure 37.6. We select **Chi-square** just to provide a test of the null hypothesis that there is no relationship between the raters in their assignment of categories to the wards, and we select **Kappa** because that is the point of this analysis.

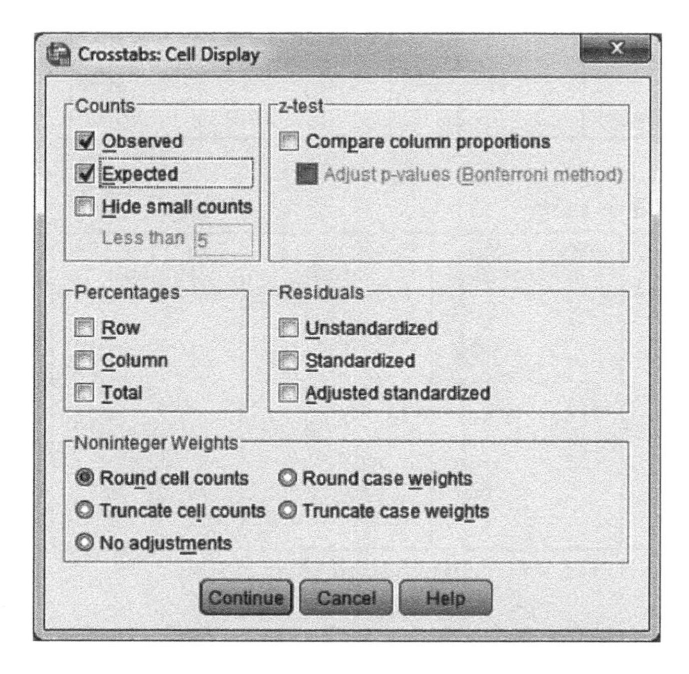

FIGURE 37.7
The **Cell Display** dialog window of **Crosstabs**.

Selecting the **Cells** pushbutton opens the **Cell Display** dialog window shown in Figure 37.7. To obtain a sense of how the raters assigned wards to categories, we check both **Observed** and **Expected** under **Counts**. These provide, respectively, the actual frequencies of rater correspondence to the category assignments and the frequencies that would be expected based on the null hypothesis. Select **Continue** to return to the main dialog window and click **OK** to perform the analysis.

37.7 ANALYSIS OUTPUT: KAPPA

The frequency **Crosstabulation** is shown in the top table of Figure 37.8. Both raters agreed on **20** wards who they believed should be kept on the grounds of the facility and should not be allowed to work in the community; both raters also agreed on **30** wards who could be allowed to leave the facility to work in the community. If we were using the simple percentage of rater agreement as our gauge, then 50 of the 56 wards were judged in the same way by these raters for a rater agreement percentage of 50/56, or about 89.3%.

The chi-square test of the null hypothesis is shown in the bottom table in Figure 37.8. Chi-square was computed to be **34.164**, and with one degree of freedom (for the 2×2 array), it is statistically significant ($p < .001$); thus, the null hypothesis of no relation in the ratings gives way to the conclusion that the ratings provided by the raters are related to each other.

rater1 * rater2 Crosstabulation

			rater2		
			1 remain on grounds	2 off-site work allowed	Total
rater1	1 remain on grounds	Count	20	4	24
		Expected Count	9.4	14.6	24.0
	2 off-site work allowed	Count	2	30	32
		Expected Count	12.6	19.4	32.0
Total		Count	22	34	56
		Expected Count	22.0	34.0	56.0

Chi-Square Tests

	Value	df	Asymp. Sig. (2-sided)	Exact Sig. (2-sided)	Exact Sig. (1-sided)
Pearson Chi-Square	34.164[a]	1	.000		
Continuity Correction[b]	31.009	1	.000		
Likelihood Ratio	38.451	1	.000		
Fisher's Exact Test				.000	.000
Linear-by-Linear Association	33.554	1	.000		
N of Valid Cases	56				

a. 0 cells (.0%) have expected count less than 5. The minimum expected count is 9.43.
b. Computed only for a 2x2 table

FIGURE 37.8 Observed and expected cell frequencies (top) and chi-square test of null hypothesis.

Symmetric Measures

		Value	Asymp. Std. Error[a]	Approx. T[b]	Approx. Sig.
Measure of Agreement	Kappa	.779	.085	5.845	.000
N of Valid Cases		56			

a. Not assuming the null hypothesis.
b. Using the asymptotic standard error assuming the null hypothesis.

FIGURE 37.9 The kappa coefficient.

How strongly the judgments of the raters are related is shown by the kappa coefficient in Figure 37.9. The kappa value is **.779**, a value noticeably lower than the straight percentage of agreement because its calculation has statistically controlled for the distribution of frequencies we would expect by chance. It is statistically significant ($p < .001$). We interpret this value as representing a reasonable amount of agreement, but the two psychologists did not evaluate the wards in complete synchrony, and some discussion between them to resolve at least some of the discrepancies should be attempted if at all possible.

PART 12

ANALYSIS OF STRUCTURE

CHAPTER 38

Principal Components and Factor Analysis

38.1 OVERVIEW OF PRINCIPAL COMPONENTS AND FACTOR ANALYSIS

Principal components and factor analysis comprise a family of exploratory data analysis procedures that are used to identify a relatively small number of dimensions or themes (the components or factors) underlying a relatively larger set of variables. These procedures represent ways to consolidate into a small number of synthesizing constructs information that is dispersed among several variables. They can be applied to a set of items on a survey (e.g., items assessing behavior on a range of health-related issues) or to a set of already-developed measures (e.g., personality characteristics). We can trace the beginnings of principal components analysis and factor analysis to Karl Pearson (1901b) and Charles Spearman (1904a), respectively. Principal components analysis was more fully developed by Harold Hotelling (1933, 1936), and factor analysis was transformed into its more modern form primarily by Louis L. Thurstone (1931, 1935).

Principal components and factor analyses each begins with the Pearson correlation matrix of the variables. Their mutual goal is to account for the variance in this matrix by fitting a series of weighted linear functions (components or factors) to the variables within the multidimensional space they occupy. The analysis is composed of two major phases: extraction and rotation.

Extraction can be accomplished through several methods. When that method is principal components, we label the analysis as the principal components analysis. Principal components is conceptually and statistically the simplest of the extraction techniques. The other methods as a set are known as factor analysis techniques; they include principal axis, maximum likelihood, and unweighted and generalized least-squares analyses. When one of these methods is used, we label the technique as a factor analysis.

In both principal components and factor analysis, the dimensions are extracted (straight line fit functions to correlate maximally with the variables) sequentially and are independent of each other (they are *orthogonal*). Thus, the first component or factor will explain what variance it can. The second component or factor must target the remaining (residual) variance to explain. The third component or factor must then address the variance not explained by the first two, and so on. Because they are orthogonal, we can add the amount of explained variance across the components or factors to speak of the cumulative amount of variance accounted for by the first d number of dimensions.

Performing Data Analysis Using IBM SPSS®, First Edition.
Lawrence S. Meyers, Glenn C. Gamst, and A. J. Guarino.

By virtue of the way components and factors are extracted, each successively extracted component or factor will explain less variance than those extracted before it. What usually happens is that the first few components or factors will cumulatively have accounted for a relatively large percentage of the to-be-explained variance, and we often quickly reach a point of diminishing returns. If we were to plot the amount of variance explained on the Y-axis as a function of components or factors extracted on the X-axis we would see a backwards J-shaped function, popularized by Raymond Cattell (1966) as a *scree plot*. The goal of the researchers in the extraction phase is to select a relatively few components or factors that cumulatively explain a fair amount of variance to bring into the rotation phase; the caveat to this is that, at the end of the rotation process, the component or factor structure needs to be meaningfully interpreted.

As was true for extraction, rotation can also be accomplished through several methods. At the end of the extraction phase, the components or factors can be conceived as intersecting at 90 degrees with respect to each other (they are orthogonal when intersecting at 90 degrees). To rotate the component or factor structure is to pivot it within the multidimensional space while the variables remain in place. This pivoting results in each component or factor being closer to (correlating more strongly with) some of the variables in that space while at the same time being further from (correlating less strongly with) some other variables. The pattern of correlations between the variables and the components or factors resulting from striving to meet this goal is termed *simple structure* after the original idea articulated by Louis L. Thurstone (1947).

One set of rotation procedures keeps the dimensions orthogonal; the most widely used of these is a varimax rotation. Another set of rotation procedures allows the components or factors to depart from 90 degrees; that is, it allows the angle between them to become oblique. These procedures are known as *oblique rotation strategies*, and intersecting obliquely results in the components or factors being correlated once the rotation is completed. One of the more commonly used oblique rotation strategies is a promax rotation.

Extraction results in the first component or factor as the strongest, the second next strongest, and so on; that is, many or most of the variables will correlate most strongly with the first component or factor, with the other components or factors falling off quickly in the order they were extracted. Because we interpret a component or factor based on the pattern of the correlations of the variables with it, and because the "strength" of the factors is a result of the algorithm used in the extraction, we do not ordinarily interpret the components or factors at the end of the extraction phase. Instead, interpretation of the components/factors resulting from the analysis is focused on the rotated factor structure. This is the case because the associations between the variables and the components or factors have been "equalized" to a large extent, providing each dimension with a viable profile of correlated variables of its own.

We will cover principal components and factor analysis in more detail as we go through the worked examples. To learn more about these procedures, readers are referred to Cudeck and MacCallum (2007), Gorsuch (2003), and Meyers et al. (2013) as the places to start with.

38.2 NUMERICAL EXAMPLE

Here, we illustrate how to perform and interpret principal components and factor analysis using a set of personality measures contained in the data file **Personality**. We will use 13 of the variables in the data file.

38.3 A STARTING PLACE

The starting place for principal components and factor analysis is the correlation matrix of the variables, and these are presented in Figure 38.1. With a few exceptions, the variables are correlated to a low or moderate extent. There are a couple of higher correlations that might suggest removal of one of the variables in the pair (because they could form the seed of what might turn out to be an artificial researcher-created factor), but we will retain all of the variables in our illustration just to have enough to show how these analyses work.

The correlation matrix shows the values of 1.000 on the diagonal of the matrix, and these are preserved in the process of performing a principal components analysis (the factor analytic procedures replace the 1's with some measure representing the relationship of each variable with the set of other variables). Principal components analysis focuses on explaining the total variance of the set of variables; in this context, the total variance is equal to the number of variables in the analysis (each variable contributes 1.000 units to the total variance). For our present example where we are analyzing 13 variables, the total variance achieves a value of 13.00. It is common practice to perform a preliminary analysis involving only the extraction phase in order to assess how we should proceed in the rest of the analysis, and this is what we do next using principal components analysis because it is the simplest of the extraction methods.

Correlation Matrix

		cntrlcpi self-control: cpi	acceptnc self-acceptance: poi	eminhib emotional inhibition: ecq	bencntrl benign control: ecq	aggcntrl aggression control: ecq	neoextra extraversion: neo	neoagree agreeableness: neo	posafect positive affect: mpq	negafect negative affect: mpq	neoneuro neuroticism: neo	depcon depression control: cecs	anxcon anxiety control: cecs	regard self-regard: poi
Correlation	cntrlcpi self-control: cpi	1.000	.124	.019	.584	.436	-.206	.393	.060	-.435	-.373	-.084	-.096	.241
	acceptnc self-acceptance: poi	.124	1.000	-.216	.183	.108	.155	.129	.266	-.404	-.432	-.218	-.101	.483
	eminhib emotional inhibition: ecq	.019	-.216	1.000	.039	.146	-.345	-.149	-.259	.128	.231	.640	.448	-.257
	bencntrl benign control: ecq	.584	.183	.039	1.000	.400	-.078	.389	.168	-.467	-.445	-.065	-.034	.319
	aggcntrl aggression control: ecq	.436	.108	.146	.400	1.000	-.068	.459	.063	-.188	-.146	.050	-.006	.098
	neoextra extraversion: neo	-.206	.155	-.345	-.078	-.068	1.000	.207	.528	-.218	-.347	-.226	-.268	.293
	neoagree agreeableness: neo	.393	.129	-.149	.389	.459	.207	1.000	.224	-.287	-.293	-.113	-.138	.240
	posafect positive affect: mpq	.060	.266	-.259	.168	.063	.528	.224	1.000	-.328	-.443	-.186	-.136	.416
	negafect negative affect: mpq	-.435	-.404	.128	-.467	-.188	-.218	-.287	-.328	1.000	.711	.168	.111	-.495
	neoneuro neuroticism: neo	-.373	-.432	.231	-.445	-.146	-.347	-.293	-.443	.711	1.000	.254	.187	-.581
	depcon depression control: cecs	-.084	-.218	.640	-.065	.050	-.226	-.113	-.186	.168	.254	1.000	.580	-.239
	anxcon anxiety control: cecs	-.096	-.101	.448	-.034	-.006	-.268	-.138	-.136	.111	.187	.580	1.000	-.151
	regard self-regard: poi	.241	.483	-.257	.319	.098	.293	.240	.416	-.495	-.581	-.239	-.151	1.000

FIGURE 38.1 Correlation matrix of the variables.

38.4 ANALYSIS SETUP: PRELIMINARY ANALYSIS

From the main menu we select **Analyze → Dimension Reduction → Factor**. This opens the main **Factor Analysis** window as shown in Figure 38.2. We move the following variables into the **Variables** panel: **cntrlcpi** (self-control), **acceptnc** (self-acceptance), **eminhib** (emotional inhibition), **bencntrl** (benign control of self), **aggcontrl** (control of one's aggressive tendencies), **neoextra** (extraversion), **neoagree** (agreeableness), **posafect** (positive affect), **negafect** (negative affect), **neoneuro** (neuroticism), **depcon** (control of depressive tendencies), **anxcon** (control of anxious tendencies), and **regard** (self-regard).

The **Descriptives** window is presented in Figure 38.3. Check both **Univariate descriptives** (this will indicate the number of cases in the analysis as a check to ensure that we have not lost too many cases due to missing values) and **Initial solution** (this will display the amount of variance explained by each component) in the **Statistics** area in the upper portion of the window.

In the **Correlation Matrix** area in the lower portion of the window shown in Figure 38.3, select **KMO and Bartlett's test of sphericity**. **KMO** stands for the Kaiser–Meyer–Olkin indicator of how adequate the correlations are for factor analysis; generally, a value of .70 or above is considered adequate (Kaiser, 1970, 1974). **Bartlett's**

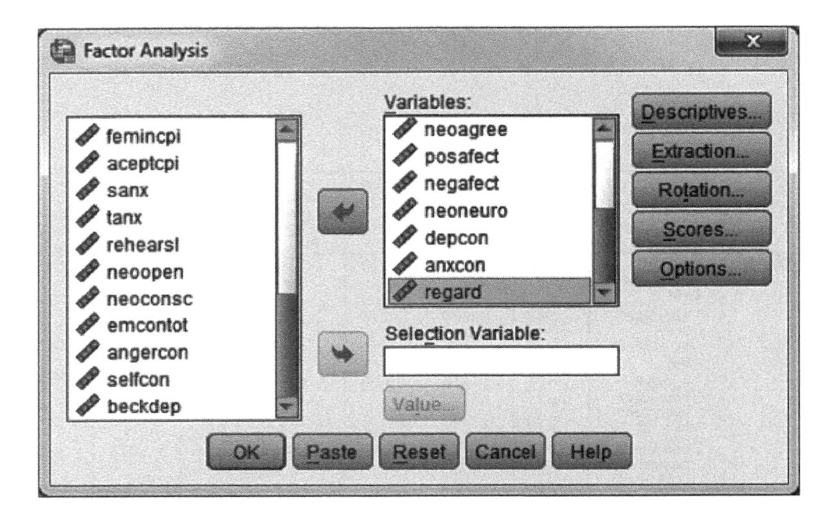

FIGURE 38.2
The main dialog window of **Factor Analysis**.

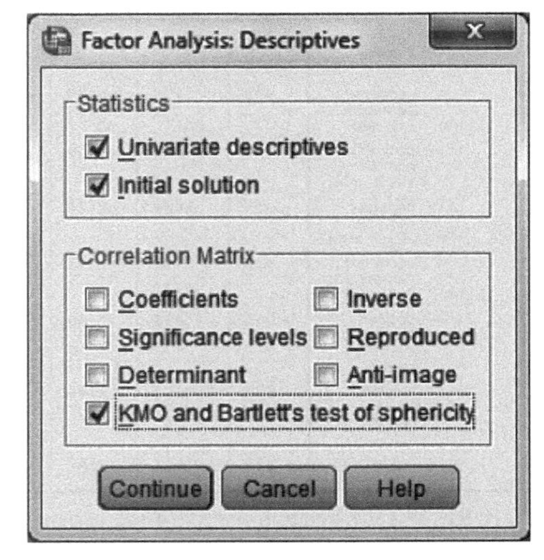

FIGURE 38.3
The **Descriptives** dialog window of **Factor Analysis**.

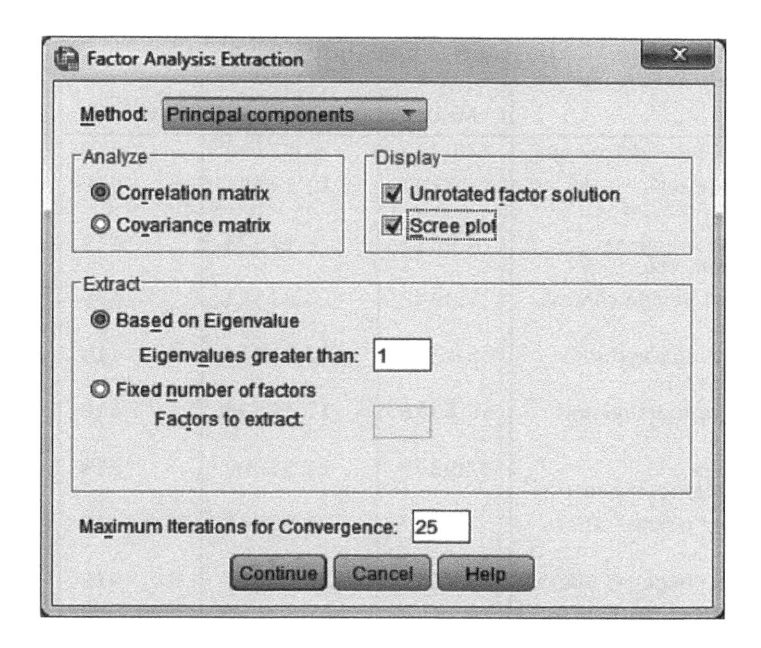

FIGURE 38.4
The **Extraction** dialog window of **Factor Analysis**.

test of sphericity tests the null hypothesis that the variables are not significantly correlated and should yield a statistically significant outcome before proceeding with the factor analysis. Click **Continue** to return to the main dialog window.

The **Extraction** window, shown in Figure 38.4, allows us to specify the method we wish to use; **Principal components** is the default **Method**, and we retain it for this analysis. We also retain the default of **Correlation matrix** as what we wish to **Analyze**. Under **Display**, we check both **Unrotated factor solution** (to obtain the correlations of the variables with the components) and **Scree plot** to obtain this classic graph. In the **Extract** area, we retain (only for this preliminary analysis) the default **Based on Eigenvalues greater than 1**. Click **Continue** to return to the main dialog window. Because the default in the **Rotation** screen is **None**, and because we will not perform a rotation in this preliminary analysis, there is no need to access this window. Click **OK** to perform the analysis.

38.5 ANALYSIS OUTPUT: PRELIMINARY ANALYSIS

The **Descriptive Statistics**, shown in Figure 38.5, indicates that we have lost only six of our original 425 cases, giving us a sample size of **419**, and so our sample seems virtually complete. With the **Kaiser–Meyer–Olkin Measure of Sampling Adequacy** exceeding .70 and **Bartlett's Test of Sphericity** being statistically significant, we have confidence of the appropriateness of the analysis.

The **Total Variance Explained** table is shown in Figure 38.6. There are two major vertical parts of the table. The **Initial Eigenvalues** portion of the table describes the initial analysis, which is always a principal components analysis (even if we specified one of the factor analysis extraction methods) and it is taken to completion. With 13 units of variance in this set of variables, it is possible to extract a maximum of 13 components. The third column under **Initial Eigenvalues** presents the **Cumulative** % of the variance that is accounted for with increasingly more components extracted. When all 13 have been extracted, they will cumulatively have explained 100% of the variance. Note that the largest chunks of explained variance are associated with the earliest extracted components; by the time we have extracted the fifth component, for example, we have

Descriptive Statistics

	Mean	Std. Deviation	Analysis N
cntrlcpi self-control: cpi	47.0723	8.92148	419
acceptnc self-acceptance: poi	53.8294	12.14851	419
eminhib emotional inhibition: ecq	4.9618	2.97573	419
bencntrl benign control: ecq	7.9642	2.81591	419
aggcntrl aggression control: ecq	8.4344	2.91217	419
neoextra extraversion: neo	55.8388	11.29538	419
neoagree agreeableness: neo	47.9305	11.96092	419
posafect positive affect: mpq	7.7017	2.91248	419
negafect negative affect: mpq	5.7613	3.71797	419
neoneuro neuroticism: neo	50.5110	11.14605	419
depcon depression control: cecs	2.1626	.64635	419
anxcon anxiety control: cecs	2.2542	.68142	419
regard self-regard: poi	55.3437	10.48484	419

KMO and Bartlett's Test

Kaiser-Meyer-Olkin Measure of Sampling Adequacy.		.798
Bartlett's Test of Sphericity	Approx. Chi-Square	2072.396
	df	78
	Sig.	.000

FIGURE 38.5

Descriptive statistics and the KMO and Bartlett's Test output.

explained almost 75% of the variance. The second column under **Initial Eigenvalues** presents the **% of Variance** that is accounted for by each component; these are simply added together to derive the **Cumulative %**.

The first column under **Initial Eigenvalues** labeled **Total** presents the *eigenvalues* associated with each component. Eigenvalues are one way to express the variance that is explained. In this analysis, there is a total of 13 units of variance. The eigenvalue associated with the first component has a value of 4.106, and 4.106/13.00 = 31.587, the percentage of variance explained by the first component. Eigenvalues are additive here because the components are orthogonal, and if we summed the column of eigenvalues, we would achieve a total of 13.00.

The set of columns under **Extraction Sums of Squared Loadings** provides the same information that we see in the **Initial Eigenvalues** columns but only for the first four components (because only these have eigenvalues of 1.00 or greater). It is the same information as in the **Initial Eigenvalues** portion of the table because the extraction procedure we specified was principal components; thus, the initial complete analysis, which is always principal components, is the same as our chosen extraction method,

Total Variance Explained

Component	Initial Eigenvalues			Extraction Sums of Squared Loadings		
	Total	% of Variance	Cumulative %	Total	% of Variance	Cumulative %
1	4.106	31.587	31.587	4.106	31.587	31.587
2	2.351	18.084	49.670	2.351	18.084	49.670
3	1.426	10.966	60.637	1.426	10.966	60.637
4	1.095	8.422	69.058	1.095	8.422	69.058
5	.758	5.829	74.888			
6	.573	4.406	79.294			
7	.537	4.131	83.424			
8	.468	3.599	87.023			
9	.414	3.187	90.211			
10	.406	3.121	93.332			
11	.318	2.447	95.779			
12	.286	2.197	97.975			
13	.263	2.025	100.000			

Extraction Method: Principal Component Analysis.

FIGURE 38.6 The **Total Variance Explained** table.

hence the outcome is identical. In summary, the first four components cumulatively accounted for 69% of the total variance.

There were only four components extracted because in the **Extraction** window, we had retained the default extraction criterion of **Based on Eigenvalues greater than 1**. Generally, researchers are very hesitant to extract (and thus subsequently rotate) components whose eigenvalue is less than 1.000, in that, at least in a principal components analysis, a single variable is worth that much variance and we would like a component to do at least as well. Even then, researchers are equally as hesitant to take into the rotation stage the full set of components whose eigenvalues exceed 1.000, as they are concerned that the set of components would be inappropriately large.

Figure 38.7 displays the scree plot; it plots the eigenvalues that are contained in the **Initial Eigenvalues** column against the components in the full principal components solution. The function appears to start leveling out at approximately the fourth or fifth component, which indicates that we want fewer components in the solution. On the basis of the values in the **Initial Eigenvalues** column and a visual inspection of them in the scree plot, it would appear that

- we make reasonably substantial gains in explained variance through three components, and thus probably want to retain at least (the first) three components in the final solution;
- we may gain enough variance to warrant including the fourth component (depending on how the rotated components are to be interpreted), but it is not clear if this will be one too many.

The **Communalities** of the variables are displayed in Figure 38.8. The column labeled **Initial** represents the values on the diagonal of the correlation matrix when the principal components method was applied and run to completion. These values are all 1.000. One way to interpret these 1's is to think of each of the variables as being "fully in" or "fully captured by" the dimensional structure; because they are fully captured by the dimensional structure, principal components attempts to explain the total amount of variance in the set of variables.

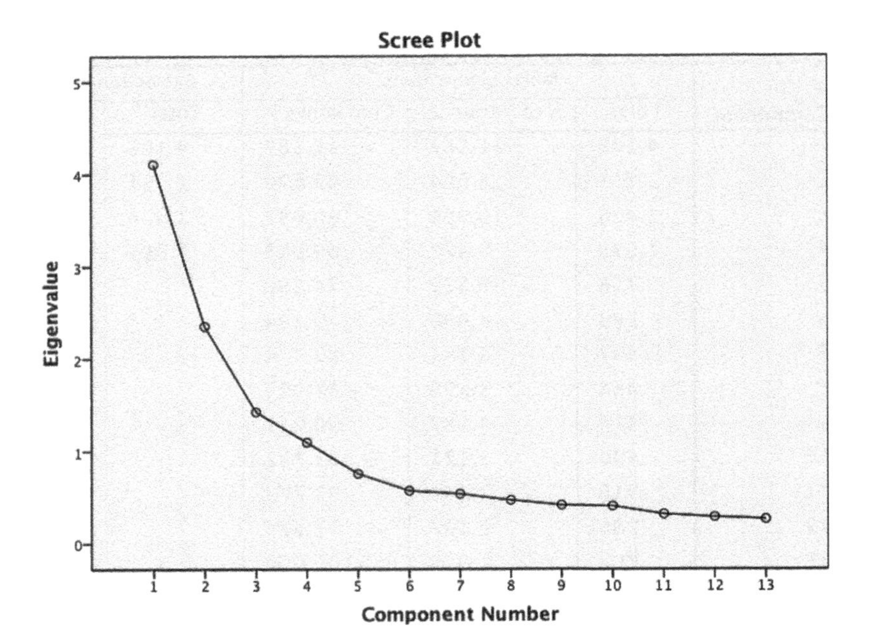

FIGURE 38.7
The **Scree Plot**.

Communalities

	Initial	Extraction
cntrlcpi self-control: cpi	1.000	.732
acceptnc self-acceptance: poi	1.000	.518
eminhib emotional inhibition: ecq	1.000	.698
bencntrl benign control: ecq	1.000	.656
aggcntrl aggression control: ecq	1.000	.647
neoextra extraversion: neo	1.000	.808
neoagree agreeableness: neo	1.000	.704
posafect positive affect: mpq	1.000	.671
negafect negative affect: mpq	1.000	.693
neoneuro neuroticism: neo	1.000	.746
depcon depression control: cecs	1.000	.797
anxcon anxiety control: cecs	1.000	.674
regard self-regard: poi	1.000	.633

Extraction Method: Principal Component Analysis.

FIGURE 38.8
Communalities of the variables.

The column labeled **Extraction** in the **Communalities** table describes the percentage of variance of each variable subsumed in the number of factors that were ultimately extracted (four in the present instance). The four extracted factors cumulatively accounted for 69% of the total variance, and so there is still unexplained variance remaining. The variable whose variance has been best captured in the four-component solution is **neoextra** with a communality of .808, and the variable whose variance has been least captured in the four-component solution is **acceptnc** with a communality of .518. Despite these differences, however, we judge that all of the variables are "participating" substantially in the four-component solution. Generally, we would hope that variables would be associated with communality values at least in the high .4's to be able to say that we have captured enough of the variance of a variable to be worthy of inclusion in the component or factor structure.

The **Component Matrix** is shown in Figure 38.9. The values in the table are the Pearson correlations between each of the variables and each of the components. Squaring the correlations and adding them provides some of the information we described earlier. The sum of squared correlations down each column is equal to the eigenvalue of that component. As is typical, the first component is more strongly associated with the variables as a set than any of the others because it accounts for more variance than any of the other components; this pattern propagates across the columns (components). The sum of squared correlations across each row is equal to the extracted communalities for each variable.

Component Matrix[a]

	Component			
	1	2	3	4
cntrlcpi self-control: cpi	.504	.586	-.338	-.139
acceptnc self-acceptance: poi	.557	-.066	.268	-.364
eminhib emotional inhibition: ecq	-.440	.633	.319	.048
bencntrl benign control: ecq	.568	.556	-.138	-.077
aggcntrl aggression control: ecq	.323	.578	-.273	.366
neoextra extraversion: neo	.435	-.516	.306	.509
neoagree agreeableness: neo	.533	.286	-.255	.522
posafect positive affect: mpq	.580	-.236	.378	.369
negafect negative affect: mpq	-.748	-.206	-.212	.216
neoneuro neuroticism: neo	-.816	-.052	-.241	.142
depcon depression control: cecs	-.471	.538	.501	.188
anxcon anxiety control: cecs	-.389	.466	.552	.001
regard self-regard: poi	.717	-.050	.289	-.180

Extraction Method: Principal Component Analysis.

a. 4 components extracted.

FIGURE 38.9

The **Component Matrix**.

38.6 OUR ANALYSIS STRATEGY FOR THE MAIN ANALYSES

There are three separate issues that need to be resolved now that we have the results of the preliminary analysis:

- whether an oblique or orthogonal rotation method should be used;
- whether we will finalize our solution using principal components or one of the factor analytic methods for our extraction;
- whether we will retain for the rotation phase three or four components or factors in the final solution.

These three issues are intertwined, in that all three elements must be specified in any single analysis that we perform. We use the following strategy to address these issues:

- We will perform an oblique (promax) rotation and examine the correlations between the components or factors that is a standard part of the output. If the correlations are in the low teens or less, we will judge the components or factors sufficiently unrelated to revert to a varimax rotation; if some of the correlations are close to .30 or higher, we will judge them sufficiently related to retain the promax rotation; if the correlations fall between these two general criteria, our bias will be to stay with the promax rotation.
- We will perform a set of analyses using principal components and, to simplify our presentation, another set of analyses using principal axis factoring (we have examined all of the major factor analytic methods and they produce very similar results; to illustrate our strategy, we limit ourselves to principal axis as representative of the factor analysis methods).
- We will examine both the three- and the four-component/factor solutions. Based on the outcome of these exploratory ventures, we will select a solution that appears to be reasonable from both the statistical and research-oriented perspectives. Because it will supply a bit more information by virtue of giving us a potentially extra component/factor, we opt to obtain the four-dimensional solution first and back-step into the three-dimensional solution, but it would be equally reasonable to perform these analyses in the reverse order. It should be pointed out here that *the three-component/factor solution is not the same as the first three components/factors of the four-component/factor rotated solution*; rather, rotating three components/factors in the multidimensional space will create a different correlation pattern between the variables and the components/factors than that is obtained when rotating four components/factors.

38.7 ANALYSIS SETUP FOR THE FOUR-FACTOR STRUCTURE

From the main menu we select **Analyze ➔ Dimension Reduction ➔ Factor** and move the variables into the **Variables** panel, as was done in the preliminary analysis (see Section 38.4).

In the **Descriptives** window presented in Figure 38.10, check only **Initial solution** in the **Statistics** area (we have already obtained the **Univariate descriptives** and **KMO and Bartlett's test of sphericity** in the preliminary analysis and do not need to generate them again). Click **Continue** to return to the main dialog window.

In the **Extraction** window, shown in Figure 38.11, we select **Principal axis factoring** from the **Method** drop-down menu. The dimensions resulting from this analysis are referred to as factors rather than as components. We will perform an identical analysis

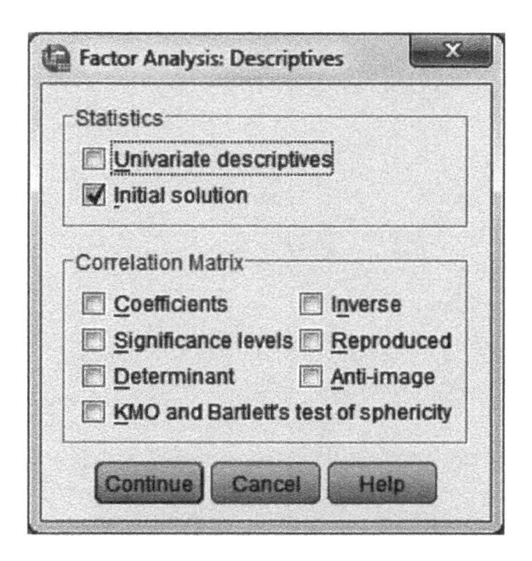

FIGURE 38.10

The **Descriptives** dialog window of **Factor Analysis**.

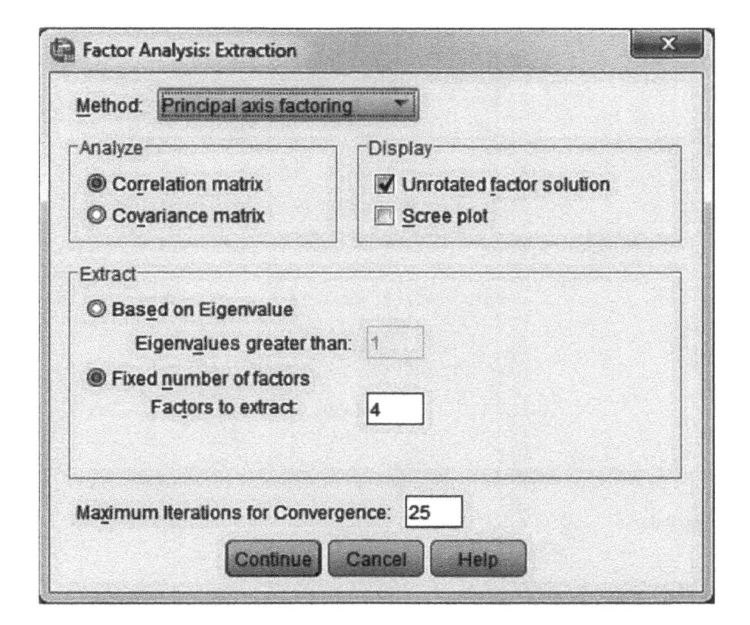

FIGURE 38.11

The **Extraction** dialog window of **Factor Analysis**.

after this, except that we will have selected **Principal components** (without showing the dialog screens). We request under **Display** the **Unrotated factor solution** but not the **Scree plot** (we have already obtained it in the preliminary analysis). In the **Extract** area, we select **Fixed number of factors** and in the panel labeled **Factors to extract** we type **4**. This specifies the number of factors that will be brought into the rotation phase. Click **Continue** to return to the main dialog window.

The **Rotation** window is shown in Figure 38.12. We select **Promax** and retain the **Kappa** value at the default of **4** (one of the steps in a promax rotation is to raise the correlations of the variables with the factors to a power labeled as *kappa* and the default is the fourth power—see Meyers et al. (2013) for a fuller description of this rotation strategy). In the **Display** area, we wish to see the **Rotated solution**. Click **Continue** to return to the main dialog window.

The **Options** window is shown in Figure 38.13. We retain the default **Missing Values** procedure of **Exclude cases listwise** (all cases must have valid values on all of the variables to be included in the analysis). Under **Coefficient Display Format**, we select **Sorted by size**; this will order the listing of the variables in the factor matrix by

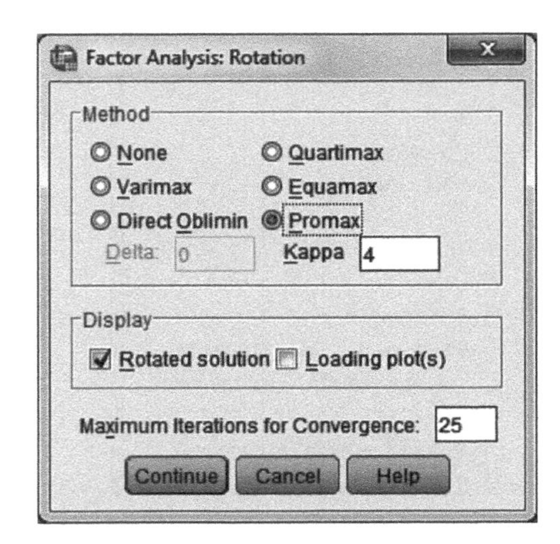

FIGURE 38.12
The **Rotation** dialog window of **Factor Analysis**.

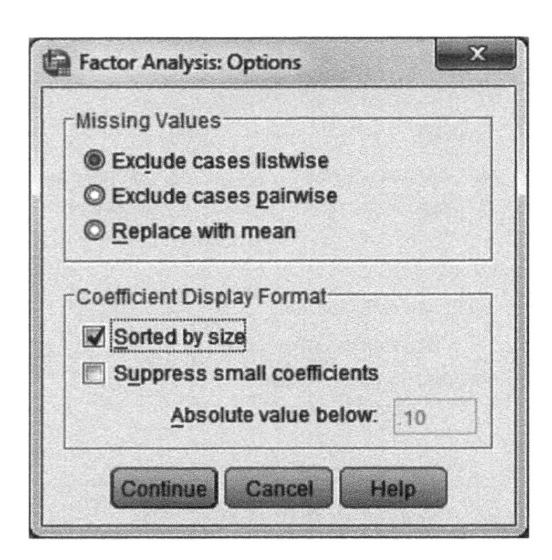

FIGURE 38.13
The **Options** dialog window of **Factor Analysis**.

the strength of their relationship to the factor, thereby making the output more convenient to read (in our judgment). Click **Continue** to return to the main dialog window and click **OK** to perform the analysis.

38.8 ANALYSIS OUTPUT FOR THE FOUR-COMPONENT/FACTOR STRUCTURE

Analysis Output for the Four-Component/Factor Structure: Total Variance Explained

The **Total Variance Explained** tables for the principal components analysis (top table) and the principal axis factor analysis (bottom table) are presented in Figure 38.14. We have already seen the principal components table except for the last column that now contains information about the rotation phase. The last column (**Rotation Sums of Squared Loadings**) displays the eigenvalues for the components after rotation, where the explained variance is distributed more evenly across the components. In an orthogonal rotation, these eigenvalues would add to the same value as the sum of the eigenvalues for the four components at the end of the extraction phase (8.978); because we have used an

Principal Components Analysis: Variance Explained

Total Variance Explained

Component	Initial Eigenvalues			Extraction Sums of Squared Loadings			Rotation Sums of Squared Loadings[a]
	Total	% of Variance	Cumulative %	Total	% of Variance	Cumulative %	Total
1	4.106	31.587	31.587	4.106	31.587	31.587	3.487
2	2.351	18.084	49.670	2.351	18.084	49.670	2.777
3	1.426	10.966	60.637	1.426	10.966	60.637	2.509
4	1.095	8.422	69.058	1.095	8.422	69.058	2.061
5	.758	5.829	74.888				
6	.573	4.406	79.294				
7	.537	4.131	83.424				
8	.468	3.599	87.023				
9	.414	3.187	90.211				
10	.406	3.121	93.332				
11	.318	2.447	95.779				
12	.286	2.197	97.975				
13	.263	2.025	100.000				

Extraction Method: Principal Component Analysis.

a. When components are correlated, sums of squared loadings cannot be added to obtain a total variance.

Principal Axis Factor Analysis: Variance Explained

Total Variance Explained

Factor	Initial Eigenvalues			Extraction Sums of Squared Loadings			Rotation Sums of Squared Loadings[a]
	Total	% of Variance	Cumulative %	Total	% of Variance	Cumulative %	Total
1	4.106	31.587	31.587	3.691	28.395	28.395	3.260
2	2.351	18.084	49.670	1.953	15.021	43.416	2.425
3	1.426	10.966	60.637	1.032	7.937	51.353	2.211
4	1.095	8.422	69.058	.666	5.123	56.475	1.700
5	.758	5.829	74.888				
6	.573	4.406	79.294				
7	.537	4.131	83.424				
8	.468	3.599	87.023				
9	.414	3.187	90.211				
10	.406	3.121	93.332				
11	.318	2.447	95.779				
12	.286	2.197	97.975				
13	.263	2.025	100.000				

Extraction Method: Principal Axis Factoring.

a. When factors are correlated, sums of squared loadings cannot be added to obtain a total variance.

FIGURE 38.14 Total Variance Explained for both the four-dimensional principal components analysis and the principal axis factor analysis.

oblique rotation strategy (where the factors are permitted to correlate), the sum of the eigenvalues based on the rotated components is a larger value (10.824).

The **Total Variance Explained** table for the principal axis factor analysis is new. We can see that the first major column labeled **Initial Eigenvalues** is identical to that in the principal components analysis (except the dimensions are now labeled as factors) because the first pass through the analysis is a principal components analysis taken to completion.

The **Extraction Sums of Squared Loadings** presents a different picture from the principal components analysis results, and this highlights a major difference between principal components analysis and the factor analysis procedures. In principal components analysis, we retain the values of 1.000 on the diagonal of the correlation matrix and thus attempt to explain the total variance. In factor analysis, the values of 1.000 are replaced by some estimate of the amount of variance each variable shares (has in common) with the other variables. The total variance is thus partitioned into variance that is common to the set of variables in the analysis and variance that is unique or specific to each variable, and the factor analysis methods attempt to explain only the common variance; that is, they identify dimensions that are descriptive of the variance common to all of the variables. As the common variance is less than the total variance, factor analysis methods will explain less variance than principal components.

In Figure 38.14, we see that the first four factors in the principal axis solution account for 56.475% of the total variance, whereas the first four components in the principal components solution account for 69.058% of the total variance. Furthermore, the fourth principal axis factor presents us with an eigenvalue substantially lower than most researchers' lower limit of 1.000, even though the fourth principal component squeaked by with an eigenvalue of 1.095. To the extent that researchers are drawn toward factor analysis over principal components (it is our view that the field is pretty evenly divided on this issue), this would probably be sufficient for those researchers to abandon the four-factor solution and examine the three-factor solution.

Analysis Output for the Four-Component/Factor Structure: Communalities

Figure 38.15 presents the **Communalities** for the variables. We have already discussed these for the principal components analysis, and we can see considerable differences in the principal axis factor analysis. The **Initial** communalities shown in the first column of the principal axis factor analysis are not the 1.000's we had in the principal components analysis but rather the R^2's between the particular variable and the others (because we focus here on the common variance). For example, if we predicted **cntrlcpi** from the other variables in a multiple regression analysis, we would obtain an R^2 value of .515. This **Initial** estimate is reestimated through an iterative process that resulted in a final estimated strength of relationship provided in the **Extraction** column. Different factor analysis procedures use somewhat different algorithms in these calculations, but communalities derived from factor analysis procedures, such as principal axis factoring, will generally be lower than those derived from principal components analysis.

Analysis Output for the Four-Component/Factor Structure: Component/Factor Correlations

One of the issues that we hoped to resolve in this set of analyses was whether an oblique rotation strategy was appropriate here. The component/factor correlation matrices shown in Figure 38.16 provide the information we use to resolve this issue. The same correlation pattern is exhibited in both tables, and so we highlight the correlations from the principal axis factor analysis. Some pairs of components/factors are not correlated (e.g., Factors

Principal Components Analysis: Communalities

Communalities

	Initial	Extraction
cntrlcpi self-control: cpi	1.000	.732
acceptnc self-acceptance: poi	1.000	.518
eminhib emotional inhibition: ecq	1.000	.698
bencntrl benign control: ecq	1.000	.656
aggcntrl aggression control: ecq	1.000	.647
neoextra extraversion: neo	1.000	.808
neoagree agreeableness: neo	1.000	.704
posafect positive affect: mpq	1.000	.671
negafect negative affect: mpq	1.000	.693
neoneuro neuroticism: neo	1.000	.746
depcon depression control: cecs	1.000	.797
anxcon anxiety control: cecs	1.000	.674
regard self-regard: poi	1.000	.633

Extraction Method: Principal Component Analysis.

Principal Axis Factor Analysis: Communalities

Communalities

	Initial	Extraction
cntrlcpi self-control: cpi	.515	.629
acceptnc self-acceptance: poi	.309	.302
eminhib emotional inhibition: ecq	.490	.554
bencntrl benign control: ecq	.470	.546
aggcntrl aggression control: ecq	.350	.450
neoextra extraversion: neo	.486	.785
neoagree agreeableness: neo	.370	.521
posafect positive affect: mpq	.388	.435
negafect negative affect: mpq	.568	.629
neoneuro neuroticism: neo	.635	.741
depcon depression control: cecs	.532	.821
anxcon anxiety control: cecs	.378	.419
regard self-regard: poi	.447	.512

Extraction Method: Principal Axis Factoring.

FIGURE 38.15 **Communalities** of the variables for both the four-dimensional principal components analysis and the principal axis factor analysis.

2 and 3 show an *r* value of −**.047**), but other pairs are relatively highly correlated (e.g., Factors 1 and 2 show an *r* value of **.517**). The presence of any correlations in the .3's or greater supports the use of an oblique rotation; thus, we move forward with the interpretation based on the obliquely rotated solution and do not pursue a varimax (orthogonal) rotated solution (although the interpretations of the varimax and promax rotations are likely to coincide).

Analysis Output for the Four-Component/Factor Structure: Rotated Component/Factor Structure Coefficients

We are finally ready to examine the rotated solutions, which are presented in a component or a factor matrix. An oblique rotation strategy yields two such matrices, a *pattern* matrix and a *structure* matrix, that present, respectively, pattern coefficients and structure coefficients.

Recall that a component or factor is a weighted linear combination of the variables in the analysis; the pattern coefficients are those weights (analogous to beta coefficients in multiple regression). If the variables are highly correlated, it is possible for these to exceed the value of 1.00. Structure coefficients are the correlation between the particular variable and the component or factor. They are true correlations and must range between ±1.00. In most analyses, these two matrices produce very similar values and yield the same interpretation. Researchers appear to be divided on which of the two matrices to

Principal Components Analysis: Component Correlations

Component Correlation Matrix

Component	1	2	3	4
1	1.000	.423	-.251	.190
2	.423	1.000	-.042	-.048
3	-.251	-.042	1.000	-.291
4	.190	-.048	-.291	1.000

Extraction Method: Principal Component Analysis.
Rotation Method: Promax with Kaiser Normalization.

Principal Axis Factor Analysis: Factor Correlations

Factor Correlation Matrix

Factor	1	2	3	4
1	1.000	.517	-.312	.221
2	.517	1.000	-.047	-.112
3	-.312	-.047	1.000	-.333
4	.221	-.112	-.333	1.000

Extraction Method: Principal Axis Factoring.
Rotation Method: Promax with Kaiser Normalization.

FIGURE 38.16

Factor/Component Correlation Matrices of the variables for both the four-dimensional principal components analysis and the principal axis factor analysis.

present, but our preference is to interpret the structure matrix and these are presented for the principal components and principal axis factor solutions in Figure 38.17 (note that an orthogonal rotation strategy such as varimax yields only a single rotated component/factor matrix because the pattern and structure coefficients take on the same values.)

It is the results of the structure (or pattern) matrix that researchers interpret; that is, we use the structure (or pattern) coefficients to interpret in plain language the dimension represented by the component or factor. As the results for the two analyses are similar, we will work through the principal axis solution.

The **Structure Matrix** for the principal axis solution shown in Figure 38.17 contains the variables as the rows and the factors as columns. Entries are the structure coefficients, that is, the correlation of each variable with each factor. Higher correlations indicate a stronger relationship between the two. To the extent that the relationship between the variable and the factor is reasonably strong, the variable can be conceived as an indicator (or aspect) of the factor.

Looking across the rows (across the factors for each variable), we see that most of the variables tend to correlate reasonably highly with (colloquially referred to as the variable "loading on") just one factor (this was the goal of simple structure). An exception is **bencntrl**, as this variable correlates at least somewhat highly with (colloquially referred to as the variable "cross-loading on") both Factors 1 and 2.

With most of the variables relatively unambiguously "loading" on one of the factors, we can proceed to interpret the factors. To interpret the factors, we work vertically down each column separately, and this is why we sorted the coefficients by size in the **Options** dialog window. The variables are listed by factor and are **sorted by size** (the value of the coefficient) within each factor. Thus, the first four variables are most highly correlated with Factor 1, the next four variables are most highly correlated with Factor 2, the next three variables are most highly correlated with Factor 3, and the last two are most highly correlated with Factor 4.

Principal Components Analysis: Component Structure Matrix

Structure Matrix

	Component			
	1	2	3	4
neoneuro neuroticism: neo	-.843	-.403	.264	-.327
negafect negative affect: mpq	-.822	-.453	.153	-.159
regard self-regard: poi	.770	.249	-.256	.334
acceptnc self-acceptance: poi	.687	.096	-.213	.151
aggcntrl aggression control: ecq	.155	.776	.088	-.047
cntrlcpi self-control: cpi	.480	.768	-.085	-.311
neoagree agreeableness: neo	.251	.747	-.182	.303
bencntrl benign control: ecq	.564	.736	-.019	-.153
depcon depression control: cecs	-.276	-.040	.886	-.184
eminhib emotional inhibition: ecq	-.236	.063	.816	-.378
anxcon anxiety control: cecs	-.132	-.110	.809	-.207
neoextra extraversion: neo	.217	-.040	-.333	.895
posafect positive affect: mpq	.451	.169	-.201	.748

Extraction Method: Principal Component Analysis.
Rotation Method: Promax with Kaiser Normalization.

Principal Axis Factor Analysis: Factor Structure Matrix

Structure Matrix

	Factor			
	1	2	3	4
neoneuro neuroticism: neo	-.854	-.426	.287	-.293
negafect negative affect: mpq	-.787	-.471	.178	-.139
regard self-regard: poi	.699	.288	-.282	.295
acceptnc self-acceptance: poi	.534	.173	-.230	.174
cntrlcpi self-control: cpi	.469	.739	-.074	-.292
bencntrl benign control: ecq	.540	.691	-.033	-.148
neoagree agreeableness: neo	.327	.650	-.170	.209
aggcntrl aggression control: ecq	.205	.648	.073	-.090
depcon depression control: cecs	-.290	-.050	.900	-.205
eminhib emotional inhibition: ecq	-.253	.042	.727	-.376
anxcon anxiety control: cecs	-.176	-.089	.640	-.232
neoextra extraversion: neo	.279	-.045	-.343	.881
posafect positive affect: mpq	.470	.169	-.245	.553

Extraction Method: Principal Axis Factoring.
Rotation Method: Promax with Kaiser Normalization.

FIGURE 38.17 **Factor/Component Structure Matrices** of the variables for both the four-dimensional principal components analysis and the principal axis factor analysis.

Researchers use the values of the correlations between the variables and the factor as a cue to factor interpretation. As a general guideline, structure coefficients of about .80 or higher are taken as very strong indicators, coefficients in the .70's are taken as relatively strong indicators, those in the .5's and .6's are taken as moderate indicators, and those in the .3's and .4's are taken as very modest indicators.

Given these guidelines, we interpret the factors based on those variables correlating most strongly with it. What those variables conceptually have in common with each other should signify the core of the factor. Based on the correlations, we interpret the factors as follows:

- Factor 1 is associated with (indicated most strongly by) lower levels of **neoneuro** and **negafect** (these are negatively correlated with the factor) and higher levels of **regard** and **acceptnc** (these are positively correlated with the factor). What construct is indicated by less neuroticism, less negative affect, more self-regard, and more acceptance of self? The answer is the interpretation of the factor, and we subjectively label it as Mental Health.
- Factor 2 is indicated by higher values of **cntrlcpi**, **bencntrl**, **neoagree**, and **aggcontrl** and appears to represent Emotional Control.
- Factor 3 is indicated by higher levels of **depcon**, **eminhib**, and **anxcon** and appears to represent Inhibition of Depression.
- Factor 4 is indicated by higher levels of **neoextra** and **posafect** and appears to represent Positive Outgoingness.

Analysis Output for the Four-Component/Factor Structure: Evaluation

The four-factor solution was interpretable, and the values of the largest structure coefficients for each variable were acceptable. This outcome therefore appears to be viable. That said, the solution has four relative weaknesses to it:

- Using the four-component solution means accepting all four components whose eigenvalues exceeded 1.00, something that researchers are reluctant to do; in the principal axis factor analysis, the fourth factor has an uncomfortably low eigenvalue.
- One pair of factors correlated (in the principal axis analysis) in excess of .50; this is a somewhat higher factor correlation than that would be preferred (correlations in excess of the .4's should cause some concerns), and at some point, perhaps in the range .70, factors correlating too strongly must be judged as redundant.
- One variable (**bencntrl**) had a rather substantial "cross-loading."
- One factor (**Factor 4**) had only two variables as indicators; even with only 13 variables in the analysis and using a four-component/factor solution, we would like to have at least three indicator variables for a factor.

Some of these weaknesses may potentially be overcome in the three-component/factor solutions. If this comes to pass and the interpretation is at least equally viable, then the three-dimensional solution may be preferred.

38.9 THE THREE-COMPONENT/FACTOR STRUCTURE

We set up the analyses exactly as described in Section 38.8 except that we type **3** in the panel labeled **Factors to extract**.

The **Total Variance Explained** tables for the two analyses are presented in Figure 38.18. Consistent with the four-dimensional solution, the first three principal components cumulatively explain 60.637% of the total variance. The principal axis factor solution has accounted for less variance because it was targeting the common rather than the total variance; the first three principal axis factors cumulatively explain 49.977% of the total variance. Furthermore, the third principal axis factor is associated with an eigenvalue just under the general criterion of 1.00; however, that may be acceptable if the factor structure lends itself to a reasonable interpretation.

Figure 38.19 presents the **Communalities** for the variables. In the principal components analysis, most of the variables have **Extraction** communality values in excess of the high .4's, partly because the analysis explains more than 60% of the variance, whereas in the principal axis factor analysis, several variables have relatively low **Extraction** communality values.

The component/factor correlation matrices are shown in Figure 38.20. Components/Factors 2 and 3 are uncorrelated but the two other pairs are sufficiently correlated to warrant an oblique rotation but not sufficiently correlated so that we judge them to represent the same underlying construct.

The structure matrices from the principal components and principal axis factor analyses resulting from the promax rotations are shown in Figure 38.21. They present the same dimensional structure, but the correlations from the principal components analysis are substantially stronger; this should not come as a surprise because this analysis accounted for more variance and, as a result, yielded larger communality values for the variables. We focus on the principal axis factor structure (for explication purposes) and interpret the factors as follows:

Principal Components Analysis: Variance Explained

Total Variance Explained

Component	Initial Eigenvalues			Extraction Sums of Squared Loadings			Rotation Sums of Squared Loadings[a]
	Total	% of Variance	Cumulative %	Total	% of Variance	Cumulative %	Total
1	4.106	31.587	31.587	4.106	31.587	31.587	3.602
2	2.351	18.084	49.670	2.351	18.084	49.670	2.914
3	1.426	10.966	60.637	1.426	10.966	60.637	2.523
4	1.095	8.422	69.058				
5	.758	5.829	74.888				
6	.573	4.406	79.294				
7	.537	4.131	83.424				
8	.468	3.599	87.023				
9	.414	3.187	90.211				
10	.406	3.121	93.332				
11	.318	2.447	95.779				
12	.286	2.197	97.975				
13	.263	2.025	100.000				

Extraction Method: Principal Component Analysis.

a. When components are correlated, sums of squared loadings cannot be added to obtain a total variance.

Principal Axis Factor Analysis: Variance Explained

Total Variance Explained

Factor	Initial Eigenvalues			Extraction Sums of Squared Loadings			Rotation Sums of Squared Loadings[a]
	Total	% of Variance	Cumulative %	Total	% of Variance	Cumulative %	Total
1	4.106	31.587	31.587	3.640	28.003	28.003	3.202
2	2.351	18.084	49.670	1.894	14.567	42.570	2.549
3	1.426	10.966	60.637	.963	7.407	49.977	2.159
4	1.095	8.422	69.058				
5	.758	5.829	74.888				
6	.573	4.406	79.294				
7	.537	4.131	83.424				
8	.468	3.599	87.023				
9	.414	3.187	90.211				
10	.406	3.121	93.332				
11	.318	2.447	95.779				
12	.286	2.197	97.975				
13	.263	2.025	100.000				

Extraction Method: Principal Axis Factoring.

a. When factors are correlated, sums of squared loadings cannot be added to obtain a total variance.

FIGURE 38.18 Total Variance Explained for both the three-dimensional principal components analysis and the principal axis factor analysis.

Principal Components Analysis: Communalities

Communalities

	Initial	Extraction
cntrlcpi self-control: cpi	1.000	.713
acceptnc self-acceptance: poi	1.000	.386
eminhib emotional inhibition: ecq	1.000	.696
bencntrl benign control: ecq	1.000	.650
aggcntrl aggression control: ecq	1.000	.513
neoextra extraversion: neo	1.000	.549
neoagree agreeableness: neo	1.000	.432
posafect positive affect: mpq	1.000	.535
negafect negative affect: mpq	1.000	.646
neoneuro neuroticism: neo	1.000	.726
depcon depression control: cecs	1.000	.762
anxcon anxiety control: cecs	1.000	.674
regard self-regard: poi	1.000	.600

Extraction Method: Principal Component Analysis.

Principal Axis Factor Analysis: Communalities

Communalities

	Initial	Extraction
cntrlcpi self-control: cpi	.515	.687
acceptnc self-acceptance: poi	.309	.265
eminhib emotional inhibition: ecq	.490	.576
bencntrl benign control: ecq	.470	.564
aggcntrl aggression control: ecq	.350	.317
neoextra extraversion: neo	.486	.433
neoagree agreeableness: neo	.370	.282
posafect positive affect: mpq	.388	.407
negafect negative affect: mpq	.568	.580
neoneuro neuroticism: neo	.635	.709
depcon depression control: cecs	.532	.744
anxcon anxiety control: cecs	.378	.433
regard self-regard: poi	.447	.500

Extraction Method: Principal Axis Factoring.

FIGURE 38.19 **Communalities** of the variables for both the three-dimensional principal components analysis and the principal axis factor analysis.

- Factor 1 is associated with (indicated most strongly by) lower levels of **neoneuro** and **negafect** (these are negatively correlated with the factor) and higher levels of **regard**, **posafect**, **neoextra**, and **acceptnc** (these are positively correlated with the factor). Note that this factor is a combination of Factors 1 and 4 in the four-factor solution; that is, the first factor in the three-factor solution splits into two in the four-factor solution. We can imagine people who have higher values on this factor to be stable, confident, and positive in their attitude; if we had to provide a label to this construct, one possibility might be Positive Life Adjustment.
- Factor 2 is indicated by higher values of **cntrlcpi**, **bencntrl**, **aggcontrl**, and **neoagree** and appears to represent Emotional Control.
- Factor 3 is indicated by higher levels of **depcon**, **eminhib**, and **anxcon** and appears to represent Inhibition of Depression.

38.10 DETERMINING WHICH SOLUTION TO ACCEPT

Overall, the three-component/factor structure has overcome most of the weaknesses of the four-component/factor structure while supporting a viable interpretation of the dimensions. The one remaining weakness is a little "cross-loading" for **negafect** in both analyses and for **neoagree** in the principal axis factor solution, but there was some "cross-loading" in the four-component/factor solution anyway. We therefore are very comfortable in accepting the three-component/factor solution over the four-component/factor solution.

Principal Components Analysis: Component Correlations

Component Correlation Matrix

Component	1	2	3
1	1.000	.297	-.339
2	.297	1.000	.025
3	-.339	.025	1.000

Extraction Method: Principal Component Analysis.
Rotation Method: Promax with Kaiser Normalization.

Principal Axis Factor Analysis: Factor Correlations

Factor Correlation Matrix

Factor	1	2	3
1	1.000	.375	-.394
2	.375	1.000	.009
3	-.394	.009	1.000

Extraction Method: Principal Axis Factoring.
Rotation Method: Promax with Kaiser Normalization.

FIGURE 38.20

Factor/Component Correlation Matrices of the variables for both the three-dimensional principal components analysis and the principal axis factor analysis.

Principal Components Analysis: Component Structure Matrix

Structure Matrix

	Component		
	1	2	3
neoneuro neuroticism: neo	-.822	-.460	.242
regard self-regard: poi	.770	.311	-.237
negafect negative affect: mpq	-.723	-.539	.117
posafect positive affect: mpq	.715	.064	-.249
acceptnc self-acceptance: poi	.620	.208	-.179
neoextra extraversion: neo	.601	-.205	-.407
cntrlcpi self-control: cpi	.215	.841	-.028
bencntrl benign control: ecq	.360	.794	.026
aggcntrl aggression control: ecq	.081	.703	.091
neoagree agreeableness: neo	.325	.615	-.204
depcon depression control: cecs	-.304	-.056	.869
eminhib emotional inhibition: ecq	-.367	.089	.822
anxcon anxiety control: cecs	-.198	-.074	.807

Extraction Method: Principal Component Analysis.
Rotation Method: Promax with Kaiser Normalization.

Principal Axis Factor Analysis Factor Structure Matrix

Structure Matrix

	Factor		
	1	2	3
neoneuro neuroticism: neo	-.816	-.498	.264
regard self-regard: poi	.702	.342	-.264
negafect negative affect: mpq	-.687	-.553	.151
posafect positive affect: mpq	.625	.117	-.258
neoextra extraversion: neo	.539	-.135	-.365
acceptnc self-acceptance: poi	.512	.235	-.210
cntrlcpi self-control: cpi	.233	.821	-.037
bencntrl benign control: ecq	.363	.745	.003
aggcntrl aggression control: ecq	.120	.553	.082
neoagree agreeableness: neo	.339	.494	-.159
depcon depression control: cecs	-.331	-.065	.859
eminhib emotional inhibition: ecq	-.380	.066	.745
anxcon anxiety control: cecs	-.243	-.061	.653

Extraction Method: Principal Axis Factoring.
Rotation Method: Promax with Kaiser Normalization.

FIGURE 38.21 **Factor/Component Structure Matrices** of the variables for both the three-dimensional principal components analysis and the principal axis factor analysis.

The researchers in this example would still need to decide which extraction method to report in disseminating the results of the research. Because the principal components solution provides for somewhat stronger structure coefficients, less "cross-loading," and communality values that have achieved our general criterion of the high .4's or better, we would select the principal components solution. However, the apparent increment in explained variance (about 60% for principal components compared to about 50% for principal axis factoring) is really not a tangible gain, as principal axis factoring was aimed at explaining only that portion of the variance common to the set of variables rather than the total amount of variance targeted by principal components. Thus, other researchers may very well opt to report the results of the principal axis analysis; ultimately, such a decision will reflect the professional preferences of each research team. However, the choice of which extraction procedure to report does not affect the interpretation of the dimensional structure of the variables.

CHAPTER 39

Confirmatory Factor Analysis

39.1 OVERVIEW

Principal components and factor analysis, covered in Chapter 38, are considered to be exploratory procedures in the sense that the components/factors are allowed to emerge from the data analysis. Researchers thus approach such an analysis from an inductive or bottom-up perspective even if they have some informal expectations about the nature of the dimensional structure they expect to find. Consistent with such an inductive approach, it is not unusual for researchers using these procedures to examine a small set of possible solutions (e.g., three, four, and five components/factors) in deciding on the one that they will finally select.

Confirmatory factor analysis represents a deductive or top-down approach. Here, researchers hypothesize an explicit factor structure (a model) based on a theoretical or an empirical framework. The analysis imposes the hypothesized model on the data and assesses the extent to which that formulation fits the data. Following are two of the major differences between exploratory and confirmatory techniques:

- In exploratory principal components and factor analyses, every variable is associated with every component or factor; the goal is to have each variable associated strongly with only one component or factor and thus weakly associated with the others. In confirmatory factor analysis, researchers ordinarily specify each variable to be associated with only one factor.
- Rotated components and factors in the exploratory technique are either all required to be uncorrelated (resulting from an orthogonal rotation) or are all allowed to correlate (resulting from an oblique rotation). In confirmatory factor analysis, researchers specify which factors, if any, are to be correlated; thus, in a single analysis, researchers may hypothesize that some factors are correlated and that other factors are orthogonal.

Confirmatory factor analysis is carried out in specialized structural equation modeling (SEM) software; in this book, we use the IBM SPSS® Amos add-on for that purpose. Using the software, we configure the model in diagram form. An example of a simple, generic confirmatory factor analysis model is presented in Figure 39.1. The general features of such a model are as follows:

Performing Data Analysis Using IBM SPSS®, First Edition.
Lawrence S. Meyers, Glenn C. Gamst, and A. J. Guarino.

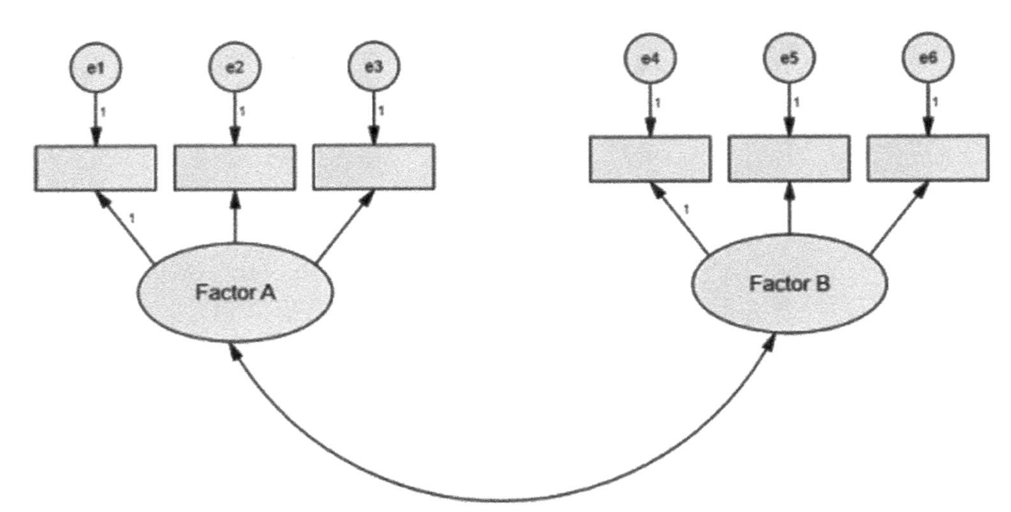

FIGURE 39.1 A generic example of a confirmatory factor analysis model.

- The two large ovals depict the hypothesized factors named **Factor A** and **Factor B**. They are drawn as circles/ovals to represent them as *latent* (unobserved) variables or constructs; latent variables are always drawn as circles/ovals. There can be any number of factors in the analysis; we have chosen two factors for illustration purposes.
- The double-headed arrow (path) between the factors represents a correlation between them; if researchers believed the factors to be orthogonal (independent), they would not draw such a path between the factors. In the final model, the value of the correlation between the factors derived from the analysis is written on that path.
- The rectangles represent the observed or measured variables in the data file; measured variables are always drawn as squares/rectangles. Three of the measured variables are hypothesized to be associated with one factor, and the other three measured variables are hypothesized to be associated with the other factor.
- Single-headed arrows (paths) emerge from the factor pointing to each associated measured variable. This represents the direction of "causal" or predictive flow. In this context, the measured variables are conceived of as *indicators* (minirepresentations) of their associated factor. In the final model, it is common practice to place on these paths the value of the standardized (path/regression) coefficients estimated from the data analysis.
- Because the latent variables require a scale or metric, the first drawn of the indicator (measured) variables for each factor is selected and the path from the latent variable to that measured variable is assigned (constrained to) the value 1 (unity). This constraining of the path to unity applies the scale from the measure to the latent variable. These path coefficients are the "loadings" of the variable of the factor and are ordinarily the pattern/structure coefficients resulting from the prediction of the measured variables based on the latent variable (the factor). As part of the analysis, the standardized values will be estimated from the data so that the constraint placed on the first indicator variable will not adversely impact the eventual solution.
- Each indicator variable is associated with its own latent variable, with the path pointing from each latent variable to the respective measured variable. These latent variables represent *measurement error* and are named **e1** through **e6** inclusive in Figure 39.1. The error components are needed to specify the model because an SEM analysis is a variation of a multiple regression analysis where, unlike

a regression analysis where the dependent variable is the *predicted* Y value, the variable predicted in an SEM analysis is the *actual* Y value (here, Y is the measured variable). Thus, the structural equation must include the unexplained (error or residual) variance of Y, which is the error term; that is, whatever variance of a given measured variable that is not explained by the factor is attributed to the error term so that explained variance plus error variance is equal to the total variance of the measured variable.

- Error terms are constrained to unity to apply the scale from the measure to the error term. These path coefficients will also be reestimated in the process of performing the analysis.
- Researchers may choose to hypothesize that some error terms associated with the indicator variables are correlated. In the model we have drawn in Figure 39.1, the error terms are hypothesized to be uncorrelated. Assuming uncorrelated errors, however, is often an oversimplification that adversely affects the fit of the model to the data. Error variables may be related to each other for several reasons. For example, two indicator variables may measure a common extraneous construct or, if they are items on an inventory, the items with which they are associated may be worded in a similar manner (e.g., reverse worded) or may share a common reading level that is somewhat different from the other items (Brown & Moore, 2012; Kline, 2011; Wang & Wang, 2012). For such reasons, the errors would be hypothesized as being correlated. If the researchers choose not to hypothesize any correlated errors at the outset of the analysis, they may wish to revise that assumption based on the results of the analysis (IBM SPSS Amos provides suggested modifications—additional paths—that can be added to the model to improve model fit) and then perform a more exploratory analysis on a revised model that specifies certain correlated errors. Our example will illustrate this.

39.2 NUMERICAL EXAMPLE

The purpose of this fictitious study was to assess the internal structure of a potential clinical assessment instrument that researchers hoped was able to differentiate grief from depression. Grief is composed of three measured variables: **no_energy**, **lonely**, and **past** (a focus on the past); depression is also composed of three measured variables: **guilt**, **pessimism**, and **self_critical**. The data file is named **grief and depression**.

We note that there are no missing values associated with the variables that we use in the analysis. IBM SPSS Amos is fussy about this; it will estimate means and intercepts for those missing values before evaluating the model but will not produce other important output (e.g., the **Modification indices**) if there are missing values in the data. We most strongly suggest that missing values be replaced by estimated values (using a single imputation method such as regression or using more elaborate multiple imputation procedures) through the Missing Values Analysis module (see Meyers et al. (2013)) or that cases with missing values be removed from the data file before performing any analysis in IBM SPSS Amos.

39.3 DRAWING THE MODEL

Drawing the Model: Some Useful Command Functions

We open the **grief and depression** data file and from the main menu we select **Analyze ➔ IBM SPSS Amos**. The initial IBM SPSS Amos window is shown in Figure 39.2, where the following portions of the screen are identified in our annotation:

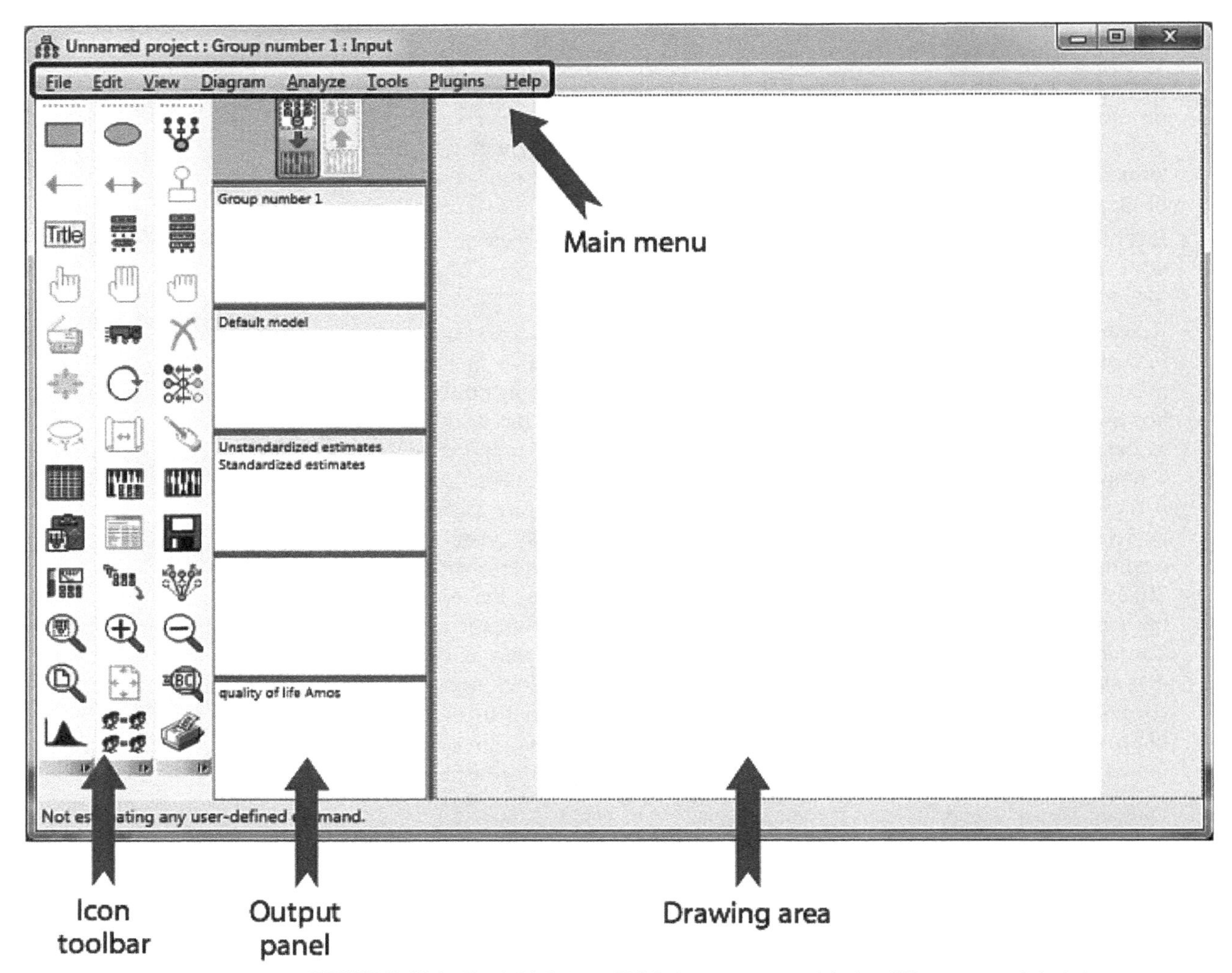

FIGURE 39.2 The initial IBM SPSS Amos screen, with the different areas labeled.

- *The icon toolbar* from which we select various functions such as drawing certain objects (e.g., latent variables) or performing certain operations (e.g., erasing objects).
- *The output panel* where we can select what portion of the output to display once the analysis has been performed.
- *The drawing area* where we configure the diagram of our model.
- *The main menu* from which we can configure our analysis.

The commands on the icon toolbar are activated with a single click. Some helpful functions from the middle of the icon toolbar are shown in Figure 39.3. To be used, some object or objects need to have been drawn in the drawing area. Generally, we select the icon (or select a function from the main **Edit** menu). This action changes our cursor to that icon. We then place the cursor on the object and click or drag as appropriate. To discontinue the function, we move the cursor to an empty portion of the drawing area and right-click. Here is how to work with some of the functions:

- To select an object or objects, we select the **Select one object at a time** icon and then click on our target objects one at a time.

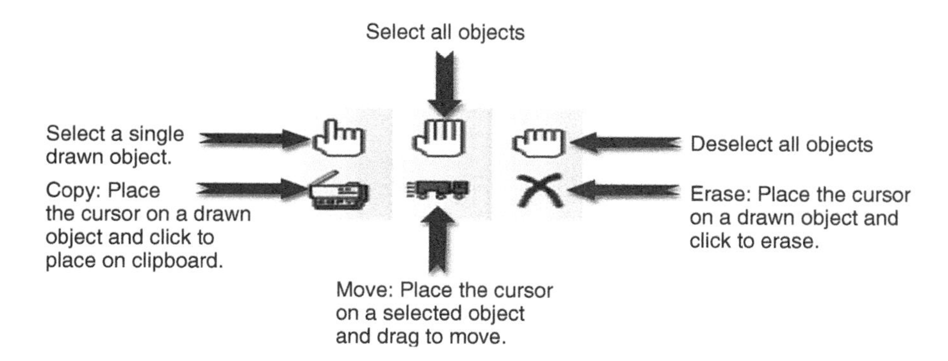

FIGURE 39.3
Some useful command functions from the icon toolbar.

- To select all objects in the drawing area, we select the **Select all objects** icon.
- To deselect all objects that have been selected, we select the **Deselect all objects** icon.
- To move an object, we select the object, select the **Move objects** (truck icon), click on our target object in the drawing area, and drag it elsewhere.
- To move a set of objects (e.g., a factor with its indicator variables and their associated errors), we select all of the separate elements, select the **Move objects** (truck icon), and then drag the set of objects to another location on the screen.
- To erase an object, we select the **Erase** (stylized X) icon and click on the target object.
- To copy and paste an object or set of objects, we select our object or objects, select **Edit ➔ Copy to clipboard** from the main menu, and then select **Edit ➔ Paste**. The copied item or items will rest on top of the original and can simply be moved by selecting the **Move** icon and dragging the object or objects.

Drawing the Model: the Drawing Tools

Because confirmatory models usually take a fair amount of horizontal space to draw, we opt to switch from the default portrait (long) orientation to landscape (wide) orientation. To change the orientation, select **View ➔ Interface Properties** from the main menu. This action opens the **Interface Properties** window as seen in Figure 39.4. Select **Landscape-Legal** under **Paper Size**, select **Apply**, and close the window.

From the icon toolbar, select the icon that represents **Draw a latent variable or add an indicator to a latent variable** (the circle with squares and more circles), as shown in Figure 39.5. We use this tool to draw our latent variables. After activating this function by selecting it, place the cursor (which has now taken on the icon) toward the right center of the drawing area (see Figure 39.6) and click to draw the factor. The drawn object will take the shape dictated by the movement of the cursor and, with a little practice, the desired shapes can be readily drawn. Then with the cursor inside the circle/oval, click one more time to draw the first indicator variable and its associated error; the size of the rectangle will be proportional to the shape of the factor. Note that both paths have been constrained to the value **1** to scale each latent variable (the factor and the error) to the measured variable.

The results of the above actions are shown in Figure 39.6. In this illustration, we have intentionally drawn the factor too small (to illustrate what can be done to remedy the situation), and as a result, the rectangle for the indicator variables is a bit too small to contain the names we will assign them. However, we can easily resize/reshape these figures. With the cursor inside the oval for the factor we right-click and select **Shape of Object** from the pop-up menu (also shown in Figure 39.6). We then place our cursor

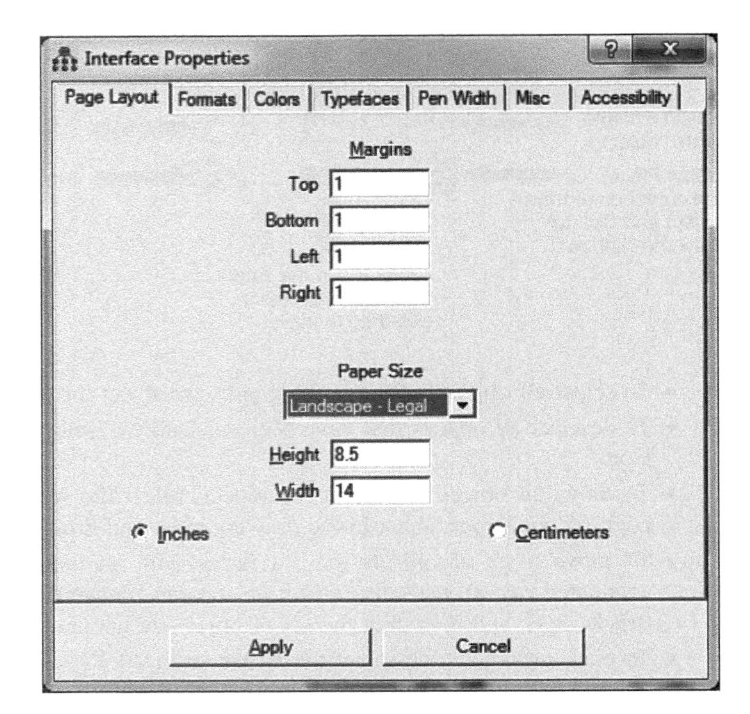

FIGURE 39.4
The **Interface Properties** window where we select **Landscape-Legal** page orientation.

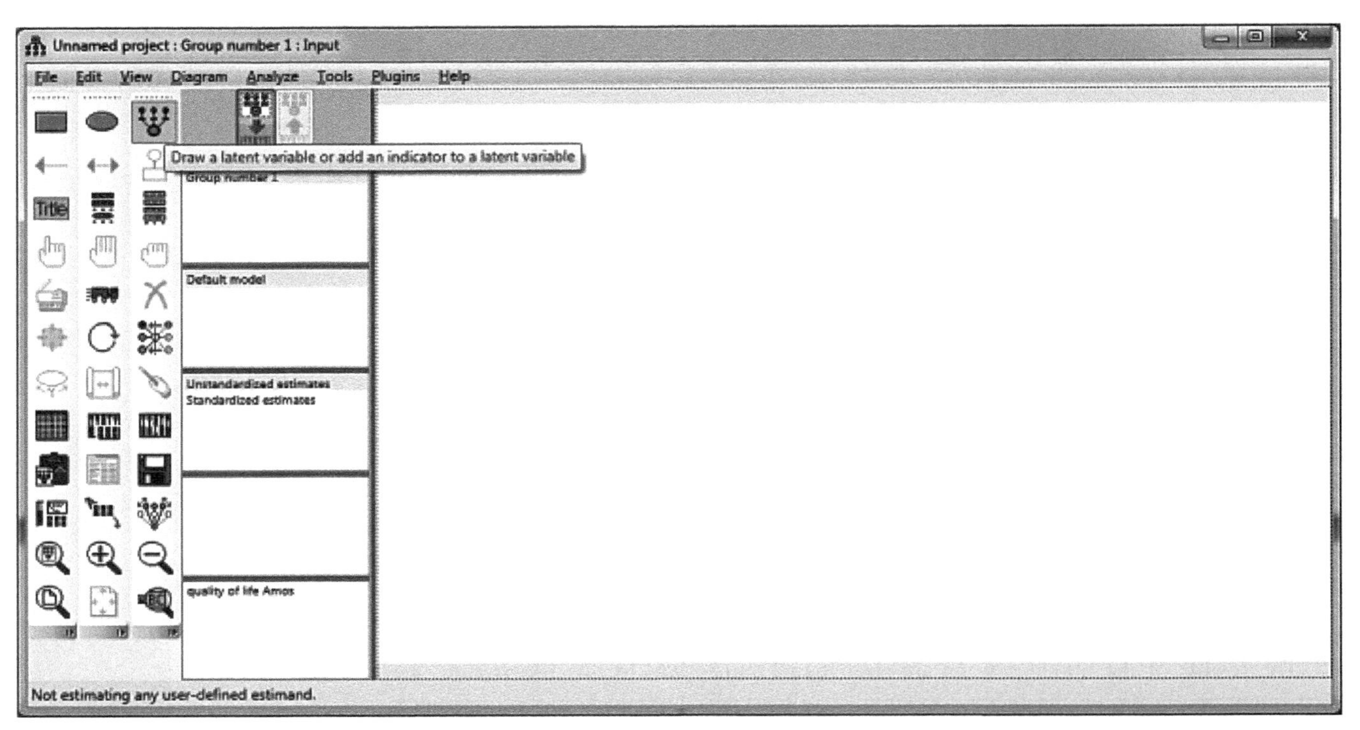

FIGURE 39.5 Selecting the **Draw a latent variable or add an indicator to a latent variable** function.

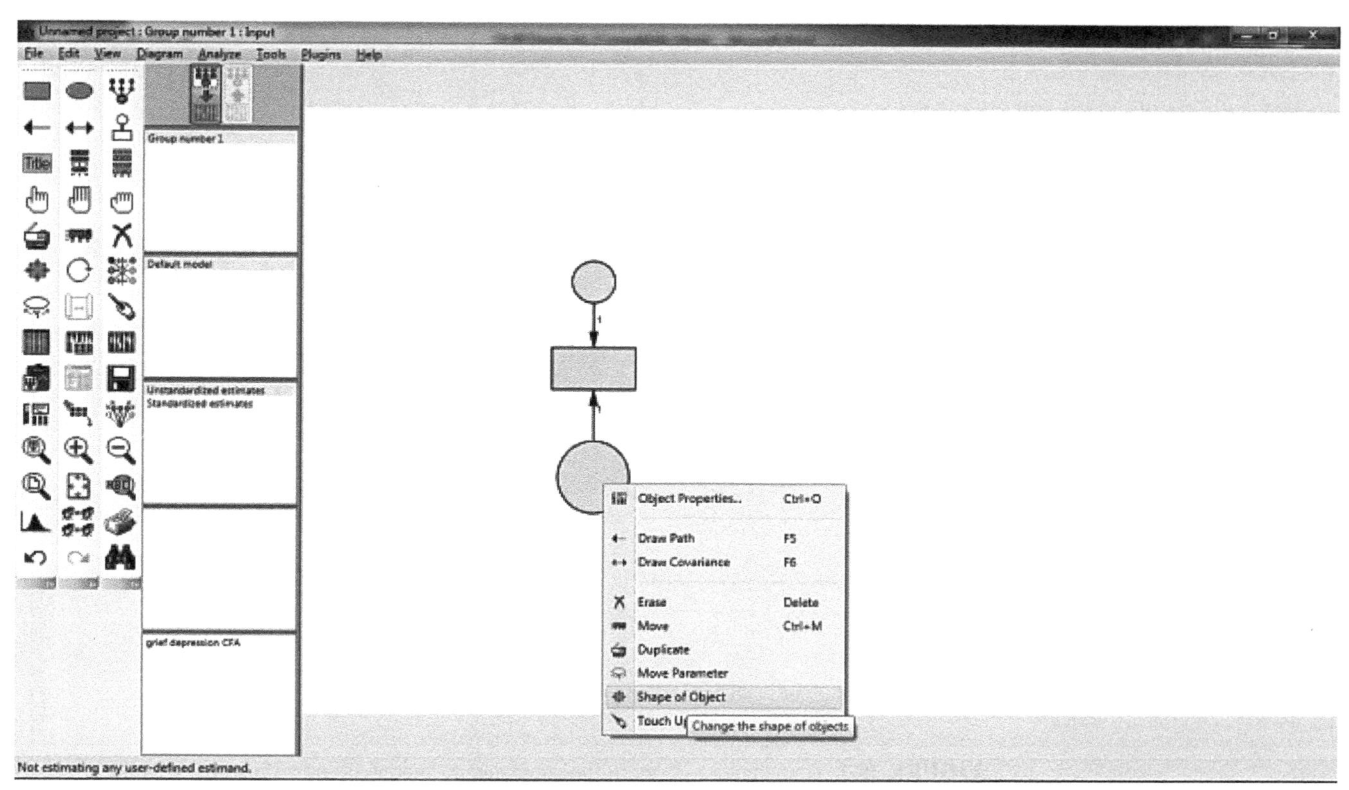

FIGURE 39.6 The first factor with its first indicator variable is drawn. We will modify the shape of these figures before drawing the rest of the confirmatory model.

inside each shape and stretch (drag) each to suit our aesthetic preferences. Once we resize these shapes to meet our preferences, we deselect the **Shape of Object** function by moving the cursor to a blank area of the screen and right-click. Then we again select the **Draw a latent variable or add an indicator to a latent variable** icon, place the icon inside the oval representing the factor, and click twice to generate two more indicator variables (these additions will match our new shape). We then repeat the process to draw the second factor. The final product is shown in Figure 39.7.

Our next task is to associate each measured variable with a variable from our data file. From the main menu, select **View ➔ Variables in Dataset**. This opens the **Variables in Dataset** window pictured in Figure 39.8. Select each variable in turn and drag it to the appropriate rectangle so that all observed variables are properly configured as shown in Figure 39.9. When finished, close the **Variables in Dataset** window.

We next draw the correlation (the covariance) path between the two factor variables because we hypothesize that the two factors are correlated. From the icon toolbar, select the **Draw covariances (double headed arrow)** as shown in Figure 39.10. **Covariance** is the (unstandardized) correlation between variables (Norman & Streiner, 2008) whose value reflects the particular metric associated with the variables (and thus can take any numeric value), whereas a correlation coefficient such as the Pearson *r* represents a standardized correlation value that is constrained to vary between ±1.00. With the **Covariance** tool activated, place the cursor on the left border of the second factor and click and drag to the first factor to draw the covariance (correlation) path as shown in Figure 39.11.

We also need to name the factors. Double-clicking inside a factor activates the **Object Properties** window shown in Figure 39.12. We type **grief** in the **Variable name** panel

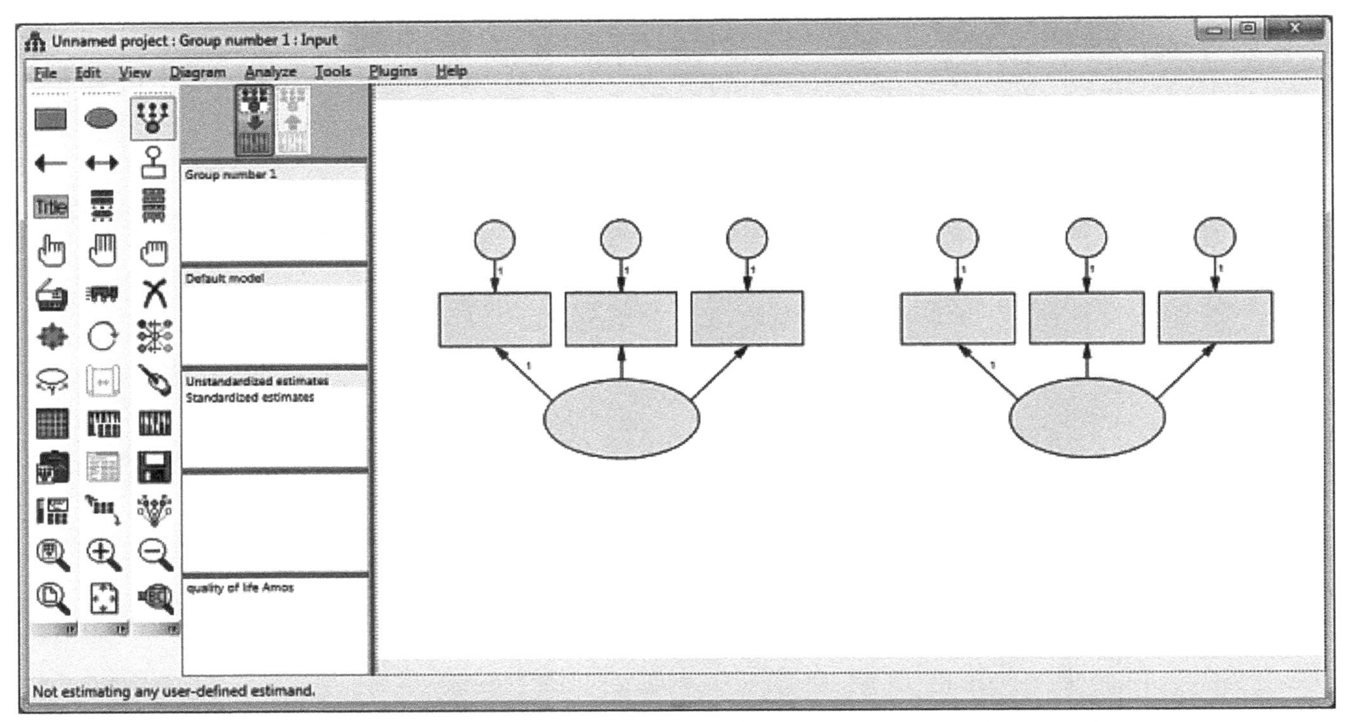

FIGURE 39.7 The latent variables in the confirmatory model are now drawn.

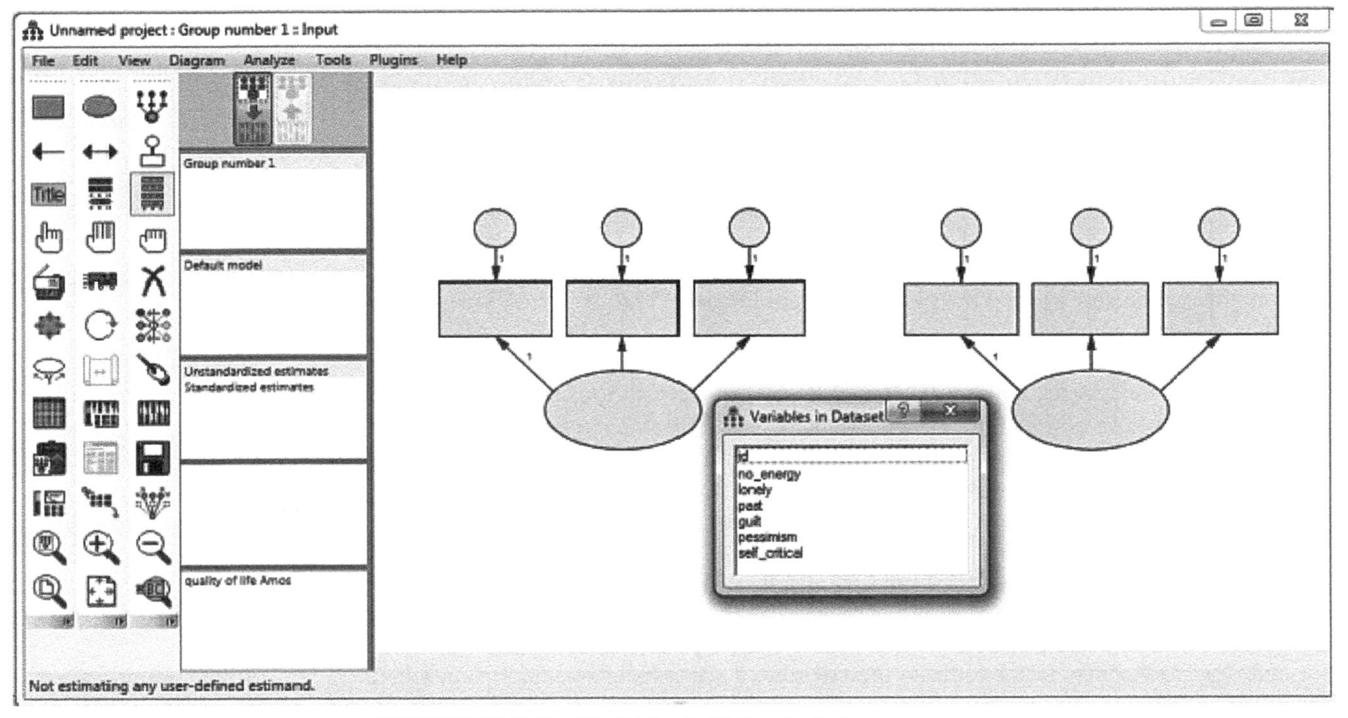

FIGURE 39.8 The **Variables in Dataset** window.

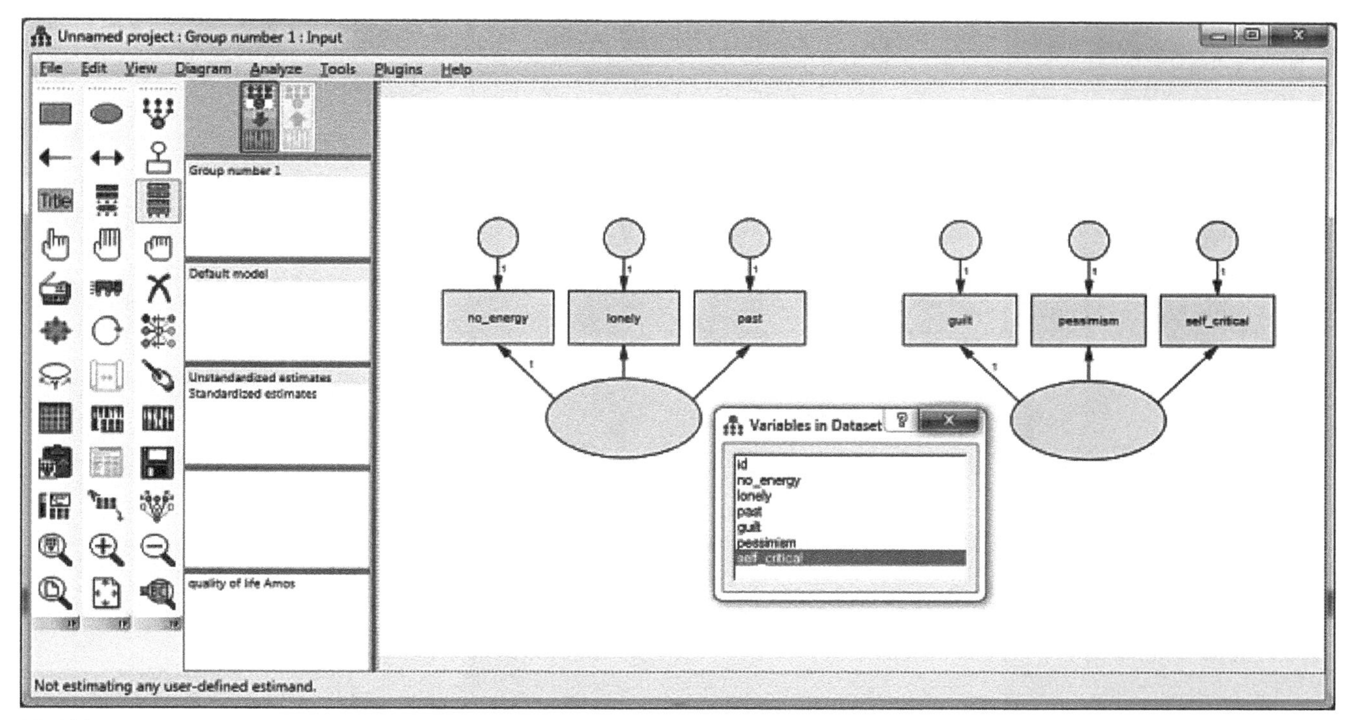

FIGURE 39.9 All of the indicator variables have been associated with the proper variables in the data file.

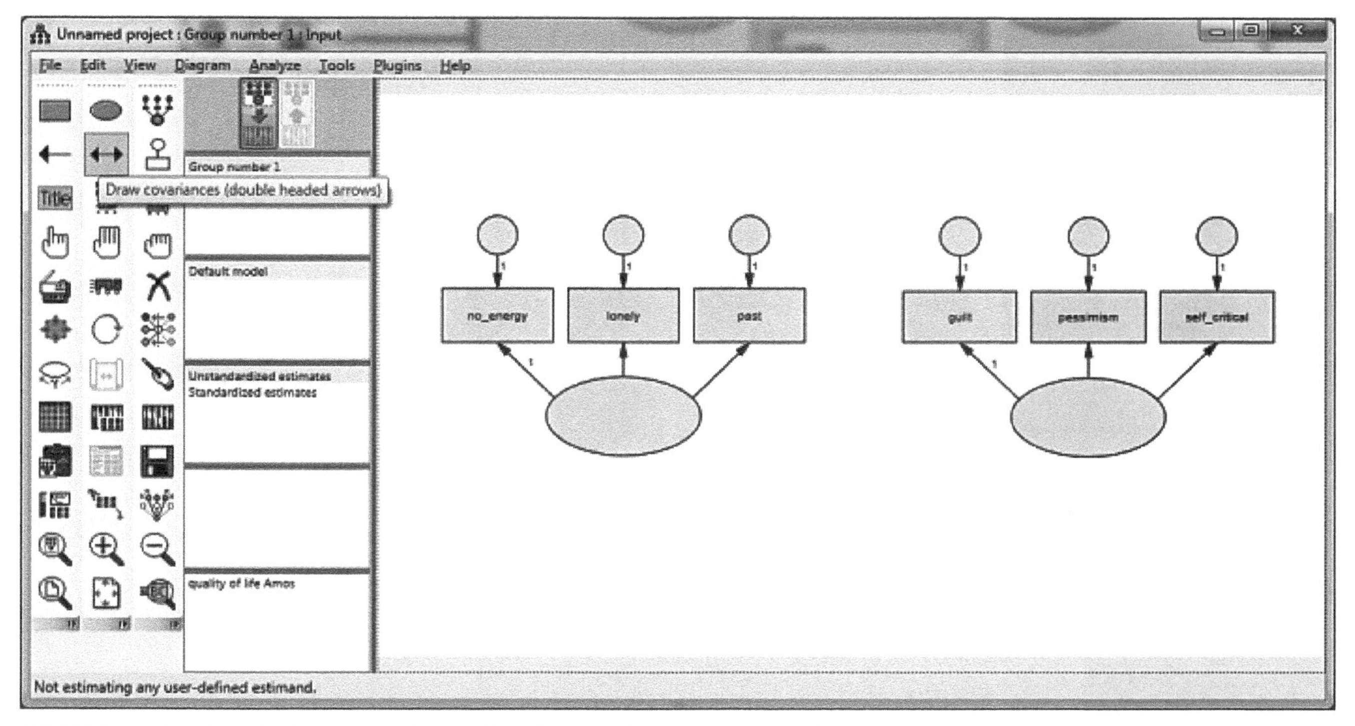

FIGURE 39.10 Select the **Draw covariances (double headed arrow)** from the icon toolbar.

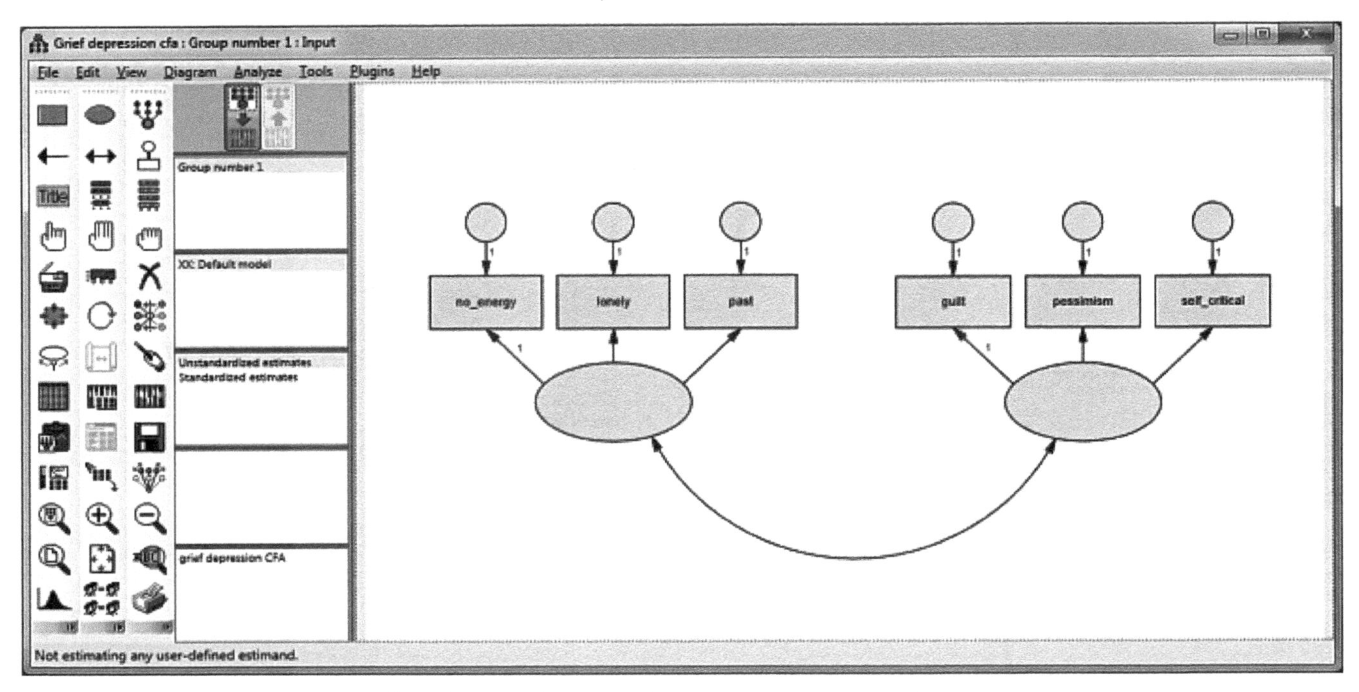

FIGURE 39.11 The correlation between the factors is now drawn.

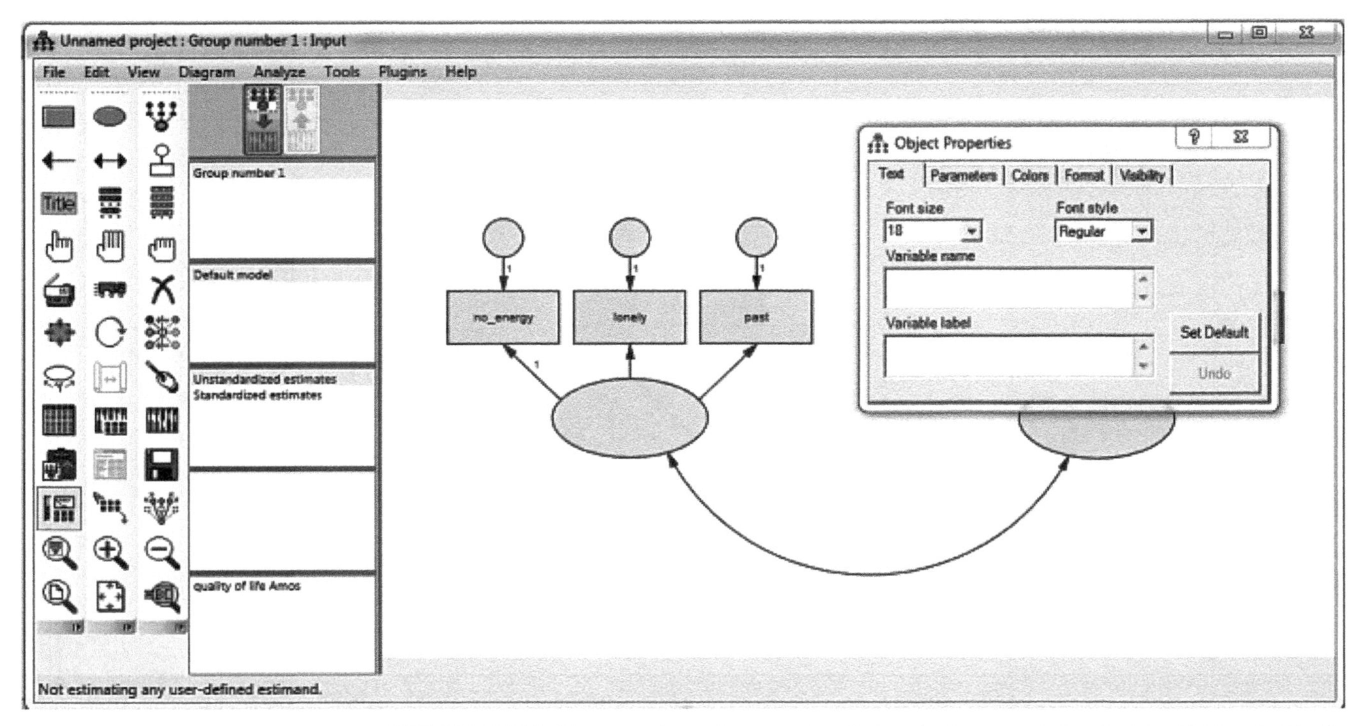

FIGURE 39.12 The covariance between the factors is drawn and, by double-clicking inside the first factor, we have activated the **Object Properties** window.

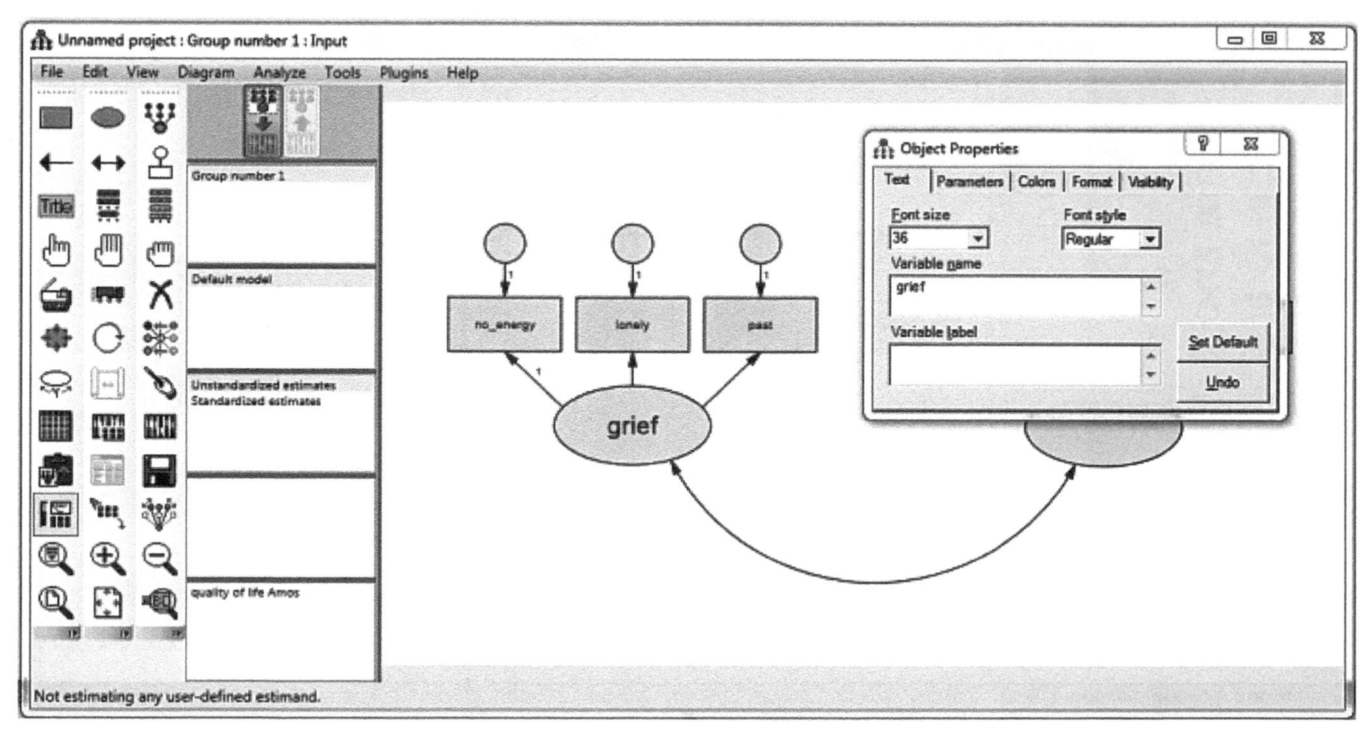

FIGURE 39.13 The first factor is now named.

and (for aesthetic reasons given the size of our ovals) increase the font size to 36; this name shows up in the diagram as shown in Figure 39.13. Repeating this process labels the second factor as **depression** (see Figure 39.14); we then close the **Object Properties** window.

We now need to name the latent error variables (called **unique variables** by IBM SPSS Amos). IBM SPSS Amos will name these generically. From the main menu, select **Plugins ➔ Name Unobserved Variables** (see Figure 39.15). On selecting this command, the **unique variables** will be labeled with an "**e**" (for error term) and given a sequential number (see Figure 39.16) resulting in the variable names **e1** through **e6**.

39.4 ANALYSIS SETUP

With the confirmatory factor analysis diagram completed, our next task is to configure the analysis. From the main menu we select **View ➔ Analysis Properties** to open the **Analysis Properties** window as shown in Figure 39.17. Select the **Output** tab. **Minimization history** is checked by default. Check **Standardized estimates** (these produce the standardized regression coefficients for the prediction of the measured variables) and **Modification indices** (these produce IBM SPSS Amos "suggestions" on how to improve the model fit).

To execute the analysis, select **Analyze ➔ Calculate Estimates** from the main menu. As this is our first analysis, we are presented with the operating system's **Save As** dialog window (see Figure 39.18). We select a location to save the file, provide a file name, and **Save**.

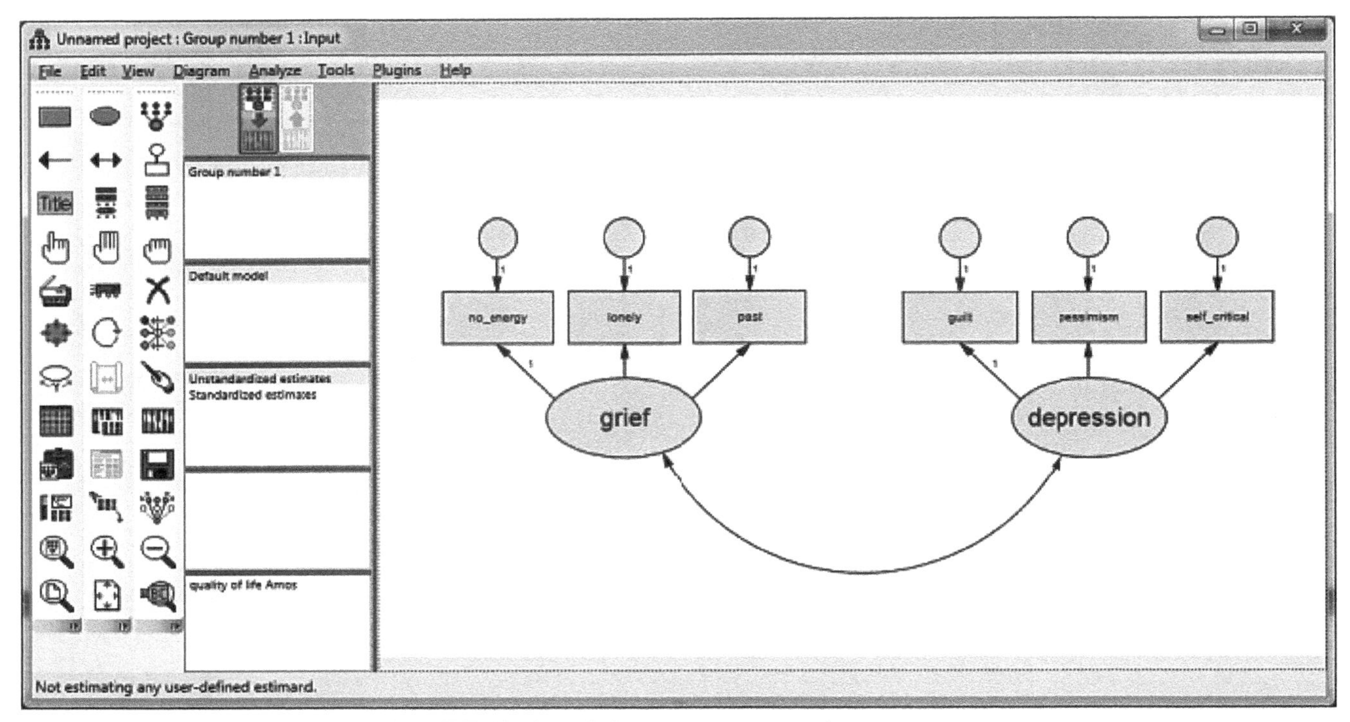

FIGURE 39.14 Both factors are now named.

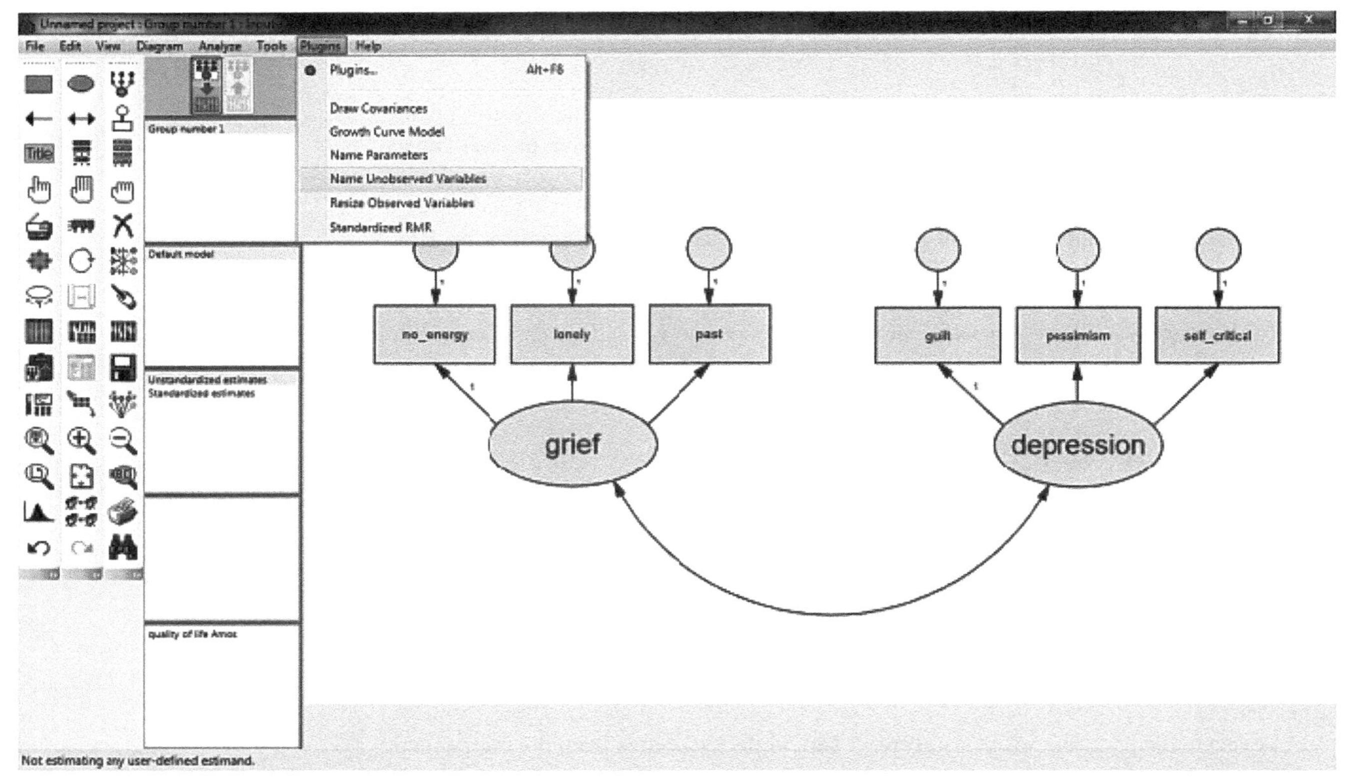

FIGURE 39.15 To name the latent error variables, we select **Plugins ➔ Name Unobserved Variables** from the main menu.

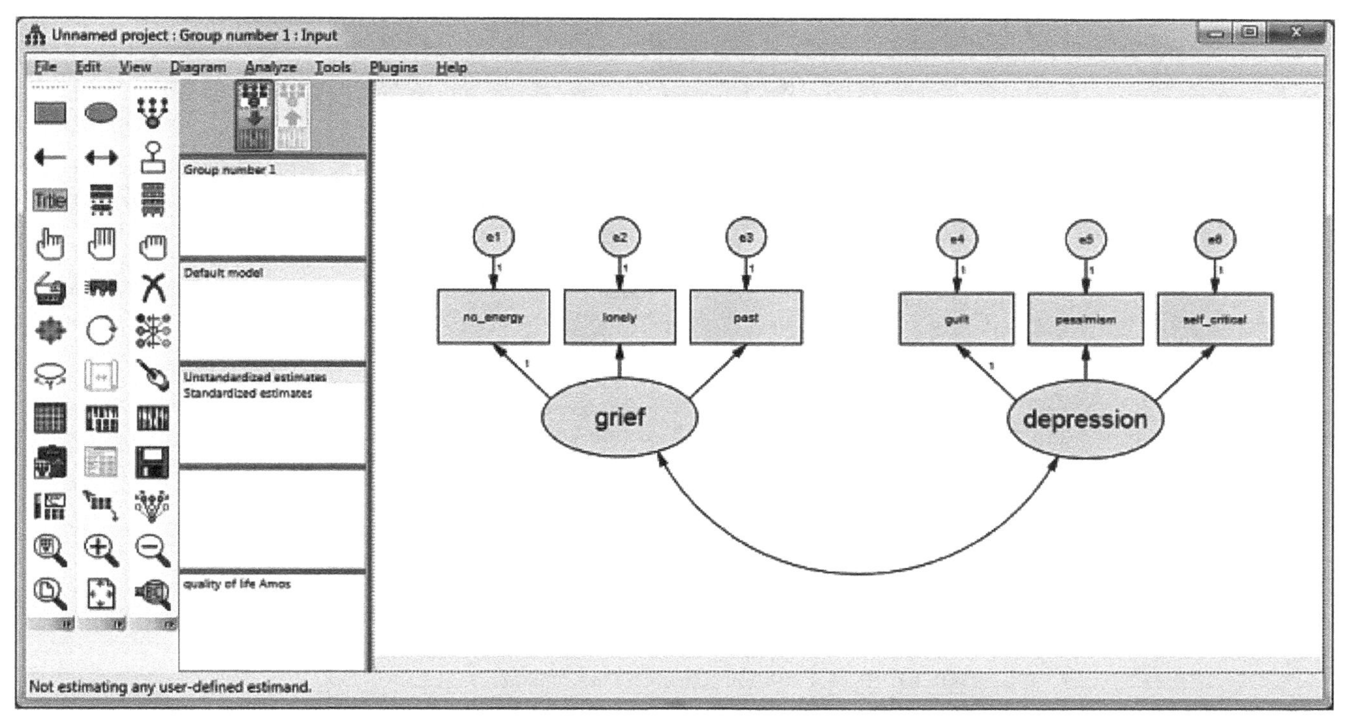

FIGURE 39.16 On selecting x, the latent error variables are labeled **e1** through **e6**.

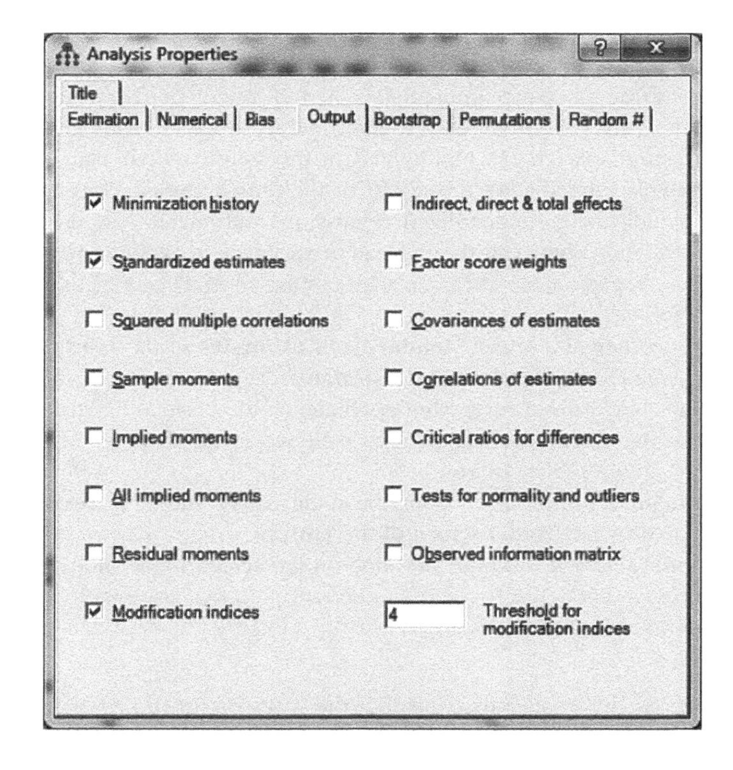

FIGURE 39.17
The **Analysis Properties** window.

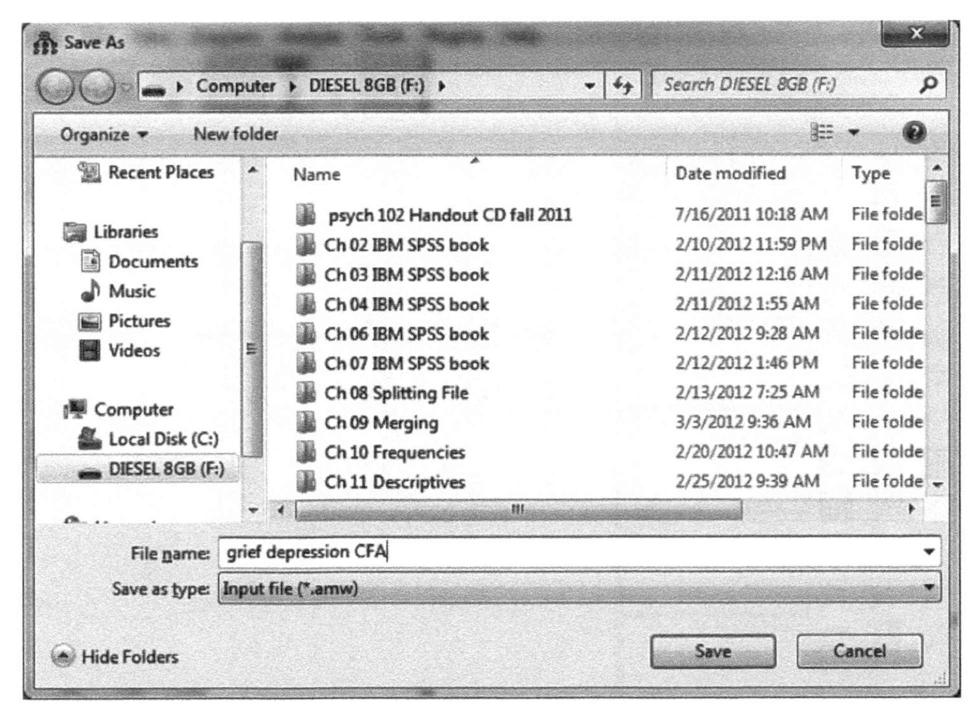

FIGURE 39.18 Performing the **Save As** function.

39.5 ANALYSIS OUTPUT

Analysis Output: Initial Screens

To access the results of the analysis, select the **View the output path diagram** icon (the one on the right) in the top output panel immediately to the left of the drawing area as shown in Figure 39.19. This action causes the confirmatory diagram with the **Unstandardized estimates** to be displayed (as highlighted in the fourth output panel from the top in the stack of panels immediately to the left of the drawing area). These **Unstandardized estimates** include the unstandardized regression coefficients and the covariance (unstandardized correlation) between the two latent variables, as well as the variances of the latent variables. These values also appear in tabular form as part of the **Text Output** as described below.

To view the standardized coefficients, select **Standardized estimates** in the fourth output panel from the top just under **Unstandardized estimates**. These are shown in Figure 39.20. In addition to the standardized regression coefficients, the correlation (the standardized covariance) of the two latent variables is also included in the display. All of these values are available in output tables as well.

To obtain the output in tabular form, select from the main menu **View → Text Output**. This action opens the **Notes for Model** screen of the **Output** window shown in Figure 39.21. The **Default model** referred to in the first line on the screen is the model that was hypothesized by the researchers. The two most important pieces of information we learn from this very global overview are as follows:

- The degrees of freedom are positive, indicating that the model is *identified* and thus viable (Meyers et al., 2013). We discuss this in a little more detail in the context of our SEM analysis in Section 43.8.

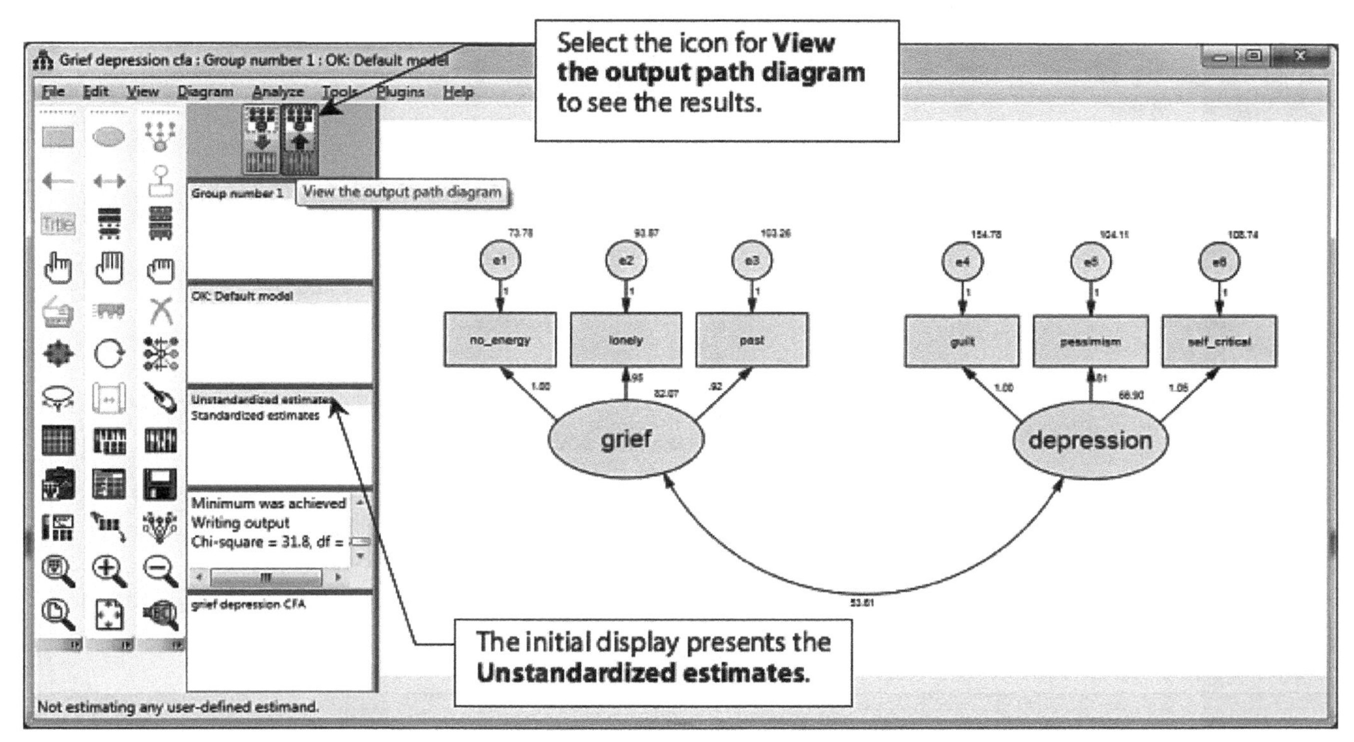

FIGURE 39.19 The **Unstandardized estimates** are displayed on selecting **View the output path diagram**.

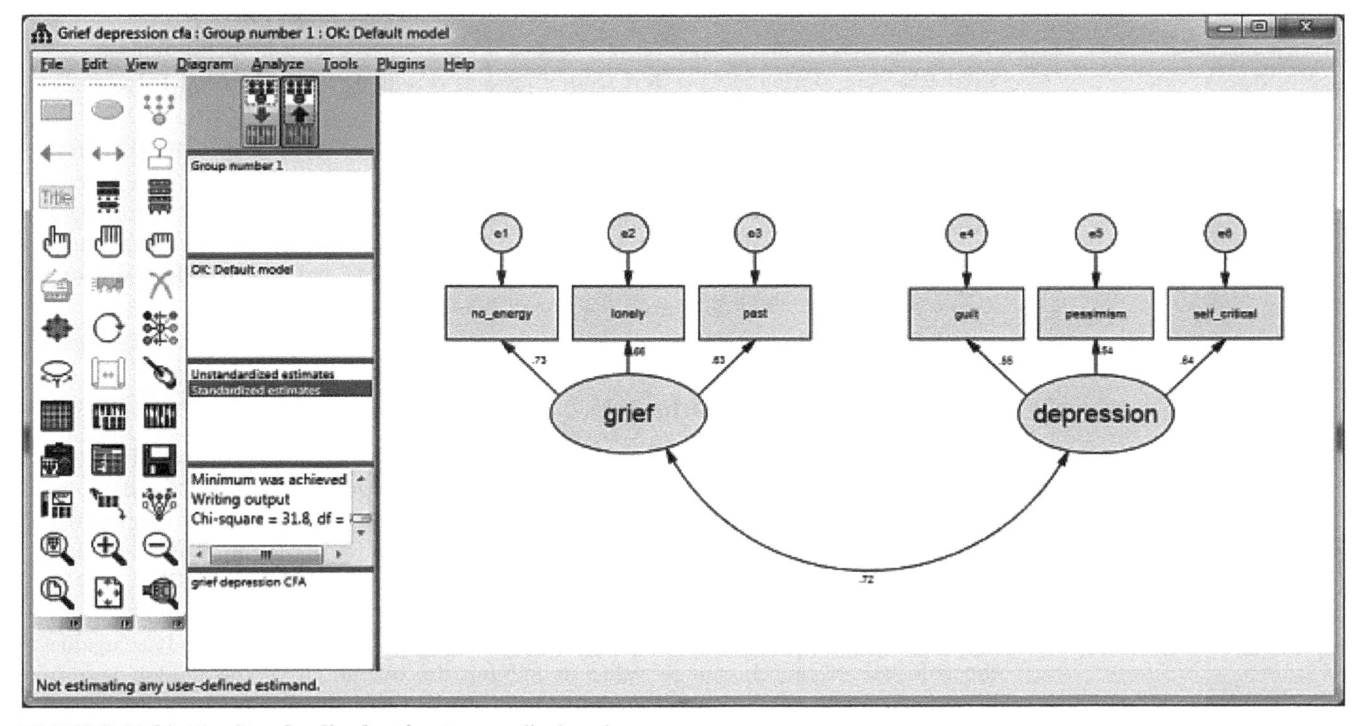

FIGURE 39.20 The **Standardized estimates** are displayed.

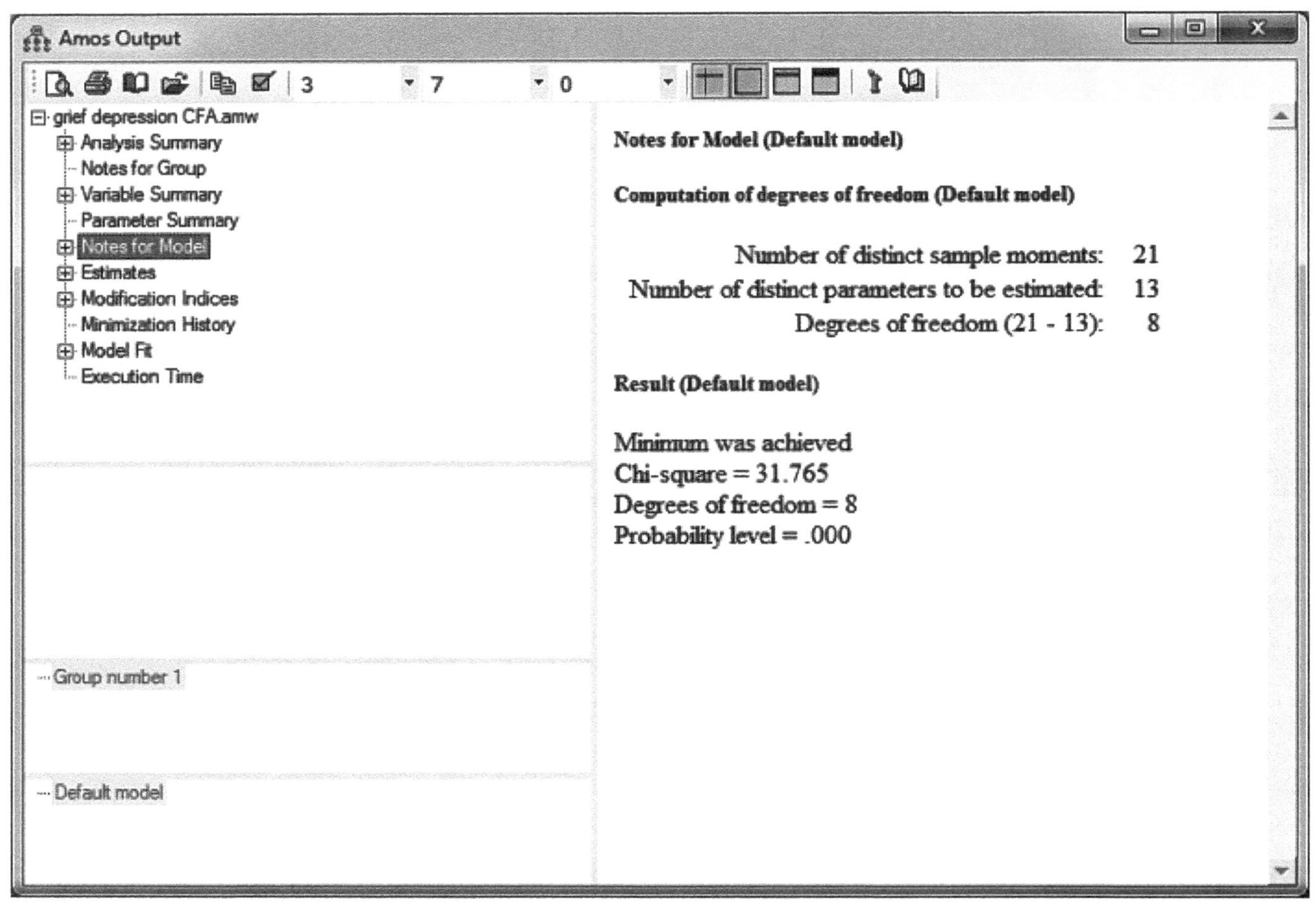

FIGURE 39.21 Having selected **Text Output** presents us with the **Notes for Model** screen.

- The chi-square value associated with the model is **31.765** and, with **8** degrees of freedom, is statistically significant ($p < .001$). This indicates that the expected values based on the model differ significantly from those represented by the data; that is, the model does not appear to adequately fit the data. That said, most researchers rely on other fit measures in addition to chi-square before drawing conclusions about how well the model fits the data.

Analysis Output: Model Fit

One of the very important aspects of the output of an SEM analysis is the set of indexes of model fit. Such indexes assess how well the hypothesized structure fits the data; the better the fit, the more the support for the model. There are quite a few fit indexes that have been developed over the years, and there are some differing opinions that are more appropriate to apply in any given research study (e.g., Hu & Bentler, 1999; Kline, 2011; West, Taylor, & Wu, 2012).

Most writers, such as Jöreskog and Sörbom (1996) and Bentler (1990), advise against the sole use of the chi-square value in judging the overall fit of the model because of the sensitivity of the chi-square to sample size. Larger sample sizes are associated with increased power, and with large sample sizes, the chi-square test can detect small unimportant discrepancies between the observed and predicted covariances and suggest that the model does not fit the data. Meyers et al. (2013) recommend reporting four fit measures in addition to chi-square. These are as follows:

- *Goodness-of-Fit Index (GFI).* The GFI is conceptually similar to the R^2 in multiple regression (Kline, 2011) and represents the proportion of variance in the sample correlation/covariance accounted for by the predicted model.
- *Normed Fit Index (NFI).* The NFI, also known as the Bentler–Bonett normed fit index, equals the difference between the chi-square of the null model and the chi-square of hypothesized model divided by the chi-square of the null model. For example, an NFI of .95 indicates the model of interest improves the fit by 95% relative to the null or independence model.
- *Comparative Fit Index (CFI).* The CFI (Bentler, 1990) roughly represents the extent to which the hypothesized model is a better fit than the independence or null model (where no correlations are assumed to exist between the variables).
- *Root-Mean-Square Error of Approximation (RMSEA).* The RMSEA is the average of the residuals between the observed correlation/covariance from the sample and the expected model estimated for the population (Browne & Cudeck, 1989, 1993).

Several sets of authors have proposed guidelines for interpretation of these fit indexes (e.g., Meyers et al., 2013; Schreiber, Nora, Stage, Barlow, & King, 2006; West et al., 2012). Generally, the present consensus is that a value of .95 for the GFI, NFI, and CFI would be taken to represent a good fit of the model to the data, although blindly applying these "lines in the sand" is ill-advised (Kline, 2011). For the RMSEA, values of .06, .08, and greater than .10 are considered to represent a good, adequate, and poor fit of the model to the data, respectively.

Selecting **Model Fit** in the left panel displays the several fit indexes produced by IBM SPSS Amos; these are presented in Figure 39.22, Figure 39.23, and Figure 39.24 (we could not show them all in a single screenshot). The **CMIN** table shown at the top in Figure 39.22 presents the *minimum discrepancy* value in the **Default model** row, which is the chi-square goodness-of-fit test that repeats the information we saw in the **Model Notes** screen. The **CMIN/DF**, also shown in Figure 39.22, is the chi-square divided by the degrees of freedom (**31.765/8** = **3.971**). Some researchers suggest values as high as five for an acceptable fit, while others maintain relative chi-square be two or less. Our value falls between these two guidelines.

The **GFI** is shown in the middle table in Figure 39.22 where we see its value of **.967** exceeds our guideline value of .95, suggesting an acceptable level of model fit. The **AGFI** (adjusted goodness-of-fit index) and the **PGFI** (parsimony goodness-of-fit index) represent adjustments to the **GFI** for degrees of freedom and number of parameters, respectively; guidelines for acceptable model fit are .95 for the **AGFI** and .50 or better for the **PGFI**. The **RMR** (root-mean-square residual) is a measure of the average size of the residuals between actual covariance and the proposed model covariance, with smaller **RMRs** indicating a better fit.

The **NFI** and the **CFI** are shown in the bottom table of Figure 39.22 where we see that their values of **.922** and **.939** fall short of our guideline value of .95, suggesting a less than acceptable level of model fit. The **RMSEA** is presented in the top table of Figure 39.24. Its value of **.095** also suggests that the model does not fit the data all that well.

In summary, the chi-square, NFI, CFI, and RMSEA indexes paint a picture of a model that falls sufficiently short of fitting the data to meet our criteria of a good fit. We will examine the regression coefficients to learn the details of the results, but we probably want to modify the model to improve the fit if the **Modification Indices** suggest any reasonable options.

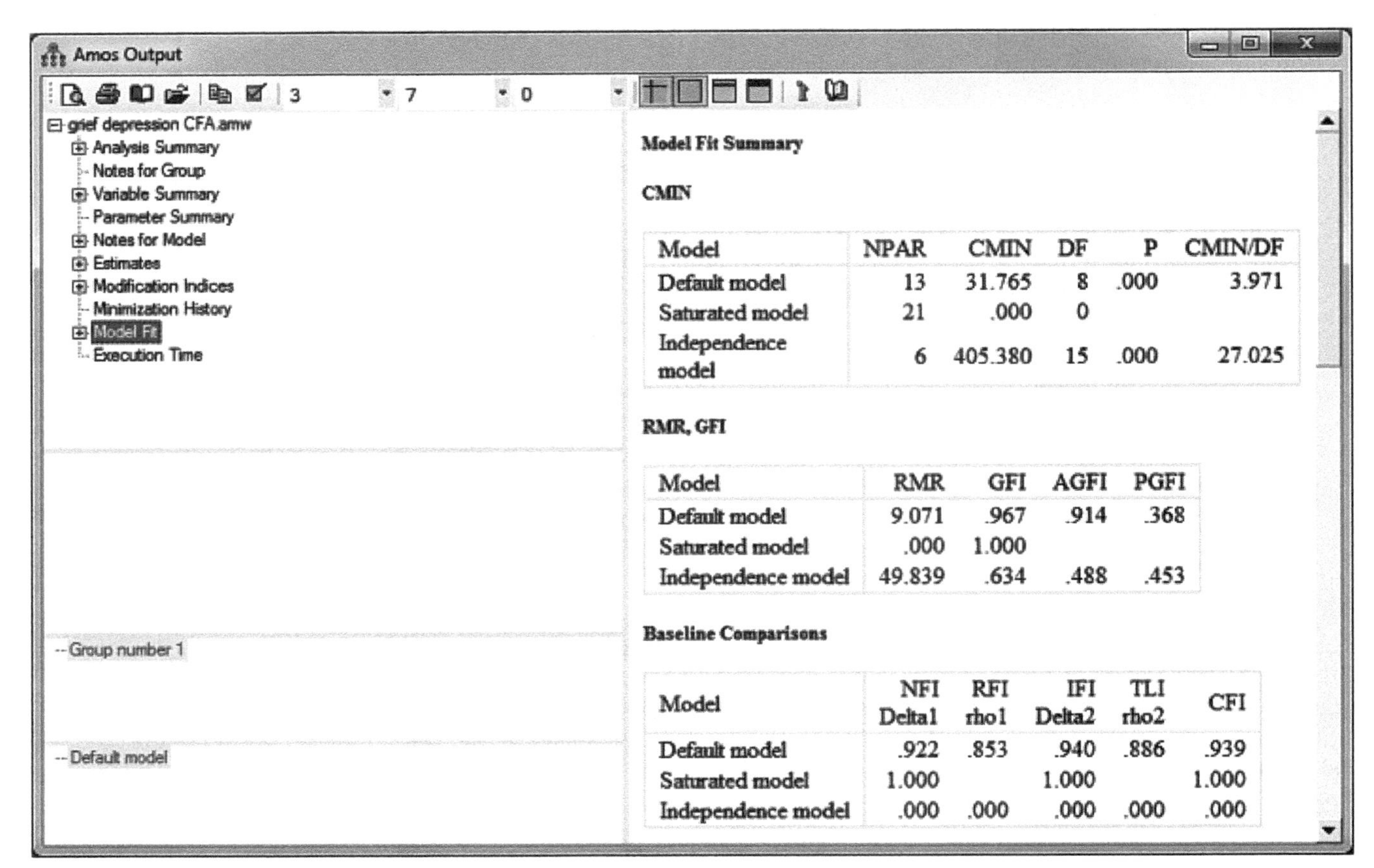

Model Fit Summary

CMIN

Model	NPAR	CMIN	DF	P	CMIN/DF
Default model	13	31.765	8	.000	3.971
Saturated model	21	.000	0		
Independence model	6	405.380	15	.000	27.025

RMR, GFI

Model	RMR	GFI	AGFI	PGFI
Default model	9.071	.967	.914	.368
Saturated model	.000	1.000		
Independence model	49.839	.634	.488	.453

Baseline Comparisons

Model	NFI Delta1	RFI rho1	IFI Delta2	TLI rho2	CFI
Default model	.922	.853	.940	.886	.939
Saturated model	1.000		1.000		1.000
Independence model	.000	.000	.000	.000	.000

FIGURE 39.22 Model fit indexes (first of three screenshots).

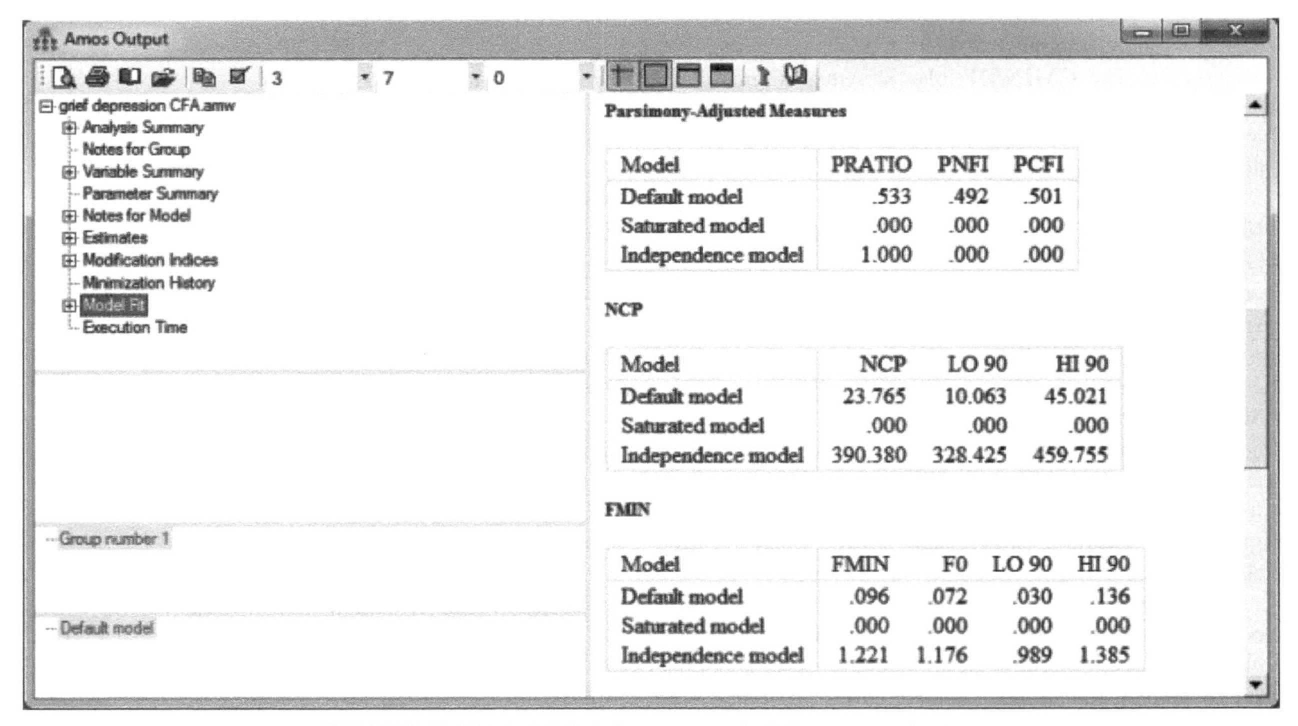

Parsimony-Adjusted Measures

Model	PRATIO	PNFI	PCFI
Default model	.533	.492	.501
Saturated model	.000	.000	.000
Independence model	1.000	.000	.000

NCP

Model	NCP	LO 90	HI 90
Default model	23.765	10.063	45.021
Saturated model	.000	.000	.000
Independence model	390.380	328.425	459.755

FMIN

Model	FMIN	F0	LO 90	HI 90
Default model	.096	.072	.030	.136
Saturated model	.000	.000	.000	.000
Independence model	1.221	1.176	.989	1.385

FIGURE 39.23 Model fit indexes (second of three screenshots).

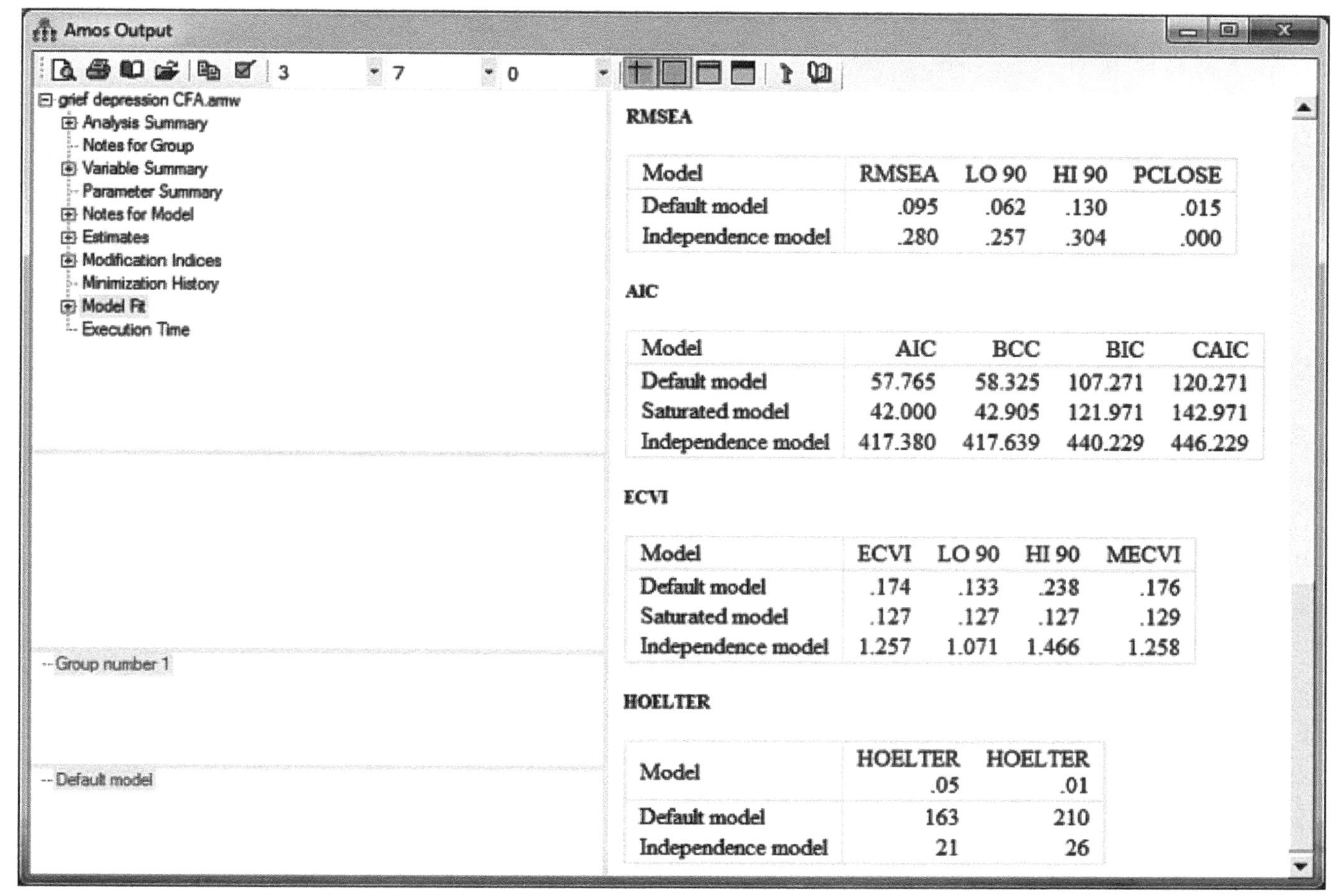

RMSEA

Model	RMSEA	LO 90	HI 90	PCLOSE
Default model	.095	.062	.130	.015
Independence model	.280	.257	.304	.000

AIC

Model	AIC	BCC	BIC	CAIC
Default model	57.765	58.325	107.271	120.271
Saturated model	42.000	42.905	121.971	142.971
Independence model	417.380	417.639	440.229	446.229

ECVI

Model	ECVI	LO 90	HI 90	MECVI
Default model	.174	.133	.238	.176
Saturated model	.127	.127	.127	.129
Independence model	1.257	1.071	1.466	1.258

HOELTER

Model	HOELTER .05	HOELTER .01
Default model	163	210
Independence model	21	26

FIGURE 39.24 Model fit indexes (third of three screenshots).

Analysis Output: Estimated Parameters

Selecting **Estimates** in the left panel displays the estimated parameters generated by the model, the first portion of which is shown in Figure 39.25. The **Regression Weights Table** in the top table of Figure 39.25 presents the unstandardized regression coefficients in the **Estimate** column together with their standard errors (**S.E.**). These are the regression weights estimated using the relevant latent variable to predict the value of the measured variable, except for the two paths that were initially constrained in specifying the model. For example, the unstandardized regression weight estimated for **lonely** as predicted by **grief** is **.948**, with a standard error of **.105**.

Each estimated regression coefficient is tested for statistical significance, the result of which is shown in the **P** column; a triple asterisk (***) indicates a probability level of $p < .001$. All unconstrained regression coefficients were statistically significant.

The third column (**C.R.**) reports the critical ratio, a value obtained by dividing the unstandardized regression coefficient by its standard error. The **C.R.** operates as a z statistic with values of 1.96 or greater indicating statistical significance.

The second table in Figure 39.25 presents the standardized regression coefficients (often referred to informally as "factor loadings") associated with each path. Note that the two constrained paths whose unstandardized values were retained as **1.000** in the **Regression Weights Table** do have estimated standardized values associated with them. In reporting the results of the model in diagram form, these are the values that are

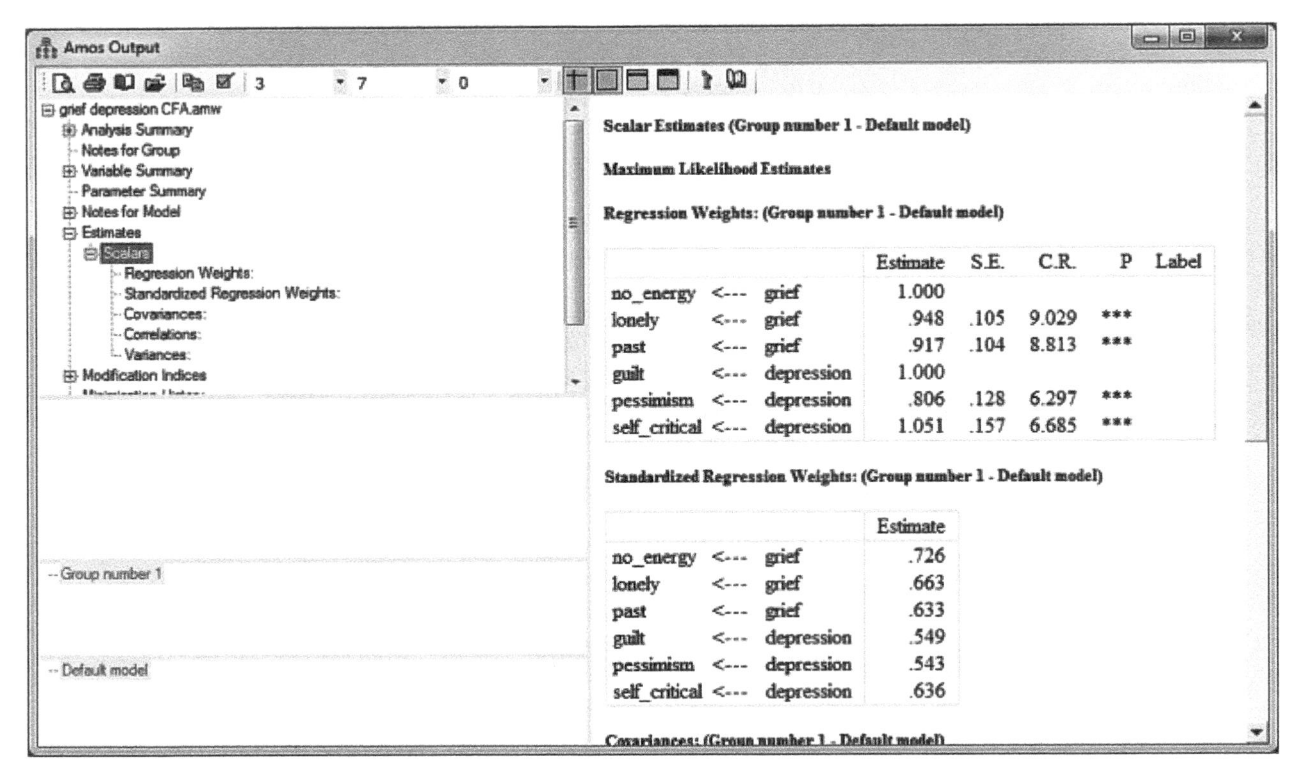

Scalar Estimates (Group number 1 - Default model)

Maximum Likelihood Estimates

Regression Weights: (Group number 1 - Default model)

			Estimate	S.E.	C.R.	P	Label
no_energy	<---	grief	1.000				
lonely	<---	grief	.948	.105	9.029	***	
past	<---	grief	.917	.104	8.813	***	
guilt	<---	depression	1.000				
pessimism	<---	depression	.806	.128	6.297	***	
self_critical	<---	depression	1.051	.157	6.685	***	

Standardized Regression Weights: (Group number 1 - Default model)

			Estimate
no_energy	<---	grief	.726
lonely	<---	grief	.663
past	<---	grief	.633
guilt	<---	depression	.549
pessimism	<---	depression	.543
self_critical	<---	depression	.636

Covariances: (Group number 1 - Default model)

FIGURE 39.25 The first portion of the **Estimates** screen.

ordinarily placed on the paths. All of these standardized regression coefficients appear to be of adequate value for us to judge that the measured variables appear to be reasonable indicators of their latent construct.

Figure 39.26 presents the remaining estimated parameters. The top two tables display the correlation between the factors in unstandardized form (**Covariances**) and standardized form (**Correlations**). As may be seen, the correlation between the two factors is **.723**. Such a correlation is quite high, suggesting that the two factors may not be independent enough of each other to justify a two-factor structure.

Analysis Output: Modification Indices

The suggested modifications offered by IBM SPSS Amos deal with adding parameters to the model on the assumption that researchers can figure out for themselves if they wish to remove any paths that were not statistically significant. It must be emphasized that these suggestions are generated purely statistically without any relationship to the theoretical, empirical, and/or common sense world in which the researchers actually live; thus, these suggested modifications need to be treated with more than a grain of salt as they may make no sense either theoretically, empirically, or both. Furthermore, by making any modifications to the model, the orientation of the analysis shifts from a confirmatory approach to more of an exploratory approach; by definition, the model fit will improve when any of these post hoc modifications are made to the model, so researchers should not interpret a model fit improvement as strong support for their modified model.

Given that the model appears to fall short of an acceptable fit to the data and that the two factors are correlated to an uncomfortable extent (i.e., the correlation in excess of .7 raises the possibility that the two factors are not able to be effectively distinguished), it is probably worthwhile to examine the suggestions offered by IBM SPSS Amos to improve

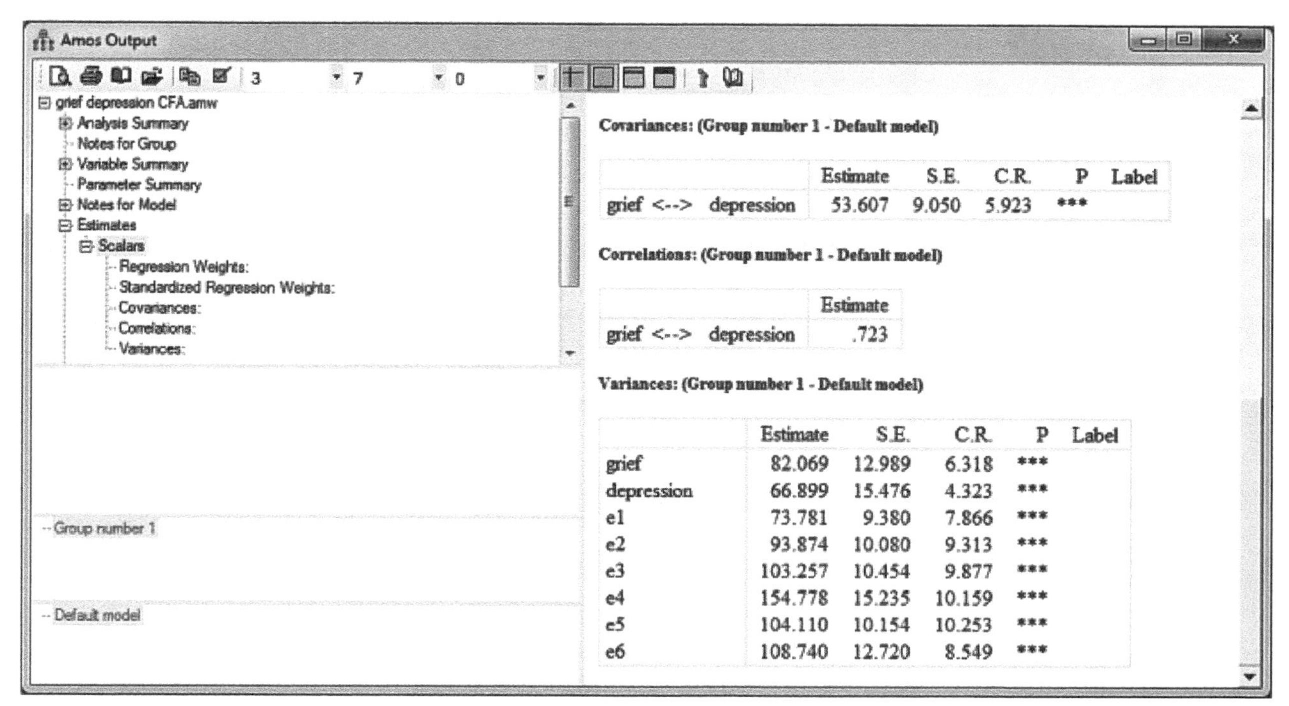

Covariances: (Group number 1 - Default model)

	Estimate	S.E.	C.R.	P	Label
grief <--> depression	53.607	9.050	5.923	***	

Correlations: (Group number 1 - Default model)

	Estimate
grief <--> depression	.723

Variances: (Group number 1 - Default model)

	Estimate	S.E.	C.R.	P	Label
grief	82.069	12.989	6.318	***	
depression	66.899	15.476	4.323	***	
e1	73.781	9.380	7.866	***	
e2	93.874	10.080	9.313	***	
e3	103.257	10.454	9.877	***	
e4	154.778	15.235	10.159	***	
e5	104.110	10.154	10.253	***	
e6	108.740	12.720	8.549	***	

FIGURE 39.26 The last portion of the **Estimates** screen.

model fit. We can view these suggested modifications by selecting **Modification Indices** in the left panel as shown in Figure 39.27.

The top table in Figure 39.27 labeled **Covariances** shows suggested additions of **Covariances** (correlations). The first column indicates the suggested parameter to add, and the **M.I.** column indicates the improvement in the chi-square value should that single change be made. The only two theoretically justifiable tweaks that appear in the table are those that deal with correlating the errors of sets of indicator variables that are each associated with the same factor. This is because it is possible that the errors may tap into a common extraneous construct or that the items may be similarly worded in some manner. These two suggested modifications are as follows:

- Adding a correlation between **e1** and **e2** (the errors associated with **no_energy** and **lonely** indicators of **grief**).
- Adding a correlation between **e5** and **e6** (the errors associated with **pessimism** and **self_critical** indicators of **depression**).

We therefore accept these two suggestions and implement the two modifications in a modified model.

The bottom table in Figure 39.27 labeled **Regression Weights** shows suggested additions of paths. As none suggest adding additional paths from the factors to indicator variables, and because it makes no theoretical sense to tie the indicator variables together in a causal chain, we do not accept any of these ideas.

39.6 ANALYSIS SETUP: MODIFIED MODEL

Select the window showing the model diagram to make it the active window. Then select the **View the output path diagram** icon to the left in the top output panel next to the

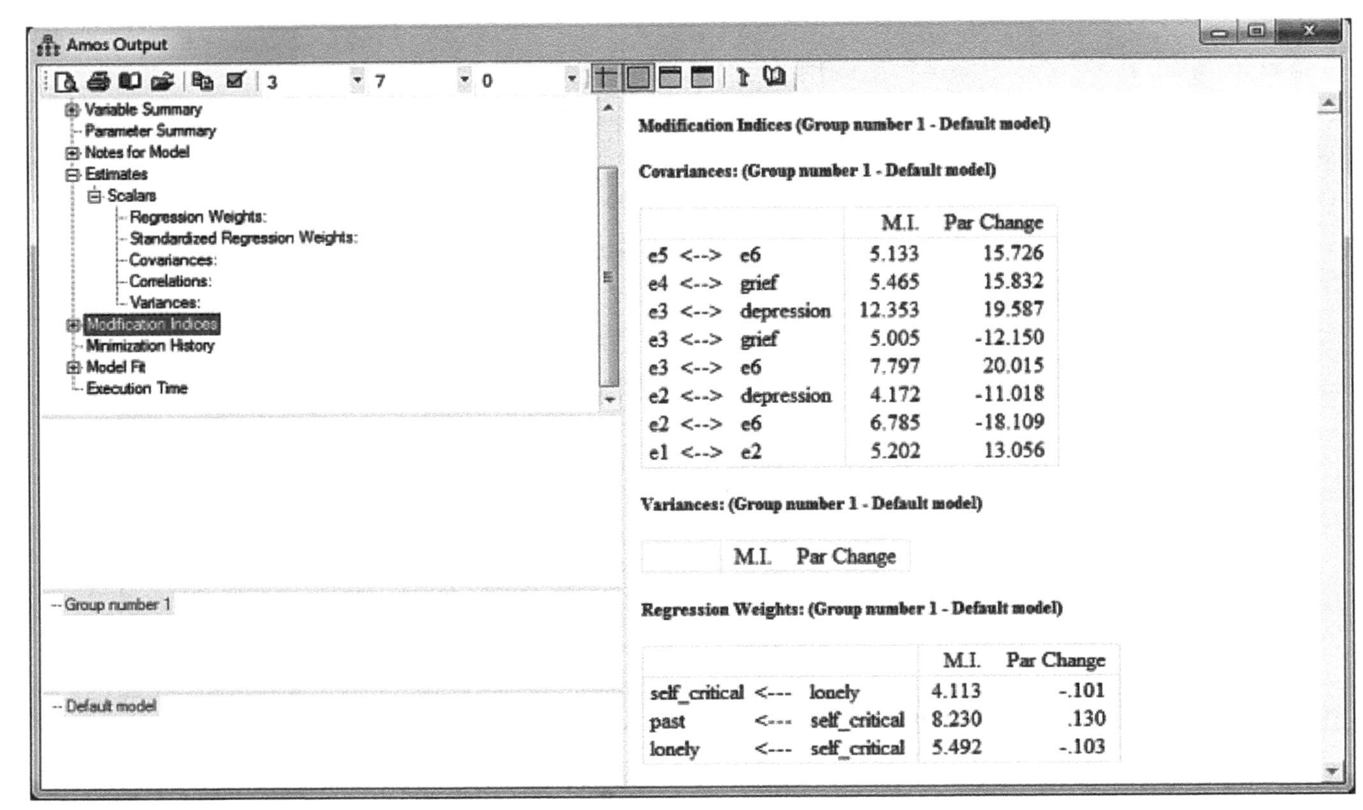

FIGURE 39.27 The **Modification Indices**.

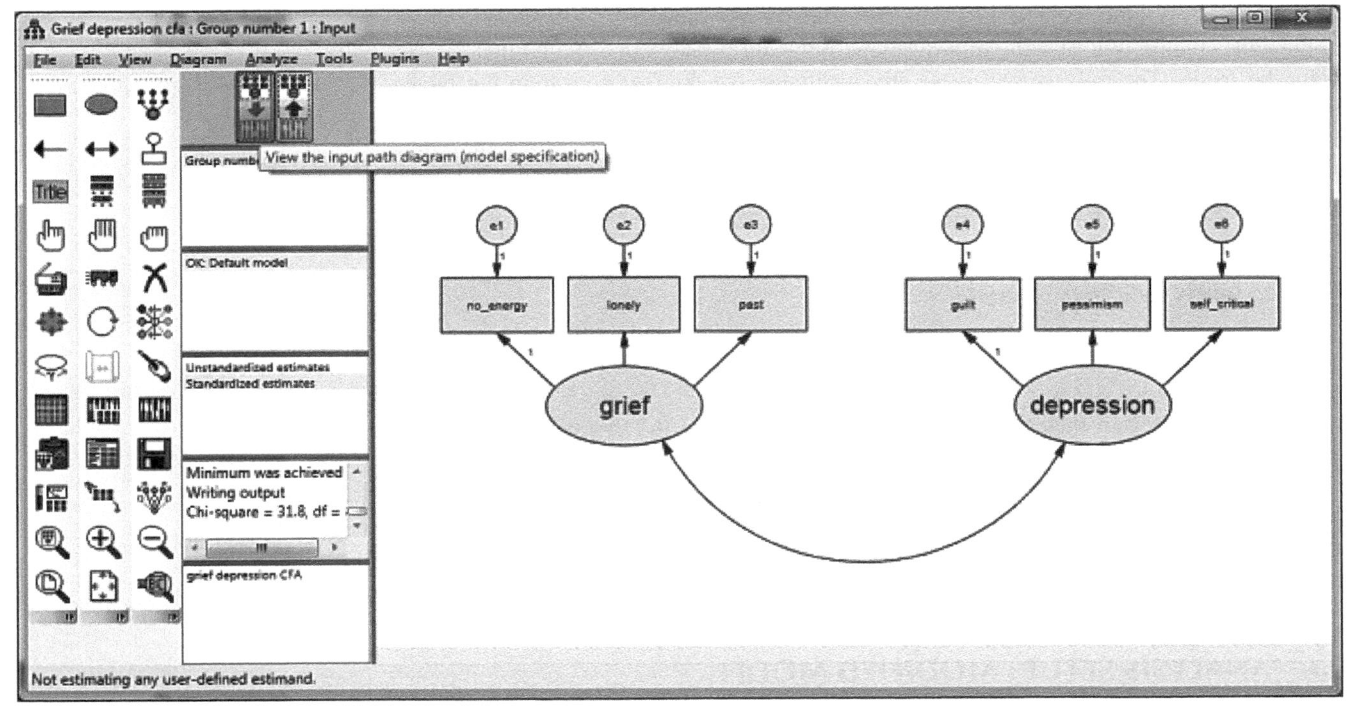

FIGURE 39.28 Select the **View the output path diagram** icon in the top panel next to the diagram.

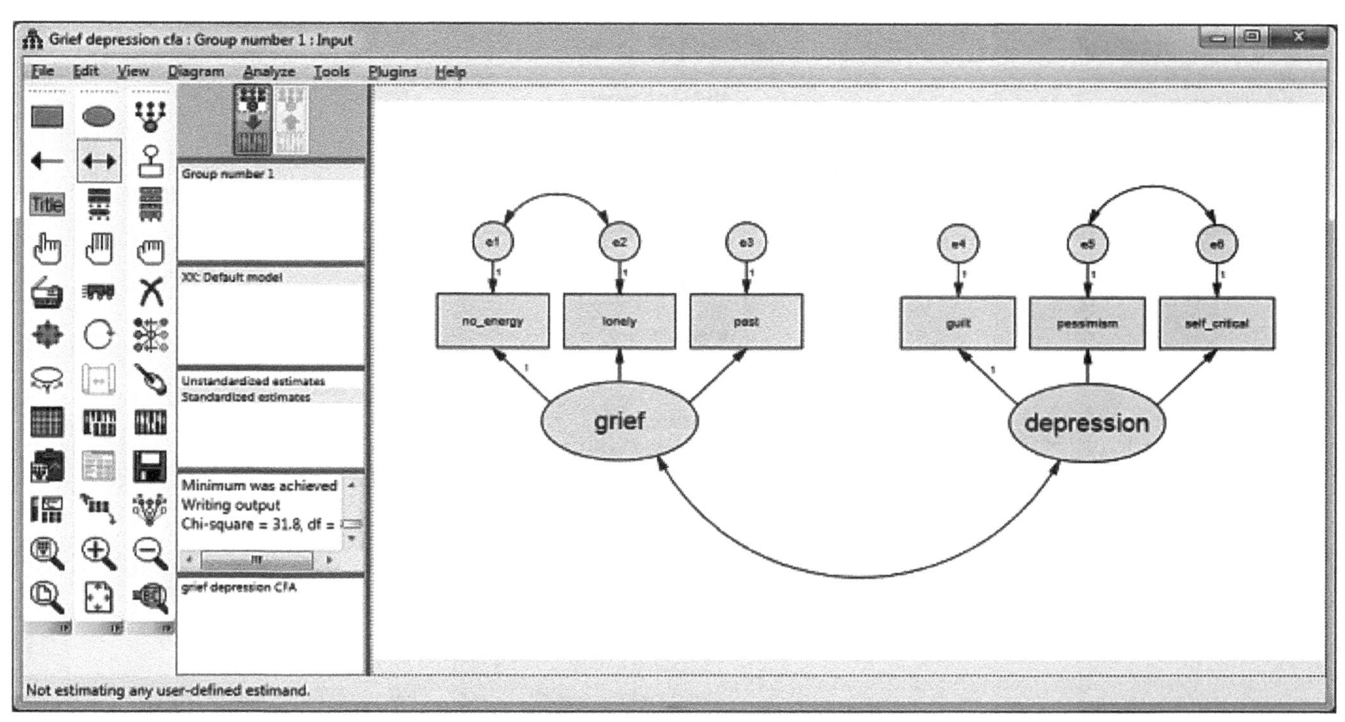

FIGURE 39.29 Select the **Draw covariances (double headed arrow)** from the icon toolbar.

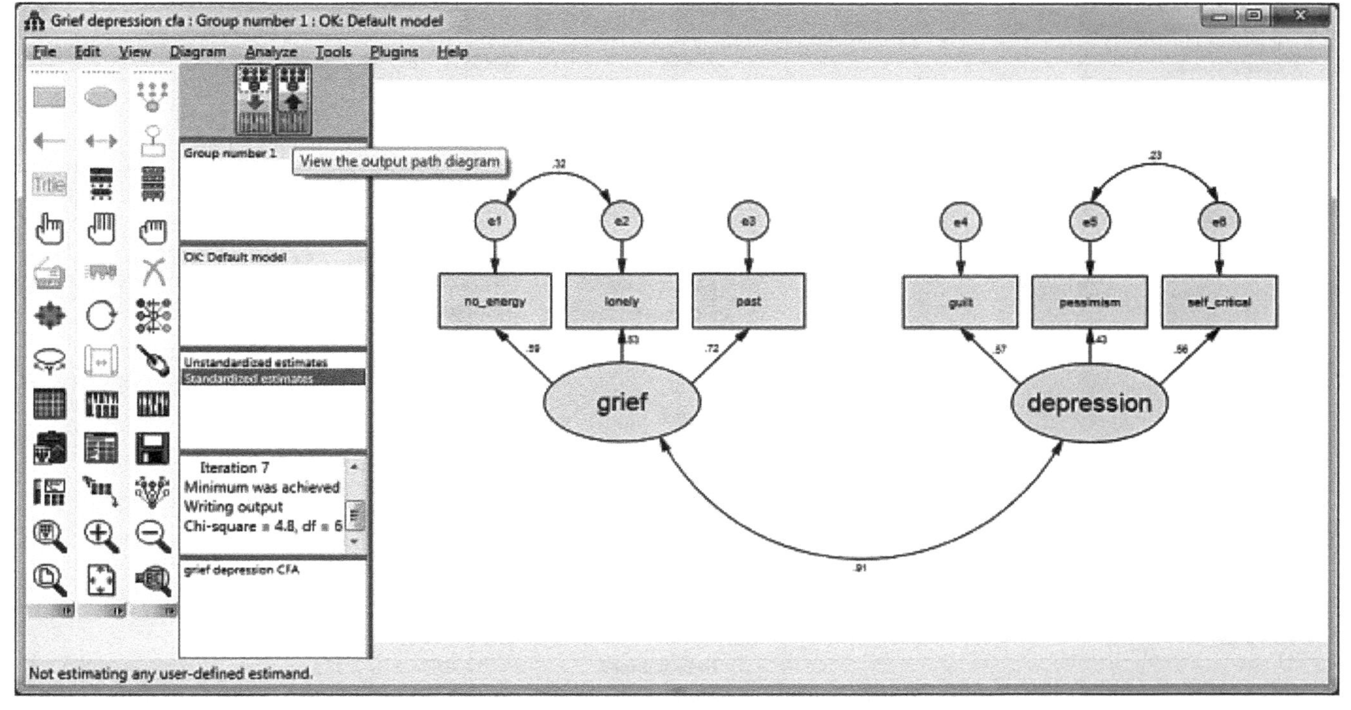

FIGURE 39.30 The **Standardized estimates** for the modified model.

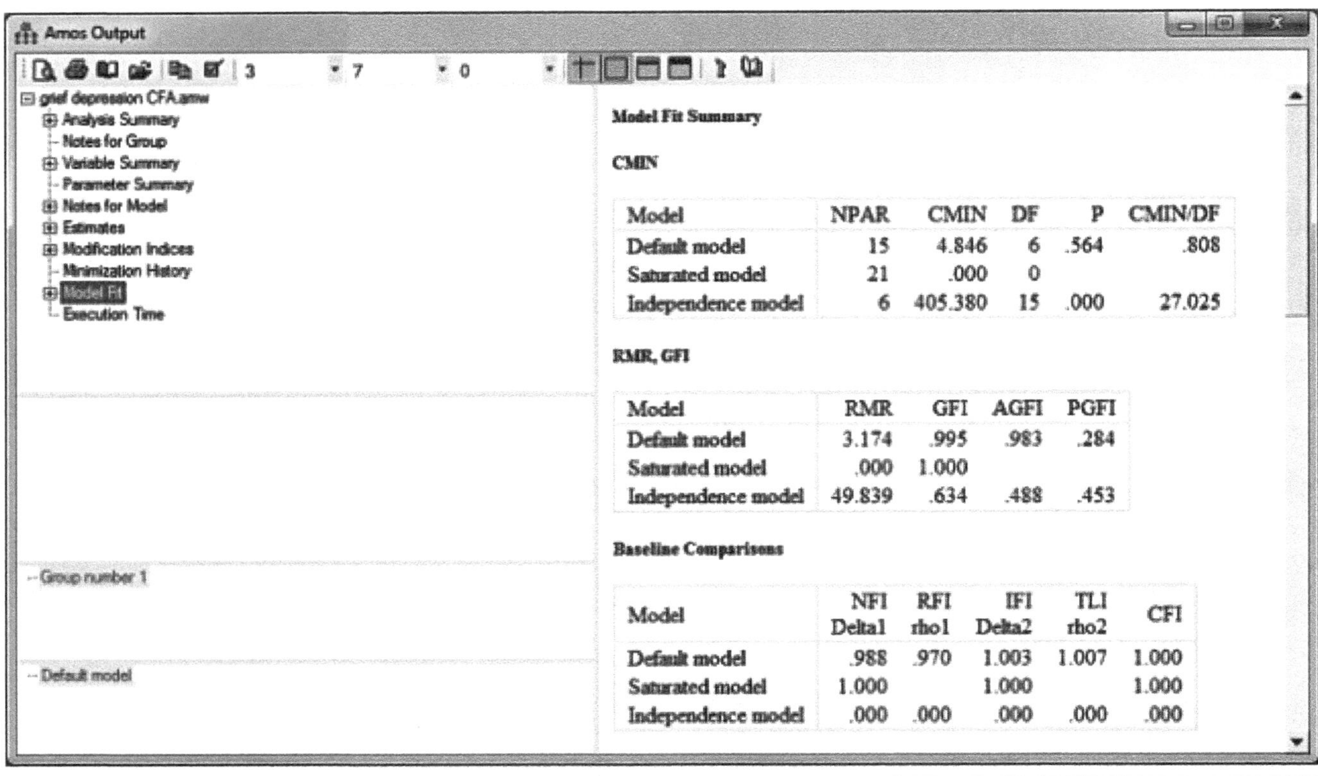

Amos Output

Model Fit Summary

CMIN

Model	NPAR	CMIN	DF	P	CMIN/DF
Default model	15	4.846	6	.564	.808
Saturated model	21	.000	0		
Independence model	6	405.380	15	.000	27.025

RMR, GFI

Model	RMR	GFI	AGFI	PGFI
Default model	3.174	.995	.983	.284
Saturated model	.000	1.000		
Independence model	49.839	.634	.488	.453

Baseline Comparisons

Model	NFI Delta1	RFI rho1	IFI Delta2	TLI rho2	CFI
Default model	.988	.970	1.003	1.007	1.000
Saturated model	1.000		1.000		1.000
Independence model	.000	.000	.000	.000	.000

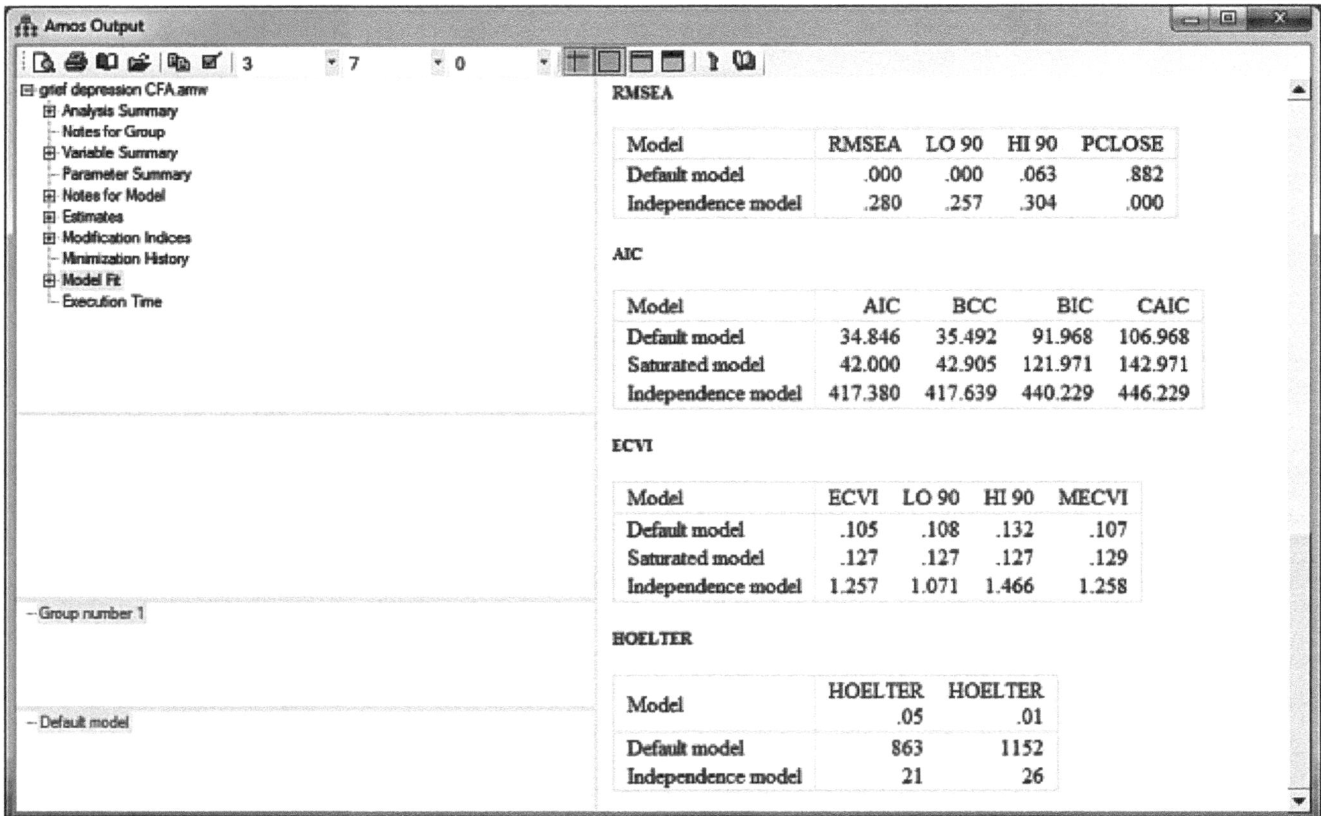

Amos Output

RMSEA

Model	RMSEA	LO 90	HI 90	PCLOSE
Default model	.000	.000	.063	.882
Independence model	.280	.257	.304	.000

AIC

Model	AIC	BCC	BIC	CAIC
Default model	34.846	35.492	91.968	106.968
Saturated model	42.000	42.905	121.971	142.971
Independence model	417.380	417.639	440.229	446.229

ECVI

Model	ECVI	LO 90	HI 90	MECVI
Default model	.105	.108	.132	.107
Saturated model	.127	.127	.127	.129
Independence model	1.257	1.071	1.466	1.258

HOELTER

Model	HOELTER .05	HOELTER .01
Default model	863	1152
Independence model	21	26

FIGURE 39.31 The model fit windows showing the fit indexes of interest to us.

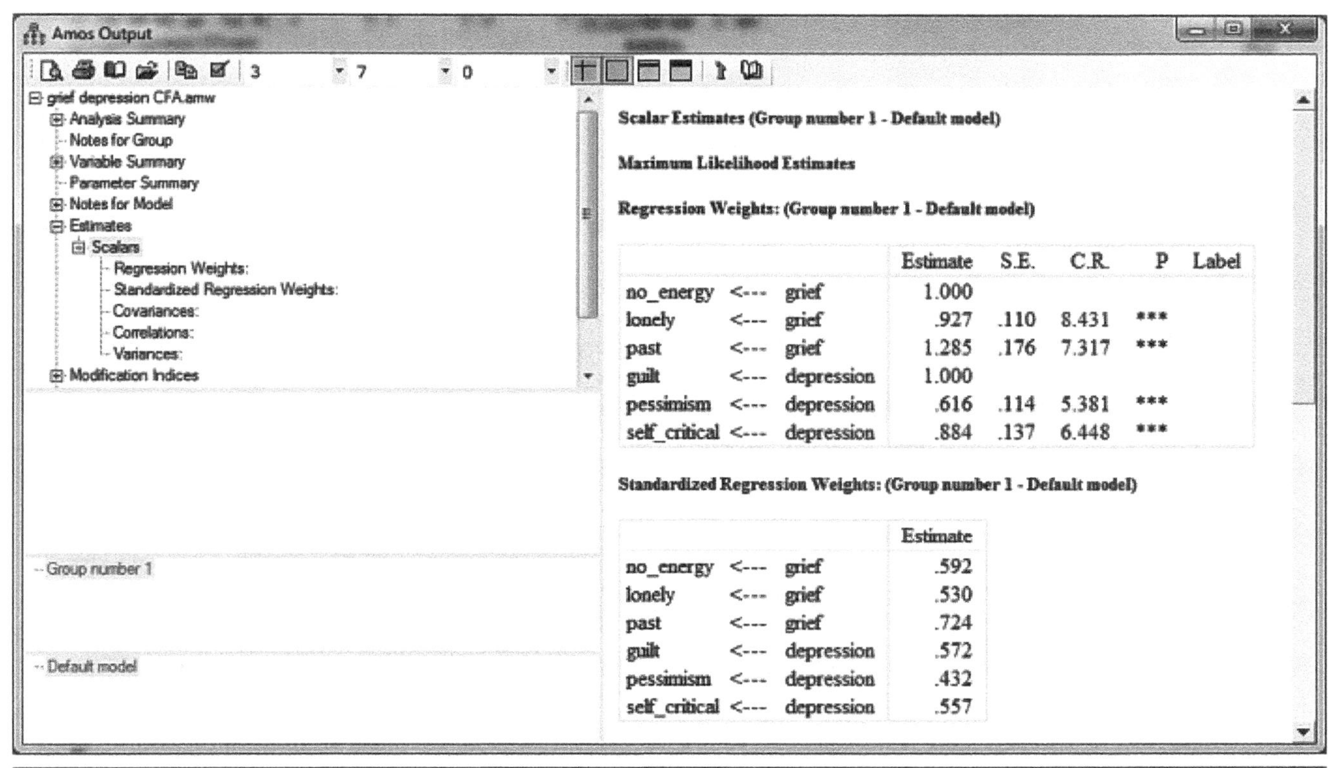

Scalar Estimates (Group number 1 - Default model)

Maximum Likelihood Estimates

Regression Weights: (Group number 1 - Default model)

			Estimate	S.E.	C.R.	P	Label
no_energy	<---	grief	1.000				
lonely	<---	grief	.927	.110	8.431	***	
past	<---	grief	1.285	.176	7.317	***	
guilt	<---	depression	1.000				
pessimism	<---	depression	.616	.114	5.381	***	
self_critical	<---	depression	.884	.137	6.448	***	

Standardized Regression Weights: (Group number 1 - Default model)

			Estimate
no_energy	<---	grief	.592
lonely	<---	grief	.530
past	<---	grief	.724
guilt	<---	depression	.572
pessimism	<---	depression	.432
self_critical	<---	depression	.557

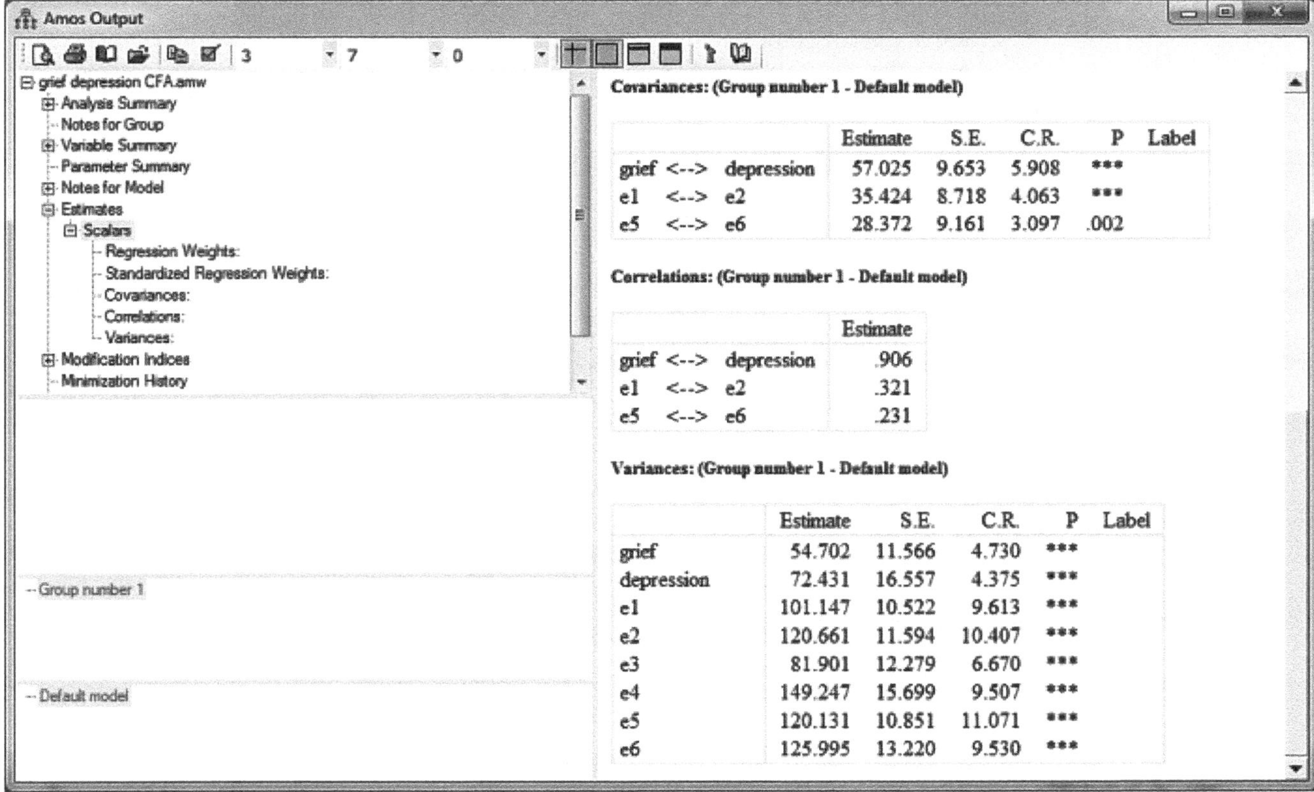

Covariances: (Group number 1 - Default model)

			Estimate	S.E.	C.R.	P	Label
grief	<-->	depression	57.025	9.653	5.908	***	
e1	<-->	e2	35.424	8.718	4.063	***	
e5	<-->	e6	28.372	9.161	3.097	.002	

Correlations: (Group number 1 - Default model)

			Estimate
grief	<-->	depression	.906
e1	<-->	e2	.321
e5	<-->	e6	.231

Variances: (Group number 1 - Default model)

	Estimate	S.E.	C.R.	P	Label
grief	54.702	11.566	4.730	***	
depression	72.431	16.557	4.375	***	
e1	101.147	10.522	9.613	***	
e2	120.661	11.594	10.407	***	
e3	81.901	12.279	6.670	***	
e4	149.247	15.699	9.507	***	
e5	120.131	10.851	11.071	***	
e6	125.995	13.220	9.530	***	

FIGURE 39.32 The **Estimates** windows.

diagram to reach the screen shown in Figure 39.28. Select the double-headed path icon **Draw covariances (double headed arrow)** from the icon toolbar, and draw in the two covariances needed, as has been done in Figure 39.29.

39.7 ANALYSIS OUTPUT: MODIFIED MODEL

From the main menu, select **Analyze ➔ Calculate Estimates**. Click the **View the output path diagram** icon. This action produces the **Standardized estimates** (see Figure 39.30). Select from the main menu **View ➔ Text Output**. This action opens the **Notes for Model** screen of the **Output** window (not shown).

Select **Model Fit** from the panel on the right side of the screen. The two screens showing the model fit indexes of interest to us are presented in Figure 39.31. The chi-square (**CMIN**) value dropped to a statistically nonsignificant value of **4.846** (p = **.564**), indicating that the expected values based on the model were relatively closely matched by the data. We can also see that the **GFI** improved to **.995**, the **NFI** went to **.988**, the **CFI** topped out at **1.000**, and the **RMSEA** bottomed out at **.000**. This model—not surprisingly—fits the data better, now that we have hypothesized that the two pairs of errors are correlated; in fact, the model represents a superb (post hoc) fit to the data.

Selecting **Estimates** from the left panel displays the parameter estimates shown in Figure 39.32. All paths are associated with statistically significant coefficients, and the standardized path coefficients appear to be respectable.

Of much greater interest is the correlation between the factors. In this revised model, the factors now correlate to **.906**. Such a very strong correlation is unacceptable, signifying that the factors lack a sufficient individual identity to justify both of them. We would therefore conclude that, despite the superb fit of the model to the data, there is not a viable two-factor structure describing these indicator variables; in fact, we would guess that the six indicator variables are likely to represent a single latent construct (this idea can be explored by constructing and testing a single-factor confirmatory model).

PART 13

EVALUATING CAUSAL (PREDICTIVE) MODELS

CHAPTER 40

Simple Mediation

40.1 OVERVIEW

Simple mediation analysis is a conceptual extension of regression analysis. Instead of just specifying a set of predictor variables, in a mediation analysis, the variables are arranged in a *predictive* (sometimes referred to as *causal*) path model to assess the dynamics of their interplay. The model represents the hypothesis of the researchers concerning how the variables interrelate. Mediation analysis was popularized by Baron and Kenny (1986) and is described in several more recent sources (e.g., MacKinnon, 2008; MacKinnon, Fairchild, & Fritz, 2007; Preacher & Hayes, 2008).

Simple mediation analysis is the least complex type of path model and is shown in generic form in the top portion of Figure 40.1. It relates just three variables to each other: a dependent or outcome variable Y, an independent or predictor variable X, and a potential mediator M. The issue addressed in such an analysis is whether the relationship between X and Y (the prediction of Y based on X) is attenuated when M is included together with X as a predictor of Y.

In the model presented in the top part of Figure 40.1, X is hypothesized to affect or influence Y in two ways. One type of influence is *direct* and is represented by the path (the single-headed arrow) leading from X directly to Y. The other type of influence is *indirect* and is represented by the path leading from X through M to Y. The key issue in a simple mediation analysis is the extent to which the presence of the mediator variable M changes the degree of prediction we would otherwise observe for X predicting Y in isolation.

To justify applying a mediation analysis, the following conditions should be met (Meyers et al., 2013):

- X must significantly predict Y in isolation.
- X must significantly predict M in isolation.
- M must significantly predict Y in the mediation model.

When these conditions are satisfied, we can determine the effect of M on the direct relationship between X and Y. There are four logically possible effects that may be

Performing Data Analysis Using IBM SPSS®, First Edition.
Lawrence S. Meyers, Glenn C. Gamst, and A. J. Guarino.

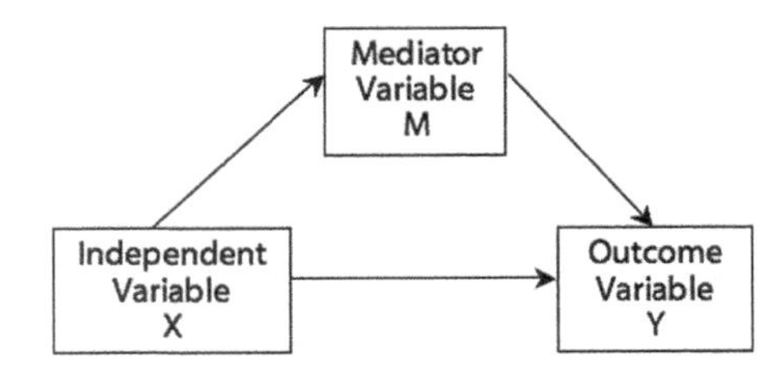

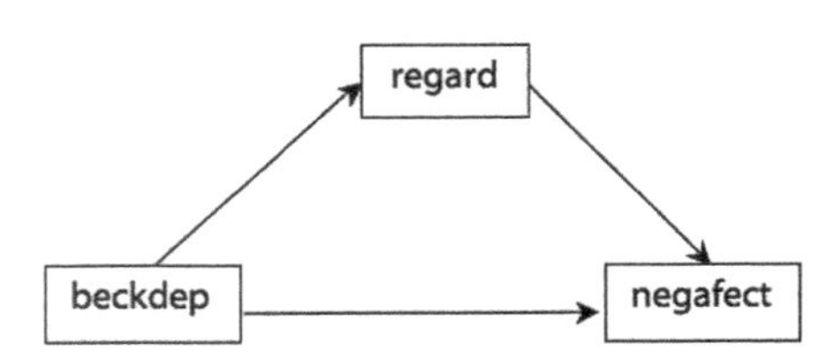

FIGURE 40.1

The general mediation model (top path diagram) and the mediation model that will be tested, with **regard** mediating the effect of **beckdep** on **negafect** (bottom path diagram).

observed in the mediated model:

- The direct relationship between X and Y is fully attenuated (X does not significantly predict Y with M in the model). With this result, we say that we have observed *full mediation*.
- The direct relationship between X and Y is less strong with M in the model than it was in isolation, but is still statistically significant. With this result, we say that we have observed *partial mediation*.
- The direct relationship between X and Y is as strong with M in the model as it was in isolation. With this result, we say that we have observed no mediated effect of M.
- The direct relationship between X and Y is significantly stronger with M in the model than it was in isolation. With this result, we say that M has acted as a *suppressor variable*; that is, the presence of M has enhanced the prediction of X on Y. This suppression effect can occur because M has accounted for some of the variance in Y not predictable by X (Darlington, 1990; Pedhazur, 1997); in some sense, the presence of M helps "purify" the relationship between X and Y (Meyers et al., 2013).

40.2 NUMERICAL EXAMPLE

The data we use for our example are extracted from a study of personality variables on 425 university students. Data are contained in the data file named **Personality**. We focus on three of the personality variables: **beckdep** as a measure of depression, **regard** as a measure of self-regard, and **negafect** as a measure of the amount of negative affect generally experienced by respondents.

40.3 ANALYSIS STRATEGY

The mediation model is shown in the bottom portion of Figure 40.1 as a path diagram. In this illustrative example, we have hypothesized that **regard** mediates the relationship between **beckdep** and **negafect**. To fully evaluate this model, we will perform a total of three linear regression analyses:

- We will use the independent variable **beckdep** to directly predict the mediator variable **regard**.
- We will use **beckdep** and **regard** to predict the outcome variable of **negafect**.
- We will generate the unmediated model with **beckdep** predicting **negafect** in isolation.

Our goal is to obtain the standardized (beta) and unstandardized (raw score) regression coefficients as well as the SEs associated with the unstandardized regression coefficients. We will then use that information to test the statistical significance of the indirect effect (the mediated effect of **beckdep** acting through **regard** to affect **negafect**) by performing the Aroian test. We will also evaluate the difference in the path coefficients between **beckdep** and **negafect** in the unmediated and mediated models by performing the Freedman–Schatzkin test. Finally, we will determine the relative strength of the mediated effect.

40.4 THE INDEPENDENT VARIABLE PREDICTING THE MEDIATOR VARIABLE

From the main menu, we select **Analyze ➔ Regression ➔ Linear**. This opens the main **Linear Regression** window shown in Figure 40.2. We move **regard** into the **Dependent** panel and **beckdep** into the **Independent(s)** panel.

In the **Statistics** dialog window shown in Figure 40.3, we check **Estimates** (to obtain the regression coefficients), **Model fit** (to obtain the R^2 and adjusted R^2), **R squared change**, **Descriptives** (to obtain the descriptive statistics), and **Part and partial correlations** (to obtain the zero-order, partial, and semipartial correlations). Click **Continue** to return to the main dialog window and click **OK** to perform the analysis.

The results of the analysis are shown in Figure 40.4. We have obtained a statistically significant amount of prediction; the unstandardized regression coefficient for **beckdep**

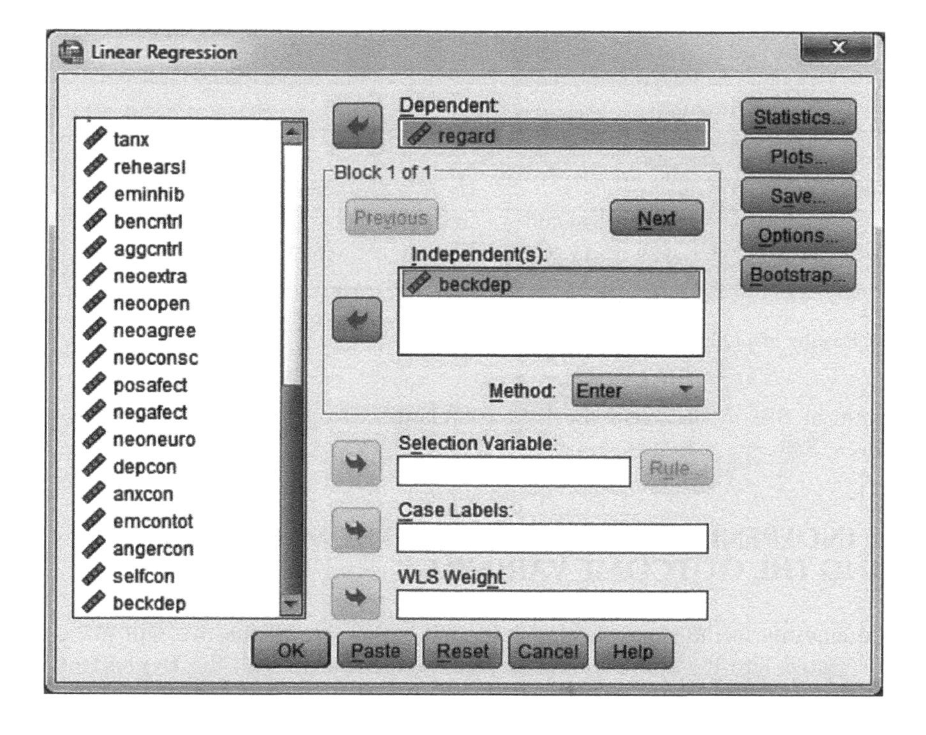

FIGURE 40.2

The main **Linear Regression** dialog window.

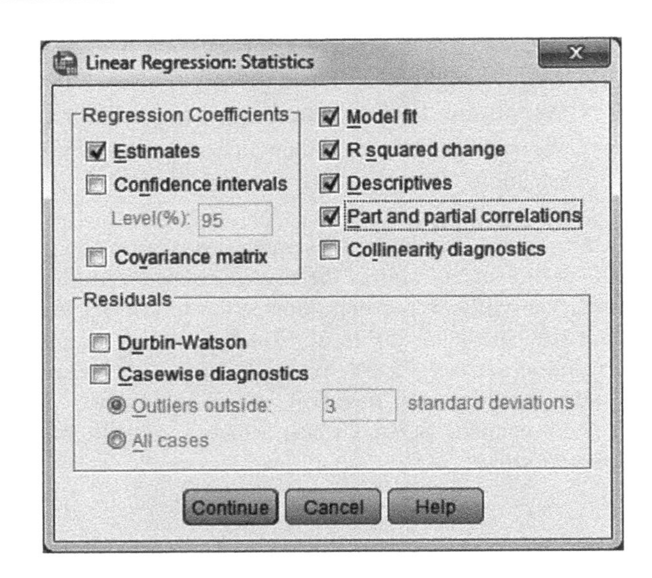

FIGURE 40.3
The **Statistics** screen of **Linear Regression**.

Model Summary

Model	R	R Square	Adjusted R Square	Std. Error of the Estimate	Change Statistics				
					R Square Change	F Change	df1	df2	Sig. F Change
1	.539[a]	.290	.288	8.90652	.290	171.187	1	419	.000

a. Predictors: (Constant), beckdep Beck depression: bdi

ANOVA[b]

Model		Sum of Squares	df	Mean Square	F	Sig.
1	Regression	13579.621	1	13579.621	171.187	.000[a]
	Residual	33237.626	419	79.326		
	Total	46817.247	420			

a. Predictors: (Constant), beckdep Beck depression: bdi
b. Dependent Variable: regard self-regard: poi

Coefficients[a]

Model		Unstandardized Coefficients		Standardized Coefficients	t	Sig.	Correlations		
		B	Std. Error	Beta			Zero-order	Partial	Part
1	(Constant)	61.316	.636		96.440	.000			
	beckdep Beck depression: bdi	-.814	.062	-.539	-13.084	.000	-.539	-.539	-.539

a. Dependent Variable: regard self-regard: poi

FIGURE 40.4 Results of predicting **regard** with **beckdep**.

was −**.814** with an SE of **.062**, and the beta (standardized) coefficient (which is the Pearson *r*) was −**.539**.

40.5 THE INDEPENDENT VARIABLE AND THE MEDIATOR PREDICTING THE OUTCOME VARIABLE

We set up the analysis in the same way as described earlier except that we will place **beckdep** and **regard** into the **Independent(s)** panel and **negafect** into the **Dependent** panel.

Model Summary

Model	R	R Square	Adjusted R Square	Std. Error of the Estimate	Change Statistics				
					R Square Change	F Change	df1	df2	Sig. F Change
1	.579[a]	.335	.331	3.05553	.335	105.130	2	418	.000

a. Predictors: (Constant), regard self-regard: poi, beckdep Beck depression: bdi

ANOVA[b]

Model		Sum of Squares	df	Mean Square	F	Sig.
1	Regression	1963.048	2	981.524	105.130	.000[a]
	Residual	3902.558	418	9.336		
	Total	5865.606	420			

a. Predictors: (Constant), regard self-regard: poi, beckdep Beck depression: bdi
b. Dependent Variable: negafect negative affect: mpq

Coefficients[a]

Model		Unstandardized Coefficients		Standardized Coefficients	t	Sig.	Correlations		
		B	Std. Error	Beta			Zero-order	Partial	Part
1	(Constant)	10.500	1.051		9.995	.000			
	beckdep Beck depression: bdi	.186	.025	.347	7.338	.000	.515	.338	.293
	regard self-regard: poi	-.110	.017	-.312	-6.586	.000	-.499	-.307	-.263

a. Dependent Variable: negafect negative affect: mpq

FIGURE 40.5 Results of predicting **negafect** with **beckdep** and **regard**.

The results of the analysis are shown in Figure 40.5. As can be seen in the output, both the independent variable **beckdep** and the mediator variable **regard** are statistically significant predictors of **negafect**.

40.6 THE UNMEDIATED MODEL WITH THE INDEPENDENT VARIABLE PREDICTING THE DEPENDENT VARIABLE

We set up the analysis such that **beckdep** is placed into the **Independent(s)** panel and **negafect** into the **Dependent** panel. The results of the analysis are shown in Figure 40.6. As can be seen in the output, the independent variable **beckdep** is a statistically significant predictor of **negafect**.

40.7 CONSOLIDATING THE RESULTS OF THE MEDIATION MODEL

We have drawn in Figure 40.7 the results obtained for the mediation analysis. Each path is associated with a letter because we will use these letters to represent values in the equations to test the statistical significance of certain aspects of the mediation model. Path coefficients are associated with each path; for example, the standardized (beta) coefficient for **beckdep** predicting **regard** was −**.539**, whereas the unstandardized path coefficient was −**.814** with an SE of .062.

The primary issue in configuring a mediation model is to determine if the hypothesized mediator affected the direct relationship between the independent and outcome variables. Visual examination of the results gives us a clue as to what our tests of significance are likely to verify. The unstandardized path coefficient from **beckdep** to **negafect**

Model Summary

Model	R	R Square	Adjusted R Square	Std. Error of the Estimate	Change Statistics				
					R Square Change	F Change	df1	df2	Sig. F Change
1	.515[a]	.266	.264	3.20632	.266	151.557	1	419	.000

a. Predictors: (Constant), beckdep Beck depression: bdi

ANOVA[b]

Model		Sum of Squares	df	Mean Square	F	Sig.
1	Regression	1558.084	1	1558.084	151.557	.000[a]
	Residual	4307.522	419	10.280		
	Total	5865.606	420			

a. Predictors: (Constant), beckdep Beck depression: bdi
b. Dependent Variable: negafect negative affect: mpq

Coefficients[a]

Model		Unstandardized Coefficients		Standardized Coefficients	t	Sig.	Correlations		
		B	Std. Error	Beta			Zero-order	Partial	Part
1	(Constant)	3.732	.229		16.306	.000			
	beckdep Beck depression: bdi	.276	.022	.515	12.311	.000	.515	.515	.515

a. Dependent Variable: negafect negative affect: mpq

FIGURE 40.6 Results of the unmediated model with **beckdep** predicting **negafect**.

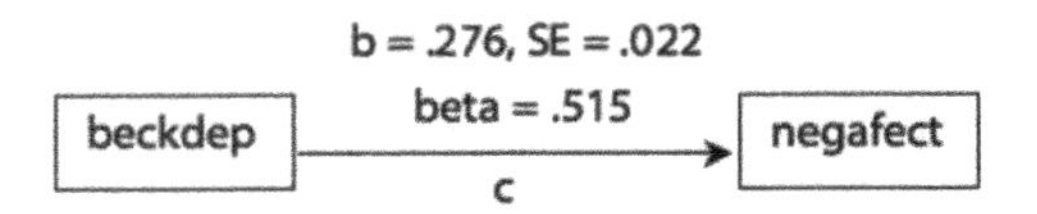

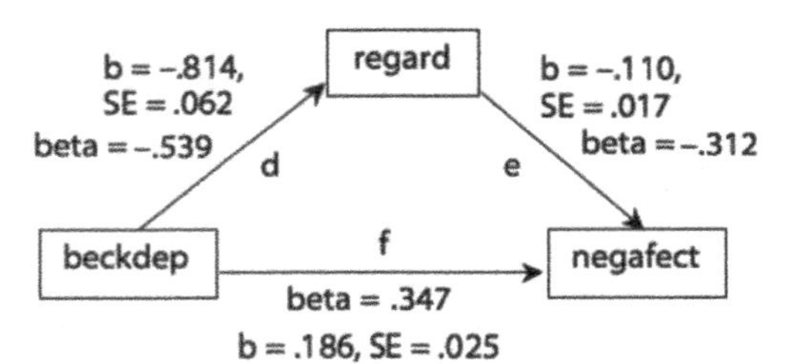

FIGURE 40.7
Results of the path model.

in the unmediated model was **.276** and was reduced to **.186** in the mediated model (the corresponding standardized coefficients were **.515** and **.347**). Because both these coefficients were statistically significant, and because visual inspection informs us that the path coefficient is lower in the mediated model, we are led to guess that we have observed a partial mediation effect. We now use a series of steps to statistically test that guess.

40.8 TESTING THE STATISTICAL SIGNIFICANCE OF THE INDIRECT EFFECT

One aspect of evaluating statistical significance in the mediation model concerns the indirect effect of **beckdep** through **regard** on **negafect**. Probably, the most widely known and most frequently used approach to address statistical significance of the indirect effect is the Sobel test (Sobel, 1982, 1986) and its variants, the Aroian test (Aroian, 1947) and the Goodman test (Goodman, 1960). Partly because the sampling distribution of the indirect effect tends to be somewhat skewed rather than meeting the assumption that it is normal (causing the tests to lose some statistical power), these tests have received some criticism in the professional community (e.g., Hayes, 2009; MacKinnon, Lockwood, Hoffman, West, & Sheets, 2002; Preacher & Hayes, 2004). Nonetheless, these tests are reported extensively in the research literature and we will apply one of the members of this family, the Aroian test, here, in that it was the variation of the Sobel family of tests popularized by Baron and Kenny (1986).

The equation for the Aroian test is presented in Figure 40.8 together with a summary of the calculations. All three tests comprising the Sobel evaluations are computed in a similar manner and differ only in how the expression ($SE_d^2 * SE_e^2$) at the end of the square root expression in the denominator of the equation is treated:

- The Aroian test adds the expression to the other terms under the square root sign in the denominator.
- The Goodman test subtracts the expression from the other terms under the square root sign in the denominator.
- The Sobel test does not include the expression under the square root sign in the denominator.

In the equation shown in Figure 40.8, the letters represent the unstandardized regression coefficients associated with the paths shown in Figure 40.7, and the SE values of the unstandardized regression weights are shown, with the subscripts indicating the reference coefficient. The outcome of the equation is a z value with an alpha level of .05 indicated by a value of 1.96 or better. Here, we obtained a value of approximately 5.79 with a hand calculator. It is also possible to use the calculator available on Kristopher J. Preacher's web site http://quantpsy.org/sobel/sobel.htm, which yields the same approximate value. The value of 5.79 is sufficient to reject the null hypothesis that the indirect path is not different from zero; instead, we conclude that the indirect effect of **beckdep** through **regard** on **negafect** is statistically significant.

40.9 TESTING THE STATISTICAL SIGNIFICANCE OF THE DIFFERENCE BETWEEN THE DIRECT PATHS IN THE UNMEDIATED AND THE MEDIATED MODELS

The Freedman–Schatzkin test (Freedman & Schatzkin, 1992) compares the relative strengths of the paths from the independent variable to the outcome variable in the unmediated model versus the mediated model. The equation is shown in Figure 40.9 together with a summary of the calculations; it produces a t value tested against a Student

$$\text{The Aroian test: } z = \frac{(d*e)}{\sqrt{(e^2 * SE_d^2 + d^2 * SE_e^2 + (SE_d^2 * SE_e^2))}}$$

$$\text{Aroian } z = \frac{(-.814)*(-.110)}{\sqrt{[(-.110)^2 * (.062)^2] + [(-.814)^2 * (.017)^2] + [(.062)^2 * (.017)^2]}}$$

$$\text{Aroian } z = 5.7918704$$

FIGURE 40.8
The Aroian test.

$$\text{The Freedman-Schatzkin test: } t = \frac{c - f}{\sqrt{(SE_c^2 + SE_f^2) - (2 * SE_c * SE_f * \sqrt{(1 - r^2_{XM})})}}$$

$$\text{Freedman-Schatzkin } t = \frac{(.276) - (.186)}{\sqrt{[(.022)^2 + (.025)^2] - [(2) * (.022) * (.025) * \sqrt{(1 - .539^2)}]}}$$

$$\text{Freedman-Schatzkin } t == 6.6621265$$

FIGURE 40.9

The Freedman–Schatzkin test.

t distribution with $N - 2$ degrees of freedom. Here, we obtained a value of approximately 6.66 with a hand calculator. With 419 degrees of freedom (our N was 421), it is sufficient to reject the null hypothesis that the coefficients are not significantly different ($p < .001$). We therefore conclude that the mediated path coefficient is significantly lower than the unmediated path coefficient, indicating that we have obtained a partial mediation effect.

40.10 DETERMINING THE RELATIVE STRENGTH OF THE MEDIATED EFFECT

We now calculate the relative strength of the mediated effect using the beta coefficients that are associated with the paths in our mediation model. This is computed as the ratio of the strength of the indirect effect to the strength of the direct effect and is calculated as follows:

- The strength of the indirect effect is the product of the beta coefficients associated with paths **beckdep** to **regard** and **regard** to **negafect** (paths **d** and **e**) in the mediated model. Here it is equal to (−**.539**) * (−**.312**), or .168.
- The strength of the isolated direct effect is the beta coefficient in the unmediated model (with a value of **.515**) where **beckdep** is the single predictor of **negafect**. It can also be calculated as the sum of the indirect effect and the beta coefficient for **beckdep** predicting **negafect** in the mediated model (path **f**): .168 + .347, or .515.
- The relative strength of the mediated effect is equal to the indirect effect divided by the direct effect. Here it is equal to .168/.515, or .326.

We may then conclude that about a third (32.6%) of the effect of **beckdep** on **negafect** is mediated through **regard**.

CHAPTER 41

Path Analysis Using Multiple Regression

41.1 OVERVIEW

In a multiple regression analysis, we assess only the direct effects of the predictors on the dependent variable. However, variables can, and often do, have an indirect (mediated) effect as well as, or instead of having, a direct effect on an outcome. Indirect or mediated effects can be analyzed by configuring the variables into a path structure. Simple mediation, described in Chapter 40, is the simplest form of path analysis. We extend that treatment in this chapter to a somewhat more complex illustration.

Sewall Wright (1920) introduced path analysis in the context of examining phylogenetic models (Lleras, 2005). The procedure achieved popularity in the 1960s (e.g., Blau & Duncan, 1967) in the field of sociology but has since steadily gained in research applications. Path analysis, sometimes called causal modeling, typically involves more than three variables and therefore subsumes more interrelationships than we see in simple mediation. The path diagram represents the model that the researchers have hypothesized weaving the measured variables together in a theoretically or empirically meaningfully way to explain the variance of the outcome variable. Thus, path analysis is used in a confirmatory manner to evaluate the structure hypothesized by the researchers rather than in an exploratory manner such as is done in multiple regression (where the variables are weighted in a linear or "flat" configuration to statistically maximize prediction).

Because there are more interconnections between variables, and because some variables take on both roles (in different portions of the analysis), the terms "independent variable" and "dependent variable" as used in multiple regression are less precise. This dual-role possibility is illustrated by considering the path model diagrammed in Figure 41.1. The variables are represented by rectangles to signify that they are observed or measured variables (as opposed to latent variables that are represented by circles or ovals). Arrows are used to depict causal or explanatory flow and are referred to as paths.

In Figure 41.1, **age** and socioeconomic status (**SES**) are hypothesized to predict the mediating variable of **optimism**, and **optimism** together with **age** is proposed to predict **quality of life**. Thus, **optimism** takes the role of a dependent variable in one portion of the analysis (it is predicted by **age** and **SES**) but takes the role of an independent variable in another portion of the analysis when it is one of the predictors of **quality of**

Performing Data Analysis Using IBM SPSS®, First Edition.
Lawrence S. Meyers, Glenn C. Gamst, and A. J. Guarino.

life. The terms that are used to better characterize the roles of the variables in the path analysis context are as follows:

- *Exogenous variables* are exclusive predictors, never serving as dependent variables. These variables are the primary or first-order predictors behind the explanatory ability of the model, in that there are no variables in the model that are presumed to "cause" or influence them; that is, no attempt is made to explain exogenous variables within the context of the path structure. In the path diagram, the causal arrows always emanate from but never point to exogenous variables. A double-arrowed curved line indicates the correlation between exogenous variables. **Age** and **SES** are exogenous variables in the model shown in Figure 41.1.
- *Endogenous variables* are predicted (explained) by other variables. The explanatory variables may be either exogenous variables (e.g., the endogenous variable of **optimism** is hypothesized as being explained by the exogenous variables of **age** and **SES**) or other endogenous variables (e.g., the endogenous variable of **quality of life** is hypothesized as being partially explained by the endogenous variable of **optimism**). In the path diagram, arrows will always point to endogenous variables; it is also possible for arrows to emanate from endogenous variables. **Optimism** and **quality of life** are endogenous variables in Figure 41.1. Each endogenous variable will serve as a dependent variable in a separate regression analysis.
- The *outcome* variable **(quality of life** in the example in Figure 41.1) is the ultimate focus of what the model is hypothesized to predict. It is always an endogenous variable.

There are two separate statistical techniques available to perform a path analysis in IBM SPSS®, multiple regression and SEM. We describe the multiple regression approach to path analysis in this chapter and treat path analysis using SEM in Chapter 42.

Using multiple regression, we perform a series of multiple or simple regression analyses, one analysis for each endogenous variable. For the model drawn in Figure 41.1, for example, we have two endogenous variables and therefore would require two separate multiple regression analyses—one analysis involves **age** and **SES** predicting **optimism** and the other analysis involves **age** and **optimism** predicting **quality of life**. The output from these regression analyses is limited to the variance explained (R^2) for each endogenous variable as well as the unstandardized and standardized path (regression) coefficients. This output may be sufficient to assess those relationships of theoretical interest, but regression analyses do not supply any estimate of the model's fit to the data.

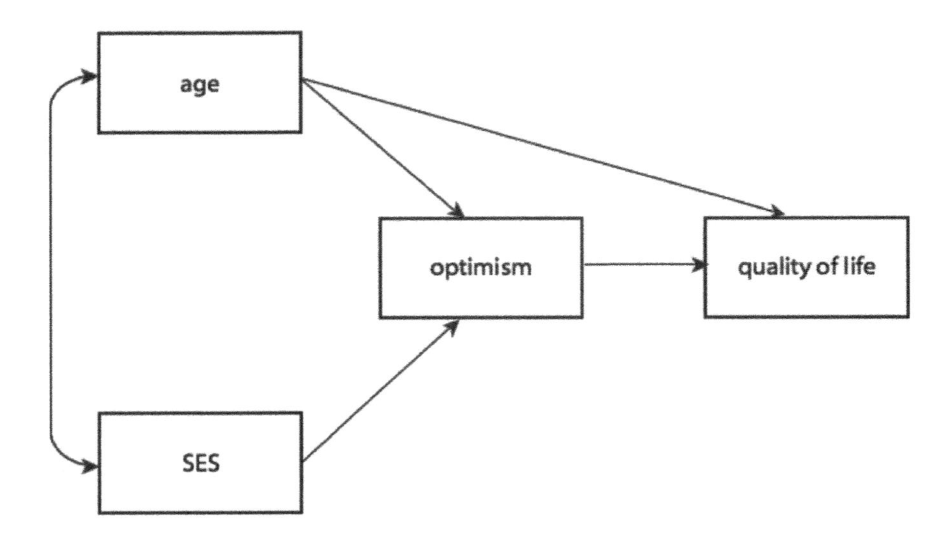

FIGURE 41.1
An example path model.

41.2 NUMERICAL EXAMPLE

The fictional data we use for our example represent a medical study of 244 patients faced with challenging medical conditions. The goal was to explain how satisfied they were with the quality of their lives (**quality_of_life**); values on **quality_of_life** could range between 0 and 20, with higher values signifying more positive judgments. **Age** and **SES** (categorized in seven steps with higher values indicating higher **SES**) serve as exogenous variables in the model, and **optimism** (scored between 30 and 90, with higher values indicating greater **optimism**) serves as a mediator variable in the model. Data are present in the data file named **quality of life**.

41.3 ANALYSIS STRATEGY

We first perform and briefly present the results of a standard ("flat") multiple regression analysis using **age**, **SES**, and **optimism** to predict the outcome variable of **quality_of_life**; this will serve as a baseline for the path analysis. We then perform a path analysis using two multiple regression procedures to generate all of the path coefficients; within that context we evaluate the simple mediation role played by **optimism**.

41.4 THE "FLAT" MULTIPLE REGRESSION ANALYSIS: SETUP

Open the **quality of life** data file and from the main menu, we select **Analyze ➔ Regression ➔ Linear**. Briefly, we move **quality_of_life** into the **Dependent** panel and all of the other variables into the **Independent(s)** panel (see Figure 41.2). We conceive of this analysis as "flat," in the sense that all of the predictors are of equal status in the model; that is, all of the predictor variables are treated as correlated exogenous variables.

In the **Statistics** screen shown in Figure 41.3, we check **Estimates** (to obtain the regression coefficients), **Model fit** (to obtain the R^2 and adjusted R^2), **R squared change** (to show this output for illustration purposes), **Descriptives** (to obtain the descriptive statistics including the correlation matrix), and **Part and partial correlations** (to obtain the zero-order, partial, and semipartial correlations). Click **Continue** to return to the main dialog window and click **OK** to perform the analysis.

41.5 THE "FLAT" MULTIPLE REGRESSION ANALYSIS: OUTPUT

The **Correlations** between the pairs of variables are shown in Figure 41.4. **quality_of_life** is substantially correlated with all three predictors. With the predictors all correlated as well, the possibility of a viable path structure might be worth considering if it is theoretically justified. The correlation between the two exogenous variables (in the path model) **age** and **SES** is −**.29**, a value that we will place in the final version of the path diagram.

The **Model Summary** presented in Figure 41.5 informs us that approximately 33% of the variance (adjusted $R^2 = .334$) of **quality_of_life** was explained by the model. The **Coefficients** associated with the individual predictors are shown in Figure 41.6. Only **optimism** was statistically significant in the model; its squared semipartial correlation of approximately .15 ($.389^2 = .151321$) suggests that about 15% of the variance of **quality_of_life** was uniquely explained by **optimism**.

From these results, some researchers may be inclined to believe that **age** and **SES** are irrelevant to the explanation (are expendable in any account of the prediction) of **quality_of_life**. In fact, these two variables would be excluded from a model based on

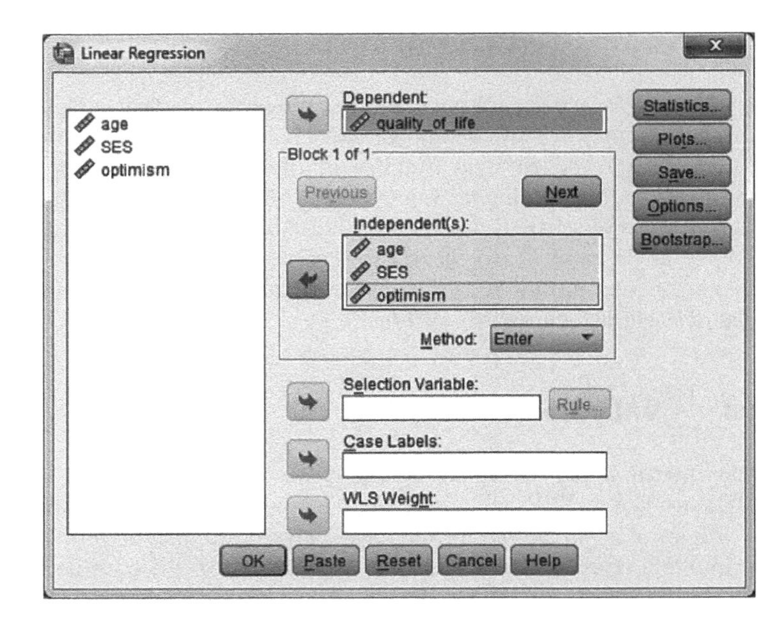

FIGURE 41.2
The main **Linear Regression** dialog window.

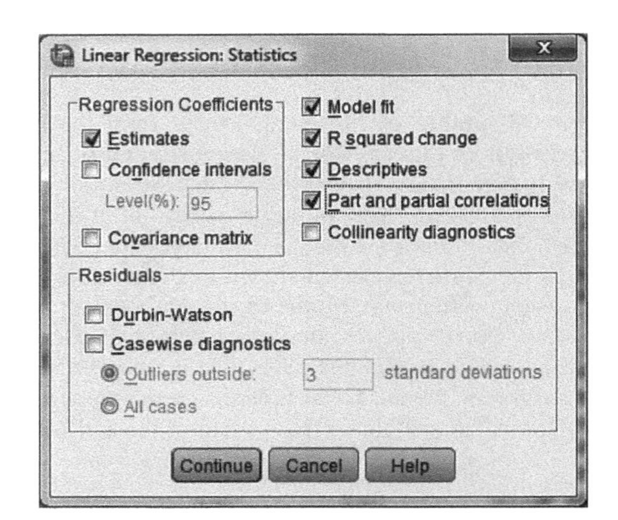

FIGURE 41.3
The **Statistics** screen of **Linear Regression**.

a step method (e.g., stepwise multiple regression). But the story may be more complicated (i.e., theoretically relevant and interesting) if something other than a "flat" analytic strategy is used to structure the variables. The opportunity to use a more complex configuration of the variables is the reason why path modeling has gained popularity in the past couple of decades.

41.6 PATH ANALYSIS USING MULTIPLE REGRESSION: ANALYSIS 1

We have presented the path model in Figure 41.1. Because there are two endogenous variables, it is necessary to perform two multiple regression analyses to obtain the full set of path coefficients, each predicting one of the endogenous variables. In the first regression analysis, we use **age** and **optimism** to predict **quality_of_life**. From the main menu, we select **Analyze ➔ Regression ➔ Linear**. Briefly (without showing the dialog

Correlations

		quality_of_life	age	SES	optimism
Pearson Correlation	quality_of_life	1.000	-.332	.369	.573
	age	-.332	1.000	-.290	-.431
	SES	.369	-.290	1.000	.520
	optimism	.573	-.431	.520	1.000
Sig. (1-tailed)	quality_of_life	.	.000	.000	.000
	age	.000	.	.000	.000
	SES	.000	.000	.	.000
	optimism	.000	.000	.000	.
N	quality_of_life	244	244	244	244
	age	244	244	244	244
	SES	244	244	244	244
	optimism	244	244	244	244

FIGURE 41.4 **Correlations** between the variables.

Model Summary

Model	R	R Square	Adjusted R Square	Std. Error of the Estimate	Change Statistics				
					R Square Change	F Change	df1	df2	Sig. F Change
1	.585[a]	.343	.334	3.647	.343	41.684	3	240	.000

a. Predictors: (Constant), optimism, age, SES

FIGURE 41.5 The **Model Summary** table.

Coefficients[a]

Model		Unstandardized Coefficients		Standardized Coefficients	t	Sig.	Correlations		
		B	Std. Error	Beta			Zero-order	Partial	Part
1	(Constant)	3.773	2.131		1.770	.078			
	age	-.073	.044	-.097	-1.675	.095	-.332	-.107	-.088
	SES	.236	.164	.089	1.441	.151	.369	.093	.075
	optimism	.175	.023	.485	7.433	.000	.573	.433	.389

a. Dependent Variable: quality_of_life

FIGURE 41.6 The **Coefficients** table.

screens), we move **quality_of_life** into the **Dependent** panel and **age** and **optimism** into the **Independent(s)** panel, retaining the above setup for the rest of the analysis.

The main results are presented in Figure 41.7. **Age** and **optimism** together accounted for approximately 33% of the variance of **quality_of_life**. Only **optimism** was associated with a statistically significant partial regression coefficient. The standardized (beta) coefficients associated with **age** and **optimism** were −**.105** and **.528**, respectively; these values are to be placed into the path diagram as path coefficients for these variables.

Model Summary

Model	R	R Square	Adjusted R Square	Std. Error of the Estimate	Change Statistics				
					R Square Change	F Change	df1	df2	Sig. F Change
1	.580[a]	.337	.331	3.655	.337	61.213	2	241	.000

a. Predictors: (Constant), optimism, age

Coefficients[a]

Model		Unstandardized Coefficients		Standardized Coefficients	t	Sig.	Correlations		
		B	Std. Error	Beta			Zero-order	Partial	Part
1	(Constant)	4.151	2.120		1.958	.051			
	age	-.078	.043	-.105	-1.801	.073	-.332	-.115	-.094
	optimism	.190	.021	.528	9.078	.000	.573	.505	.476

a. Dependent Variable: quality_of_life

FIGURE 41.7 The **Model Summary** and **Coefficients** tables when predicting **quality_of_life**.

Model Summary

Model	R	R Square	Adjusted R Square	Std. Error of the Estimate	Change Statistics				
					R Square Change	F Change	df1	df2	Sig. F Change
1	.596[a]	.356	.350	9.999	.356	66.496	2	241	.000

a. Predictors: (Constant), SES, age

Coefficients[a]

Model		Unstandardized Coefficients		Standardized Coefficients	t	Sig.	Correlations		
		B	Std. Error	Beta			Zero-order	Partial	Part
1	(Constant)	61.251	4.311		14.208	.000			
	age	-.634	.112	-.306	-5.657	.000	-.431	-.342	-.293
	SES	3.187	.399	.431	7.979	.000	.520	.457	.413

a. Dependent Variable: optimism

FIGURE 41.8 The **Model Summary** and **Coefficients** tables when predicting **optimism**.

41.7 PATH ANALYSIS USING MULTIPLE REGRESSION: ANALYSIS 2

In the second analysis, **age** and **SES** are used to predict **optimism**. From the main menu, we select **Analyze ➔ Regression ➔ Linear**. Without showing the dialog screens, we move **optimism** into the **Dependent** panel and **age** and **SES** into the **Independent(s)** panel, retaining the statistical setup for the rest of the analysis that was already described in Section 41.4.

The main results are presented in Figure 41.8. **Age** and **SES** together accounted for approximately 35% of the variance of **optimism**. Each predictor was associated with a statistically significant partial regression coefficient. The standardized (beta) coefficients associated with **age** and **SES** were **−.306** and **.431**, respectively; these values are to be placed into the path diagram as path coefficients for these variables.

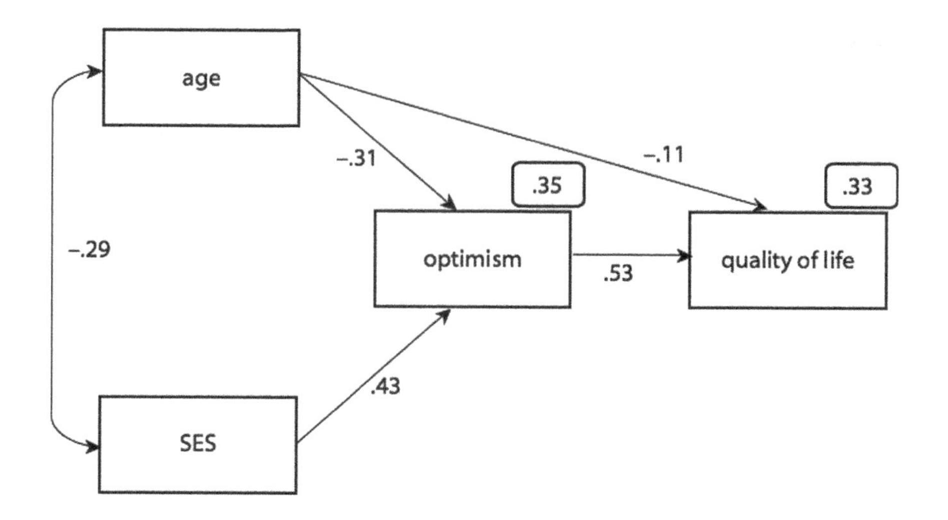

FIGURE 41.9
The path model with the path coefficients and the R^2 values in round-edged rectangles to the upper right part of each endogenous variable.

41.8 PATH ANALYSIS USING MULTIPLE REGRESSION: SYNTHESIS

The completed path model is shown in Figure 41.9. The correlation between **age** and **SES** as well as the standardized path coefficients are placed in their respective paths, and the adjusted R^2 values for the outcome variable **quality_of_life** and the endogenous variable **optimism** are shown in round-edged rectangles near their upper right corners. Based on the multiple regression results, researchers might be inclined to trim this model by removing the nonsignificant path from **age** to **quality_of_life** should it be theoretically reasonable to do so. Such an action stretches the use of path analysis from confirmatory to that borders on exploratory, but some researchers would take that action based on the principle of parsimony.

41.9 ASSESSING THE STATISTICAL SIGNIFICANCE OF THE INDIRECT EFFECTS

Our visual inspection of the path coefficients suggests that both **age** and **SES** exerted an indirect effect on **quality_of_life** through **optimism** as a mediating variable. The statistical significance of the indirect effects can be evaluated by the Aroian (1947) test as described in Chapter 40:

- Using the raw partial regression coefficients and SEs associated with the paths **age** to **optimism** ($b = -$**.634**, SE = **.112**) and **optimism** to **quality_of_life** ($b =$ **.190**, SE = **.021**), we obtain (based on Kristopher J. Preacher's web site http://quantpsy.org/sobel/sobel.htm) an Aroian z value of approximately −4.78, $p < .001$.
- Using the raw partial regression coefficients and SEs associated with the paths **SES** to **optimism** ($b =$ **3.187**, SE = **.399**) and **optimism** to **quality_of_life** ($b =$ **.190**, SE = **.021**), we obtain (based on Kristopher J. Preacher's web site http://quantpsy.org/sobel/sobel.htm) an Aroian z value of approximately 5.97, $p < .001$.

We may therefore conclude that the indirect effects of **age** and **SES** through **optimism** in explaining **quality_of_life** are statistically significant.

41.10 ASSESSING THE STRENGTH OF EACH INDIRECT EFFECT

We can assess the strength of each indirect effect by multiplying the standardized (beta) coefficients associated with each segment of each path. Thus, the absolute value of the indirect effect of **age** through **optimism** on **quality_of_life** is $-$**.31** * **.53** = .16, and the value of the indirect effect of **age** through **optimism** on **quality_of_life** is **.43** * **.53** = .23. Both of these values are quite substantial for indirect effects and should be treated as being of practical worth. Our general conclusion from this analysis is that **age** and **SES** indirectly influence **quality_of_life** through the mediated influence of **optimism**.

41.11 EVALUATING THE POSSIBILITY OF MEDIATION

Although the path from **age** to **quality_of_life** did not achieve statistical significance in the model, it does raise the interesting question of whether **age** can predict **quality_of_life** in isolation. If that is the case, then we may have observed full mediation with the inclusion of **optimism** in the model. To test that possibility, we have performed an additional regression analysis using **age** in isolation to predict **quality_of_life**. The results, shown in Figure 41.10, indicate that **age** on its own does predict **quality_of_life** ($p < .001$); the unstandardized regression coefficient is $-$**.248** (with SE = **.045**) and the standardized regression coefficient is $-$**.332**.

41.12 TESTING THE STATISTICAL SIGNIFICANCE OF THE DIFFERENCE BETWEEN THE DIRECT PATHS IN THE UNMEDIATED AND THE MEDIATED MODELS

Although it may be obvious what the result will be, it is instructive to perform the Freedman–Schatzkin test, as described in Chapter 40, to compare the relative strengths of the paths from **age** to **quality_of_life** in the unmediated and the mediated models. When we solve the Freedman–Schatzkin equation, we obtain a t value of 8.699. Evaluated against a Student t distribution with $N-2$ degrees of freedom (here we have **244** $-$ 2, or 242, degrees of freedom), our result indicates that the coefficients are significantly different ($p < .001$). We may therefore conclude that we have obtained a full mediation effect (given that the path in the mediated model is not statistically significant).

Model Summary

					Change Statistics				
Model	R	R Square	Adjusted R Square	Std. Error of the Estimate	R Square Change	F Change	df1	df2	Sig. F Change
1	.332[a]	.110	.106	4.225	.110	29.943	1	242	.000

a. Predictors: (Constant), age

Coefficients[a]

		Unstandardized Coefficients		Standardized Coefficients			Correlations		
Model		B	Std. Error	Beta	t	Sig.	Zero-order	Partial	Part
1	(Constant)	20.098	1.372		14.649	.000			
	age	-.248	.045	-.332	-5.472	.000	-.332	-.332	-.332

a. Dependent Variable: quality_of_life

FIGURE 41.10 The **Model Summary** and **Coefficients** tables using **age** to predict **quality_of_life**.

CHAPTER 42

Path Analysis Using Structural Equation Modeling

42.1 OVERVIEW

The IBM SPSS® Amos module that we used in Chapter 39 to perform a confirmatory factor analysis can also be used to perform a path analysis. As was the case for the multiple regression approach to path analysis (discussed in Chapter 41), we obtain in an SEM analysis the path (regression) coefficients and R^2 for the endogenous variables. However, because it uses SEM, IBM SPSS Amos also provides fit indexes informing us how well the proposed model fits the data. These fit indexes allow us to compare different plausible models to help determine which theoretical approach has more empirical support.

Some of the differences between the multiple regression and SEM approaches to path analysis are as follows:

- The SEM analysis is completed as a whole rather than in portions. Thus, the path coefficients are estimated simultaneously based on all of the hypothesized interrelationships among the variables rather than based only on those involved in separate portions of the analysis as is done in the multiple regression approach.
- By performing the analysis as a whole using SEM, we obtain fit indexes that inform us how well the hypothesized model fits the data; multiple regression does not supply these fit indexes.
- The multiple regression approach involves a weighted linear composite of the predictors as the variate and the predicted value of Y as the dependent variable. In SEM, the predictor side of the model includes an error term as a latent variable (in an SEM model diagram, each endogenous variable will have an associated error term as an explanatory variable); this difference means that the dependent variable in a structural model is the actual Y value rather than the predicted Y value.

We analyze the same path model evaluated by multiple regression analysis in Chapter 41. As indicated in Chapter 39, the data used in an SEM analysis should contain no missing values, and our data meet this criterion.

Performing Data Analysis Using IBM SPSS®, First Edition.
Lawrence S. Meyers, Glenn C. Gamst, and A. J. Guarino.

42.2 PATH ANALYSIS BASED ON SEM: DRAWING THE MODEL

We open the **quality of life** data file and from the main menu, we select **Analyze ➔ IBM SPSS Amos**. The initial IBM SPSS Amos window is shown in Figure 42.1 where the model is to be drawn in the large pane on the right and frequently used commands are located in the icon toolbar on the left. Readers are referred to Section 39.3 for additional details on some commonly used tools on the icon toolbar.

As we have done in Chapter 39, we opt to switch from the default portrait (long) orientation to landscape (wide) orientation. To change the orientation, select **View ➔ Interface properties** from the main menu. This action opens the **Interface Properties** window as seen in Figure 42.2. Select **Landscape-Legal** under **Paper Size**, select **Apply**, and close the window.

From the icon toolbar menu, we select the **Draw observed variables** icon (the rectangle in the upper left corner) as shown in Figure 42.3 and draw our first measured variable for the path model (shown in Figure 42.4). If any drawn object looks less than aesthetically pleasing or if the drawn object is not what is desired, either **Erase** the object and draw it again or right-click to select change the **Shape of object**.

We then continue this to draw the other three measured variables of **SES, optimism**, and **quality of life** as shown in Figure 42.5. These will represent our measured variables.

Our next task is to associate each depiction of a measured variable with the proper variable in the data file. From the main menu, we select **View ➔ Variables in Dataset**. This opens the **Variables in Dataset** window pictured in Figure 42.6. Select each variable

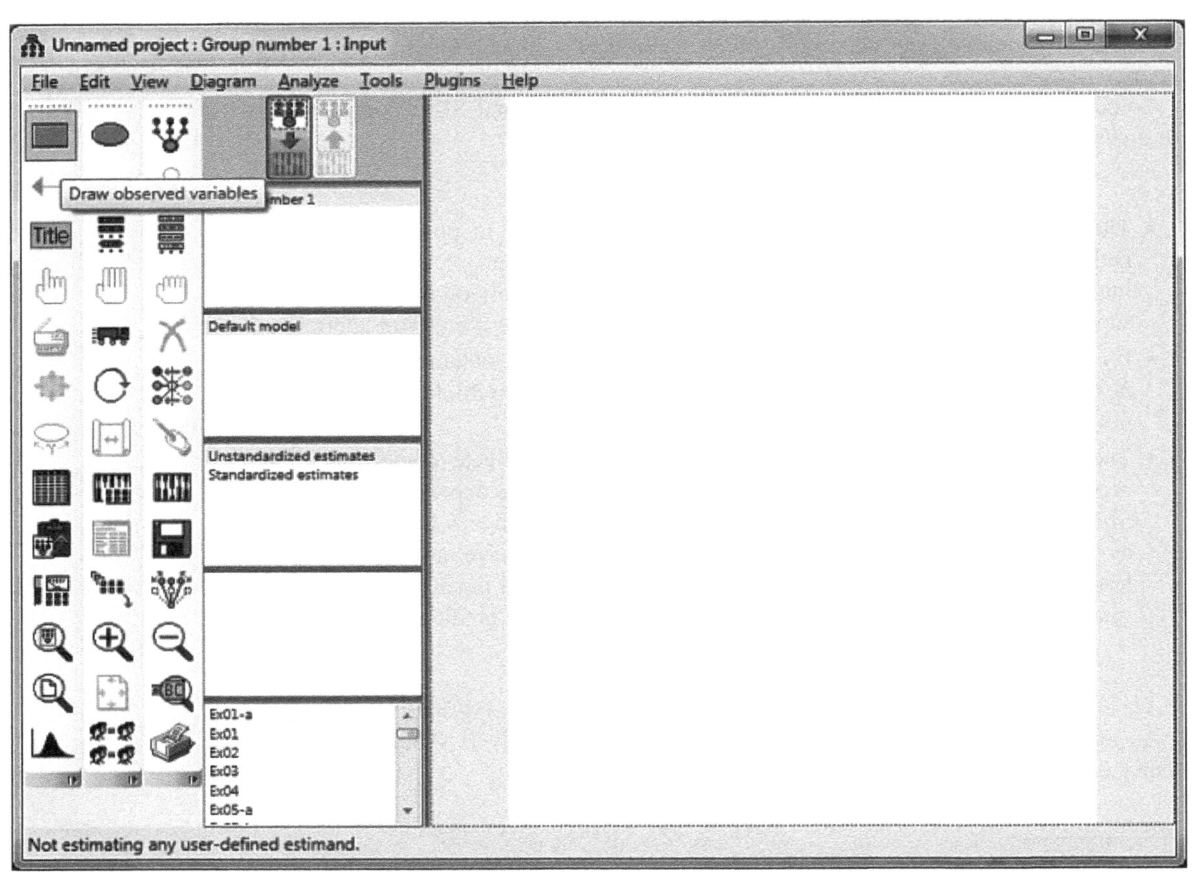

FIGURE 42.1 The initial IBM SPSS Amos screen.

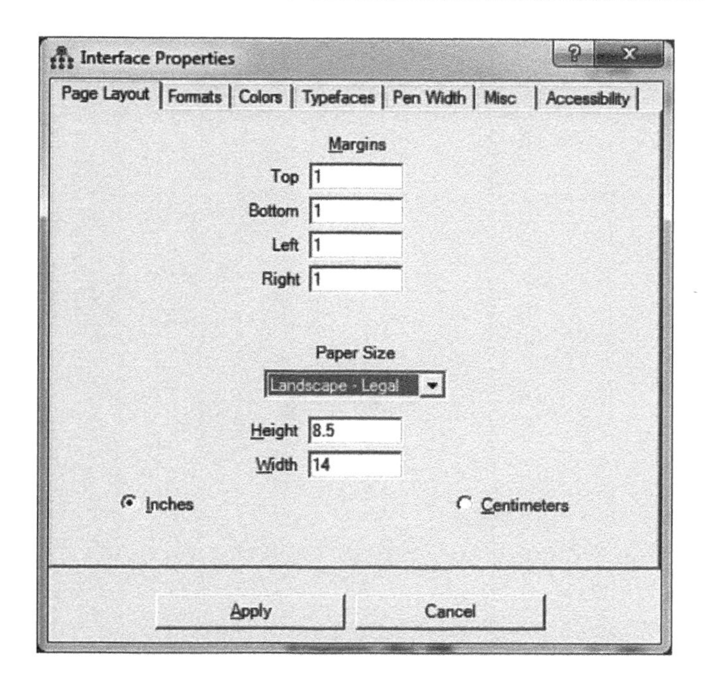

FIGURE 42.2
The **Interface Properties** window where we select **Landscape-Legal** page orientation.

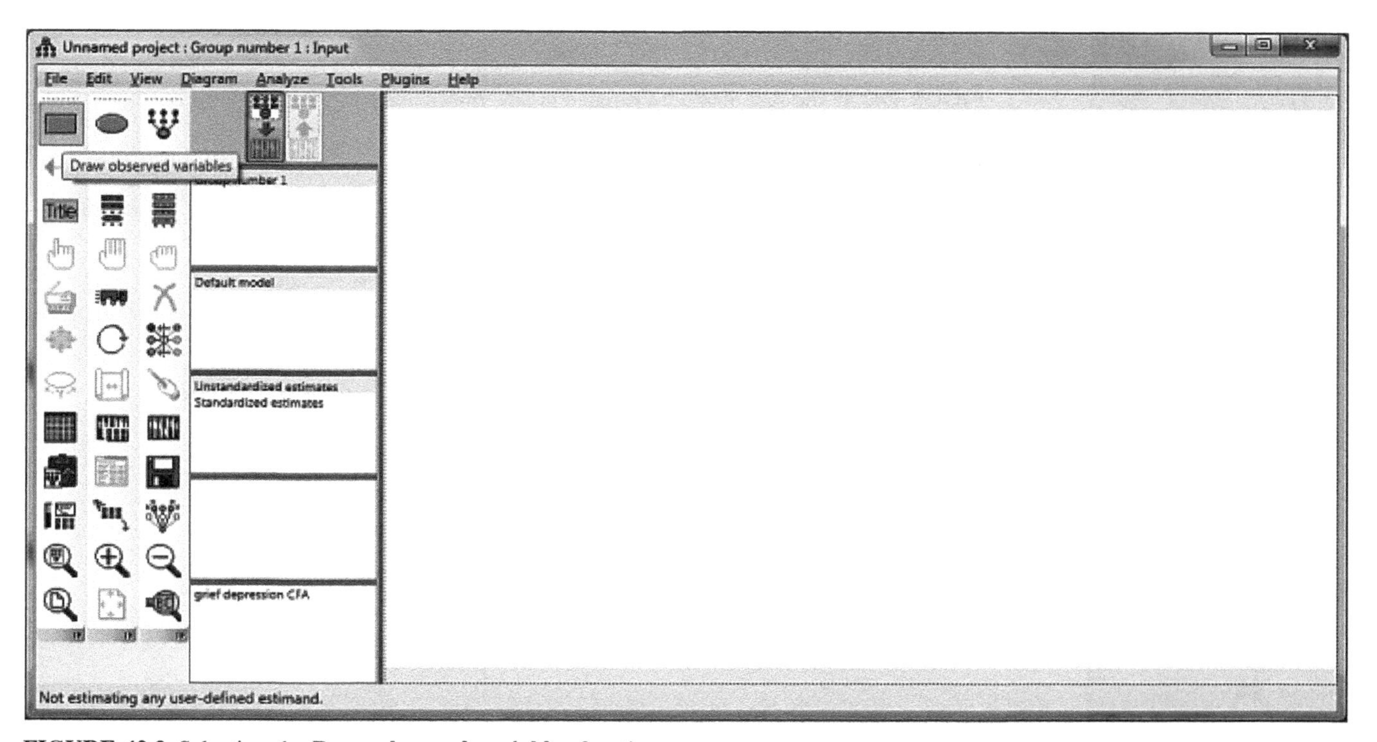

FIGURE 42.3 Selecting the **Draw observed variables** function.

in turn and drag it to the appropriate figure so that all observed variables are properly configured (see Figure 42.7). When finished, close the **Variables in Dataset** window.

We now need to draw the paths. From the icon toolbar, select the **Draw paths (single headed arrow)** as shown in Figure 42.8. With this tool activated, place the cursor on the right border of the **age** rectangle and click and drag to the **optimism** rectangle. Use this

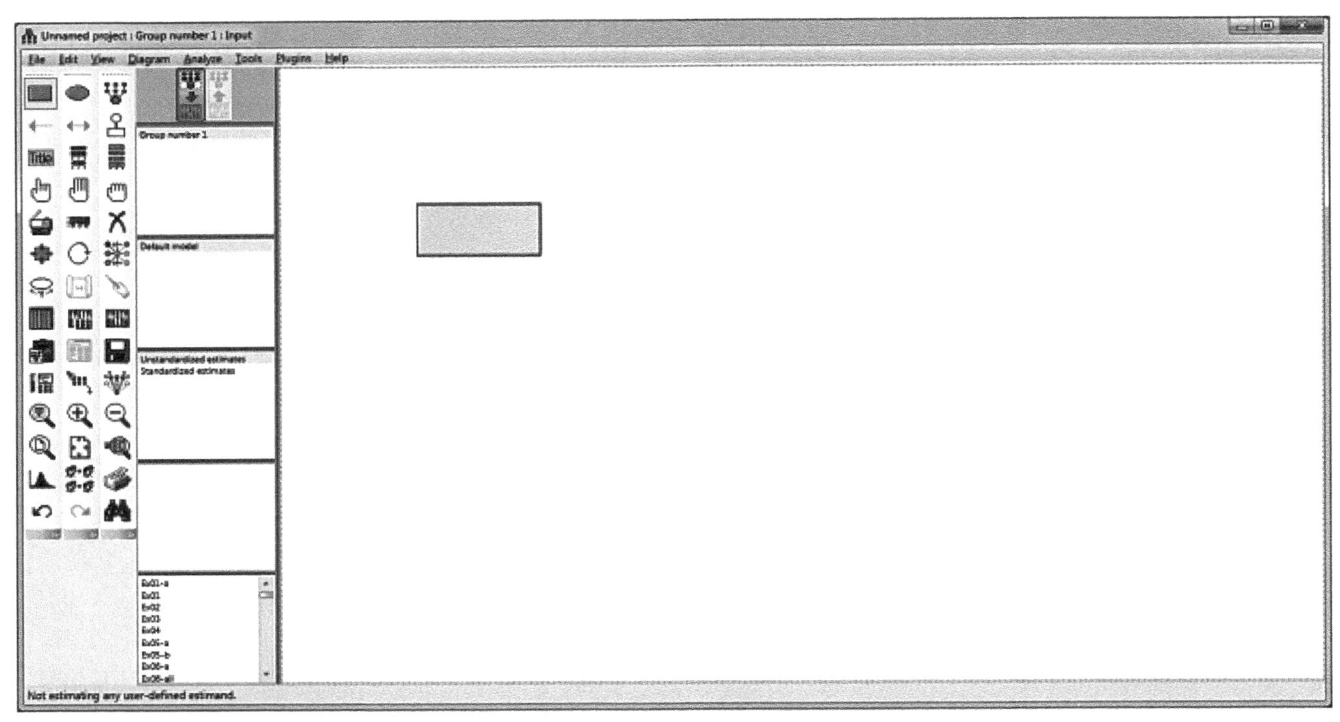

FIGURE 42.4 The first measured variable is now drawn.

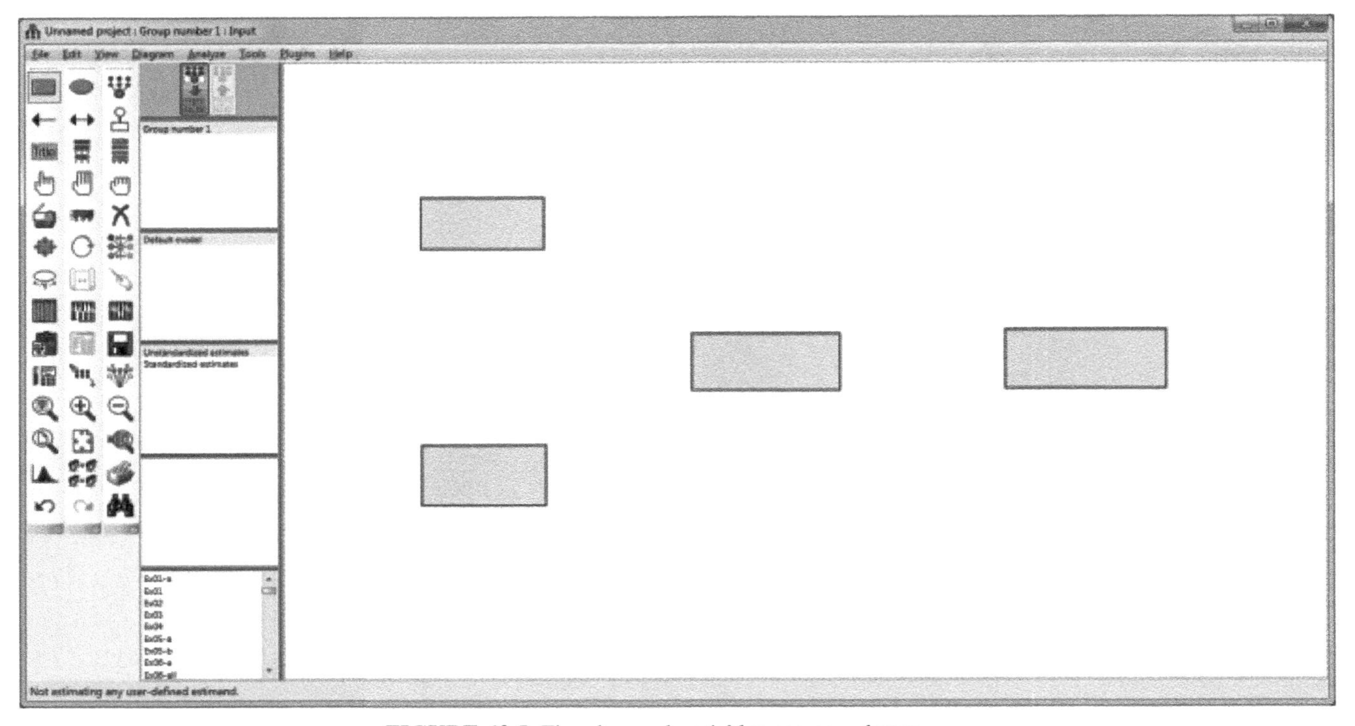

FIGURE 42.5 The observed variables are now drawn.

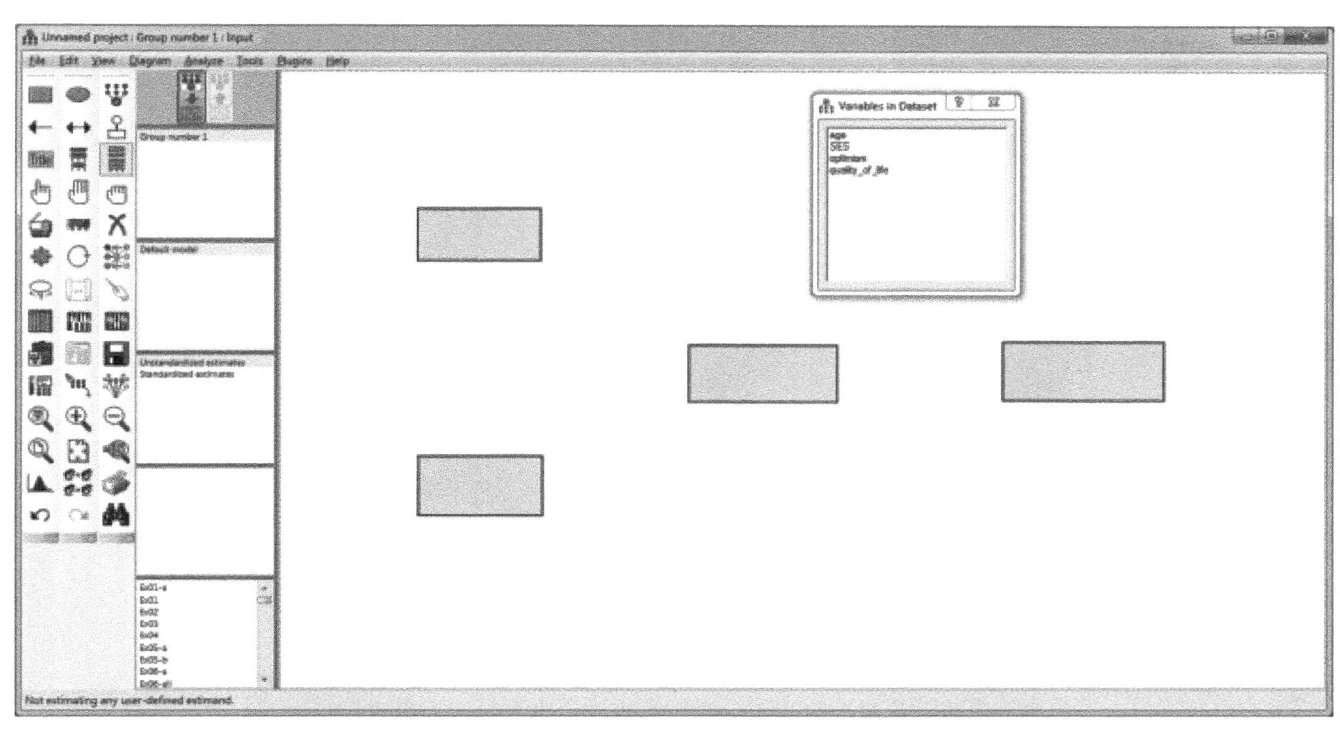

FIGURE 42.6 The **Variables in Dataset** window.

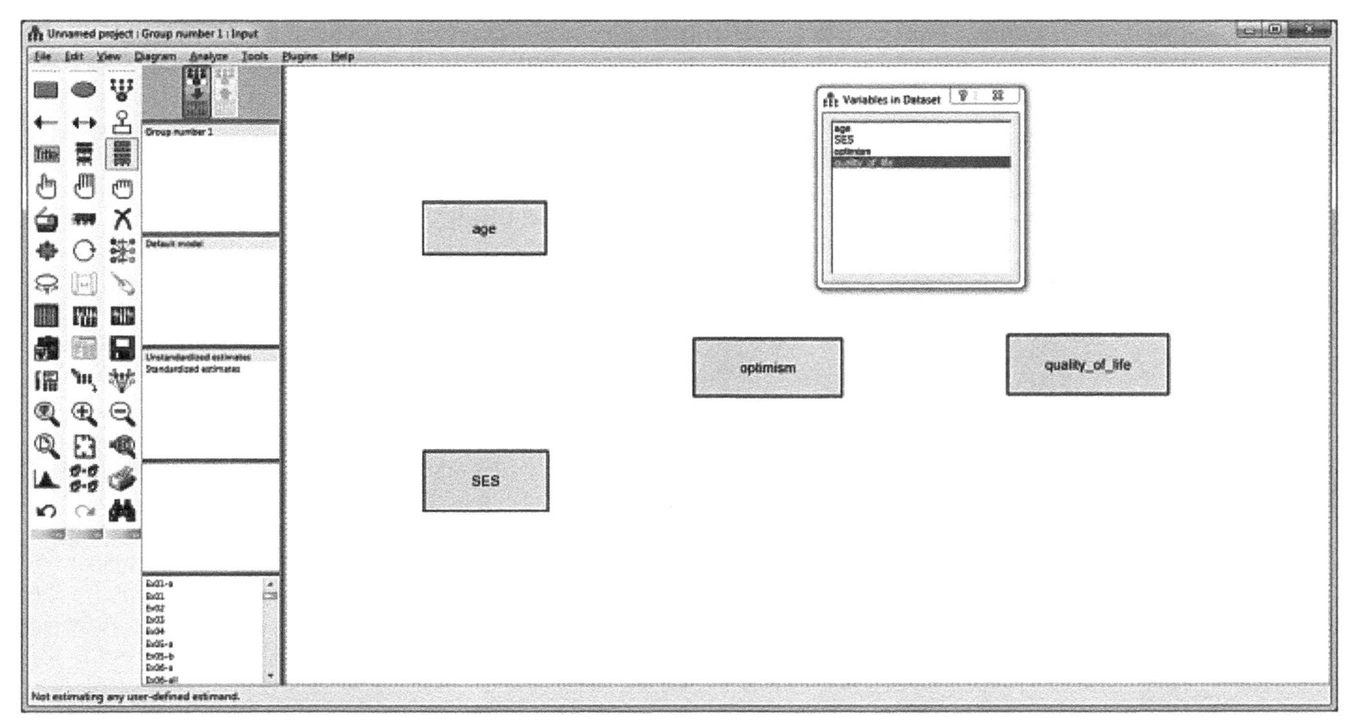

FIGURE 42.7 All of the rectangles have been associated with the proper observed variables.

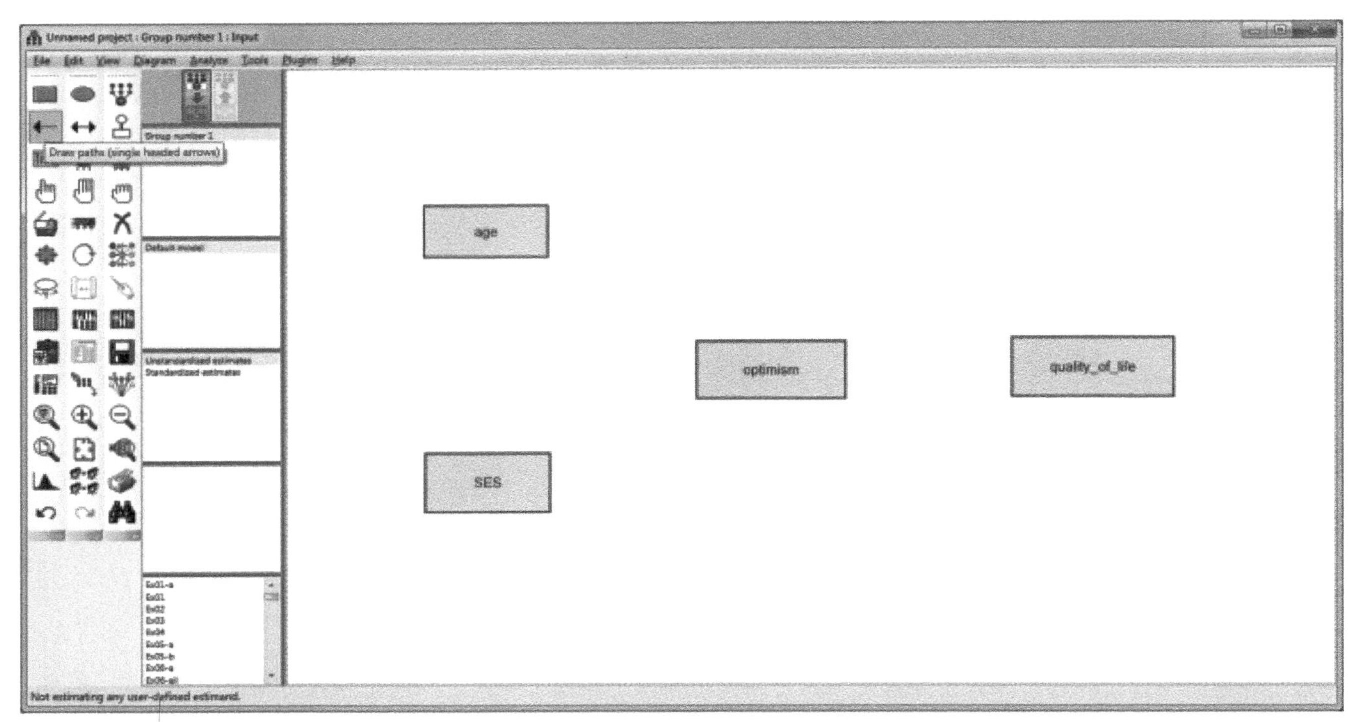

FIGURE 42.8 Select the **Draw paths (single headed arrow)** from the icon toolbar.

process to draw each of the other paths in the model so that the structure mirrors what is in Figure 42.9.

We next draw the correlation (the covariance) path between the two exogenous variables, as we hypothesize that they are correlated. From the icon toolbar, select the **Draw covariances (double headed arrow)** as shown in Figure 42.10. With this tool activated, place the cursor on the left border of the **SES** rectangle and click and drag to the **age** rectangle to draw the covariance path as shown in Figure 42.11.

In the SEM approach to path analysis, and even with all of the variables as measured (observed) variables, it is necessary to explicitly include latent error variables (called **unique variables** by IBM SPSS Amos) that are to be associated with each endogenous variable (because the dependent variables in the structural equation are the actual rather than the predicted Y values). These **unique variables** are not specifically measured and are thus represented as latent variables, to be depicted by circles or ovals. To add these latent variables to the model, we select from the icon toolbar the **Add a unique variable to an existing variable** as shown in Figure 42.12.

With this tool activated, place the cursor inside the **optimism** figure and click. This results in a latent variable (with its path initially constrained to a value of **1**) being placed above the **optimism** rectangle as shown in Figure 42.13. The standardized value of this path coefficient will be reestimated as a part of the main analysis. Because the latent error variable is in a diagrammatically awkward position, it is best to shift its position. Keeping the cursor inside the **optimism** rectangle and clicking repeatedly rotates the **unique variables** clockwise. Sufficient clicking fixes it in a more aesthetically pleasing place (see Figure 42.14). Applying this same procedure to **quality_of_life** yields the configuration shown in Figure 42.15.

We now need to name our **unique variables**, although IBM SPSS Amos will do so rather generically. From the main menu, we select **Plugins ➔ Name Unobserved Variables** (see Figure 42.16). On selecting this command, the **unique variables** will be

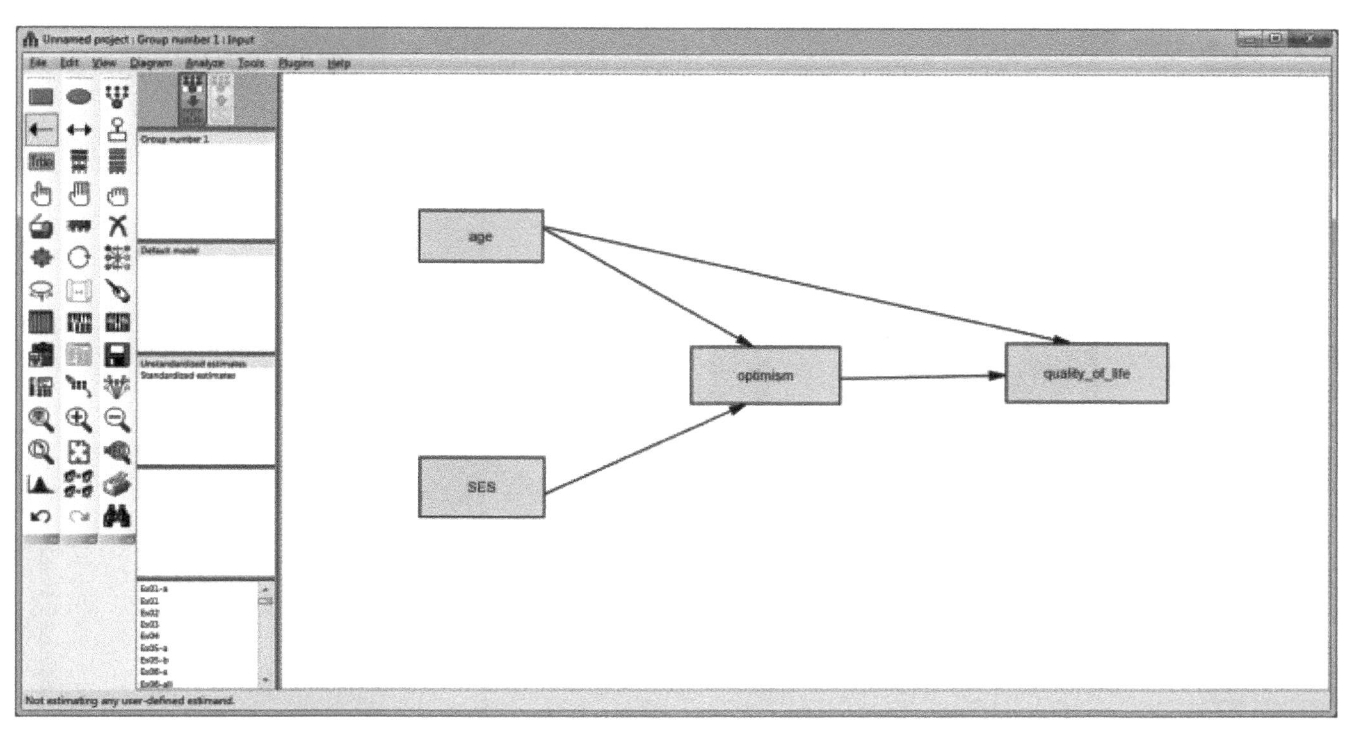

FIGURE 42.9 The paths are now drawn.

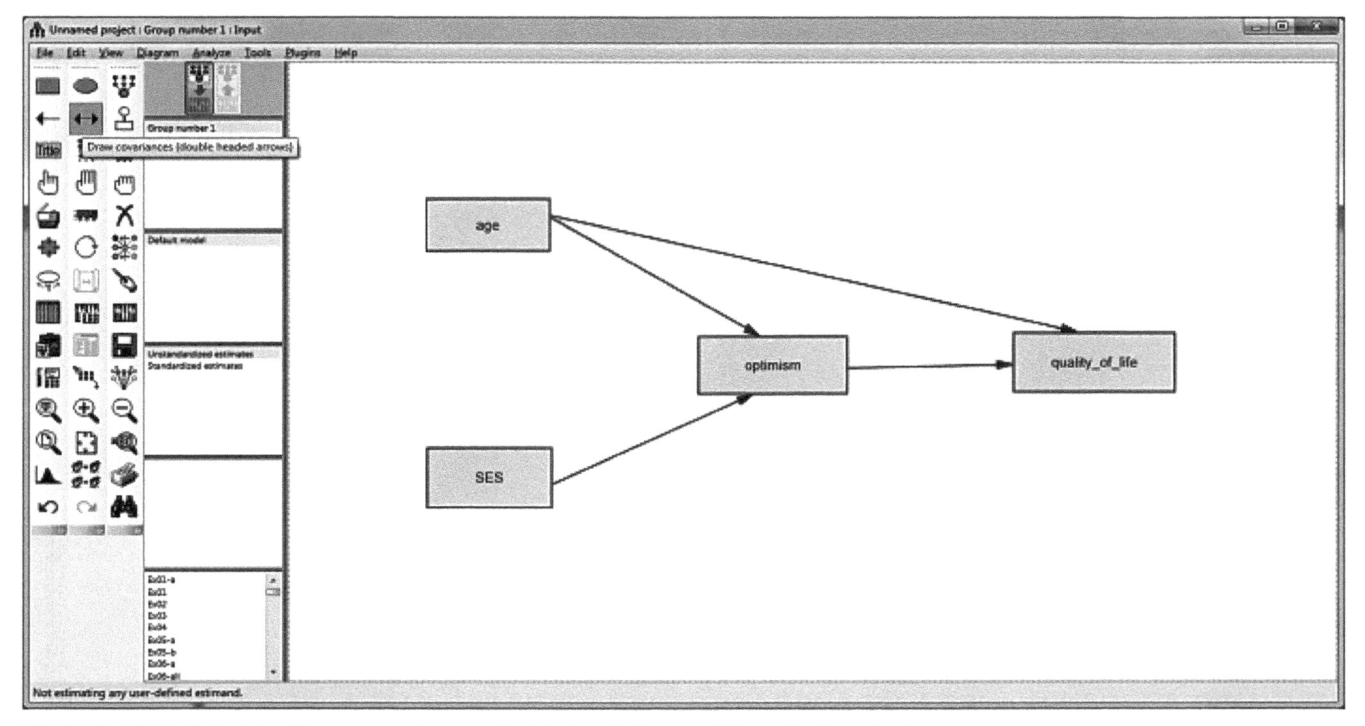

FIGURE 42.10 Select the **Draw covariances (double headed arrow)** from the icon toolbar.

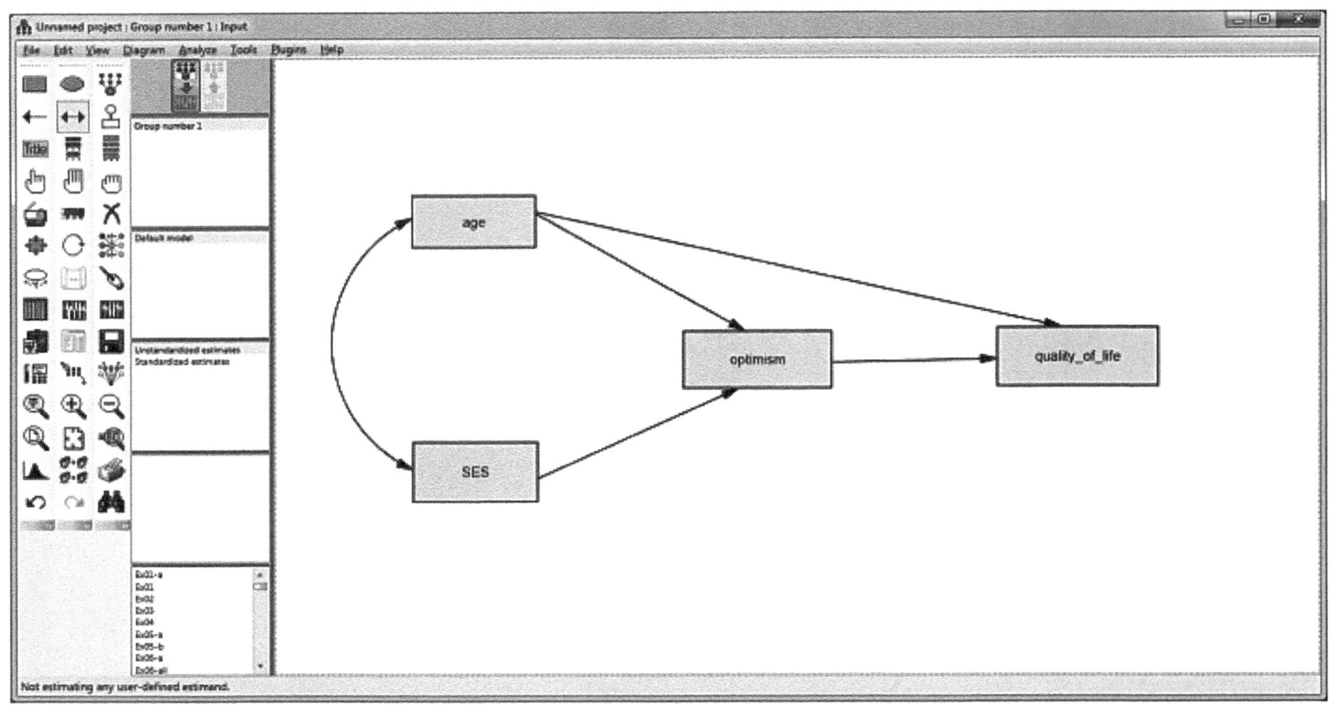

FIGURE 42.11 The covariance path is now drawn.

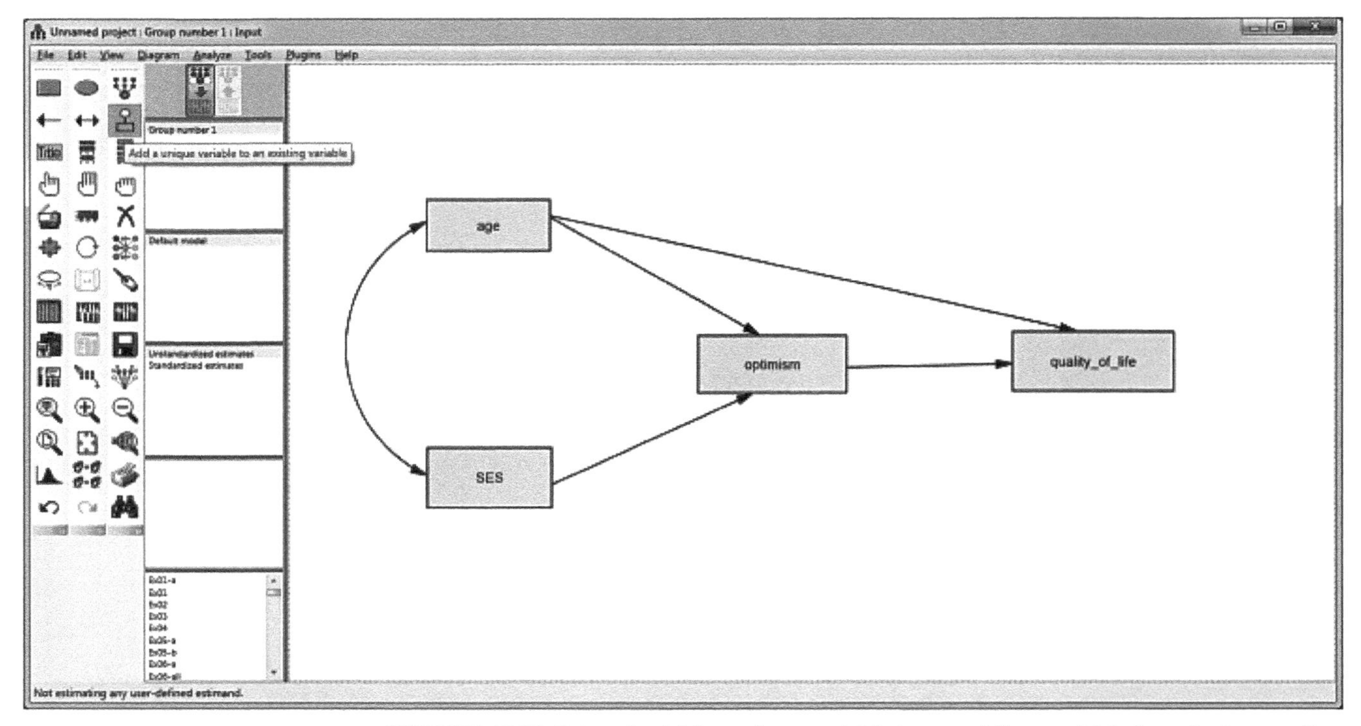

FIGURE 42.12 Select the **Add a unique variable to an existing variable** from the icon toolbar.

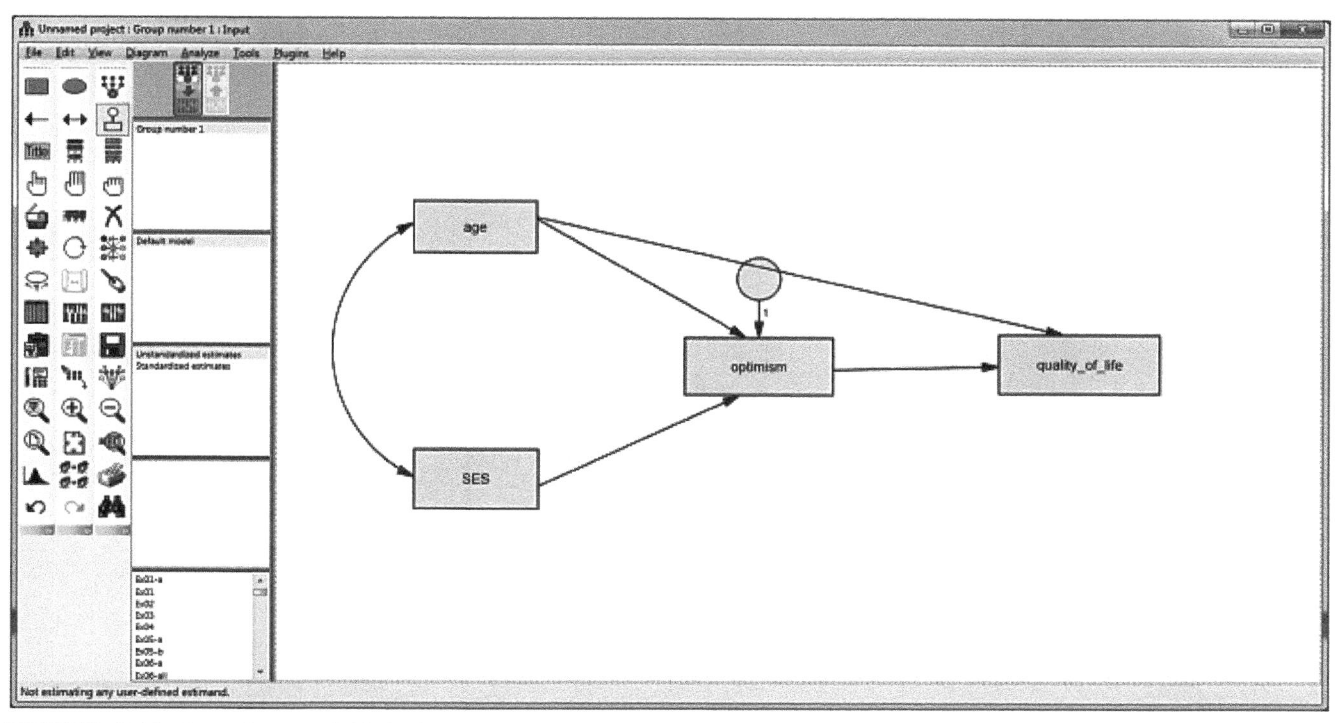

FIGURE 42.13 The **unique variable** is drawn on top of the **optimism** rectangle.

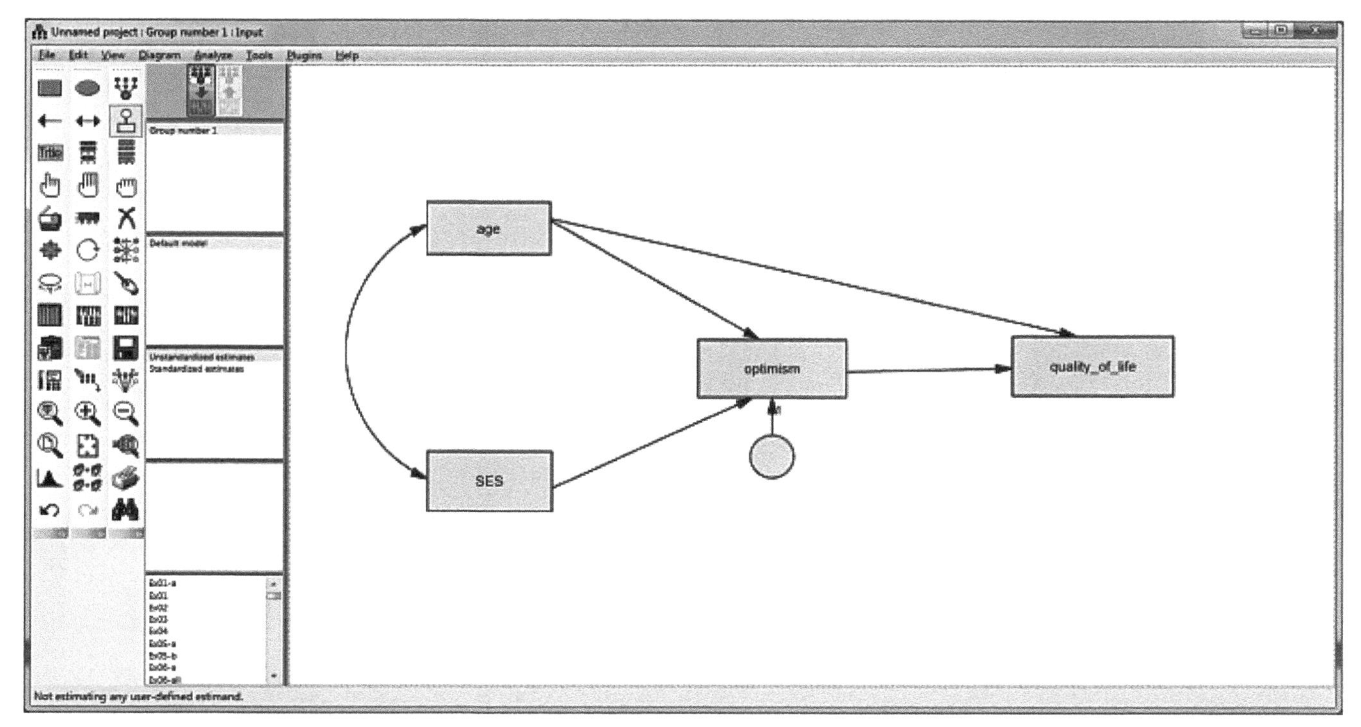

FIGURE 42.14 The **unique variable** has been shifted to a better spatial position.

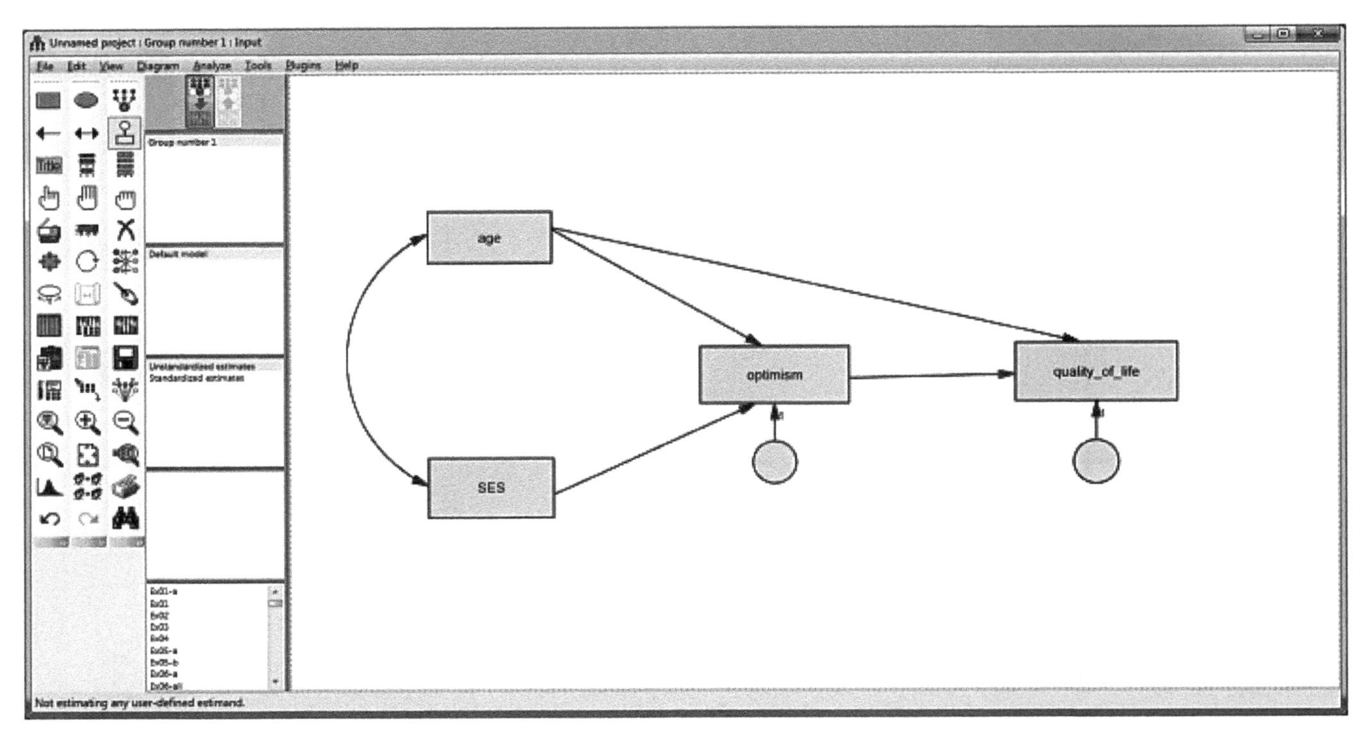

FIGURE 42.15 Both **unique variables** for the endogenous variables are now in place.

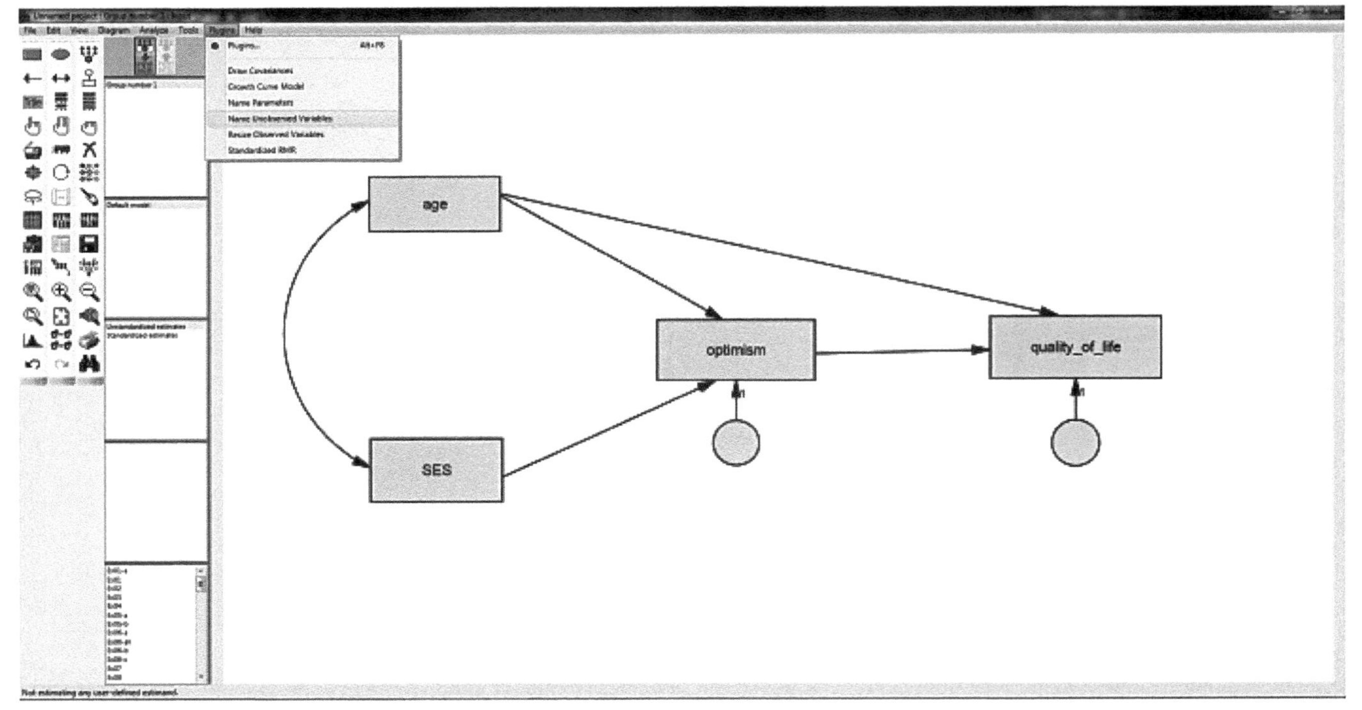

FIGURE 42.16 Select **Plugins ➔ Name Unobserved Variables** to name the **unique variables**.

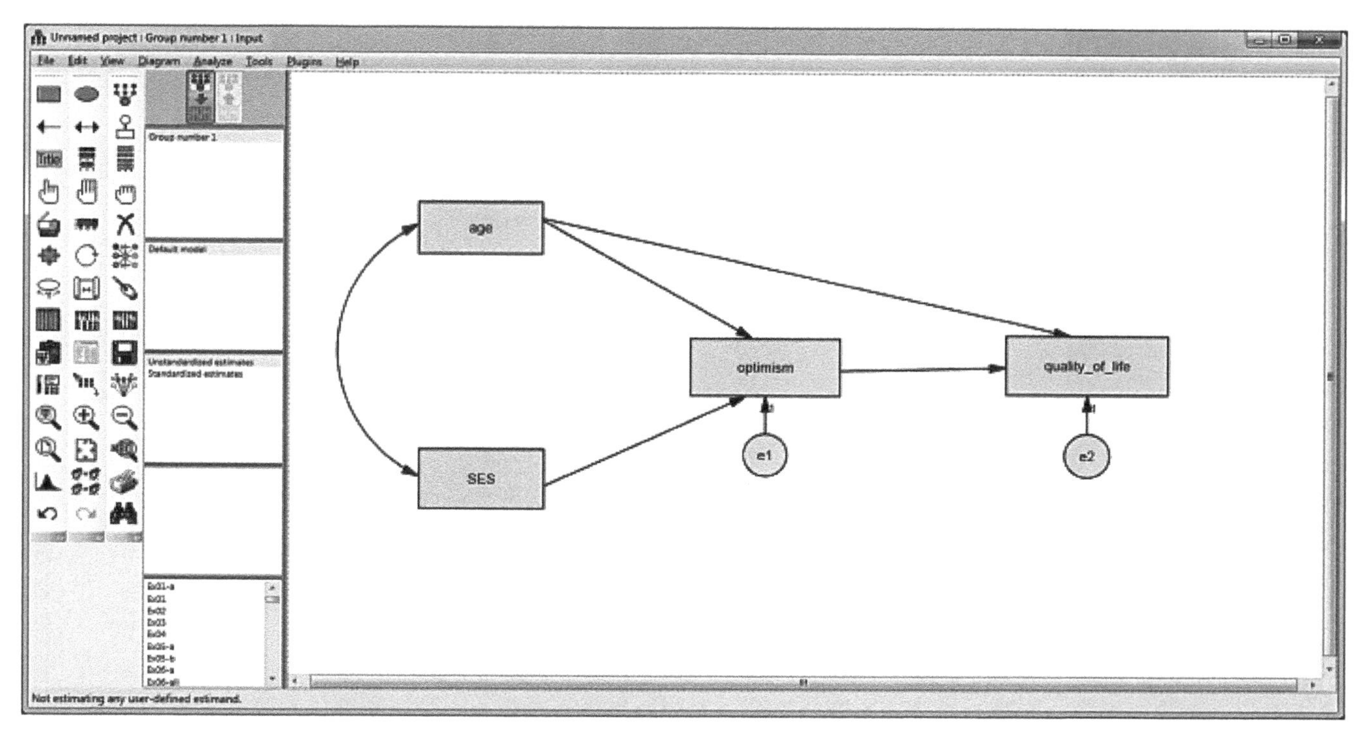

FIGURE 42.17 The **unique variables** have now been labeled as "**e**" for error term, using sequential numbers.

labeled as "**e**" (for error term) and given the sequential numbers **e1** and **e2** as shown in Figure 42.17.

42.3 PATH ANALYSIS BASED ON SEM: ANALYSIS SETUP

With the path diagram completed, we next configure the path analysis. From the main menu, we select **View ➔ Analysis Properties** to open the **Analysis Properties** window as shown in Figure 42.18. Select the **Output** tab and check **Standardized estimates**, **Squared multiple correlations**, and **Indirect, direct & total effects**. Because of the simplicity of our model, we do not request **Modification indices** to have IBM SPSS Amos offer "suggestions" on how to improve the model fit.

To execute the analysis, select **Analyze ➔ Calculate Estimates** from the main menu. As this is our first analysis, we are presented with the operating system's **Save As** dialog window (see Figure 42.19). We select a location to save the file, provide a file name, and **Save**.

42.4 PATH ANALYSIS BASED ON SEM: ANALYSIS OUTPUT

To access the results of the analysis, select the **View the output path diagram** icon (the one on the right) in the top output panel immediately to the left of the drawing area as shown in Figure 42.20. What is displayed is the path diagram with the **Unstandardized estimates** (as highlighted in the fourth output panel from the top). These values, the unstandardized regression coefficients and the covariance between the exogenous variables of **age** and **SES**, also appear in tabular form (we show these shortly).

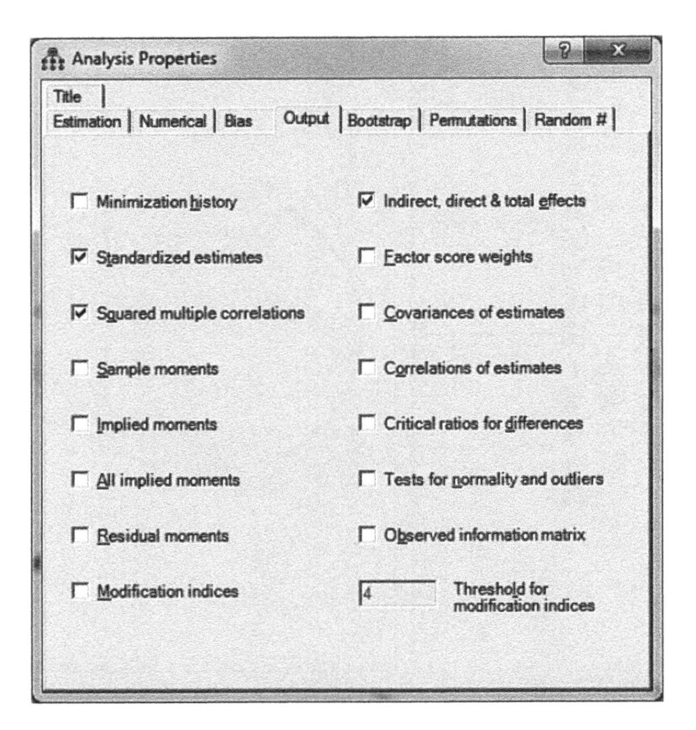

FIGURE 42.18
The configured **Output** screen in the **Analysis Properties** window.

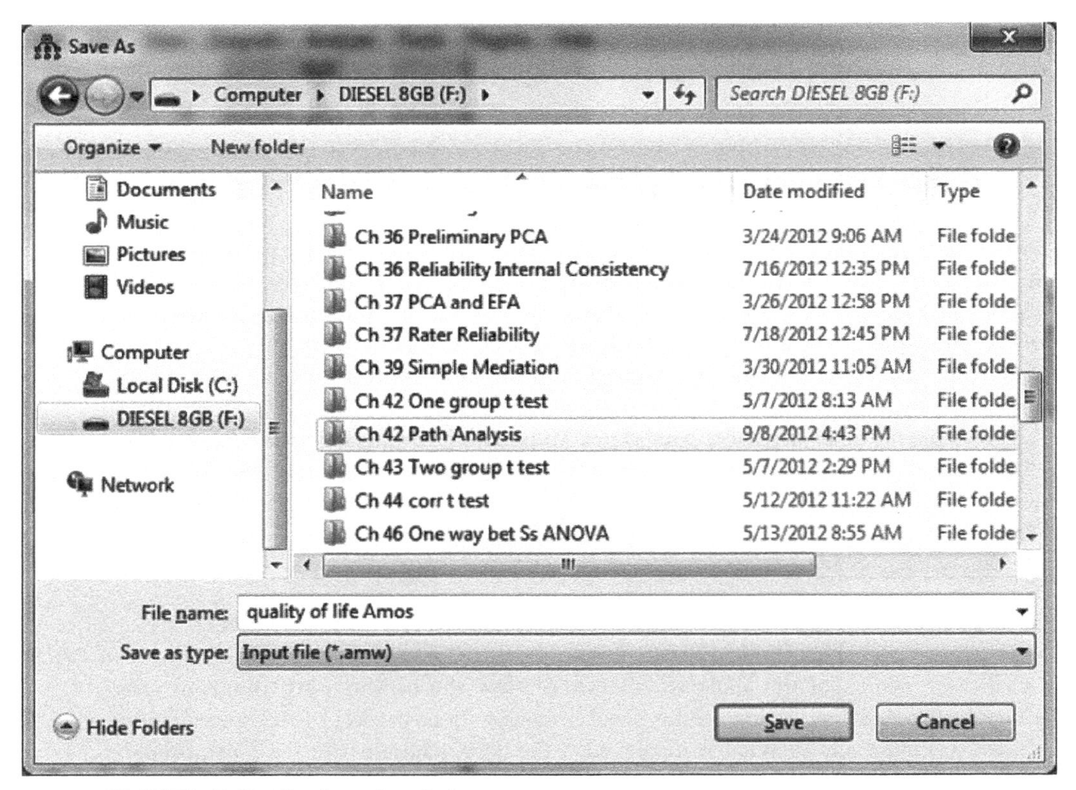

FIGURE 42.19 The **Save As** window.

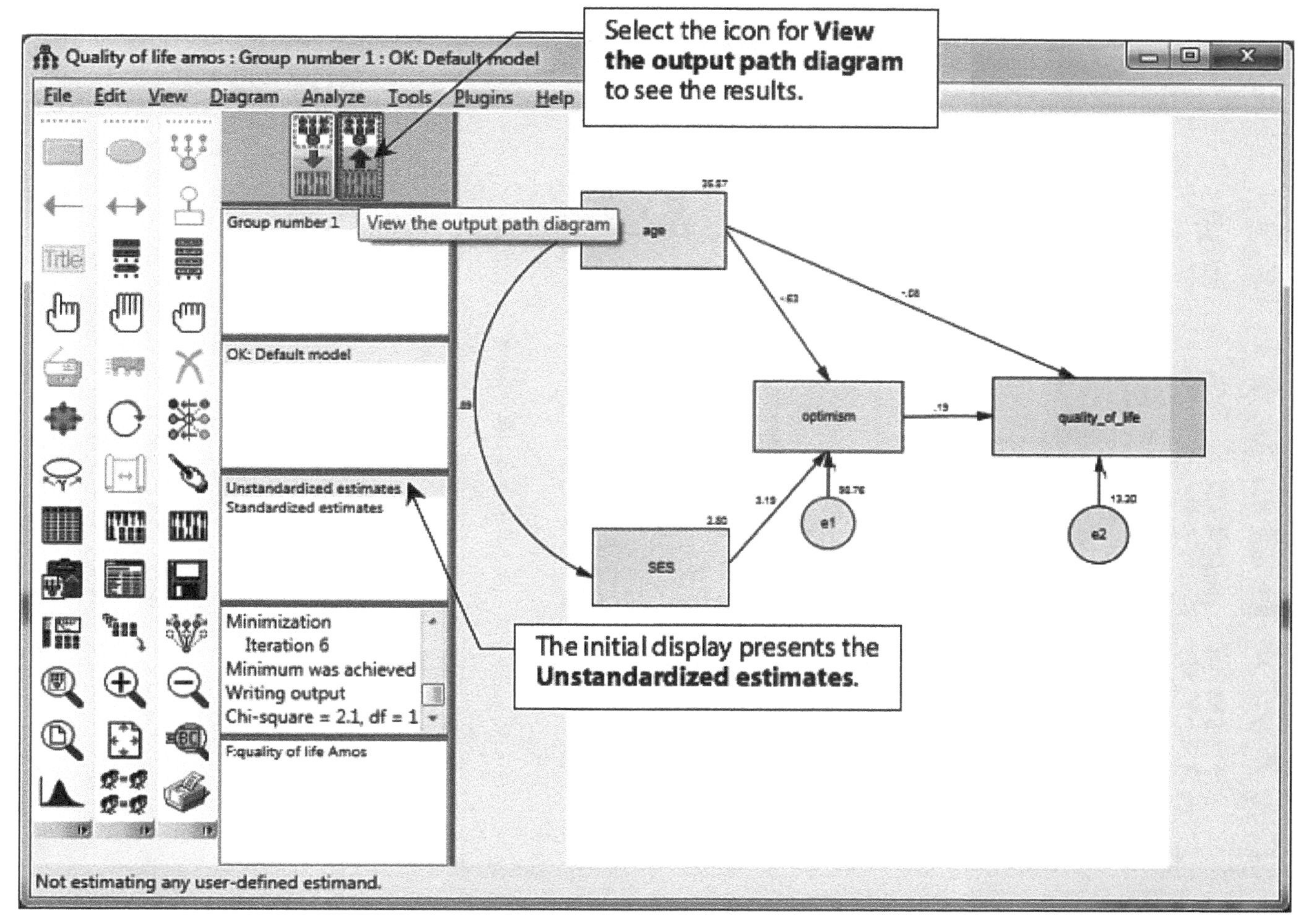

FIGURE 42.20 To access the results, select the **View the output path diagram** icon (the one on the right) in the top panel immediately to the left of the drawing area.

To view the standardized path coefficients in the path diagram, select **Standardized estimates** in the fourth output panel from the top just under **Unstandardized estimates**. These are shown in Figure 42.21. In addition to the unstandardized regression coefficients, the R^2 for each endogenous variable and the correlation of the two exogenous variables are also included in the display. All of these values are available in output tables as well.

To obtain the output in tabular form, select from the main menu **View → Text Output**. This action opens the **Notes for Model** screen shown in Figure 42.22. The **Default model** referred to in the first line on the screen is the model that we configured. The chi-square value associated with the model is **2.094** and, with **1** degree of freedom, is not statistically significant ($p = .148$). This indicates that the expected values based on the model do not differ significantly from those represented by the data; that is, the model fits the data. This is what researchers hope to find.

The model fit indexes, accessed by selecting **Model Fit** from the left panel, are shown in Figure 42.23, Figure 42.24, and Figure 42.25. As described in Chapter 39, we focus on a subset of the indexes that are presented:

- The **CMIN** index (minimal discrepancy as assessed by chi-square) is a statistically nonsignificant value of **2.094** (see top table of Figure 42.23), indicating that the

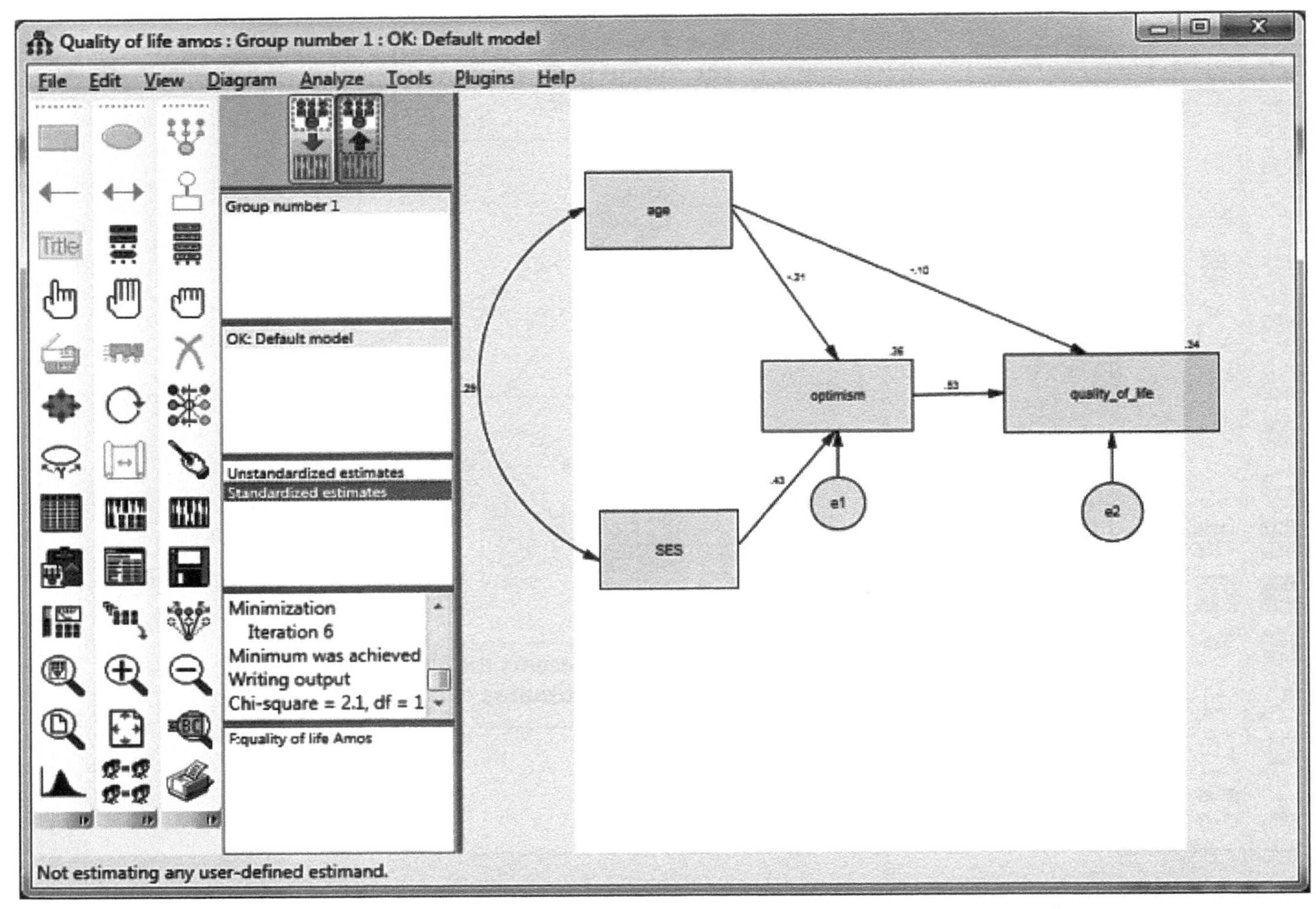

FIGURE 42.21 The **Standardized estimates** are shown in the path diagram.

model fits the data. We have already seen this result in the **Notes for Model** screen.

- The **GFI** is **.996** (see middle table of Figure 42.23) and indicates an excellent fit (the goal is to exceed .95).
- The **NFI** is **.991** (see bottom table of Figure 42.23) and indicates an excellent fit (the goal is to exceed .95).
- The **CFI** is **.995** (see bottom table of Figure 42.23) and indicates an excellent fit (the goal is to exceed .95).
- The **RMSEA** is **.067** (see top table of Figure 42.25) and indicates a reasonably good fit (we exceeded the benchmark of .06 indicating a very good fit but succeeded in achieving a value below .08 indicating an adequate fit).

Selecting **Estimates** in the output panel changes the display to a set of tables, the first screenshot of which is shown in Figure 42.26. The top table presents the unstandardized regression coefficients (labeled as **Regression Weights**), the standard error associated with the coefficients (**S.E.**), and the critical ratio (**C.R.**). The CR is the coefficient divided by the standard error and operates something as a *z* statistic with absolute values of 1.96 or greater indicating statistical significance based on an alpha level of .05.

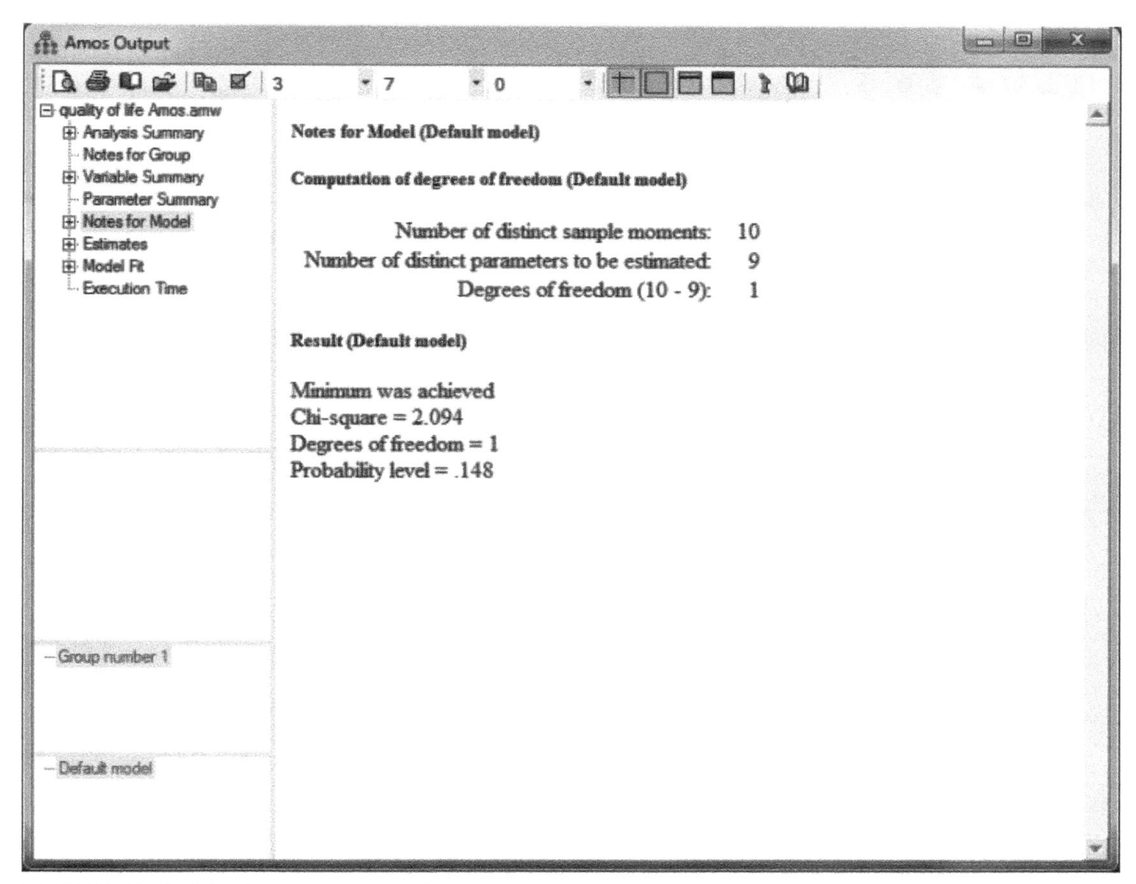

FIGURE 42.22 Selecting **Text Output** from the main menu opens the **Notes for Model** screen.

Each coefficient is tested for statistical significance (reported under the column labeled **P**). The triple asterisks (***) represent a probability level of $p < .001$. Three of the four regression coefficients reached statistical significance ($p < .001$); the exception was the path from **age** to **quality_of_life** that yielded a trend toward statistical significance ($p =$ **.071**) but fell short of an alpha level of .05. It is worth noting that these unstandardized regression coefficients produced by maximum likelihood estimation have achieved the same values as those produced by the ordinary least-squares multiple regression procedure, thanks in part to the stability of the model based on our very large sample size.

The middle table of Figure 42.26 presents the standardized regression coefficients. Not surprisingly, given the match of the unstandardized coefficients, these also match the values obtained by the multiple regression analyses. The bottom table of Figure 42.26 presents the covariance between **age** and **SES**. **Covariance** is a different way to express correlation; it is based on the original measurement units in the computation rather than being standardized to produce values (that we ordinarily see in reporting correlations) between ±1.00 (Norman & Streiner, 2008).

Figure 42.27 presents the next portion of the **Estimates** display. The top table shows the correlation between **age** and **SES** and is the translation of the covariance into the standardized form. The correlation of −**.29** is what we had obtained in the multiple regression analysis. Variances of the predictor variables are shown in the middle table

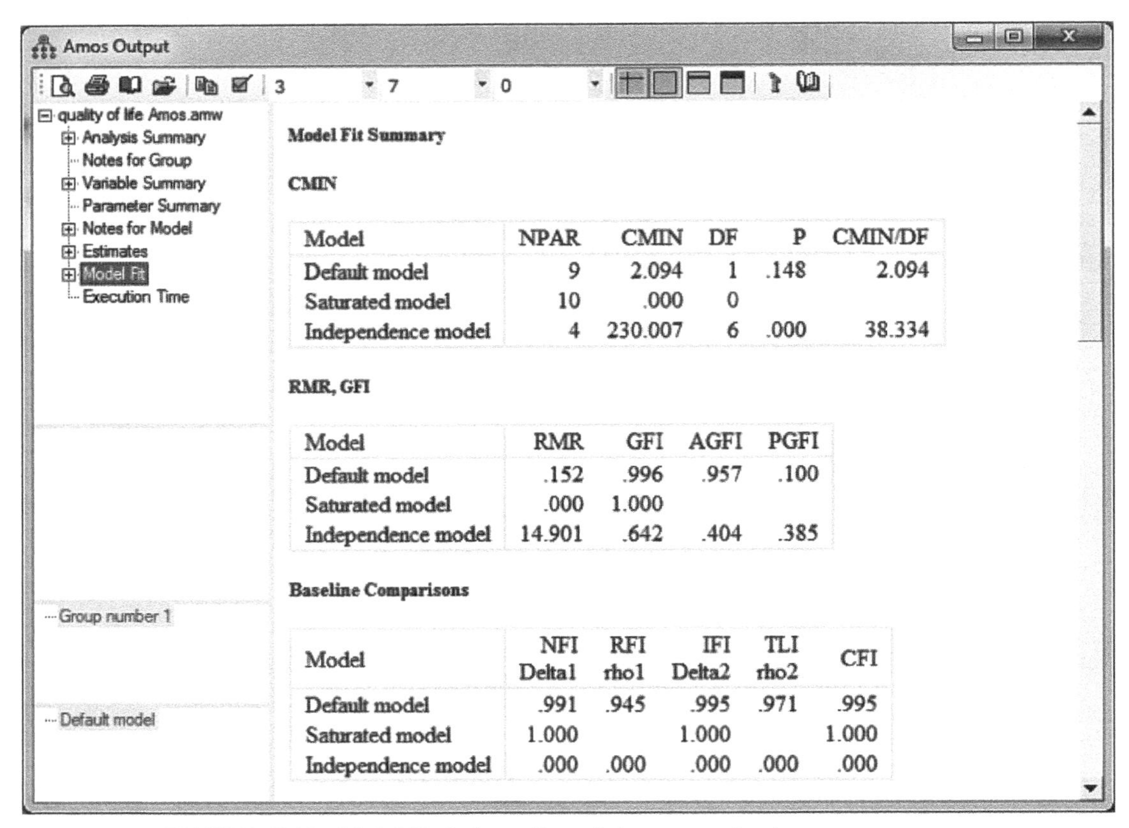

Model Fit Summary

CMIN

Model	NPAR	CMIN	DF	P	CMIN/DF
Default model	9	2.094	1	.148	2.094
Saturated model	10	.000	0		
Independence model	4	230.007	6	.000	38.334

RMR, GFI

Model	RMR	GFI	AGFI	PGFI
Default model	.152	.996	.957	.100
Saturated model	.000	1.000		
Independence model	14.901	.642	.404	.385

Baseline Comparisons

Model	NFI Delta1	RFI rho1	IFI Delta2	TLI rho2	CFI
Default model	.991	.945	.995	.971	.995
Saturated model	1.000		1.000		1.000
Independence model	.000	.000	.000	.000	.000

FIGURE 42.23 **Model Fit** indexes (first of three screenshots).

and the R^2 (**Squared Multiple Correlations**) values for **optimism** and **quality_of_life** are shown in the bottom table of Figure 42.27. The R^2 value of **.337** for **quality_of_life** matches exactly what was obtained from the multiple regression analysis; the R^2 value of **.356** for **optimism** is close to but not exactly the same as that we had obtained from the multiple regression analysis.

Figure 42.28 shows the results of the calculations for the indirect effects. The bottom table presents the standardized results; the values of **.227** for **SES** through **optimism** and **−.161** for **age** through **optimism** are equivalent to those computed (by hand) from the multiple regression analyses.

The substantive conclusion based on the SEM analysis did not change from what resulted from the series of multiple regression analyses, but we were able to obtain fit measures informing us that the model was an excellent fit to the data. Such an outcome strongly supports the structure of the variables hypothesized by the researchers with the one possible exception of the hypothesized causal path of **age** predicting **quality_of_life** that did not reach the .05 level of statistical significance.

Although we will not repeat the calculations here because we are dealing with the same values and therefore would achieve the same outcomes, it would be appropriate to engage in the following procedures at this point in our explication of the model results:

- We would want to perform the Aroian tests to evaluate the statistical significance of the indirect effects.

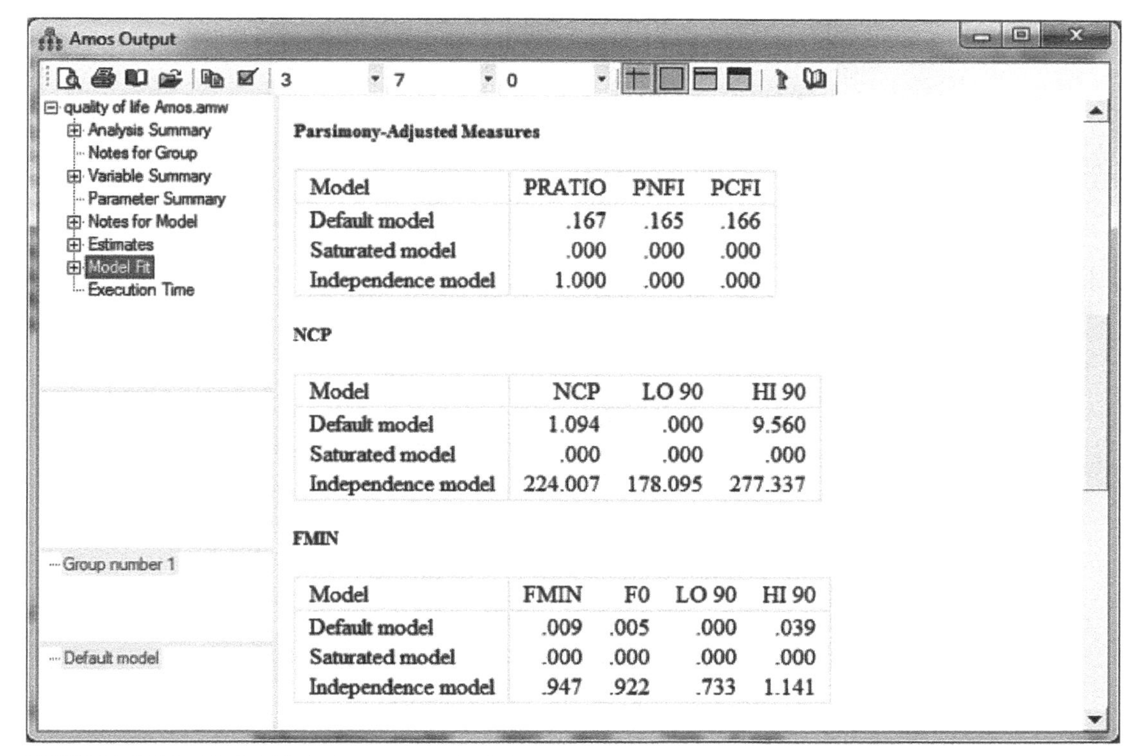

Parsimony-Adjusted Measures

Model	PRATIO	PNFI	PCFI
Default model	.167	.165	.166
Saturated model	.000	.000	.000
Independence model	1.000	.000	.000

NCP

Model	NCP	LO 90	HI 90
Default model	1.094	.000	9.560
Saturated model	.000	.000	.000
Independence model	224.007	178.095	277.337

FMIN

Model	FMIN	F0	LO 90	HI 90
Default model	.009	.005	.000	.039
Saturated model	.000	.000	.000	.000
Independence model	.947	.922	.733	1.141

FIGURE 42.24 **Model Fit** indexes (second of three screenshots).

- We would want to examine the possibility of mediation by creating a model with only **age** predicting **quality_of_life**.
- Having determined that **age** predicted **quality_of_life**, we would want to perform the Freedman–Schatzkin test to compare the relative strengths of the paths from **age** to **quality_of_life** in the unmediated model and the mediated models.

The remaining issue is how to treat the nonsignificant path between **age** and **quality_of_life**. One option open to researchers, if they believed that it is theoretically defensible, is to consider removing the path between **age** and **quality_of_life**. In a multiple regression approach, the researchers would have to make a decision about removing the path based on no more information than its failure to achieve a statistically significant outcome. Within the context of SEM, researchers have one additional avenue of investigation open to them: they can perform a new analysis excluding the path between **age** and **quality_of_life** to determine how much the model fit is changed by the omission of that path. We choose that option here.

42.5 PATH ANALYSIS BASED ON SEM: MODIFIED MODEL OUTPUT

Without showing the screenshots, we have returned to the path drawing by selecting the icon just to the left of the **View the output path diagram** icon at the top of the output panel, removed the path between **age** and **quality_of_life** by activating the **Erase**

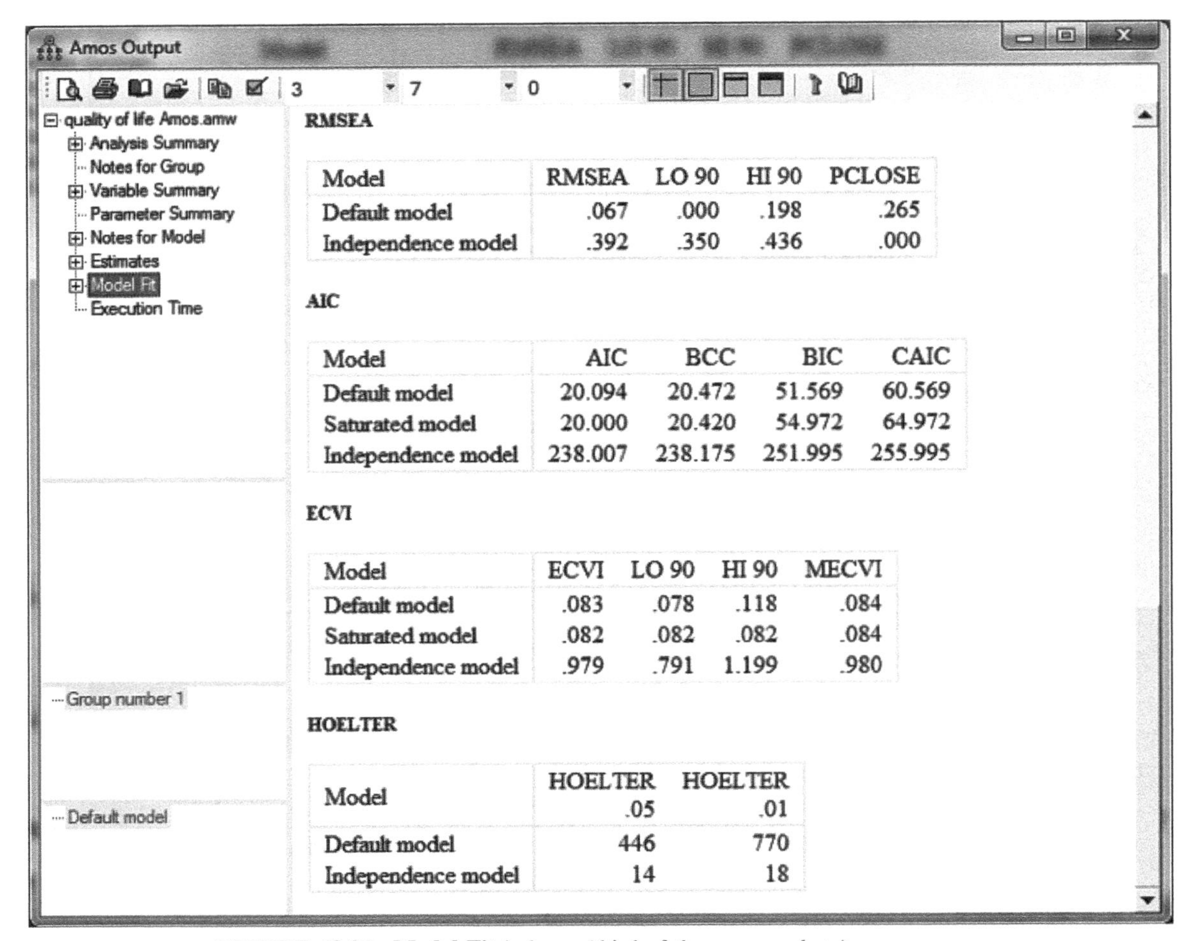
Amos Output

quality of life Amos.amw
Analysis Summary
Notes for Group
Variable Summary
Parameter Summary
Notes for Model
Estimates
Model Fit
Execution Time

Group number 1

Default model

RMSEA

Model	RMSEA	LO 90	HI 90	PCLOSE
Default model	.067	.000	.198	.265
Independence model	.392	.350	.436	.000

AIC

Model	AIC	BCC	BIC	CAIC
Default model	20.094	20.472	51.569	60.569
Saturated model	20.000	20.420	54.972	64.972
Independence model	238.007	238.175	251.995	255.995

ECVI

Model	ECVI	LO 90	HI 90	MECVI
Default model	.083	.078	.118	.084
Saturated model	.082	.082	.082	.084
Independence model	.979	.791	1.199	.980

HOELTER

Model	HOELTER .05	HOELTER .01
Default model	446	770
Independence model	14	18

FIGURE 42.25 **Model Fit** indexes (third of three screenshots).

function and clicking on the path, and performed a new analysis mirroring the setup of the original. The results of the modified model (which should be treated as an exploratory analysis) are briefly summarized in the following.

The **Model Fit** screens of relevance are presented in Figure 42.29. It is interesting that the chi-square value for the modified model of **5.343**, while still not statistically significant ($p = .069$), is greater than the value of **2.094** obtained in the original analysis. To determine if this model represents a statistically poorer fit than the initial model, a *chi-square difference test* can be performed. To conduct a chi-square difference test, we simply subtract the smaller chi-square value from the larger chi-square value. This chi-square difference is evaluated with degrees of freedom equal to the difference in degrees of freedom between the two models.

In this example, the chi-square difference is 3.249 (**5.343** − **2.094** = 3.249). The degrees of freedom for the original and modified models were **1** and **2**, respectively, giving us a difference of one degree of freedom. Based on the chi-square critical values table in Appendix A, we require a chi-square value of **3.841** to be statistically significant at a .05 alpha level for one degree of freedom (a value of **2.706** is required to meet an alpha level of .10). It therefore appears that our modified model does not represent a statistically worse fit than the original model based on a .05 alpha level. That said, the

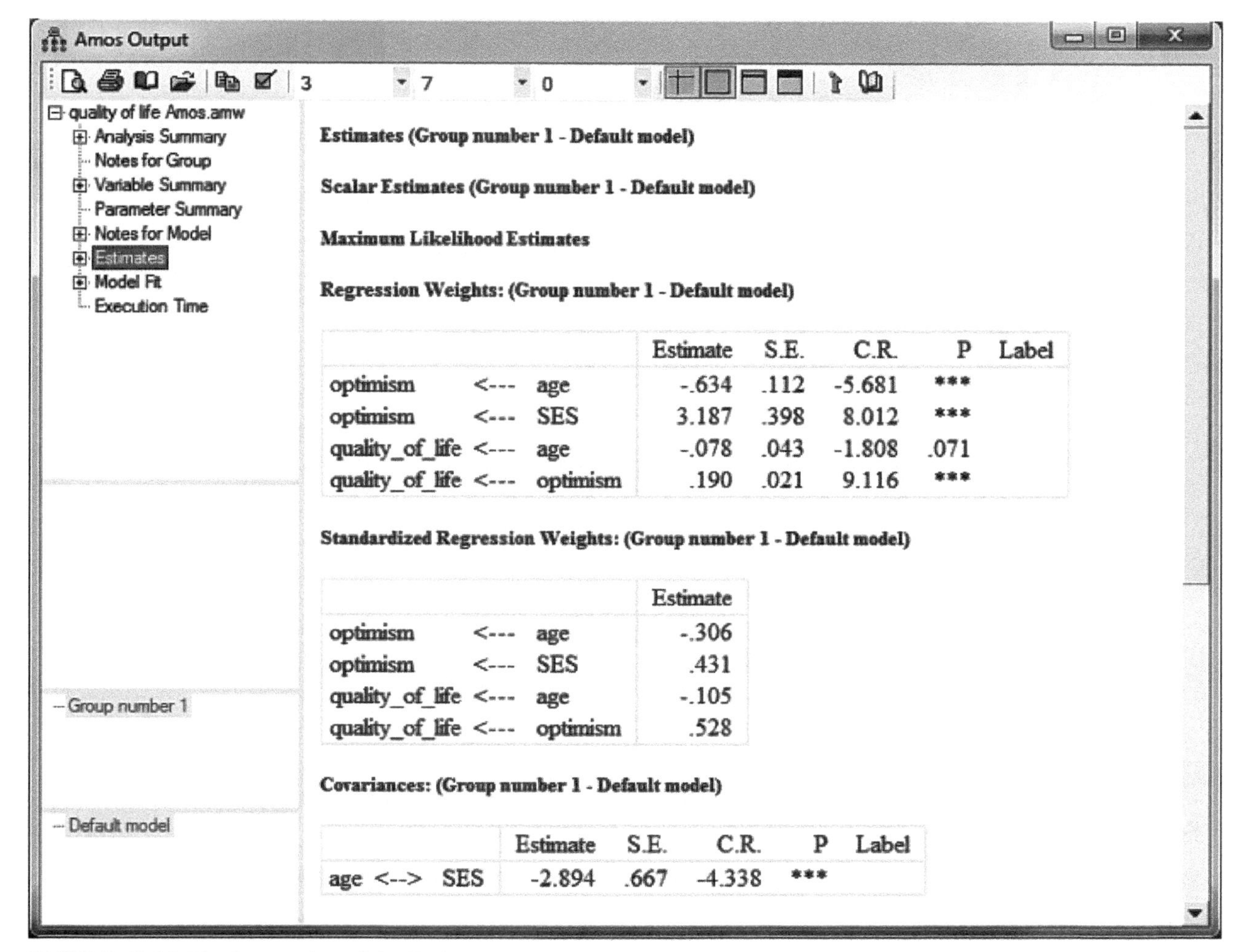

Amos Output

quality of life Amos.amw
Analysis Summary
Notes for Group
Variable Summary
Parameter Summary
Notes for Model
Estimates
Model Fit
Execution Time

Group number 1

Default model

Estimates (Group number 1 - Default model)

Scalar Estimates (Group number 1 - Default model)

Maximum Likelihood Estimates

Regression Weights: (Group number 1 - Default model)

			Estimate	S.E.	C.R.	P	Label
optimism	<---	age	-.634	.112	-5.681	***	
optimism	<---	SES	3.187	.398	8.012	***	
quality_of_life	<---	age	-.078	.043	-1.808	.071	
quality_of_life	<---	optimism	.190	.021	9.116	***	

Standardized Regression Weights: (Group number 1 - Default model)

			Estimate
optimism	<---	age	-.306
optimism	<---	SES	.431
quality_of_life	<---	age	-.105
quality_of_life	<---	optimism	.528

Covariances: (Group number 1 - Default model)

	Estimate	S.E.	C.R.	P	Label
age <--> SES	-2.894	.667	-4.338	***	

FIGURE 42.26 The first portion of the **Estimates** screen.

obtained chi-square value for this trimmed model does suggest a somewhat poorer fit than what we obtained for the original model.

The picture painted by the chi-square analysis is reinforced when we examine the other fit indexes.

- The **GFI** was **.996** in the original analysis; it has dropped to **.989** in the new analysis.
- The **NFI** was **.991** in the original analysis; it has dropped to **.977** in the new analysis.
- The **CFI** was **.995** in the original analysis; it has dropped to **.985** in the new analysis.
- The **RMSEA** was **.067** in the original analysis; it has dropped to **.083** (suggesting not quite an adequate fit) in the new analysis.

The **Estimates** resulting from the modified model are shown in Figure 42.30. These values changed minimally from the original analysis.

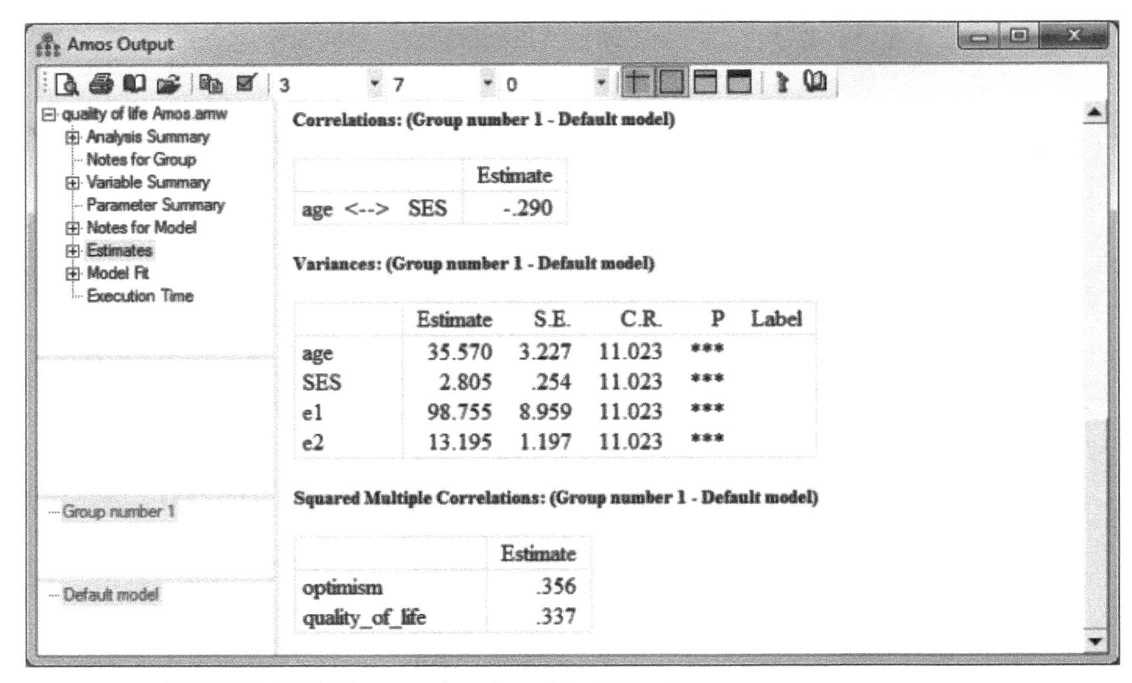

FIGURE 42.27 The second portion of the **Estimates** screen.

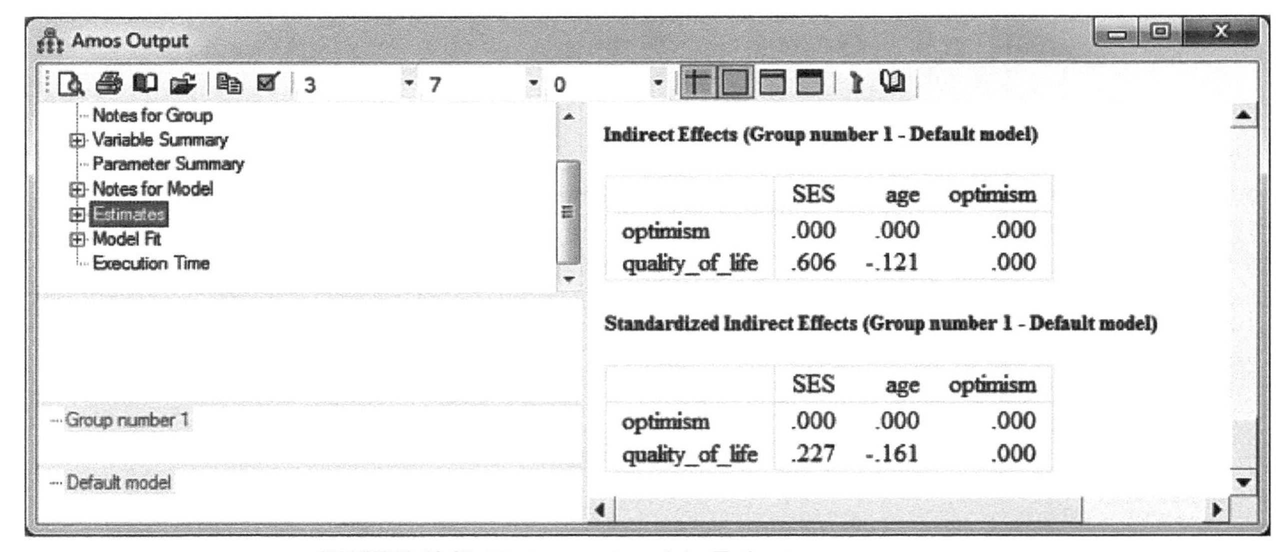

FIGURE 42.28 The last portion of the **Estimates** screen.

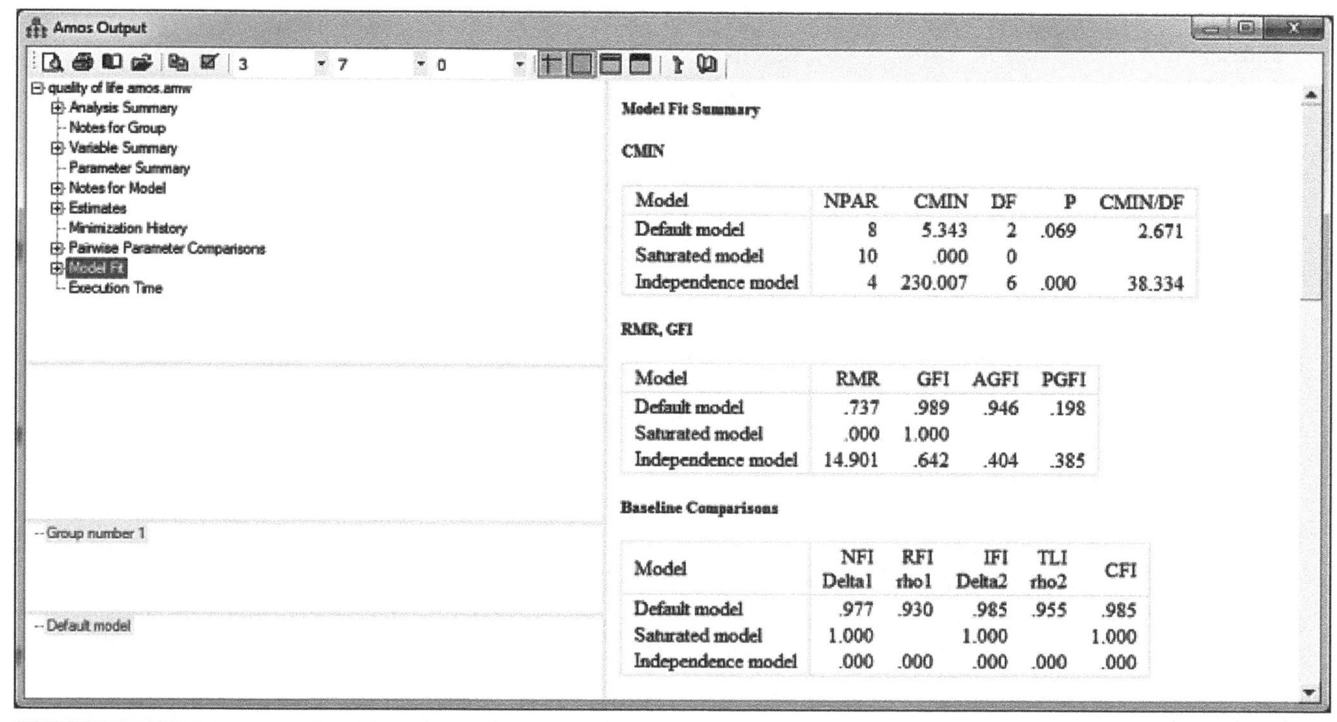

Model Fit Summary

CMIN

Model	NPAR	CMIN	DF	P	CMIN/DF
Default model	8	5.343	2	.069	2.671
Saturated model	10	.000	0		
Independence model	4	230.007	6	.000	38.334

RMR, GFI

Model	RMR	GFI	AGFI	PGFI
Default model	.737	.989	.946	.198
Saturated model	.000	1.000		
Independence model	14.901	.642	.404	.385

Baseline Comparisons

Model	NFI Delta1	RFI rho1	IFI Delta2	TLI rho2	CFI
Default model	.977	.930	.985	.955	.985
Saturated model	1.000		1.000		1.000
Independence model	.000	.000	.000	.000	.000

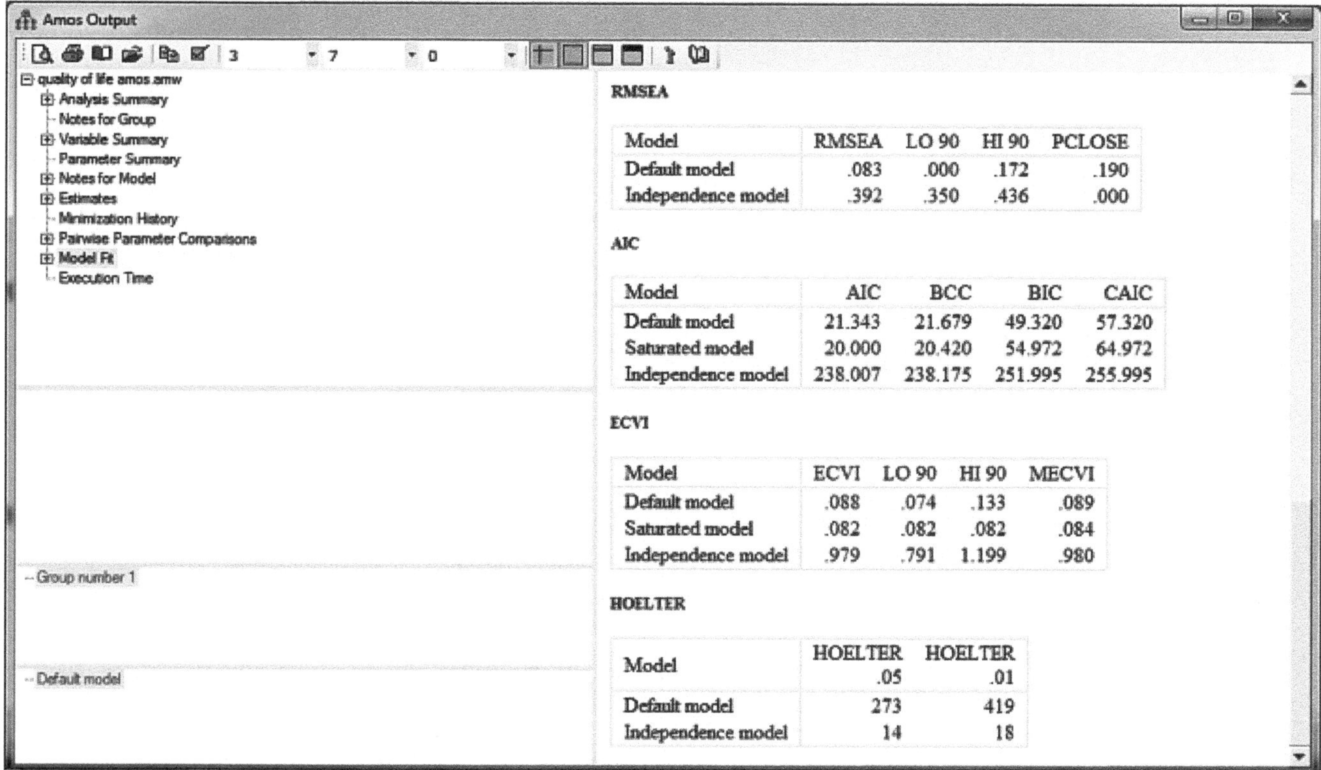

RMSEA

Model	RMSEA	LO 90	HI 90	PCLOSE
Default model	.083	.000	.172	.190
Independence model	.392	.350	.436	.000

AIC

Model	AIC	BCC	BIC	CAIC
Default model	21.343	21.679	49.320	57.320
Saturated model	20.000	20.420	54.972	64.972
Independence model	238.007	238.175	251.995	255.995

ECVI

Model	ECVI	LO 90	HI 90	MECVI
Default model	.088	.074	.133	.089
Saturated model	.082	.082	.082	.084
Independence model	.979	.791	1.199	.980

HOELTER

Model	HOELTER .05	HOELTER .01
Default model	273	419
Independence model	14	18

FIGURE 42.29 The **Model Fit** indexes.

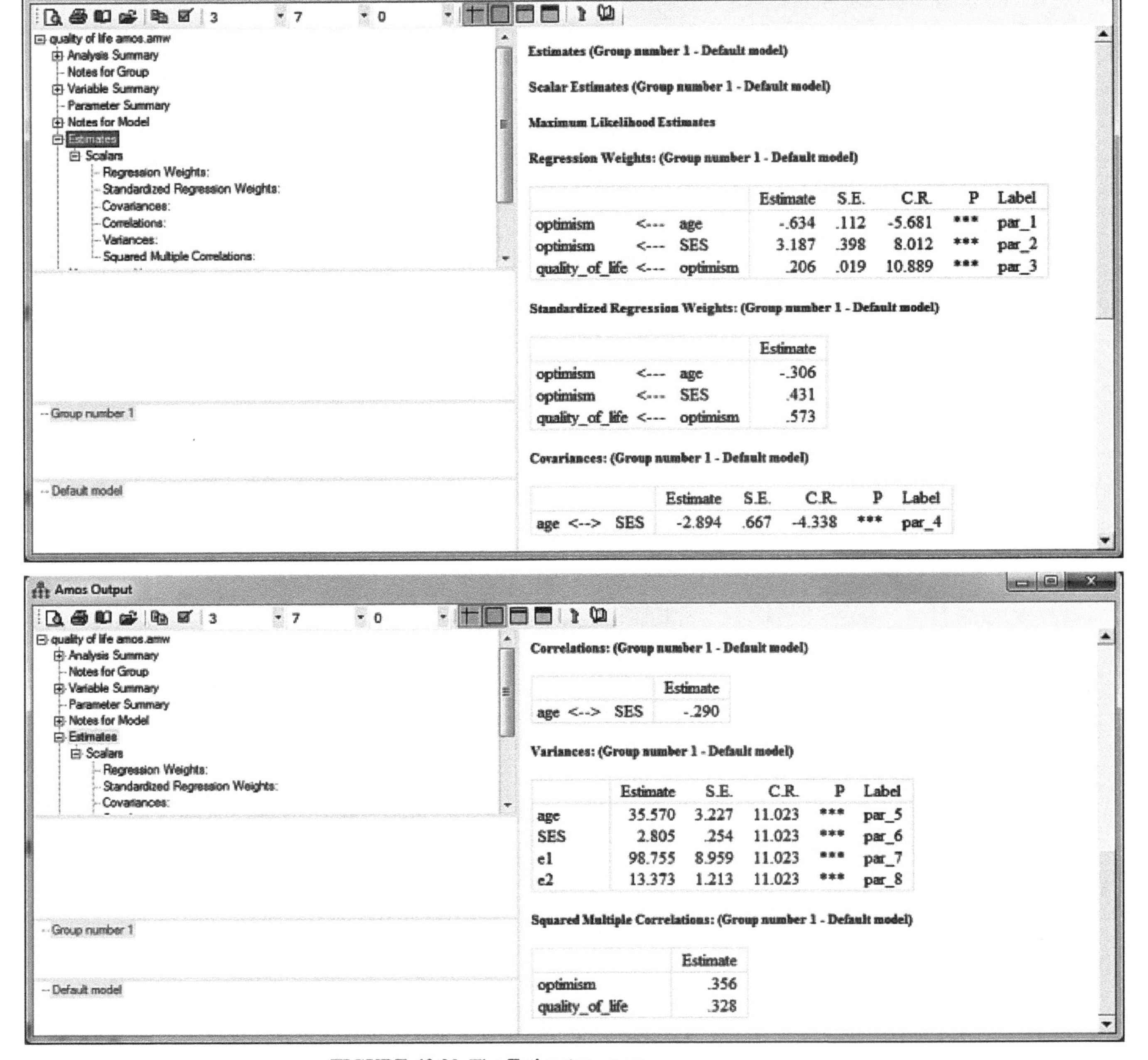

FIGURE 42.30 The **Estimates** screens.

In summary, researchers might be inclined to remove the nonsignificant path from the model based on multiple regression analyses; however, the SEM analysis suggests that the original model should be retained, as removing the path between **age** and **quality_of_life** results in a somewhat (although not a statistically significant) poorer fit of the model to the data. Thus, our conclusion is that the original model is as good a description of the data as we have, even with the inclusion of a path that does not quite reach the .05 alpha level of statistical significance.

CHAPTER 43

Structural Equation Modeling

43.1 OVERVIEW

SEM extends what we have discussed in Chapter 42 with respect to path analysis—where only measured variables are included in the model—to causal or predictive models that incorporate latent variables. We broached the concept of latent variables in Chapter 39 in the context of confirmatory factor analysis, where the factors were presented as latent variables. We treat the idea of latent variables in the same way in this chapter—latent variables are factors or constructs that are not directly observed but rather are indicated or reflected (Edwards & Bagozzi, 2000) by measured (observed) variables; a much richer discussion of latent variables may be found in Bollen (2002). Because latent variables (and their accompanying measured variables) are included in the model, SEM thus subsumes confirmatory factor analysis.

An example of a structural model is presented in Figure 43.1. Latent variables A and B are the exogenous variables in the model; they are hypothesized to be correlated as indicated by the double-headed arrow between them. The outcome variable in the model is the latent variable D; it is hypothesized to be driven by both exogenous variables in both directly and indirectly through the latent variable C. As endogenous variables, latent variables C and D have associated error terms. Each of the latent variables is indicated by two measured variables (for illustration purposes). Structural models always contain latent variables; these models can also include measured variables as either exogenous or endogenous variables (we use only latent variables in our example).

SEM is composed of two major and sequentially dependent components. These components are both contained within the SEM analysis. In our example, we will treat them separately in an effort to make these components clear to readers. The two components are as follows:

- The *measurement model* represents the degree to which the indicator variables relate to their respective latent variables, given the hypothesized relationships among the latent variables. This measurement model is assessed by performing a confirmatory factor analysis as part of the analysis. The presumption is that there is sufficient internal validity evidence associated with the measurement model to support (act as the foundation for) the structural (predictive) model; if the measurement model does not have sufficient integrity, then there is little point in even evaluating the structural model.

Performing Data Analysis Using IBM SPSS®, First Edition.
Lawrence S. Meyers, Glenn C. Gamst, and A. J. Guarino.

- The *structural model* represents the causal or predictive relationships between the exogenous and endogenous variables as specified by the researchers. It is conceptually akin to a path analysis in terms of both analysis setup and statistical output but includes latent variables that are weaved together in a causal configuration.

43.2 NUMERICAL EXAMPLE

The fictional data we use for our example are present in the data file named **statistics achievement** and represent a study of 196 students who were evaluated on their achievement in a statistics class (**statistics**). The model that integrates these variables is shown in Figure 43.2. The latent constructs of anxiety in academic contexts (**anxiety**) and academic self-efficacy (**efficacy**) are hypothesized to directly predict both science

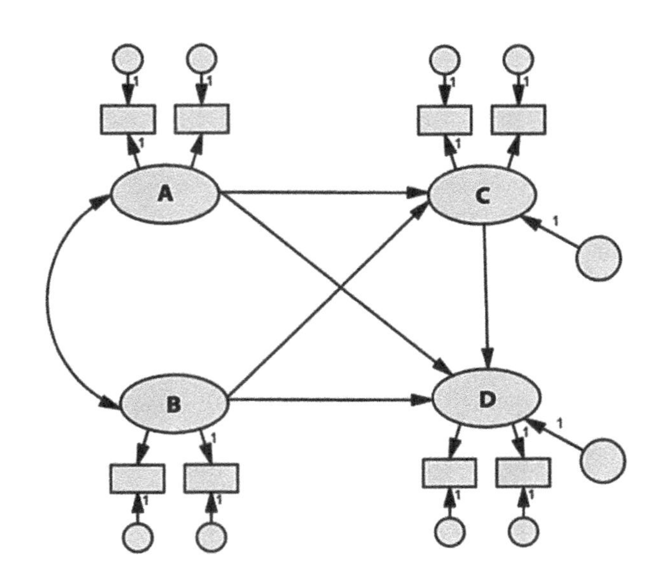

FIGURE 43.1
An illustration of a generic structural model.

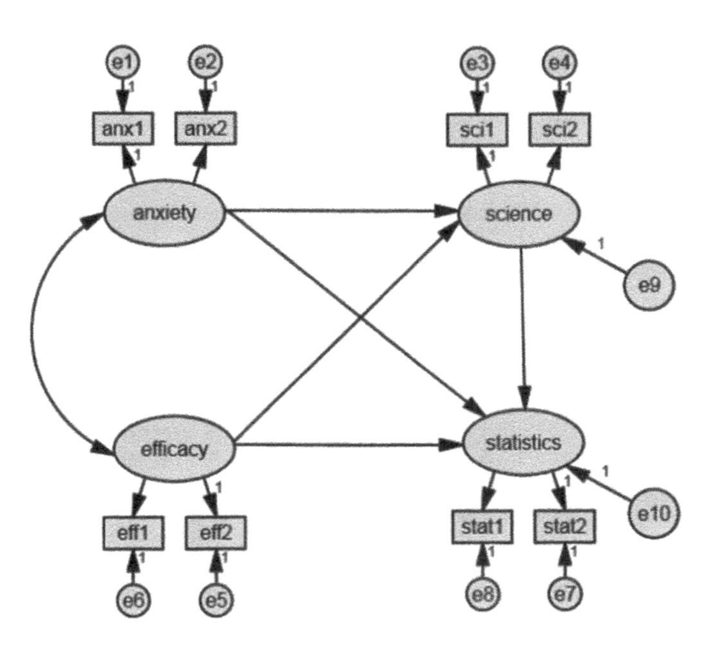

FIGURE 43.2
The hypothesized model.

aptitude (**science**) and statistics achievement (**statistics**), with **science** placed as a mediator variable between the two exogenous variables and the outcome variable.

The two indicator variables (associated measured variables) of **statistics** (**stat1** and **stat2**) could vary from 40 (low ability) to 100 (high ability). The indicators of **anxiety** (**anx1** and **anx2**) could vary from 0 (low anxiety) to 12 (high anxiety), and the indicators of **efficacy** could vary from 10 (low efficacy) to 30 (high efficacy). Finally, **sci1** and **sci2** as indicators of **science** could vary from 20 (low aptitude) to 60 (high aptitude).

43.3 ANALYSIS STRATEGY

When the structural model is specified and executed in IBM SPSS® Amos, results pertaining to the measurement model and to the structural model components are both contained in the output. To make the two components clear to readers, we perform the analysis in two phases corresponding to the two components of the overall model. First, we configure the confirmatory analysis to evaluate the measurement model. Assuming that the measurement model is acceptable, we then evaluate the causal (structural) model. The output for the structural model contains the results reflecting the measurement model; we show where in the output these results are located when discussing the output of the second analysis. Readers are also referred to Section 39.2 where we have discussed the need to have no missing values, a condition that is met by our data set.

43.4 EVALUATING THE MEASUREMENT MODEL: DRAWING THE MODEL

Much of what is covered here has already been described in Chapters 39 and 42, and so we will go into a little less detail in this section; readers are encouraged to consult those two chapters for additional details if needed.

We open the **statistics achievement** data file and from the main menu we select **Analyze → IBM SPSS Amos**. As we have done in the previous chapters, we opt to switch from the default portrait (long) orientation (opened by default in IBM SPSS Amos) to landscape (wide) orientation. To change the orientation, we select **View → Interface properties** from the main menu. This action opens the **Interface Properties** window as seen in Figure 43.3. Select **Landscape-Legal** under **Paper Size**, select **Apply**, and close the window.

The measurement model is evaluated by performing a confirmatory factor analysis. From the icon toolbar menu, select the icon (the circle with squares and more circles) representing **Draw a latent variable or add an indicator to a latent variable** as shown in Figure 43.4. We use this tool to draw our confirmatory model. After activating this function by selecting it, click to draw the first factor toward the left of the drawing area. Place the cursor inside the circle/oval. Then click twice to draw the two indicator variables and their associated errors (changing the shape of the objects if necessary as described in Section 39.3.2). Draw the next three latent variables down the line resulting in the configuration shown in Figure 43.5. Note that all the paths from the errors and the path from each factor to its first indicator variable have been automatically constrained to the value of **1** to scale each latent variable (the factor and the error) to the measured variables. Right-click in a blank portion of the drawing area to end that function (and return to the operating system cursor).

We next draw the correlation (the covariance) paths between the factors based on the hypothesized structural model shown in Figure 43.2. Each latent variable in the model has a path from itself to every other latent variable; most of these paths are causal (predictive) paths (single-headed arrows) but there is also one covariance (correlation) path between

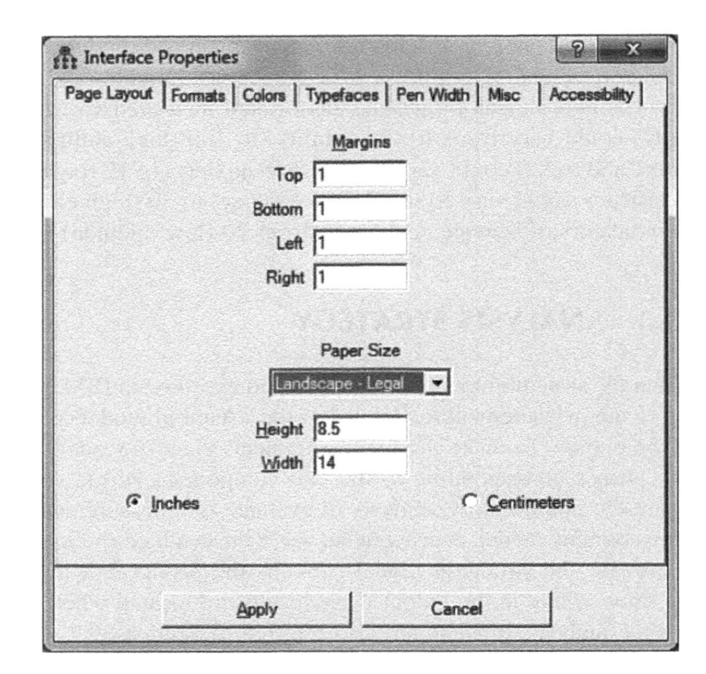

FIGURE 43.3

The **Interface Properties** window where we select **Landscape-Legal** page orientation.

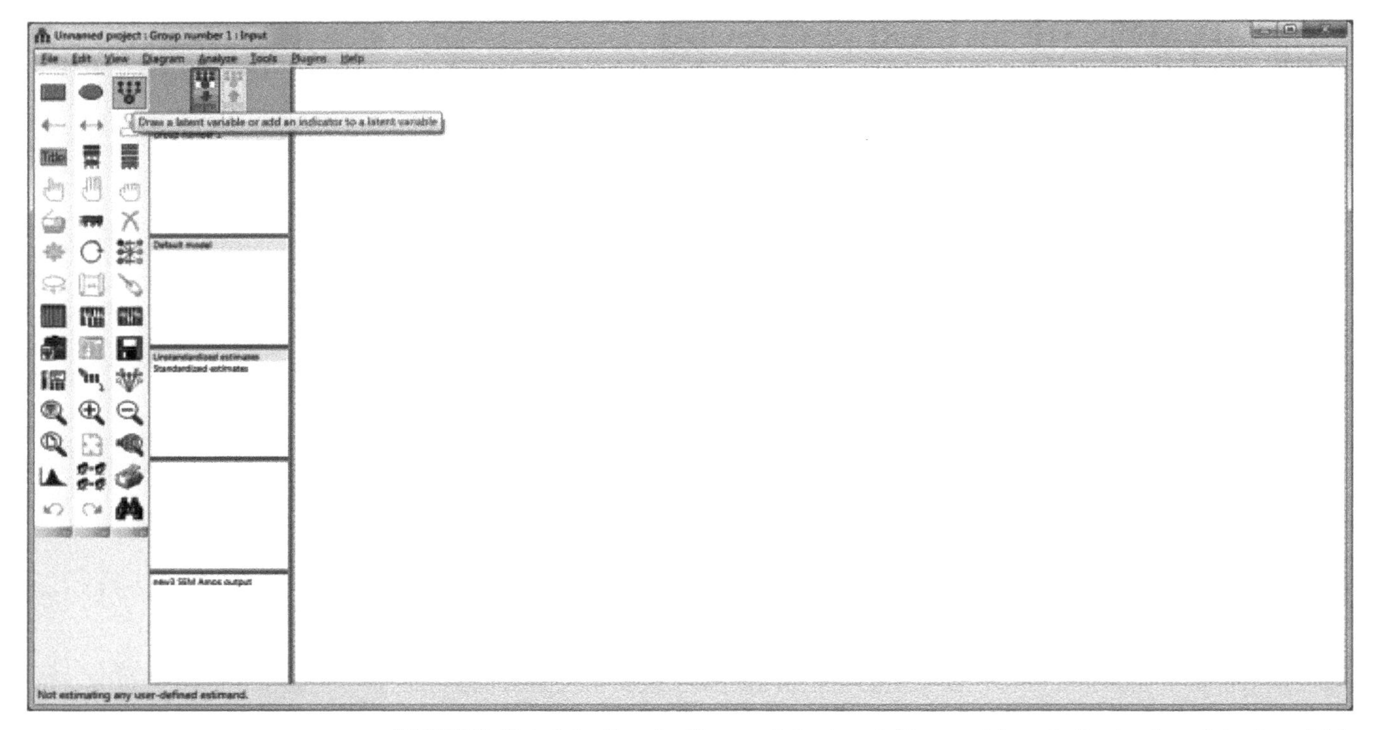

FIGURE 43.4 Selecting the **Draw a latent variable or add an indicator to a latent variable** function.

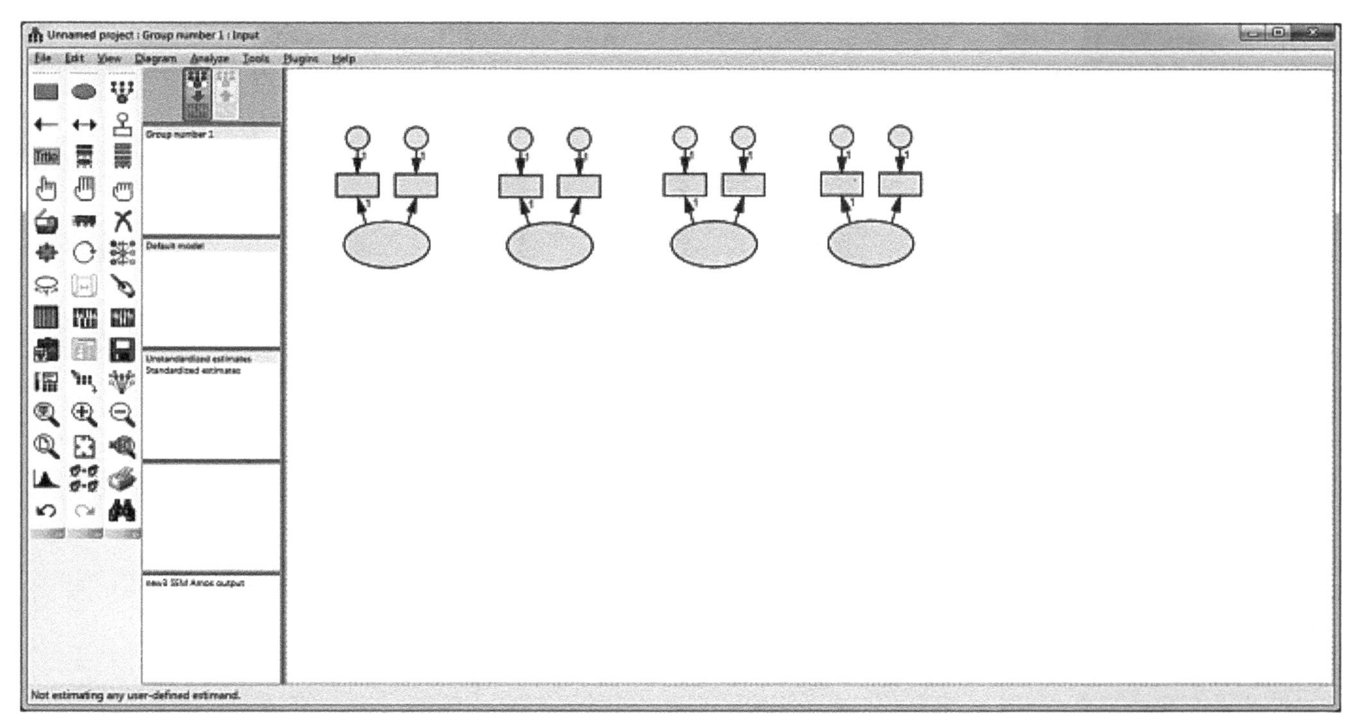

FIGURE 43.5 The latent variables in the confirmatory model are now drawn.

anxiety and **efficacy**; these relationships need to be captured in our confirmatory factor analysis. From the icon toolbar, we select the **Draw covariances (double headed arrow)** and, working from right to left, draw all of the covariances to create a structure mirroring that shown in Figure 43.6.

Our next task is to associate each measured (indicator) variable with a variable from our data file. From the main menu, we select **View Variables in Dataset**. This opens the **Variables in Dataset** window pictured in Figure 43.7. Select each variable in turn and drag it to the appropriate rectangle so that all observed variables are properly configured as shown in Figure 43.8. When finished, close the **Variables in Dataset** window.

To name the factors, we double-click inside a factor to activate the **Object Properties** window shown in Figure 43.9. We type the name we wish to use for each latent variable in turn; we use the names **anxiety**, **efficacy**, **science**, and **statistics**. The result of this activity is shown in Figure 43.10.

The final task in drawing the model is to name the latent error variables (called **unique variables** by IBM SPSS Amos). From the main menu, we select **Plugins ➔ Name Unobserved Variables** (see Figure 43.11). On selecting this command, the **unique variables** will be labeled as "**e**" (for error term) and given a sequential number (see Figure 43.12), resulting in the variable names **e1** through **e8**.

43.5 EVALUATING THE MEASUREMENT MODEL: ANALYSIS SETUP

With the confirmatory factor analysis diagram completed, our next job is to configure the analysis. From the main menu, we select **View ➔ Analysis Properties** to open the

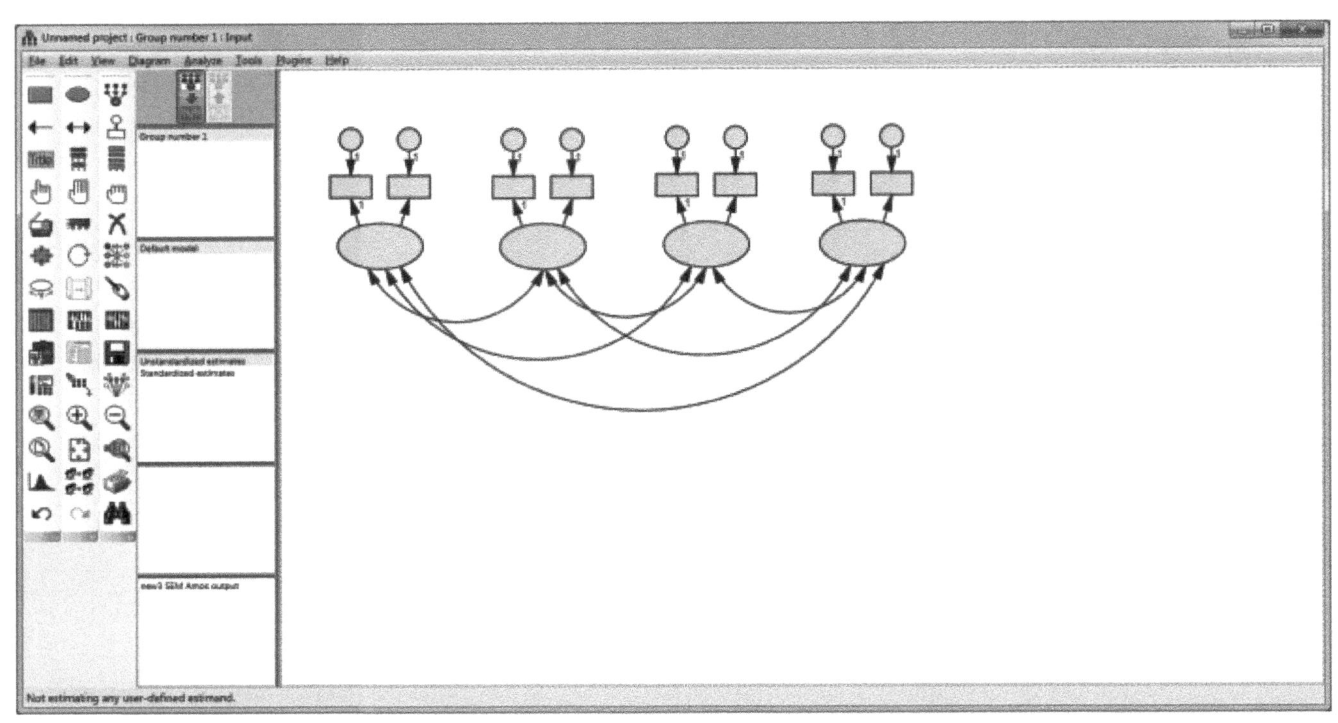

FIGURE 43.6 The correlations between the factors are now drawn.

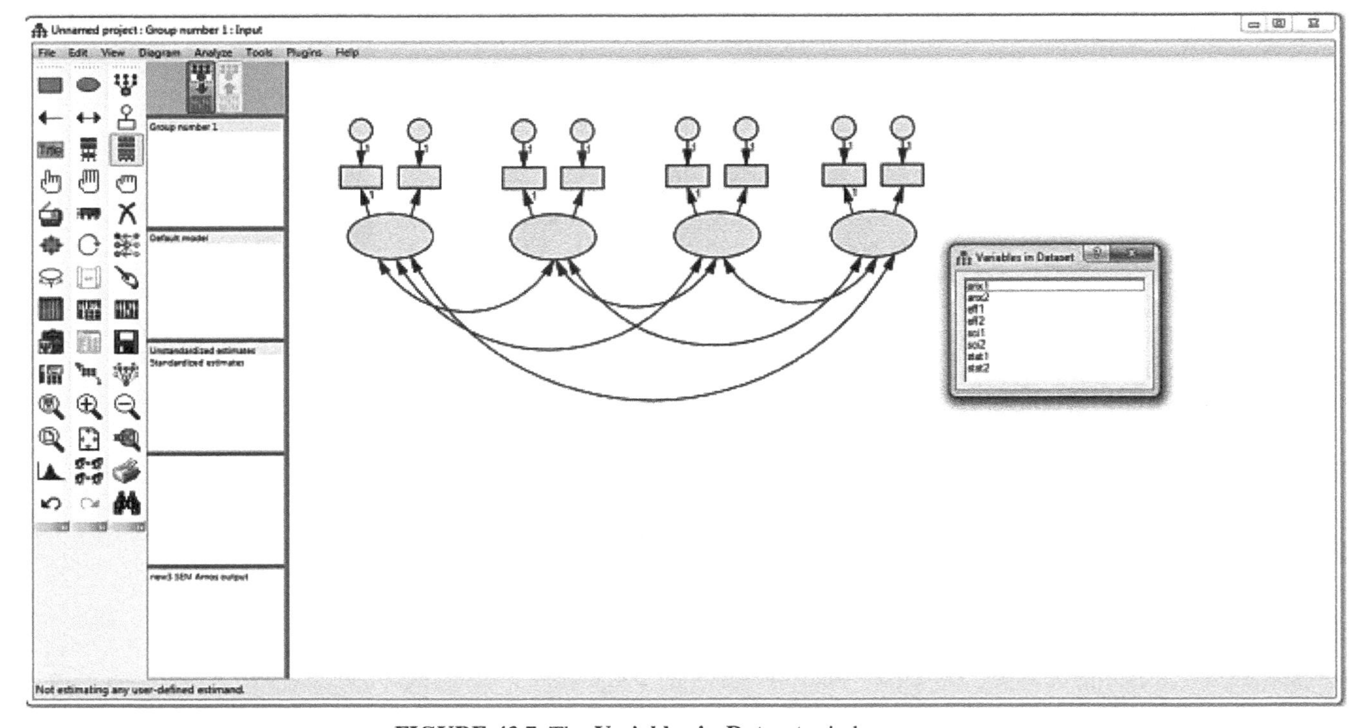

FIGURE 43.7 The **Variables in Dataset** window.

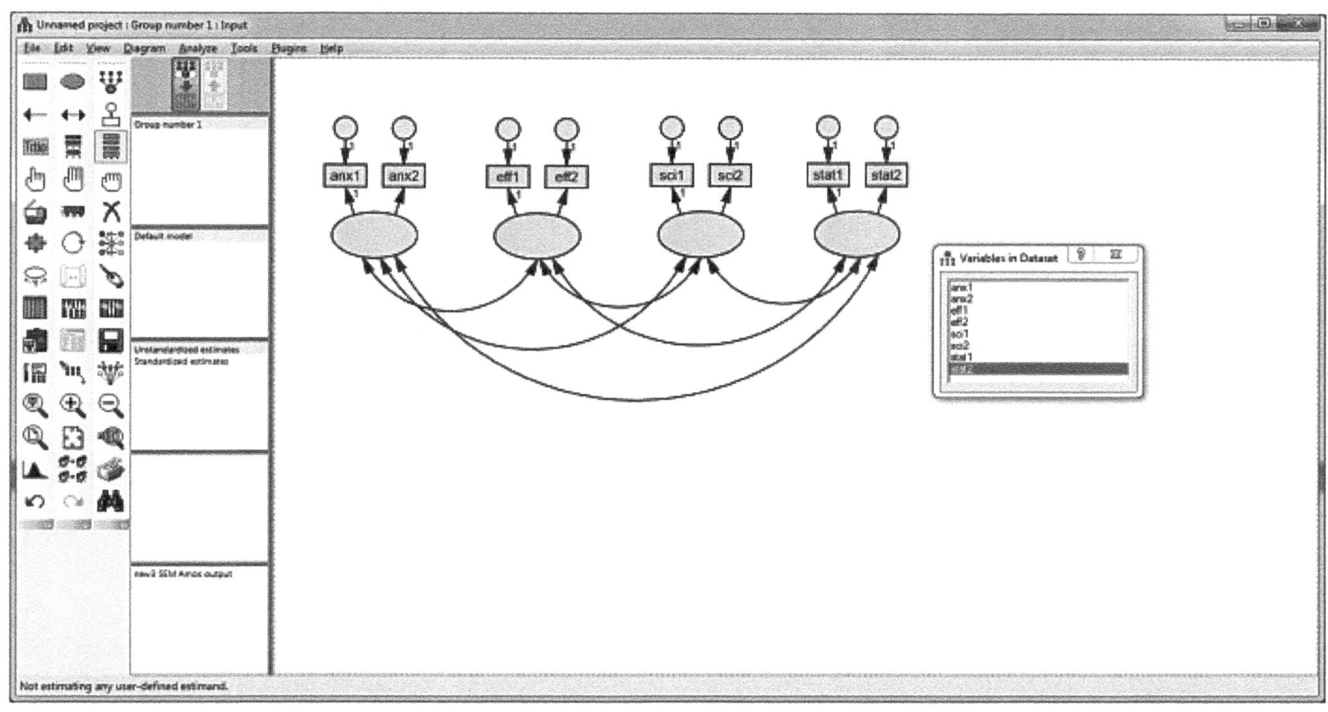

FIGURE 43.8 All of the indicator variables have now been associated with the proper variables in the data file.

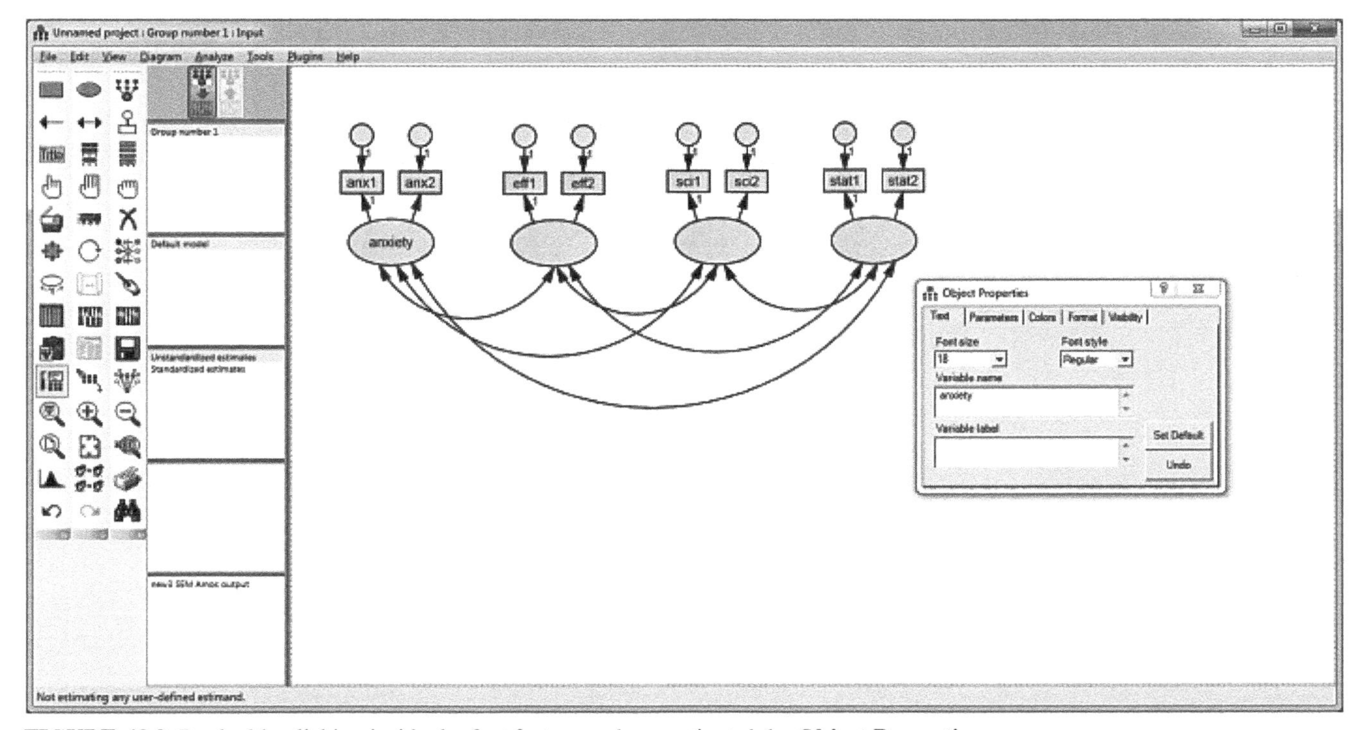

FIGURE 43.9 By double-clicking inside the first factor, we have activated the **Object Properties** window.

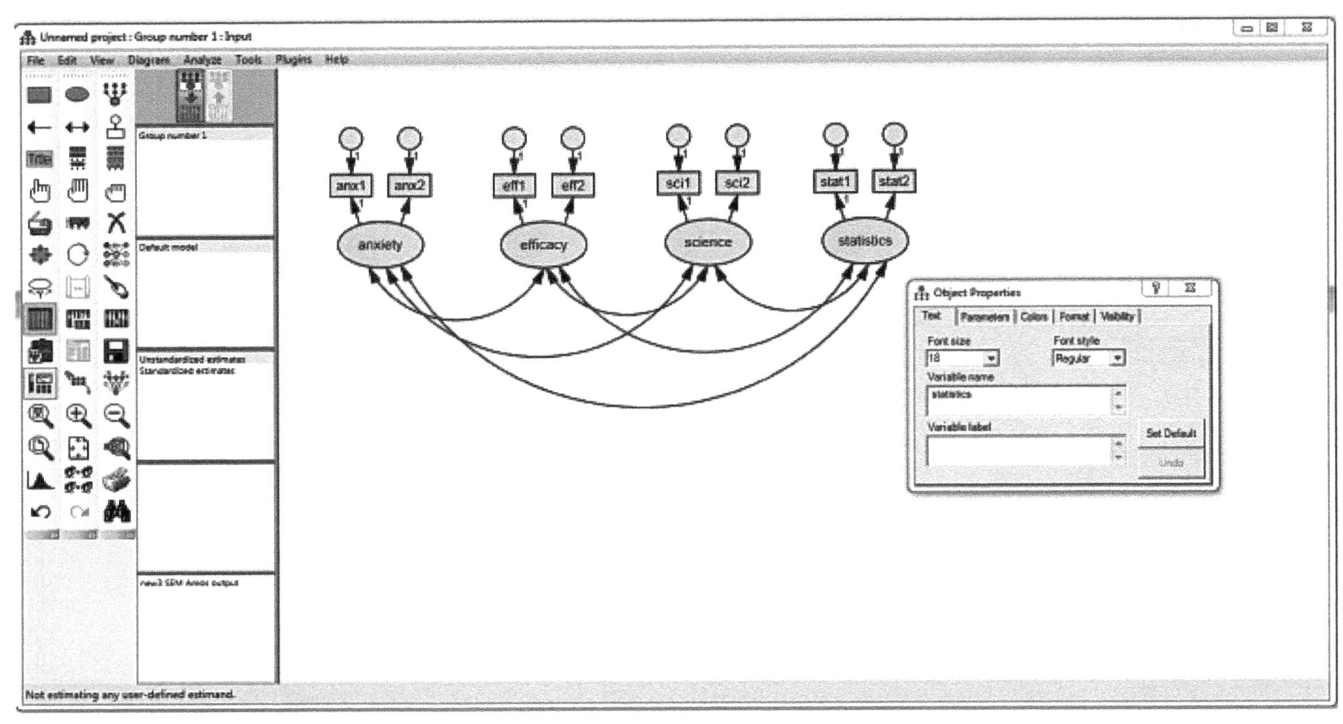

FIGURE 43.10 All of the factors are now named.

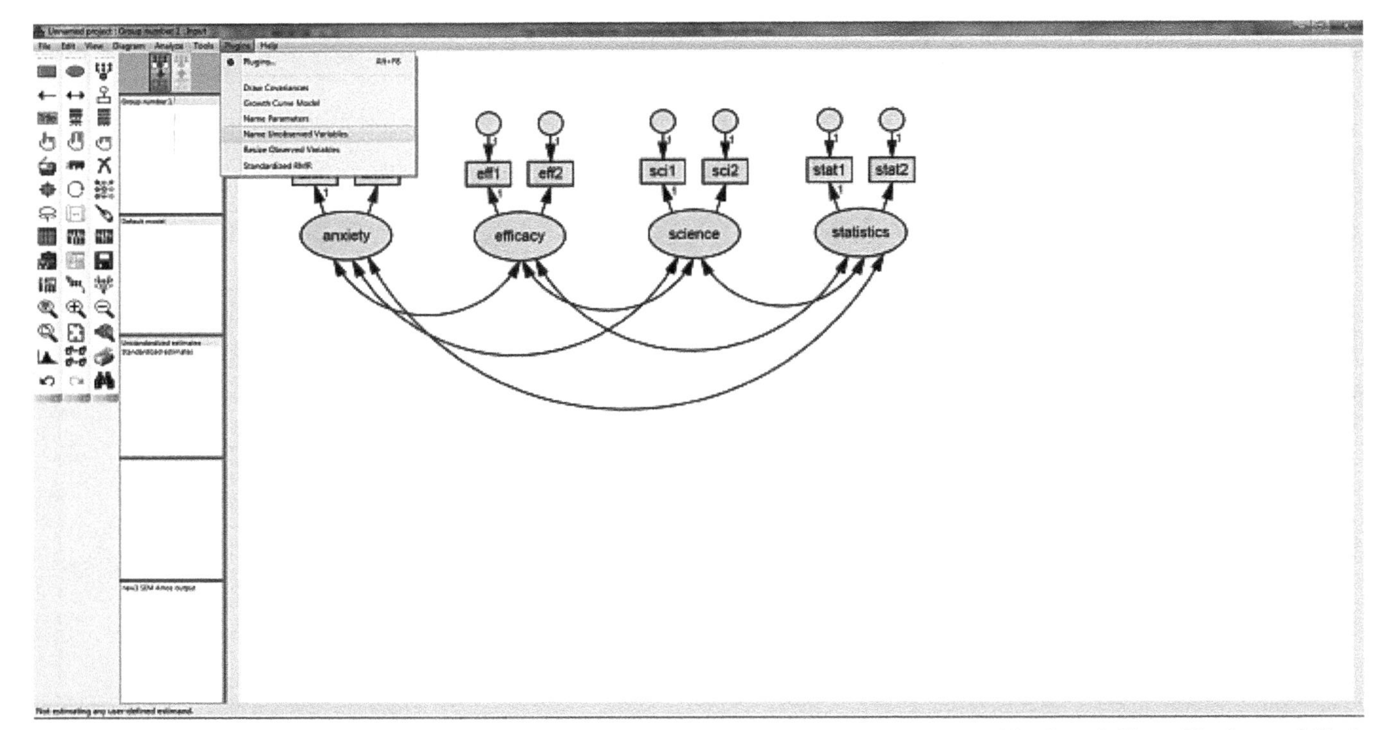

FIGURE 43.11 To name the latent error variables, we select **Plugins ➔ Name Unobserved Variables** from the main menu.

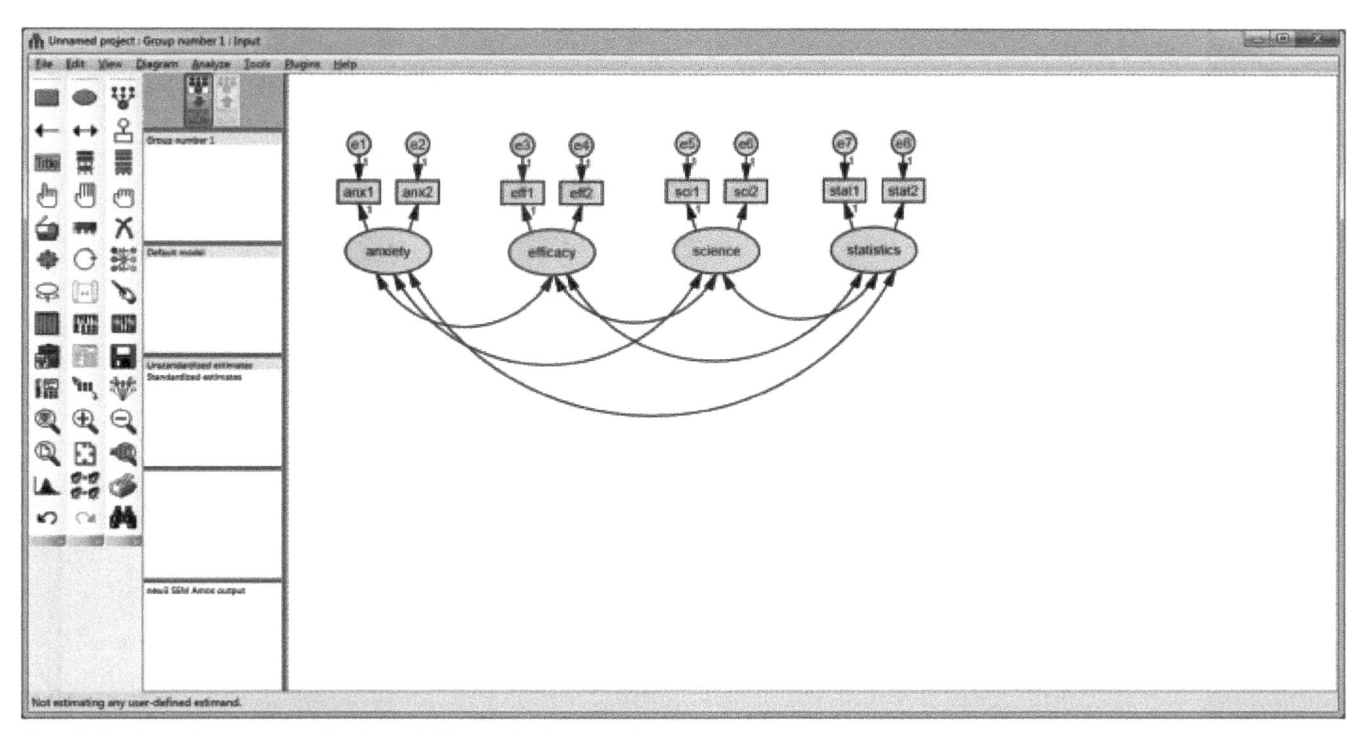

FIGURE 43.12 On selecting **Plugins ➔ Name Unobserved Variables**, the latent error variables are labeled **e1** through **e8**.

Analysis Properties window as shown in Figure 43.13. Select the **Output** tab. Check **Standardized estimates** (these produce the standardized regression coefficients for the prediction of the measured variables) and **Modification indices** (these produce IBM SPSS Amos "suggestions" on how to improve the model fit). To execute the analysis, select **Analyze ➔ Calculate Estimates** from the main menu, saving the output file as needed to an appropriate location.

43.6 EVALUATING THE MEASUREMENT MODEL: ANALYSIS OUTPUT

To access the results of the analysis, select the **View the output path diagram** icon (the one on the right) in the top output panel immediately to the left of the drawing area as shown in Figure 43.14. What is displayed is the confirmatory diagram with the **Unstandardized estimates** displayed.

To obtain the output in tabular form, select from the main menu **View ➔ Text Output**. This action opens the **Notes for Model** screen of the **Output** window shown in Figure 43.15. The **Default model** referred to in the first line on the screen is the confirmatory model shown in Figure 43.12. It may be seen that the chi-square value associated with the model is **24.695** and, with **14** degrees of freedom, is statistically significant (p = **.038**). This suggests that the expected values based on the model differ significantly from those represented by the data, but we use other fit measures as well, in that the chi-square test is overly powerful and can detect small discrepancies between the observed and predicted covariances, suggesting (sometimes inappropriately) that the model does not fit the data (Bentler, 1990; Jöreskog & Sörbom 1996).

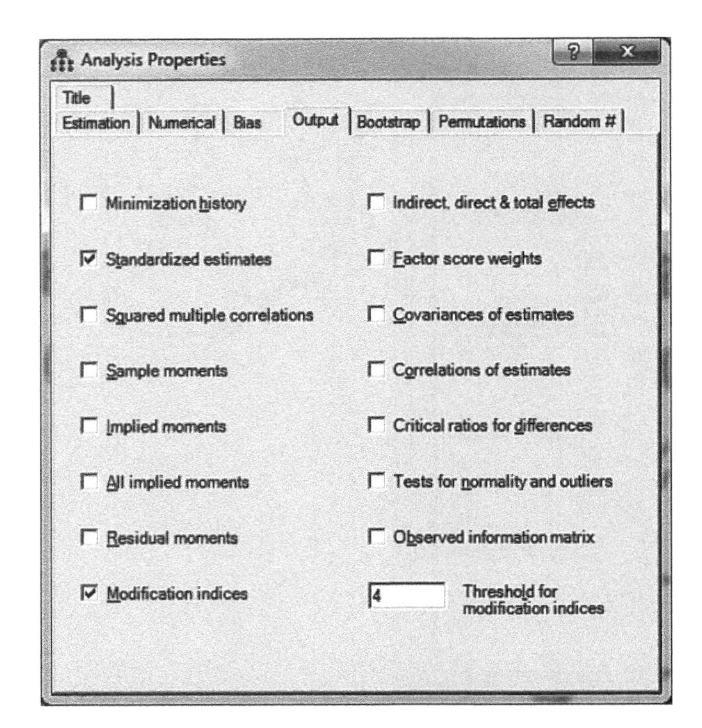

FIGURE 43.13
The **Analysis Properties** window.

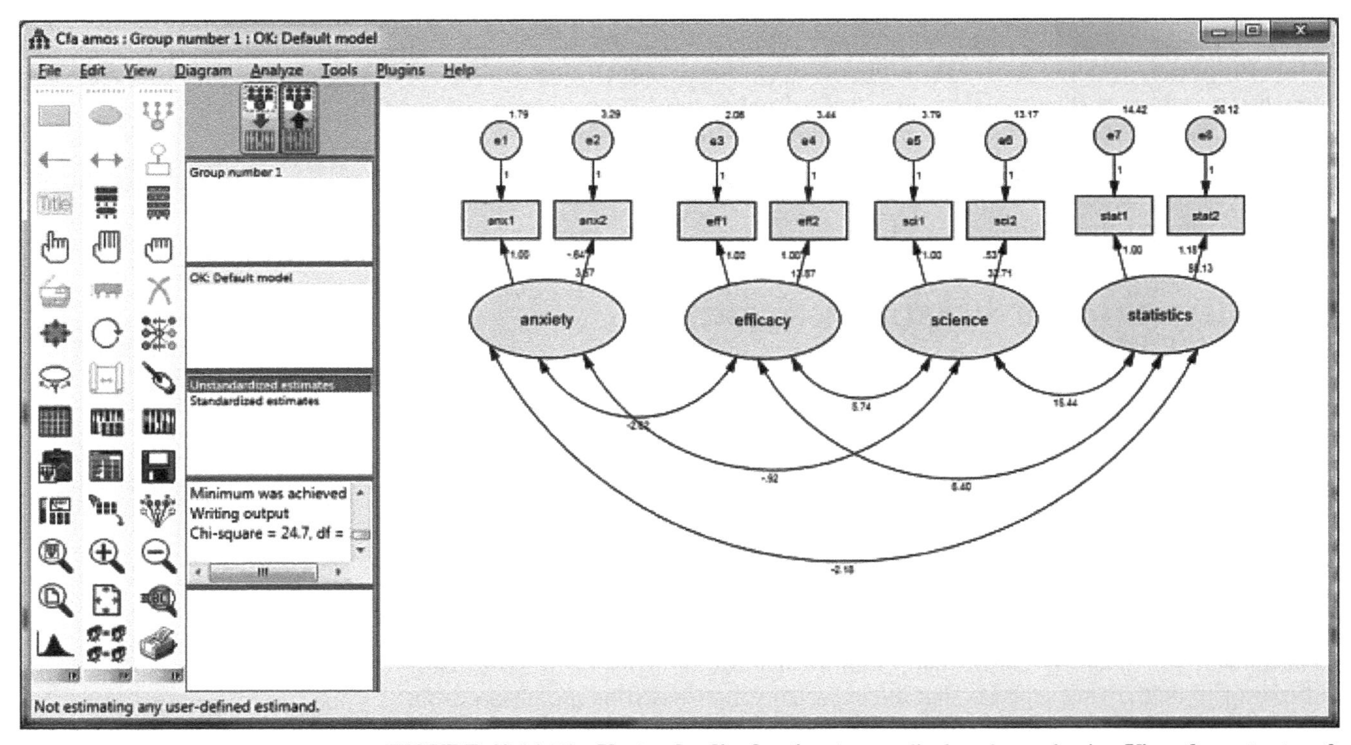

FIGURE 43.14 The **Unstandardized estimates** are displayed on selecting **View the output path diagram**.

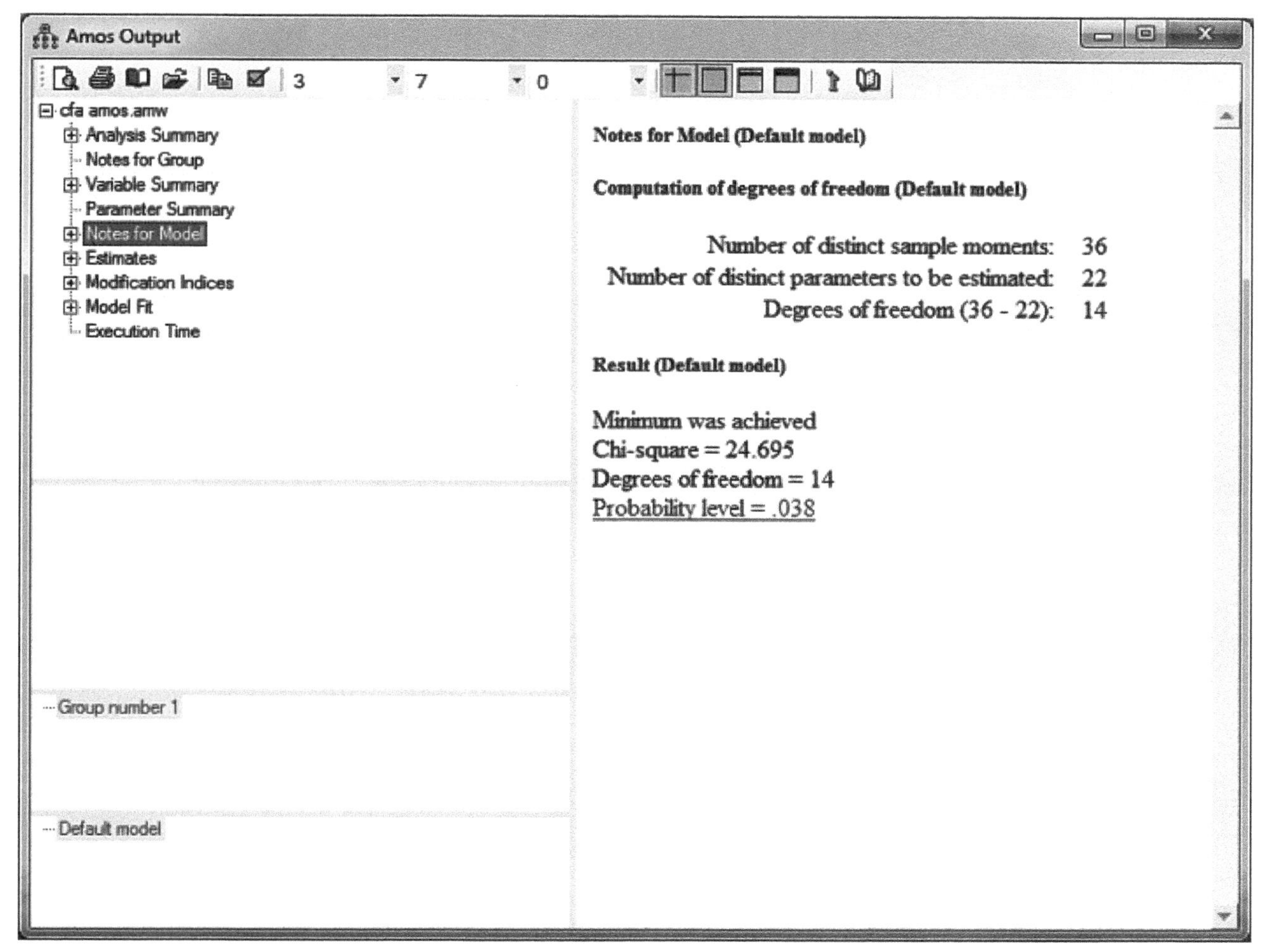

FIGURE 43.15 Selecting **Text Output** presents us with the **Notes for Model** screen.

Selecting **Model Fit** in the left panel displays the several fit indexes produced by IBM SPSS Amos; the ones on which we focus are presented in Figure 43.16. The **CMIN** table shown at the top in Figure 43.16 presents the *minimum discrepancy* value in the **Default model** row, which is the chi-square goodness-of-fit test that repeats the information we saw in the **Model Notes** screen. The **GFI** is shown in the next table in Figure 43.16 where we see its value of **.973** exceeds our guideline value of .95, suggesting an acceptable level of model fit. The **NFI** and the **CFI** are shown in the bottom table of the top screenshot in Figure 43.16 where we see their respective values of **.962** and **.983** also meet our guideline value of .95, suggesting an acceptable level of model fit. The **RMSEA** table is presented in the bottom screenshot of Figure 43.16. Its value of **.063** also suggests that the model fits the data fairly well.

In summary, the NFI, CFI, and RMSEA indexes suggest that our confirmatory model is a reasonably good fit to the data despite an associated chi-square that was statistically significant at an alpha level of .05.

Selecting **Estimates** in the left panel displays the estimated parameters generated by the model, the factor coefficients portion of which is shown in Figure 43.17. The **Regression Weights Table** Figure 43.17 presents the unstandardized regression coefficients and the bottom table presents the same information in the standardized form (which is easier to read). All coefficients that were estimated were statistically significant, and the

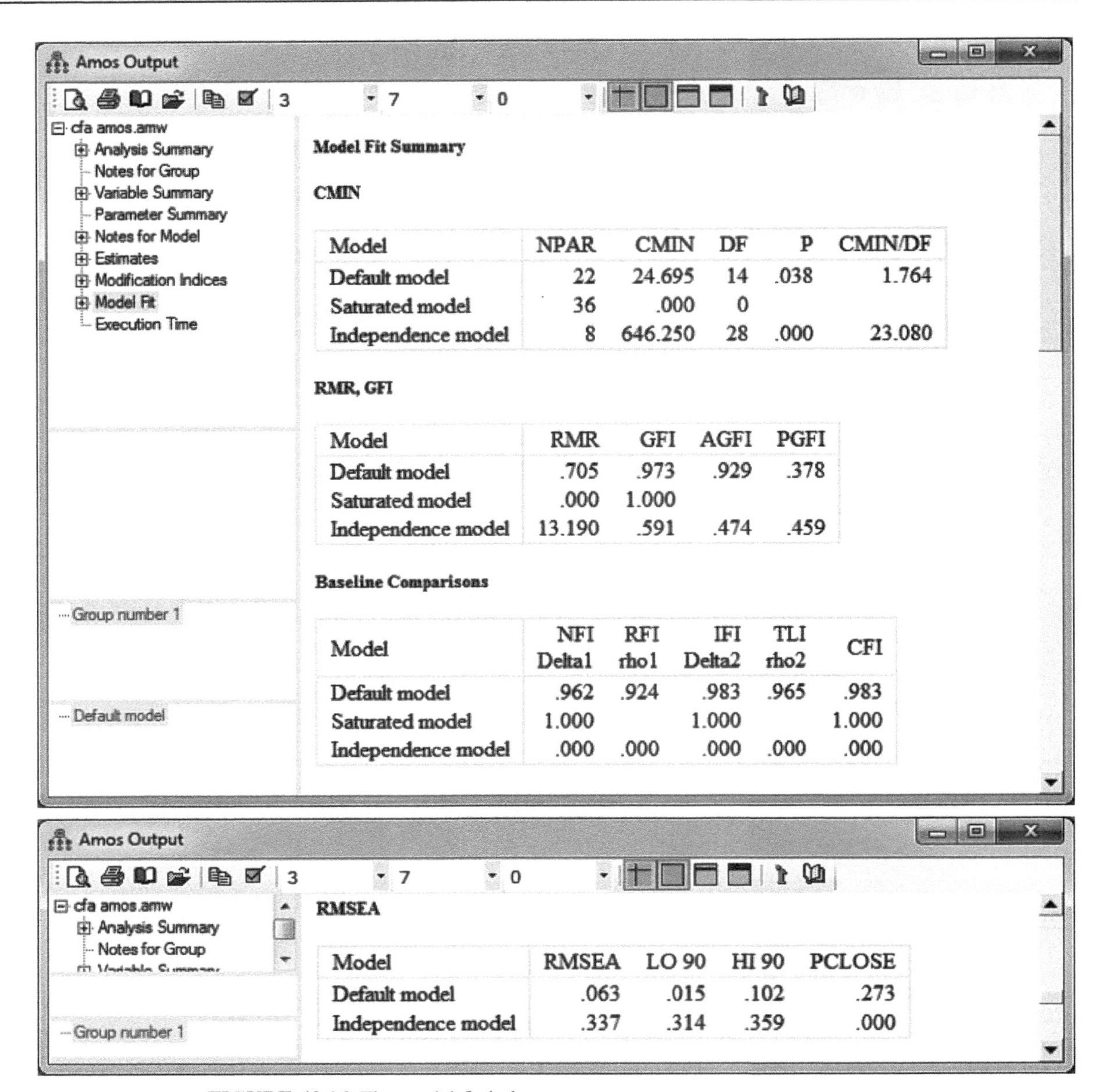

Model Fit Summary

CMIN

Model	NPAR	CMIN	DF	P	CMIN/DF
Default model	22	24.695	14	.038	1.764
Saturated model	36	.000	0		
Independence model	8	646.250	28	.000	23.080

RMR, GFI

Model	RMR	GFI	AGFI	PGFI
Default model	.705	.973	.929	.378
Saturated model	.000	1.000		
Independence model	13.190	.591	.474	.459

Baseline Comparisons

Model	NFI Delta1	RFI rho1	IFI Delta2	TLI rho2	CFI
Default model	.962	.924	.983	.965	.983
Saturated model	1.000		1.000		1.000
Independence model	.000	.000	.000	.000	.000

RMSEA

Model	RMSEA	LO 90	HI 90	PCLOSE
Default model	.063	.015	.102	.273
Independence model	.337	.314	.359	.000

FIGURE 43.16 The model fit indexes.

standardized coefficients suggest that the measured variables are good indicators of their respective factors.

Figure 43.18 presents the **Covariances** in the top table (unstandardized correlations) and the **Correlations** between the factors in the bottom table. The factors are indeed correlated, but the strengths of those relationships are at most in the moderate range and are thus perfectly acceptable for a confirmatory model (we would not want the factors correlated to such an extent that they were indistinguishable from one another, as was obtained in the example for Chapter 39).

The **Modification Indices** are shown in Figure 43.19. There appear to be no theoretically viable suggestions here, and so they can be ignored.

As an overall conclusion with respect to this component of the analysis, it appears that the measurement model is acceptable; that is, we take the hypothesized structure to be a sufficiently good fit to the data to accept the viability of the measurement model.

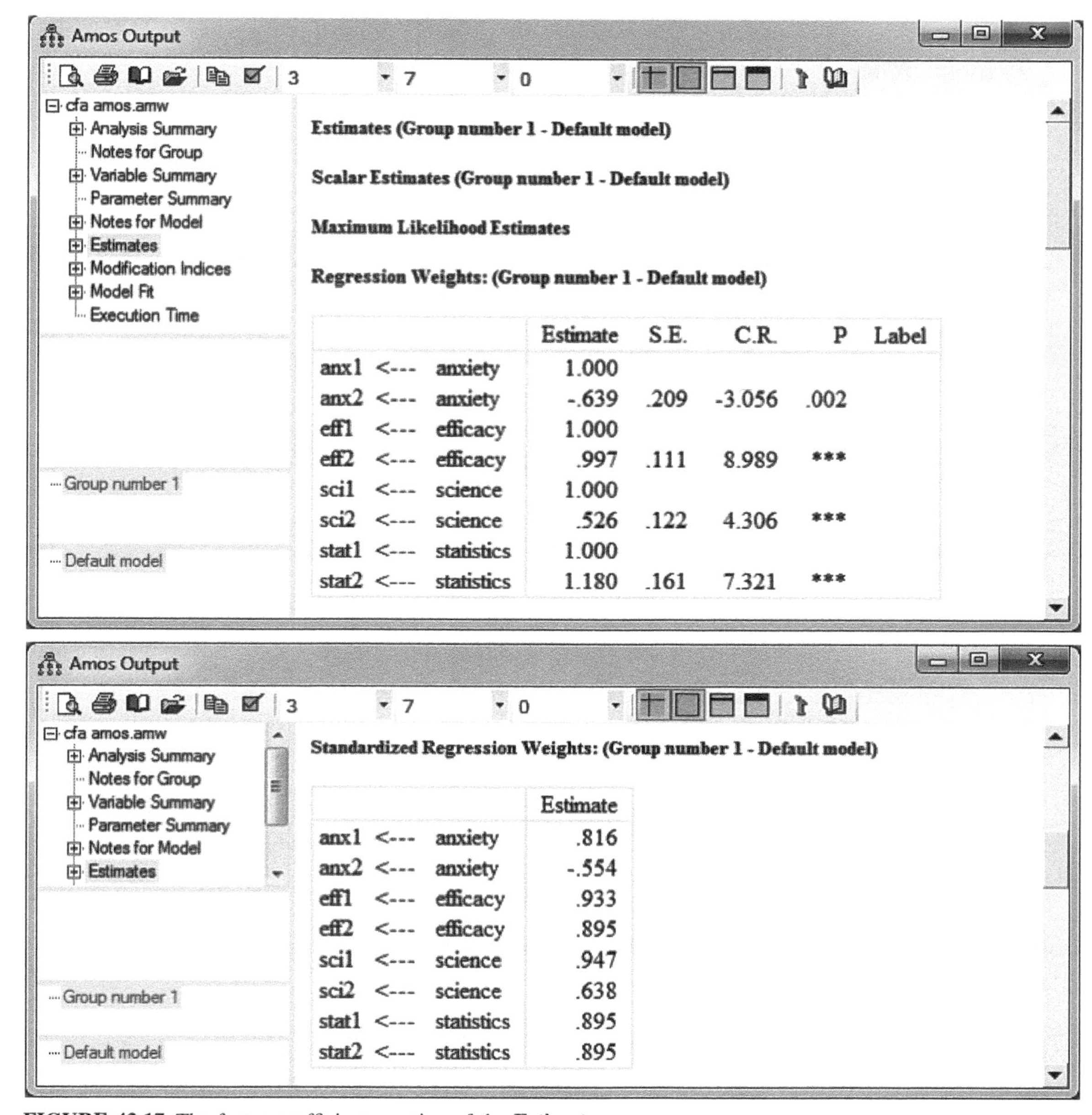

Estimates (Group number 1 - Default model)

Scalar Estimates (Group number 1 - Default model)

Maximum Likelihood Estimates

Regression Weights: (Group number 1 - Default model)

			Estimate	S.E.	C.R.	P	Label
anx1	<---	anxiety	1.000				
anx2	<---	anxiety	-.639	.209	-3.056	.002	
eff1	<---	efficacy	1.000				
eff2	<---	efficacy	.997	.111	8.989	***	
sci1	<---	science	1.000				
sci2	<---	science	.526	.122	4.306	***	
stat1	<---	statistics	1.000				
stat2	<---	statistics	1.180	.161	7.321	***	

Standardized Regression Weights: (Group number 1 - Default model)

			Estimate
anx1	<---	anxiety	.816
anx2	<---	anxiety	-.554
eff1	<---	efficacy	.933
eff2	<---	efficacy	.895
sci1	<---	science	.947
sci2	<---	science	.638
stat1	<---	statistics	.895
stat2	<---	statistics	.895

FIGURE 43.17 The factor coefficients portion of the **Estimates** screen.

Given this conclusion, we can now proceed with specifying and evaluating the structural model proper.

43.7 EVALUATING THE STRUCTURAL MODEL: DRAWING THE MODEL

We start drawing the structural model in exactly the same way described in Section 43.4, in that we wish to draw the four latent variables and their indicators. Briefly, we open the **statistics achievement** data file, select **Analyze → IBM SPSS Amos** from the main menu, select **View → Interface properties** from the main menu, choose

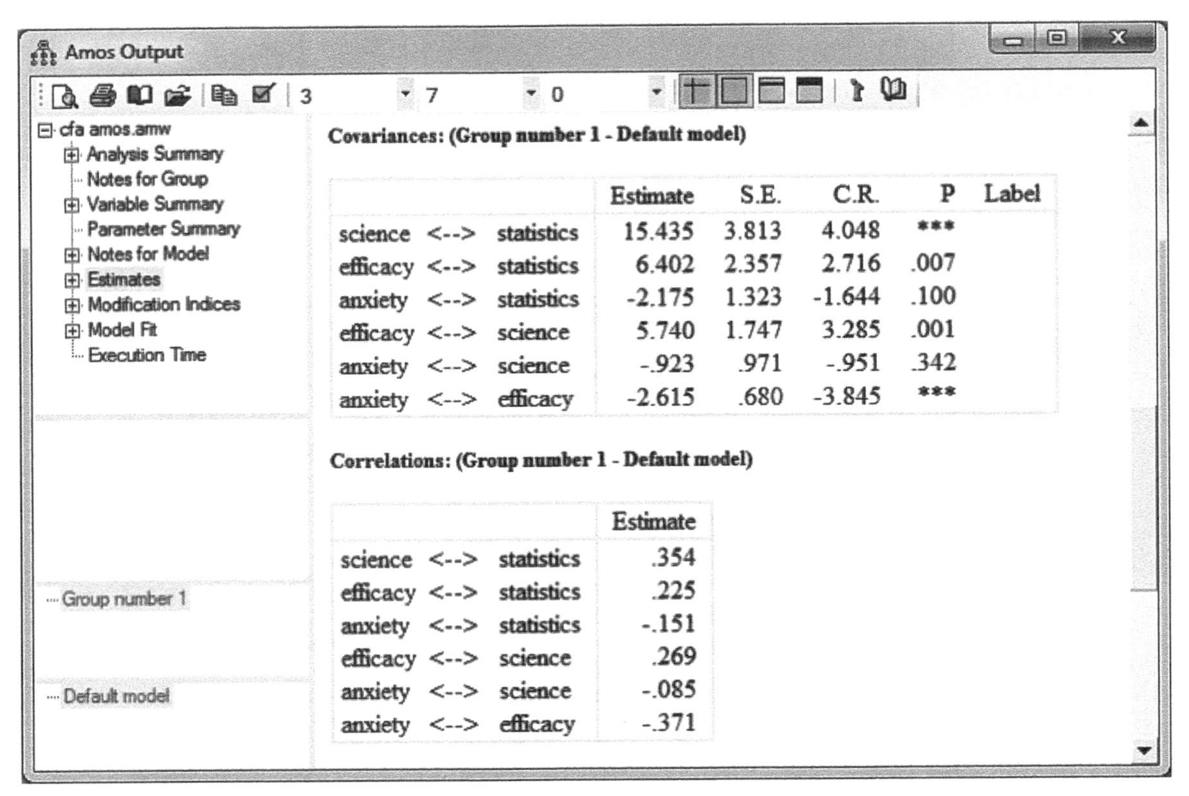

Covariances: (Group number 1 - Default model)

			Estimate	S.E.	C.R.	P	Label
science	<-->	statistics	15.435	3.813	4.048	***	
efficacy	<-->	statistics	6.402	2.357	2.716	.007	
anxiety	<-->	statistics	-2.175	1.323	-1.644	.100	
efficacy	<-->	science	5.740	1.747	3.285	.001	
anxiety	<-->	science	-.923	.971	-.951	.342	
anxiety	<-->	efficacy	-2.615	.680	-3.845	***	

Correlations: (Group number 1 - Default model)

			Estimate
science	<-->	statistics	.354
efficacy	<-->	statistics	.225
anxiety	<-->	statistics	-.151
efficacy	<-->	science	.269
anxiety	<-->	science	-.085
anxiety	<-->	efficacy	-.371

FIGURE 43.18 Covariances and correlations.

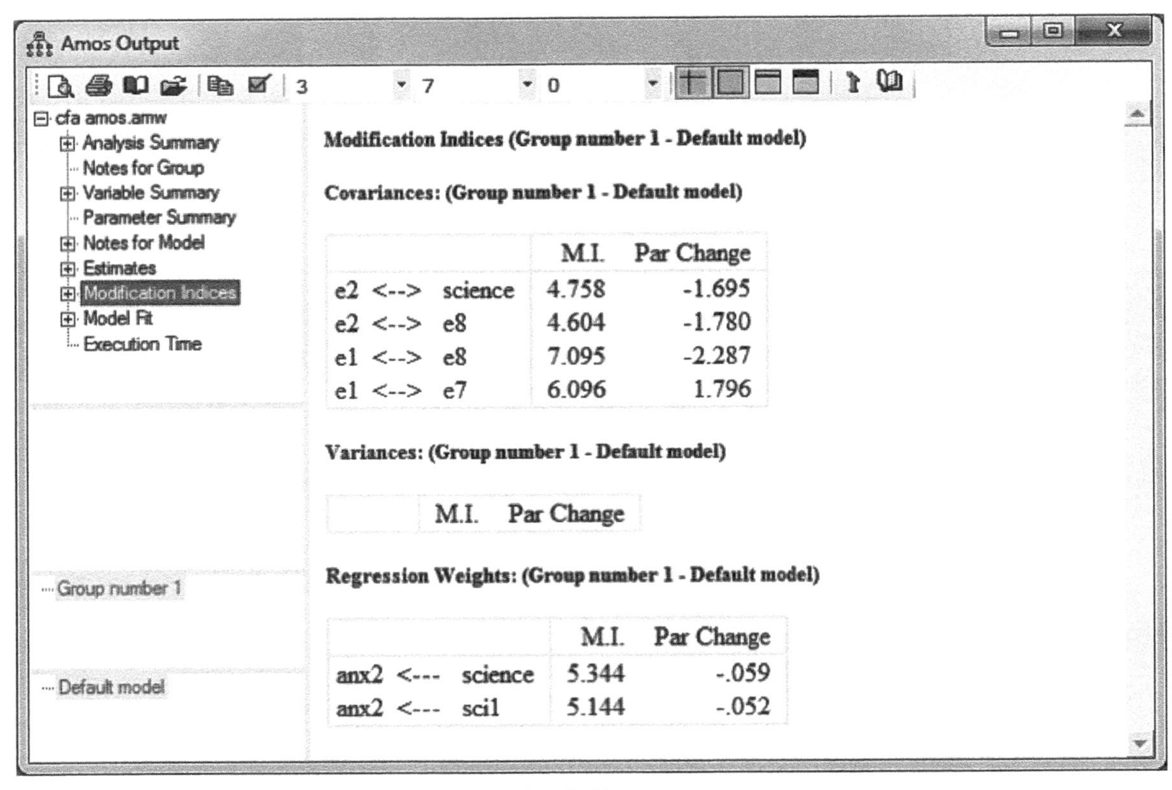

Modification Indices (Group number 1 - Default model)

Covariances: (Group number 1 - Default model)

			M.I.	Par Change
e2	<-->	science	4.758	-1.695
e2	<-->	e8	4.604	-1.780
e1	<-->	e8	7.095	-2.287
e1	<-->	e7	6.096	1.796

Variances: (Group number 1 - Default model)

	M.I.	Par Change

Regression Weights: (Group number 1 - Default model)

			M.I.	Par Change
anx2	<---	science	5.344	-.059
anx2	<---	sci1	5.144	-.052

FIGURE 43.19 The **Modification Indices**.

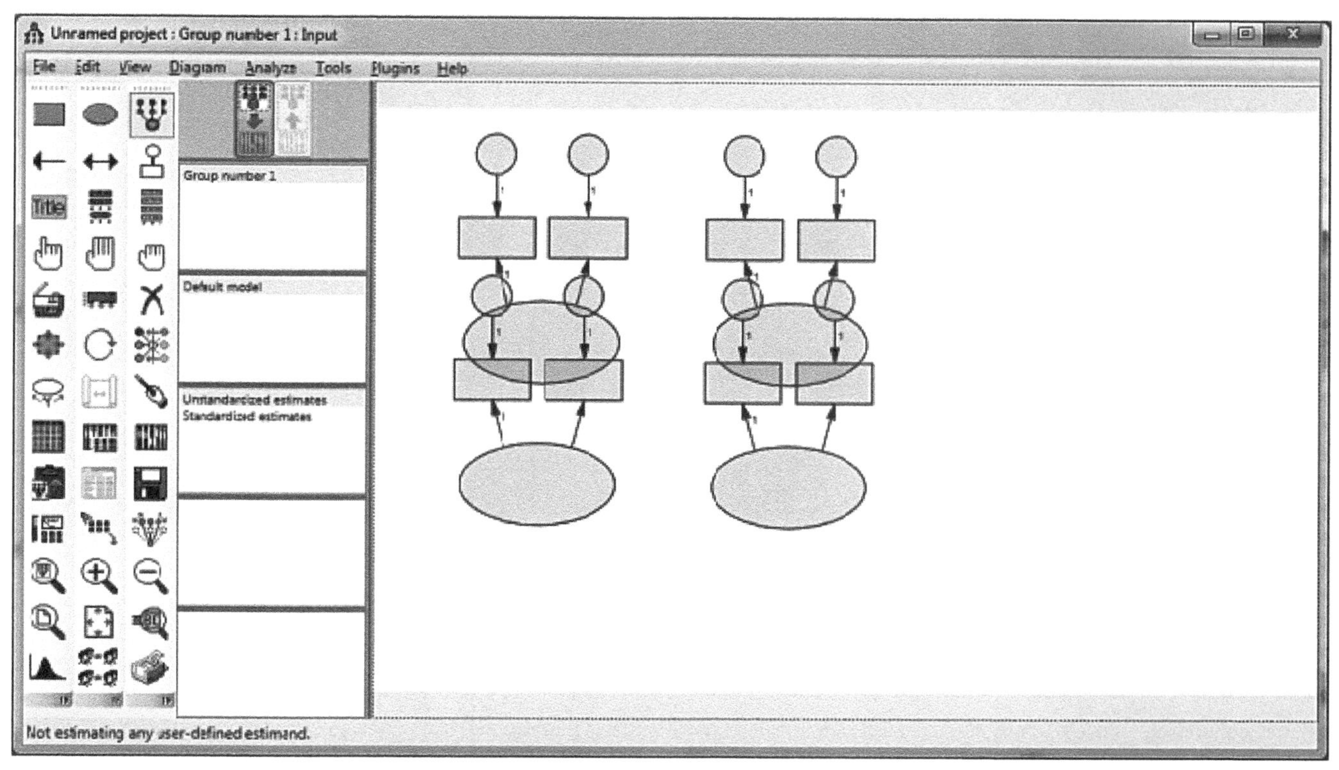

FIGURE 43.20 The first pass at drawing the model.

Landscape-Legal under **Paper Size** to switch to landscape orientation, and select the icon (the circle with squares and more circles) representing **Draw a latent variable or add an indicator to a latent variable** from the icon toolbar menu. Configure the factors and their indicator variables in a two-row by two-column array as shown in Figure 43.20.

On our screen, the second row drawings interfere with the first row drawings, but even if they did not (on a larger monitor), we would want to have the indicator variables in the second row facing down rather than up (for aesthetic reasons). To accomplish this, place the cursor inside one of the ovals in the second row and right-click. From the pop-up menu, select **Rotate** (see Figure 43.21). The cursor now includes the word **rotate**. Click twice. This changes the orientation of the indicator variables. With the **Rotate** function still in effect, repeat the operation for the next latent variable to achieve the result shown in Figure 43.22. Place the cursor in a blank area and right-click to end the **Rotate** function.

Drawing the details of the structural model uses the same procedures as described in Section 42.2. Select the **Draw paths (single headed arrow)** from the icon toolbar as shown in Figure 43.23. With this tool activated, draw each of the paths in the model so that the structure mirrors that in Figure 43.24.

We next draw the correlation (the covariance) between the two exogenous latent variables. From the icon toolbar, select the **Draw covariances (double headed arrow)** as shown in Figure 43.25. With this function activated, place the cursor on the left border of the lower left latent variable and click and drag to the upper latent variable as shown in Figure 43.26.

Each endogenous variable needs to be associated with an error term. We select from the icon toolbar the **Add a unique variable to an existing variable**, as shown in Figure 43.27. With this tool activated, place the cursor inside one of the latent variables in the second column and click thrice (once to create the **unique variable** and twice more

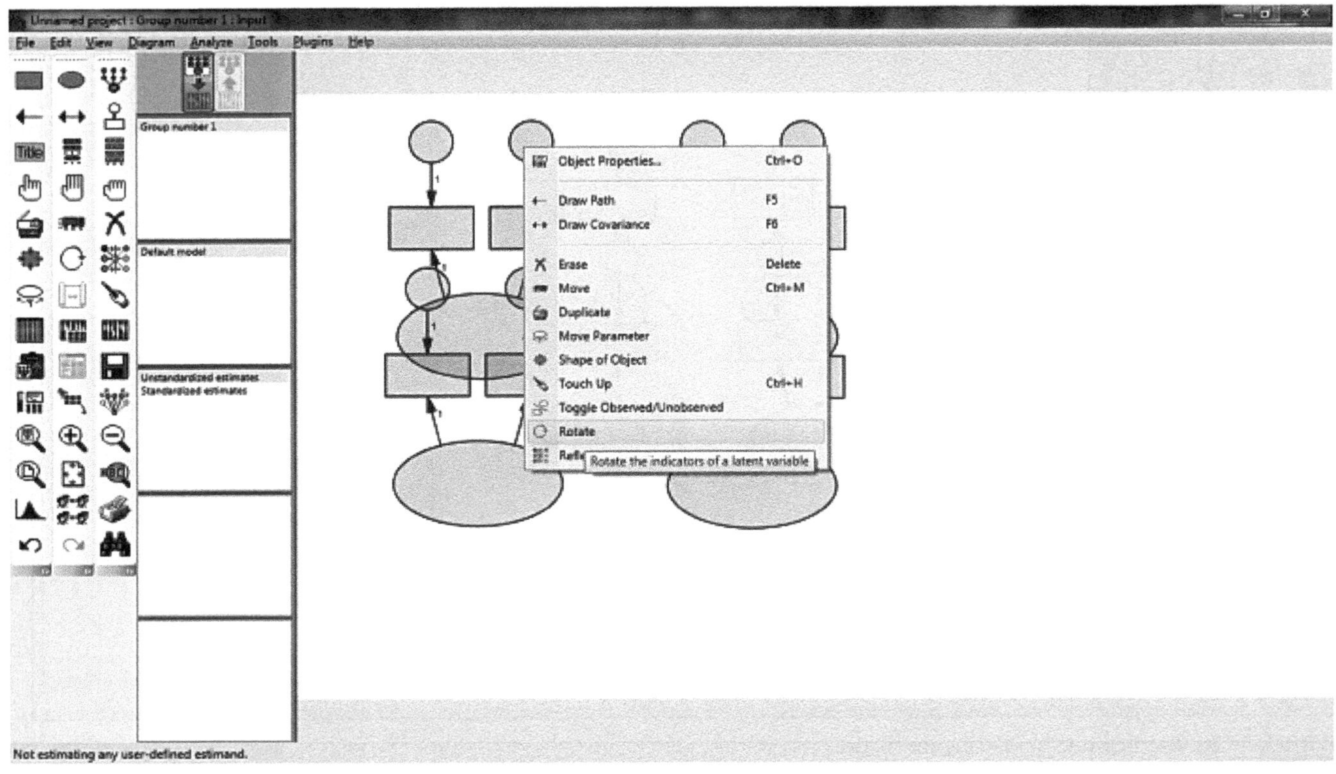

FIGURE 43.21 By rotating the indicator variables and their error terms in the second row, we can make the drawing much clearer.

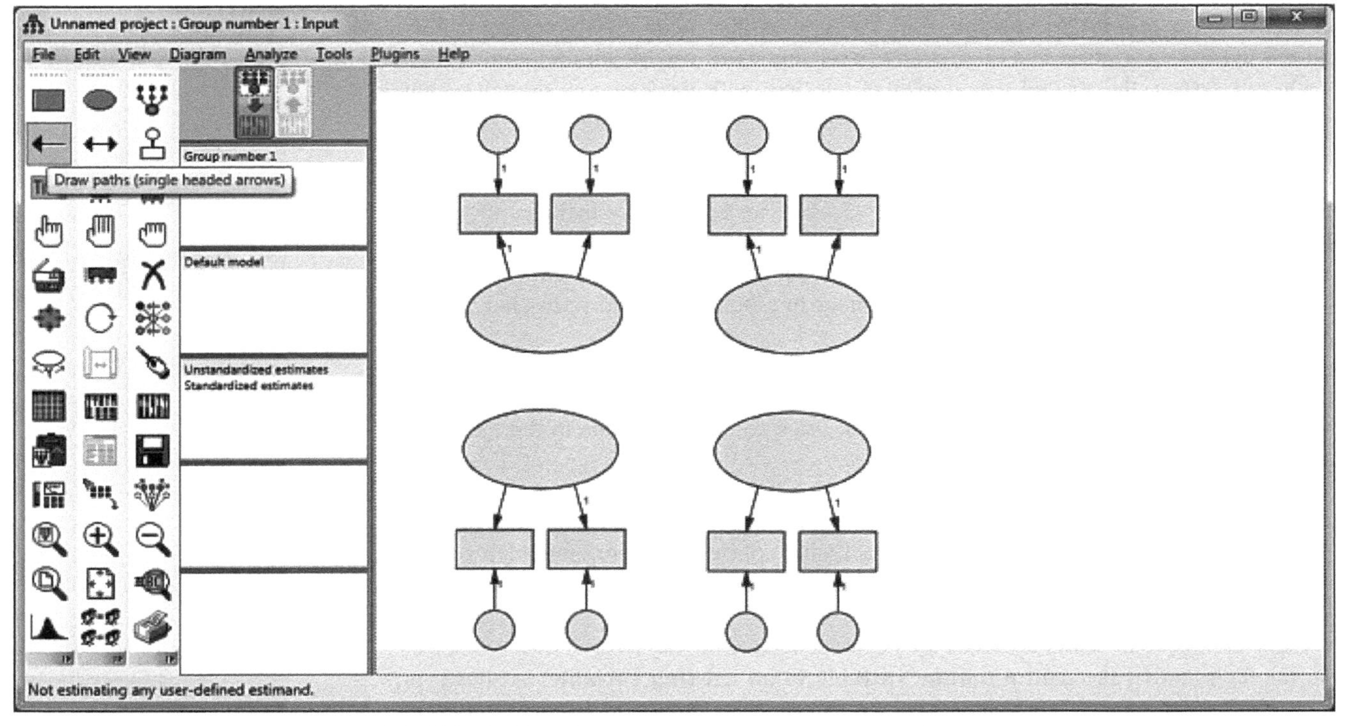

FIGURE 43.22 After the rotation, the diagram is much clearer.

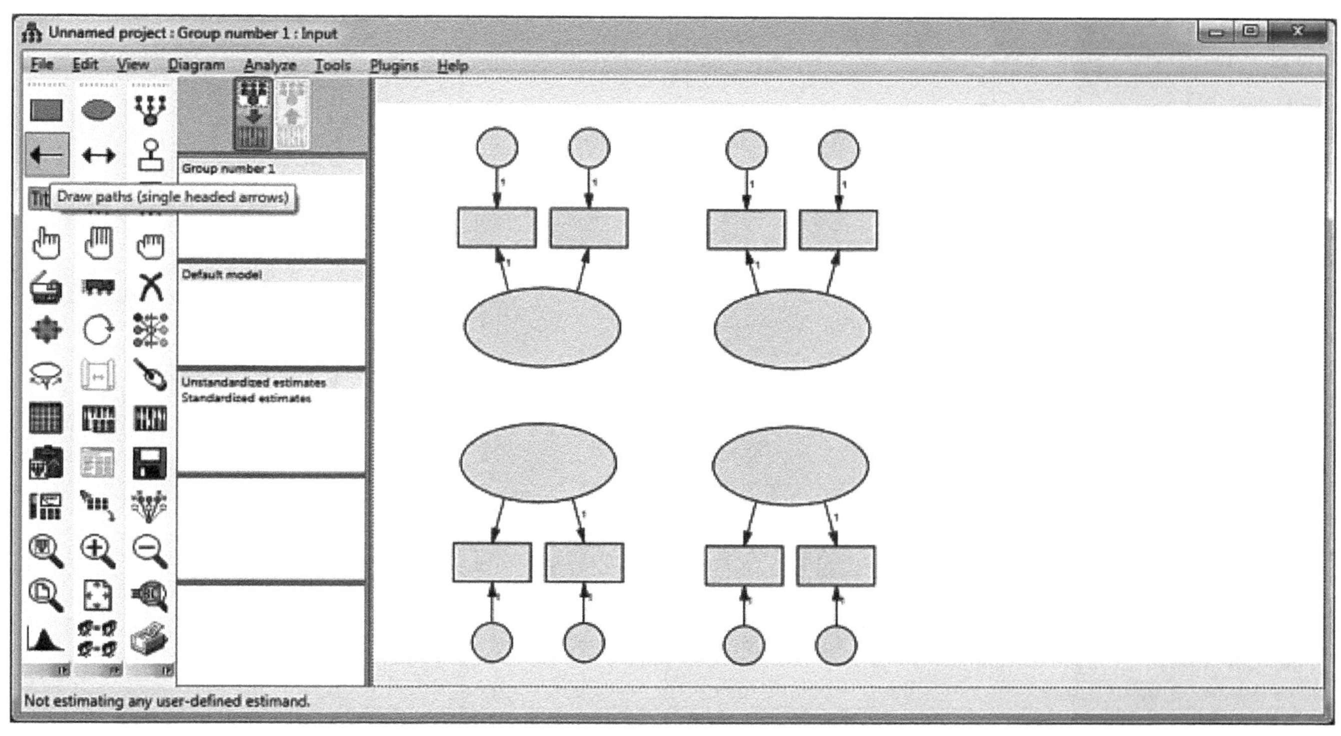

FIGURE 43.23 Select the **Draw paths (single headed arrow)** from the icon toolbar.

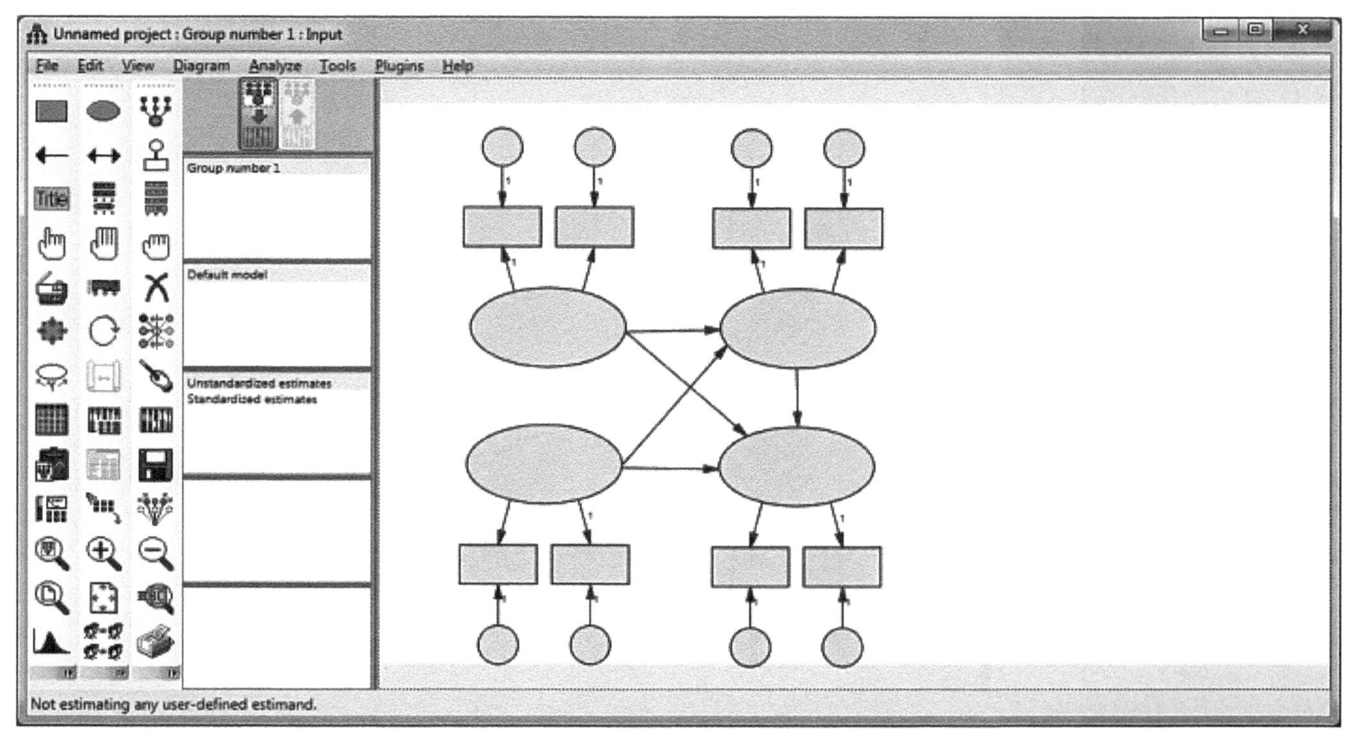

FIGURE 43.24 The paths are now drawn.

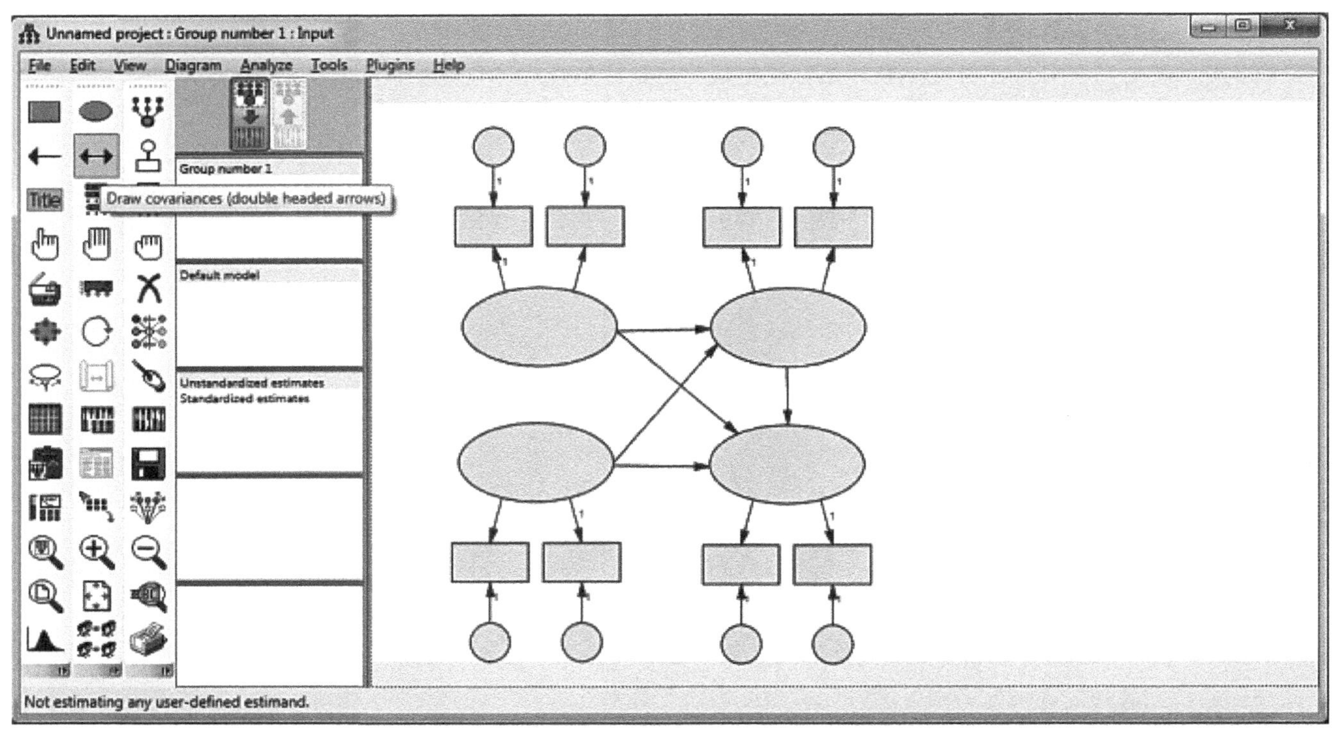

FIGURE 43.25 Select the **Draw covariances (double headed arrow)** from the icon toolbar.

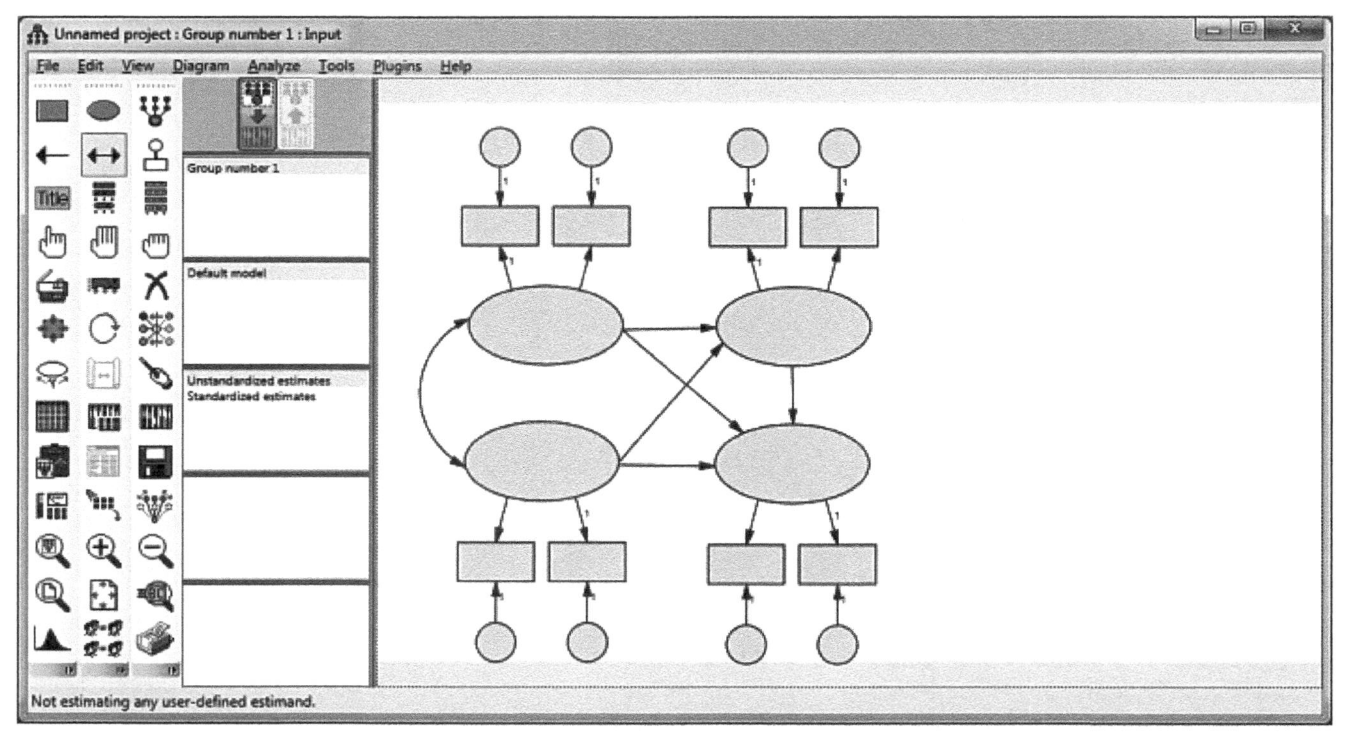

FIGURE 43.26 The covariance path between the two latent variables is now drawn.

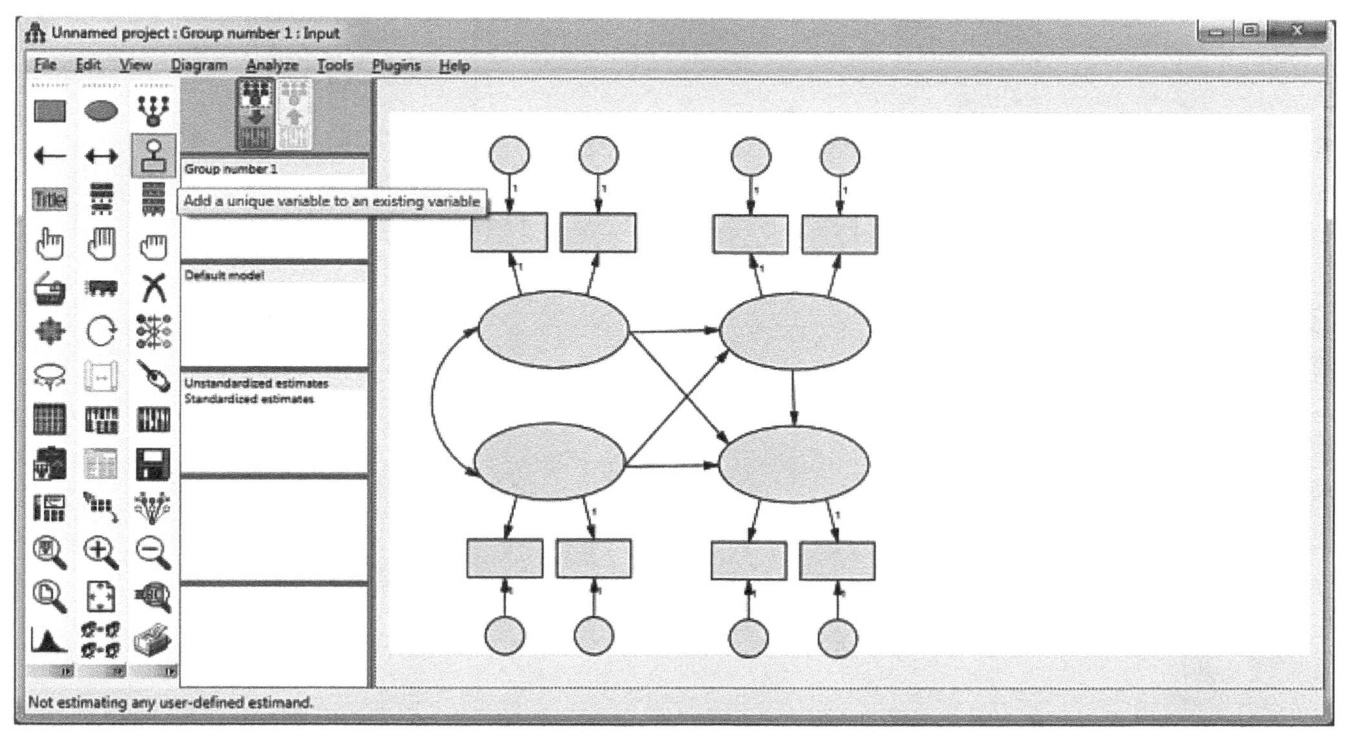

FIGURE 43.27 Select the **Add a unique variable to an existing variable** from the icon toolbar.

to rotate it out of the way). Doing the same for the other endogenous latent variable produces the configuration shown in Figure 43.28.

Our next task is to associate each measured variable with the proper variable in the data file. From the main menu, select **View ➔ Variables in Dataset**. This opens the **Variables in Dataset** window. Select each variable in turn and drag it to the appropriate figure so that all observed variables are properly named as in Figure 43.29. When finished, close the **Variables in Dataset** window.

We name the factors by double-clicking inside each. This activates the **Object Properties** window. After typing each name in turn, we obtain the result shown in Figure 43.30.

Naming the **unique variables** is the last step in drawing the model. It is accomplished by selecting **Plugins ➔ Name Unobserved Variables** from the main menu (see Figure 43.31). On selecting this command, the **unique variables** will be given the sequential numbers **e1** through **e10**, as shown in Figure 43.32.

43.8 EVALUATING THE STRUCTURAL MODEL: ANALYSIS SETUP

From the main menu, we select **View ➔ Analysis Properties** to open the **Analysis Properties** window, as shown in Figure 43.33. Select the **Output** tab and check **Standardized estimates**, **Squared multiple correlations**, **Modification indices**, and **Indirect, direct & total effects**. To execute the analysis, select **Analyze ➔ Calculate Estimates** from the main menu and **Save** as needed.

In order for IBM SPSS Amos to perform the analysis, the model must be *identified* (Bollen, 1989; Byrne, 2010; Meyers et al., 2013; Raykov & Marcoulides, 2008). Briefly, there must be more known or nonredundant elements (e.g., covariances and variances of

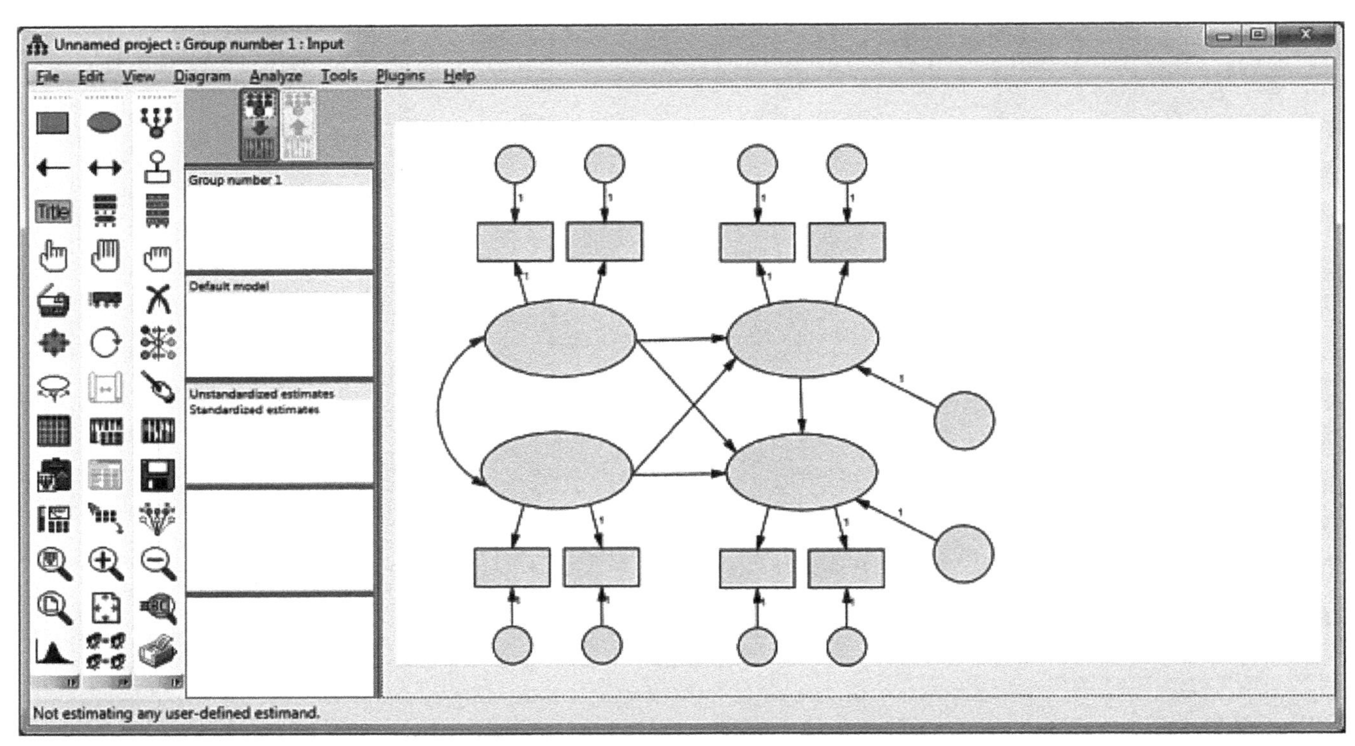

FIGURE 43.28 Both **unique variables** for the endogenous latent variables are now in place.

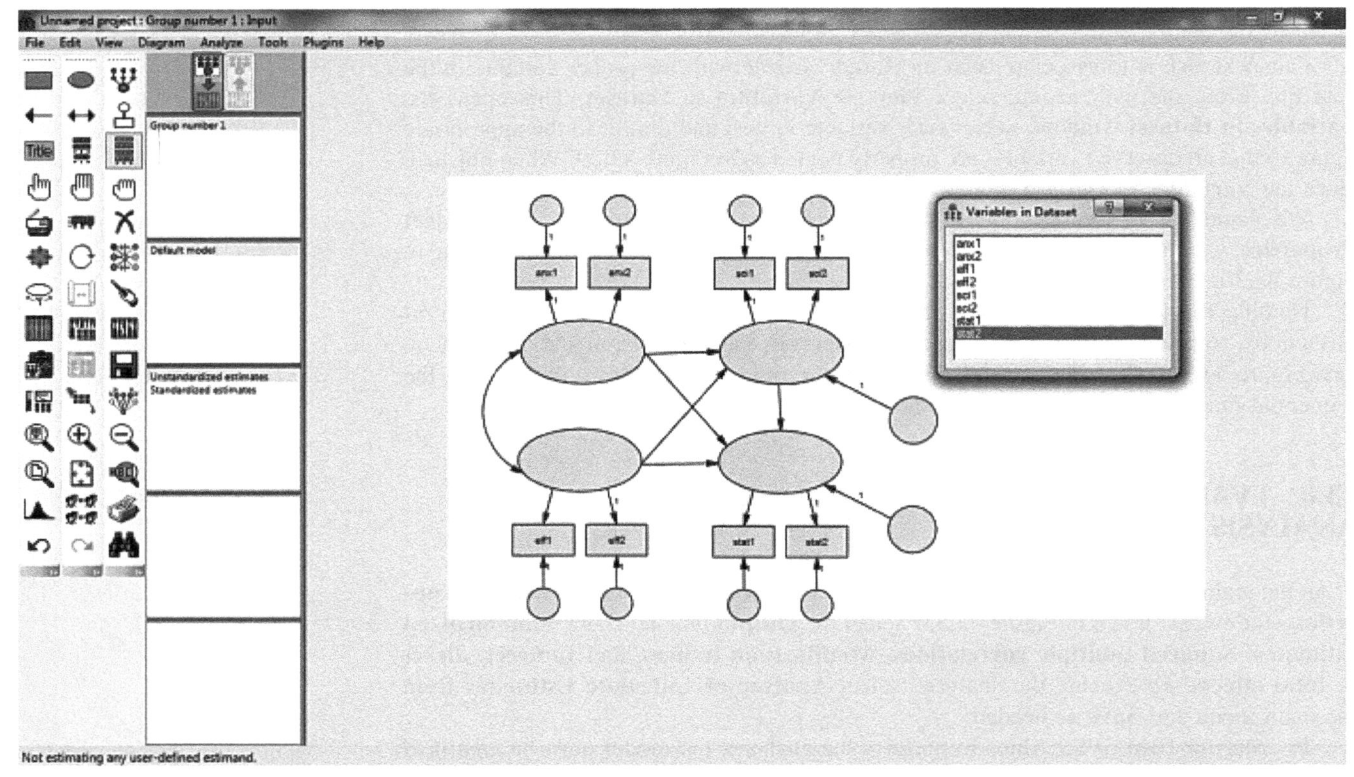

FIGURE 43.29 All of the measured variables in the model have been associated with the proper observed variables.

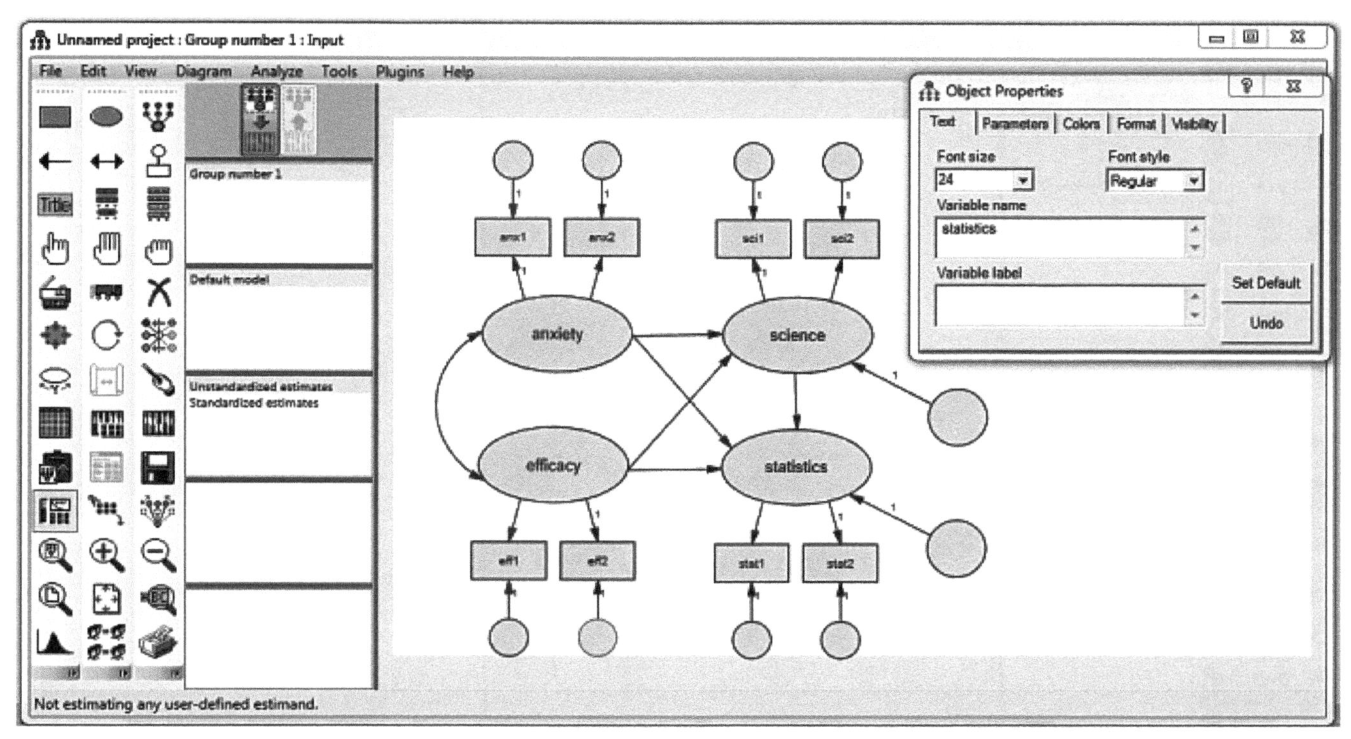

FIGURE 43.30 All of the latent variables are now named.

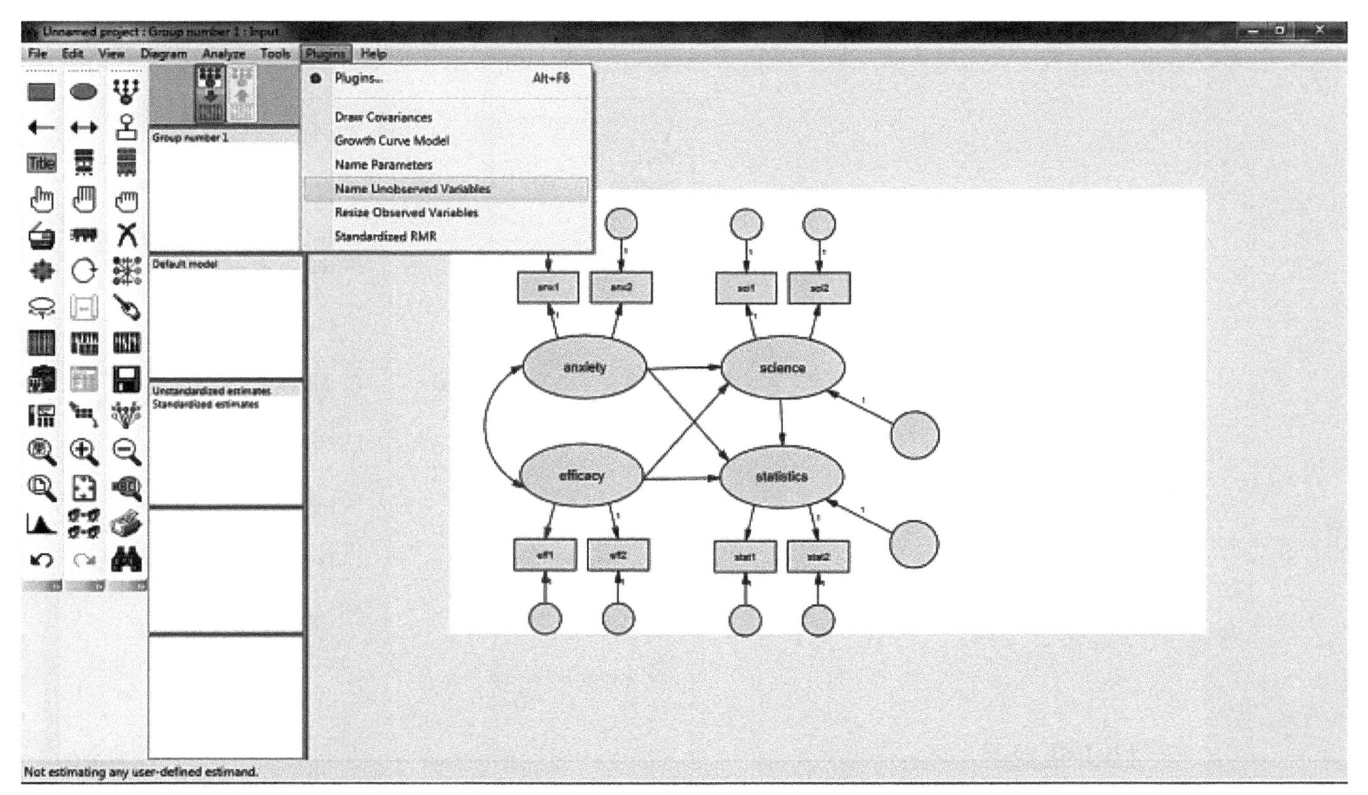

FIGURE 43.31 To name the latent error variables, we select **Plugins ➔ Name Unobserved Variables** from the main menu.

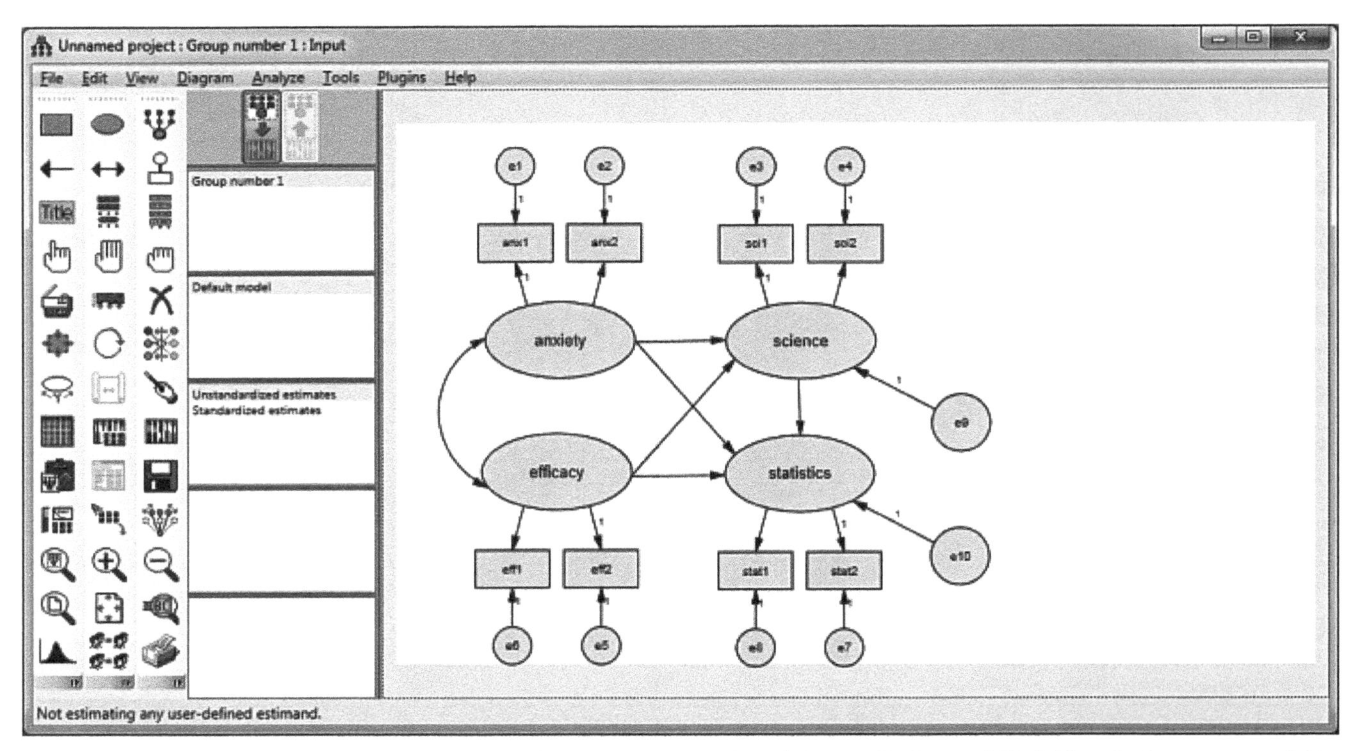

FIGURE 43.32 On selecting **Plugins** ➔ **Name Unobserved Variables**, the latent error variables are labeled **e1** through **e10**.

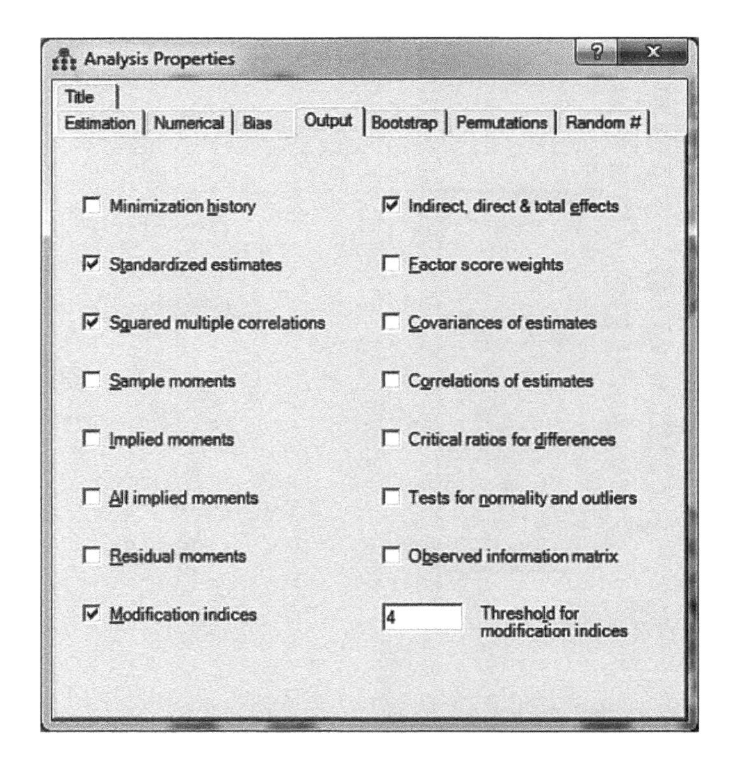

FIGURE 43.33
The **Analysis Properties** window.

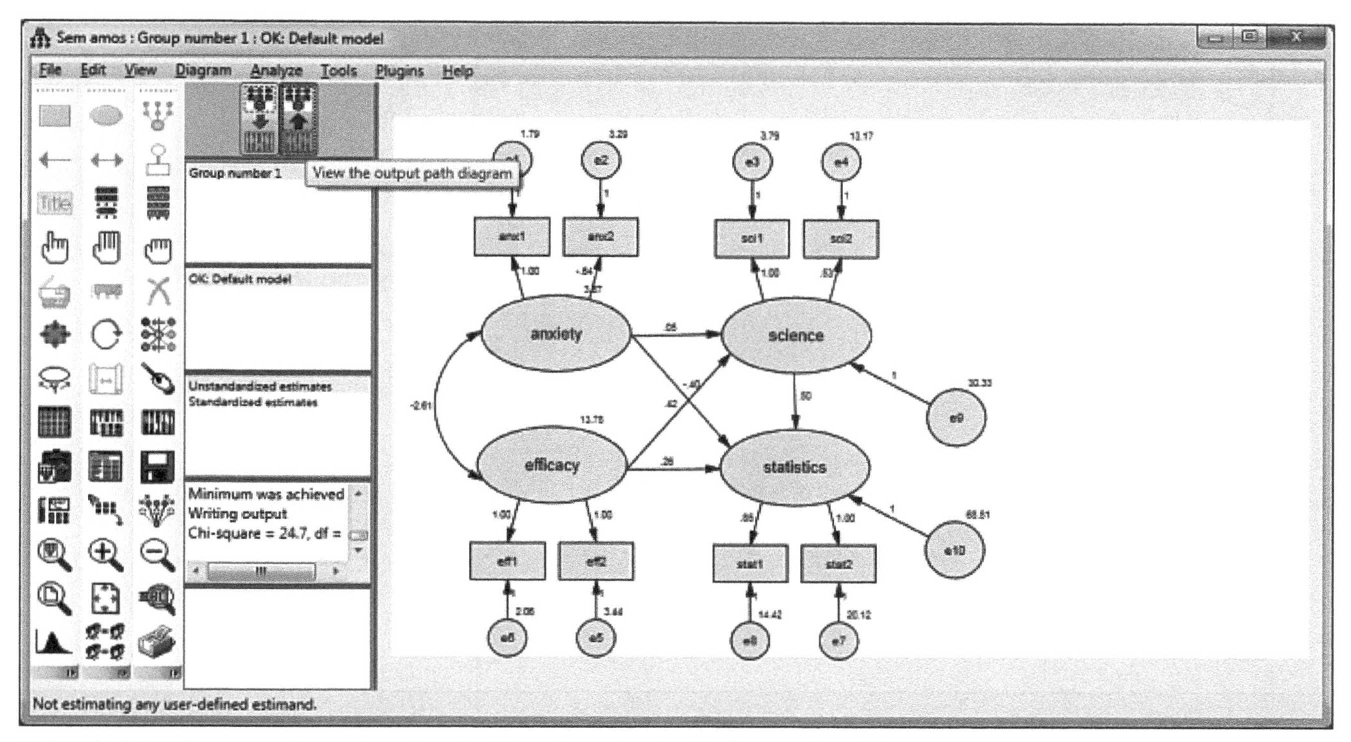

FIGURE 43.34 To access the results, select the **View the output path diagram** icon (the one on the right) in the top panel immediately to the left of the drawing area.

the measured variables) rather than parameters to be estimated (e.g., variances of the latent variables, pattern/structure coefficients from the latent variable to the indicator variables) associated with the model. Subtracting the number of parameters to be estimated from the number of known elements yields the *degrees of freedom* of the model. The goal is for the model to be *overidentified*, that is, for its degrees of freedom to be greater than zero (i.e., have a positive value).

If the model is not properly identified, IBM SPSS Amos will request that one or more additional paths be constrained (thereby reducing the number of parameters to be estimated). If researchers receive such a message, they should determine which path they wish to constrain, place the cursor on that path, double-click to produce the **Object Properties** pop-up window, select the **Parameters** tab, and in the **Regression weight** panel type in the value **1**. Finally, close the dialog pop-up window and attempt to execute the analysis again.

43.9 EVALUATING THE STRUCTURAL MODEL: ANALYSIS OUTPUT

To access the results of the analysis, select the **View the output path diagram** icon (the one on the right) in the top output panel immediately to the left of the drawing area as shown in Figure 43.34. What is displayed is the path diagram with the **Unstandardized estimates**.

To obtain the output in tabular form, select from the main menu **View → Text Output**. This action opens the **Notes for Model** screen shown in Figure 43.35. The chi-square results shown in this screen are identical to those obtained in the confirmatory factor analysis, as the model fit output addresses the measurement model. We show the other fit indexes of relevance in Figure 43.36 (they duplicate our confirmatory factor

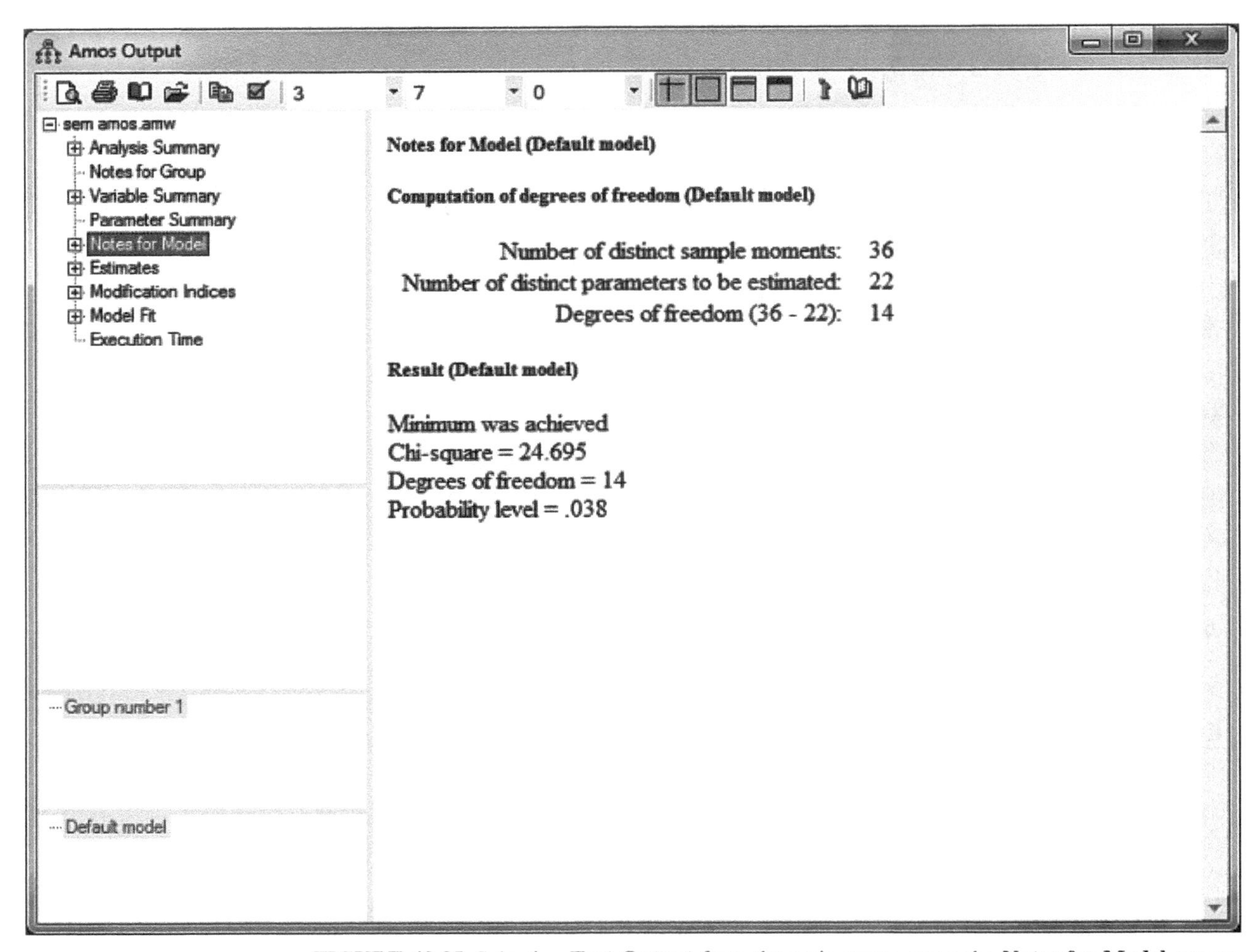

FIGURE 43.35 Selecting **Text Output** from the main menu opens the **Notes for Model** screen.

analysis results; we performed the two separate analyses to make it clear that the structural model was dependent on the integrity of the measurement model).

Selecting **Estimates** in the left panel presents the unstandardized regression coefficients (labeled as **Regression Weights**), the SE associated with the coefficients (**S.E.**), and the **C.R.**, as shown in Figure 43.37. The critical ratio is the coefficient divided by the SE and operates something as a z statistic with absolute values of 1.96 or greater, indicating statistical significance based on an alpha level of .05. The standardized regression coefficients are shown in Figure 43.38.

Each coefficient is tested for statistical significance (reported under the column labeled **P** in Figure 43.36). The triple asterisks (***) represent a probability level of $p < .001$. The lower set of coefficients in both the unstandardized and standardized tables duplicate what we obtained in the confirmatory analysis. What is new in the present analysis is the upper set of coefficients relating the latent variables to each other. Although the fit indexes, with the exception of chi-square, suggested a good measurement model fit, only two of the five hypothesized paths were associated with statistically significant coefficients, neither of which involved **anxiety**.

- The path from **efficacy** to **science** yielded an unstandardized coefficient of **.425** and a standardized coefficient of **.276** ($p =$ **.001**).

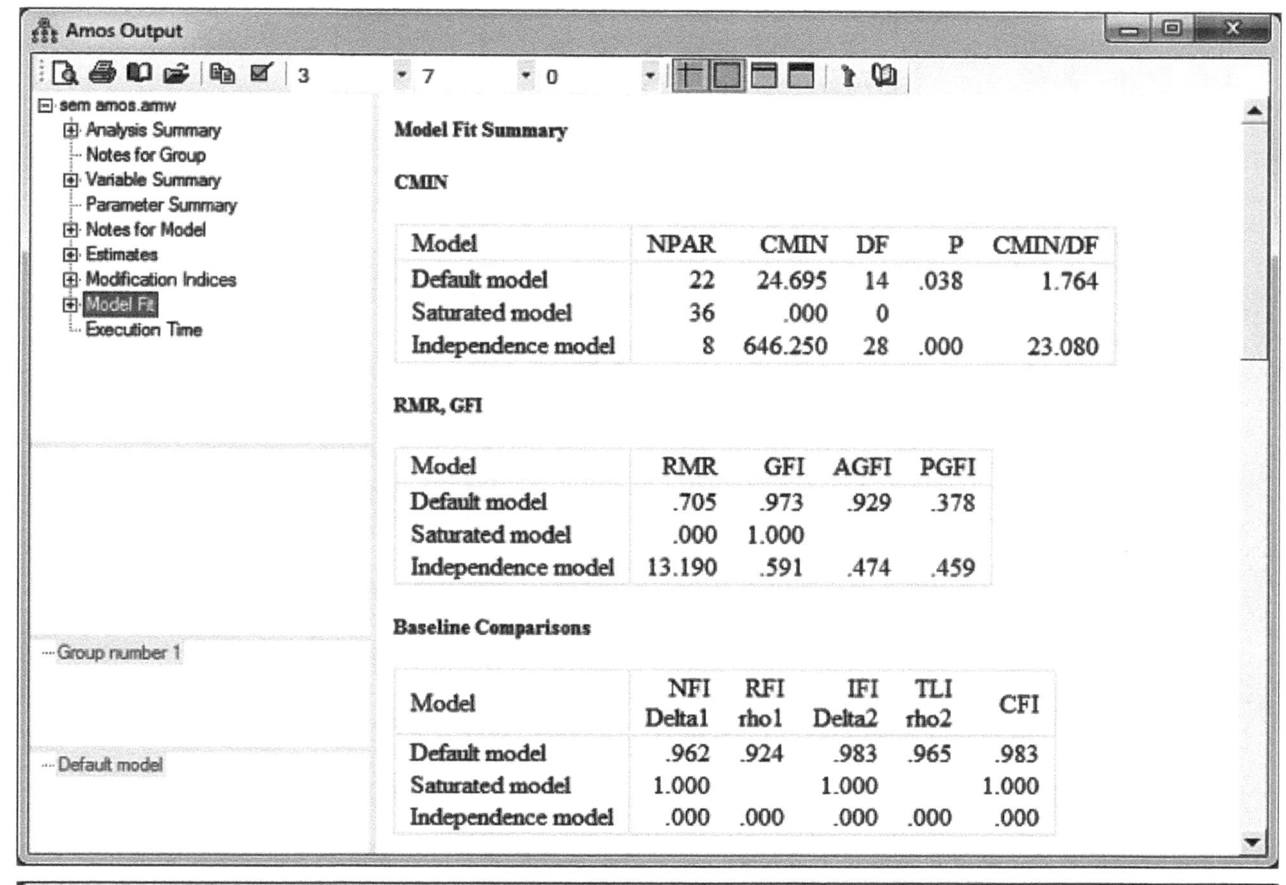

Model Fit Summary

CMIN

Model	NPAR	CMIN	DF	P	CMIN/DF
Default model	22	24.695	14	.038	1.764
Saturated model	36	.000	0		
Independence model	8	646.250	28	.000	23.080

RMR, GFI

Model	RMR	GFI	AGFI	PGFI
Default model	.705	.973	.929	.378
Saturated model	.000	1.000		
Independence model	13.190	.591	.474	.459

Baseline Comparisons

Model	NFI Delta1	RFI rho1	IFI Delta2	TLI rho2	CFI
Default model	.962	.924	.983	.965	.983
Saturated model	1.000		1.000		1.000
Independence model	.000	.000	.000	.000	.000

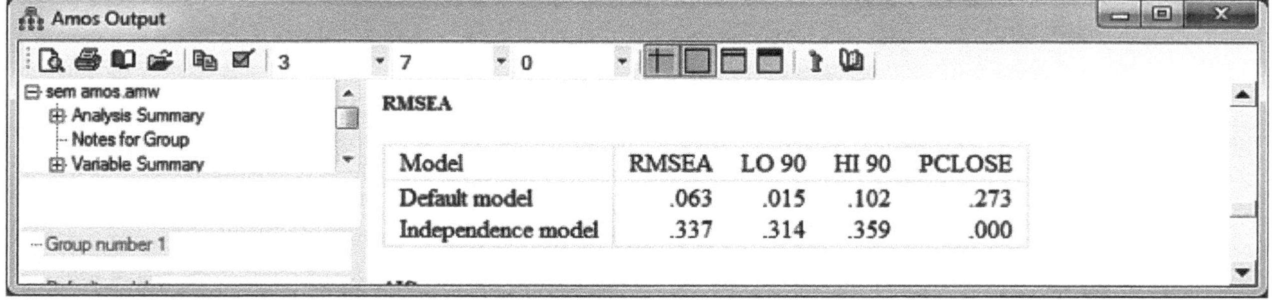

RMSEA

Model	RMSEA	LO 90	HI 90	PCLOSE
Default model	.063	.015	.102	.273
Independence model	.337	.314	.359	.000

FIGURE 43.36 The **Model Fit** indexes of relevance.

- The path from **science** to **statistics** yielded an unstandardized coefficient of **.499** and a standardized coefficient of **.317** (p = **.003**).

Figure 43.39 presents the **Covariance** and **Correlation** between the two exogenous latent variables of **anxiety** and **efficacy**. The two were correlated to −**.371**.

Figure 43.40 presents the **Squared Multiple Correlations** for all endogenous variables. Of relevance are the two latent variables, **science** and **statistics**, shown in the first two rows. The model could explain approximately 7% and 15% of the variance of these two variables, respectively.

Figure 43.41 shows the results of the calculations for the indirect effects based on the standardized coefficients. Given the statistical significance of the set of paths, the only relevant indirect path is from **efficacy** through **science** to **statistics**. Hand-calculating

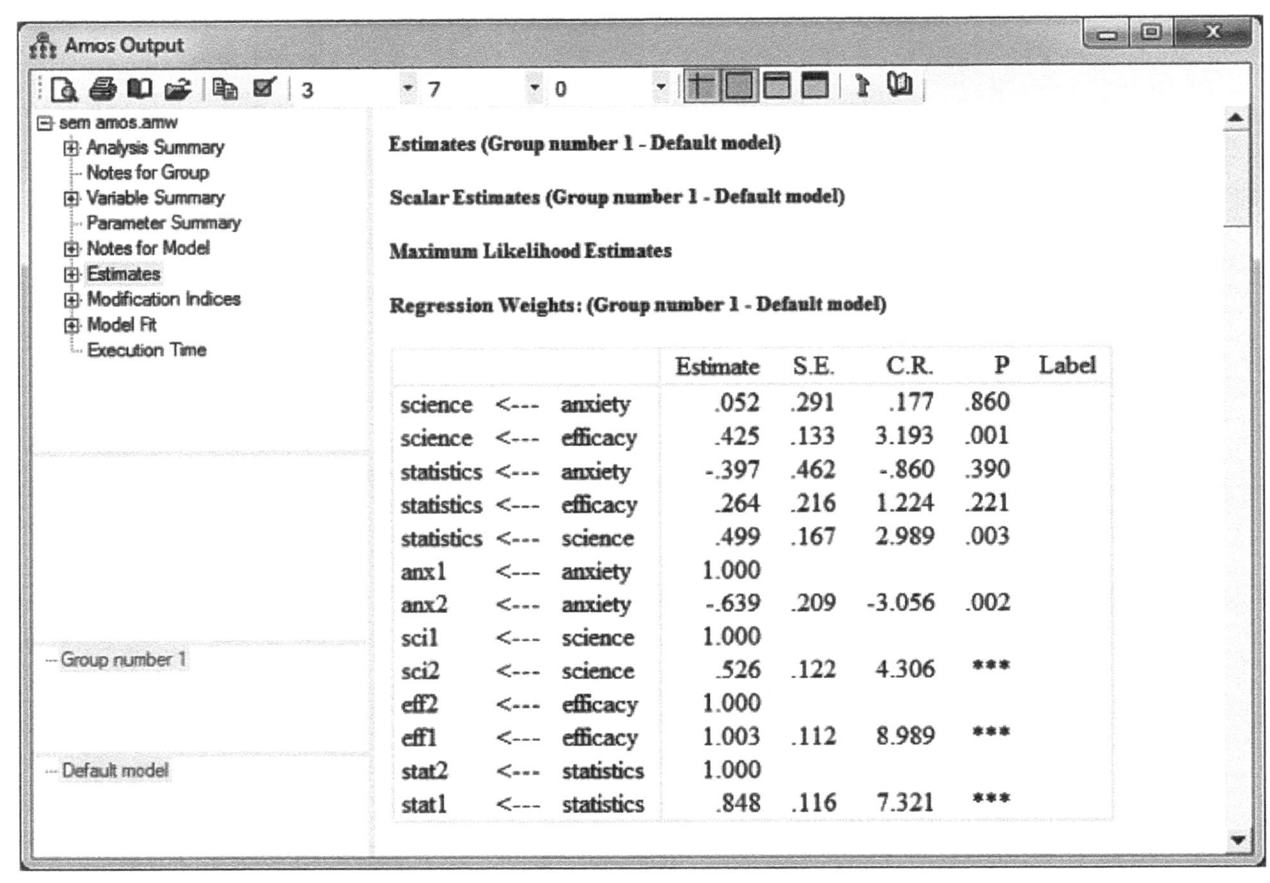

Estimates (Group number 1 - Default model)

Scalar Estimates (Group number 1 - Default model)

Maximum Likelihood Estimates

Regression Weights: (Group number 1 - Default model)

			Estimate	S.E.	C.R.	P	Label
science	<---	anxiety	.052	.291	.177	.860	
science	<---	efficacy	.425	.133	3.193	.001	
statistics	<---	anxiety	-.397	.462	-.860	.390	
statistics	<---	efficacy	.264	.216	1.224	.221	
statistics	<---	science	.499	.167	2.989	.003	
anx1	<---	anxiety	1.000				
anx2	<---	anxiety	-.639	.209	-3.056	.002	
sci1	<---	science	1.000				
sci2	<---	science	.526	.122	4.306	***	
eff2	<---	efficacy	1.000				
eff1	<---	efficacy	1.003	.112	8.989	***	
stat2	<---	statistics	1.000				
stat1	<---	statistics	.848	.116	7.321	***	

FIGURE 43.37 The unstandardized regression coefficients.

this value (multiplying the two path coefficients) yields **.276** * **.317** or .087 (on our hand calculator). The more precise value of **.088** is found in the table at the coordinate of **efficacy** and **statistics** (the first entry in the second row).

The **Modification Indices** are shown in Figure 43.42. None of these additions to the model appear theoretically justifiable and so we will not act on them.

43.10 EVALUATING THE STRUCTURAL MODEL: SYNTHESIS

Although the measurement model appeared to represent an acceptable fit to the data, three of the five hypothesized paths weaving the latent variables together into a coherent structure were not statistically significant. **Anxiety** was not associated with any statistically significant path (at least with **efficacy** in the model), and **efficacy** appeared to influence **statistics** only indirectly through **science** (at least with **anxiety** in the model).

Most researchers would likely take two steps at this juncture to productively move forward in their research program: (a) they would probably trim the model down in a thoughtful manner and rerun the procedure as an exploratory analysis and (b) then they might rethink the dynamics underlying the relationships between the variables, perhaps consider other variables as predictors, collect new data on a revised model that they might envision, and test the fit of their new model on the newly collected data. We take the first of these two steps here.

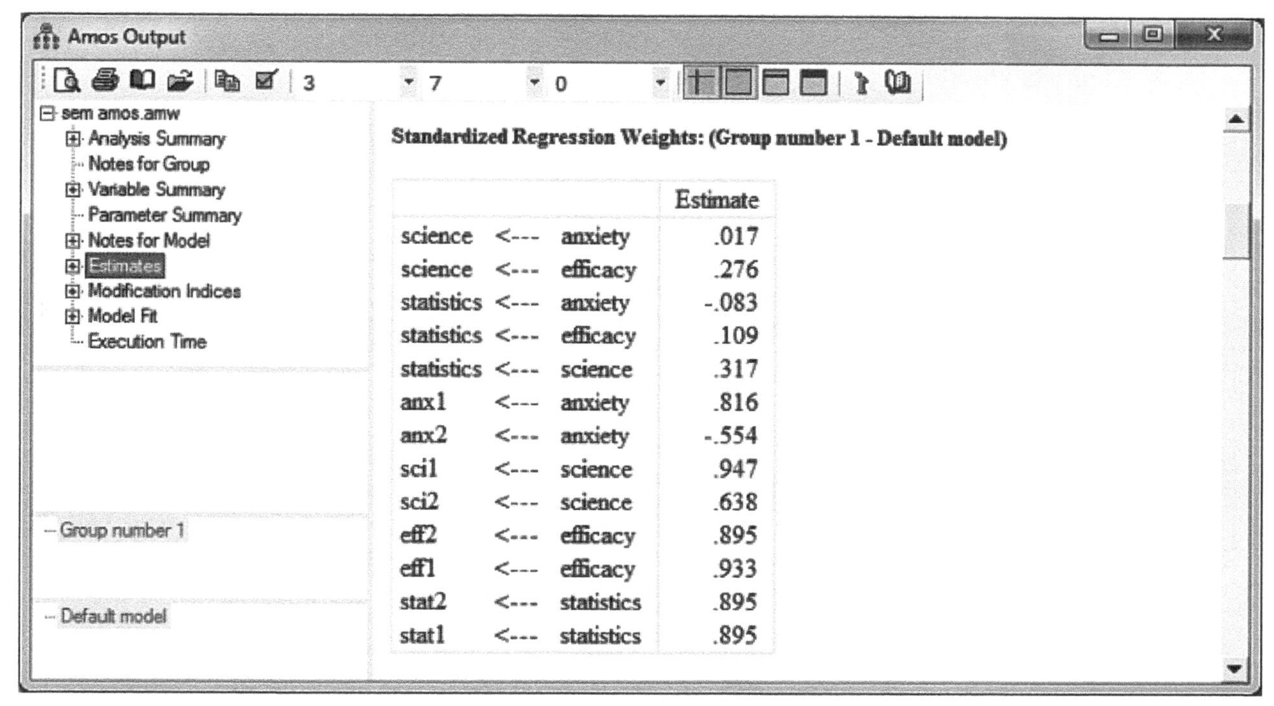

Standardized Regression Weights: (Group number 1 - Default model)

			Estimate
science	<---	anxiety	.017
science	<---	efficacy	.276
statistics	<---	anxiety	-.083
statistics	<---	efficacy	.109
statistics	<---	science	.317
anx1	<---	anxiety	.816
anx2	<---	anxiety	-.554
sci1	<---	science	.947
sci2	<---	science	.638
eff2	<---	efficacy	.895
eff1	<---	efficacy	.933
stat2	<---	statistics	.895
stat1	<---	statistics	.895

FIGURE 43.38 The standardized regression coefficients.

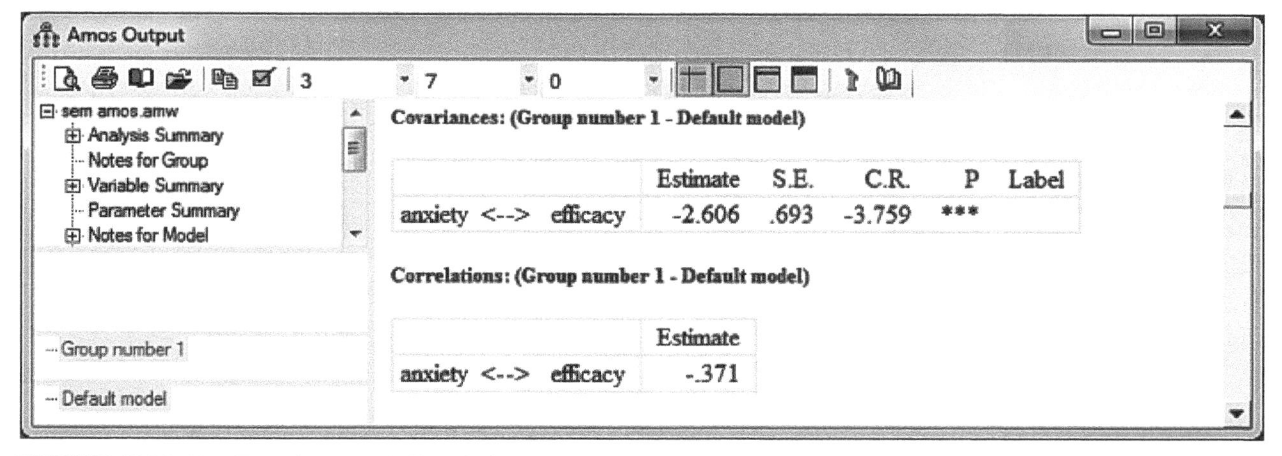

Covariances: (Group number 1 - Default model)

			Estimate	S.E.	C.R.	P	Label
anxiety	<-->	efficacy	-2.606	.693	-3.759	***	

Correlations: (Group number 1 - Default model)

			Estimate
anxiety	<-->	efficacy	-.371

FIGURE 43.39 The **Covariance** and **Correlation** between the two exogenous latent variables **anxiety** and **efficacy**.

43.11 THE STRATEGY TO CONFIGURE AND ANALYZE A TRIMMED MODEL

With no statistically significant path connecting **anxiety** to either endogenous latent variable, we make the decision to remove it from the model. But a model is a holistic entity and removing one of its parts can change the nature of the relationships that remain in a reconfigured model. When **anxiety** was in the model, the only statistically significant path leading from **efficacy** was to **science**—its direct effect on **statistics** was not statistically significant. Let us presume at this point that the indirect effect of **efficacy** through **science** on **statistics** will remain viable, with **anxiety** removed from the model.

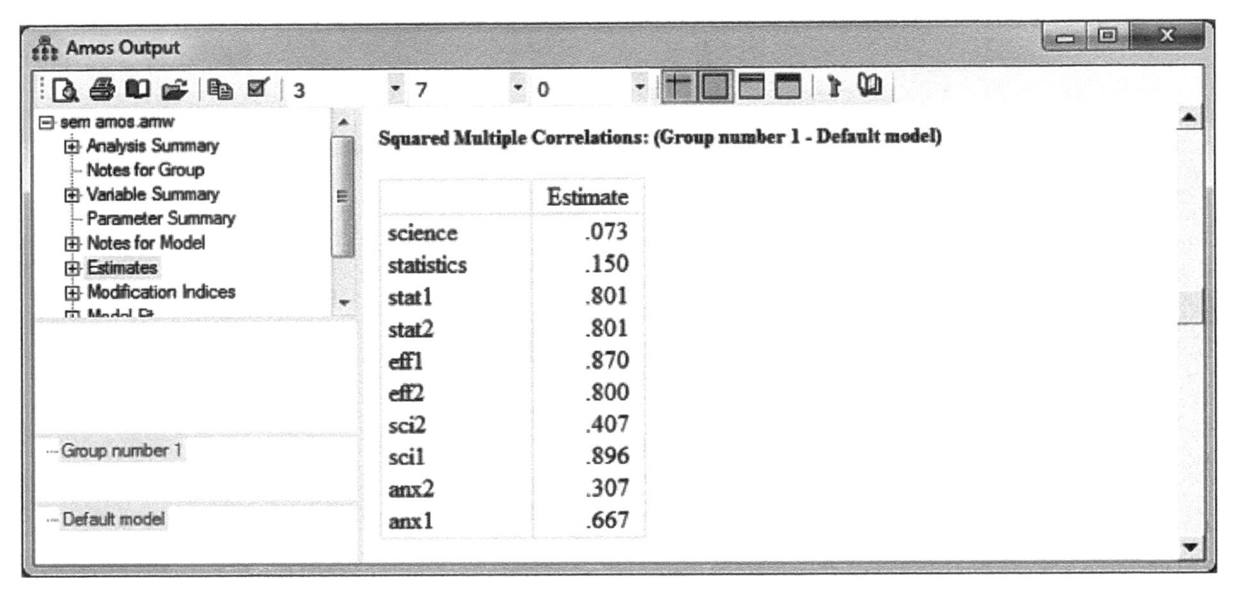

Squared Multiple Correlations: (Group number 1 - Default model)

	Estimate
science	.073
statistics	.150
stat1	.801
stat2	.801
eff1	.870
eff2	.800
sci2	.407
sci1	.896
anx2	.307
anx1	.667

FIGURE 43.40 The **Squared Multiple Correlations** for all endogenous variables.

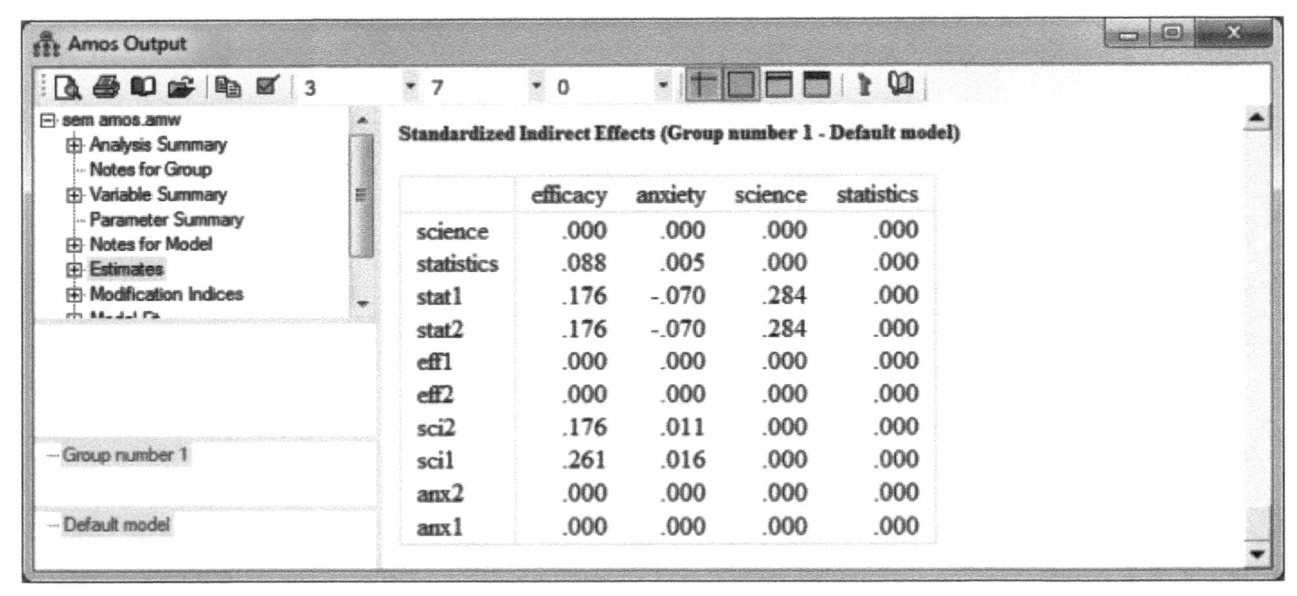

Standardized Indirect Effects (Group number 1 - Default model)

	efficacy	anxiety	science	statistics
science	.000	.000	.000	.000
statistics	.088	.005	.000	.000
stat1	.176	-.070	.284	.000
stat2	.176	-.070	.284	.000
eff1	.000	.000	.000	.000
eff2	.000	.000	.000	.000
sci2	.176	.011	.000	.000
sci1	.261	.016	.000	.000
anx2	.000	.000	.000	.000
anx1	.000	.000	.000	.000

FIGURE 43.41 The **Standardized Indirect Effects**.

With the removal of **anxiety** in a revised model, it is possible that the direct effect of **efficacy** on **statistics** could now be worth hypothesizing. The extent to which this option is reasonable hinges on whether, in isolation, **efficacy** is significantly predictive of **statistics**. If **efficacy** significantly predicts **statistics** in isolation, then we may have a situation where the effect of **efficacy** on **statistics** is mediated either partially or completely by **science**.

This line of thought suggests a strategy to use in exploring a trimmed model that is analogous to the strategy described in Chapter 40 for simple mediation and Chapter 41 for path analysis. First, we will perform a simple SEM analysis (assuming that any SEM analysis can be called simple) with **efficacy** hypothesized to predict **statistics** in isolation

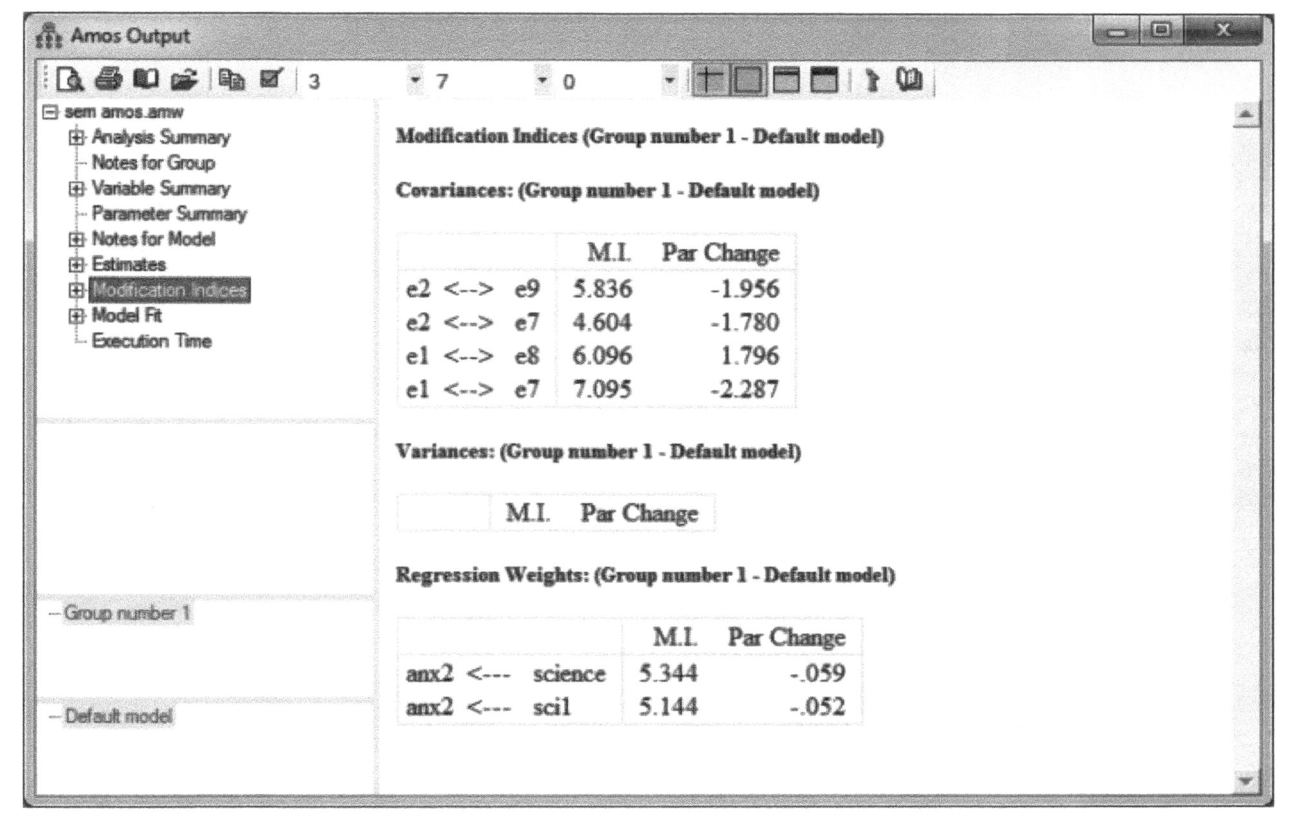

FIGURE 43.42 The **Modification Indices**.

(i.e., we will have just two latent variables in the model). We would then follow one of these two scenarios depending on the outcome of that analysis:

- If **efficacy** is a statistically significant predictor of **statistics** in isolation, then we will perform a mediation analysis akin to the path structure described in Chapter 40; that is, we will include in the trimmed model the direct path from **efficacy** to **statistics** in addition to the indirect path from **efficacy** through **science** to **statistics**.
- If **efficacy** does not significantly predict **statistics** in isolation, then we will specify the trimmed model with only the indirect path from **efficacy** through **science** to **statistics**.

43.12 EXAMINING THE DIRECT EFFECT OF EFFICACY ON STATISTICS IN ISOLATION

With the **statistics achievement** data file open and IBM SPSS Amos in landscape orientation, we draw the model in which **efficacy** directly predicts **statistics** as shown in Figure 43.43. Then from the main menu, we select **View ➔ Analysis Properties** to open the **Analysis Properties** window, and in the window of the **Output** tab, we check **Standardized estimates** and **Squared multiple correlations**. To execute the analysis, select **Analyze ➔ Calculate Estimates** from the main menu.

To access the results of the analysis, select the **View the output path diagram** icon (the one on the right) in the top output panel immediately to the left of the drawing area. The path diagram with the **Unstandardized estimates** is shown in the top screenshot

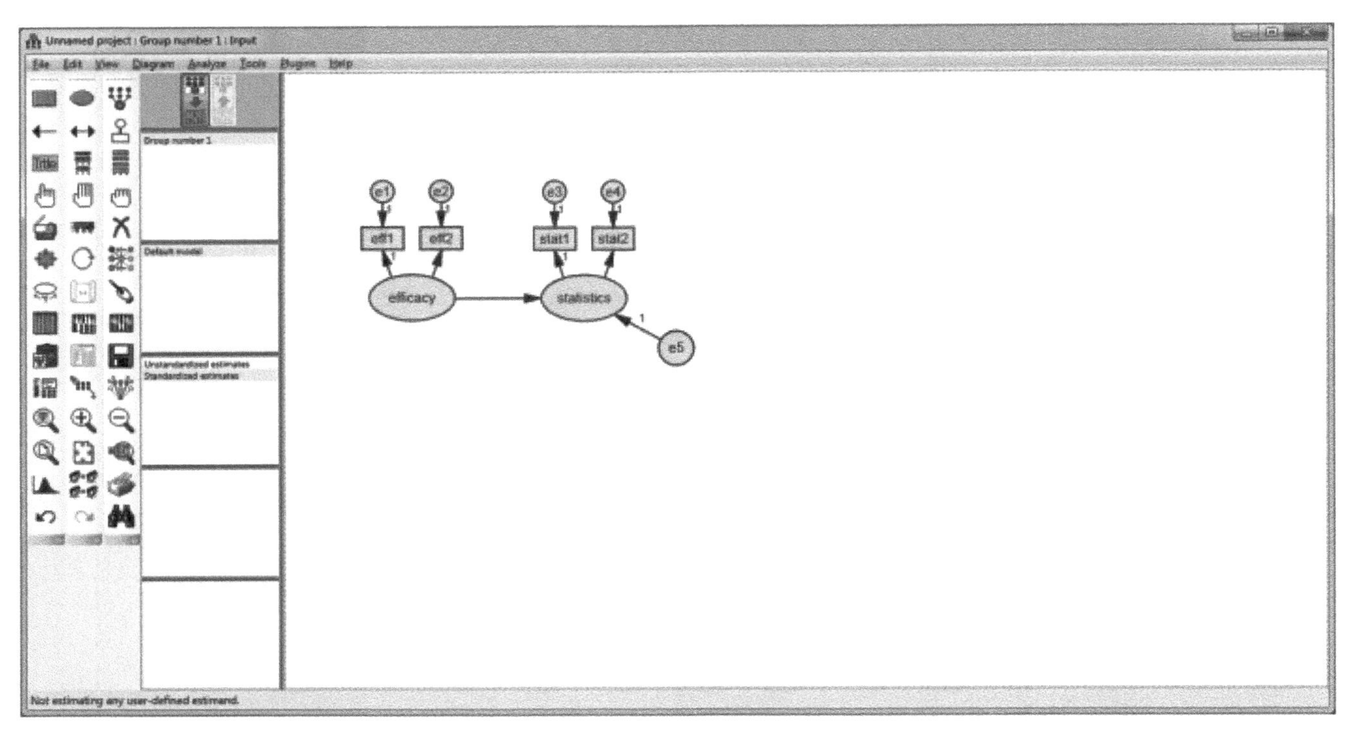

FIGURE 43.43 The simple model with **efficacy** predicting **statistics**.

in Figure 43.44; we also show the **Standardized estimates** in the bottom screenshot in Figure 43.44.

To obtain the output in tabular form, select from the main menu **View → Text Output** and select **Model Fit** in the left panel. The fit indexes of relevance are shown in Figure 43.45. The chi-square value is **1.342**; with **1** degree of freedom, it is not statistically significant ($p = .247$), suggesting concordance between the observed and expected values. The **GFI**, **NFI**, **CFI**, and **RMSEA** yielded values of **.997**, **.997**, **.999**, and **.042**, respectively, indicating a very good fit of the measurement model.

The **Estimates** screenshots are presented in Figure 43.46. The top screenshot informs us that the path from **efficacy** to **statistics** was statistically significant ($p = .006$), yielding an unstandardized coefficient of **.501** (with an SE of **.181**) and a standardized coefficient of **.222**. The bottom screenshot in Figure 43.46 shows us that **efficacy** explained almost 5% of the variance of **statistics** (a **Squared Multiple Correlation** of **.049**). Thus, with **efficacy** significantly predicting **statistics** in isolation, we can proceed to configure a mediation structure as our trimmed model in the next analysis.

43.13 EXAMINING THE MEDIATED EFFECT OF EFFICACY ON STATISTICS THROUGH SCIENCE

The mediation model with **efficacy** hypothesized as exerting both a direct effect on **statistics** as well as an indirect effect on **statistics** through **science** is shown in Figure 43.47. After performing the analysis, the **Unstandardized estimates** and the **Standardized estimates** are presented in the top and bottom screenshots, respectively, of Figure 43.48.

Figure 43.49 presents the relevant information from the **Model Fit Summary**. The chi-square value is **4.017**; with **6** degrees of freedom, it is not statistically significant ($p = .674$), suggesting close agreement between the observed and the expected values

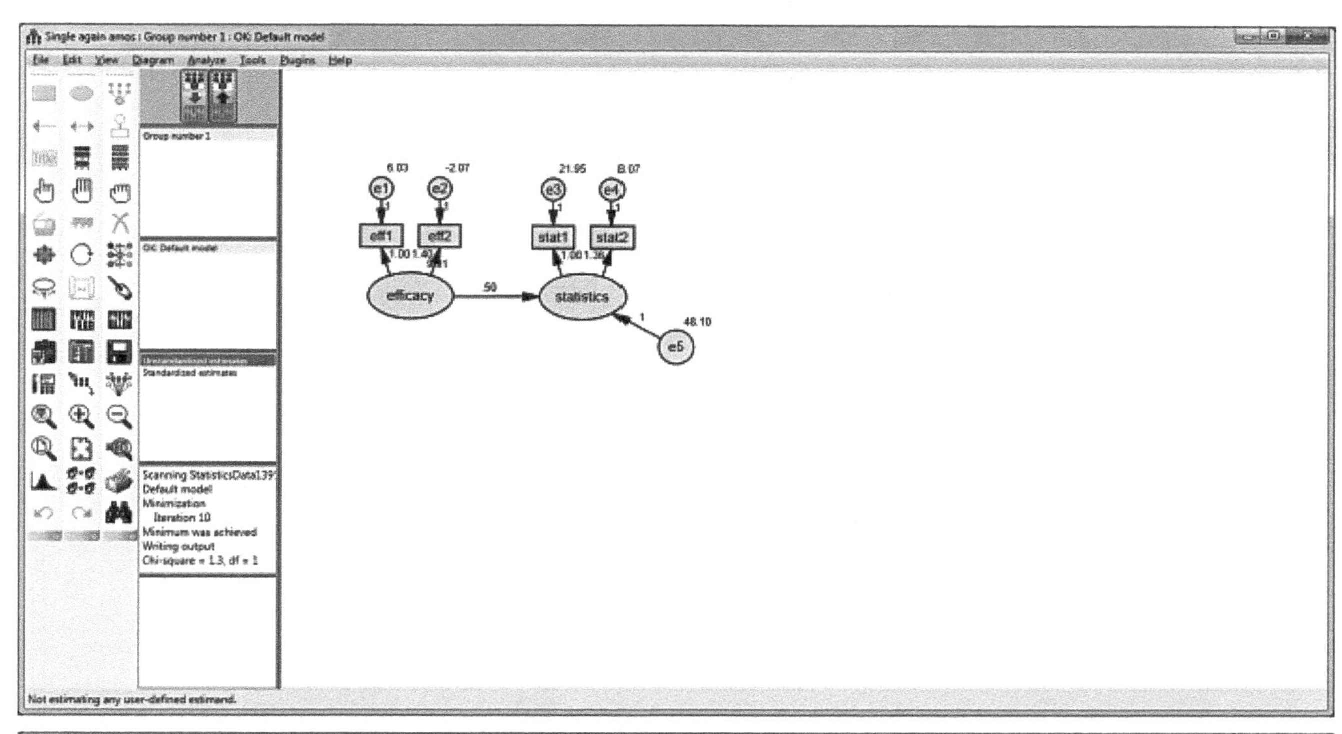

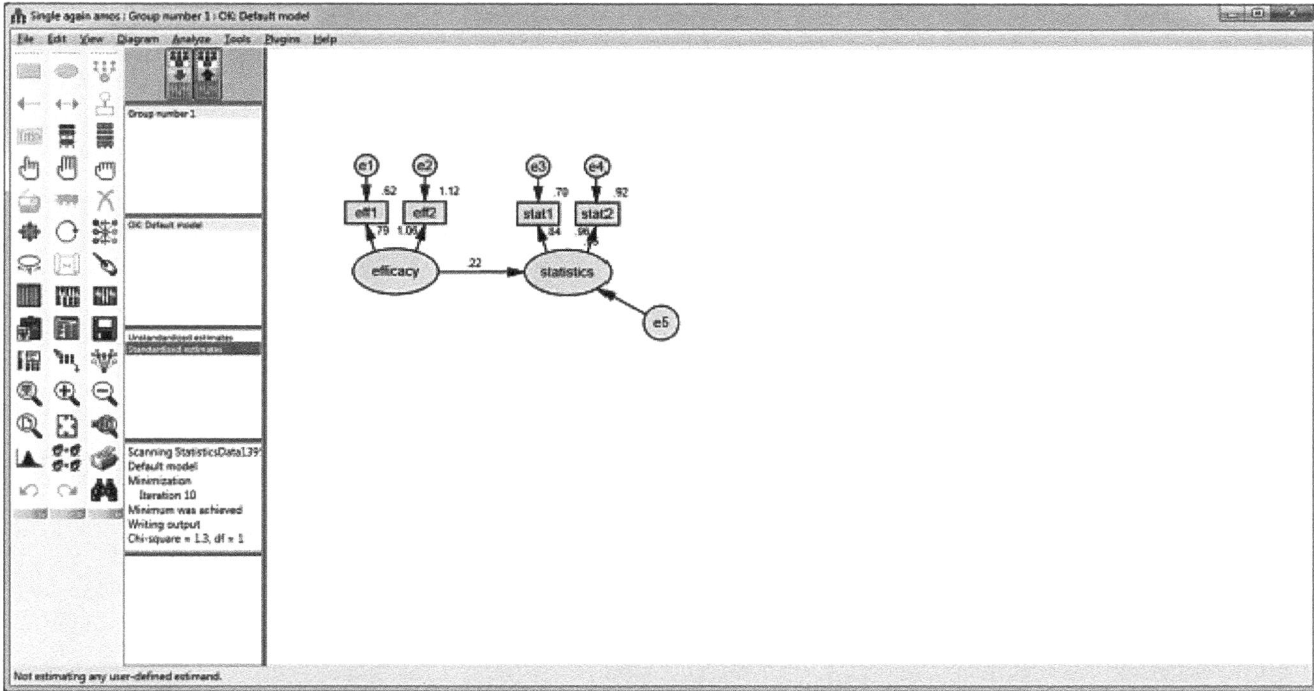

FIGURE 43.44 The unstandardized and standardized coefficients for the model.

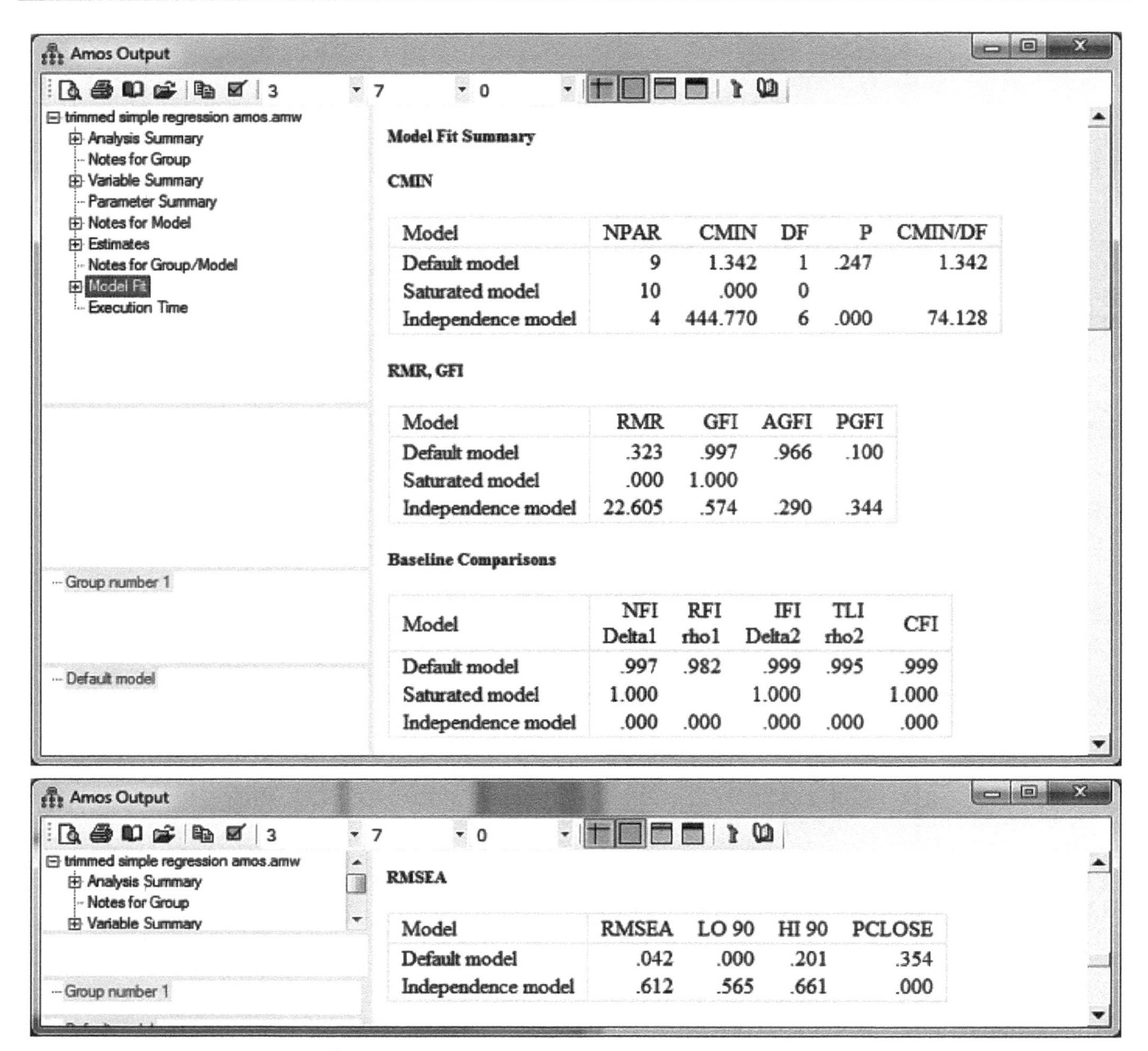

Model Fit Summary

CMIN

Model	NPAR	CMIN	DF	P	CMIN/DF
Default model	9	1.342	1	.247	1.342
Saturated model	10	.000	0		
Independence model	4	444.770	6	.000	74.128

RMR, GFI

Model	RMR	GFI	AGFI	PGFI
Default model	.323	.997	.966	.100
Saturated model	.000	1.000		
Independence model	22.605	.574	.290	.344

Baseline Comparisons

Model	NFI Delta1	RFI rho1	IFI Delta2	TLI rho2	CFI
Default model	.997	.982	.999	.995	.999
Saturated model	1.000		1.000		1.000
Independence model	.000	.000	.000	.000	.000

RMSEA

Model	RMSEA	LO 90	HI 90	PCLOSE
Default model	.042	.000	.201	.354
Independence model	.612	.565	.661	.000

FIGURE 43.45 The model fit indexes for the simple model.

based on the model. The **GFI**, **NFI**, **CFI**, and **RMSEA** yielded values of **.993**, **.993**, **1.000**, and **.000**, respectively, indicating an exceptionally good fit of the measurement model to the data.

The **Estimates** screenshots are presented in Figure 43.50. From the top screenshot, we learn that all three paths in the mediation model are statistically significant:

- The path from **efficacy** to **science** yielded an unstandardized coefficient of **.442** (with an SE of **.127**) and a standardized coefficient of **.268** ($p < .001$).
- The path from **efficacy** to **statistics** yielded an unstandardized coefficient of **.350** (with an SE of **.176**) and a standardized coefficient of **.156** ($p = .047$).

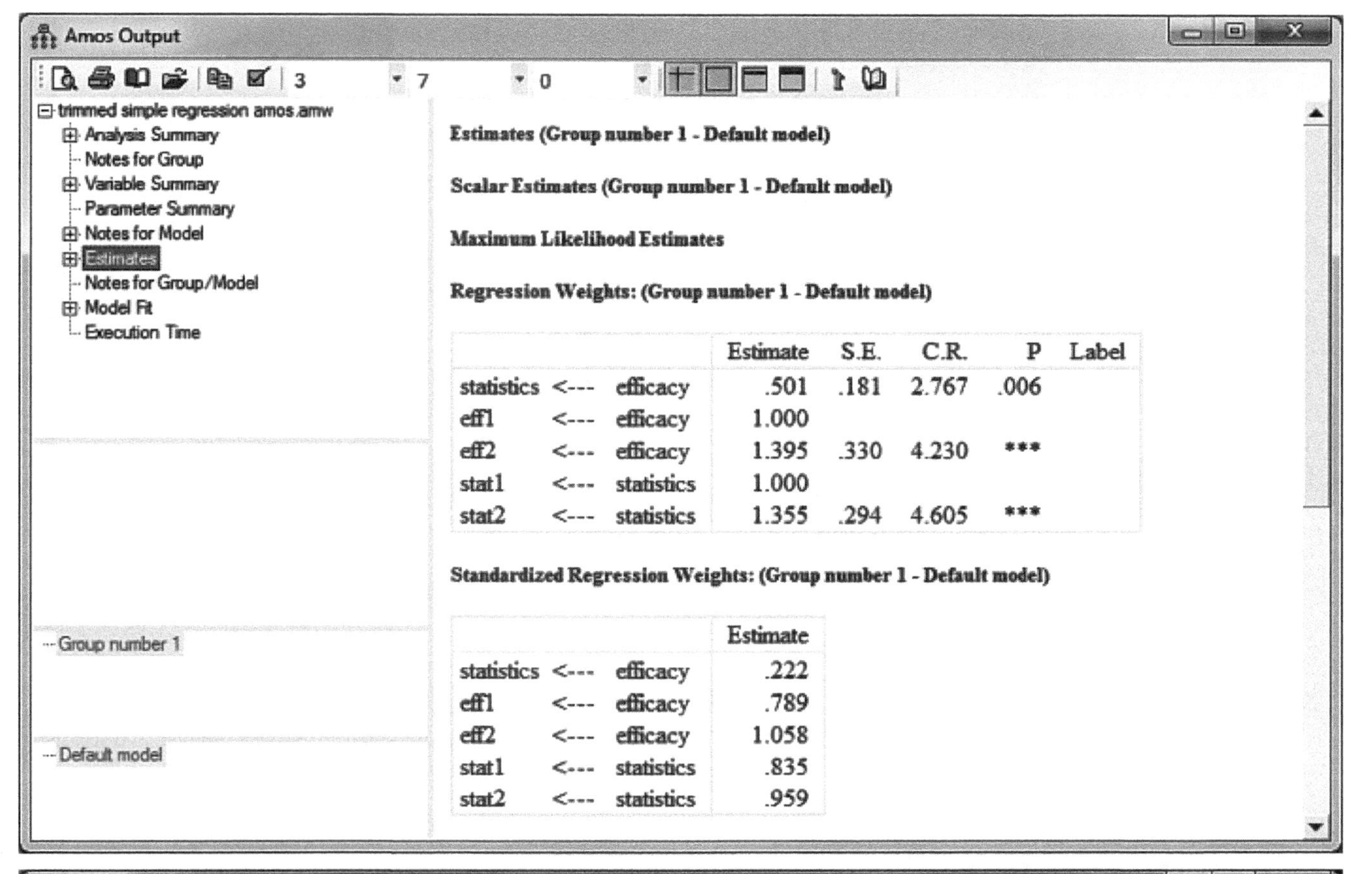

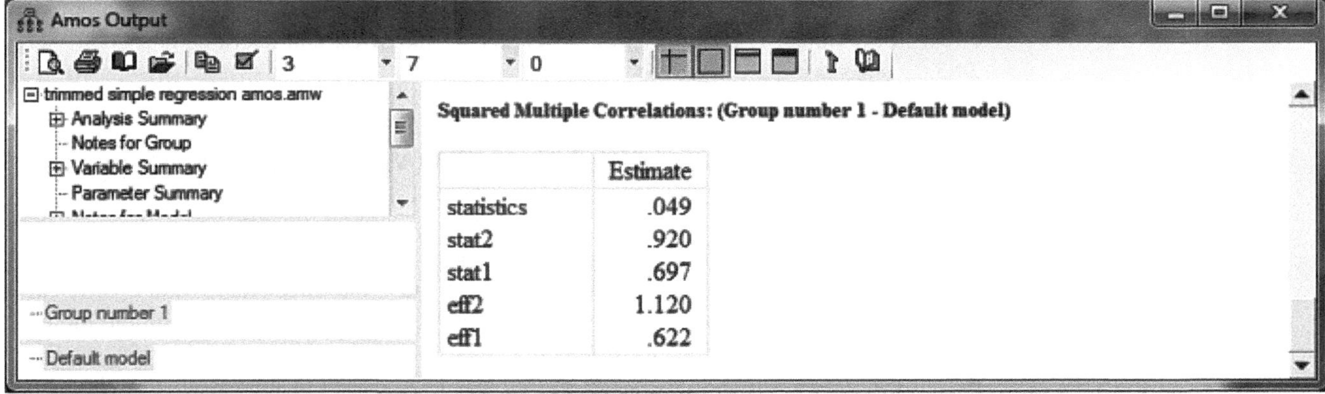

FIGURE 43.46 The **Unstandardized** and **Standardized Regression Weights**, as well as the **Squared Multiple Correlations** for the simple model.

- The path from **science** to **statistics** yielded an unstandardized coefficient of **.430** (with an SE of **.142**) and a standardized coefficient of **.316** (p = **.003**).

The bottom screenshot shows us that **efficacy** explained approximately 7% of the variance of **science** (a **Squared Multiple Correlation** of **.072**) and that **efficacy** in combination with **science** explained approximately 15% of the variance of **statistics** (a **Squared Multiple Correlation** of **.151**).

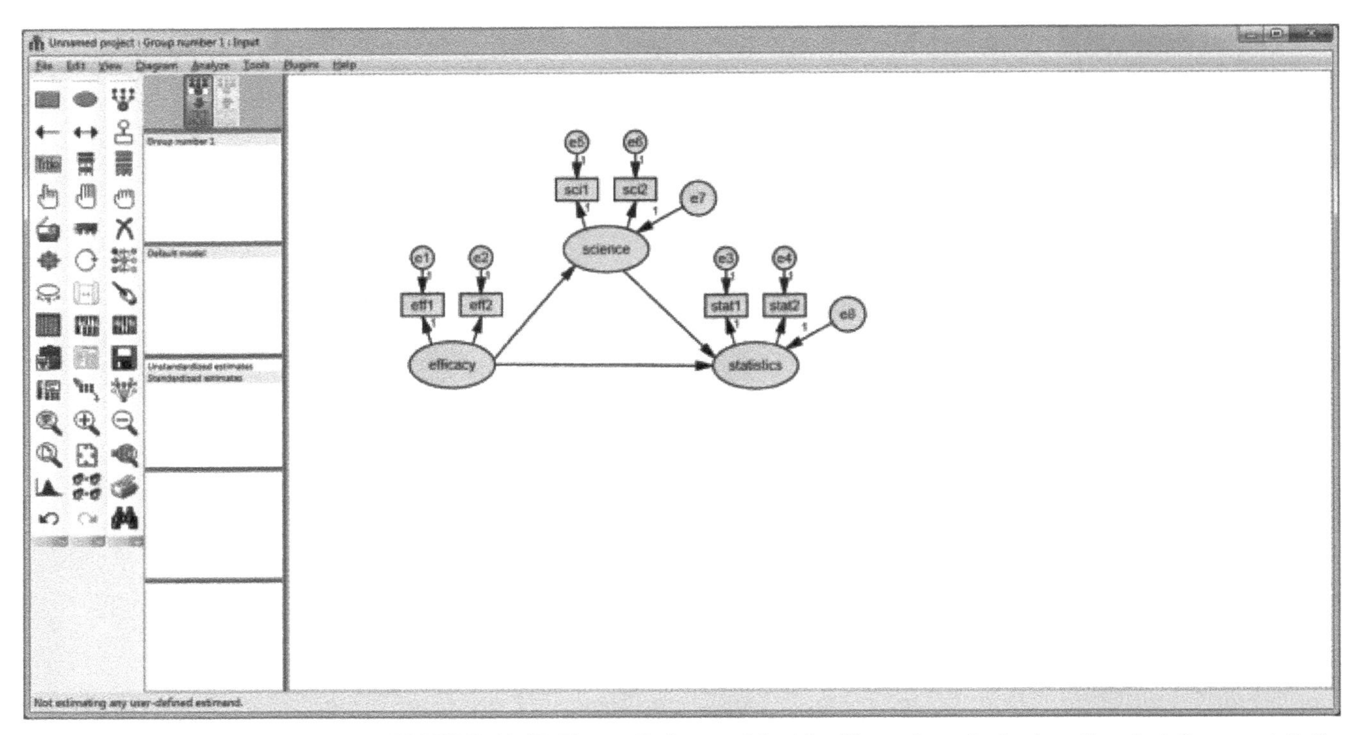

FIGURE 43.47 The mediation model with **efficacy** hypothesized to directly influence **statistics** as well as indirectly influence **statistics** through **science**.

The **Standardized Indirect Effects** are displayed in Figure 43.51. The only indirect effect in the model is **efficacy** through **science** to **statistics**, and its value is a nontrivial **.085**.

43.14 SYNTHESIS OF THE TRIMMED (MEDIATED) MODEL RESULTS

Based on visual inspection of the results, it appears that the direct effect of **efficacy** on **statistics** is mediated by **science**. Furthermore, because the direct regression coefficient from **efficacy** to **statistics** in the mediated model is lower than that in the simple model (but still statistically significant), it would appear that we have observed a partial mediation effect. These impressions based on the visual inspection of the results now need to be evaluated statistically so that we can draw our conclusions from the data analysis.

43.15 STATISTICAL SIGNIFICANCE OF THE INDIRECT EFFECT: THE AROIAN TEST

As described in Section 40.8, we use the Aroian test (Aroian, 1947) as a representative of the Sobel test family to evaluate the statistical significance of the indirect effect. Based on the unstandardized partial regression coefficients and SEs associated with the paths **efficacy** to **science** (b = **.442**, SE = **.127**) and **science** to **statistics** (b = **.430**, SE = **.142**), we obtain (using the calculator available on Kristopher J. Preacher's web site http://quantpsy.org/sobel/sobel.htm) an Aroian z value of approximately 2.23, $p = .023$.

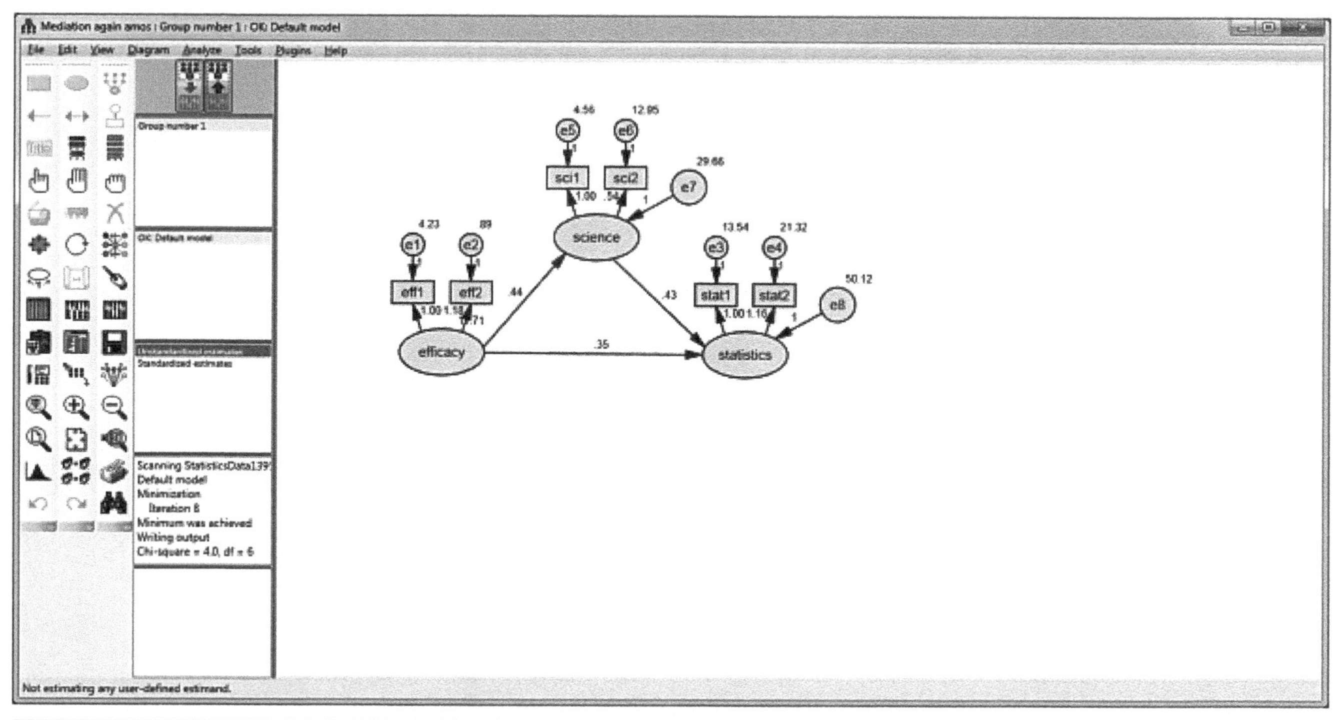

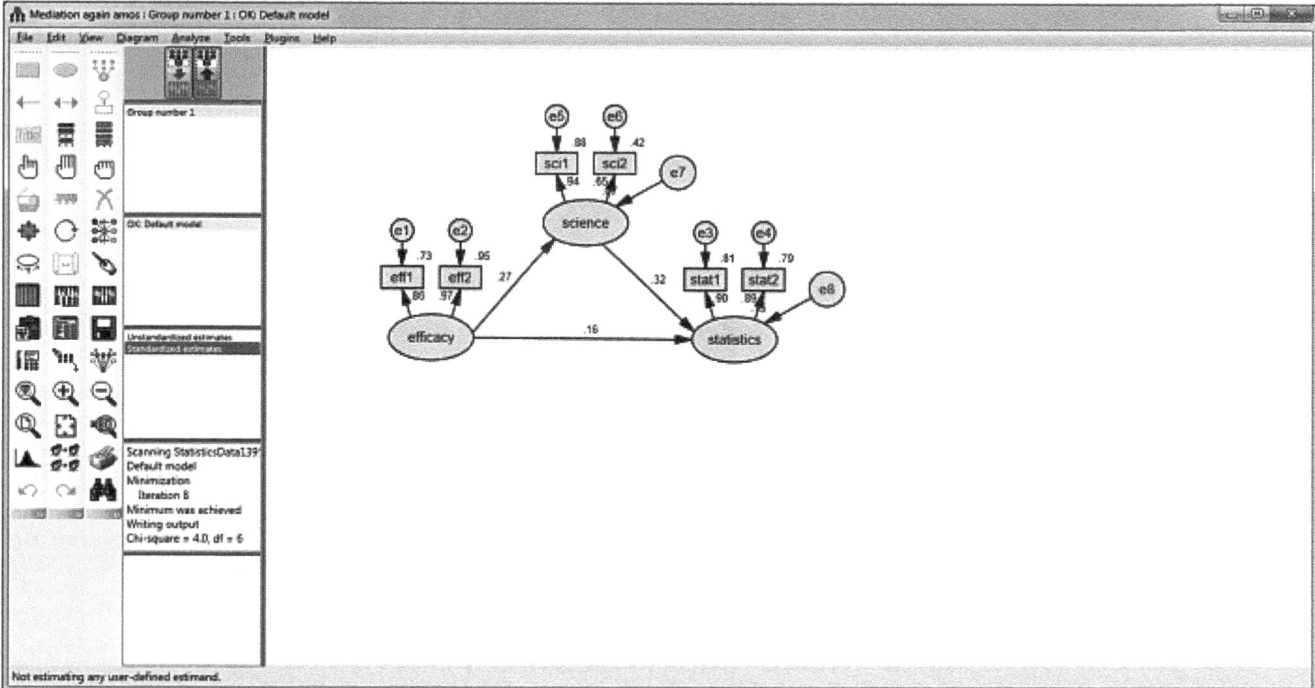

FIGURE 43.48 The unstandardized and standardized coefficients for the mediation model.

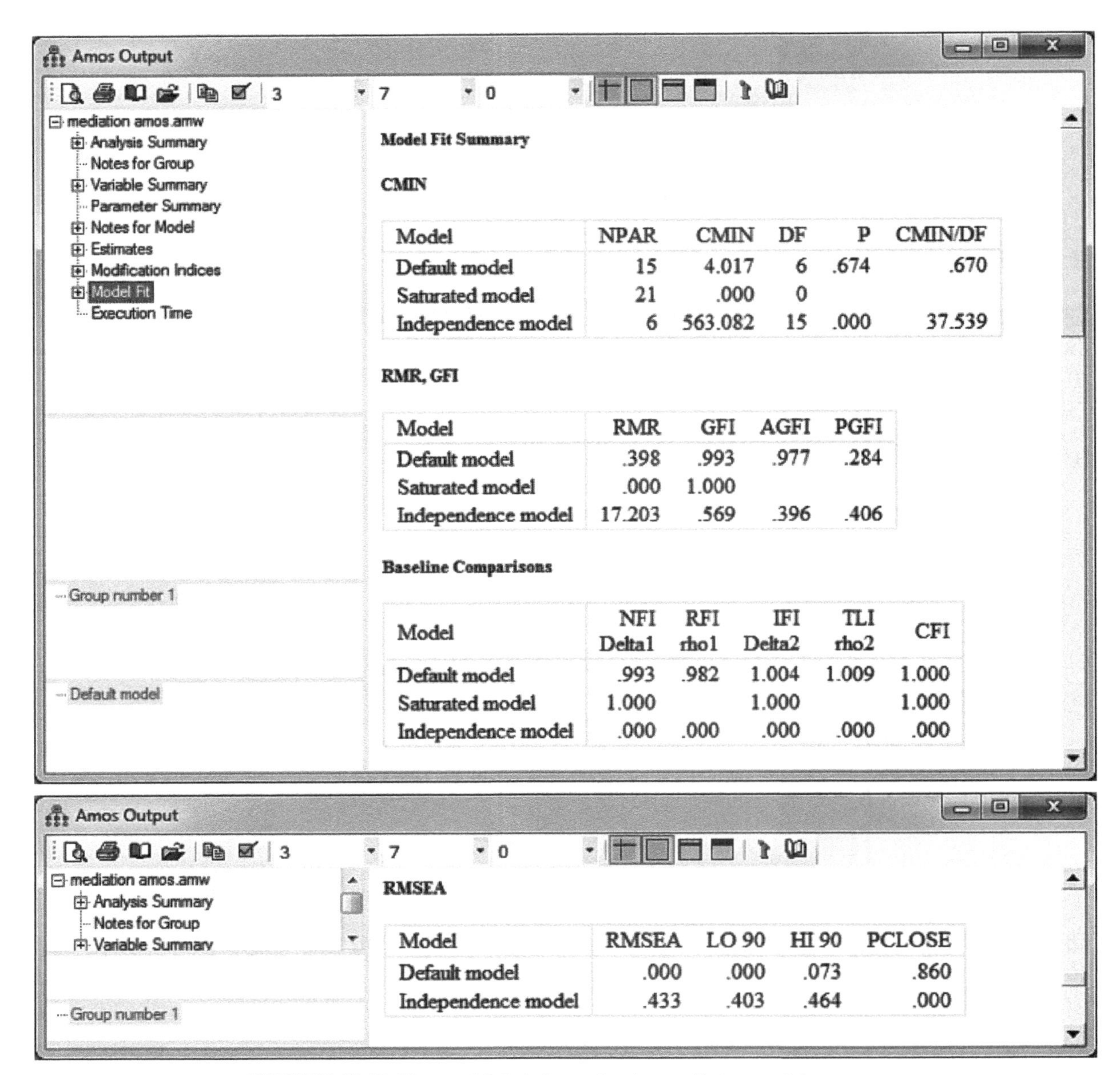

Model Fit Summary

CMIN

Model	NPAR	CMIN	DF	P	CMIN/DF
Default model	15	4.017	6	.674	.670
Saturated model	21	.000	0		
Independence model	6	563.082	15	.000	37.539

RMR, GFI

Model	RMR	GFI	AGFI	PGFI
Default model	.398	.993	.977	.284
Saturated model	.000	1.000		
Independence model	17.203	.569	.396	.406

Baseline Comparisons

Model	NFI Delta1	RFI rho1	IFI Delta2	TLI rho2	CFI
Default model	.993	.982	1.004	1.009	1.000
Saturated model	1.000		1.000		1.000
Independence model	.000	.000	.000	.000	.000

RMSEA

Model	RMSEA	LO 90	HI 90	PCLOSE
Default model	.000	.000	.073	.860
Independence model	.433	.403	.464	.000

FIGURE 43.49 The model fit indexes for the mediation model.

We therefore conclude that the indirect path from **efficacy** through **science** to **statistics** is statistically significant.

43.16 COMPARING THE DIRECT EFFECTS OF EFFICACY ON STATISTICS IN THE SIMPLE MODEL AND THE MEDIATED MODEL: THE FREEDMAN-SCHATZKIN TEST

In the simple model, the unstandardized regression coefficient from **efficacy** to **statistics** was **.501**, with an SE of **.181**; in the mediated model, the unstandardized regression coefficient from **efficacy** to **statistics** was **.350**, with an SE of **.176**. Based on the Freedman–Schatzkin test (Freedman & Schatzkin, 1992), by comparing the relative

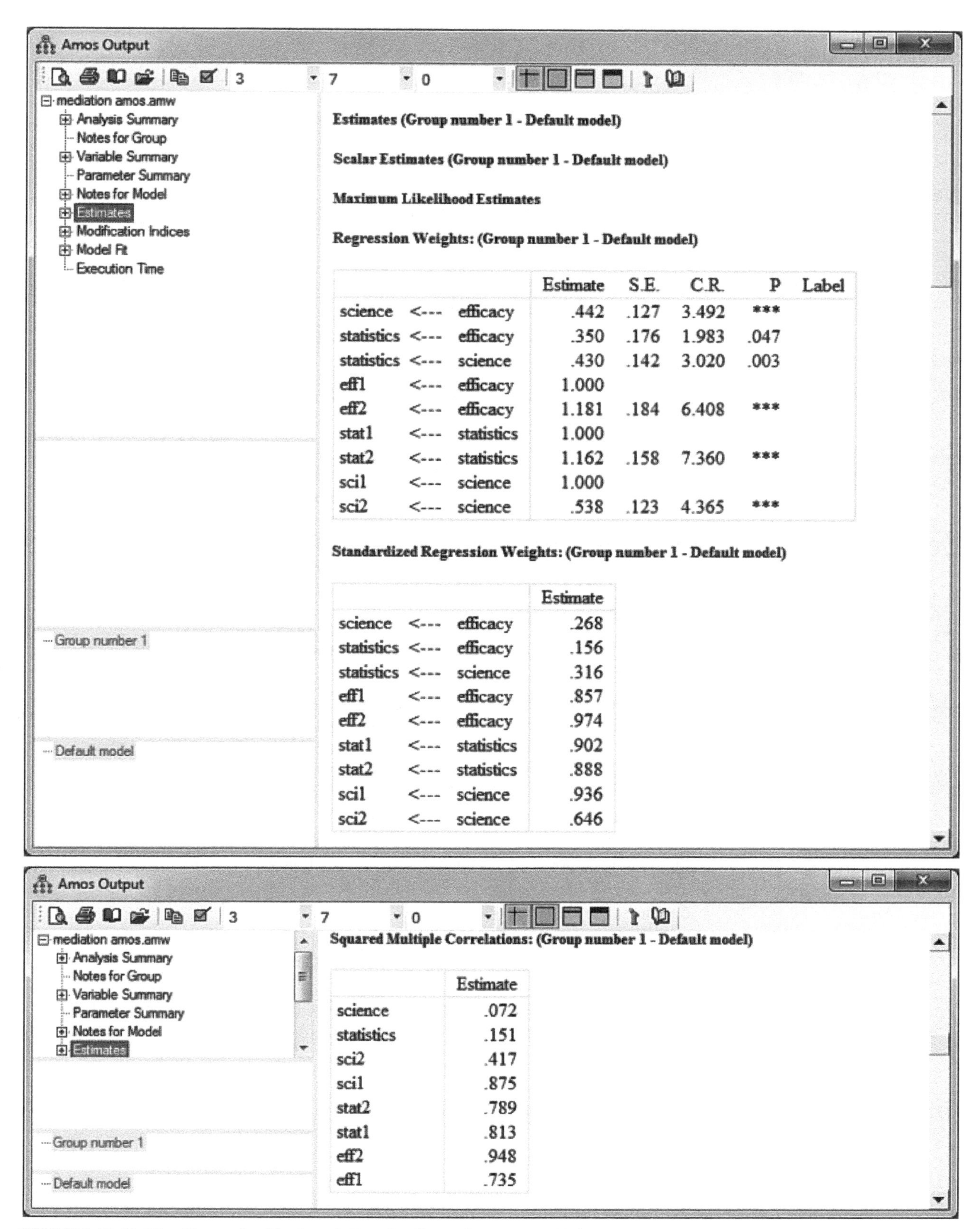

Estimates (Group number 1 - Default model)

Scalar Estimates (Group number 1 - Default model)

Maximum Likelihood Estimates

Regression Weights: (Group number 1 - Default model)

			Estimate	S.E.	C.R.	P	Label
science	<---	efficacy	.442	.127	3.492	***	
statistics	<---	efficacy	.350	.176	1.983	.047	
statistics	<---	science	.430	.142	3.020	.003	
eff1	<---	efficacy	1.000				
eff2	<---	efficacy	1.181	.184	6.408	***	
stat1	<---	statistics	1.000				
stat2	<---	statistics	1.162	.158	7.360	***	
sci1	<---	science	1.000				
sci2	<---	science	.538	.123	4.365	***	

Standardized Regression Weights: (Group number 1 - Default model)

			Estimate
science	<---	efficacy	.268
statistics	<---	efficacy	.156
statistics	<---	science	.316
eff1	<---	efficacy	.857
eff2	<---	efficacy	.974
stat1	<---	statistics	.902
stat2	<---	statistics	.888
sci1	<---	science	.936
sci2	<---	science	.646

Squared Multiple Correlations: (Group number 1 - Default model)

	Estimate
science	.072
statistics	.151
sci2	.417
sci1	.875
stat2	.789
stat1	.813
eff2	.948
eff1	.735

FIGURE 43.50 The **Unstandardized** and **Standardized Regression Weights** as well as the **Squared Multiple Correlations** for the mediation model.

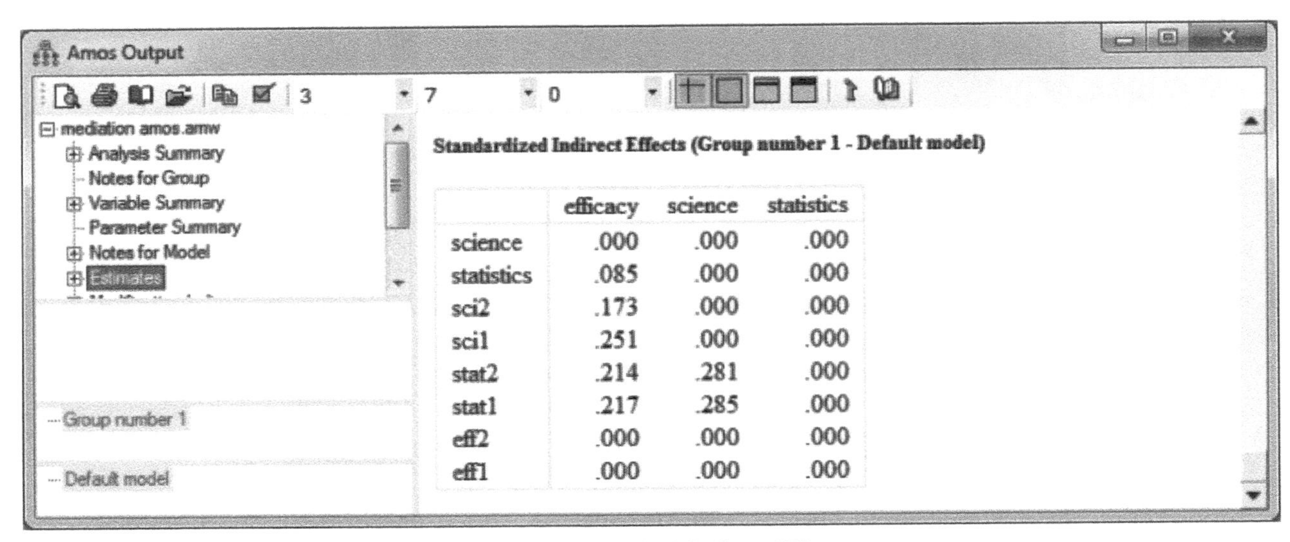

Standardized Indirect Effects (Group number 1 - Default model)

	efficacy	science	statistics
science	.000	.000	.000
statistics	.085	.000	.000
sci2	.173	.000	.000
sci1	.251	.000	.000
stat2	.214	.281	.000
stat1	.217	.285	.000
eff2	.000	.000	.000
eff1	.000	.000	.000

FIGURE 43.51 The **Standardized Indirect Effects**.

strengths of the paths from the exogenous variable to the outcome variable in the unmediated model and the mediated model as described in Section 40.9, we obtain a t value of approximately 3.76. With a sample size of 196 cases, the degrees of freedom are 194 and the obtained t value is statistically significant ($p < .001$). We therefore conclude that the direct path from **efficacy** to **statistics** is less significant (but still statistically significant) in the mediated model, indicating that we have observed a partial mediation effect.

43.17 THE RELATIVE STRENGTH OF THE MEDIATED EFFECT

We now calculate the relative strength of the mediation effect using the standardized coefficients that are associated with the paths in our mediation model (as described in Section 40.10). This is computed as the ratio of the strength of the indirect effect to the strength of the direct effect and is calculated as follows:

- The strength of the indirect effect is indexed by the product of the standardized coefficients associated with paths **efficacy** to **science** and **science** to **statistics** in the mediated model. As seen in Figure 43.51, this value is **.085**.
- The strength of the isolated direct effect is indexed by the value of the standardized coefficient in the unmediated model; its value is **.222**.
- The relative strength of the mediated effect is equal to the indirect effect divided by the direct effect. Here it is equal to **.085/.222** or approximately .383.

We may then conclude that about 38% (38.3%) of the effect of **efficacy** on **statistics** is mediated through **science**.

PART 14

t TEST

CHAPTER 44

One-Sample *t* Test

44.1 OVERVIEW

William Gosset was a chemist and mathematician who worked for the Guinness Brewing Company and developed the *t* test in an effort to track the quality of the beer that was produced. When Gosset wished to publish a description of this statistical technique, he had to do so under a pseudonym to protect proprietary trade secrets. With the help of his colleague Karl Pearson, he published under the name of *Student* in *Biometrika* in 1908 (Student, 1908). The letter *t* was selected from the last letter in Student to depict the new procedure and its distribution (Salsburg, 2001).

There are three types of *t* test procedures: *one-sample t test, independent-samples t test*, and the *paired-samples t test*. The one-sample *t* test described in this chapter is used when we have data on a dependent variable from a single group or sample and wish to determine whether the sample mean is statistically different from the known (or assumed) population mean. The general strategy is as follows:

- The standard error of the sample mean is computed.
- A confidence interval that corresponds to a particular alpha level (e.g., .05, or 95% confidence interval) is computed from the standard error.
- We determine where the population parameter falls with respect to the confidence interval.

If the estimate of the population parameter falls inside the confidence interval or range, we would judge the sample mean and population to be not significantly different; if the population parameter falls outside the confidence interval, we would judge the sample mean and population parameter to be significantly different. The statistical assessment is made by means of a *t* test, with the null hypothesis assuming equivalence between the sample mean and the population parameter.

44.2 NUMERICAL EXAMPLE

To illustrate the one-sample *t* test, we use the hypothetical mental health data set, a portion of which is shown in Figure 44.1. A total of 25 mental health consumers were given

Performing Data Analysis Using IBM SPSS®, First Edition.
Lawrence S. Meyers, Glenn C. Gamst, and A. J. Guarino.

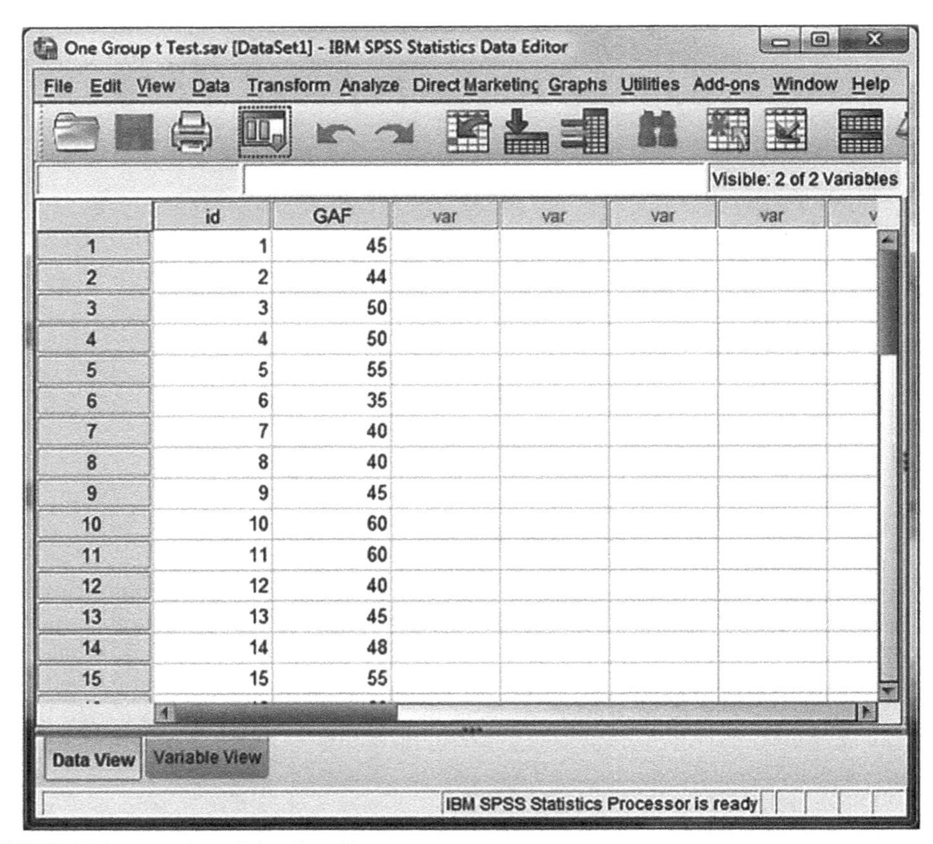

FIGURE 44.1 A portion of the data file.

intake evaluation interviews at a new mental health facility. The assessment included evaluating each consumer on their GAF Axis V rating of the Diagnostic and Statistical Manual of Mental Disorders, IV-TR (DSM-IV-TR; American Psychiatric Association, 2000). The GAF scale values can range from 1 (*severe impairment*) to 100 (*good general functioning*). The question being investigated is whether the consumer GAF sample mean is comparable to the consumer GAF statewide average of 55. The data can be found in the file named **One Group t Test**.

44.3 ANALYSIS SETUP

Open the IBM SPSS® save file named **One Group t Test** and from the main menu select **Analyze → Compare Means → One-Sample T Test**. This produces the **One-Sample T Test** dialog window shown in Figure 44.2. We have moved the **GAF** variable to the **Test Variable(s)** panel and typed **55** in the **Test Value** panel.

Clicking the **Options** pushbutton produces the **Options** dialog window in Figure 44.3. This dialog box allows us to set the **Confidence Interval Percentage**; in this example, we have retained the default value of **95%**. The **Missing Values** option has also been left at its default setting of **Exclude cases analysis by analysis**. Clicking **Continue** takes us back to the main dialog window, and clicking **OK** generates the analysis.

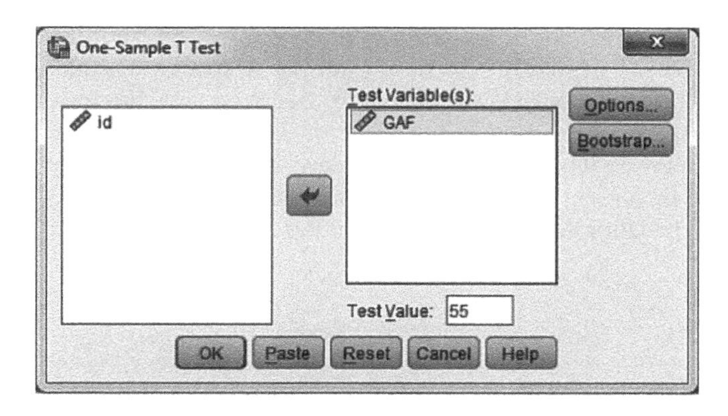

FIGURE 44.2
The main **One-Sample T Test** dialog window.

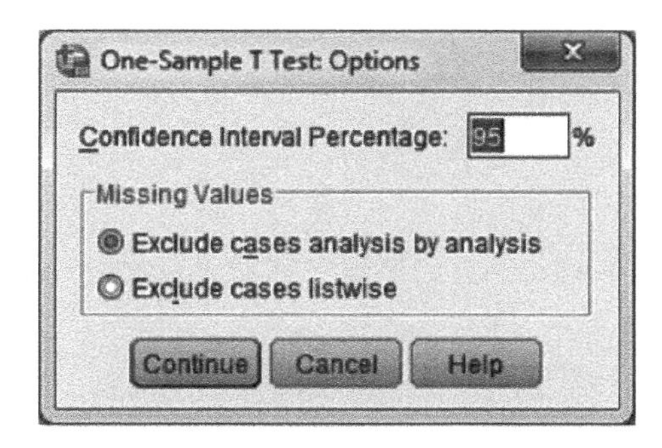

FIGURE 44.3
The **Options** dialog window of the **One-Sample T Test**.

One-Sample Statistics

	N	Mean	Std. Deviation	Std. Error Mean
GAF Global Assessment of Functioning	25	47.44	7.703	1.541

One-Sample Test

	Test Value = 55					
					95% Confidence Interval of the Difference	
	t	df	Sig. (2-tailed)	Mean Difference	Lower	Upper
GAF Global Assessment of Functioning	-4.907	24	.000	-7.560	-10.74	-4.38

FIGURE 44.4 Output of the **One-Sample T Test**.

44.4 ANALYSIS OUTPUT

The output for the analysis is presented in Figure 44.4. The top table, **One-Sample Statistics**, provides the sample size, the sample mean and standard deviation, and the standard error of the mean. The bottom table, **One-Sample Test**, contains the presumed population mean (labeled **Test Value** at the top of the table), the *t* statistic, its degrees of freedom, significance level, mean difference, and the 95% lower and upper bounds of the confidence interval of the difference.

From the output, we can see that the value of the *t* statistic is −**4.907**. It is negative for the following reason: the mean difference was calculated as the sample mean (47.44) minus the population mean (55.00), resulting in the mean difference of **−7.56**, and thus the t value will be negative. With **24** degrees of freedom (number of cases − 1), the t value was statistically significant (significantly lower than zero) based on an alpha level of .05 (IBM SPSS displays a **Sig. (2-tailed)** probability value of **.000**, which is treated as $p < .001$). The significance level is labeled "2-tailed" because it sums the areas of both tails of the t distribution. The present data indicate that the sample mean of **47.75** is significantly lower than the population mean of 55.

CHAPTER 45

Independent-Samples *t* Test

45.1 OVERVIEW

The independent-samples *t* test assesses whether the means of two independent groups (measured as a nonmetric or categorical independent variable) significantly differ on a metric or quantitatively measured dependent variable. A one-way between-subjects ANOVA, as treated in Chapter 47, is the general case of the independent-samples *t* test and is often used in its stead.

The sampling distribution of *t* is leptokurtic (a bit compressed toward the center with slightly raised tails compared to the normal distribution). It is symmetrically distributed around a mean difference of zero, which is the null hypothesis. With increasingly larger sample sizes, the *t* distribution becomes more normal, and by the time we reach *N*s of triple figures, it is very close to a normal distribution.

Interpreting the results of a *t* test rests on meeting the following assumptions:

- The observations are independent. This assumption is generally achieved through explicit randomization procedures, where participants are randomly and independently assigned to treatment conditions or levels of the independent variable (Gamst et al., 2008).
- The dependent variable is normally distributed. Normality can be assessed by a graphical examination of the distribution of dependent variable scores or through special tests of univariate normality (see Chapter 20).
- The variance of the dependent variable is comparable across levels of the independent variable. This assumption of homogeneity of variance is addressed by default directly within the IBM SPSS® *t* test procedure, and we discuss it later in more detail.

The equality of variance assumption, known as *homogeneity of variance*, is that the two groups have comparable variability on the dependent variable. This assumption is tested in IBM SPSS by the Levene test; if the test is statistically significant, then homogeneity of variance has been violated. To the extent that this homogeneity assumption is violated, use of the standard *t* test is not appropriate, as it will generally increase the chances of making a Type I error; instead, the alternative *t* test procedure introduced by Welch (1937) and modified by Satterthwaite (1946), provided in the default IBM SPSS

Performing Data Analysis Using IBM SPSS®, First Edition.
Lawrence S. Meyers, Glenn C. Gamst, and A. J. Guarino.

output (Moser & Stevens, 1992), is to be used instead. We present two examples to demonstrate how to work with homogeneity of variances; the first is an instance where the assumption is met by the data set and the second is an instance where there is a violation of the assumption.

45.2 NUMERICAL EXAMPLE: MEETING THE HOMOGENEITY OF VARIANCE ASSUMPTION

Our example concerns performance in a history test (**history_test**) by male (**sex** coded as **1**) and female (**sex** coded as **2**) middle-school students. We are interested in whether or not there is a reliable (statistically different) test score difference between males and females. The data can be found in the data file **history test**, and a screenshot of part of the data file can be seen in Figure 45.1.

45.3 ANALYSIS SETUP: MEETING THE HOMOGENEITY OF VARIANCE ASSUMPTION

Open the file named **history test** and from the main menu select **Analyze → Compare Means → Independent-Samples T Test**. This produces the main **Independent-Samples T Test** dialog window shown in Figure 45.2. We have moved the **history_test** variable to the **Test Variable(s)** panel; this will be the dependent variable in the analysis.

We next move **sex** into the **Grouping Variable** panel to represent our independent variable. Note in Figure 45.2 the question marks in parentheses in the panel next to **sex**. This is a prompt from IBM SPSS asking us to indicate the values (codes) of the **Grouping Variable**. We accomplish this by clicking the **Define Groups** pushbutton, which displays the **Define Groups** dialog window seen in Figure 45.3. The two group

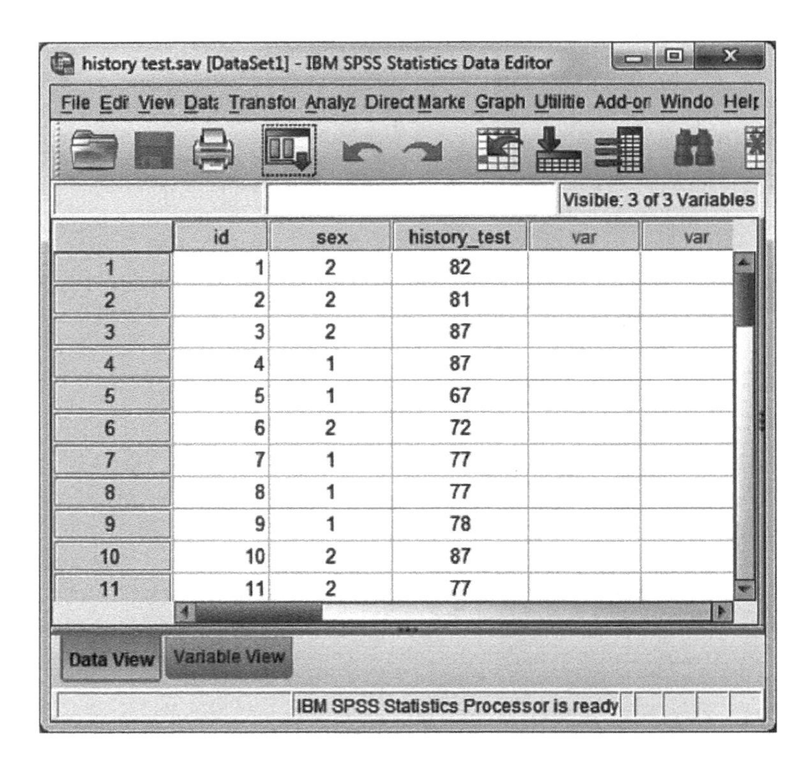

	id	sex	history_test	var	var
1	1	2	82		
2	2	2	81		
3	3	2	87		
4	4	1	87		
5	5	1	67		
6	6	2	72		
7	7	1	77		
8	8	1	77		
9	9	1	78		
10	10	2	87		
11	11	2	77		

FIGURE 45.1

A portion of the data file.

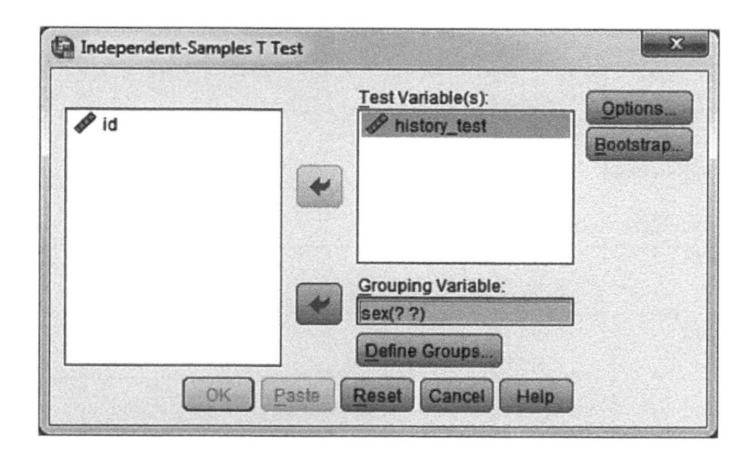

FIGURE 45.2

The main **Independent-Samples T Test** dialog window.

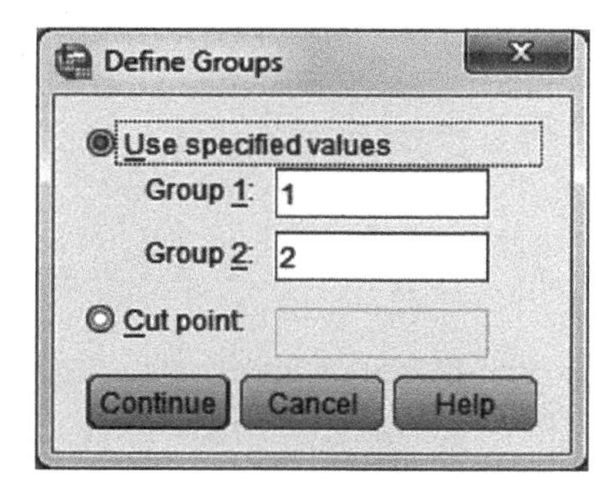

FIGURE 45.3

The **Define Groups** screen of the **Independent-Samples T Test** dialog window.

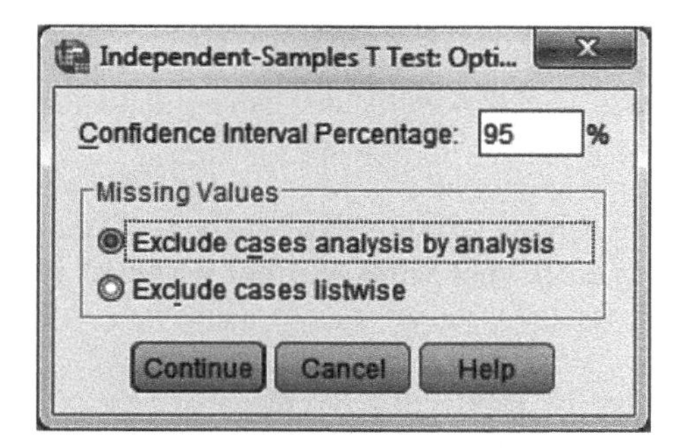

FIGURE 45.4

The options window of the **Independent-Samples T Test**.

codes are **1** and **2** and we have filled them in. Clicking **Continue** brings us back to the main **Independent-Samples T Test** dialog window.

Clicking the **Options** pushbutton produces the **Options** dialog window shown in Figure 45.4. We can set the **Confidence Interval Percentage** to whatever level we wish, but we have retained the default value of **95%**. The **Missing Values** box has also been left at its default setting of **Exclude cases analysis by analysis**. Clicking **Continue** takes us back to the main dialog window, and clicking **OK** generates the analysis.

45.4 ANALYSIS OUTPUT: MEETING THE HOMOGENEITY OF VARIANCE ASSUMPTION

The output for the analysis is presented in Figure 45.5. The **Group Statistics** table provides the group size, mean, standard deviation, and standard error of the mean for

Group Statistics

	sex	N	Mean	Std. Deviation	Std. Error Mean
history_test	1 Male	12	75.58	5.485	1.583
	2 Female	13	83.46	5.911	1.639

Independent Samples Test

		Levene's Test for Equality of Variances		t-test for Equality of Means						
									95% Confidence Interval of the Difference	
		F	Sig.	t	df	Sig. (2-tailed)	Mean Difference	Std. Error Difference	Lower	Upper
history_test	Equal variances assumed	.141	.711	-3.446	23	.002	-7.878	2.286	-12.608	-3.149
	Equal variances not assumed			-3.457	22.998	.002	-7.878	2.279	-12.593	-3.163

FIGURE 45.5 Output from the **Independent-Samples T Test**.

each level of the categorical independent variable. Males and females averaged **75.58** and **83.46**, respectively, in the history test.

The **Independent Samples Test** table provides several pieces of information. The first two columns of the statistical output show the results for Levene's equality of variances test, tested with an *F* ratio (see Chapter 47). The Levene test evaluates the ratio of the group variances (larger variance divided by smaller variance), and a statistically significant result indicates unequal variance across groups. In the present example, the ***F*** of **.141** yielded a probability of occurrence (**Sig.**) of **.711**, given the null hypothesis (of no difference in variances) is true. The *F* ratio was therefore not statistically significant, indicating that the assumption of equal variances was met.

To the right of the Levene test results, the table splits into two rows, and the one we use depends on whether the assumption of homogeneity of variances was met (**Equal variances assumed** in the top row) or was violated (**Equal variances not assumed** in the bottom row). Because the assumption holds for the data set we are using here, we examine the output in the first row.

The **Mean Difference** of **−7.878** is computed by subtracting the mean of Group 2 (**83.46**) from the mean of Group 1 (**75.58**), and the **Std. Error Difference** (**2.286**) is the denominator of the *t* ratio. The **95% Confidence Interval of the Difference** is centered on the mean difference and is based on the **Std. Error Difference**.

The *t* test represented for **Equal variances assumed** is the standard *t* test. The *t* statistic of **−3.446** was negative, informing us that the group defined as **1** in the **Define Groups** dialog window had a smaller mean than of the group defined as **2** (the mean difference is computed as mean of Group 1 minus mean of Group 2). Had these codes been reversed (the coding in a two-group *t* test is arbitrary), the *t* value would have been positive. Note that whether positive or negative, the probability indicates the distance in the sampling distribution between zero and this absolute value. Here, with **23** degrees of freedom (total degrees of freedom is $N - 1$, which is $25 - 1$, or 24, and we subtract from that one additional degree of freedom for the two-group independent variable), the *t* value of **−3.446** was statistically significant; its probability of occurrence if the null hypothesis is true (**Sig. (2-tailed)**) is **.002**, which is less than our (presumed) alpha level of .05. Viewing the means allows us to interpret this result to indicate that females performed significantly better than males in the history test.

45.5 MAGNITUDE OF THE MEAN DIFFERENCE

With a statistically significant mean difference, it is appropriate to assess the magnitude of the effect. There are two related and not mutually exclusive ways to make such a determination:

- *Strength of Effect.* This approach is quite general and applies to most procedures based on the general linear model (e.g., t tests, ANOVA, least-squares regression); its index is the squared correlation coefficient known as eta square.
- *Effect Size.* This approach converts the mean difference between two conditions into standard deviation units; its index is Cohen's d statistic.

Strength of Effect

Strength of effect is indexed by eta square and represents the amount of variance of the dependent variable that is explained by the independent variable (Meyers et al., 2013); it is akin to R^2 in ordinary least-squares regression (Cohen et al., 2003). In the absence of any context, the general guideline suggested by Cohen (1988, pp. 285–288) for interpreting the strength of effect is that eta square values of .01, .06, and .14 can be considered to be small, medium, and large, respectively. For the two-group t test, eta square can be computed as follows (Hays, 1981):

$$\text{eta square} = t^2/(t^2 + \text{degrees of freedom})$$

In the present example, $t = -3.446$, $t^2 = 11.875$, and degrees of freedom $= 23$; eta square thus computes to .043. We would therefore consider the superior performance of the females to represent a medium effect (with no context within which to frame the importance of this difference); that is, test-taker sex explained about 4% of the variance in history test scores.

Effect Size

Effect size is a concept rather strongly identified by Jacob Cohen (1988) and is symbolized by d. It provides a criterion for interpreting the separation of two means by determining the distance between them in standard deviation units. Cohen's d is computed by dividing the mean difference by the (weighted) average standard deviation of the groups. For example, a d value of .50 informs us that the means are half a standard deviation apart. In the absence of any context, effect sizes of .20, .50, and .80 are considered to be small, medium, and large, respectively (Cohen, 1988, pp. 24–27).

In the present example, the weighted average of the standard deviations is computed as follows: $((\mathbf{5.484} * 12) + (\mathbf{5.911} * 13))/25 = 5.706$. With a mean difference of -7.878, Cohen's $d = -7.878/5.706 = -1.38$. Thus, males scored approximately one and a third standard deviation units lower than females; generally, a mean difference exceeding one standard deviation would be considered a relatively large effect size.

45.6 NUMERICAL EXAMPLE: VIOLATING THE HOMOGENEITY OF VARIANCE ASSUMPTION

The present example concerns performance in a test of aptitude for science (**science_aptitude**) by male (**sex** coded as **1**) and female (**sex** coded as **2**) elementary-school students. We are interested in whether or not there are reliable (statistically different) test

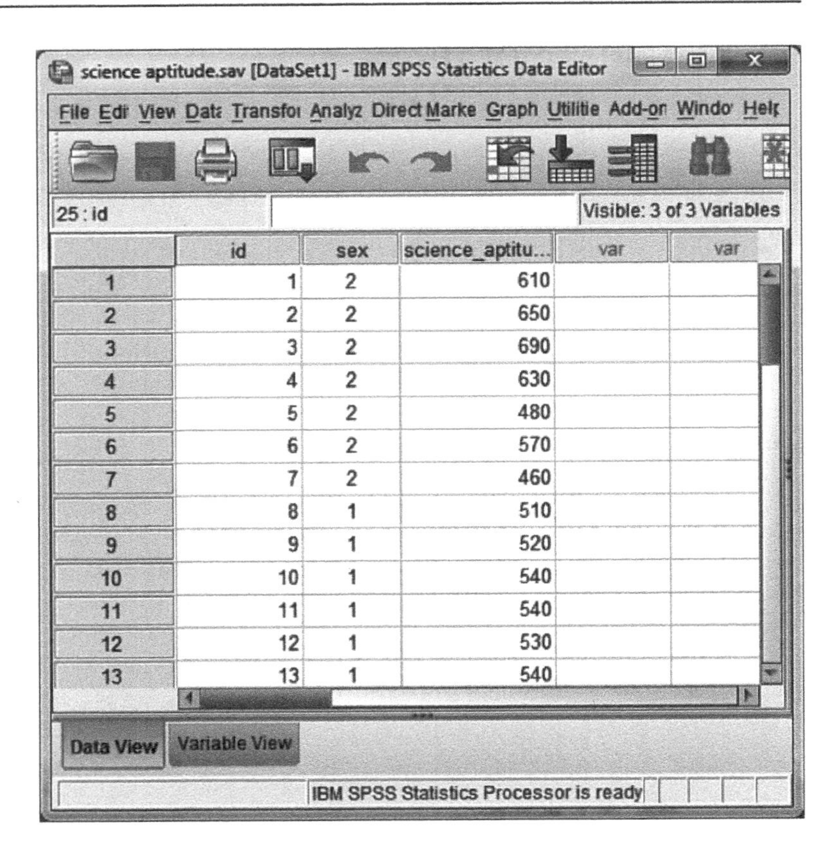

FIGURE 45.6
A portion of the data file.

score differences between males and females. The data can be found in the file named **science aptitude**, and a screenshot of a part of the data file can be seen in Figure 45.6.

45.7 ANALYSIS SETUP: VIOLATING THE HOMOGENEITY OF VARIANCE ASSUMPTION

Open the IBM SPSS file named **science aptitude** and from the main menu select **Analyze ➔ Compare Means ➔ Independent-Samples T Test**. This produces the **Independent-Samples T Test** dialog window shown in Figure 45.7. We have moved the **science_aptitude** variable to the **Test Variable(s)** panel to serve as the dependent variable in the analysis.

We move **sex** into the **Grouping Variable** panel to represent our independent variable and click the **Define Groups** pushbutton where we supply the two group codes **1** and **2**. Clicking **Continue** brings us back to the main **Independent-Samples T Test** dialog window. As we do not wish to change any of the defaults in the **Options** dialog window, we click **OK** to generate the analysis.

45.8 ANALYSIS OUTPUT: VIOLATING THE HOMOGENEITY OF VARIANCE ASSUMPTION

The output for the analysis is presented in Figure 45.8. The **Group Statistics** table shows us that males and females averaged **510.77** and **584.29**, respectively, in the aptitude assessment. The **Independent Samples Test** table provides the results for Levene's equality of variances test. In the present example, the F of **11.024** yielded a probability of

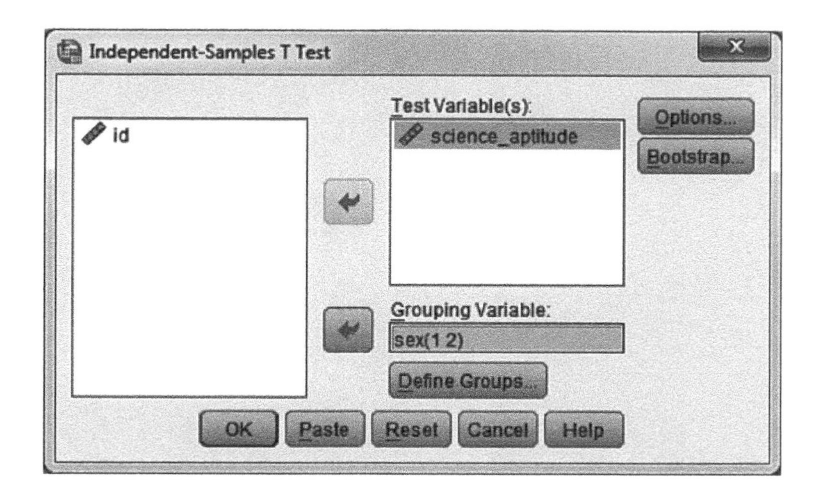

FIGURE 45.7

The main **Independent-Samples T Test** dialog window.

Group Statistics

	sex	N	Mean	Std. Deviation	Std. Error Mean
science_aptitude	1 Male	13	510.77	31.215	8.657
	2 Female	7	584.29	86.382	32.649

Independent Samples Test

		Levene's Test for Equality of Variances		t-test for Equality of Means						
									95% Confidence Interval of the Difference	
		F	Sig.	t	df	Sig. (2-tailed)	Mean Difference	Std. Error Difference	Lower	Upper
science_aptitude	Equal variances assumed	11.024	.004	-2.800	18	.012	-73.516	26.257	-128.680	-18.353
	Equal variances not assumed			-2.176	6.856	.067	-73.516	33.778	-153.728	6.695

FIGURE 45.8 Output from the **Independent-Samples T Test**.

occurrence (**Sig.**) of **.004**, given the null hypothesis (of no difference in variances) is true. In this instance, the *F* ratio was statistically significant, indicating that the assumption of equal variances was violated.

To the right of the Levene test results, the table splits into two rows, and because the assumption of equal variances was not met for the data set we are using here, we examine the output in the second row (**Equal variances not assumed**).

The **Mean Difference** of **−73.516** is computed by subtracting the mean of Group 2 (**584.29**) from the mean of Group 1 (**510.77**), and the **Std. Error Difference** (**33.778**) is the denominator of the *t* ratio. The **95% Confidence Interval of the Difference** is centered on the mean difference and is based on the **Std. Error Difference**.

The *t* test represented for **Equal variances not assumed** is the Welch–Satterthwaite *t* test. The *t* statistic of **−2.176** is not statistically significant ($p = .067$) based on an adjusted degrees of freedom value of **6.856**. Had we used the standard *t* test (**Equal variances assumed**), our *t* value of **−2.800** would have been based on **18** degrees of freedom and with an expected probability of occurrence of **.012** would have reached an alpha level of .05. Had we taken the standard *t* test as our result, we would have probably committed a Type I error (claiming a difference when the group means are not really different).

CHAPTER 46

Paired-Samples *t* Test

46.1 OVERVIEW

The *paired-samples t test* is a somewhat different way to analyze data arrayed in a Pearson correlation (see Chapter 22) or simple regression (see Chapter 24) design. Each case is associated with two scores that may represent either different variables or the same variable measured at two different points in time. We invoke the *t* test when we are interested in the differences between the means of the two measures. Such an approach uses a paired-samples *t* test because the two measures are linked or paired for each case.

Probably the more common application of the paired-samples design is the pretest–posttest design to assess any meaningful change that may have occurred in the dependent variable between the two time periods. This is known as a repeated measures or within-subject design, and the one-way within-subjects ANOVA is the more general procedure of this design (see Chapter 51). The paired-samples *t* test explicitly makes use of the Pearson correlation between the measures.

46.2 NUMERICAL EXAMPLE

Our present example examines the degree of change in mental health of clients following 6 months of mental health treatment. The dependent variable is performance on the GAF described in Chapter 44. Clients were assessed at intake (pretest named **GAF_time1**) and again after 6 months (posttest named **GAF_time2**). The question of interest is whether a statistically significant change occurred between the two time periods. The data can be found in the file **GAF time1 time2**. A portion of the data file is presented in Figure 46.1, and as can be seen, the data structure is identical to that of a correlation design.

46.3 ANALYSIS SETUP

Open the file named **GAF time1 time2**. From the main menu select **Analyze→ Compare Means→ Paired-Samples T Test**, which produces the main **Paired-Samples T Test** dialog window shown in Figure 46.2. From the variable list on the left side of the dialog window, we have moved **GAF_time1** to the **Variable 1** panel and **GAF_time2** to the **Variable 2** panel.

Performing Data Analysis Using IBM SPSS®, First Edition.
Lawrence S. Meyers, Glenn C. Gamst, and A. J. Guarino.

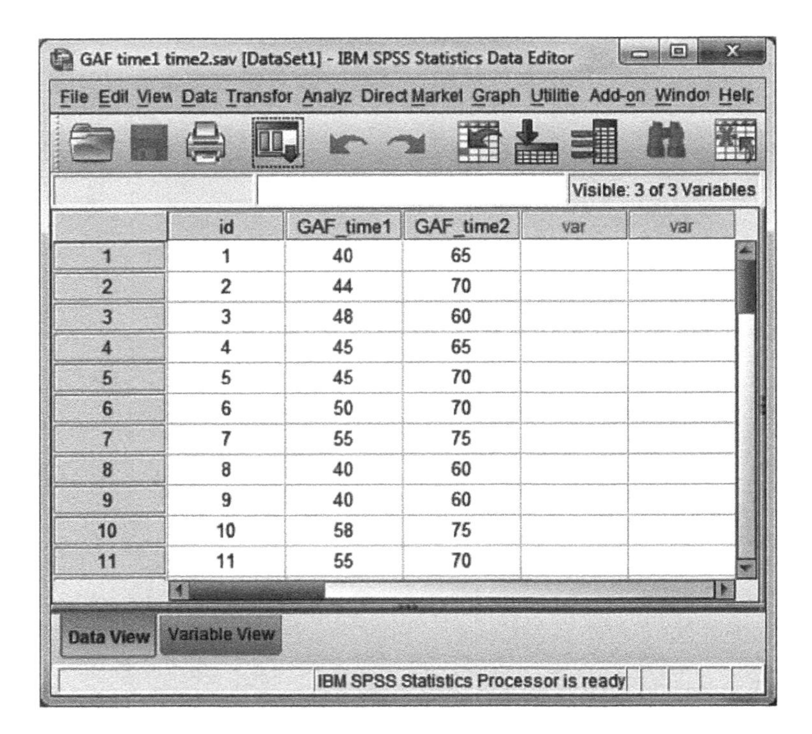

	id	GAF_time1	GAF_time2	var	var
1	1	40	65		
2	2	44	70		
3	3	48	60		
4	4	45	65		
5	5	45	70		
6	6	50	70		
7	7	55	75		
8	8	40	60		
9	9	40	60		
10	10	58	75		
11	11	55	70		

FIGURE 46.1
A part of the data file.

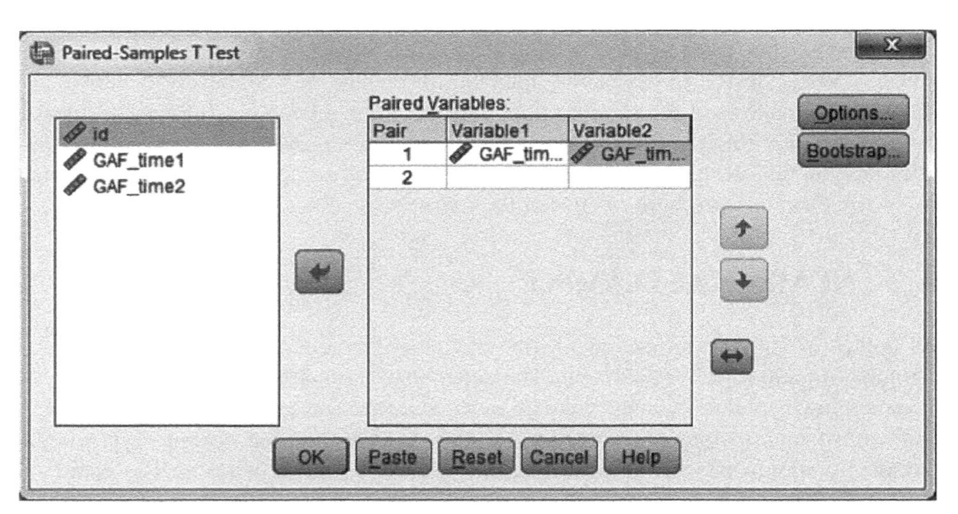

FIGURE 46.2 The main dialog window of **Paired-Samples T Test**.

The **Options** dialog window is the same as we have seen in Chapter 45; we can set the **Confidence Interval Percentage** to whatever level we wish and deal with missing values in a couple of different ways. We retain the defaults of **95% Confidence Interval Percentage** and **Exclude cases analysis by analysis** (see Figure 46.3). Clicking **Continue** takes us back to the main dialog window, and clicking **OK** generates the analysis.

46.4 ANALYSIS OUTPUT

The output for the analysis is presented in Figure 46.4. The top table (**Paired Samples Statistics**) provides descriptive statistics for each measure. It appears that the mean for

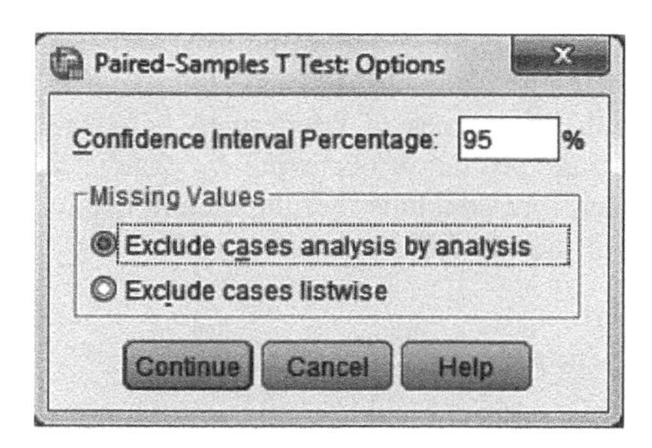

FIGURE 46.3
The **Options** dialog window of **Paired-Samples T Test**.

Paired Samples Statistics

		Mean	N	Std. Deviation	Std. Error Mean
Pair 1	GAF_time1 Global Assessment of Functioning Time 1	46.08	25	5.852	1.170
	GAF_time2 Global Assessment of Functioning Time 2	65.52	25	6.232	1.246

Paired Samples Correlations

		N	Correlation	Sig.
Pair 1	GAF_time1 Global Assessment of Functioning Time 1 & GAF_time2 Global Assessment of Functioning Time 2	25	.801	.000

Paired Samples Test

		Paired Differences					t	df	Sig. (2-tailed)
					95% Confidence Interval of the Difference				
		Mean	Std. Deviation	Std. Error Mean	Lower	Upper			
Pair 1	GAF_time1 Global Assessment of Functioning Time 1 - GAF_time2 Global Assessment of Functioning Time 2	-19.440	3.831	.766	-21.021	-17.859	-25.375	24	.000

FIGURE 46.4 Output from the **Paired-Samples T Test**.

GAF_time2 (**65.52**) is higher than the mean for **GAF_time1** (**46.08**); as higher scores on the GAF represent healthier functioning, it appears that the treatment might have been effective.

The middle table of Figure 46.4 (**Paired Samples Correlations**) informs us that the Pearson correlation between the GAF scores is **.801**. The **Sig**. value of **.000** indicates that the chances of this correlation occurring by chance if the null hypothesis is true is less than .001, and we would take this to be statistically significant.

The bottom table of Figure 46.4 shows the results of the *t* test. The mean difference **−19.440** (mean of **GAF_time1** − mean of **GAF_time2**) is negative because the posttest scores are higher than the pretest scores. The standard deviation of the differences between pairs of variables is **3.831**, and the correlated difference standard error of the mean is **.766**. The value of the t statistic is **−25.375**; with **24** degrees of freedom ($N - 1$), the **Sig**. value of **.000** indicates that the chances of this correlation occurring by chance if the null hypothesis is true is less than .001. We therefore conclude that the mental health treatment significantly improved the GAF score of the clients.

46.5 MAGNITUDE OF THE MEAN DIFFERENCE

As was true for the independent-groups t test, we can evaluate the magnitude of the mean difference as a squared correlation coefficient; in this context, it is indexed by computing the squared Pearson r. The IBM SPSS® output has provided the Pearson r of **.801**; squaring this value yields .64. We may then say that the pretest and posttest scores share 64% of their variance.

Cohen's d is computed as the mean difference divided by the average standard deviation. The mean difference is **−19.440** and the average standard deviation is 6.042 ((**6.232** + **5.852**)/2) = 6.042); Cohen's d thus computes to 3.217. With the mean difference spanning more than three standard deviation units, we would judge the effect size to be very large.

PART 15

UNIVARIATE GROUP DIFFERENCES: ANOVA AND ANCOVA

CHAPTER 47

One-Way Between-Subjects ANOVA

47.1 OVERVIEW

A one-way between-subjects ANOVA is the generalized form of an independent-groups t test. The t test involves the comparison of just two groups representing a single independent variable; the ANOVA permits us to compare two or more groups. The name of the ANOVA design is derived from the following considerations:

- It is a *one-way* design in that there is only one independent variable, although any number of groups or levels representing that independent variable can be subsumed.
- It is a *between-subjects* design in that the cases in each group must be independent of each other, that is, they must be different entities.

The origins of ANOVA can be traced to Sir Ronald Aylmer Fisher when he worked at the Rothamsted Agricultural Experimental Station from 1919 to 1933. The staff researchers at the station in evaluating the effectiveness of different fertilizers were using only one fertilizer each year, by comparing them across the years despite differences in temperature and rainfall. Fisher (1921a) showed this methodology to be flawed. Instead, he started using multiple fertilizers each year by randomly assigning them to different plots within the field to control for a host of environmental conditions and developed the family of ANOVA techniques to analyze the data (Fisher, 1921b, 1925, 1935a; Fisher & Eden, 1927; Fisher & Mackenzie, 1923).

The statistic that is computed and tested for statistical significance in the ANOVA procedure is an F ratio. The letter F was first used by George W. Snedecor at the Iowa State University in the first edition of his book *Statistical Methods* (Snedecor, 1934) as a way to honor Fisher who he knew personally and very much respected. A more complete history can be found in Salsburg (2001), and a more comprehensive treatment of a wide range of ANOVA designs can be found in Gamst et al. (2008).

The statistical strategy underlying the ANOVA is to partition (analyze) the total variance of the dependent variance into its constituent sources of variance. This total variance is defined in terms of the difference between the grand or overall mean of the entire sample and each score associated with each case.

Different ANOVA designs partition the variance of the dependent variable into somewhat different sources of variance. In a one-way between-subjects design, the two sources

Performing Data Analysis Using IBM SPSS®, First Edition.
Lawrence S. Meyers, Glenn C. Gamst, and A. J. Guarino.

of variance into which the total variance of the dependent variable is partitioned are as follows:

- *Between-Groups Variance.* This reflects the differences in means between the groups and is defined in terms of the differences between the group means and the grand mean. It represents the effect of the independent variable. However, if there is any systematic experimental error (confounding) resulting from some other source covarying with the treatment effect, the mean differences will also represent the source of experimental error (Keppel & Wickens, 2004). The variance associated with the between-groups variance is the numerator of the F ratio.
- *Within-Groups Variance.* This reflects the variability within each group and is defined in terms of the difference between the group mean and each score within that group. It represents the error of measurement in the study. The variance associated with this source of variance is the denominator of the F ratio.

47.2 NUMERICAL EXAMPLE

The data we use for our example are fictional. The Vehicle Testing Corporation has compared three types of vehicles (**vehicle_type**), namely, **mid-sized cars** (coded as **1** in the data file), **full-sized cars** (coded as **2** in the data file), and **cross-over SUVs** (coded as **3** in the data file), based on their risk to drivers in case of an accident. Using crash-test dummies, the researchers have estimated the level of injury that may be expected (**expected_injury**) from a 40-mph collision with a solid and immovable object. Higher **expected_injury** scores reflect a greater amount of anticipated injury. The data file is named **vehicle safety**, and a screenshot of part of the data file is shown in Figure 47.1.

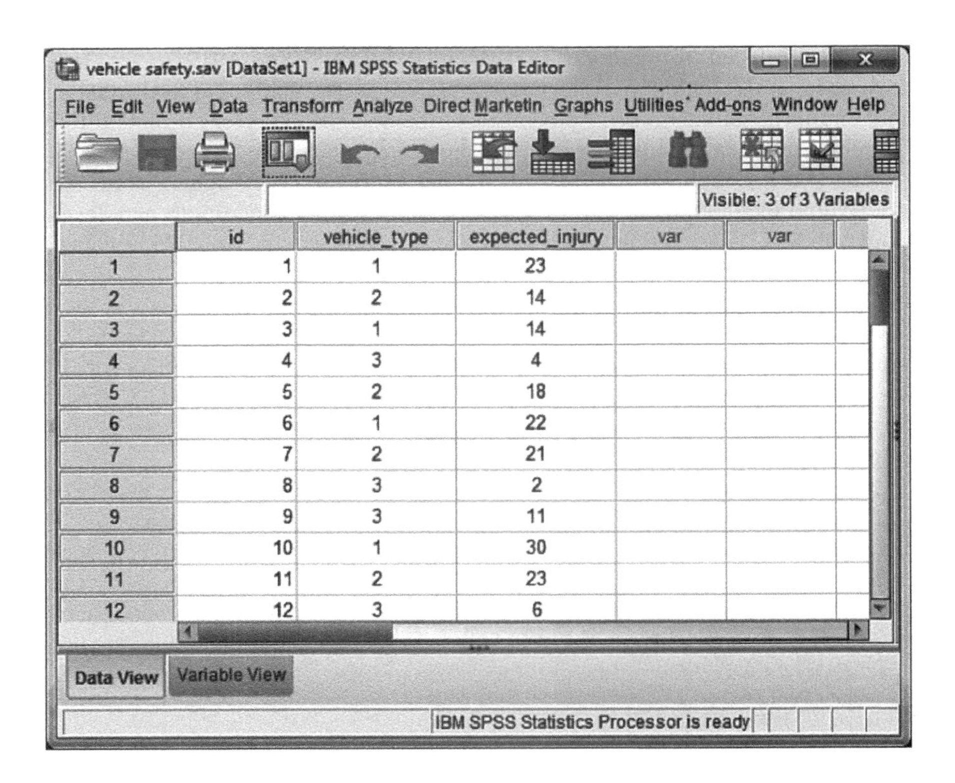

	id	vehicle_type	expected_injury	var	var
1	1	1	23		
2	2	2	14		
3	3	1	14		
4	4	3	4		
5	5	2	18		
6	6	1	22		
7	7	2	21		
8	8	3	2		
9	9	3	11		
10	10	1	30		
11	11	2	23		
12	12	3	6		

FIGURE 47.1
A portion of the data file.

47.3 ANALYSIS STRATEGY

IBM SPSS® has several procedures available that we can use to perform an ANOVA. We use the **General Linear Model** in our ANOVA chapters in this book, as this module can also be applied to the range of designs that we discuss in the subsequent chapters; the one exception to this will be in Chapter 48 where the analysis of polynomial trends is conveniently performed with the **One-Way ANOVA** procedure.

If the effect of the independent variable is statistically significant (in what is called the *omnibus* or overall analysis), then it is necessary to perform additional statistical tests to determine which pairs of means are statistically significant (assuming that there are more than two groups in the analysis); these procedures are generically called *multiple comparison tests* or *tests of simple effects*. There are many types of multiple comparison tests available as described in detail in Gamst et al. (2008), but we will illustrate the easiest to apply of the procedures here: the post hoc test. Of those post hoc tests that appear to be a good compromise for statistical power and also protect against alpha level inflation, we will use the Studentized Range variation of the Ryan–Enoit–Gabriel–Welsch test. For convenience, we will perform this test in the setup of the omnibus analysis, but not examine its outcome unless the between-groups F ratio is statistically significant.

47.4 ANALYSIS SETUP

We open the data file **vehicle safety** and from the main menu, we select **Analyze ➔ General Linear Model ➔ Univariate** (because we have just a single dependent variable in the analysis). This opens the main **Univariate** window as shown in Figure 47.2. We move **expected_injury** into the **Dependent Variable** panel and **vehicle_type** into the **Fixed Factor(s)** panel.

The **Options** dialog window shown in Figure 47.3 is divided into two major areas, the upper **Estimated Marginal Means** area and the lower **Display** area. For our one-way between-groups design, we do not work with **Estimated Marginal Means**. In the **Display** area, we check **Descriptive statistics** and **Homogeneity tests** (to obtain the Levine test of equal group variances). Click **Continue** to return to the main dialog window.

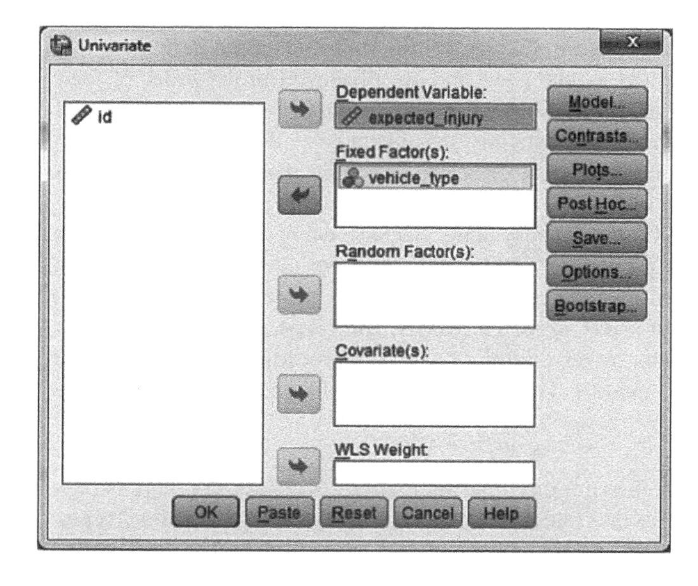

FIGURE 47.2

The main **Univariate** dialog window in the **General Linear Model**.

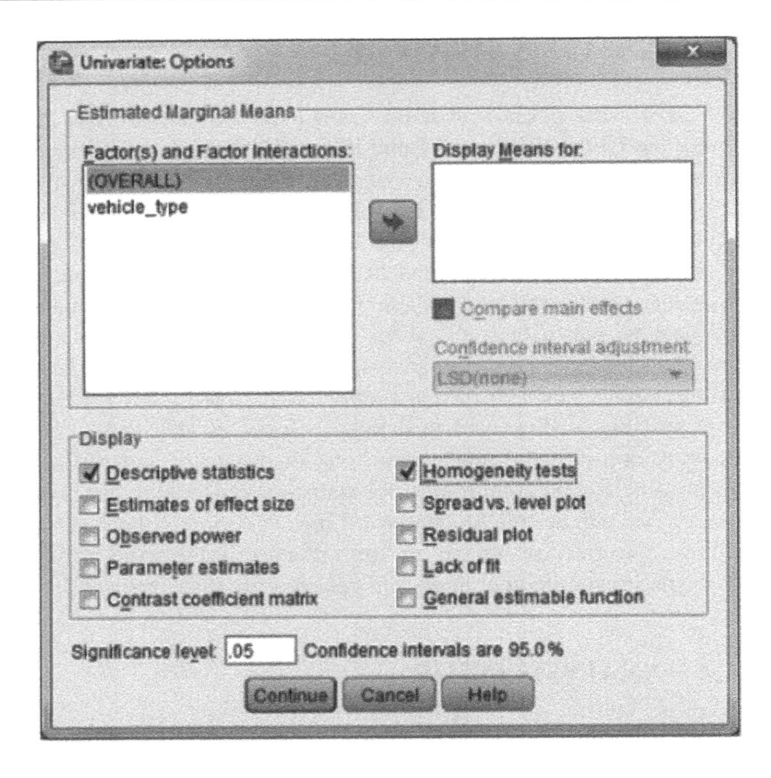

FIGURE 47.3
The **Options** dialog window of **Univariate**.

The **Post Hoc** dialog window is shown in Figure 47.4. In the upper portion of the window, we move **vehicle_type** into the **Post Hoc Tests for** panel and check **R-E-G-W-Q** to obtain the Ryan–Enoit–Gabriel–Welsch Studentized Range test; this test assumes equal variances and, if the Levene test returned a statistically significant result (indicating that we failed to meet this assumption), we would perform the analysis again with the **Tamhane's T2** test. Click **Continue** to return to the main dialog window and click **OK** to perform the analysis.

47.5 ANALYSIS OUTPUT

The **Descriptive Statistics** table in Figure 47.5 shows the mean, standard deviation, and group sizes. Groups are represented by the rows and are displayed in the order of their codes. The bottom table in Figure 47.5 shows Levene's statistic for homogeneity of variances. The F ratio of **.545** was not statistically significant (p = **.586**) when evaluated with **2** and **30** degrees of freedom; we may therefore conclude that the assumption of equal group variances was not violated.

Figure 47.6 displays the ANOVA summary table. Because we performed this analysis in the **General Linear Model** procedure, IBM SPSS presents all of the output from the general linear model analysis. Three of the rows include the output from the full regression model that we will not use (**Corrected Model**, **Intercept**, and **Total**), whereas other rows are comparable to what other (simpler) ANOVA procedures ordinarily yield (a *corrected*, *partial*, or *reduced* model). The sums of squares for the rows represent the following:

- **Corrected Model**. This is the sum of all between-subjects effects. As there is only one between-subjects effect in a one-way between-subjects design (**vehicle_type**), this sum of squares is equal to that of the effect of **vehicle_type**. It is called

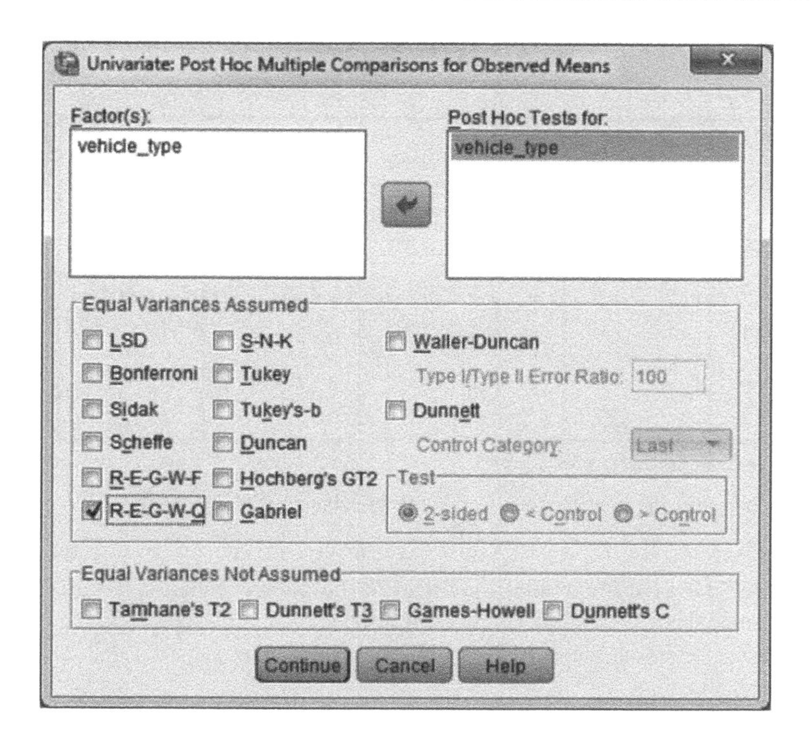

FIGURE 47.4

The **Post Hoc** dialog window of **Univariate**.

Descriptive Statistics

Dependent Variable:expected_injury

vehicle_type	Mean	Std. Deviation	N
1 mid-size car	23.91	8.288	11
2 full-size car	19.82	7.441	11
3 cross-over SUV	9.64	6.249	11
Total	17.79	9.387	33

Levene's Test of Equality of Error Variances[a]

Dependent Variable:expected_injury

F	df1	df2	Sig.
.545	2	30	.586

Tests the null hypothesis that the error variance of the dependent variable is equal across groups.

a. Design: Intercept + vehicle_type

FIGURE 47.5

Descriptive statistics and the results of the Levene test.

Tests of Between-Subjects Effects

Dependent Variable:expected_injury

Source	Type III Sum of Squares	df	Mean Square	F	Sig.
Corrected Model	1188.424[a]	2	594.212	10.929	.000
Intercept	10441.485	1	10441.485	192.046	.000
vehicle_type	1188.424	2	594.212	10.929	.000
Error	1631.091	30	54.370		
Total	13261.000	33			
Corrected Total	2819.515	32			

a. R Squared = .421 (Adjusted R Squared = .383)

FIGURE 47.6 Summary table for the omnibus one-way between-subjects ANOVA.

"**Corrected Model**" because it excludes the effect of the intercept that is specific to the regression analysis.

- **Intercept**. This is a regression-specific effect representing the value of the Y intercept in the model.
- **vehicle_type**. This is the effect of the independent variable and our focus of interest.
- **Error**. This is the within-groups effect and is an estimate of the measurement error.
- **Total**. This is the total sum of squares including the regression-specific **Intercept** effect.
- **Corrected Total**. This is the total for the reduced model including only the effects of the independent variable and the error source of variance. When computing the value of eta square as an estimate of the strength of effect for the independent variable, this is the proper denominator of that ratio.

We focus on the reduced model. The F ratio is computed as the **Mean Square** (the variance) associated with the effect for the independent variable (computed as sum of squares for **vehicle_type** divided by its degrees of freedom shown under **df**) divided by the **Mean Square** (the variance) associated with the **Error** effect (computed as sum of squares for **Error** divided by its degrees of freedom). With **2** and **30** degrees of freedom, the F ratio for the effect of the independent variable of **10.929** is statistically significant ($p < .001$). We note that attributing the effect to the independent variable presumes that there is no systematic error component covarying with the treatment effect (Keppel & Wickens, 2004).

The eta square value associated with the effect is computed as the sum of squares associated with **vehicle_type** divided by the sum of squares associated with **Corrected Total**. In the present example, we have **1188.424/2819.515** or .421; this value is shown in the footnote for the table as **R Squared** together with a value for the adjusted R^2 (it is traditional in the context of ANOVA to report R^2 rather than the adjusted R^2 and to refer to it as an eta square value).

With the effect of the independent variable statistically significant, we are confident that some group means (or some combination of group means) differ from some others (or some combination of group means). The Ryan–Enoit–Gabriel–Welsch post hoc test addresses this issue, and its results are shown in Figure 47.7. The groups, shown in the

Post Hoc Tests
vechicle_type
Homogeneous Subsets

expected_injury

Ryan-Einot-Gabriel-Welsch Range[a]

vehicle_type	N	Subset 1	Subset 2
3 cross-over SUV	11	9.64	
2 full-size car	11		19.82
1 mid-size car	11		23.91
Sig.		1.000	.203

Means for groups in homogeneous subsets are displayed.
Based on observed means.
The error term is Mean Square(Error) = 54.370.

a. Alpha =

FIGURE 47.7
Results of the Ryan–Enoit–Gabriel–Welsch post hoc test.

rows, are ordered by the magnitude of their means; the columns represent subsets of means whose values do not significantly differ under an alpha level of .05.

In the present instance, there are two **Homogeneous Subsets** of means. Means within a given **Subset** column do not statistically differ; means in different **Subset** columns do statistically differ. The interpretation of the output is therefore as follows:

- The **cross-over SUV** mean of **9.64** differs significantly from (it is significantly lower than) the other two means, as it is in its own **Subset** column.
- The **full-size car** mean of **19.82** and the **mid-size car** mean of **23.91** do not significantly differ ($p = .203$), as they are together in a **Subset** column.

We may therefore conclude that crossover SUVs are projected by the Vehicle Testing Corporation to result in less serious injury following a 40-mph crash than either full-sized or mid-sized cars.

We can also compute Cohen's d value for each significant mean difference. For **cross-over SUVs** and **full-size cars**, the mean difference is 10.18 and the average standard deviation is 6.825; Cohen's d thus computes to approximately 1.49. For **cross-over SUVs** and **mid-size cars**, the mean difference is 14.27 and the average standard deviation is 7.269; Cohen's d thus computes to approximately 1.96. Both of these represent substantial effect sizes.

CHAPTER 48

Polynomial Trend Analysis

48.1 OVERVIEW

A polynomial trend analysis is a specialized application of a one-way between-subjects ANOVA. It examines the shape of the function (in terms of polynomials) relating three or more groups where the groups are represented on the X-axis and the value of the means on the dependent variable are represented on the Y-axis. The data points are connected by lines to show the pattern. The types of possible polynomial relationships are a function of the number of groups.

- With two groups, we are limited to a linear relationship. Thus, the dependent variable is predictable only from a linear function: Y is a function of X.
- With three groups, we may have up to a quadratic relationship. Thus, the dependent variable is potentially predictable from a linear and quadratic function: Y is a function of X and X^2.
- With four groups, we may have up to a cubic relationship. Thus, the dependent variable is potentially predictable from a linear, quadratic, and cubic function: Y is a function of X, X^2, and X^3.
- With five groups, we may have up to a quartic relationship. Thus, the dependent variable is potentially predictable from a linear, quadratic, cubic, and quartic function: Y is a function of X, X^2, X^3, and X^4.

The primary condition to be met in a polynomial trend analysis is that the groups are able to be ordered in magnitude based on at least an interval-level measurement variable. This condition is necessary because the "spacing" between the groups on the X-axis must be fixed in order for the shape of the function to make sense (if we could place the groups anywhere on the axis, then there could be no "true" shape to the function). Examples of groups that would meet this requirement include:

- different exam preparation times (e.g., 1 week, 2 weeks, and 3 weeks);
- different drug dosages (50, 100, 150 mg).

Just because a relationship is possible does not mean that it will be exhibited in the data of every study. Some of the polynomial functions may not be statistically significant,

Performing Data Analysis Using IBM SPSS®, First Edition.
Lawrence S. Meyers, Glenn C. Gamst, and A. J. Guarino.

and more than one can sometimes describe the data. For example, with three groups, the means can be related in a linear and/or quadratic manner:

- *Fully Linear.* The means can fall on a straight line; the quadratic component will not be significant.
- *Fully Quadratic.* The means can be arrayed in a "V" shape or an inverted "V" shape; the linear component will not be significant.
- *Partially Linear and Quadratic.* The means can show an increase or a decrease and then level off; both the linear and quadratic components may be statistically significant.

48.2 NUMERICAL EXAMPLE

The data we use for our example are fictional. The Good Food Market has a large number of stores across the country selling a variety of food and drink products, including wine. Management was interested in determining the effect on wine sales when different numbers of wineries were represented on the shelves of the stores. A total of 56 of their stores participated in this study, 14 in each of 4 groups. Stores in each group stocked vintages from a different number of wineries, 20, 40, 60, or 80 different wineries; this was the independent variable named **n_wineries** in the data file. We note here that number of wineries qualifies as an interval-level variable, thus meeting our requirement for performing a polynomial trend analysis. The dependent variable named **sales_dollars** in the data file was the sales volume on average each day measured in hundreds of dollars of wine sales. The data file is named **Wine Sales**.

48.3 ANALYSIS STRATEGY

A polynomial trend analysis is a one-way between-subjects ANOVA with one extra piece: the sum of squares associated with the between-groups variance (the effect of the independent variable) is partitioned into polynomial components. The analysis thus entails two levels of partitioning of the variance of the dependent variable.

- At the highest level, the variance of the dependent variable is partitioned into between-groups variance and error or within-groups variance. The F ratio (mean square associated with between-groups variance divided by mean square associated with error variance) assesses the effect of the independent variable (number of different wineries stocked). With a statistically significant effect, we can perform our multiple comparison tests.
- At a more microscopic level, the between-groups variance is itself partitioned into polynomial components. Each polynomial component will be associated with (will explain) a certain amount of variance and will thus be associated with its own F ratio. We can examine a plot of the means and interpret each of the statistically significant polynomial components.

48.4 ANALYSIS SETUP

We open the data file **Wine Sales** and from the main menu, we select **Analyze ➔ Compare Means ➔ One-Way ANOVA**. This opens the main **One-Way ANOVA** window as shown in Figure 48.1. The **One-Way ANOVA** is a specialized procedure that allows us

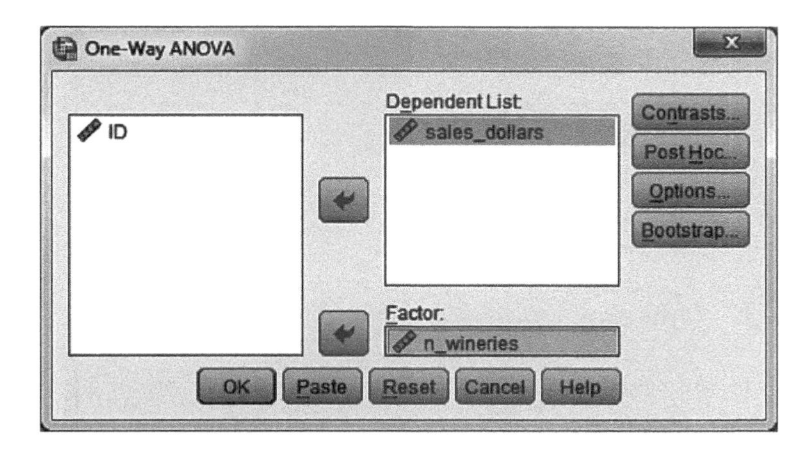

FIGURE 48.1

The main **One-Way ANOVA** dialog window

to perform several types of mean contrasts (see Gamst et al. (2008)) including polynomial contrasts. We move **sales_dollars** into the **Dependent List** panel and **n_wineries** into the **Factor** panel.

In the **Options** dialog window shown in Figure 48.2, we select **Descriptives**, **Homogeneity of variance test**, and **Means plot** (to obtain a visual representation of the function, a very useful part of the output for a polynomial trend analysis), retaining the default of **Exclude cases analysis by analysis**.

The **One-Way ANOVA** procedure also has options for us to perform ANOVA when the assumption of equal variances among the groups is violated. Both the Brown–Forsythe and the Welch tests accommodate for unequal group variances. We do not select these now because the polynomial partitioning is based on the ordinary ANOVA; if our data violated the equal variance assumption, we would evaluate the *F* ratio using a more stringent alpha level. Click **Continue** to return to the main dialog window.

In the **Post Hoc** dialog window shown in Figure 48.3, we check **R-E-G-W Q** to obtain the Ryan–Enoit–Gabriel–Welsch Studentized Range test; this test assumes equal variances and, if the Levene test returned a statistically significant result (indicating that we failed to meet this assumption), we would perform the analysis again with **Tamhane's T2** test. Click **Continue** to return to the main dialog window.

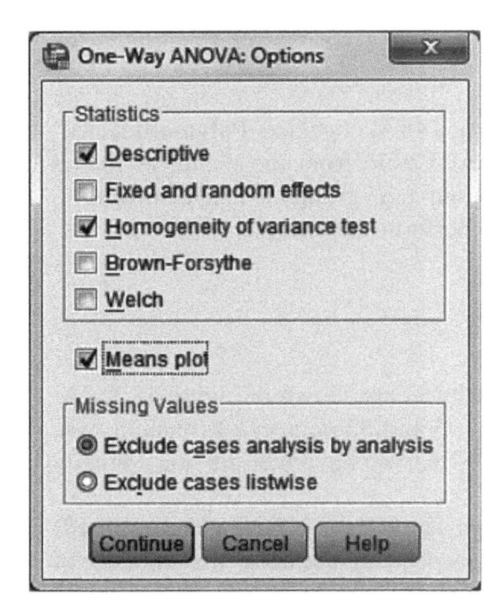

FIGURE 48.2

The **Options** dialog window in **One-Way ANOVA**.

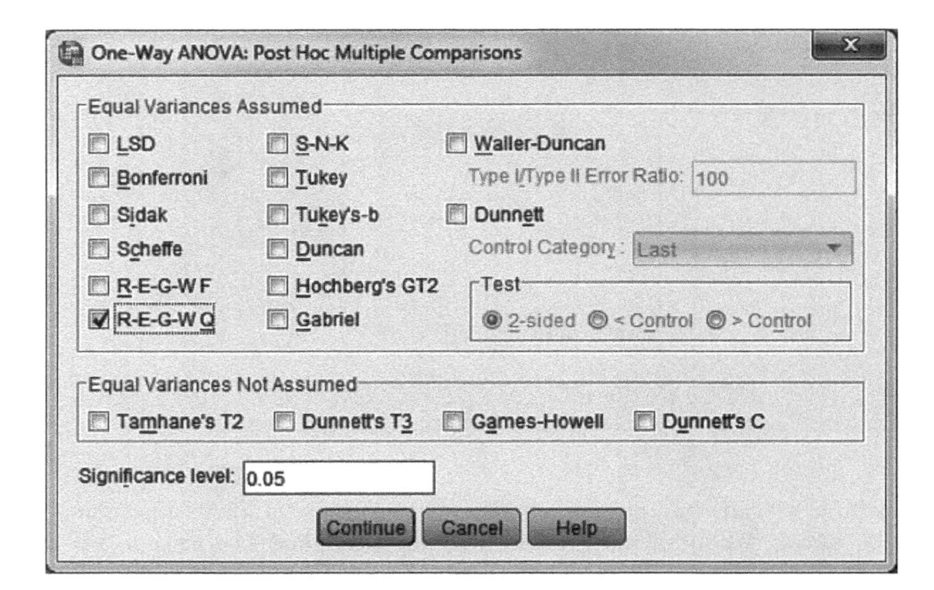

FIGURE 48.3

The **Post Hoc** dialog window in **One-Way ANOVA**.

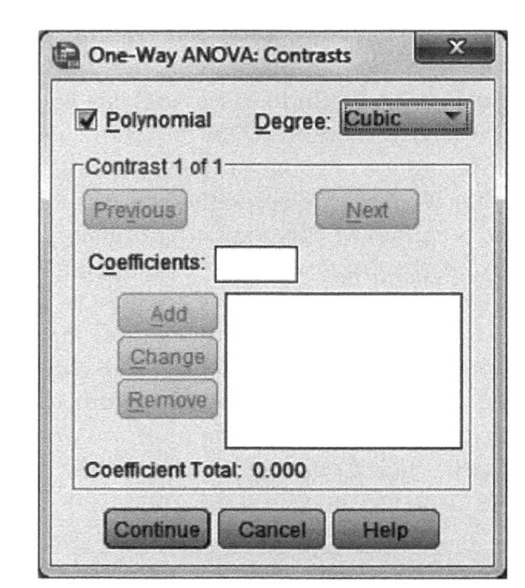

FIGURE 48.4

The **Contrasts** dialog window in **One-Way ANOVA**.

In the **Contrasts** dialog window shown in Figure 48.4, checking **Polynomial** activates the **Degree** drop-down menu next to it. Select **Cubic** from the menu, as this is the highest polynomial contrast we can obtain with our four groups. Click **Continue** to return to the main dialog window and click **OK** to perform the analysis.

48.5 ANALYSIS OUTPUT

The descriptive statistics and the results of the Levene test are shown in Figure 48.5. The **Levene Statistic** was computed to be **.795** and, with **3** and **52** degrees of freedom, was not statistically significant ($p = .502$). It thus appears that we have met the homogeneity of variance assumption.

Figure 48.6 displays the ANOVA summary table. The **Between Groups (combined)** source of variance represents the effect of the independent variable, the number of wineries stocked by the stores. It specifies **(combined)** because this is the effect of

Descriptives

sales_dollars hundreds dollars per day

	N	Mean	Std. Deviation	Std. Error	95% Confidence Interval for Mean		Minimum	Maximum
					Lower Bound	Upper Bound		
20	14	30.93	4.582	1.225	28.28	33.57	24	39
40	14	32.00	4.114	1.099	29.62	34.38	26	38
60	14	46.14	3.634	.971	44.04	48.24	40	51
80	14	39.43	4.972	1.329	36.56	42.30	32	47
Total	56	37.13	7.513	1.004	35.11	39.14	24	51

Test of Homogeneity of Variances

sales_dollars hundreds dollars per day

Levene Statistic	df1	df2	Sig.
.795	3	52	.502

FIGURE 48.5 The descriptive statistics and Levene's test of homogeneity of variances.

ANOVA

sales_dollars hundreds dollars per day

			Sum of Squares	df	Mean Square	F	Sig.
Between Groups	(Combined)		2118.054	3	706.018	37.232	.000
	Linear Term	Contrast	1100.089	1	1100.089	58.013	.000
		Deviation	1017.964	2	508.982	26.841	.000
	Quadratic Term	Contrast	212.161	1	212.161	11.188	.002
		Deviation	805.804	1	805.804	42.494	.000
	Cubic Term	Contrast	805.804	1	805.804	42.494	.000
Within Groups			986.071	52	18.963		
Total			3104.125	55			

FIGURE 48.6 The summary table for ANOVA.

the independent variable as a whole (without considering the polynomial components). The F ratio is **37.232**, which, with **3** (number of groups – 1) and **52** degrees of freedom, is statistically significant ($p < .001$). The eta square value associated with this effect can be computed as the sum of squares for the between-groups variance divided by the total variance (**2118.054/3104.125**), or .68; thus, the differences between the groups explained about 68% of the variance of the dependent variable.

The evaluation of the group differences is seen in Figure 48.7. The results of the Ryan–Enoit–Gabriel–Welsch Studentized Range test inform us (in order of the magnitude of the means) that there is no reliable wine sales difference when the stores stocked 20 or 40 different wineries, but that stocking 80 different wineries produced a higher sales volume from the 20–40 winery level, and that stocking 60 different wineries produced the highest sales volume of all the groups.

Thus far, we have described the outcome of an ordinary one-way ANOVA. But this was a polynomial trend analysis, and we can now address this portion of the analysis.

Post Hoc Tests
Homogeneous Subsets

sales_dollars hundreds dollars per day

Ryan-Einot-Gabriel-Welsch Range[a]

n_wineries n different wineries available in store	N	Subset for alpha = 0.05		
		1	2	3
20	14	30.93		
40	14	32.00		
80	14		39.43	
60	14			46.14
Sig.		.768	1.000	1.000

Means for groups in homogeneous subsets are displayed.

a. Critical values are not monotonic for these data. Substitutions have been made to ensure monotonicity. Type I error is therefore smaller.

FIGURE 48.7

Results of the Ryan–Enoit–Gabriel–Welsch Studentized Range test.

Means Plots

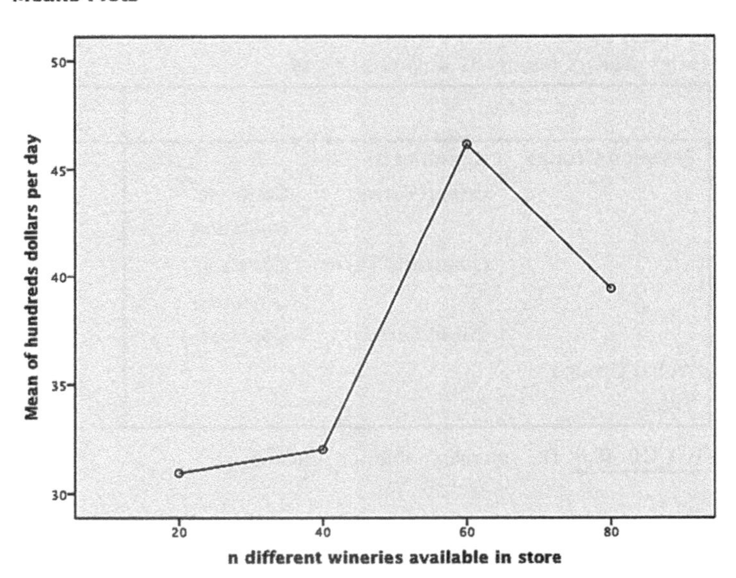

FIGURE 48.8

The plot of the group means.

In the summary table (Figure 48.6) we have the polynomial results. Under the row for **Between Groups (combined)** is a set of rows that document the polynomial components of the between-groups variance. We treat each in order.

A polynomial trend analysis is as much a statistical as a visual analytic enterprise, and it is useful to view the plot of means as we discuss each polynomial component of the variance. The means are plotted in Figure 48.8. It appears, generally, that the function increases to a certain extent from 20 to 40 wineries, sharply increases to 60 wineries, and then sharply drops to 80 wineries.

In the summary table, the row labeled **Linear Term** has two parts to it: the row labeled **Contrast** and the row labeled **Deviation**. The **Contrast** row assesses the linear component. Its **Sum of Squares** is **1100.089**, and the *F* ratio of **58.013** is statistically significant. This can be seen in the plot of means shown in Figure 48.8. There is a general trend of the function to rise from 20 to 80, even though it is rather "bumpy"; this general rise in the function is the linear component in that we could fit a straight line through the data points that would angle upward from 20 to 80.

The **Deviation** row under the **Linear Contrast** in the summary table in Figure 48.6 assesses the remaining nonlinear components—whatever is left after the linear component of the variance has been explained (which in the present example is the combination of the quadratic and cubic components). Its **Sum of Squares** is **1017.964**, and the *F* ratio of **26.841** is statistically significant. Thus, there is a statistically significant nonlinear component to the means. This can be seen in the plot of means shown in Figure 48.8—it is the "bumpiness" in the function that was not explained by the linear component. Note that the **Sums of Squares** of the **Linear Contrast** (**1100.089**) and the **Linear Deviation** (**1017.964**) sum to the total **Between Groups (combined) Sum of Squares** (**2118.054**).

We can compute the eta square value associated with the linear component (or any polynomial component) with respect to either the between-subjects variance or the total variance. With respect to the between-subject variance, the eta square would be **1100.089/2118.054**, or about .519; with respect to the total variance, the eta square would be **1100.089/3104.125**, or about .354. Thus, the linear component accounted for approximately 52% of the between-subject variance and approximately 35% of the total variance.

The row labeled **Quadratic Term** in the summary table in Figure 48.6 also has two parts to it: the row labeled **Contrast** and the row labeled **Deviation**. The **Contrast** row assesses the quadratic component. Its **Sum of Squares** is **212.161**, and the *F* ratio of **11.188** is statistically significant. Thus, there is a statistically significant quadratic component to the means; its eta square with respect to the between-groups effect is about 10% (**212.161/2118.054**) and with respect to the total variance is about 7% (**212.161/3104.125**). This can be seen in the plot of means shown in Figure 48.8. Although far from symmetric (its eta square value is much less than that associated with the linear component), the function does appear to somewhat resemble an inverted "V" indicative of a quadratic relationship.

The **Deviation** row under the **Quadratic Contrast** in the summary table in Figure 48.6 assesses the remaining component when the linear and quadratic components of the variance have been explained. Its **Sum of Squares** is **805.804**, and the *F* ratio of **42.494** is statistically significant. Thus, there is a statistically significant amount of variance yet to be explained by the remaining polynomial component(s); in the present example, there is only one such component remaining, and it is the cubic component.

The row labeled **Cubic Term** in the summary table in Figure 48.6 has only one part to it, as it is the last polynomial component possible to be assessed in our four-group example. Its **Sum of Squares** is **805.804**, and the *F* ratio of **42.494** is statistically significant. This is equivalent to the row above it in that only the cubic contrast remaining after the quadratic component was assessed. Thus, there is a statistically significant cubic component to the means; its eta square with respect to the between-groups effect is about 38% (**805.804/2118.054**) and with respect to the total variance is about 26% (**805.80/3104.125**). This can be seen in the plot of means shown in Figure 48.8. There appear to be two distinct inflexion points in the function, one at 40 and another at 60, and it is these slope changes that are likely driving the cubic component.

Overall, although there is a major linear component in the results, suggesting at a general level that stocking more wineries is associated with more sales volume, the cubic function combined with the post hoc test results indicate that the optimal level of wineries carried by the Good Food Market appears to be 60. Increasing the number of wineries on their shelves beyond that optimal level appears to actually adversely affect wine sales.

CHAPTER 49

One-Way Between-Subjects ANCOVA

49.1 OVERVIEW

Imagine the following scenario. We have k levels of an independent variable that we would like to associate with differences in dependent variable scores. But there is another (quantitative) variable that could also potentially affect the performance of the dependent variable, and so any differences we observe between the k groups may be at least partly attributable to the other variable. One procedure that is available to deal with this possible confounding issue is to use a one-way between-subjects analysis of covariance (ANCOVA) design in this situation in an attempt to "purge" the scores of the dependent variable of this other quantitative variable (by statistically controlling for it) so that we can evaluate a "purer" effect of the independent variable.

We can statistically control for any number of quantitatively measured variables, and these controlled-for variables are assigned the role of *covariates* in the ANCOVA; we will restrict our coverage here to a single quantitative covariate. We can also apply ANCOVA to ANOVA designs subsuming any number of independent variables, but we will restrict our coverage here to the one-way between-subjects design.

In the first stage of a one-way between-subjects ANCOVA design, the relationship (correlation) between the covariate and the dependent variable is determined for the sample as a whole. As a result of this evaluation, the values of the dependent measure are *adjusted* (in much the same way as is done in a multiple regression analysis) such that the variance associated with the covariate is removed from the dependent variable scores. These adjusted values (representing the residual variance of the dependent variable) become the "new" or "revised" scores of the dependent variable and are then evaluated by ANOVA to assess the effect of the independent variable. If the effect of the independent variable is statistically significant, the k groups differ *in the means of the adjusted scores* (and not necessarily in the raw or observed means).

All of the assumptions underlying ANOVA (normal distribution, independence, and homogeneity of variance) apply to ANCOVA, but ANCOVA adds two additional assumptions:

- *Linearity of Regression.* It is assumed that there is a linear relationship between the covariate variable and the dependent variable. This is important because the adjustment process is based on an ordinary least-squares (linear) regression procedure.

Performing Data Analysis Using IBM SPSS®, First Edition.
Lawrence S. Meyers, Glenn C. Gamst, and A. J. Guarino.

- *Homogeneity of Regression.* It is assumed that the slope of the regression function, with the covariate predicting the dependent variable, is equal across the k groups. This is important because the adjustment procedure is done once on the sample as a whole, and thus we need to have the relationship between the covariate and the dependent variable comparable for each group in order for the score adjustment procedure to be interpretable.

49.2 NUMERICAL EXAMPLE

The fictional data we use for our example are in the data file named **math teaching methods**. A school district wished to explore alternative methods of teaching elementary-grade math. The researchers established three instructional groups of 12 classes each under the independent variable of **teaching_method**: the **currently used** method (the standard method is coded as **1**), a **social method** where two students worked in teams to solve the problems (the social method is coded as **2**), and **computer assisted instruction** (the CAI method is coded as **3**). Believing that children with greater levels of math ability would probably do better under any teaching method, and wishing to statistically control for such ability, the researchers obtained the standardized math ability scores (based on a statewide testing program) from the district files to be used as a covariate in this study (**math_ability_cov**). The dependent variable was the average final examination scores of each class of students reflecting course performance over the period of instruction (**exam_grade_dv**). A screenshot of part of the data file is shown in Figure 49.1.

49.3 ANALYSIS STRATEGY

The analysis strategy is as follows:

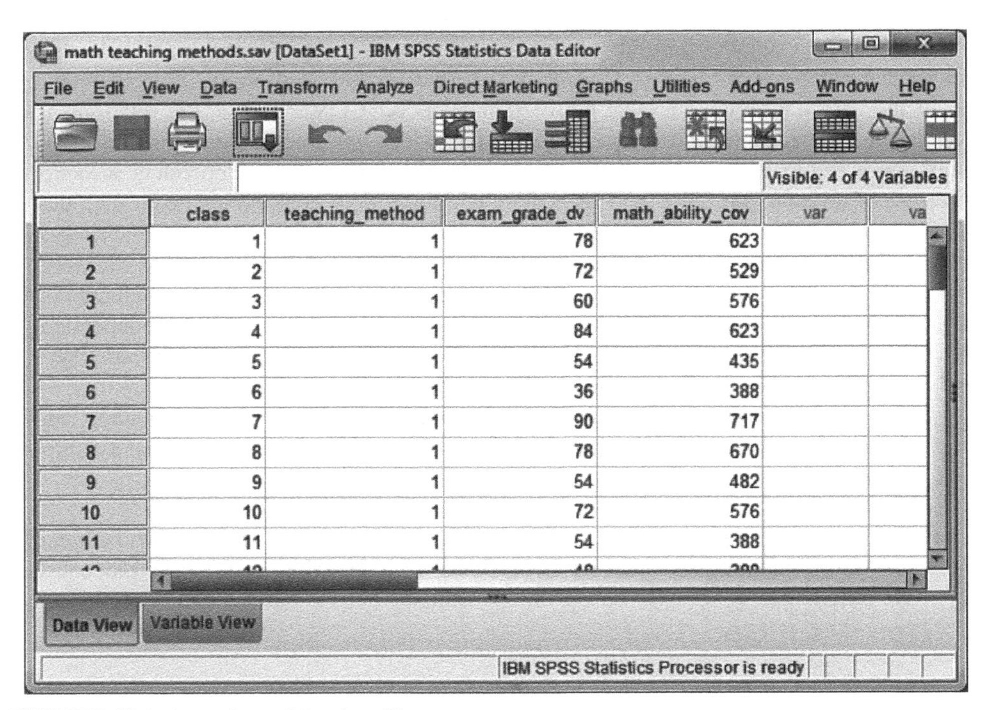

	class	teaching_method	exam_grade_dv	math_ability_cov	var
1	1	1	78	623	
2	2	1	72	529	
3	3	1	60	576	
4	4	1	84	623	
5	5	1	54	435	
6	6	1	36	388	
7	7	1	90	717	
8	8	1	78	670	
9	9	1	54	482	
10	10	1	72	576	
11	11	1	54	388	

FIGURE 49.1 A portion of the data file.

- First, we perform an ANOVA on the observed scores to show the results when the covariate is not included in the analysis.
- Second, we assess the degree to which the data set meets the extra assumptions of ANCOVA.
- Third, we perform the ANCOVA, making sure to carry out the multiple comparison tests (assuming we obtain a statistically significant effect of the independent variable) on the adjusted rather than the raw (observed) means.

49.4 ANALYSIS SETUP: ANOVA

We open the data file **math teaching methods** and from the main menu, we select **Analyze ➔ General Linear Model ➔ Univariate**. This opens the main **Univariate** window as shown in Figure 49.2. We move **exam_grade_dv** into the **Dependent Variable** panel and **teaching_method** into the **Fixed Factor(s)** panel.

In the **Options** dialog window (see Figure 49.3), we check **Descriptive statistics** and **Homogeneity tests** (to obtain the Levene test of equal group variances). Click **Continue** to return to the main dialog window.

We will also request a post hoc test, but will examine its results only if the effect of the independent variable is statistically significant. In the upper portion of the **Post Hoc** dialog window (see Figure 49.4), we move **teaching_method** into the **Post Hoc Tests for** panel and check **R-E-G-W-Q** to obtain the Ryan–Enoit–Gabriel–Welsch Studentized Range test. Click **Continue** to return to the main dialog window and click **OK** to perform the analysis.

49.5 ANALYSIS OUTPUT: ANOVA

The **Descriptive Statistics** table in Figure 49.5 shows the mean, standard deviation, and group sizes. As can be seen, the group means are very close. The middle table in Figure 49.5 shows Levene's statistic for homogeneity of variances. The F ratio of **2.202** was not statistically significant (p = **.127**) when evaluated with **2** and **33** degrees of

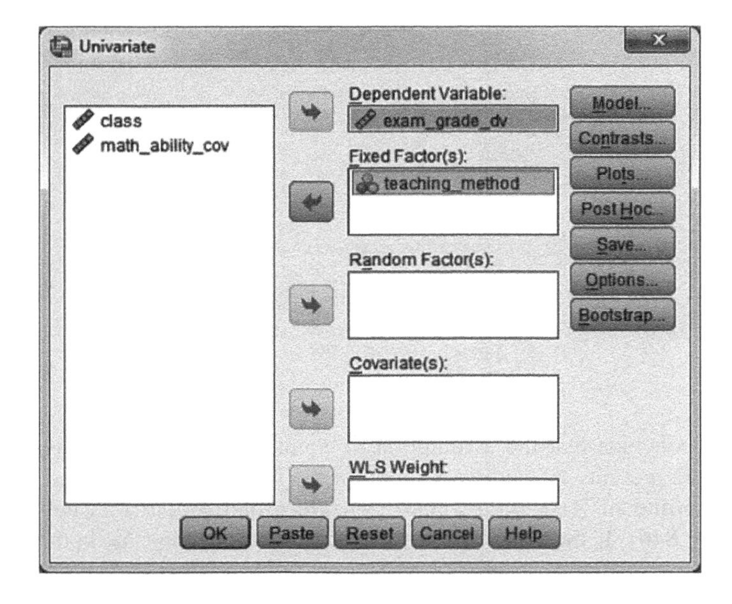

FIGURE 49.2

The main dialog window of the **Univariate** procedure of the **General Linear Model**.

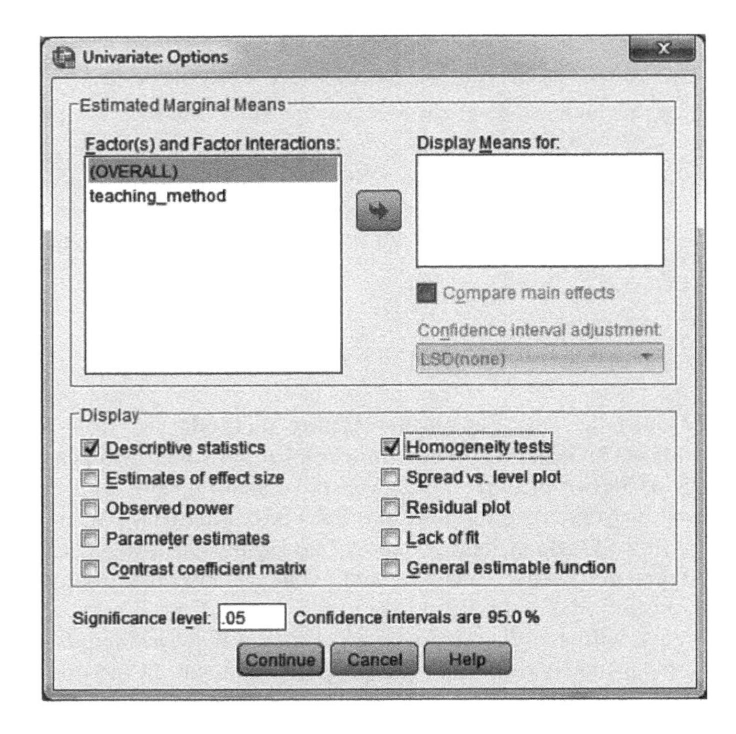

FIGURE 49.3

The **Options** dialog window of the **Univariate** procedure of the **General Linear Model**.

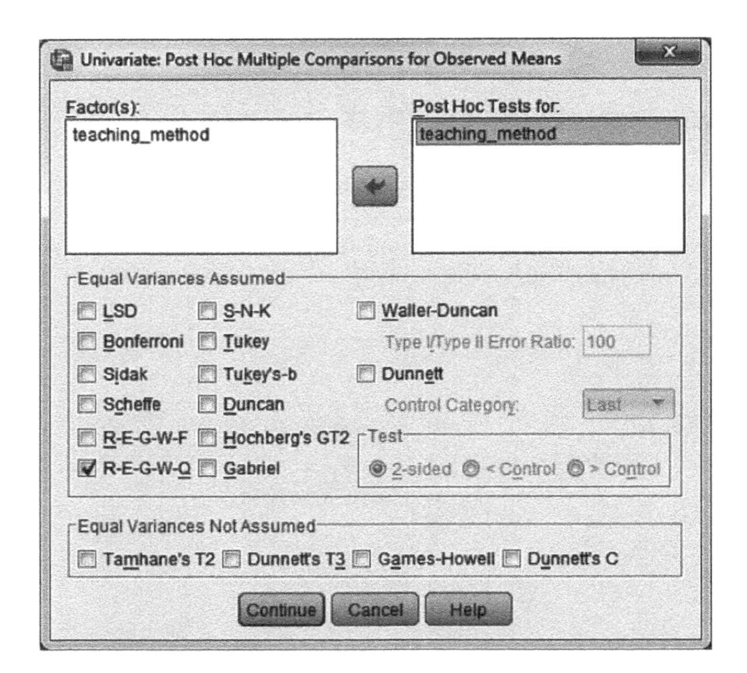

FIGURE 49.4

The **Post Hoc** dialog window of the **Univariate** procedure of the **General Linear Model**.

freedom; we may therefore conclude that the assumption of equal group variances was not violated. The bottom table presents the summary table for the omnibus analysis, and the F ratio for **teaching_method** of **.168**, with **2** and **33** degrees of freedom was not statistically significant ($p =$ **.846**). It therefore appears that the three teaching methods were equally effective.

Descriptive Statistics

Dependent Variable:exam_grade_dv

teaching_method	Mean	Std. Deviation	N
1 standard method	65.00	16.348	12
2 social method	65.50	13.406	12
3 CAI method	68.00	10.340	12
Total	66.17	13.261	36

Levene's Test of Equality of Error Variances[a]

Dependent Variable:exam_grade_dv

F	df1	df2	Sig.
2.202	2	33	.127

Tests the null hypothesis that the error variance of the dependent variable is equal across groups.

a. Design: Intercept + teaching_method

Tests of Between-Subjects Effects

Dependent Variable:exam_grade_dv

Source	Type III Sum of Squares	df	Mean Square	F	Sig.
Corrected Model	62.000[a]	2	31.000	.168	.846
Intercept	157609.000	1	157609.000	853.618	.000
teaching_method	62.000	2	31.000	.168	.846
Error	6093.000	33	184.636		
Total	163764.000	36			
Corrected Total	6155.000	35			

a. R Squared = .010 (Adjusted R Squared = -.050)

FIGURE 49.5 Descriptive statistics, Levene's Test and Omnibus Analysis Results.

49.6 EVALUATING THE ANCOVA ASSUMPTIONS

Linearity of Regression

We ordinarily evaluate linearity of regression by visually examining a scatterplot of the covariate and the dependent variables. From the main menu select **Graphs ➔ Legacy Dialogs ➔ Scatter/Dot** to open the **Scatter/Dot** window shown in Figure 49.6. Select **Simple Scatter** and click **Define**. This opens the **Simple Scatterplot** dialog window (see Figure 49.7). Click **exam_grade_dv** to the **Y Axis** panel (the dependent variable is represented on the *Y*-axis) and click **math_ability_cov** to the **X Axis** panel. Click **OK**.

The scatterplot of the dependent variable (**exam_grade_dv**) as a function of the covariate (**math_ability_cov**) result of this setup is shown in Figure 49.8, and it appears that the two variables are linearly related. To display the regression line, double-click

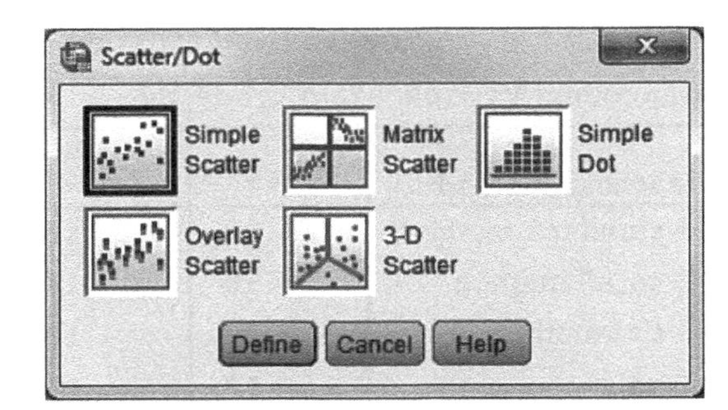

FIGURE 49.6
Selecting the simple scatterplot.

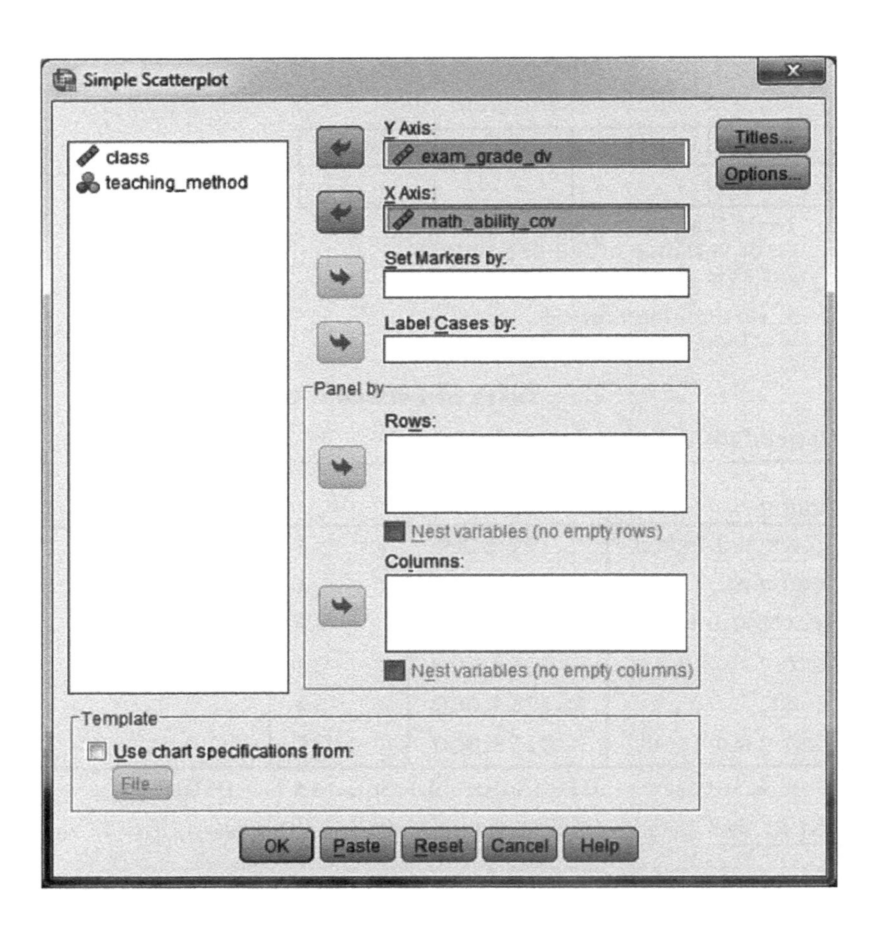

FIGURE 49.7
Configuring the scatterplot.

inside the scatterplot in the IBM SPSS® output. This gives us access to the **Chart Editor**. Select **Elements ➔ Fit Line at Total**. As soon as this is selected, the line of best fit will appear superimposed on the scatterplot. This is shown in Figure 49.9.

Homogeneity of Regression

What we are looking for when we evaluate the assumption of homogeneity of regression is whether the individual group regression functions predicting the dependent variable from the covariate are the same (they have slopes comparable to the total sample regression line shown in the scatterplot of Figure 49.8). Obtaining the Independent Variable × Covariate interaction effect allows us to test this assumption; we presume that we have conformed

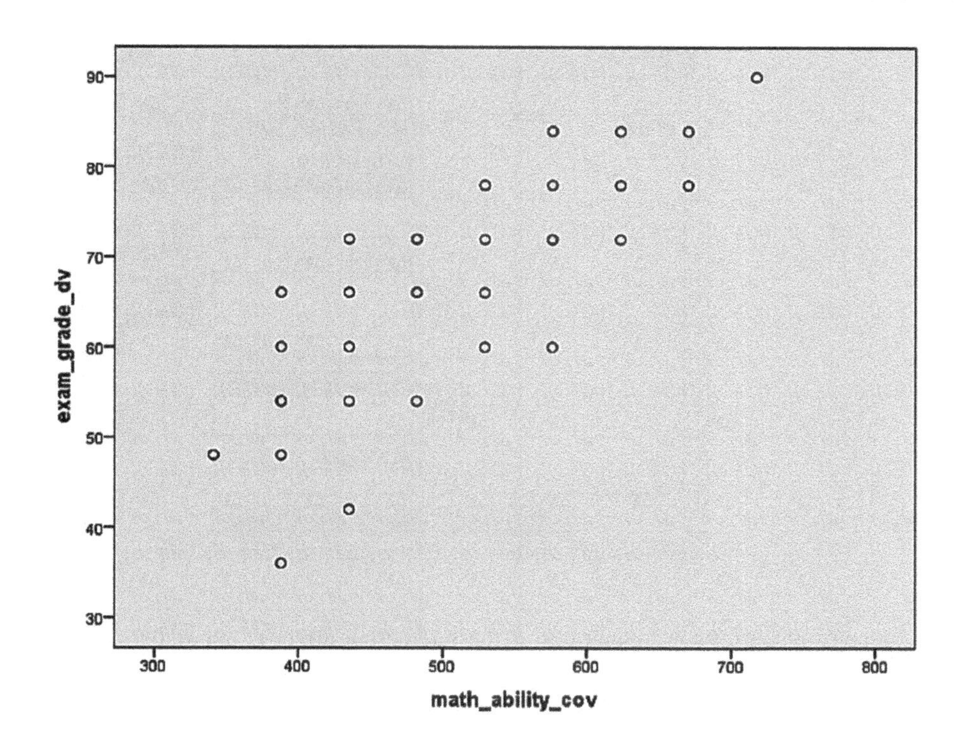

FIGURE 49.8 Output of the scatterplot.

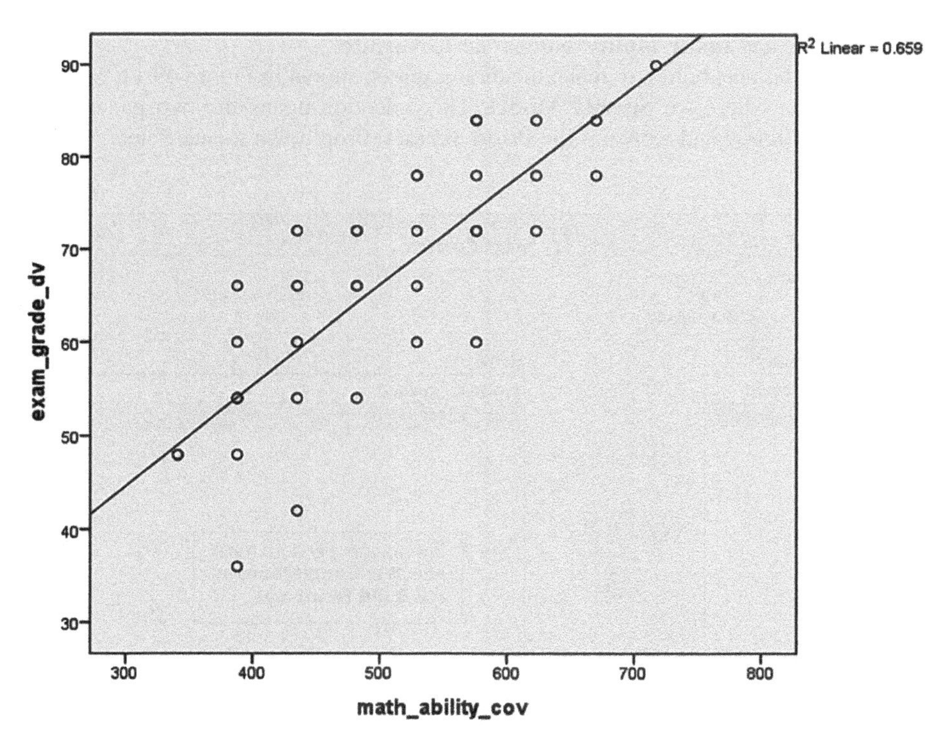

FIGURE 49.9 The scatterplot with the least-squares regression line fit.

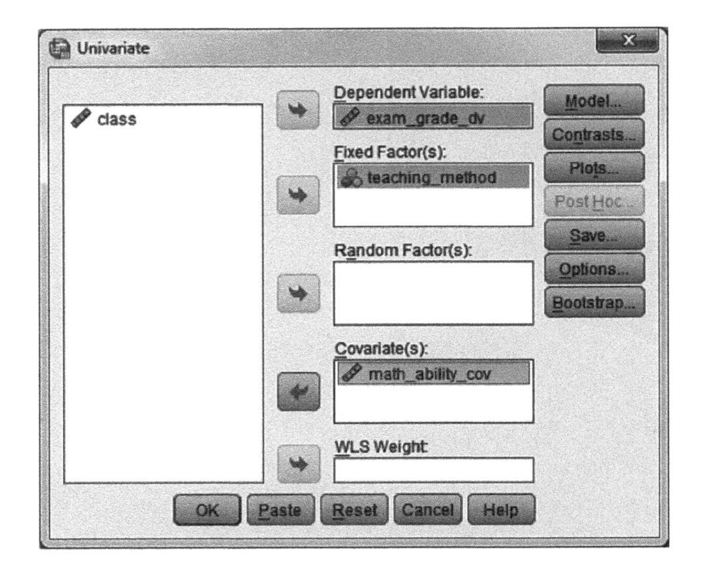

FIGURE 49.10
The main dialog window of the **Univariate** procedure of the **General Linear Model**.

to the homogeneity of regression assumption if the interaction effect is not statistically significant.

From the main IBM SPSS menu select **Analyze ➔ General Linear Model ➔ Univariate**. This opens the main **Univariate** dialog window shown in Figure 49.10. We have configured it with **teaching_method** as the **Fixed Factor**, **exam_grade_dv** as the dependent variable, and **math_ability_cov** as the **Covariate**.

Select the **Model** pushbutton to reach the dialog screen shown in Figure 49.11. Select **Custom** in the area where we **Specify Model**. This selection opens the two panels on either side of the window and activates the **Build Term(s)** drop-down menu. Select **Main**

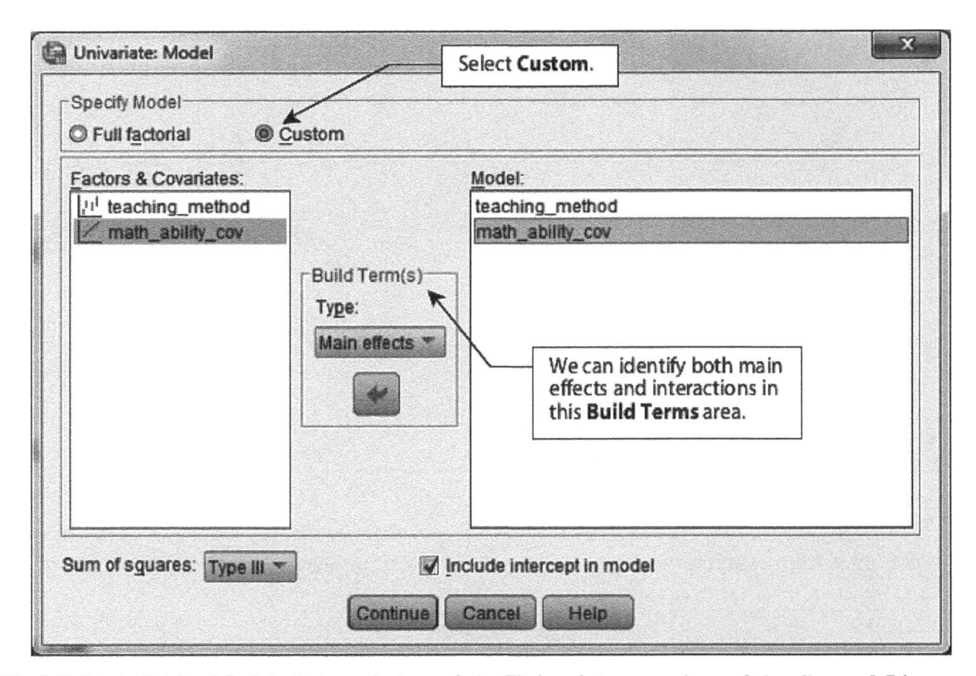

FIGURE 49.11 The **Model** dialog window of the **Univariate** procedure of the **General Linear Model** configured for main effects.

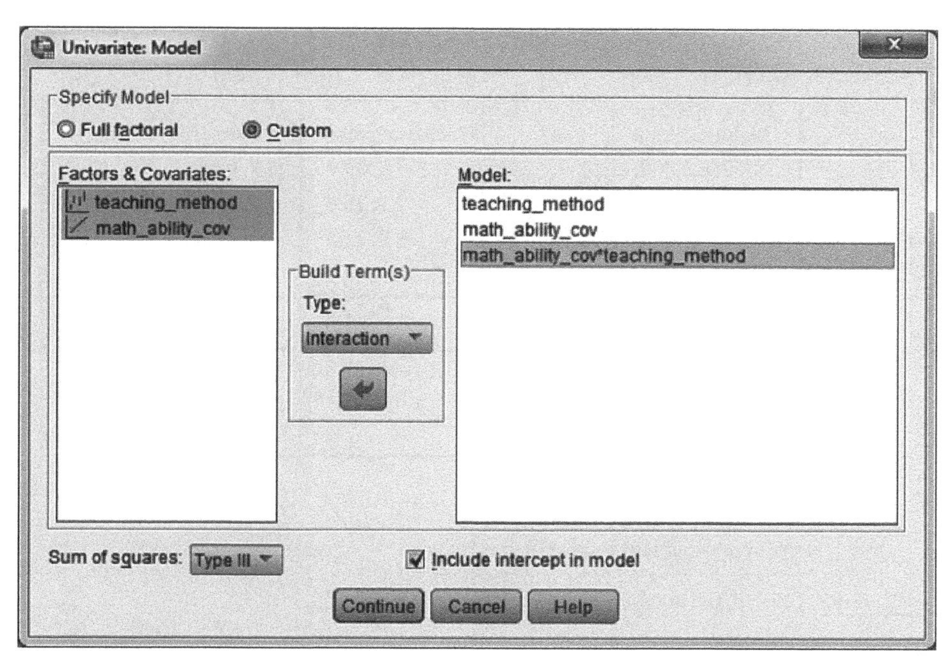

FIGURE 49.12 The **Model** dialog window of the **Univariate** procedure of the **General Linear Model** configured for main effects and the interaction.

effects from the pull-down menu under **Build Term(s)** and click **teaching_method** and **math_ability_cov** to the **Model** panel. This is shown in Figure 49.11.

Now select **Interaction** from the **Build Term(s)** pull-down menu (replacing the **Main effects** choice). Select both **teaching_method** and **math_ability_cov** (by holding down the **Ctrl** or **Shift** key while selecting the variables one at a time) and use the arrow button to click them over to the **Model** panel. The result of this is shown in Figure 49.12. Click **Continue** to return to the main dialog window and click **OK** to perform the analysis.

The only output that interests us is the test of statistical significance of the **teaching_method*math_ability_cov** interaction shown in the summary table in Figure 49.13. As can be seen in the summary table, the effect is not statistically significant with an F ratio of **.548** (p = **.584**). We thus presume that the assumption of homogeneity of regression has not been violated and proceed with the ANCOVA.

49.7 ANALYSIS SETUP: ANCOVA

Select **Analyze → General Linear Model → Univariate**. Configure the main dialog window as we did in testing the assumption of homogeneity of regression, with **teaching_method** as the **Fixed Factor**, **exam_grade_dv** as the dependent variable, and **math_ability_cov** as the **Covariate**. In the **Model** window (see Figure 49.10), set the **Specify Model** to **Full factorial**.

In the **Options** dialog window displayed in Figure 49.14, we check **Homogeneity tests** (to obtain the Levene test of equal group variances on the adjusted scores); we do not request the **Descriptive statistics** because (a) we already have them on the observed scores from our initial ANOVA and (b) the covariance analysis is performed on the adjusted and not the observed scores.

We obtain the adjusted means from the **Options** dialog window. As shown in Figure 49.14, in the **Estimated Marginal Means** area in the top half of the **Options**

Tests of Between-Subjects Effects

Dependent Variable:exam_grade_dv

Source	Type III Sum of Squares	df	Mean Square	F	Sig.
Corrected Model	4579.870[a]	5	915.974	17.446	.000
Intercept	81.156	1	81.156	1.546	.223
teaching_method	118.471	2	59.236	1.128	.337
math_ability_cov	3685.019	1	3685.019	70.185	.000
teaching_method * math_ability_cov	57.539	2	28.770	.548	.584
Error	1575.130	30	52.504		
Total	163764.000	36			
Corrected Total	6155.000	35			

a. R Squared = .744 (Adjusted R Squared = .701)

Our interest is only in the statistical significance of the interaction effect.

FIGURE 49.13 Tests of Between-Subjects Effects.

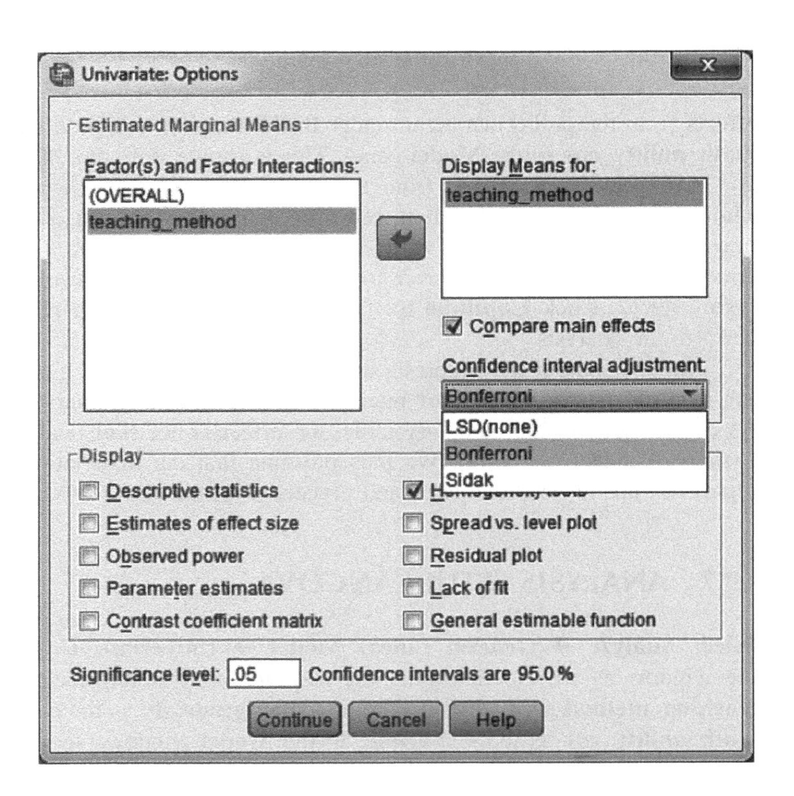

FIGURE 49.14

The **Options** dialog window of the **Univariate** procedure of the **General Linear Model** configured for obtaining the estimated marginal means and the Bonferroni corrected *t* tests on the group means.

window, we select **teaching_method** from the **Factor(s) and Factor Interactions** panel and click it over to the **Display Means for** panel. These means are labeled by IBM SPSS as *estimated marginal means* and are also known as *least-squares means*: they are unweighted means of the values (the adjusted values here) associated with the groups and are accompanied by standard errors rather than standard deviations.

We also click the checkbox under the **Display Means for** panel for **Compare main effects** (see Figure 49.14). This is how we obtain multiple comparison tests in a one-way covariance design because the **Post Hoc** tests we have used up to this point are not available for adjusted scores (the post hoc tests can only be used on raw or observed data). There are three multiple comparison tests available in the **Confidence interval adjustment** drop-down menu, all of which use a t test to assess the mean differences for each pair of means. The primary difference among them is how they control for alpha level (Type I error) inflation:

- *LSD.* This is the Least Significant Difference test and does not control for alpha level inflation. As such, it is the most powerful of the three (it will detect more "significant" differences than the other two) but the comparisons are really being evaluated at less stringent alpha levels. It may be appropriate to use when testing one or two a priori predictions but is generally not recommended for exploratory purposes.
- *Bonferroni.* Named after the mathematician Carlo Emilio Bonferroni, this set of pairwise t tests controls for alpha inflation by dividing the ordinarily used .05 level by the number of comparisons—what is called the *Bonferroni correction* to the alpha level. It is the most conservative of the three available methods and is probably the one most frequently used.
- *Sidak.* Named after the mathematician Zbynek Sidak, this set of pairwise t tests is a variation of the Bonferroni method that adds a tad more power (it is slightly less conservative than the Bonferroni correction) but is still relatively conservative.

We select the **Bonferroni** procedure from the drop-down menu. Click **Continue** to return to the main dialog window and click **OK** to perform the analysis.

49.8 ANALYSIS OUTPUT: ANCOVA

The top table in Figure 49.15 shows the results of Levene's test of the equality of error variances. The F ratio is **.105** (p = **.901**), indicating that we appear to meet the homogeneity of variance assumption. Note that this Levene's F value is based on the adjusted scores and yields a different result from that computed on the observed scores (see Figure 49.5).

The bottom table in Figure 49.15 presents the estimated marginal means together with their standard errors. Note that these means are different from the observed means shown in Figure 49.5; the estimated marginal means reflect the statistical "removal" of variance due to math ability and exhibit greater differences than did the observed means.

We show the summary table for the omnibus ANCOVA in Figure 49.16. Both the covariate of **math_ability_cov** and the effect of the independent variable of **teaching_method** are statistically significant. The eta square value for **teaching_method** is computed by dividing its sum of squares (**469.055**) by the **Corrected Total** sum of squares (**6155.000**) to yield a value of .076. The eta square value for **math_ability_cov** is computed by dividing its sum of squares (**4460.331**) by the **Corrected Total** sum of squares (**6155.000**) to yield a value of .725.

The Bonferroni corrected pairwise comparisons of the estimated marginal means are shown in Figure 49.17. The **Pairwise Comparisons** table contains some redundancy. Each major row focuses on one of the three groups and compares the other two to it. Consider the first major row focusing on the **standard method**. The difference between the estimated marginal mean for that method and the estimated marginal mean for the **social method** is **61.211** − **67.167**, or **−5.956**. This difference is not statistically significant (p = **.161**). However, the difference between the estimated marginal

Levene's Test of Equality of Error Variances[a]

Dependent Variable:exam_grade_dv

F	df1	df2	Sig.
.105	2	33	.901

Tests the null hypothesis that the error variance of the dependent variable is equal across groups.

a. Design: Intercept + math_ability_cov + teaching_method

Estimated Marginal Means teaching_method

Estimates

Dependent Variable:exam_grade_dv

teaching_method	Mean	Std. Error	95% Confidence Interval	
			Lower Bound	Upper Bound
1 standard method	61.211[a]	2.101	56.931	65.492
2 social method	67.167[a]	2.070	62.951	71.383
3 CAI method	70.122[a]	2.074	65.896	74.347

a. Covariates appearing in the model are evaluated at the following values: math_ability_cov = 500.28.

FIGURE 49.15 ANCOVA homogeneity of variance test and estimated marginal means.

Tests of Between-Subjects Effects

Dependent Variable:exam_grade_dv

Source	Type III Sum of Squares	df	Mean Square	F	Sig.
Corrected Model	4522.331[a]	3	1507.444	29.546	.000
Intercept	83.510	1	83.510	1.637	.210
math_ability_cov	4460.331	1	4460.331	87.422	.000
teaching_method	469.055	2	234.527	4.597	.018
Error	1632.669	32	51.021		
Total	163764.000	36			
Corrected Total	6155.000	35			

a. R Squared = .735 (Adjusted R Squared = .710)

FIGURE 49.16 ANCOVA summary table.

mean for the **standard method** and the estimated marginal mean for the **CAI method** is **61.211** − **70.122** or **−8.911** and is statistically significant (p = **.016**). Examining the other rows suggests that this is the only reliable difference.

Based on the ANCOVA, we would conclude that, when controlling for effects of math ability, the **social method** is not more effective than the **standard method** but the **CAI method** appeared to be better than the **standard method**. Note that this outcome

Pairwise Comparisons

Dependent Variable:exam_grade_dv

(I) teaching_method	(J) teaching_method	Mean Difference (I-J)	Std. Error	Sig.[a]	95% Confidence Interval for Difference[a]	
					Lower Bound	Upper Bound
1 standard method	2 social method	-5.956	2.974	.161	-13.469	1.557
	3 CAI method	-8.911*	2.984	.016	-16.449	-1.372
2 social method	1 standard method	5.956	2.974	.161	-1.557	13.469
	3 CAI method	-2.955	2.916	.956	-10.323	4.414
3 CAI method	1 standard method	8.911*	2.984	.016	1.372	16.449
	2 social method	2.955	2.916	.956	-4.414	10.323

Based on estimated marginal means

a. Adjustment for multiple comparisons: Bonferroni.
*. The mean difference is significant at the

FIGURE 49.17 ANCOVA multiple comparison tests.

is different from what we obtained in our original ANOVA; without taking math ability into account, the researchers might have erroneously concluded that the two alternative teaching methods were no more effective than the one they had been using all along, whereas, once we take math ability into account, the computer-based method appears to be better than the one currently being used by the school district.

CHAPTER 50

Two-Way Between-Subjects ANOVA

50.1 OVERVIEW

A two-way between-subjects ANOVA involves two between-subjects independent variables. The levels of these two variables are combined in all possible combinations (factorially combined), with each combination representing an independent group of cases. For example, with two levels of independent variable A (a_1 and a_2) and two levels of independent variable B (b_1 and b_2), there are four separate conditions (separate groups of cases) in the design (a_1b_1, a_1b_2, a_2b_1, and a_2b_2).

Evolving the design from a one-way between-subjects ANOVA to a two-way between-subjects ANOVA is both a quantitative and a qualitative advancement. It is a quantitative advancement in that in a one-way design we evaluate only the effect of the single independent variable in the study, whereas in a two-way design we can evaluate the effects of each of the independent variables (as though we implemented two one-way designs). In the two-way design, the effects of the independent variables are called *main effects*.

The two-way design is also a qualitative advancement over the one-way design because we have two independent variables combined factorially. Such a factorial combination yields a "bonus" effect that is extremely valuable—the omnibus (overall) *interaction effect*. In a two-way design, this interaction effect is known as the two-way interaction and is ordinarily referred to as the A × B interaction (in a research study A and B take on the actual names of the independent variables).

These three effects, the two main effects and the two-way interaction, represent separate partitions of the variance of the dependent variable, and each is separately tested for statistical significance. The null hypothesis in each test is that the effect accounts for none of the variance of the dependent variable (i.e., the eta square value for each effect is zero).

Probably, the primary reason why researchers are drawn to two-way designs is that they can examine the interaction effect of the two independent variables. Interactions are higher order effects than main effects (they carry more information), and if they are statistically significant, they supercede (take priority in explicating over) main effects. This idea of an interaction being a higher order effect is based on the following points:

- The omnibus interaction effect explicitly deals with the means of the separate conditions (e.g., a_1b_1, a_1b_2, a_2b_1, and a_2b_2), thus carrying considerable detailed

Performing Data Analysis Using IBM SPSS®, First Edition.
Lawrence S. Meyers, Glenn C. Gamst, and A. J. Guarino.

information. The main effects deal only with their own respective levels; for example, with two levels of A, the main effect of A addresses the difference between a_1 and a_2 (collapsing across the levels of B).

- Because the interaction effect carries more information, the information contained in the main effects can be explicitly seen in the interaction; because each main effect averages across the effect of the other, information contained in the interaction is "lost" when looking at the main effects.
- Given the above two points, the strategy that researchers typically adopt, all else equal, is to focus their attention on the omnibus interaction effect if it is statistically significant; they then perform multiple comparison tests on the means of the conditions (called *tests of simple effects* in this context) to "simplify" or explicate the omnibus effect.
- Only when the interaction is not significant will most researchers examine in detail the main effects by comparing the means of the levels within each independent variable (e.g., using post hoc tests).

A statistically significant omnibus interaction effect will be obtained when different levels of one independent variable produce a different pattern of results across the other independent variable. Under these conditions, it may be said that one of the independent variables *moderates* the effect of the other, in that the pattern across levels is not parallel. This verbal characterization is illustrated relatively simply in Figure 50.1.

The two plots on the left in Figure 50.1 depict two (of many possible) interaction effects. The top plot shows the lines representing a_1 and a_2 crossing and the bottom plot shows the lines representing a_1 and a_2 not crossing. We can recognize that both depict omnibus interaction effects because the functions for a_1 and a_2 are not parallel (they bear different relationships across the levels of the independent variable B).

Tests of simple effects compare each pair of means. In the bottom plot, for example, we would want to determine whether a_1 differed from a_2 at b_1, whether a_1 differed from a_2 at b_2, whether b_1 differed from b_2 under condition a_1, and whether b_1 differed from b_2 under condition a_2.

We can also glean some information about the main effects by examining the interaction. For example, in the bottom plot, we can reasonably guess that there is probably a significant main effect of A; we can make this guess because the function for a_2 is, on average, higher than a_1 (the mean of a_2 is reasonably higher than the mean of a_1). On the other hand, in the top plot, the mean of a_1 (the midpoint of the a_1 line) appears to be very close to the mean of a_2 (the midpoint of the a_2 line), suggesting that the main effect of A (assessing the mean difference between a_1 and a_2) is probably not statistically significant.

The single plot to the right in Figure 50.1 shows the absence of an interaction effect. The key in diagnosing the absence of a statistically significant interaction is that the functions for a_1 and a_2 are parallel (they bear the same relationship across the levels of the independent variable B). Graphed in this manner, the likely significance of the main effect of A also becomes apparent.

50.2 NUMERICAL EXAMPLE

The data we use for our example is present in the data file named **swelling treatment** are fictional. A drug for a certain type of muscle swelling has just been developed and is undergoing initial limited testing on 54 patients with the symptom. Swelling is assessed on a 25-point measure, with higher scores indicating higher levels of swelling; the dependent variable is named **swelling** in the data file. Patients with comparable levels

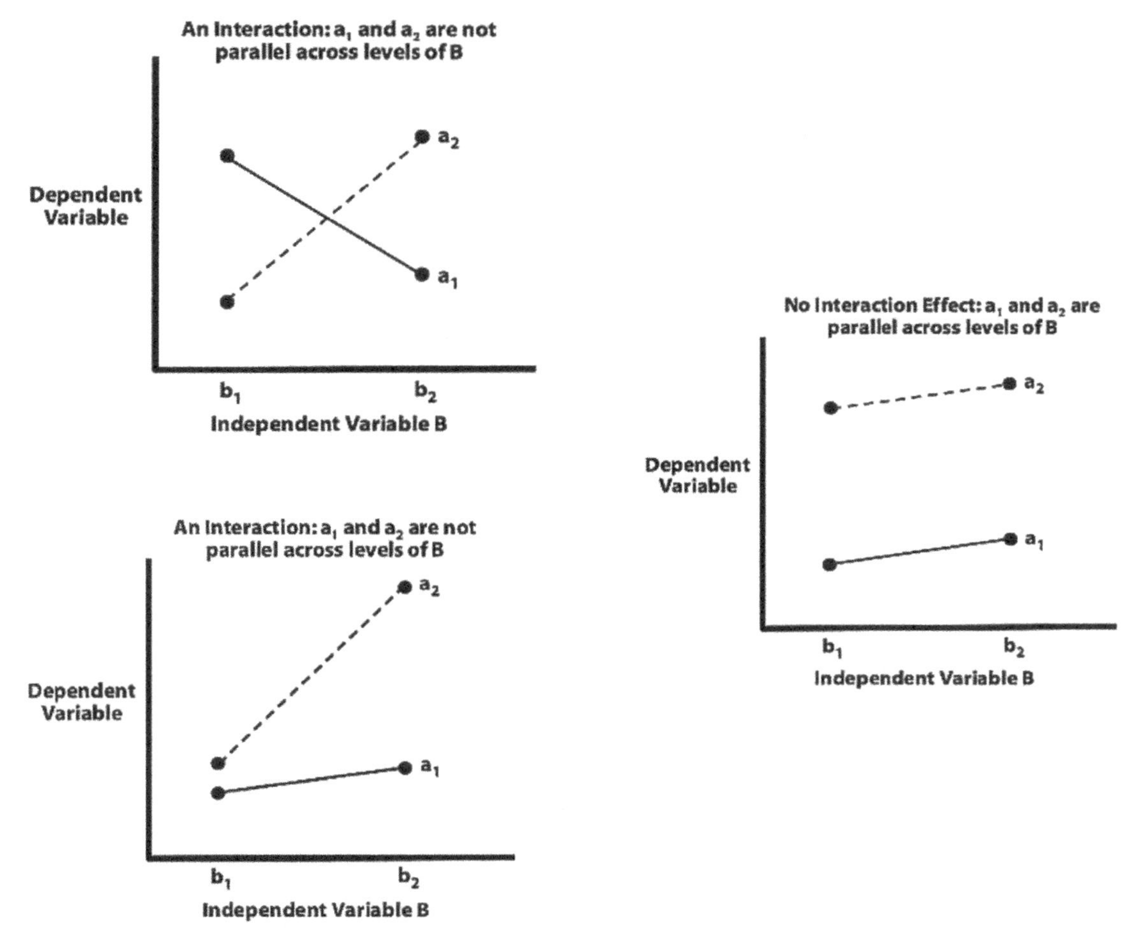

FIGURE 50.1 Two examples of an interaction and an illustration of no statistically significant interaction effect.

of swelling at the outset of the study were randomly assigned to one of three conditions under the independent variable named **treatment**: **no treatment** (coded as **1** in the data file), **placebo** (coded as **2** in the data file), or **medication** (coded as **3** in the data file). Patient **sex** is coded in the data file as **1** for females and **2** for males. A screenshot of part of the data file so that the coding may be seen is displayed in Figure 50.2.

50.3 ANALYSIS SETUP

We open the data file **swelling treatment** and from the main menu, we select **Analyze ➔ General Linear Model ➔ Univariate**. This opens the main **Univariate** window as shown in Figure 50.3. We move **swelling** into the **Dependent Variable** panel and **treatment** and **sex** into the **Fixed Factor(s)** panel.

The **Post Hoc** dialog window is shown in Figure 50.4. In the upper portion of the window, we move **treatment** into the **Post Hoc Tests for** panel, as this independent variable has three levels. If this main effect is statistically significant, and assuming we met the assumption of homogeneity of variance, we would want to compare the **treatment** means if the interaction is not statistically significant. We check **R-E-G-W-Q** to obtain the Ryan–Enoit–Gabriel–Welsch Studentized Range test but are prepared

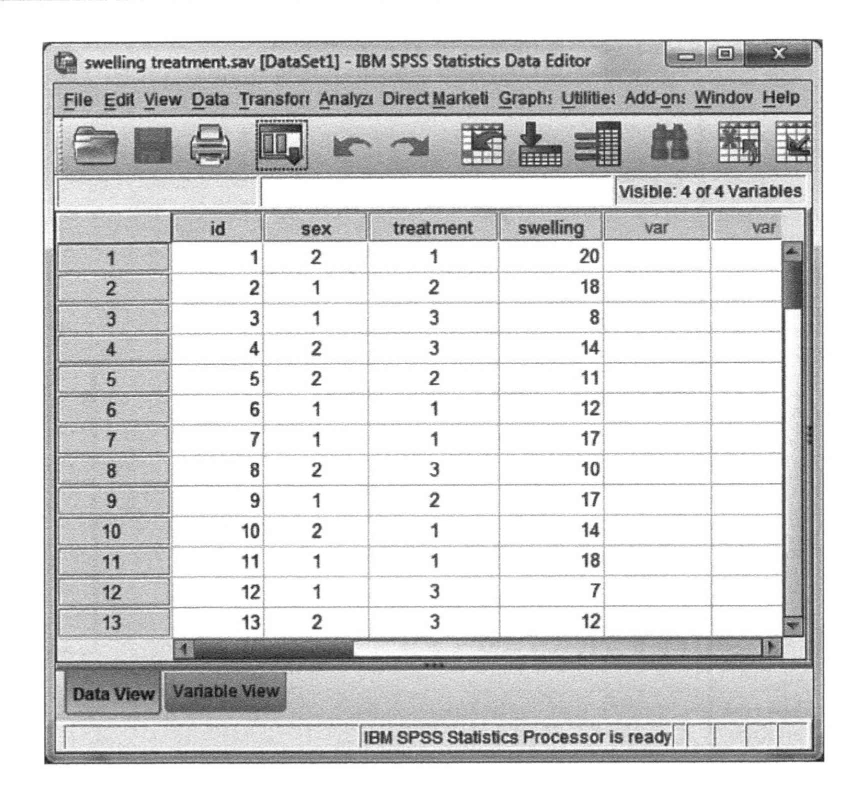

FIGURE 50.2
A portion of the data file.

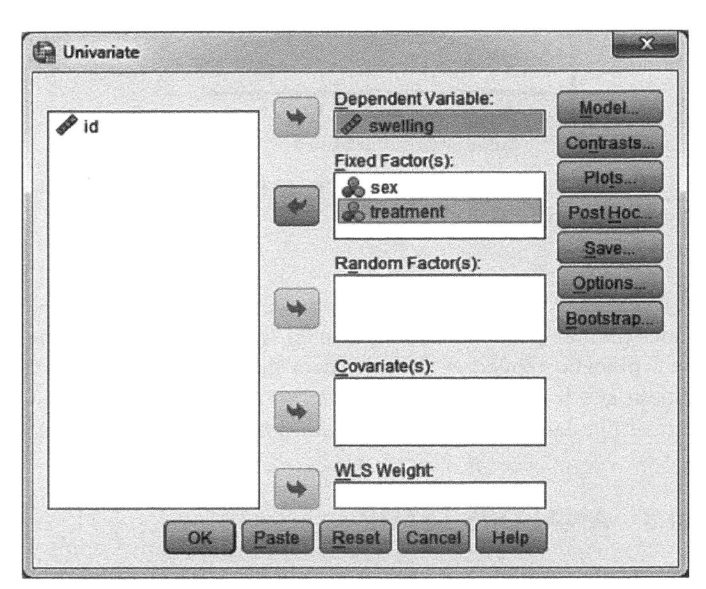

FIGURE 50.3
The main **Univariate** dialog window in **General Linear Model**.

to perform the analysis again with **Tamhane's T2** test if necessary. Click **Continue** to return to the main dialog window.

Anticipating the possibility of a statistically significant interaction effect, it is useful to obtain a plot of the means for two reasons: (a) interactions can best be understood by examining their visual depiction and (b) mapping our simple effects tests to a visual display facilitates our interpretation. Selecting the **Plots** pushbutton opens the **Profile Plots** dialog window (see Figure 50.5). As both independent variables are categorical, it is arbitrary how we structure the plot. We have therefore placed **sex** in the **Horizontal**

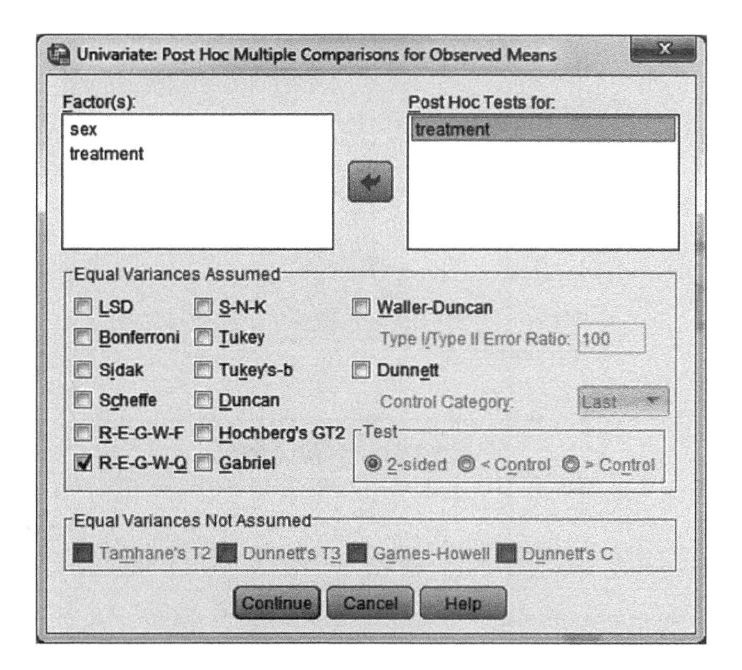

FIGURE 50.4

The **Post Hoc** dialog window of **Univariate**.

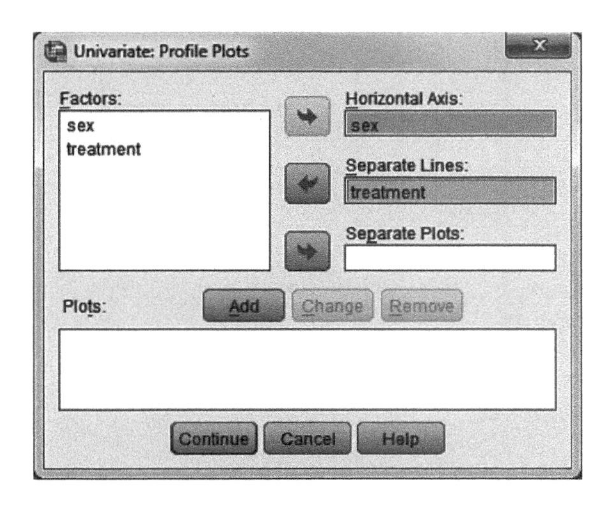

FIGURE 50.5

The **Profile Plots** dialog window of **Univariate** to set up the plot.

axis and will have **Separate Lines** for each **treatment**. Click the **Add** pushbutton to place the plot in the **Plots** panel (see Figure 50.6) and click **Continue** to return to the main dialog window.

The **Options** dialog window shown in Figure 50.7. In the **Display** area, we check **Descriptive statistics** and **Homogeneity tests** (to obtain the Levene test of equal group variances). We also begin the setup for the simple effects tests but will examine those results only if the interaction effect is statistically significant. The first step in setting up the simple effects tests is to focus on the **Estimated Marginal Means** portion of the **Options** window (the upper half of the window). We move the interaction shown as **sex*treatment** from the **Factor(s) and Factor Interactions** panel into the **Display Means for** panel, and click **Continue** to return to the main dialog window.

The next step in setting up the simple effects tests is to click **Paste**. This opens the syntax window shown in Figure 50.8 that presents the analysis setup (this is what our pointing and clicking generated and what actually drives the IBM SPSS® analysis). The

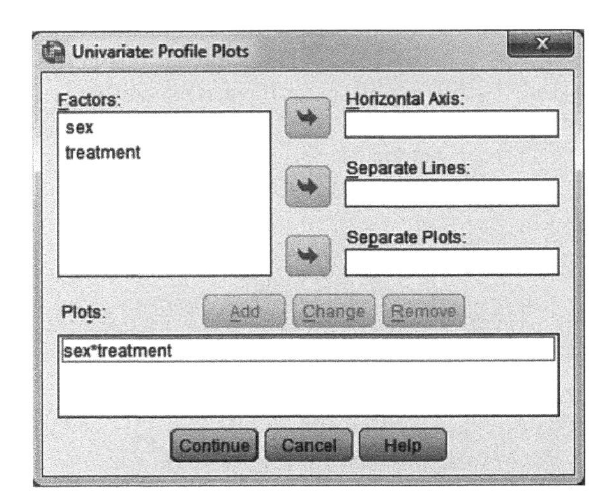

FIGURE 50.6

The **Profile Plots** dialog window of **Univariate** with the plot registered with IBM SPSS.

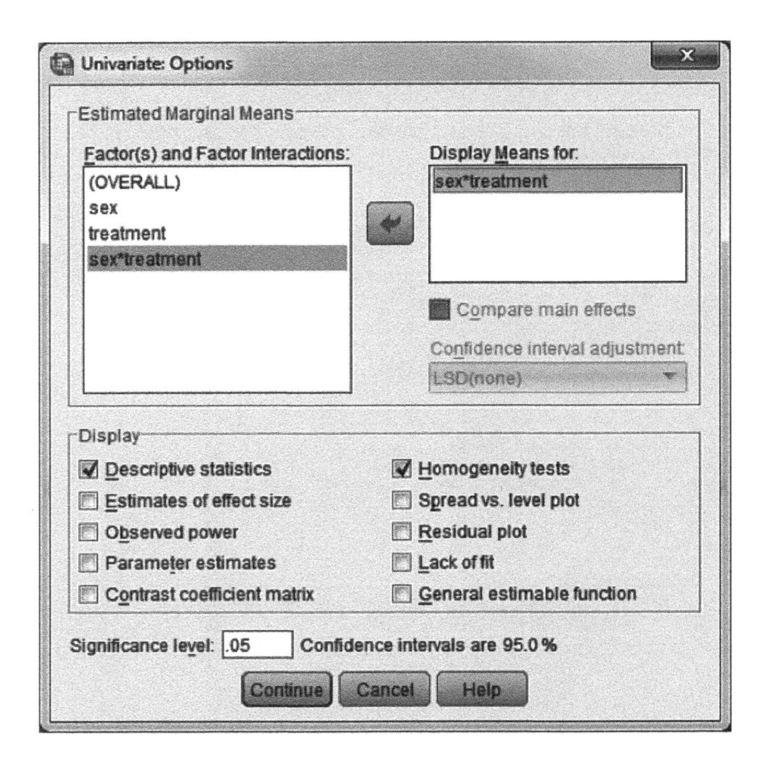

FIGURE 50.7

The **Options** dialog window of **Univariate** requesting a table of estimated marginal means for the interaction effect.

syntax in a syntax window can be edited in a word processing manner (but with a lot less power), and we will need to add a few words to this underlying syntax.

Our focus is on the subcommand **/EMMEANS = TABLES(sex*treatment)**. The translation of this syntax is roughly "create a table of estimated marginal means for the **sex*treatment** interaction." Our simple effects tests will be based on the estimated marginal means. We take the following steps to perform the full set of simple effects tests:

- Copy and paste the subcommand directly below the original making sure that it is indented as the original (see Figure 50.9).
- At the end of the original line, type **compare (sex) adj (Bonferroni)**. Do not type a period. IBM SPSS will anticipate what is being typed and will offer prompts

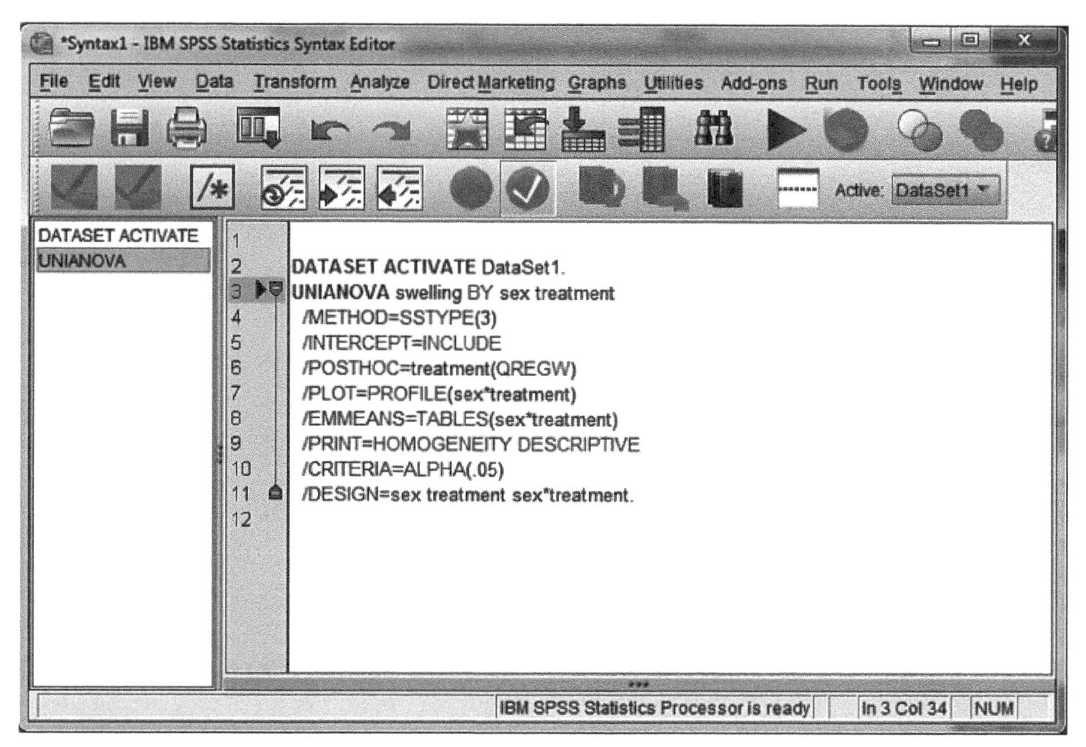

FIGURE 50.8 The **Pasted** syntax in a syntax window.

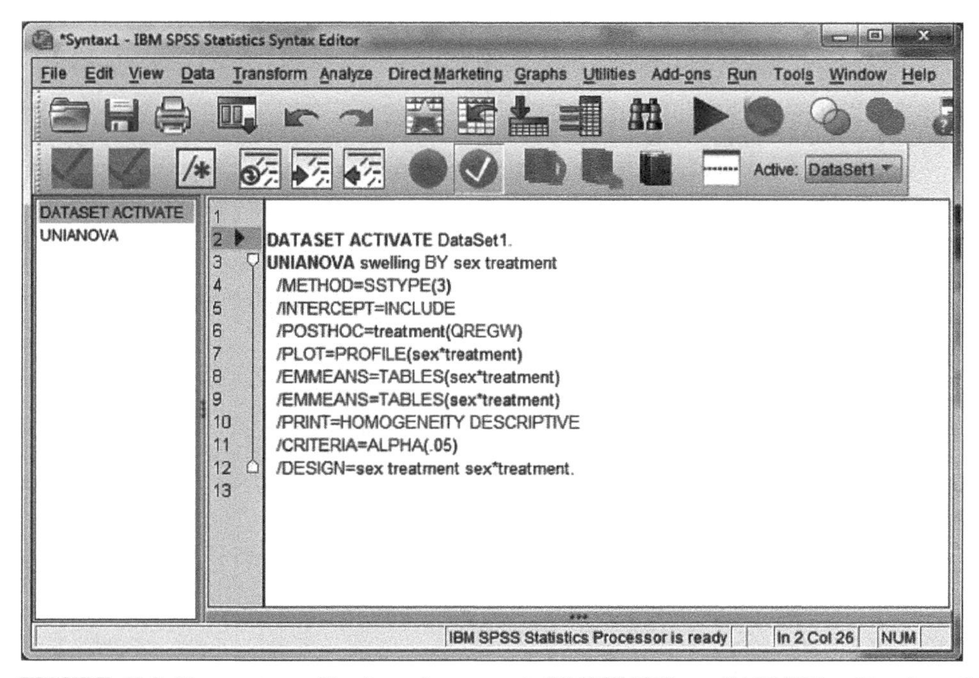

FIGURE 50.9 The syntax with the subcommand **/EMMEANS = TABLES(sex*treatment)** duplicated.

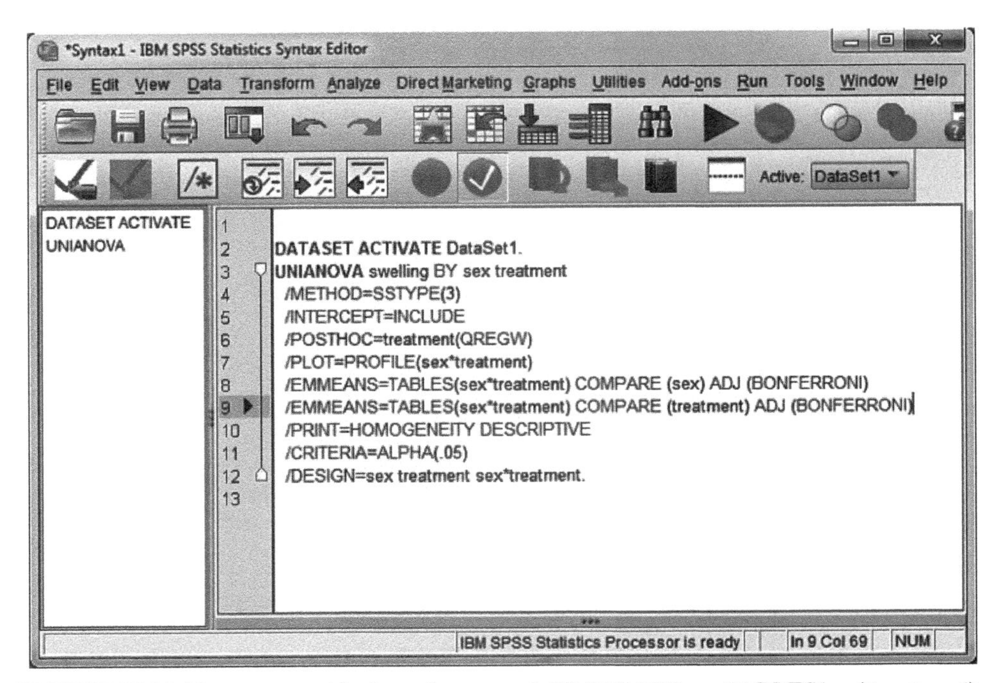

FIGURE 50.10 The syntax with the subcommand **/EMMEANS = TABLES(sex*treatment)** duplicated and the simple effects tests configured.

that are useful to accept (the screenshot in Figure 50.10 reflects the acceptance of the prompts). This additional syntax translates to "compare the levels of **sex** (females vs. males) separately for each level of the other independent variable (compare females with males for the **no treatment** condition, again for the **placebo** condition, and still again for the **medication** condition) using a Bonferroni alpha level correction."

- At the end of the line that we copied and pasted, type **compare (treatment) adj (Bonferroni)**. Do not type a period. This addition translates to "compare the levels of **treatment** (**no treatment** vs. **placebo** vs. **medication**) separately for each level of the other independent variable (compare these three groups for the female patients and compare them again for the male patients) using a Bonferroni alpha level correction" (see Figure 50.10).
- From the main menu, we select **Run ➔ All** to obtain the output.

50.4 ANALYSIS OUTPUT: OMNIBUS ANALYSIS

The **Descriptive Statistics** table in Figure 50.11 shows the observed mean, standard deviation, and group sizes for each sex by treatment condition. The bottom table in Figure 50.10 shows Levene's statistic for homogeneity of variances. The F ratio of **1.507** was not statistically significant ($p = .$**205**) when evaluated with **5** and **48** degrees of freedom; we may therefore conclude that the assumption of equal group variances was not violated.

Figure 50.12 displays the ANOVA summary table. Because we performed this analysis in the **General Linear Model** procedure, IBM SPSS presents all of the output from the general linear model analysis, as described in Chapter 47. We focus on the reduced model as described in Section 47.5. The F ratio is generally computed as the

Descriptive Statistics

Dependent Variable:swelling

sex	treatment	Mean	Std. Deviation	N
1 female	1 no treatment	15.22	2.167	9
	2 placebo	13.56	3.644	9
	3 medication	4.22	2.224	9
	Total	11.00	5.602	27
2 male	1 no treatment	16.67	2.915	9
	2 placebo	10.11	3.257	9
	3 medication	11.22	3.270	9
	Total	12.67	4.206	27
Total	1 no treatment	15.94	2.600	18
	2 placebo	11.83	3.792	18
	3 medication	7.72	4.509	18
	Total	11.83	4.978	54

Levene's Test of Equality of Error Variances[a]

Dependent Variable:swelling

F	df1	df2	Sig.
1.507	5	48	.205

Tests the null hypothesis that the error variance of the dependent variable is equal across groups.

a. Design: Intercept + sex + treatment + sex * treatment

FIGURE 50.11 Descriptive statistics and the results of the Levene test.

Tests of Between-Subjects Effects

Dependent Variable:swelling

Source	Type III Sum of Squares	df	Mean Square	F	Sig.
Corrected Model	891.722[a]	5	178.344	20.296	.000
Intercept	7561.500	1	7561.500	860.529	.000
sex	37.500	1	37.500	4.268	.044
treatment	608.444	2	304.222	34.622	.000
sex * treatment	245.778	2	122.889	13.985	.000
Error	421.778	48	8.787		
Total	8875.000	54			
Corrected Total	1313.500	53			

a. R Squared = .679 (Adjusted R Squared = .645)

FIGURE 50.12 Summary table for the omnibus two-way between-subjects ANOVA.

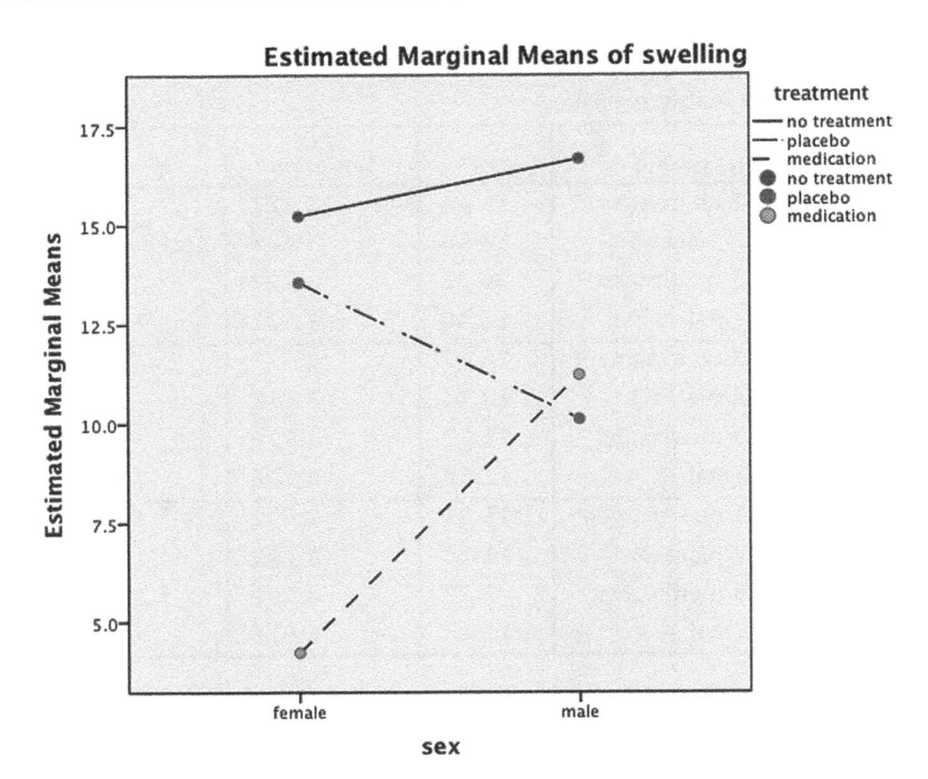

FIGURE 50.13

Plot of the interaction.

Mean Square (the variance) associated with the effect divided by the **Mean Square** (the variance) associated with the **Error** effect.

There are three effects of interest. In the order that they appear in the summary table, these effects are as follows:

- The main effect of **sex** with an F ratio of **4.268** and degrees of freedom of **1** and **48** is statistically significant (p = **.044**); its eta square value is **37.500/1313.500**, or approximately .03.
- The main effect of **treatment** with an F ratio of **34.622** and degrees of freedom of **2** and **48** is statistically significant ($p < .001$); its eta square value is **608.444/1313.500**, or approximately .46.
- The interaction of **sex** and **treatment** with an F ratio of **13.985** and degrees of freedom of **2** and **48** is statistically significant ($p < .001$); its eta square value is **245.778/1313.500**, or approximately .19.

Although the main effect of **treatment** appears to be a highly potent effect, the statistically significant interaction informs us that the treatment effect (the differences among the means of the **no treatment** vs. **placebo** vs. **medication** groups) is moderated (affected) by the **sex** of the patient. This can be seen in the plot shown in Figure 50.13. Based on visual inspection of the plot, it appears that

- the **no treatment** condition (solid line) appears to exhibit the most swelling, and the difference between females and males is small and not likely to be statistically significant;
- for the **placebo** condition (dashed–dotted line), females may not be significantly different from the **no treatment** control, whereas males appear to have less swelling than the **no treatment** control;

- for the **medication** condition (dashed line), females appear to have very little swelling, whereas males are not very different from the males in the **placebo** condition.

50.5 ANALYSIS OUTPUT: SIMPLE EFFECTS TESTS

The syntax we added at the end of the analysis setup generated two sets of simple effects, one for each subcommand line. The first set of simple effects was designed to compare **sex** (compare females and males) for each of the **treatment** levels. The results of this analysis are shown in Figure 50.14.

The **Estimates** table shows the estimated marginal means, standard deviations, and 95% confidence intervals for each of the groups. Because we are focused on the individual groups (and are not combining groups of unequal sample sizes), these estimated marginal means are equal to the observed means.

The **Pairwise Comparisons** table shows the Bonferroni corrected t test results on each pair of means. We described the structure of such a table in Chapter 49. Briefly, we read across the three major rows to learn that

Estimates

Dependent Variable:swelling

				95% Confidence Interval	
sex	treatment	Mean	Std. Error	Lower Bound	Upper Bound
1 female	1 no treatment	15.222	.988	13.236	17.209
	2 placebo	13.556	.988	11.569	15.542
	3 medication	4.222	.988	2.236	6.209
2 male	1 no treatment	16.667	.988	14.680	18.653
	2 placebo	10.111	.988	8.124	12.098
	3 medication	11.222	.988	9.236	13.209

Pairwise Comparisons

Dependent Variable:swelling

						95% Confidence Interval for Difference[a]	
treatment	(I) sex	(J) sex	Mean Difference (I-J)	Std. Error	Sig.[a]	Lower Bound	Upper Bound
1 no treatment	1 female	2 male	-1.444	1.397	.306	-4.254	1.365
	2 male	1 female	1.444	1.397	.306	-1.365	4.254
2 placebo	1 female	2 male	3.444*	1.397	.017	.635	6.254
	2 male	1 female	-3.444*	1.397	.017	-6.254	-.635
3 medication	1 female	2 male	-7.000*	1.397	.000	-9.810	-4.190
	2 male	1 female	7.000*	1.397	.000	4.190	9.810

Based on estimated marginal means

a. Adjustment for multiple comparisons: Bonferroni.

*. The mean difference is significant at the

FIGURE 50.14 Simple effects tests comparing females and males.

- for the **no treatment** condition, females and males do not differ (p = **.306**);
- for the **placebo** condition, females have significantly more swelling than males (p = **.017**);
- for the **medication** condition, females have significantly less swelling than males ($p < .001$).

The second set of simple effects was designed to compare **treatment** (compare **no treatment** vs. **placebo** vs. **medication**) for each of the **sex** levels. The results of this analysis are shown in Figure 50.15. Briefly, we read across the two major rows to learn the following.

sex * treatment

Estimates

Dependent Variable:swelling

sex	treatment	Mean	Std. Error	95% Confidence Interval	
				Lower Bound	Upper Bound
1 female	1 no treatment	15.222	.988	13.236	17.209
	2 placebo	13.556	.988	11.569	15.542
	3 medication	4.222	.988	2.236	6.209
2 male	1 no treatment	16.667	.988	14.680	18.653
	2 placebo	10.111	.988	8.124	12.098
	3 medication	11.222	.988	9.236	13.209

Pairwise Comparisons

Dependent Variable:swelling

sex	(I) treatment	(J) treatment	Mean Difference (I-J)	Std. Error	Sig.[a]	95% Confidence Interval for Difference[a]	
						Lower Bound	Upper Bound
1 female	1 no treatment	2 placebo	1.667	1.397	.717	-1.800	5.133
		3 medication	11.000*	1.397	.000	7.533	14.467
	2 placebo	1 no treatment	-1.667	1.397	.717	-5.133	1.800
		3 medication	9.333*	1.397	.000	5.867	12.800
	3 medication	1 no treatment	-11.000*	1.397	.000	-14.467	-7.533
		2 placebo	-9.333*	1.397	.000	-12.800	-5.867
2 male	1 no treatment	2 placebo	6.556*	1.397	.000	3.089	10.022
		3 medication	5.444*	1.397	.001	1.978	8.911
	2 placebo	1 no treatment	-6.556*	1.397	.000	-10.022	-3.089
		3 medication	-1.111	1.397	1.000	-4.578	2.355
	3 medication	1 no treatment	-5.444*	1.397	.001	-8.911	-1.978
		2 placebo	1.111	1.397	1.000	-2.355	4.578

Based on estimated marginal means

a. Adjustment for multiple comparisons: Bonferroni.

*. The mean difference is significant at the

FIGURE 50.15 Simple effects tests comparing treatment levels.

- For females, the **medication** condition resulted in significantly less swelling than both the **placebo** ($p < .001$) and the **no treatment** ($p < .001$) conditions; however, the **placebo** and **no treatment** conditions did not differ significantly in the amount of swelling ($p =$ **.717**).
- For males, both the **placebo** ($p < .001$) and the **medication** conditions resulted in significantly less swelling than the **no treatment** ($p < .001$) conditions; however, the **placebo** and **medication** conditions did not differ significantly in the amount of swelling ($p =$ **1.000**).

Overall, the results is that the new medication appeared to work for females but not for males; the placebo worked as well as the medication for males but females obtained no significant relief from the placebo.

CHAPTER 51

One-Way Within-Subjects ANOVA

51.1 OVERVIEW

A one-way within-subjects ANOVA design represents the generalized form of a paired-samples *t* test, with the ANOVA permitting us to compare two or more conditions. It is also known as a one-way *repeated measures* design. The name of the ANOVA design is derived from the following considerations:

- It is a *one-way* design in that there is only one independent variable, although any number of levels representing that independent variable can be used in the research. Usually, this single independent variable is a time marker where measurement of the same dependent variable is made at various points in time, that is, cases are repeatedly assessed on a given dependent variable.
- It is a *within-subjects* or *repeated measures* design in that the same cases are repeatedly measured on the same dependent variable; in this sense, the cases serve as their own control. It is often referred to as a *pretest–posttest* design when the within-subjects factor is time related.
- Although it is common to envision a repeated measure design to assess the same variable on two or more different occasions (a pretest–posttest design), a one-way within-subjects ANOVA is also capable of assessing two or more different conditions experienced by the participants. For example, an instructor could use three different types of graded requirements for a class, such as a class presentation, a term paper, and a final exam. The scores on these three components could comprise three levels of a within-subjects factor and could be statistically analyzed by a one-way within-subjects ANOVA.

Using a time-related independent variable generally involves administering one or more pretest measurements to establish a baseline level of performance on the dependent measure and then administering a "treatment" of some kind. Following the completion of the treatment, one or more posttests are made to determine the effect of the treatment and, potentially, the longevity of its effects.

Substantial drawbacks to the one-way within-subjects design are associated with the substantial threats to internal and external validity (Shadish, Cook, & Campbell, 2002). Within-subjects designs that are time-based have no control group; thus, changes in performance over time in the dependent measure may be due in part to the treatment but may

Performing Data Analysis Using IBM SPSS®, First Edition.
Lawrence S. Meyers, Glenn C. Gamst, and A. J. Guarino.

also be due to factors that covary with time during which the treatment was administered but are, by definition, neither controlled nor measured. Despite the substantial drawback of no control group being present in such a study, the design is used in situations where a set of cases is unique (no appropriate control group can be identified) and subdividing the set of cases into a "holdout" (control) subsample and a treatment subsample is not feasible.

The statistical procedure in designs containing a within-subjects variable is to partition the variance of the dependent variable into two major portions—between-subjects variance and within-subjects variance—and then subpartition those portions when possible. In a one-way within-subjects design, the partitioning is as follows:

- The between-subjects variance is not able to be partitioned further; the variance represents the overall individual differences between the cases and is of little interest to researchers (IBM SPSS® actually labels this portion of the variance as between-subjects error).
- The within-subjects variance is further partitioned into the overall effect of the independent variable—the main effect of the treatment—and within-subjects error. The error represents the differences in patterns of performance of the cases across the levels of the treatment and is often called the Treatment × Subjects interaction. The *F* ratio for the treatment effect is computed by dividing the Mean Square associated with the treatment by the Mean Square associated with the within-subjects error.

51.2 NUMERICAL EXAMPLE

The fictional data we use for our example depict an evaluation of an educational component in connection with a mandatory drug rehabilitation program for offenders younger than 21 years. The data file is named **drug rehab**. A total of 11 youth offenders were assessed on their attitude toward using illegal drugs before the start of the program (**pretest**), after completing the 2-week program (**posttest1**), and 1 month following program completion (**posttest2**). Scores on the attitude measure could range between 5 and 20, with higher scores representing more favorable attitudes toward drug use.

A screenshot of the data file is shown in Figure 51.1. Note that the levels of the within-subjects variable are variables and that the values recorded under those variables are the scores from the attitude survey. Thus, the independent variable in this within-subjects design—the treatment—is represented by multiple variables (three in this example); it is for this reason that researchers can conceptualize within-subjects (repeated measures) designs as a form of a multivariate (multiple dependent measures) design.

51.3 ANALYSIS SETUP

We open the data file **drug rehab** and from the main menu, we select **Analyze → General Linear Model → Repeated Measures**. This opens the **Define Factor(s)** window of **Repeated Measures** as shown in Figure 51.2. The **Within-Subject Factor Name** panel opens with the default of **factor1**; we change that to **drug_education** as the name we created for our independent variable, as shown in Figure 51.3. We also type **3** in the **Number of Levels** panel. This action sets **drug_education** as our within-subjects independent variable. Click **Add** to register this information with IBM SPSS (see Figure 51.3) and click **Define** to open the main dialog window.

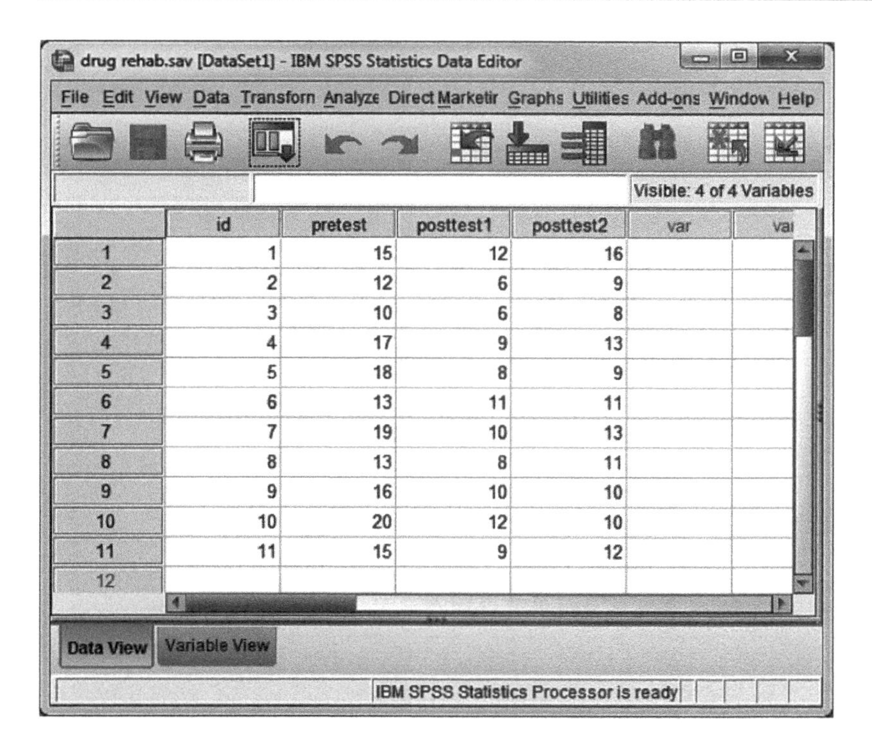

	id	pretest	posttest1	posttest2	var	var
1	1	15	12	16		
2	2	12	6	9		
3	3	10	6	8		
4	4	17	9	13		
5	5	18	8	9		
6	6	13	11	11		
7	7	19	10	13		
8	8	13	8	11		
9	9	16	10	10		
10	10	20	12	10		
11	11	15	9	12		
12						

FIGURE 51.1
The data file.

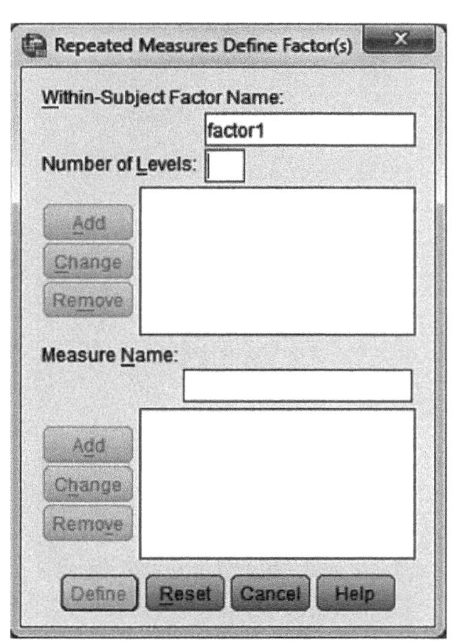

FIGURE 51.2
The **Define Factors** window of **General Linear Model Repeated Measures**.

The main **General Linear Model Repeated Measures** dialog window (see Figure 51.4) opens with three question-marked lines in the **Within-Subjects Variables** panel. Our name for the treatment effect, **drug_education**, appears just above the panel. There are three question marks because we defined the **Within-Subject Factor** as having three levels in the previous **Define Factors** window. Note that the question marks are numbered to accommodate a time-based set of variables. We move the variables in order into the **Within-Subjects Variables** panel as shown in Figure 51.5.

In the **Options** dialog window shown in Figure 51.6, we select **Descriptive statistics** in the **Display** area. We do not select **Homogeneity tests** because it applies only to

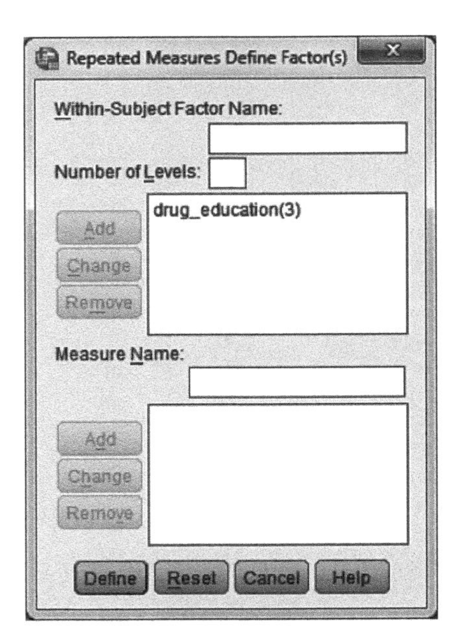

FIGURE 51.3
The **Define Factors** window of **General Linear Model Repeated Measures** with the treatment effect defined.

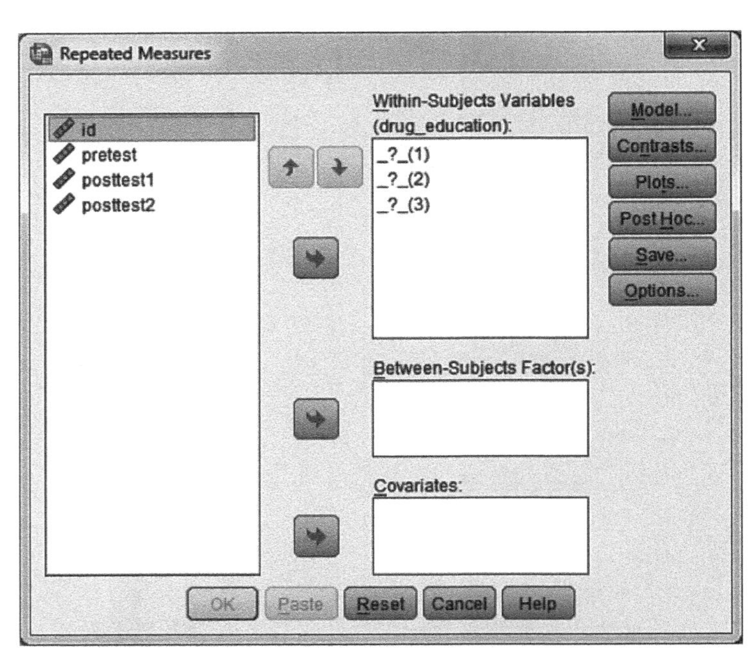

FIGURE 51.4
The main dialog window of **General Linear Model Repeated Measures**.

between-subjects independent variables; instead, the **Repeated Measures** procedure will automatically perform a *sphericity* test (the repeated measures rough analog of Levene's test) so long as we have at least three levels of the within-subjects variable.

Recall in the one-way between-subjects ANOVA, there was the assumption of homogeneity of variances (i.e., that all levels of the independent variable produced outcomes with equivalent variances). In the within-subjects ANOVA, the general sphericity assumption requires that the variances of the differences between each level are equivalent. It also assumes that each pair of levels (**pretest** and **posttest1**, **pretest** and **posttest2**, and **posttest1** and **posttest2**) is correlated to the same extent. These assumptions are tested together as sphericity. The general method of assessing these assumptions is to evaluate a variance/covariance structure known as *compound symmetry*, which specifies that both

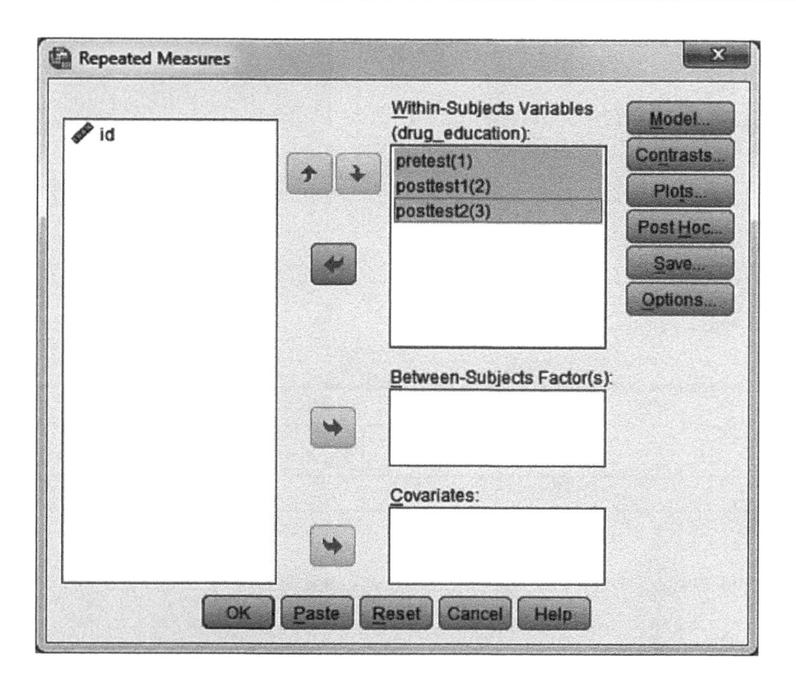

FIGURE 51.5

The main dialog window of **General Linear Model Repeated Measures** configured with the three levels of our treatment variable.

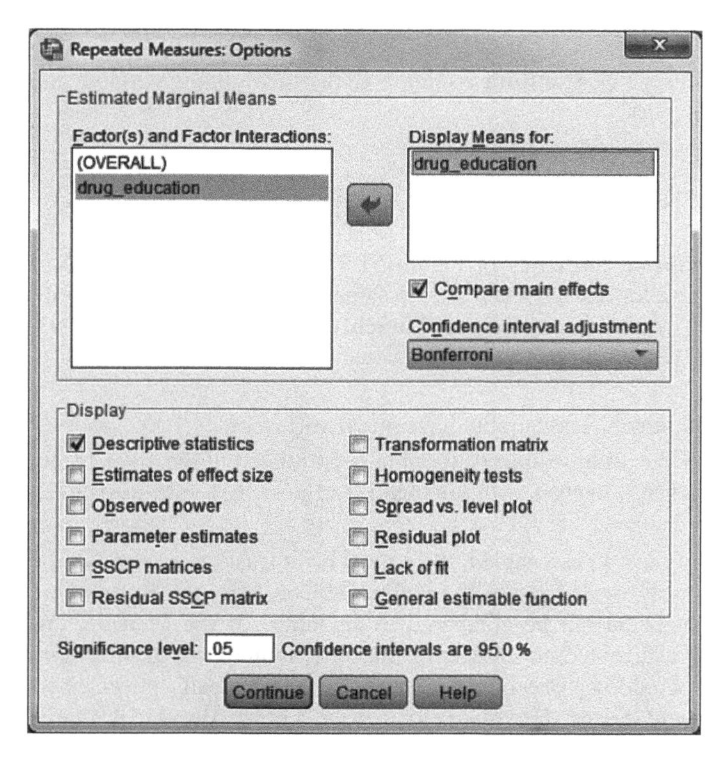

FIGURE 51.6

The **Options** window of **Repeated Measures**.

the variances and the covariances (correlations) between each level are comparable (see Chapter 52 for a more complete discussion of this topic).

As was the case with the ANCOVA (see Chapter 49), we can request our multiple comparison tests in the **Estimated Marginal Means** area. We therefore move **drug_ education** from the **Factor(s) and Factor Interactions** panel to the **Display Means for** panel, check **Compare main effects**, and select **Bonferroni** in the **Confidence interval**

Descriptive Statistics

	Mean	Std. Deviation	N
pretest pre program	15.27	3.101	11
posttest1 end of program	9.18	2.089	11
posttest2 one month folluwup	11.09	2.300	11

Mauchly's Test of Sphericity[b]

Measure:MEASURE_1

					Epsilon[a]		
Within Subjects Effect	Mauchly's W	Approx. Chi-Square	df	Sig.	Greenhouse-Geisser	Huynh-Feldt	Lower-bound
drug_education	.643	3.972	2	.137	.737	.834	.500

Tests the null hypothesis that the error covariance matrix of the orthonormalized transformed dependent variables is proportional to an identity matrix.

a. May be used to adjust the degrees of freedom for the averaged tests of significance. Corrected tests are displayed in the Tests of Within-Subjects Effects table.
b. Design: Intercept
Within Subjects Design: drug_education

FIGURE 51.7 **Descriptive Statistics** and **Mauchly's Test of Sphericity**.

adjustment drop-down menu. Click **Continue** to return to the main dialog window and click **OK** to perform the analysis.

51.4 ANALYSIS OUTPUT

The table labeled as **Descriptive Statistics** in Figure 51.7 shows the means, standard deviations, and sample sizes (even though these are the same 11 cases in each condition). The bottom table in Figure 51.7 shows the results of **Mauchly's** sphericity test. Mauchly's test evaluates the following two assumptions simultaneously:

- The levels of the within-subjects variable have equal variances.
- Each pair of levels of the within-subjects variable is correlated to the same extent (i.e., pretest with **posttest1**, **pretest** with **posttest2**, and **posttest1** with **posttest2**).

In order for the sphericity test to be enabled, there must be at least three levels of the independent variable (with only two levels of the within-subjects variable, there would be only one correlation that could not be compared to anything). If the Mauchly test returns a statistically significant outcome, indicating that the assumption of sphericity has not been met, then we would use one of the corrections automatically provided by IBM SPSS, the **Greenhouse–Geisser**, **Huynh–Feldt**, or the **Lower Bound**. All three corrections work by multiplying the *df* for the *F* ratio associated with the within-subjects treatment effect by the respective epsilon value shown in the table and evaluating the significance of *F* against these adjusted degrees of freedom.

Mauchly's W statistic is evaluated against a chi-square distribution with degrees of freedom of one less than the number of levels of our treatment effect. In the present case, **Mauchly's W** of **.643** corresponds to a chi-square value of **3.972**. With **2** degrees of freedom, this chi-square value is not statistically significant (p = **.137**), and so we judge that we have not violated the assumption of sphericity.

Tests of Within-Subjects Effects

Measure:MEASURE_1

Source		Type III Sum of Squares	df	Mean Square	F	Sig.
drug_education	Sphericity Assumed	213.515	2	106.758	30.876	.000
	Greenhouse-Geisser	213.515	1.474	144.854	30.876	.000
	Huynh-Feldt	213.515	1.667	128.072	30.876	.000
	Lower-bound	213.515	1.000	213.515	30.876	.000
Error(drug_education)	Sphericity Assumed	69.152	20	3.458		
	Greenhouse-Geisser	69.152	14.740	4.691		
	Huynh-Feldt	69.152	16.671	4.148		
	Lower-bound	69.152	10.000	6.915		

Tests of Between-Subjects Effects

Measure:MEASURE_1
Transformed Variable:Average

Source	Type III Sum of Squares	df	Mean Square	F	Sig.
Intercept	4632.758	1	4632.758	374.892	.000
Error	123.576	10	12.358		

FIGURE 51.8 Summary tables showing the omnibus ANOVA results.

Figure 51.8 displays the ANOVA summary tables. These correspond to the major partitions of the variance of the dependent variable. The **Tests of Between-Subjects Effects** table takes account of the individual differences in overall performance across the three testing occasions. In this analysis, IBM SPSS treats these individual differences as **Error** in the table (with a **Sum of Squares** of **123.576**) and is of little interest to us in this context.

The **Tests of Within-Subjects Effects** table presents the results for our treatment variable **drug_education** and its associated within-subject error term. We focus on the rows for **Sphericity Assumed** because the Mauchly test returned a nonsignificant outcome. The F ratio for the effect is **30.876** and, with **2** and **20** degrees of freedom, is statistically significant ($p < .001$). It therefore appears that there are differences in drug attitudes across the three measurements.

In a design containing a within-subjects variable, eta square is ordinarily computed separately for within-subjects effects and (if there are any) between-subjects effects. Our single treatment variable is a within-subjects effect and so the reference sum of squares is the total within-subjects sum of squares; in the example, it is **213.515** + **69.152**, or 282.667. We may then say that the **drug_education** effect explained **213.515**/282.667 or approximately 76% of the within-subjects variance.

IBM SPSS also provides another set of results evaluating the effect of the within-subjects variable, and these are shown in Figure 51.9. As we indicated in describing the structure of the data file, the three levels of the within-subjects factor are structured as separate variables (assembled under the umbrella of a within-subjects factor in the **Define Factor(s)** dialog window). These are treated by IBM SPSS as separate dependent variables in a *multivariate* ANOVA. Using this language, our one-way repeated measures ANOVA would be called a *univariate* design, in that it conceives of the repeated factor as a single dependent variable measured at three different times.

Multivariate Tests[b]

Effect		Value	F	Hypothesis df	Error df	Sig.
drug_education	Pillai's Trace	.890	36.256[a]	2.000	9.000	.000
	Wilks' lambda	.110	36.256[a]	2.000	9.000	.000
	Hotelling's Trace	8.057	36.256[a]	2.000	9.000	.000
	Roy's Largest Root	8.057	36.256[a]	2.000	9.000	.000

a. Exact statistic
b. Design: Intercept
Within Subjects Design: drug_education

FIGURE 51.9 **Multivariate tests** of statistical significance.

Tests of Within-Subjects Contrasts

Measure:MEASURE_1

Source	drug_education	Type III Sum of Squares	df	Mean Square	F	Sig.
drug_education	Linear	96.182	1	96.182	17.872	.002
	Quadratic	117.333	1	117.333	76.522	.000
Error(drug_education)	Linear	53.818	10	5.382		
	Quadratic	15.333	10	1.533		

FIGURE 51.10 Polynomial contrasts for the three levels of the within-subjects variable.

We will discuss more about the general multivariate approach in Chapter 54. For now, it is sufficient to say that this multivariate analysis tests the null hypothesis that the dependent variables have the same mean; as such, it is the same null hypothesis evaluated in the univariate ANOVA. However, the multivariate tests do not rest on the assumption of sphericity (because the levels of the within-subjects factor are treated as separate dependent variables).

The four **Multivariate Tests** each occupy a row in the table shown in Figure 51.9. Although they are calculated somewhat differently (see Meyers et al. (2013)), they are translated here into an identical F value of **36.256**. Based on **2** and **9** degrees of freedom, all are statistically significant ($p < .001$). The most commonly used of these multivariate tests is Wilks' lambda. Its value is an estimate of the unexplained variance and so subtracting that value from 1.00 yields an estimated multivariate eta square value; here, 1.00 − **.110** is .89, and we estimate that 89% of the variance is explained by the within-subjects factor.

IBM SPSS also performs polynomial contrasts (see Chapter 48) on the levels of the within-subjects factor, and the results of this portion of the analysis are shown in Figure 51.10. With only three levels in our current example, we can obtain only linear and quadratic components, and each of these is tested for statistical significance. The major row of interest in the table is labeled **drug_education**. Both the linear ($p =$ **.002**) and the quadratic ($p < .001$) components are significant.

These contrasts may be understood by examining the means of the levels. As shown in Figure 51.7, the means of the **pretest**, **posttest1**, and **posttest2** are **15.27**, **9.18**, and **11.09**, respectively. The linear trend is seen by the overall drop in positive attitudes toward drugs between the **pretest** and the **posttest2**; the quadratic trend is seen because drug attitudes dropped sharply from the **pretest** to the first posttest (**posttest1**) but showed an unfortunate (from the standpoint of the rehabilitation program) resurgence from the first to the second posttest.

Estimated Marginal Means drug_education

Estimates

Measure:MEASURE_1

drug_education	Mean	Std. Error	95% Confidence Interval Lower Bound	95% Confidence Interval Upper Bound
1	15.273	.935	13.189	17.356
2	9.182	.630	7.778	10.585
3	11.091	.694	9.546	12.636

Pairwise Comparisons

Measure:MEASURE_1

(I) drug_education	(J) drug_education	Mean Difference (I-J)	Std. Error	Sig.[a]	95% Confidence Interval for Difference[a] Lower Bound	95% Confidence Interval for Difference[a] Upper Bound
1	2	6.091*	.756	.000	3.920	8.261
	3	4.182*	.989	.005	1.343	7.021
2	1	-6.091*	.756	.000	-8.261	-3.920
	3	-1.909*	.579	.024	-3.572	-.247
3	1	-4.182*	.989	.005	-7.021	-1.343
	2	1.909*	.579	.024	.247	3.572

Based on estimated marginal means

*. The mean difference is significant at the

a. Adjustment for multiple comparisons: Bonferroni.

FIGURE 51.11 Results of the Bonferroni-corrected paired comparison t tests.

The estimated marginal means together with their standard errors and 95% confidence intervals are shown in the top table in Figure 51.11. These means are identical to the observed means.

The Bonferroni-corrected pairwise comparisons of the estimated marginal means are shown in the bottom table in Figure 51.11. Each major row focuses on one of the three conditions and compares it to the other two. The first major row focuses on the **pretest** (shown simply as **1** in the table, forcing us to remember our coding of the variable). The difference between the estimated marginal mean for that condition and the estimated marginal mean for **posttest1** (shown as **2** in the table) is **6.091** (**pretest** – **posttest1**). The difference is positive and statistically significant ($p < .001$), indicating that attitudes toward drugs were less favorable at **posttest1** than at **pretest**. The difference between the estimated marginal mean for the **pretest** (shown as **1** in the table) and the estimated marginal mean for **posttest2** (shown as **3** in the table) is **4.182** (**posttest1** – **posttest2**) and is also positive and statistically significant ($p =$ **.005**); thus, drug attitudes were still less favorable at **posttest2** than at **pretest**. Examining the other rows indicates that the difference between **posttest1** and **posttest2** is statistically significant as well ($p =$ **.024**);

examining the means informs us that attitudes toward drugs became somewhat more favorable from **posttest1** to **posttest2**.

Given the means of the three conditions and the results of the polynomial contrasts, we would conclude that the drug education component of the program was effective in lowering positive attitudes toward illegal drug use by the time it was completed. However, the offenders did regain some favorable attitude toward drug use after a month following the end of the program, although they were still less favorable at that time than they were at the start of the program. This resurgence of favorable attitude toward drugs should cause the researchers to consider what could be done to revise the program so that this effect could be reduced or eliminated.

CHAPTER 52

Repeated Measures Using Linear Mixed Models

52.1 OVERVIEW

A repeated measures ANOVA was applied in Chapter 51 to assess differences at three time points of youth offenders' attitude toward using drugs. The analysis was performed within the **General Linear Model** that requires the data to be normally distributed, have homogeneous variance across the levels of the within-subjects variable, and represent independent observations. However, the assumption of independence is automatically violated in repeated measures analyses in that the same cases are contributing multiple data points; therefore, to the extent that there are any individual differences, scores within any given case are going to be more related (correlated) than scores between cases.

In the **Linear Mixed Models** module of IBM SPSS®, the independent observations assumption is not necessary to be made in that the correlations among the scores tied to each individual case can be taken into account. In addition, the **Linear Mixed Models** is more adaptable. For example, the **General Linear Model** procedure for repeated measures requires participants to be observed at constant time points (e.g., every week), that we have complete data on all cases, and that all cases are observed at the same time points (e.g., if the schedule calls for measurements to be made every month at a specified time, observations cannot occur for some participants at a 6-week interval); the **Linear Mixed Models** is robust to these restrictions.

In the **General Linear Model**, individual differences are assumed to be *fixed effects* in the sense that all participants have the same intercept (initial score) and slope (changes in scores over time). In the **Linear Mixed Models**, individual differences are treated as a *random effect* in that cases can differ in their intercepts and slopes; such treatment of the variable tends to provide a better model of the data.

The **Linear Mixed Models** approach allows for alternative variance/covariance structures, whereas the **General Linear Model** procedure uses only a **Compound Symmetry** structure; alternative variance/covariance structures can lead to the **Linear Mixed Models** approach being statistically more powerful than the **General Linear Model** procedure.

The variance/covariance structure is contained in what is known as an **R** matrix. We have generated such a matrix for the analysis described in Chapter 51 by requesting the **Residual SSCP matrix** in the **Options** dialog window. The portion of the **Residual SSCP matrix** related to our current discussion is presented in the major **Covariance** row (the middle row) of the table shown in Figure 52.1. Such a matrix places the variances

Performing Data Analysis Using IBM SPSS®, First Edition.
Lawrence S. Meyers, Glenn C. Gamst, and A. J. Guarino.

Residual SSCP Matrix

		pretest pre program	posttest1 end of program	posttest2 one month folluwup
Sum-of-Squares and Cross-Products	pretest pre program	96.182	38.455	20.727
	posttest1 end of program	38.455	43.636	29.818
	posttest2 one month folluwup	20.727	29.818	52.909
Covariance	pretest pre program	9.618	3.845	2.073
	posttest1 end of program	3.845	4.364	2.982
	posttest2 one month folluwup	2.073	2.982	5.291
Correlation	pretest pre program	1.000	.594	.291
	posttest1 end of program	.594	1.000	.621
	posttest2 one month folluwup	.291	.621	1.000

Based on Type III Sum of Squares

FIGURE 52.1 The **Residual SSCP Matrix** with **Covariance** as the middle row.

of the measures on the diagonal. The largest variance was at **pretest (9.618)**. Following that there was a reduction of the variance in **posttest1 (4.364)**, with a slight increase at **posttest2 (5.291)**.

The entries off the diagonal (the upper and lower elements are redundant) are the covariances. Covariances are akin to correlations but use the original measurement units in the computation; that is, they are not standardized to produce values between ±1.00 (Norman & Streiner, 2008). Inspecting the covariances shown in Figure 52.1 reveals that the largest covariance occurs between **pretest** and **posttest1 (3.845)**, followed by **posttest1** and **posttest2 (2.982)** with the smallest covariance between **pretest** and **posttest2 (2.073)**. It appears that the covariances have become smaller after the **pretest**; as the intervals between time points increase, the covariances appear to decrease.

The structure of the **R** matrix is essential to the repeated measures analysis. The aim of the **Linear Mixed Models** procedure is to apply an **R** matrix that best fits the observed **R** matrix (i.e., the differences between the two matrices are minimized). In the **General Linear Model** approach, the **R** matrix is imposed on the data, whereas the **Linear Mixed Models** allows for alternative **R** matrices to be invoked. Obtaining an accurate **R** matrix is essential in calculating valid contrasts between the levels of the within-subjects factor.

In the **General Linear Model Repeated Measures ANOVA**, the observed **R** matrix is compared to the **Compound Symmetry R** matrix; this is how the assumption of *sphericity* (i.e., the variances and correlations between each level are equivalent) is tested in the **General Linear Model ANOVA** procedure. **Compound Symmetry** mandates that each of the three variances is equal and that all three covariances (correlations) are likewise equal. Although there are six parameters (three variances and three covariances) in the **R** matrix, the **Compound Symmetry** covariance structure needs to estimate only two of them (the variance and the covariance), as it proposes that within each set of three (variances and covariances), the values are equal, and this affords it relatively greater statistical power. The **Compound Symmetry R** matrix is, however, often not very realistic, and the assumption of sphericity in the **General Linear Model ANOVA** is frequently violated because of the statistical power associated with it.

An alternative to the **ANOVA** procedure in the **General Linear Model** approach is the multivariate technique that is automatically reported. This approach utilizes the **Unstructured R** matrix where there are no assumptions concerning the variances and the covariances of the elements. However, the **Unstructured** matrix is extremely complex because the variances and covariances need to be estimated at every time point (in this example, the three variances and the three covariances for a total of six parameters as opposed to only two parameters in the **Compound Symmetry** matrix). The inclusion of these additional parameters attenuates statistical power (Norman & Streiner, 2008). The **Unstructured** matrix may be applied if researchers are unable to determine the **R** matrix.

An alternative to both the **Compound Symmetry** and the **Unstructured** matrices—a viable and reasonable middle ground between the two—is to assume that the variances are either homogeneous (i.e., equivalent) or heterogeneous (i.e., different) but to consider that the correlation of each successive measurement with respect to the first time point would attenuate with each successive time point. This attenuation over time (that the covariances—correlations—between adjacent pairs decrease across the time points) is assessed by the *first-order autoregressive* covariance structure. We use here the variant of the autoregressive covariance structure that assumes homogeneity of variances. It has less statistical power than the **Compound Symmetry** covariance structure but is perhaps the approach most descriptive of the majority of such data structures, including the one shown in Figure 52.1.

52.2 NUMERICAL EXAMPLE

We use the data contained in the file named **drug rehab** that we used in Chapter 51. A total of 11 youth offenders were assessed on their attitude toward using drugs before the start of the program (**pretest**), after completing the 2-week program (**posttest1**), and 1 month following program completion (**posttest2**). Scores on the attitude measure could range between 5 and 20, with higher scores representing more favorable attitudes toward drug use.

52.3 ANALYSIS STRATEGY

The **Linear Mixed Models** procedure requires a different data structure in order to analyze data in a repeated measures design. We therefore first convert the data file from Chapter 51 into the necessary structure using the **Restructure Data Wizard**. We then perform the **Linear Mixed Models** analysis using the first-order autoregressive covariance structure. Finally, we repeat the **Linear Mixed Models** analysis two additional times, first using the **Compound Symmetry** covariance structure and then using the **Unstructured** covariance structure.

52.4 RESTRUCTURING THE DATA FILE

We open the data file **drug rehab** and from the main menu, we select **Data → Restructure**. This opens the **Restructure Data Wizard** window as shown in Figure 52.2. Our data file is currently in the *multivariate* or *wide* format, where repeated measures on the same dependent variable (e.g., attitudes toward drugs) are in separate columns on a single row for each case. The **Linear Mixed Models** procedure requires the data to be

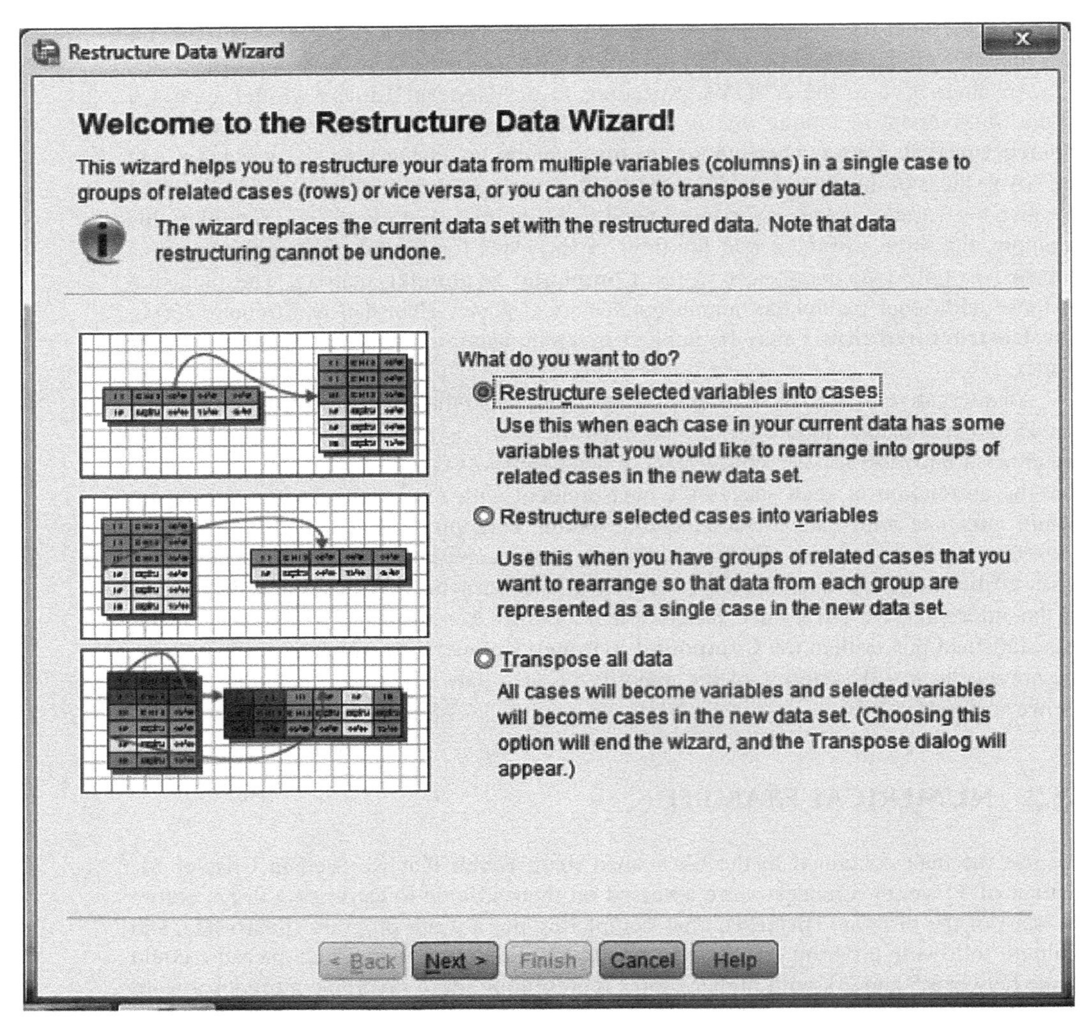

FIGURE 52.2 The initial **Restructure Data Wizard** window.

in the *univariate* or *long* format, where only one repeated measure on any one dependent variable is allowed in a single row. Thus, in our restructuring process, each case will occupy three rows (first for **pretest**, second for **posttest1**, and third for **posttest2**). The first choice of **Restructure selected variables into cases** specifies this format and we select it. Click **Next**.

Step 2 of the **Restructure Data Wizard** is to specify the **Number of Variable Groups** we have in our data set (this is the number of within-subjects variables in our study). This is shown in Figure 52.3. We select **One (for example, w1, w2, and w3)** and click **Next**.

Figure 52.4 presents the initial **Select Variables** screen, with default information filled in that we need to replace. In the **Case Group Identification** area, we select from the drop-down menu **Use selected variable** and move **id** into the panel. This variable identifies each case and will appear in three successive rows in the restructured data file (because there are three measures and one measure is placed on each row). This is shown in Figure 52.5.

In the area of the **Select Variables** screen labeled as **Variables to be Transposed**, we replace the default **trans1** in the **Target Variable** panel with some descriptive name for our measure. As we have assessed attitudes toward drugs, we have created the variable

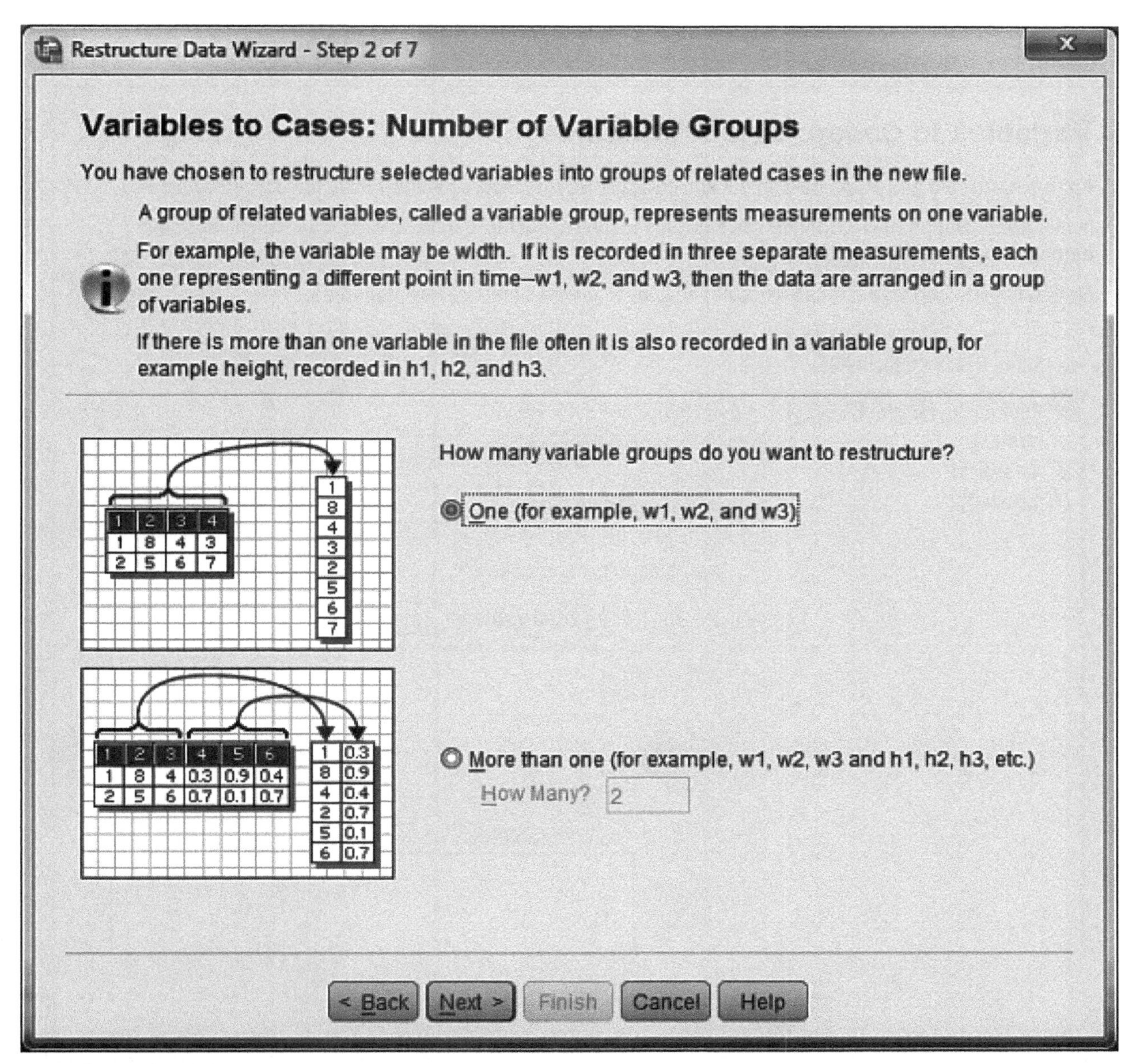

FIGURE 52.3 Here we specify the **Number of Variable Groups** (within-subjects variables) we have in our data set.

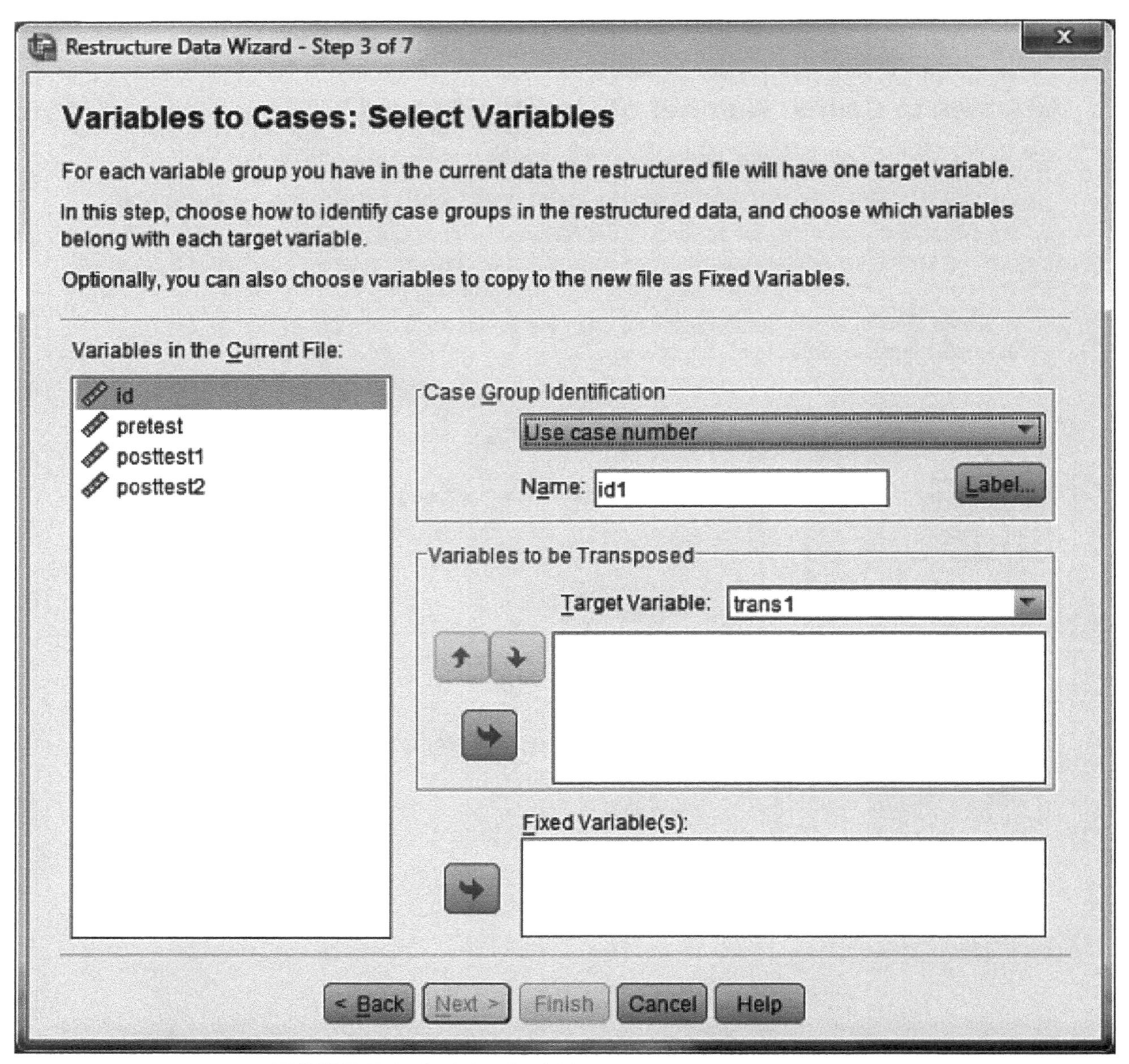

FIGURE 52.4 The **Select Variables** screen.

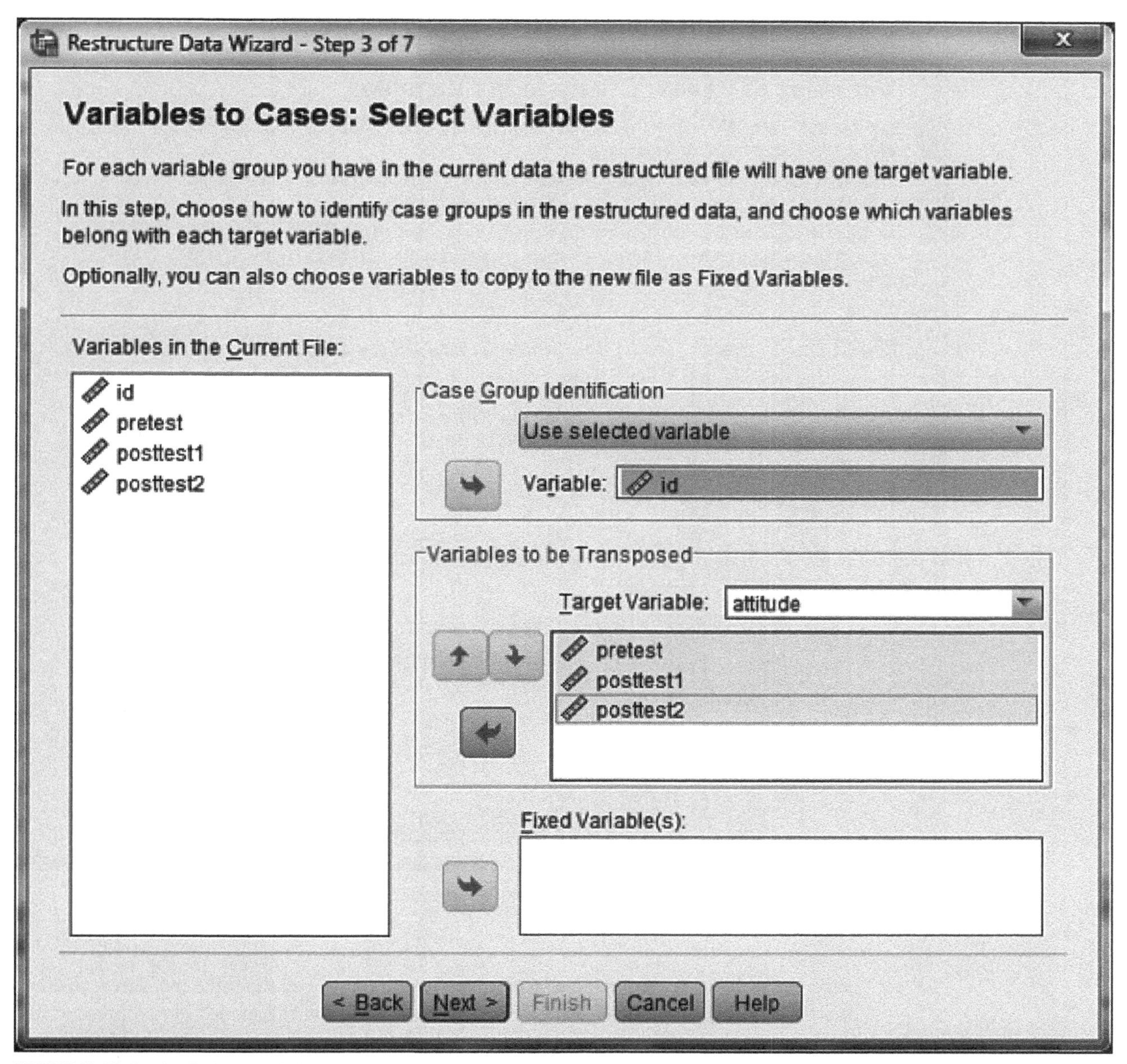

FIGURE 52.5 The **Select Variables** screen with the variables now specified.

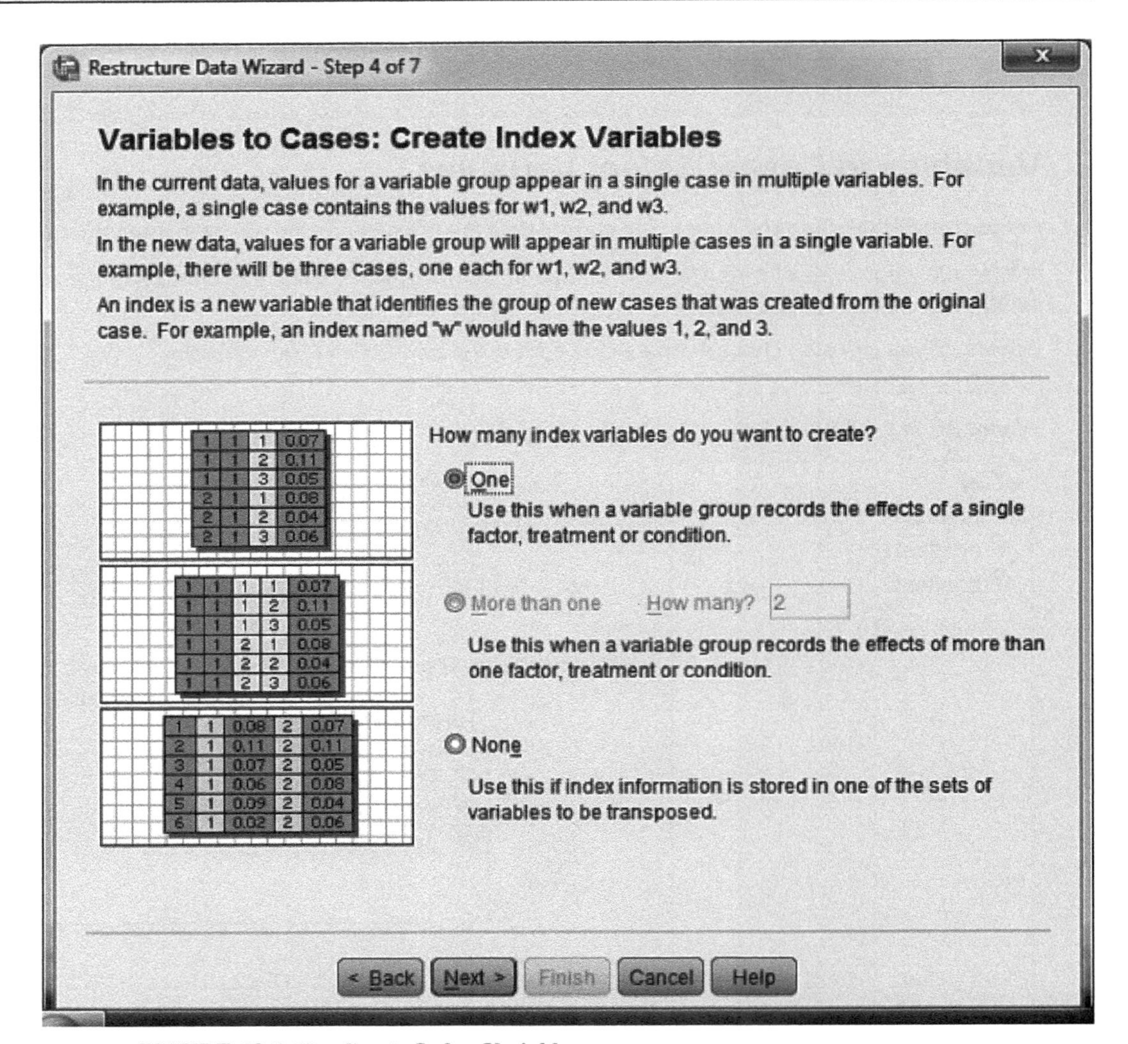

FIGURE 52.6 The **Create Index Variables** screen.

name **attitude** and typed it in the panel. We have then moved the three measures, **pretest**, **posttest1**, and **posttest2**, to **Target Variable** panel below the name. This represents the measurements made on our dependent variable and is also shown in Figure 52.5. Click **Next**.

The **Create Index Variables** screen is shown in Figure 52.6. An index variable tracks the level of the repeated measure. With our three levels, each case will have a score named **attitude** for **pretest**, **posttest1**, and **posttest2**. The index variable will code these levels as 1, 2, and 3 for the first case, 1, 2, and 3 for the second case, and so on. We have only one such variable and so we select **One**. Click **Next**.

We can now name our index variable in the **Create One Index Variable** screen (see Figure 52.7). We choose **Sequential numbers** under **What kind of index values?** We then replace the default **Index1** shown in Figure 52.7 with a name we create for this purpose. As shown in Figure 52.8, we have named our index variable **time**. Click **Next**.

The **Options** window is configured by default in the way we wish (see Figure 52.9) and so we simply click **Next**.

The **Finish** screen is shown in Figure 52.10. We wish to **Restructure the data now**. Clicking **Finish** opens a confirmation screen (see Figure 52.11) to which we click **OK**.

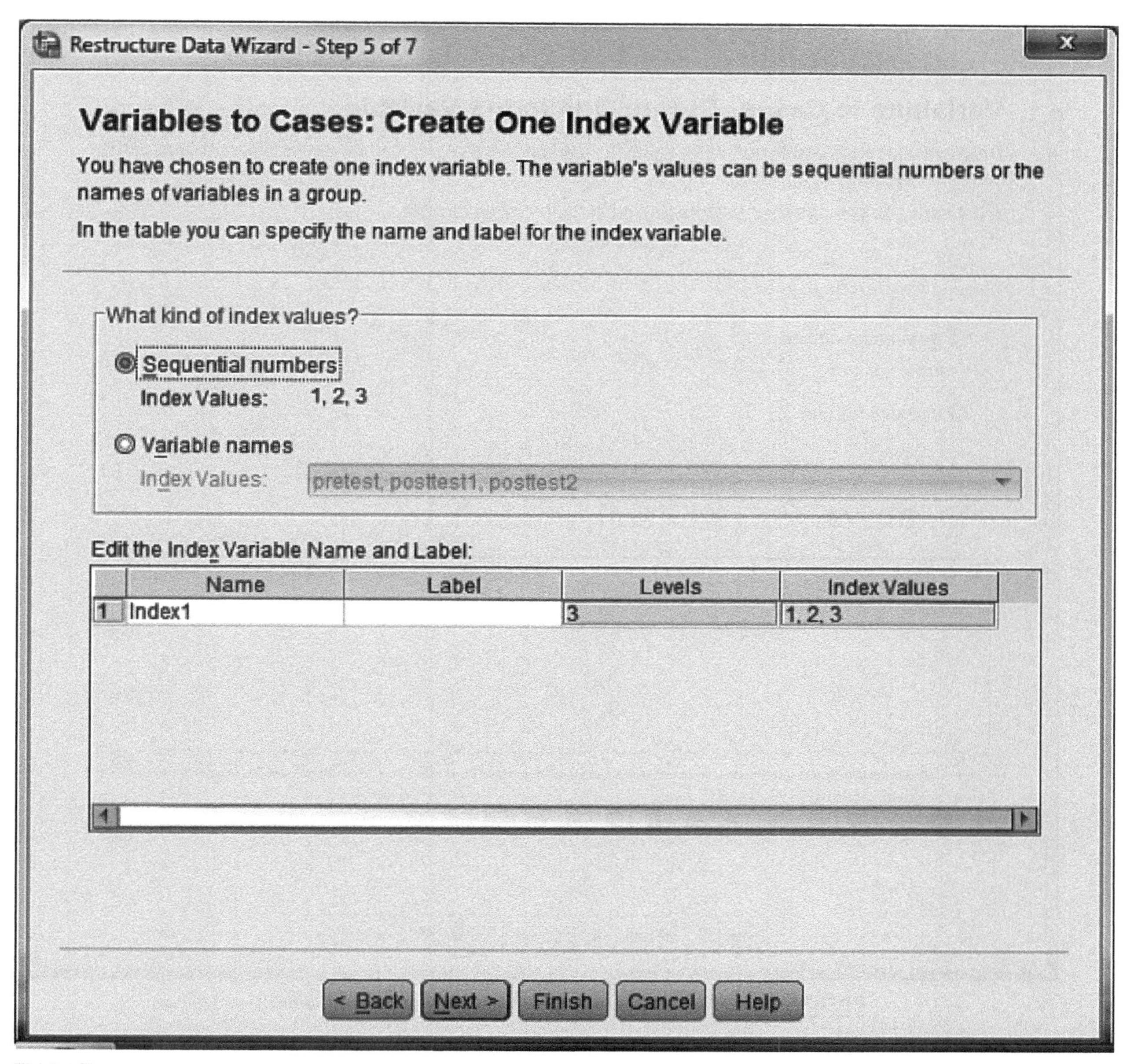

FIGURE 52.7 The **Create One Index Variable** screen.

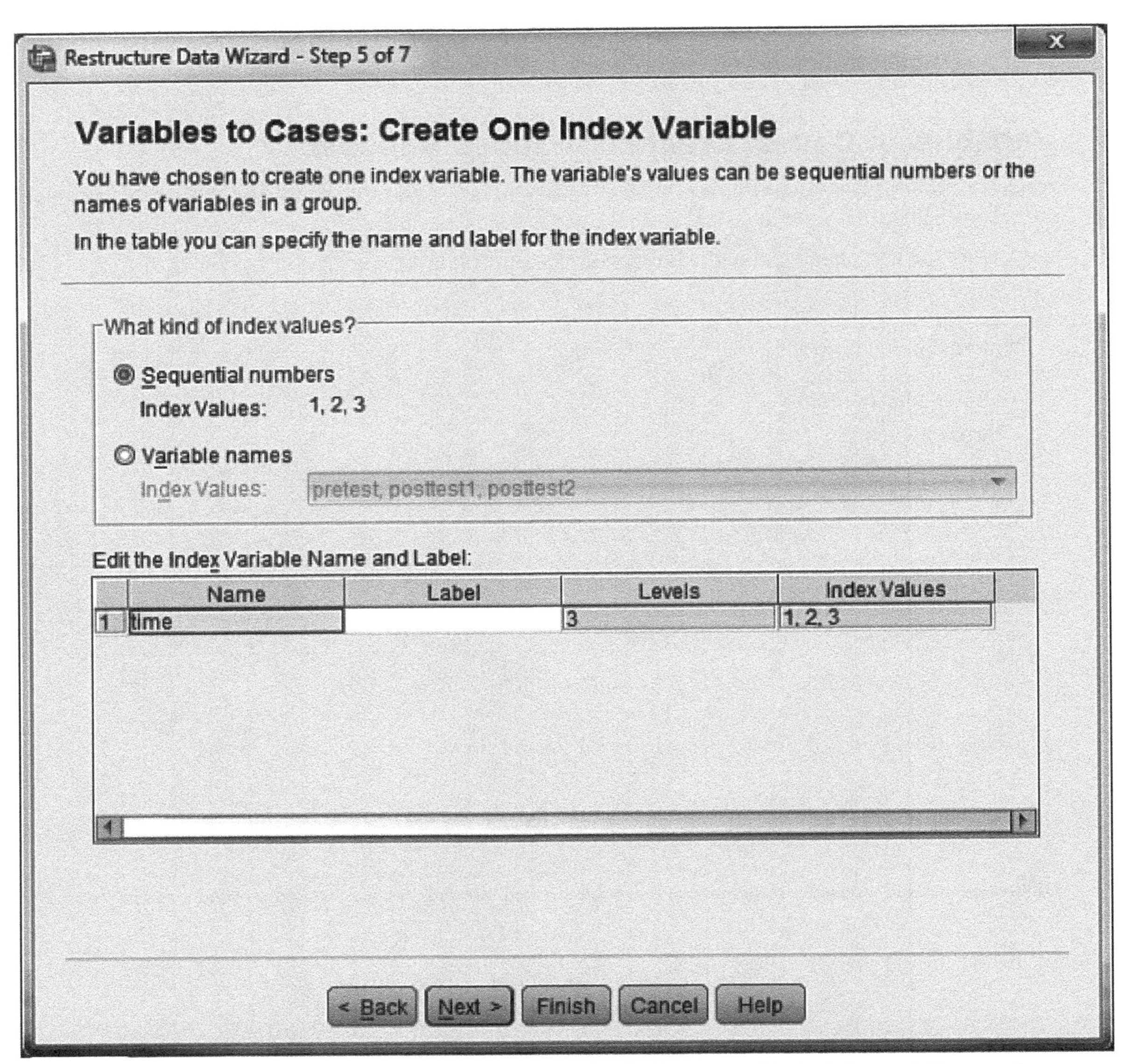

FIGURE 52.8 The **Create One Index Variable** screen with the index variable named.

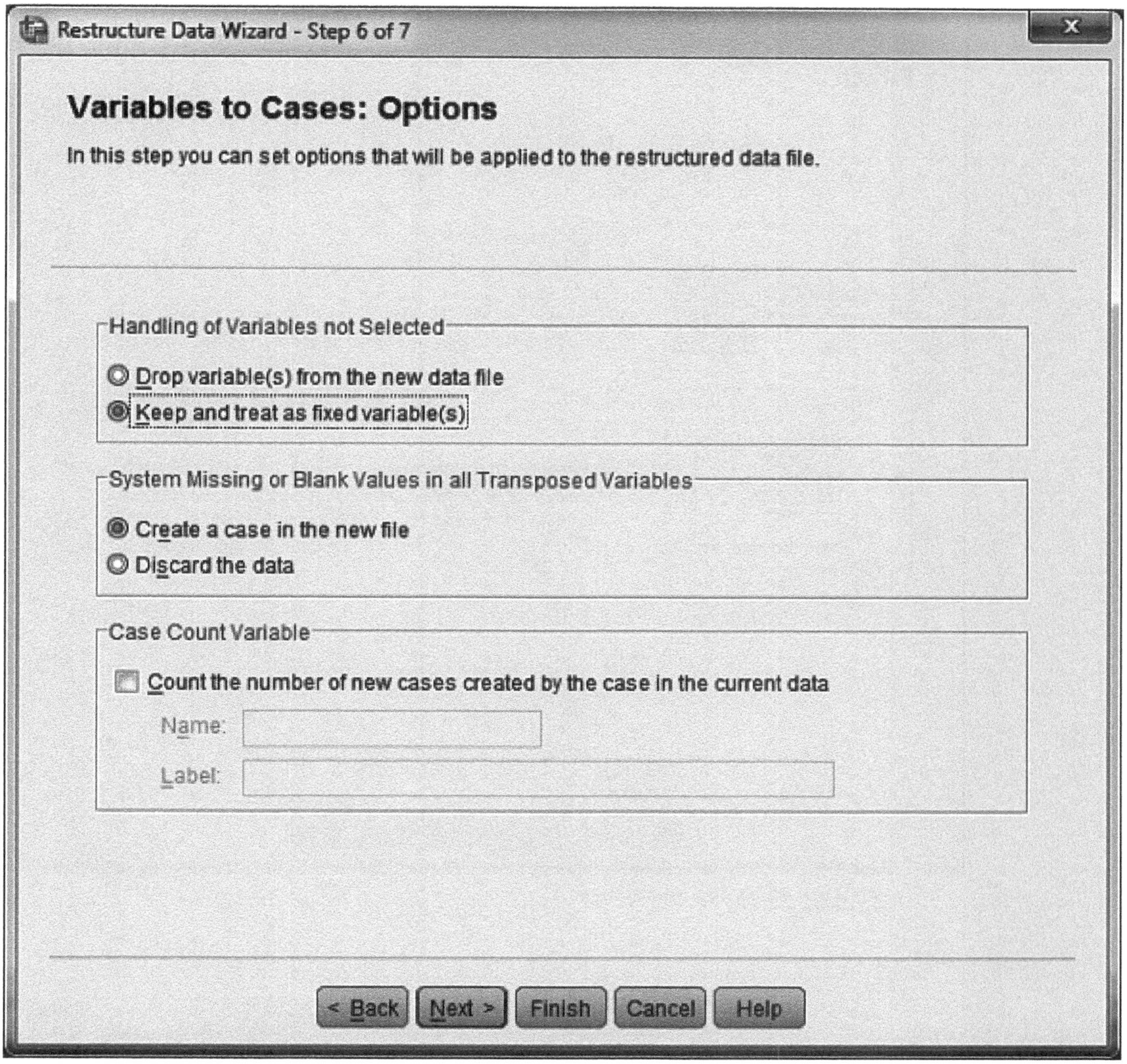

Restructure Data Wizard - Step 6 of 7

Variables to Cases: Options

In this step you can set options that will be applied to the restructured data file.

Handling of Variables not Selected

Drop variable(s) from the new data file

Keep and treat as fixed variable(s)

System Missing or Blank Values in all Transposed Variables

Create a case in the new file

Discard the data

Case Count Variable

Count the number of new cases created by the case in the current data

Name:

Label:

< Back | Next > | Finish | Cancel | Help

FIGURE 52.9 The **Options** screen.

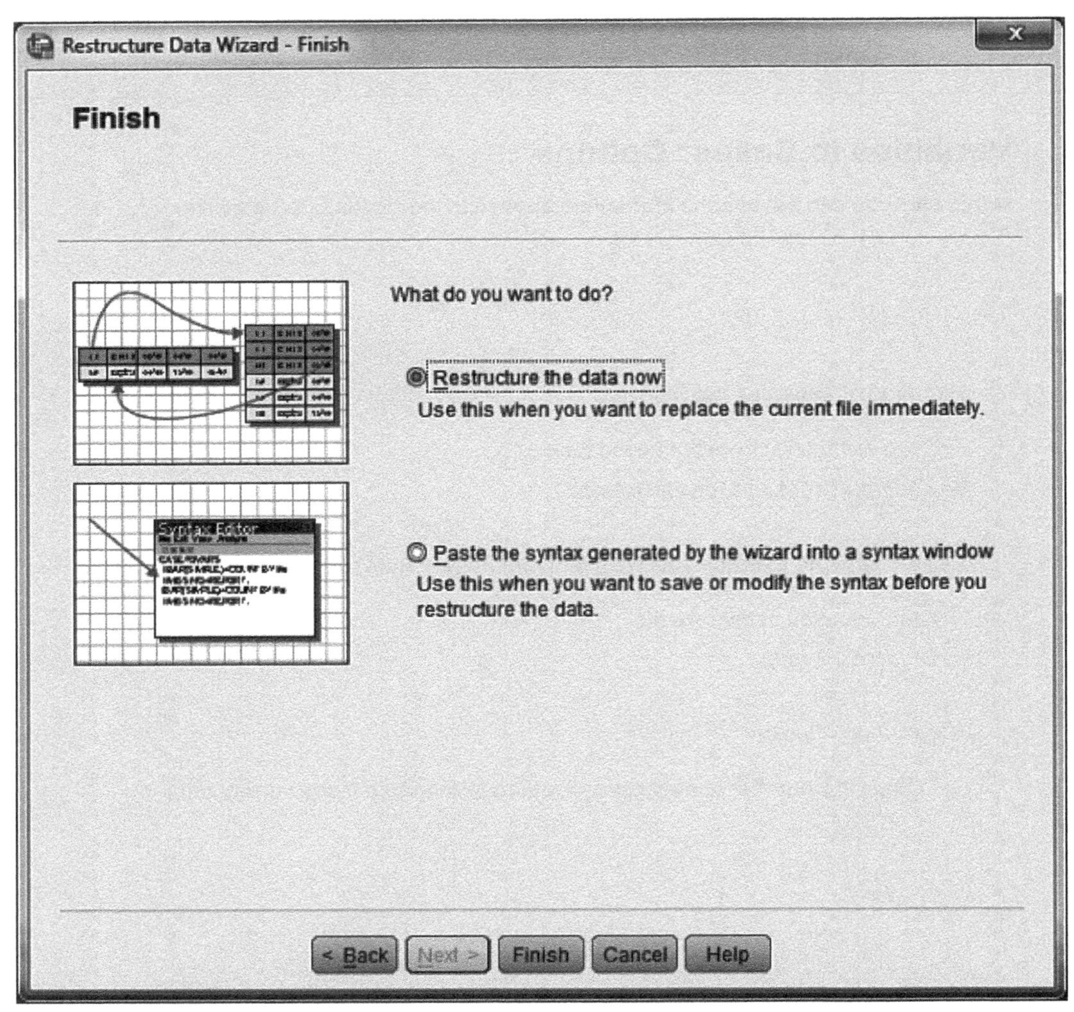

FIGURE 52.10 The **Finish** screen.

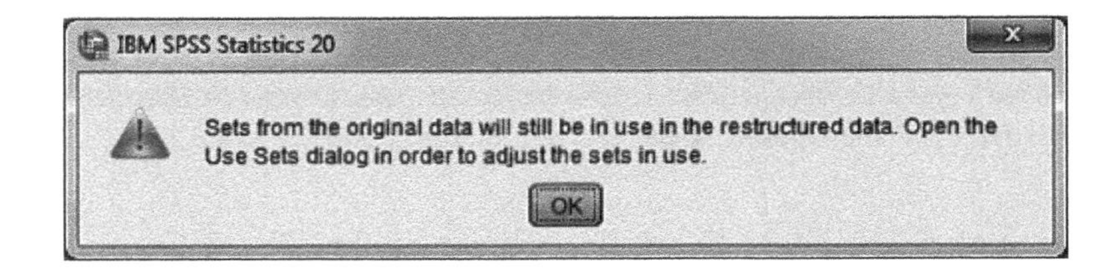

FIGURE 52.11
Accept this information and click **OK**.

A portion of the restructured data file is shown in Figure 52.12. This univariate or long format can be read as follows. The first case (**id 1**) is represented in the first three rows. The index variable of **time** represents the set of variables originally named **pretest**, **posttest1**, and **posttest2**. Under **attitude**, we see the values corresponding to **pretest**, **posttest1**, and **posttest2**. Each case is structured similarly. We save this file under the name **drug rehab long format** to preserve our original data file that is in the multivariate format.

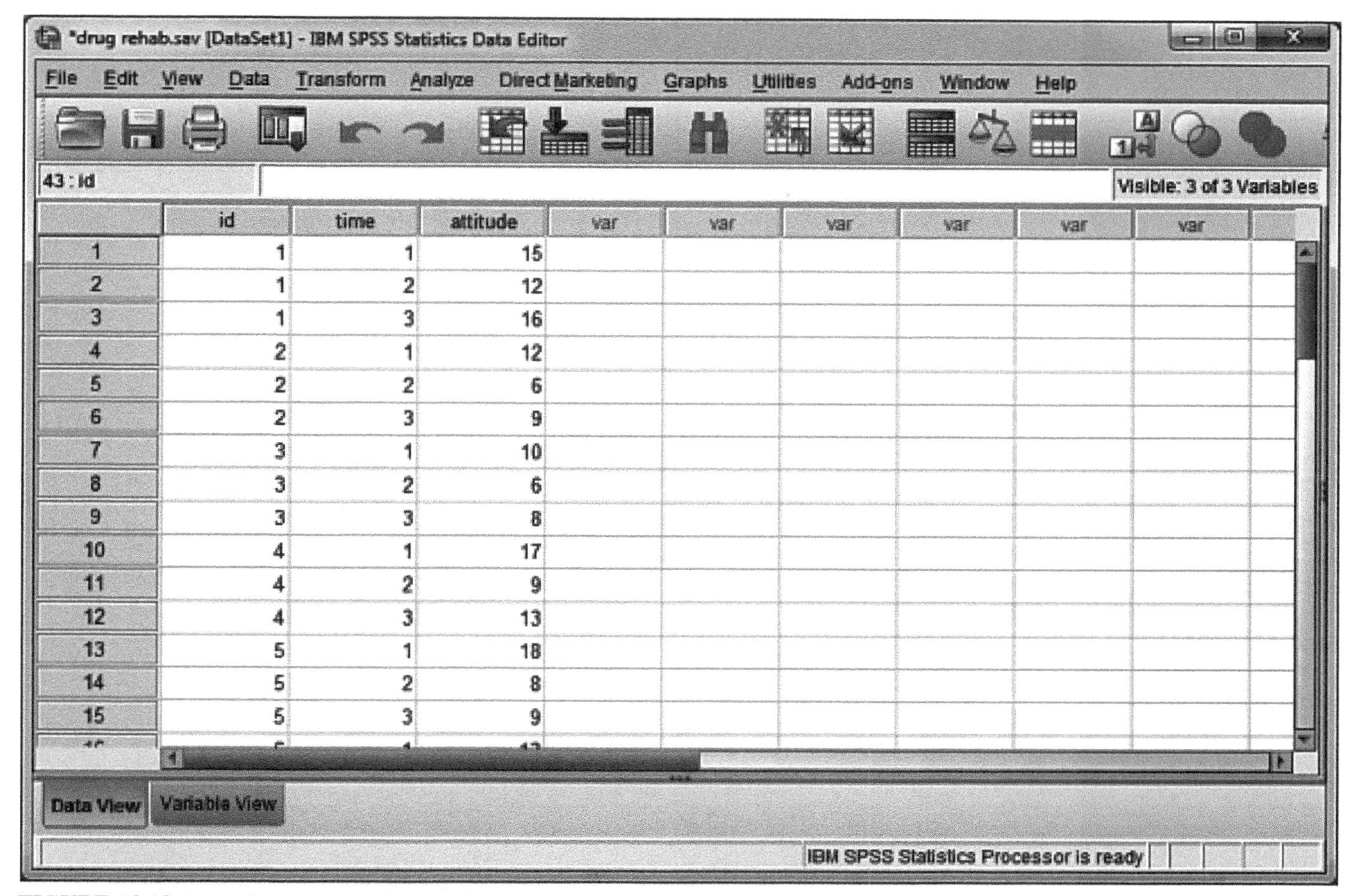

	id	time	attitude	var	var	var	var	var	var
1	1	1	15						
2	1	2	12						
3	1	3	16						
4	2	1	12						
5	2	2	6						
6	2	3	9						
7	3	1	10						
8	3	2	6						
9	3	3	8						
10	4	1	17						
11	4	2	9						
12	4	3	13						
13	5	1	18						
14	5	2	8						
15	5	3	9						

FIGURE 52.12 A portion of the restructured data file.

52.5 ANALYSIS SETUP: AUTOREGRESSIVE COVARIANCE STRUCTURE

We open the data file **drug rehab long format** and from the main menu, we select **Analyze → Mixed Models → Linear**. This opens the initial **Specify Subjects and Repeated** window shown in Figure 52.13. We move **id** into the **Subjects** panel (this variable represents the different cases) and **time** (this is our index variable that tracks the levels of the within-subjects factor) into the **Repeated** panel. From the drop-down menu for **Repeated Covariance Type**, we select **AR(1)**; this is the autoregressive structure that assumes homogeneous variances. Click **Continue** to reach the main **Linear Mixed Models** dialog window.

The main **Linear Mixed Models** window is shown in Figure 52.14. We move **attitude** into the **Dependent Variable** panel and **time** into the **Factor(s)** panel.

Selecting the **Fixed** pushbutton opens the **Fixed Effects** dialog window shown in Figure 52.15. Highlight **time** and select the **Add** button that becomes active once it is selected; this action places **time** into the **Model** panel. Click **Continue** to return to the main **Linear Mixed Models** dialog window.

Selecting the **Random** pushbutton opens the **Random Effects** dialog window shown in Figure 52.16. In the **Subject Groupings** area at the bottom of the screen, double-click **id** from the **Subjects** panel into the **Combinations** panel. We do not check the box associated with **Include intercept** (including the intercept yields a **Warning** about

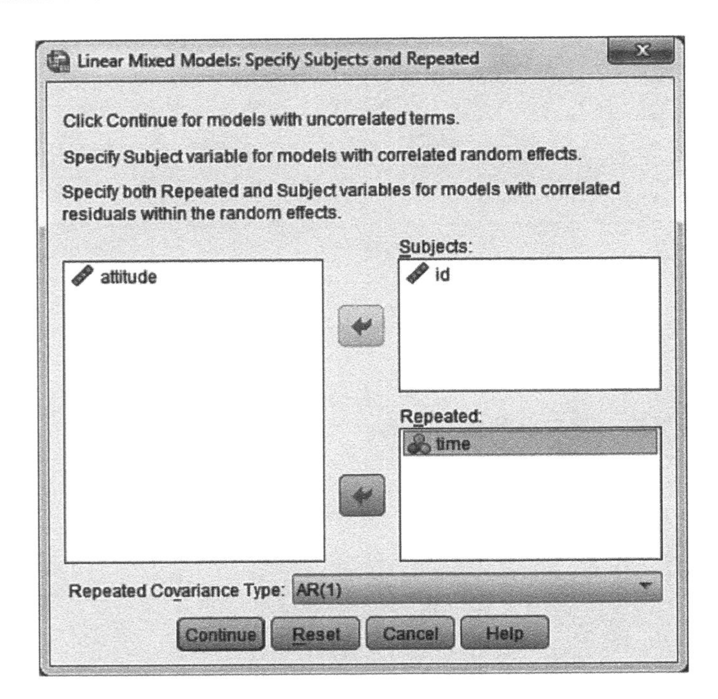

FIGURE 52.13
The initial **Specify Subjects and Repeated** window of **Linear Mixed Models**.

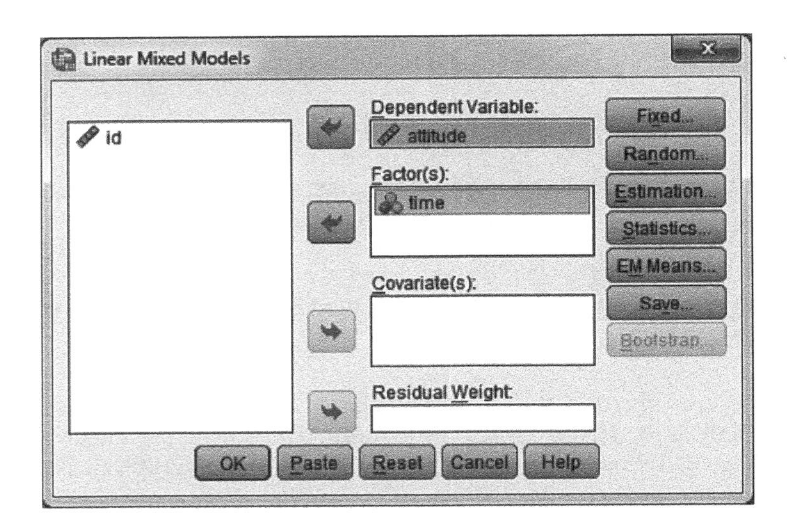

FIGURE 52.14
The main dialog window of **Linear Mixed Models**.

iteration being terminated and a footnote in the **Estimates of Covariance Parameters** table indicates that its covariance parameter is redundant and cannot be computed). Click **Continue** to return to the main **Linear Mixed Models** dialog window.

Selecting the **Statistics** pushbutton opens the **Statistics** dialog window shown in Figure 52.17. Under **Model Statistics**, check the boxes associated with **Parameter estimates, Tests for covariance parameters**, and **Covariances of residuals**. Click **Continue** to return to the main **Linear Mixed Models** dialog window.

Select the **EM Means** pushbutton to obtain the estimated marginal means. In the dialog window shown in Figure 52.18, we move **time** from the **Factors and Factor Interactions** panel to the **Display Means for** panel. To obtain the pairwise comparisons of the means of the three time measurements, we check **Compare main effects** and select **Bonferroni** from the drop-down menu, retaining **None (all pairwise)** under **Reference**

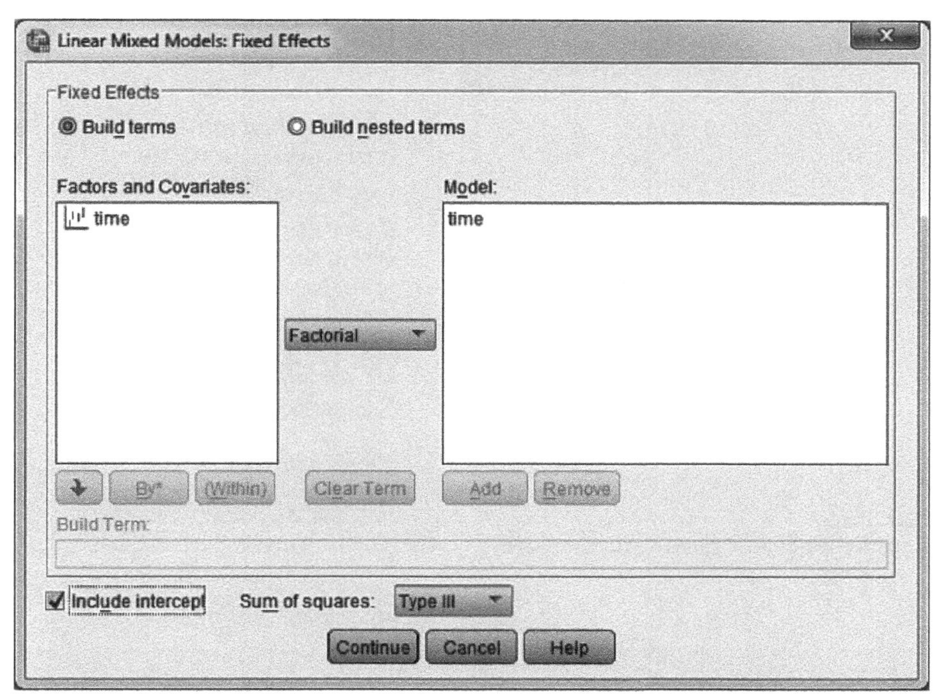

FIGURE 52.15 The **Fixed Effects** window of **Linear Mixed Models**.

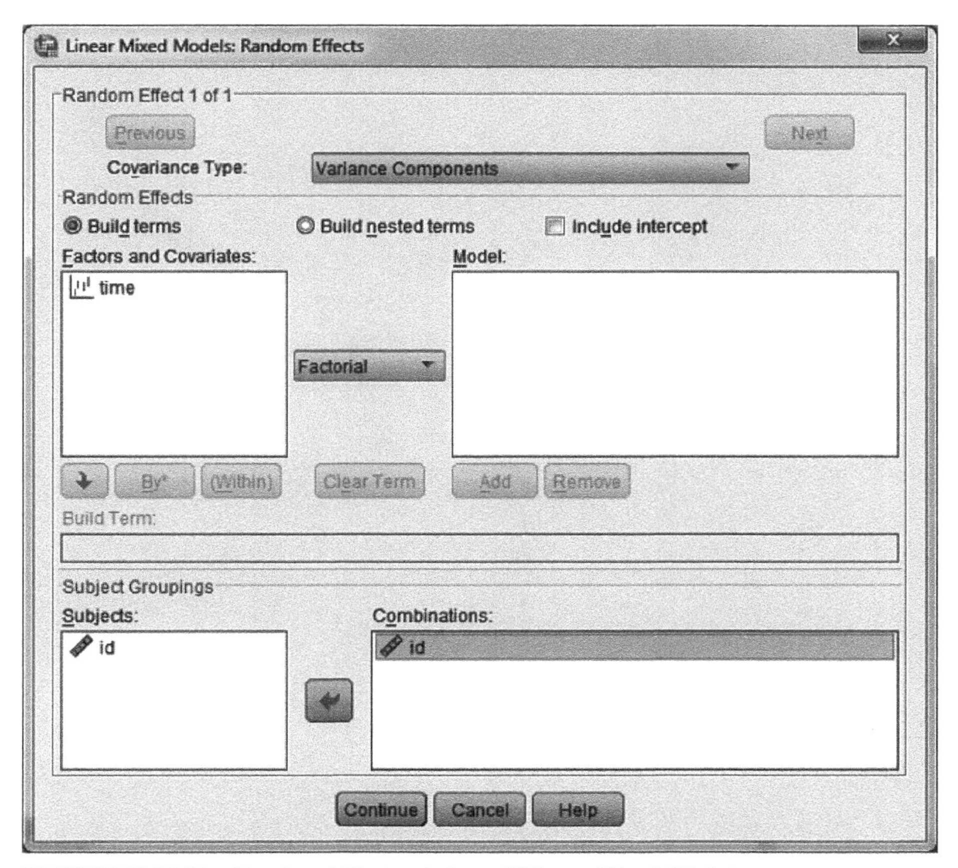

FIGURE 52.16 The **Random Effects** window of **Linear Mixed Models**.

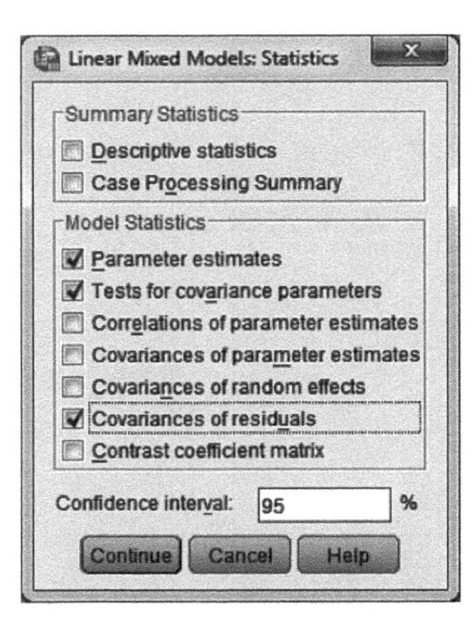

FIGURE 52.17
The **Statistics** window of **Linear Mixed Models**.

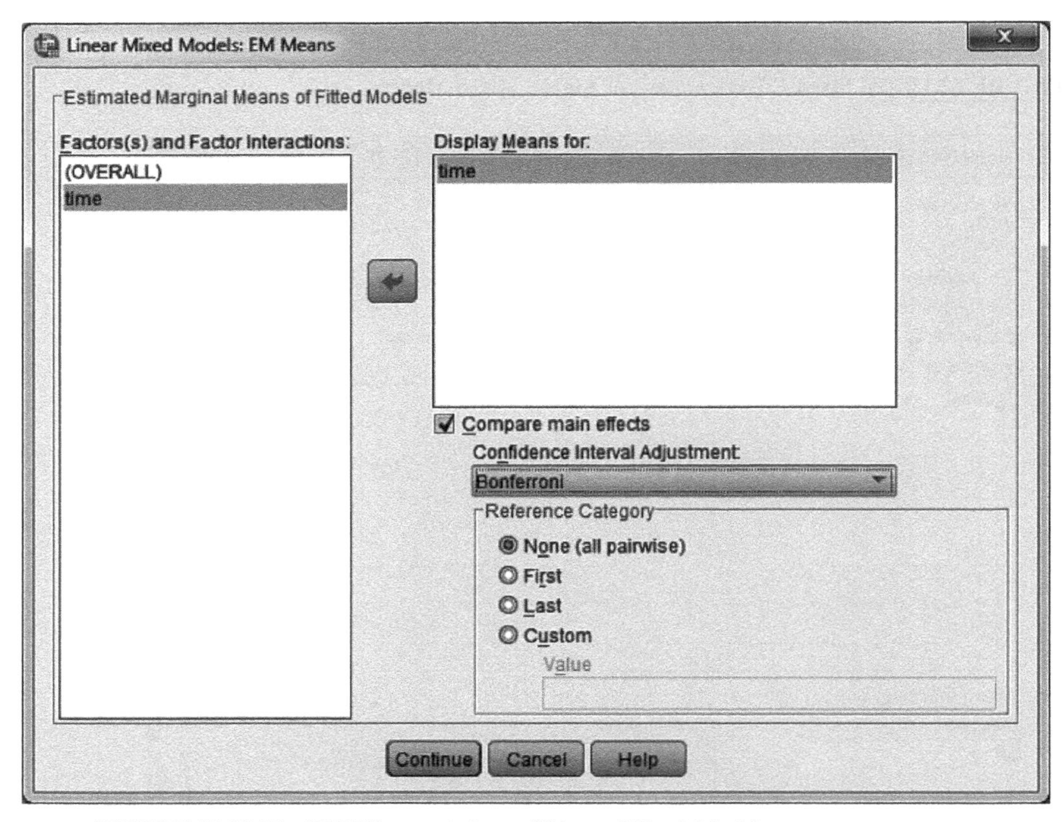

FIGURE 52.18 The **EM Means** window of **Linear Mixed Models**.

Category. Click **Continue** to return to the main **Linear Mixed Models** dialog window and click **OK** to perform the analysis.

52.6 ANALYSIS OUTPUT: AUTOREGRESSIVE COVARIANCE STRUCTURE

The **Model Dimension** table shown in Figure 52.19 provides a review of the number of parameters to be determined by the analysis. The **First-Order Autoregressive** is displayed as the **Covariance Structure** and is associated with **2** parameters to be estimated. These two parameters are as follows: (a) a single variance estimate for all three levels as we are assuming homogeneity of variances and (b) one covariance estimate that is exponentially smaller over the time measurements.

The **Information Criteria** table shown in Figure 52.20 presents how well the model fits the data. We will compare the **Autoregressive** model to the **Compound Symmetry and Unstructured** models. To assess which model best fits the data, Norusis (2008, 2012) recommends the use of the **Akaike information criterion (AIC)** and the **Bayesian information criterion (BIC)** with lower scores as being indicative of a more accurate model. Both indexes penalize the estimation of more parameters and thus increase, all else equal, with additional parameters in the model (Hox, 2010).

The **Type III Tests of Fixed Effects** table is presented in Figure 52.21. Our focus is on the **time** row where the omnibus or overall effect of the within-subjects factor is evaluated. The F ratio of **40.110** assessed with **2** and **18.673** degrees of freedom indicates that there are significant differences ($p < .001$) in the dependent variable means across the three **time** measurements.

Model Dimension[a]

		Number of Levels	Covariance Structure	Number of Parameters	Subject Variables	Number of Subjects
Fixed Effects	Intercept	1		1		
	time	3		2		
Repeated Effects	time	3	First-Order Autoregressive	2	id	11
Total		7		5		

a. Dependent Variable: attitude.

FIGURE 52.19 The **Model Dimension** table.

Information Criteria[a]

-2 Restricted Log Likelihood	139.957
Akaike's Information Criterion (AIC)	143.957
Hurvich and Tsai's Criterion (AICC)	144.401
Bozdogan's Criterion (CAIC)	148.759
Schwarz's Bayesian Criterion (BIC)	146.759

The information criteria are displayed in smaller is better forms.

a. Dependent Variable: attitude.

FIGURE 52.20 The **Information Criteria** table.

Type III Tests of Fixed Effects[a]

Source	Numerator df	Denominator df	F	Sig.
Intercept	1	9.653	332.285	.000
time	2	18.673	40.110	.000

a. Dependent Variable: attitude.

FIGURE 52.21 Omnibus assessment of the effects of the within-subjects factor.

Estimates of Fixed Effects[b]

Parameter	Estimate	Std. Error	df	t	Sig.	95% Confidence Interval	
						Lower Bound	Upper Bound
Intercept	11.090909	.781749	18.058	14.187	.000	9.448895	12.732924
[time=1]	4.181818	.871037	25.744	4.801	.000	2.390508	5.973128
[time=2]	-1.909091	.685231	18.525	-2.786	.012	-3.345791	-.472391
[time=3]	0[a]	0	.	.	.	.	.

a. This parameter is set to zero because it is redundant.
b. Dependent Variable: attitude.

FIGURE 52.22 The **Estimates of Fixed Effects** table.

The **Estimates of Fixed Effects** table is shown in Figure 52.22. These estimates focus on the differences among the three test time points. Referring back to the results of the repeated measures ANOVA discussed in Chapter 51, we may recall that the means (shown in Figure 51.1) of the **pretest**, **posttest1**, and **posttest2** conditions were **15.273**, **9.182**, and **11.091**, respectively.

The **Intercept** in the **Estimates of Fixed Effects** table is the mean score at **posttest2** (taken to six decimal places), with the remaining rows presenting deviations between **posttest2** and each level of the within-subjects factor. The row labeled as **[time = 1]** displays the deviation score between the **posttest2** and the **pretest**; by adding 4.18 to 11.09, we obtain the **pretest** mean score of **15.27**. Similarly, for the **posttest1** labeled as **[time = 2]**, the deviation score is 11.09 – 1.90, or 9.19. The row labeled as **[time = 3]** displays the deviation score between **posttest2** and the **posttest2**, which is **0**.

These mean differences are tested for statistical significance, and the results indicate that each difference suggests a significant difference. However, these comparisons neither represent the full set of pairwise comparisons nor have they been subjected to the Bonferroni correction.

In the **Estimates of Covariance Parameters** table in Figure 52.23, the **AR1 diagonal** row deals with the estimated variances. This refers to the diagonal of the **R** matrix. There is only one such estimate because this procedure assumes homogeneity of variance (i.e., all three variances are equal). The estimated variance is **6.722444**.

Estimates of Covariance Parameters[a]

Parameter		Estimate	Std. Error	Wald Z	Sig.	95% Confidence Interval	
						Lower Bound	Upper Bound
Repeated Measures	AR1 diagonal	6.722444	2.237206	3.005	.003	3.501469	12.906368
	AR1 rho	.615842	.158729	3.880	.000	.213750	.839495

a. Dependent Variable: attitude.

FIGURE 52.23 The **Estimates of Covariance Parameters** table.

Residual Covariance (R) Matrix[a]

	[time = 1]	[time = 2]	[time = 3]
[time = 1]	6.722444	4.139963	2.549562
[time = 2]	4.139963	6.722444	4.139963
[time = 3]	2.549562	4.139963	6.722444

First-Order Autoregressive

a. Dependent Variable: attitude.

FIGURE 52.24
The **Residual Covariance (R) Matrix**.

The second parameter that was estimated is the correlation between successive measurements, and this is shown in the row labeled as **AR1 rho**. Although there are three pairs of correlations (**pretest** and **posttest1**, **posttest1** and **posttest2**, and **pretest** and **posttest2**), the way in which the procedure is associated with only a single parameter is to estimate a single value for the correlation between adjacent pairs (pairs that are "next to" each other in the design; here this is the correlation between **pretest** and **posttest1** and between **posttest1** and **posttest2**). This correlation was estimated to be **.615842** and was determined by the Wald statistic to be statistically significant ($p < .001$).

The correlation between the pair of time measurements that are two steps removed from each other (**pretest** and **posttest2**) is estimated by squaring the parameter (because the covariance is assumed to become exponentially smaller as intervals increase). Thus, the correlation between **pretest** and **posttest2** is estimated as approximately $.615^2$ or approximately .378. These results indicate that the time points are statistically and practically significantly correlated with score change.

The **Residual Covariance (R) Matrix** is presented in Figure 52.24. These are the parameters that have been estimated under the **First-Order Autoregressive** procedure. As we have just described, the estimated variance of **6.722444** is placed on each diagonal element. The covariances are simply unstandardized variations of the correlations described earlier.

The **Estimated Marginal Means** and the **Pairwise Comparisons** are presented in Figure 52.25. The pairwise comparisons results lead to the same conclusion that we drew in the repeated measures ANOVA in Chapter 51, but there are a couple of subtle differences in the probability levels. Specifically, the comparison between **pretest** and **posttest1** is still significant at **.000**, the comparison between **pretest** and **posttest2** yielded a probability of **.005** in the ANOVA but here the probability is **.000**, and the comparison between **posttest1** and **posttest2** yielded a probability of **.024** in the ANOVA but here the probability is **.036**.

52.7 ANALYSIS: COMPOUND SYMMETRY

We repeat the analysis exactly as described earlier with the following exception: in the initial **Specify Subjects and Repeated** window, we select **Compound Symmetry** from the drop-down menu for **Repeated Covariance Type**. This is shown in Figure 52.26.

The **Model Dimension** table in Figure 52.27 indicates that only two parameters are being estimated: the variance (assumed to be equivalent) and the correlations (also assumed to be equivalent) between all pairs of time measures. The bottom table in Figure 52.27 shows the **Information Criteria** associated with this analysis. Comparison to Figure 52.20 reveals that the AIC and BIC indexes are higher for the **Compound Symmetry** structure than for the **Autoregressive** structure, suggesting that the **Autoregressive** structure is a better fit for the data.

Estimated Marginal Means time

Estimates[a]

time	Mean	Std. Error	df	95% Confidence Interval	
				Lower Bound	Upper Bound
1	15.273	.782	18.058	13.631	16.915
2	9.182	.782	18.058	7.540	10.824
3	11.091	.782	18.058	9.449	12.733

a. Dependent Variable: attitude.

Pairwise Comparisons[b]

(I) time	(J) time	Mean Difference (I-J)	Std. Error	df	Sig.[a]	95% Confidence Interval for Difference[a]	
						Lower Bound	Upper Bound
1	2	6.091*	.685	18.525	.000	4.288	7.894
	3	4.182*	.871	25.744	.000	1.951	6.412
2	1	-6.091*	.685	18.525	.000	-7.894	-4.288
	3	-1.909*	.685	18.525	.036	-3.712	-.106
3	1	-4.182*	.871	25.744	.000	-6.412	-1.951
	2	1.909*	.685	18.525	.036	.106	3.712

Based on estimated marginal means

*. The mean difference is significant at the .05 level.
a. Adjustment for multiple comparisons: Bonferroni.
b. Dependent Variable: attitude.

FIGURE 52.25 The **Estimated Marginal Means** and the **Pairwise Comparisons**.

The **Estimates of Fixed Effects** table is presented in Figure 52.28. The *F* ratio of **30.876** duplicates the result obtained from the repeated measures ANOVA described in Chapter 51 (see Figure 51.8) as both are based on **Compound Symmetry**.

Compound Symmetry estimates just two parameters as shown in the **Residual Covariance (R) Matrix** in the bottom table of Figure 52.28. The estimated variance of **6.424242** in this analysis is placed on each diagonal element, and the estimated covariance for each pair of time measures, irrespective of the number of steps removed from each other, is **2.966667**.

Figure 52.29 presents the **Pairwise Comparisons** based on **Compound Symmetry**. In this analysis, unlike the results for either the repeated measures ANOVA or the **Autoregressive** structure, the difference between **posttest1** and **posttest2** was not statistically significant ($p = .077$).

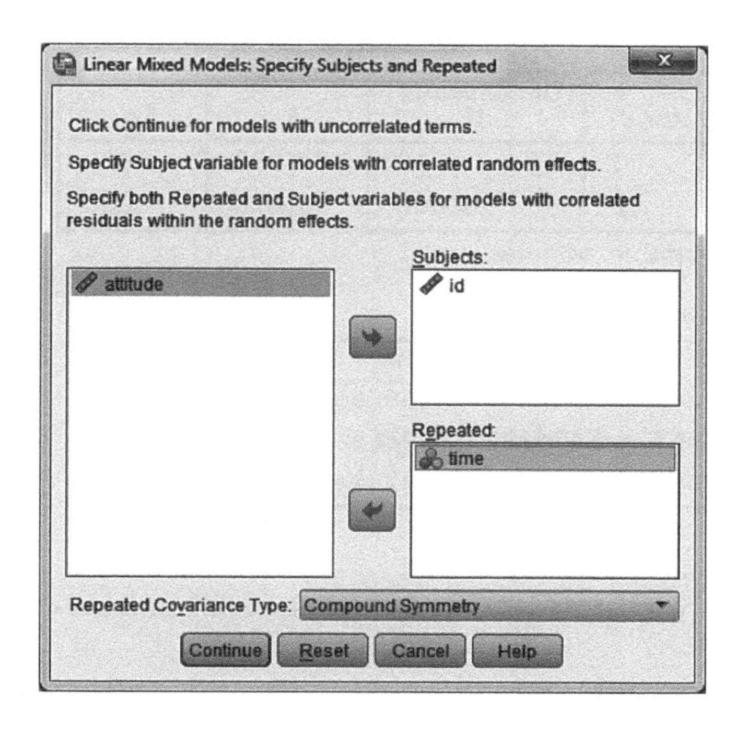

FIGURE 52.26 The initial **Specify Subjects and Repeated** window of **Linear Mixed Models**, with **Compound Symmetry** selected as the **Repeated Covariance Type**.

Model Dimension[a]

		Number of Levels	Covariance Structure	Number of Parameters	Subject Variables	Number of Subjects
Fixed Effects	Intercept	1		1		
	time	3		2		
Repeated Effects	time	3	Compound Symmetry	2	id	11
Total		7		5		

a. Dependent Variable: attitude.

Information Criteria[a]

-2 Restricted Log Likelihood	142.284
Akaike's Information Criterion (AIC)	146.284
Hurvich and Tsai's Criterion (AICC)	146.728
Bozdogan's Criterion (CAIC)	151.086
Schwarz's Bayesian Criterion (BIC)	149.086

The information criteria are displayed in smaller-is-better forms.

a. Dependent Variable: attitude.

FIGURE 52.27 The **Model Dimension** and **Information Criteria** tables based on **Compound Symmetry**.

Type III Tests of Fixed Effects[a]

Source	Numerator df	Denominator df	F	Sig.
Intercept	1	10	374.892	.000
time	2	20.000	30.876	.000

a. Dependent Variable: attitude.

Residual Covariance (R) Matrix[a]

	[time = 1]	[time = 2]	[time = 3]
[time = 1]	6.424242	2.966667	2.966667
[time = 2]	2.966667	6.424242	2.966667
[time = 3]	2.966667	2.966667	6.424242

Compound Symmetry

a. Dependent Variable: attitude.

FIGURE 52.28 The **Estimates of Fixed Effects** and **Residual Covariance (R) Matrix** tables based on **Compound Symmetry**.

Pairwise Comparisons[b]

						95% Confidence Interval for Difference[a]	
(I) time	(J) time	Mean Difference (I-J)	Std. Error	df	Sig.[a]	Lower Bound	Upper Bound
1	2	6.091*	.793	20.000	.000	4.019	8.162
	3	4.182*	.793	20.000	.000	2.110	6.253
2	1	-6.091*	.793	20.000	.000	-8.162	-4.019
	3	-1.909	.793	20.000	.077	-3.981	.162
3	1	-4.182*	.793	20.000	.000	-6.253	-2.110
	2	1.909	.793	20.000	.077	-.162	3.981

Based on estimated marginal means

*. The mean difference is significant at the .05 level.
a. Adjustment for multiple comparisons: Bonferroni.
b. Dependent Variable: attitude.

FIGURE 52.29 The **Pairwise Comparisons** based on **Compound Symmetry**.

52.8 ANALYSIS: UNSTRUCTURED COVARIANCE

We repeat the analysis one more time but here we select **Unstructured** from the drop-down menu for **Repeated Covariance Type** in the initial **Specify Subjects and Repeated** window This is shown in Figure 52.30.

The **Model Dimension** table shown in Figure 52.31 indicates that six parameters are being estimated. These are the three variances and the three correlations between all pairs of **time** measures. The bottom table in Figure 52.31 shows the **Information Criteria** associated with this analysis. Comparison to Figure 52.20 and Figure 52.27 reveals that

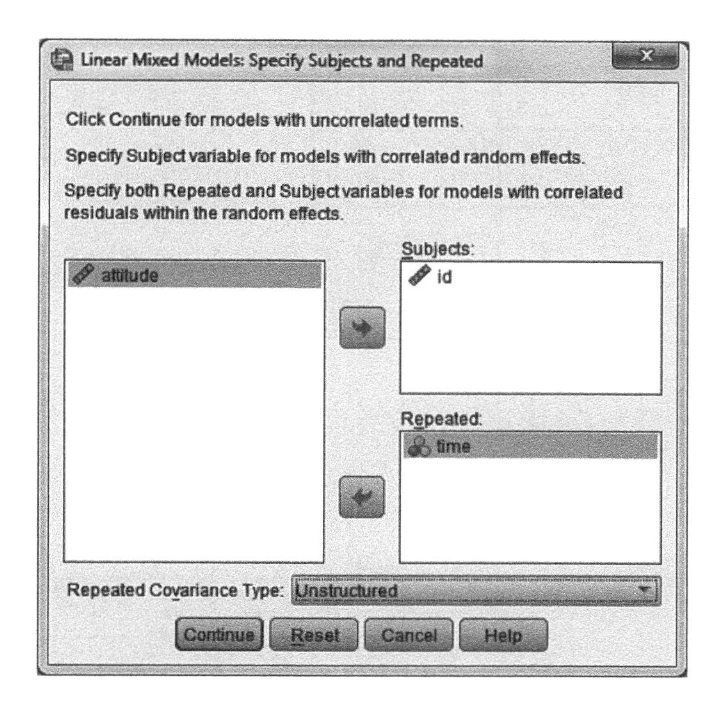

FIGURE 52.30 The initial **Specify Subjects and Repeated** window of the **Linear Mixed Models**, with **Unstructured** selected as the **Repeated Covariance Type**.

Model Dimension[a]

		Number of Levels	Covariance Structure	Number of Parameters	Subject Variables	Number of Subjects
Fixed Effects	Intercept	1		1		
	time	3		2		
Repeated Effects	time	3	Unstructured	6	id	11
Total		7		9		

a. Dependent Variable: attitude.

Information Criteria[a]

-2 Restricted Log Likelihood	136.999
Akaike's Information Criterion (AIC)	148.999
Hurvich and Tsai's Criterion (AICC)	152.652
Bozdogan's Criterion (CAIC)	163.407
Schwarz's Bayesian Criterion (BIC)	157.407

The information criteria are displayed in smaller-is-better forms.

a. Dependent Variable: attitude.

FIGURE 52.31 The **Model Dimension** and **Information Criteria** tables based on **Unstructured Covariance**.

Type III Tests of Fixed Effects[a]

Source	Numerator df	Denominator df	F	Sig.
Intercept	1	10.000	374.892	.000
time	2	10	40.285	.000

a. Dependent Variable: attitude.

Residual Covariance (R) Matrix[a]

	[time = 1]	[time = 2]	[time = 3]
[time = 1]	9.618182	3.845455	2.072727
[time = 2]	3.845455	4.363636	2.981818
[time = 3]	2.072727	2.981818	5.290909

Unstructured

a. Dependent Variable: attitude.

Pairwise Comparisons[b]

(I) time	(J) time	Mean Difference (I-J)	Std. Error	df	Sig.[a]	95% Confidence Interval for Difference[a]	
						Lower Bound	Upper Bound
1	2	6.091*	.756	10	.000	3.920	8.261
	3	4.182*	.989	10	.005	1.343	7.021
2	1	-6.091*	.756	10	.000	-8.261	-3.920
	3	-1.909*	.579	10	.024	-3.572	-.247
3	1	-4.182*	.989	10	.005	-7.021	-1.343
	2	1.909*	.579	10	.024	.247	3.572

Based on estimated marginal means

*. The mean difference is significant at the .05 level.
a. Adjustment for multiple comparisons: Bonferroni.
b. Dependent Variable: attitude.

FIGURE 52.32 The **Estimates of Fixed Effects**, the **Residual Covariance (R) Matrix**, and the **Pairwise Comparisons** tables based on **Unstructured Covariance**.

the AIC and BIC indexes are higher for the **Unstructured** covariance than for the other two, reinforcing our earlier supposition that the **Autoregressive** structure is a best fit of the three approaches for the data.

The **Estimates of Fixed Effects** table, the **Residual Covariance (R) Matrix**, and the **Pairwise Comparisons** based on **Unstructured Covariance** are shown in Figure 52.32. The *F* ratio is somewhat higher from what we have seen earlier, every parameter takes on a different value in the **Residual Covariance (R) Matrix**, and the probability levels for the **Pairwise Comparisons** are the same as those obtained in the repeated measures ANOVA.

CHAPTER 53

Two-Way Mixed ANOVA

53.1 OVERVIEW

A mixed ANOVA design contains at least one between-subjects variable and at least one within-subjects variable. A two-way mixed ANOVA must, by definition, contain one independent variable of each type and is often called a *simple mixed* design because it is the minimal (or simplest) mixed ANOVA design that is possible; an overview of complex mixed designs is presented in Gamst et al. (2008). From an ANOVA standpoint, a two-way mixed ANOVA is a repeated measures design with the addition of a between-subjects factor. We thus partition the variance in the same overall way as we did in the one-way within-subjects design: into between-subjects and within-subjects variance. However, both major partitions can be further partitioned in the following way in a two-way mixed design:

- The between-subjects variance is further partitioned into variance attributed to the between-subjects independent variable and error variance. The effect associated with the independent variable is evaluated for statistical significance against this between-subjects error (the *F* ratio is computed by dividing the mean square associated with the independent variable by the mean square associated with the error term).
- The within-subjects variance is further partitioned into the overall effect of the within-subjects independent variable, the interaction of the between-subjects and within-subjects variables, and the within-subjects error. As was true for the one-way within-subjects design, the error represents the differences in patterns of performance of the cases across the levels of the within-subjects independent variable. The *F* ratios for both the within-subjects independent variable effect and the interaction are computed by dividing the Mean Square associated with each of these by the Mean Square associated with the within-subjects error.

The two-way mixed design allows us to track the performance of two or more independent groups over time on a dependent measure or to compare the performance of two or more independent groups on different measures. By plotting the performance

Performing Data Analysis Using IBM SPSS®, First Edition.
Lawrence S. Meyers, Glenn C. Gamst, and A. J. Guarino.

of the groups as separate lines across the set of measures represented on the X-axis, we can immediately discern any differences in the pattern of performance among the groups. Examining these pattern differences has often been referred to as a form of *profile analysis* (Fitzmaurice, Laird, & Ware, 2011; Kim & Neil, 2007).

53.2 NUMERICAL EXAMPLE

The fictional data we use for our example track the effectiveness of an advertising campaign for a new widget. There are two levels of the between-subjects variable we have named **ad_campaign**. The 20 sample markets serving as cases for this study were randomly assigned to one of the two conditions: 10 were assigned to a **media blitz** advertisement campaign (coded as **1**) and 10 were assigned to a **targeted marketing** advertisement campaign (coded as **2**). The dependent variable is thousands of sales dollars and was measured before the start of the campaigns (**pretest**) and 1 (**post1**), 2 (**post2**), and 3 (**post3**) weeks after the start of the advertisement campaign. The data file is named **advertising campaign**.

A screenshot of a portion of the data file is shown in Figure 53.1. Note that the values of **ad_campaign** in the data file show the codes **1** and **2** for the two types of marketing strategies because it is a between-subjects variable. The levels of the within-subjects variable measure the sales dollars at four points in time, and each level is a variable in the data file. These levels will be linked together in the **Define Factors** dialog window to form the within-subjects variable with four levels.

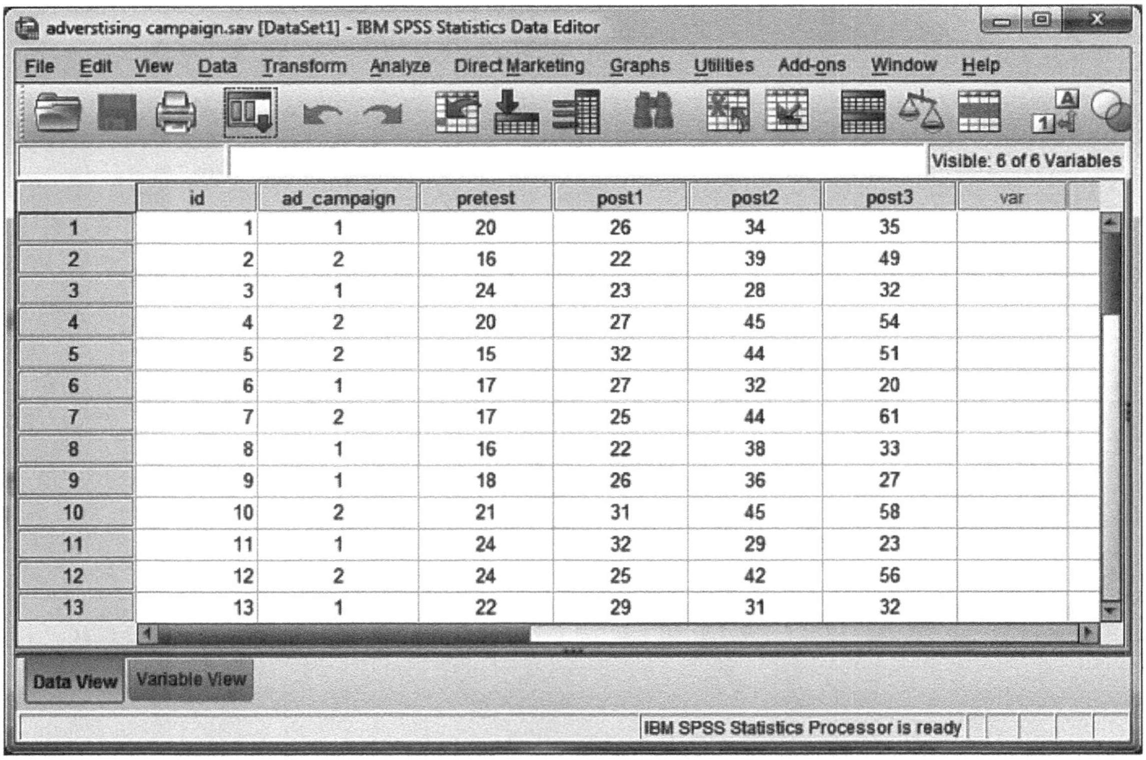

adverstising campaign.sav [DataSet1] - IBM SPSS Statistics Data Editor

File Edit View Data Transform Analyze Direct Marketing Graphs Utilities Add-ons Window Help

Visible: 6 of 6 Variables

	id	ad_campaign	pretest	post1	post2	post3	var
1	1	1	20	26	34	35	
2	2	2	16	22	39	49	
3	3	1	24	23	28	32	
4	4	2	20	27	45	54	
5	5	2	15	32	44	51	
6	6	1	17	27	32	20	
7	7	2	17	25	44	61	
8	8	1	16	22	38	33	
9	9	1	18	26	36	27	
10	10	2	21	31	45	58	
11	11	1	24	32	29	23	
12	12	2	24	25	42	56	
13	13	1	22	29	31	32	

Data View Variable View

IBM SPSS Statistics Processor is ready

FIGURE 53.1 A portion of the data file.

53.3 ANALYSIS SETUP

We open the data file **advertising campaign** and from the main menu, we select **Analyze → General Linear Model → Repeated Measures**. This opens the **Define Factor(s)** window of **Repeated Measures** as shown in Figure 53.2. The **Within-Subject Factor Name** panel opens with the default of **factor1**; we change that to **sales** as the name we create for our within-subjects independent variable. We also type **4** in the **Number of Levels** panel. Click **Add** to register this information with IBM SPSS® and click **Define** to open the main dialog window.

The main **General Linear Model Repeated Measures** dialog window opens with four question-marked lines in the **Within-Subjects Variables** panel. Our name for the treatment effect, **sales**, appears just above the panel. As described in Chapter 51, we move the variables **pretest**, **post1**, **post2**, and **post3** in order into the **Within-Subjects Variables** panel as shown in Figure 53.3. We also move **ad_campaign** into the **Between-Subjects Factor(s)** panel.

Anticipating the possibility of a statistically significant interaction effect, we will generate a plot. Selecting the **Plots** pushbutton opens the **Profile Plots** dialog window (see Figure 53.4). Because the within-subjects variable of **sales** represents a time-related factor, it almost always facilitates interpretation to place such a variable on the horizontal axis; this also simplifies the figure in that there will be only two functions (one for each level of **ad_campaign**). We therefore place **sales** on the **Horizontal axis** and will have **Separate Lines** for each level of **ad_campaign**. Click the **Add** pushbutton to place the plot in the **Plots** panel and click **Continue** to return to the main dialog window.

In the **Options** dialog window shown already configured in Figure 53.5, we select **Descriptive statistics** and **Homogeneity tests** (because we have a between-subjects variable in our design) in the **Display** area; the **Repeated Measures** procedure will automatically perform a *sphericity* test for the levels of the **sales** variable.

As was the case with the one-way within-subjects design (see Chapter 51), we can request our multiple comparison tests for the main effect of the within-subjects variable in the **Estimated Marginal Means** area should we need it. We therefore move **sales** from the **Factor(s) and Factor Interactions** panel to the **Display Means for** panel,

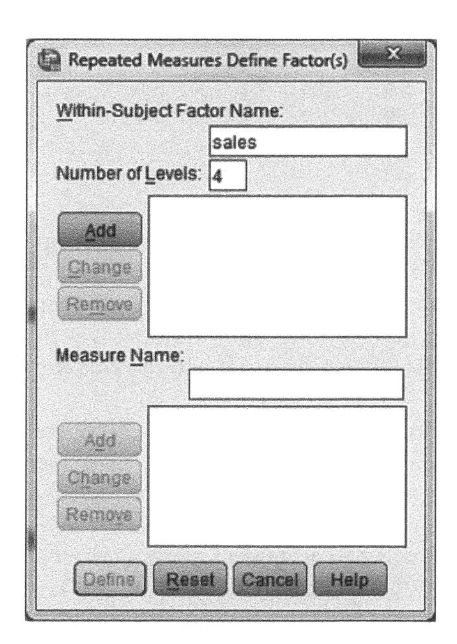

FIGURE 53.2

The **Define Factors** window of **General Linear Model Repeated Measures**.

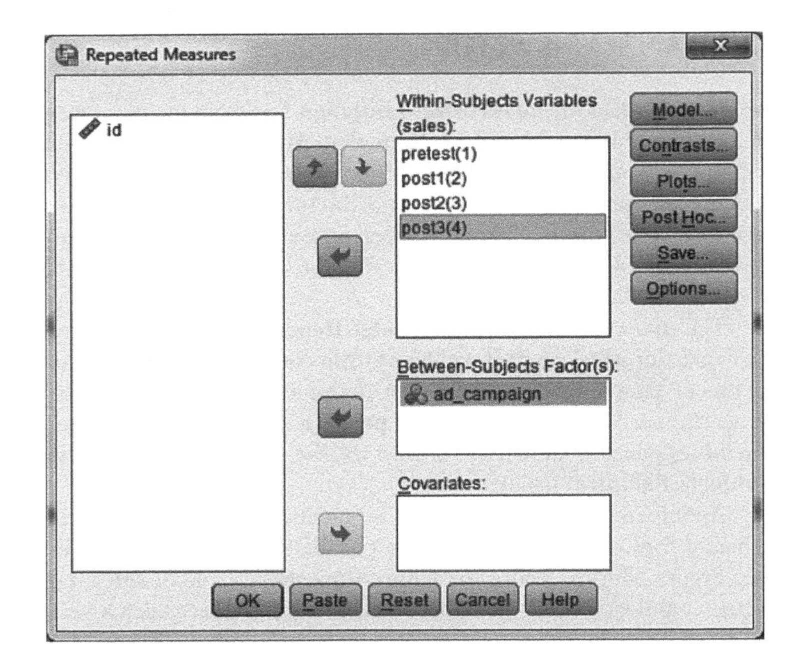

FIGURE 53.3
The main dialog window of **General Linear Model Repeated Measures**.

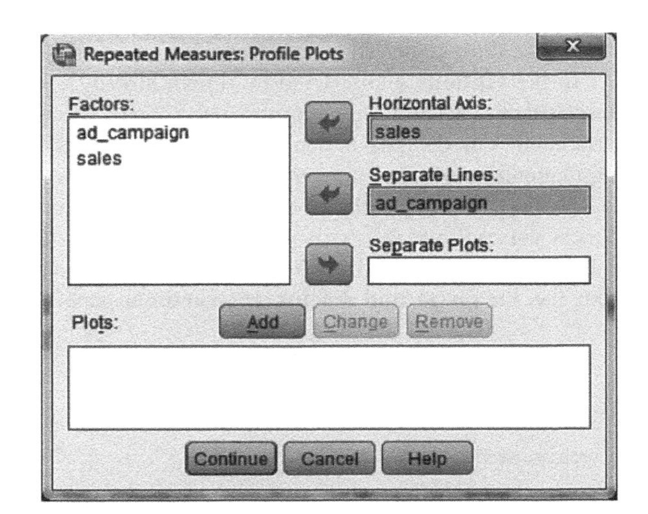

FIGURE 53.4
The **Profile Plots** dialog window of **Univariate** to set up the plot of the interaction.

check **Compare main effects**, and select **Bonferroni** from the **Confidence interval adjustment** drop-down menu.

Remaining in the **Options** window, we can also set up for the simple effects tests of the interaction should that prove needed by moving the interaction term shown as **ad_campaign*sales** from the **Factor(s) and Factor Interactions** panel into the **Display Means for** panel.

Click **Continue** to return to the main dialog window, and click **Paste** to copy the underlying syntax to a syntax window (see Figure 53.6). Note that the **/EMMEANS = TABLES(sales)** is already configured for the Bonferroni-corrected *t* tests of the main effect of **sales** (it does not mention the variable **sales** in parentheses after **Compare** as this is the only variable in the **TABLES(sales)** portion of the syntax and IBM SPSS recognizes this fact). This syntax resulted from us having checked **Compare main effects** and selected **Bonferroni** from the **Confidence interval**

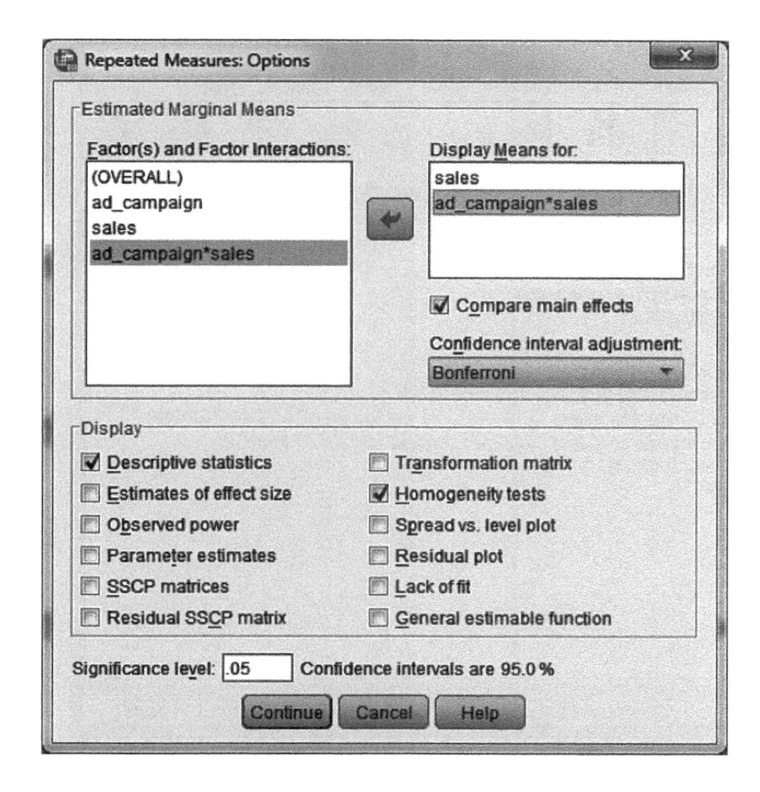

FIGURE 53.5
The **Options** window of **Repeated Measures**.

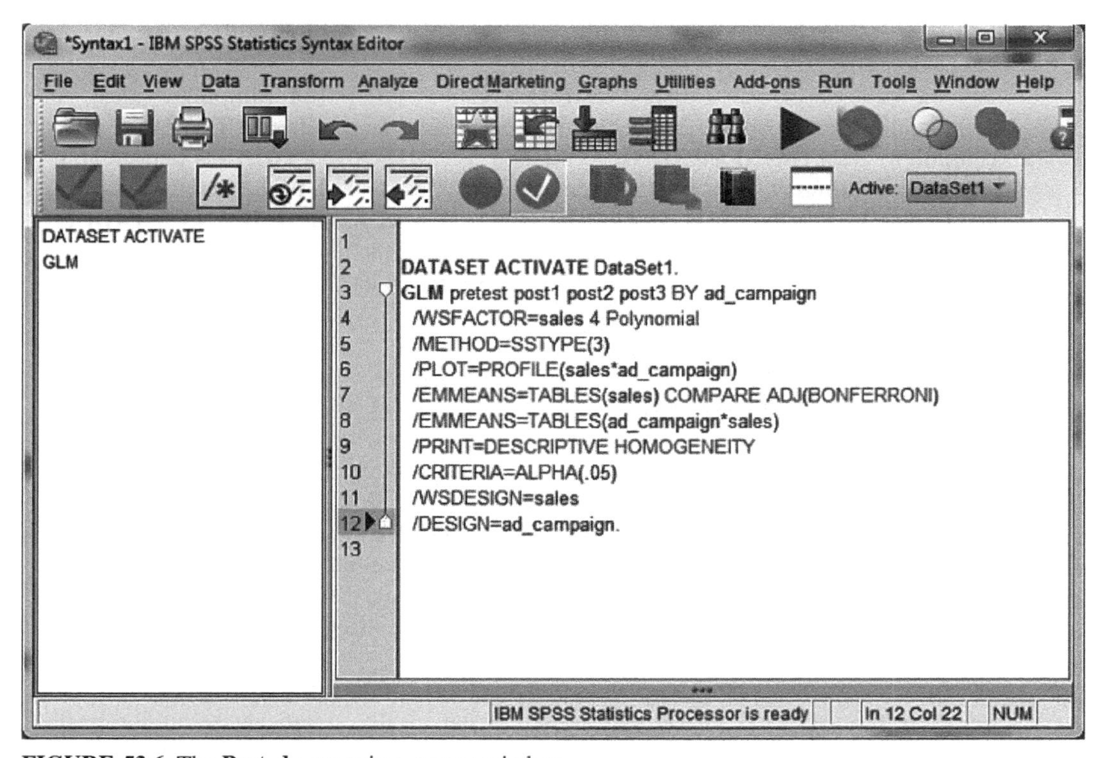

FIGURE 53.6 The **Pasted** syntax in a syntax window.

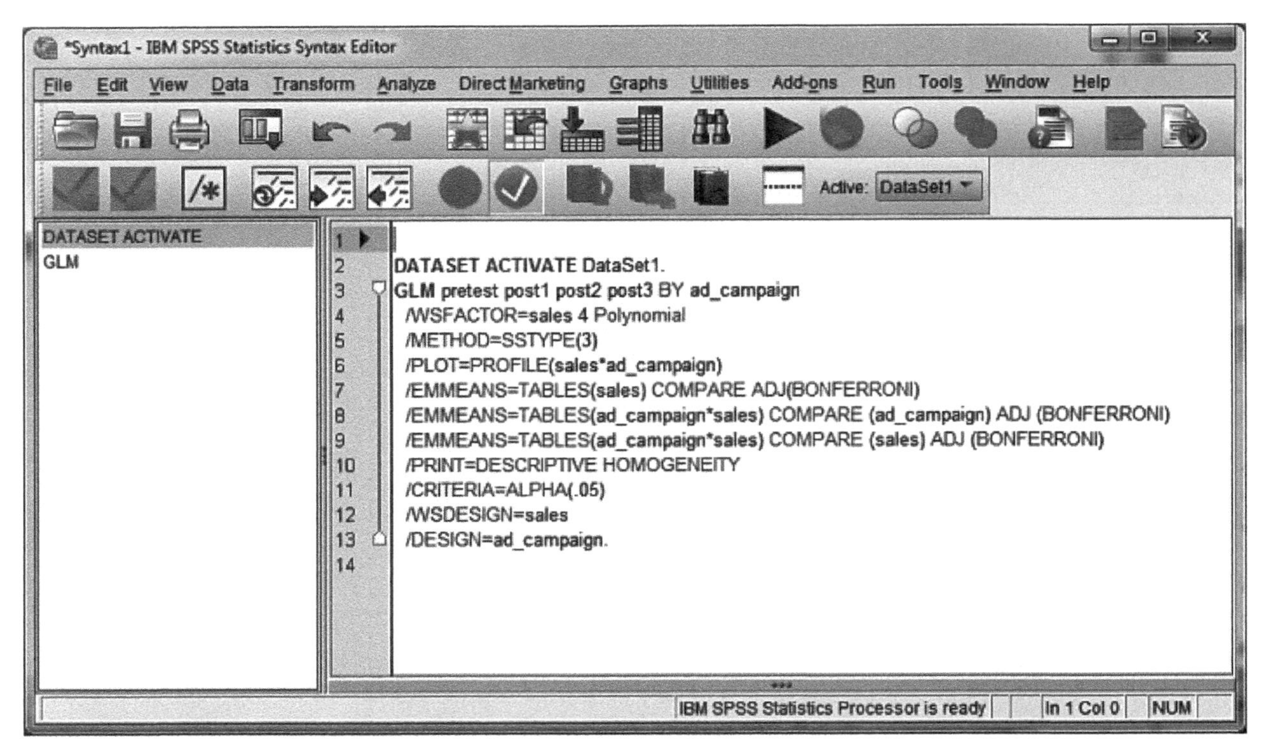

FIGURE 53.7 The syntax with the full set of simple effects tests configured.

adjustment drop-down menu. We now need to configure the simple effects tests for the interaction effect.

We focus on the subcommand **/EMMEANS = TABLES(ad_campaign*sales)**. As described in Chapter 50, we duplicate this subcommand line, and add **COMPARE (ad_campaign) ADJ (BONFERRONI)** to the first **/EMMEANS = TABLES** line (to compare the two different types of advertisement campaigns for each of the four time periods) and **COMPARE (sales) ADJ (BONFERRONI)** to the second **/EMMEANS = TABLES** line (to compare the four different time periods once for the blitz campaign and again for the targeted campaign). The completed syntax is shown in Figure 53.7. From the main menu select **Run → All** to obtain the output.

53.4 ANALYSIS OUTPUT

The **Descriptive Statistics** with the means, standard deviations, and sample sizes are shown in Figure 53.8. Figure 53.9 shows the results of **Box's** test, **Mauchly's** sphericity test, and **Levene's** test. **Box's Test of Equality of Covariance Matrices** tests the assumption that the dependent variable covariance matrices are equal across the levels of the independent variable (**ad_campaign**); in this data set, the assumption has been met (Box's M = **11.245**, p = **.582**). The data also appear to meet the sphericity assumption (Mauchly's W = **11.245**, p = **.464**).

Levene's test of homogeneity of variance returned statistically nonsignificant results for **pretest**, **post1**, and **post2**, but showed a statistically significant result for **post3** (p = **.041**). However, even with just four evaluations, there is a risk of alpha inflation here; a more stringent Bonferroni-corrected alpha level (.05/number of statistical tests) would give us a corrected alpha level of .05/4, or .0125, and this would render the

Descriptive Statistics

	ad_campaign	Mean	Std. Deviation	N
pretest before advertising	1 media blitz	19.60	3.169	10
	2 targeted marketing	19.90	3.665	10
	Total	19.75	3.338	20
post1 one week after	1 media blitz	26.40	3.406	10
	2 targeted marketing	27.50	4.035	10
	Total	26.95	3.677	20
post2 two weeks after	1 media blitz	31.60	4.648	10
	2 targeted marketing	41.60	4.274	10
	Total	36.60	6.723	20
post3 three weeks after	1 media blitz	27.60	5.700	10
	2 targeted marketing	54.80	3.676	10
	Total	41.20	14.713	20

FIGURE 53.8 The **Descriptive Statistics** output.

Box's Test of Equality of Covariance Matrices[a]

Box's M	11.245
F	.849
df1	10
df2	1549.004
Sig.	.582

Tests the null hypothesis that the observed covariance matrices of the dependent variables are equal across groups.

a. Design: Intercept + ad_campaign
Within Subjects Design: sales

Levene's Test of Equality of Error Variances[a]

	F	df1	df2	Sig.
pretest before advertising	.525	1	18	.478
post1 one week after	.735	1	18	.403
post2 two weeks after	.099	1	18	.757
post3 three weeks after	4.865	1	18	.041

Tests the null hypothesis that the error variance of the dependent variable is equal across groups.

a. Design: Intercept + ad_campaign
Within Subjects Design: sales

Mauchly's Test of Sphericity[b]

Measure:MEASURE_1

					Epsilon[a]		
Within Subjects Effect	Mauchly's W	Approx. Chi-Square	df	Sig.	Greenhouse-Geisser	Huynh-Feldt	Lower-bound
sales	.758	4.627	5	.464	.875	1.000	.333

Tests the null hypothesis that the error covariance matrix of the orthonormalized transformed dependent variables is proportional to an identity matrix.

a. May be used to adjust the degrees of freedom for the averaged tests of significance. Corrected tests are displayed in the Tests of Within-Subjects Effects table.
b. Design: Intercept + ad_campaign
Within Subjects Design: sales

FIGURE 53.9 Tests of homogeneity and sphericity.

Tests of Within-Subjects Effects

Measure:MEASURE_1

Source		Type III Sum of Squares	df	Mean Square	F	Sig.
sales	Sphericity Assumed	5566.050	3	1855.350	119.386	.000
	Greenhouse-Geisser	5566.050	2.624	2121.046	119.386	.000
	Huynh-Feldt	5566.050	3.000	1855.350	119.386	.000
	Lower-bound	5566.050	1.000	5566.050	119.386	.000
sales * ad_campaign	Sphericity Assumed	2343.250	3	781.083	50.260	.000
	Greenhouse-Geisser	2343.250	2.624	892.939	50.260	.000
	Huynh-Feldt	2343.250	3.000	781.083	50.260	.000
	Lower-bound	2343.250	1.000	2343.250	50.260	.000
Error(sales)	Sphericity Assumed	839.200	54	15.541		
	Greenhouse-Geisser	839.200	47.236	17.766		
	Huynh-Feldt	839.200	54.000	15.541		
	Lower-bound	839.200	18.000	46.622		

Tests of Between-Subjects Effects

Measure:MEASURE_1
Transformed Variable:Average

Source	Type III Sum of Squares	df	Mean Square	F	Sig.
Intercept	77501.250	1	77501.250	3524.564	.000
ad_campaign	1862.450	1	1862.450	84.700	.000
Error	395.800	18	21.989		

FIGURE 53.10 Within-subjects and between-subjects summary tables.

Levene test for **post3** nonsignificant. Thus, we should probably be cautious concerning any within-subjects effect whose significance levels hover around the .05 mark.

Figure 53.10 displays the ANOVA summary tables; the **Repeated Measures** procedure of IBM SPSS reports only the reduced model, thus simplifying the output. All three effects are statistically significant ($p < .001$). Because interaction effects supercede main effects, we focus on that effect. The total sum of squares for the within-subjects partition of the variance is **5566.050** + **2343.250** + **839.200**, or 8748.500. Thus, the eta square value for the interaction is **2343.250**/8748.500, or .268, indicating that the interaction effect explained about 27% of the within-subjects variance.

The plot of the interaction effect is presented in Figure 53.11, and a visual inspection of the difference in the pattern or group profiles tells much of the general story. Both advertisement campaigns appeared to work early but the media blitz (solid line) was losing ground to the targeted campaign by the second week (**sales3** on the X-axis) and had pretty much run its course while the targeted campaign continued to generate increased sales.

We can also see the main effect of **sales** in the plot, as both functions rise from the pretest (**sales1**) to the final measure at **sales4**, but it is clear that the rise shows a different rate or pattern for each group and reinforces the idea that the interaction contains more detailed information in it than the main effect.

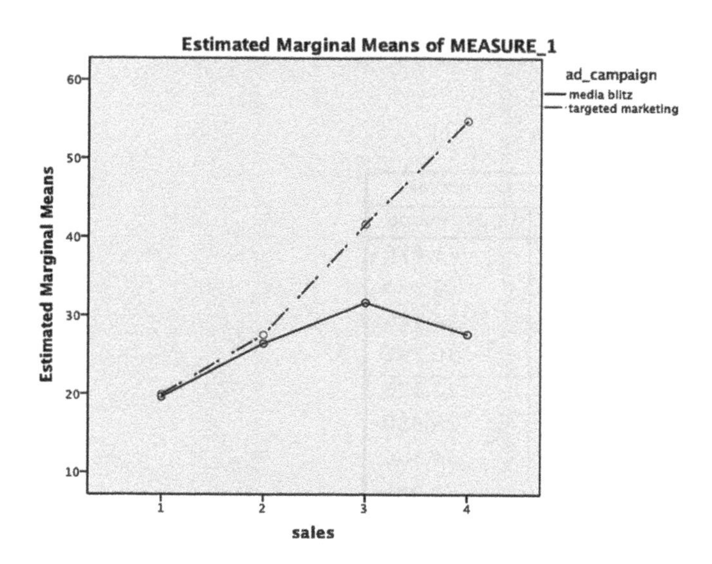

FIGURE 53.11
Plot of the interaction effect.

1. ad_campaign * sales

Estimates

Measure:MEASURE_1

ad_campaign	sales	Mean	Std. Error	95% Confidence Interval	
				Lower Bound	Upper Bound
1 media blitz	1	19.600	1.083	17.324	21.876
	2	26.400	1.181	23.920	28.880
	3	31.600	1.412	28.634	34.566
	4	27.600	1.517	24.414	30.786
2 targeted marketing	1	19.900	1.083	17.624	22.176
	2	27.500	1.181	25.020	29.980
	3	41.600	1.412	38.634	44.566
	4	54.800	1.517	51.614	57.986

Pairwise Comparisons

Measure:MEASURE_1

sales	(I) ad_campaign	(J) ad_campaign	Mean Difference (I-J)	Std. Error	Sig.[a]	95% Confidence Interval for Difference[a]	
						Lower Bound	Upper Bound
1	1 media blitz	2 targeted marketing	-.300	1.532	.847	-3.519	2.919
	2 targeted marketing	1 media blitz	.300	1.532	.847	-2.919	3.519
2	1 media blitz	2 targeted marketing	-1.100	1.670	.518	-4.608	2.408
	2 targeted marketing	1 media blitz	1.100	1.670	.518	-2.408	4.608
3	1 media blitz	2 targeted marketing	-10.000*	1.997	.000	-14.195	-5.805
	2 targeted marketing	1 media blitz	10.000*	1.997	.000	5.805	14.195
4	1 media blitz	2 targeted marketing	-27.200*	2.145	.000	-31.706	-22.694
	2 targeted marketing	1 media blitz	27.200*	2.145	.000	22.694	31.706

Based on estimated marginal means

a. Adjustment for multiple comparisons: Bonferroni.
*. The mean difference is significant at the

FIGURE 53.12 Simple effects tests comparing the two levels of **ad_campaign** for each time period.

2. ad_campaign * sales

Estimates

Measure:MEASURE_1

ad_campaign	sales	Mean	Std. Error	95% Confidence Interval	
				Lower Bound	Upper Bound
1 media blitz	1	19.600	1.083	17.324	21.876
	2	26.400	1.181	23.920	28.880
	3	31.600	1.412	28.634	34.566
	4	27.600	1.517	24.414	30.786
2 targeted marketing	1	19.900	1.083	17.624	22.176
	2	27.500	1.181	25.020	29.980
	3	41.600	1.412	38.634	44.566
	4	54.800	1.517	51.614	57.986

Pairwise Comparisons

Measure:MEASURE_1

ad_campaign	(I) sales	(J) sales	Mean Difference (I-J)	Std. Error	Sig.[a]	95% Confidence Interval for Difference[a]	
						Lower Bound	Upper Bound
1 media blitz	1	2	-6.800*	1.553	.002	-11.400	-2.200
		3	-12.000*	2.055	.000	-18.088	-5.912
		4	-8.000*	1.638	.001	-12.853	-3.147
	2	1	6.800*	1.553	.002	2.200	11.400
		3	-5.200*	1.710	.042	-10.267	-.133
		4	-1.200	1.891	1.000	-6.803	4.403
	3	1	12.000*	2.055	.000	5.912	18.088
		2	5.200*	1.710	.042	.133	10.267
		4	4.000	1.683	.172	-.985	8.985
	4	1	8.000*	1.638	.001	3.147	12.853
		2	1.200	1.891	1.000	-4.403	6.803
		3	-4.000	1.683	.172	-8.985	.985
2 targeted marketing	1	2	-7.600*	1.553	.001	-12.200	-3.000
		3	-21.700*	2.055	.000	-27.788	-15.612
		4	-34.900*	1.638	.000	-39.753	-30.047
	2	1	7.600*	1.553	.001	3.000	12.200
		3	-14.100*	1.710	.000	-19.167	-9.033
		4	-27.300*	1.891	.000	-32.903	-21.697
	3	1	21.700*	2.055	.000	15.612	27.788
		2	14.100*	1.710	.000	9.033	19.167
		4	-13.200*	1.683	.000	-18.185	-8.215
	4	1	34.900*	1.638	.000	30.047	39.753
		2	27.300*	1.891	.000	21.697	32.903
		3	13.200*	1.683	.000	8.215	18.185

Based on estimated marginal means

*. The mean difference is significant at the

a. Adjustment for multiple comparisons: Bonferroni.

FIGURE 53.13 Simple effects tests comparing the four levels of sales for each level of **ad_campaign**.

The details of what we were able to see in the plot of the means are revealed in the simple effects tests. The first set of simple effects tests (the first **/EMMEANS = TABLES** subcommand in the syntax) called for the two levels of **ad_campaign** to be compared for each time period, and these results are shown in Figure 53.12. As can be seen from the **Pairwise Comparisons** table (see Chapter 49 for how to read the **Pairwise Comparisons** table), the two sets of widget markets were comparable in sales during the **pretest** (**sales1** in the table) period ($p =$ **.847**); thus, the random assignment was effective in generating two comparable groups. At the end of the first week (**sales2** in the table), the groups were still comparable in widget sales ($p =$ **.518**). However, at the final two measurement times, the **targeted marketing** campaign resulted in significantly more widget sales than the **media blitz** campaign.

The other half of the story suggested by the plot of the interaction is conveyed in the second set of simple effects analyses shown in Figure 53.13 where the four time periods are compared for each type of advertisement campaign. For the **media blitz** campaign, sales increased from the pretest (**sales1** in the table) to **post2** (**sales3** in the table) but leveled off at a point where **post2** and **post3** (**sales3** and **sales4**, respectively, in the table) do not differ significantly. However, for the **targeted marketing** campaign, sales of widgets significantly increased at every measurement period. When all is said and done, we may conclude that the **targeted marketing** campaign was the better of the two in that its effect extended over a longer time period than did the one associated with the **media blitz** campaign.

PART 16

MULTIVARIATE GROUP DIFFERENCES: MANOVA AND DISCRIMINANT FUNCTION ANALYSIS

CHAPTER 54

One-Way Between-Subjects MANOVA

54.1 OVERVIEW

ANOVA is defined as a univariate procedure in that there is only one dependent variable in the analysis. But in many research contexts, the cases in the study are affected in more than one manner by the independent variables, and when this is a possibility, it is useful to assess the cases on multiple dependent variables. When we compare the cases in each independent variable level on their performance on two or more dependent variables, the design that we use is a multivariate ANOVA (MANOVA). The key feature of a MANOVA is that, at least in the initial portion of the analysis, the dependent variables are *treated together as a set* rather than analyzed in isolation from each other. We consider in this chapter the one-way between-subjects design, where we have a single independent variable with k levels and two or more dependent variables.

The strategy we use in a MANOVA is intimately tied to discriminant function analysis, a topic that is covered in Chapter 55. The first step in the strategy is typically to determine if there are multivariate differences between the groups by examining the omnibus MANOVA results. If there are, then researchers ordinarily engage in two additional analyses in either order depending on the research focus. One analysis is to determine the latent dimensions along which the groups differ; such a multivariate approach is performed with discriminant function analysis. Another analysis is to perform univariate ANOVAs on each dependent variable, examining the effects as justified from the MANOVA results.

The statistical objective in a MANOVA is to form a weighted linear composite (a variate) of the dependent variables analogous to what takes place in multiple regression. In MANOVA, however, the weights are generated such that the groups are maximally differentiated. The value that results from solving the discriminant function (the weighted linear composite of the dependent variables) is a discriminant score (i.e., a discriminant score is computed for each case), and the weights assigned to each dependent variable achieves the goal of maximally separating (differentiating) the groups based on their average discriminant score. The groups are then compared using an ANOVA but with the discriminant score (representing the variate or "super variable") as the dependent variable in the analysis (Meyers et al., 2013).

To most effectively use MANOVA, it is best that the dependent variables are modestly to moderately correlated. Variables very highly correlated are redundant, and it is pointless (and perhaps a bit distorting) to claim they are separate, stand-alone measures.

Performing Data Analysis Using IBM SPSS®, First Edition.
Lawrence S. Meyers, Glenn C. Gamst, and A. J. Guarino.

Variables that are barely correlated can best be directly analyzed in separate univariate ANOVAs, as joining them together into a variate makes little theoretical sense given that they are not related to each other. Generally, it is preferable to have correlations between the dependent variables in the approximate range .30–.60 (Stevens, 2009; Tabachnick & Fidell, 2007).

There are four multivariate tests of the effect of the independent variable (Pillai's Trace, Wilks' lambda, Hotelling's Trace, and Roy's Largest Root) produced by IBM SPSS®. Each is converted to a multivariate *F* ratio and tested for statistical significance.

If the focus is on comparing the means of the groups (as it is in this chapter) rather than on identifying the latent dimensions underlying the group differences and the multivariate tests indicate that the groups differ significantly on their discriminant scores, we then examine the results of the univariate ANOVAs where each dependent variable is analyzed separately; these univariate ANOVAs are conveniently produced by IBM SPSS as part of the default output. Researchers will often apply an alpha level inflation correction when examining the ANOVAs, such as the Bonferroni correction where .05 is divided by the number of dependent variables, to establish a revised alpha level. Each statistically significant effect of the independent variable (using the corrected alpha level) is then subjected to a multiple comparison test (if there are more than two levels) to determine which groups differ significantly from which others.

54.2 NUMERICAL EXAMPLE

The fictional study we use for our example examines some personality characteristics of employees in companies that are experiencing one of three types of business cycles. The between-subjects independent variable is named **business_cycle**, and its levels are **bankruptcy** (coded as **1** in the data file), **steady state** (coded as **2** in the data file), and **rapid expansion** (coded as **3** in the data file). The dependent variables are the Big-5 personality dimensions of neuroticism (**neoneuro**), extraversion (**neoextra**), openness (**neoopen**), agreeableness (**neoagree**), and conscientiousness (**neoconsc**) and represent scores of employees sampled from each type of **business_cycle**. The data file is named **business climate**.

54.3 CORRELATION ANALYSIS

We will perform a correlation analysis on the dependent variables as a preliminary to our MANOVA. We open the data file **business climate** and from the main menu, we select **Analyze ➔ Correlate ➔ Bivariate**. This opens the main **Bivariate Correlation** window (the general dialog window can be seen in Chapter 22). We move **neoneuro**, **neoextra**, **neoopen**, **neoagree**, and **neoconsc** into the **Variables** panel, and click **OK** to perform the analysis.

The **Correlations** output is shown in Figure 54.1. Correlations ranged between absolute values of approximately .01 and approximately .38 and would be considered quite modest in the present context. We can proceed directly to the MANOVA.

54.4 ANALYSIS SETUP: MANOVA

We select **Analyze ➔ General Linear Model ➔ Multivariate**. This opens the main **Multivariate** window shown in Figure 54.2. We move **business_cycle** into the **Fixed Factor(s)** panel and **neoneuro**, **neoextra**, **neoopen**, **neoagree**, and **neoconsc** into the **Dependent Variables** panel.

Correlations

		neoneuro	neoextra	neoopen	neoagree	neoconsc
neoneuro	Pearson Correlation	1	-.375**	-.149**	-.297**	-.375**
	Sig. (2-tailed)		.000	.000	.000	.000
	N	634	634	634	634	634
neoextra	Pearson Correlation	-.375**	1	.096*	.204**	.262**
	Sig. (2-tailed)	.000		.016	.000	.000
	N	634	634	634	634	634
neoopen	Pearson Correlation	-.149**	.096*	1	.207**	-.006
	Sig. (2-tailed)	.000	.016		.000	.880
	N	634	634	634	634	634
neoagree	Pearson Correlation	-.297**	.204**	.207**	1	.119**
	Sig. (2-tailed)	.000	.000	.000		.003
	N	634	634	634	634	634
neoconsc	Pearson Correlation	-.375**	.262**	-.006	.119**	1
	Sig. (2-tailed)	.000	.000	.880	.003	
	N	634	634	634	634	634

**. Correlation is significant at the 0.01 level (2-tailed).
*. Correlation is significant at the 0.05 level (2-tailed).

FIGURE 54.1 **Correlations** of the dependent variables.

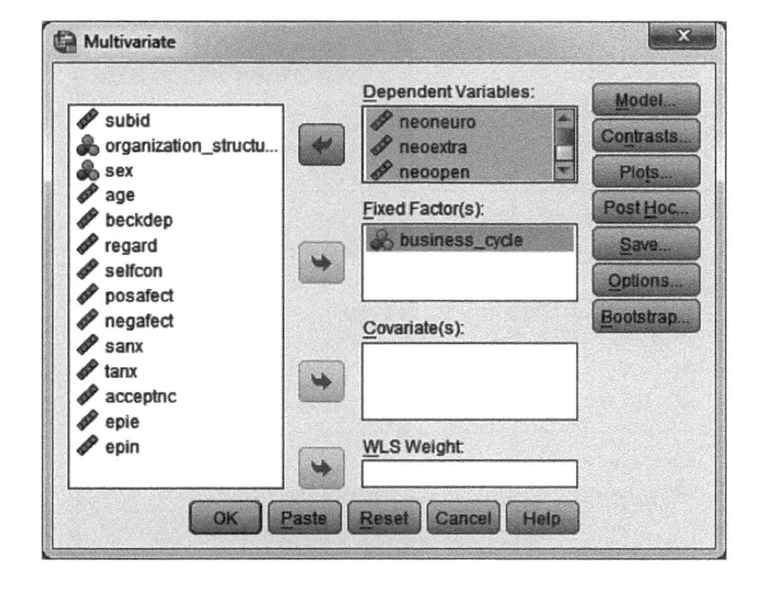

FIGURE 54.2
The main dialog window of **General Linear Model Multivariate**.

Selecting the **Options** pushbutton opens the **Options** dialog window shown in Figure 54.3. In the **Display** area, check **Descriptive statistics**, **Residual SSCP Matrix** (to obtain Bartlett's test of sphericity evaluating whether or not there is statistically significant correlation between the dependent variables), and **Homogeneity tests** (to obtain Box's *M* and the Levene tests). Click **Continue** to return to the main dialog window.

We will anticipate the possibility of a statistically significant main effect and thus select the **Post Hoc** pushbutton. In the **Post Hoc** window (see Figure 54.4), we move

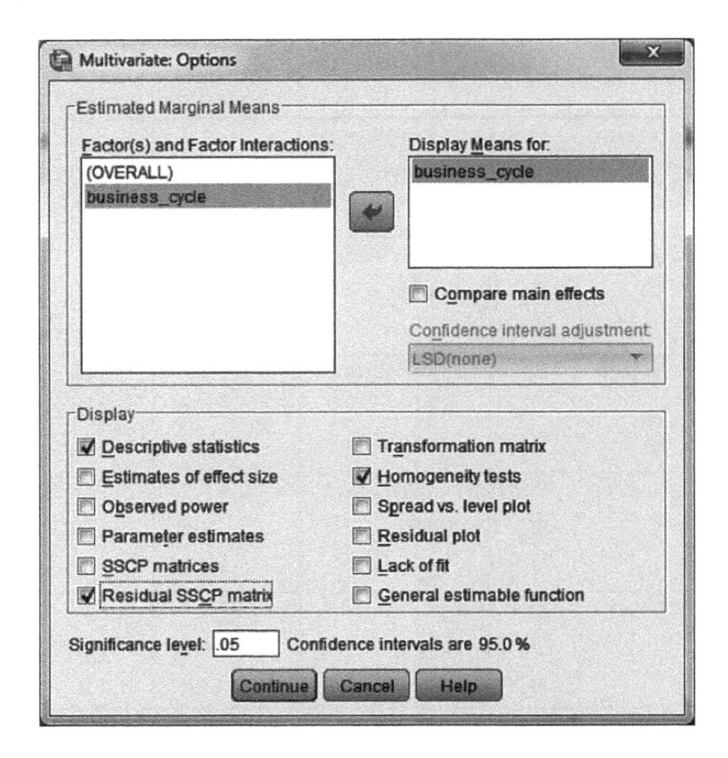

FIGURE 54.3
The **Options** dialog window of **General Linear Model Multivariate**.

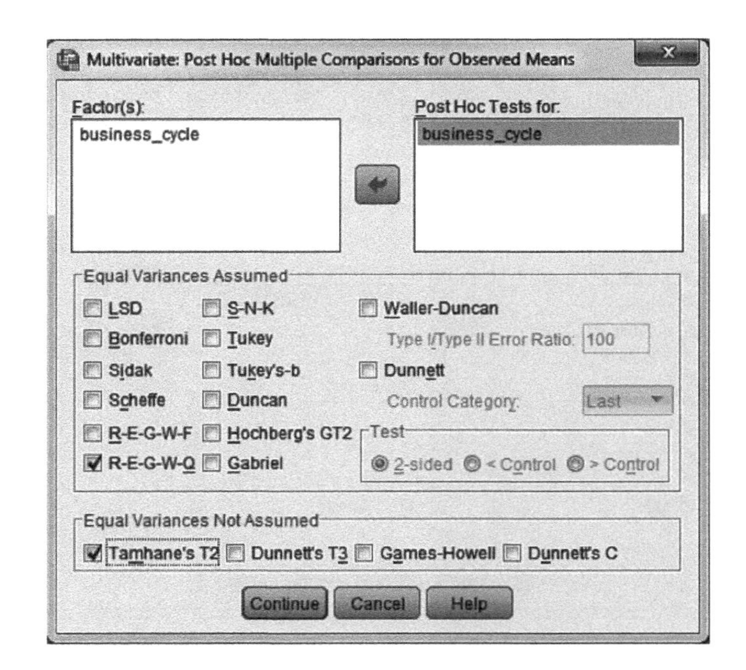

FIGURE 54.4
The **Post Hoc** dialog window of **General Linear Model Multivariate**.

business_cycle into the panel labeled **Post Hoc Tests for**. It is possible that we may have unequal variances, and so we select **R-E-G-W-Q** under **Equal Variances Assumed** and **Tamhane's T2** under **Equal Variances Not Assumed**, planning to use whichever is the more appropriate with respect to each dependent variable. Click **Continue** to return to the main dialog window and click **OK** to perform the analysis.

54.5 ANALYSIS OUTPUT: MANOVA

The **Descriptive Statistics** with the mean, standard deviation, and sample sizes are shown in Figure 54.5. Figure 54.6 shows the results of **Box's** test and **Bartlett's** test. **Box's Test of Equality of Covariance Matrices** tests the assumption that the dependent variable covariance matrices are equal across the levels of the independent variable (**business_cycle**); in this data set, the assumption has apparently not been met (Box's $M = \mathbf{166.018}$, $p < .001$). However, Box's M statistic has a good deal of statistical power when the sample size is large (Norman & Streiner, 2008), and so we will still evaluate our results but exercise caution; because Box's M was statistically significant, we use Pillai's Trace rather than Wilks' lambda. **Bartlett's Test of Sphericity** determines if the correlations between the dependent variables are sufficiently strong to support the MANOVA. A statistically significant outcome indicates that the correlations are sufficient and that is what was obtained (approximate chi-square = **275.796**, $p < .001$).

Figure 54.7 displays the results of the four multivariate tests of statistical significance. Wilks' lambda and Pillai's Trace are the two most commonly used criteria, and with the statistical significance of Box's M, we should be focused on and should report the outcome based on Pillai's Trace, but the effect for **business_cycle** is statistically significant ($p < .001$) based on all four indexes. Wilks' lambda represents the amount

Descriptive Statistics

	business_cycle	Mean	Std. Deviation	N
neoneuro	1 bankruptcy	66.3253	7.65660	145
	2 steady state	53.5585	6.81634	229
	3 rapid expansion	43.3233	7.51283	254
	Total	52.3665	11.48228	628
neoextra	1 bankruptcy	44.0397	12.24871	145
	2 steady state	53.3882	11.96425	229
	3 rapid expansion	57.2991	10.80371	254
	Total	52.8115	12.63648	628
neoopen	1 bankruptcy	52.7634	12.54515	145
	2 steady state	53.9871	11.05835	229
	3 rapid expansion	57.3912	11.17124	254
	Total	55.0814	11.61080	628
neoagree	1 bankruptcy	40.5809	13.50701	145
	2 steady state	45.6679	10.10016	229
	3 rapid expansion	50.6869	11.33657	254
	Total	46.5234	12.09587	628
neoconsc	1 bankruptcy	38.7657	10.18847	145
	2 steady state	44.3662	12.21000	229
	3 rapid expansion	46.2805	9.98008	254
	Total	43.8473	11.25663	628

FIGURE 54.5 The **Descriptive Statistics** output.

Box's Test of Equality of Covariance Matrices[a]

Box's M	166.018
F	5.466
df1	30
df2	800776.681
Sig.	.000

Tests the null hypothesis that the observed covariance matrices of the dependent variables are equal across groups.

a. Design: Intercept + business_cycle

Bartlett's Test of Sphericity[a]

Likelihood Ratio	.000
Approx. Chi-Square	275.796
df	14
Sig.	.000

Tests the null hypothesis that the residual covariance matrix is proportional to an identity matrix.

a. Design: Intercept + business_cycle

FIGURE 54.6
Results of Box's and Bartlett's tests.

Multivariate Tests[c]

Effect		Value	F	Hypothesis df	Error df	Sig.
Intercept	Pillai's Trace	.993	18419.663[a]	5.000	621.000	.000
	Wilks' Lambda	.007	18419.663[a]	5.000	621.000	.000
	Hotelling's Trace	148.306	18419.663[a]	5.000	621.000	.000
	Roy's Largest Root	148.306	18419.663[a]	5.000	621.000	.000
business_cycle	Pillai's Trace	.635	57.933	10.000	1244.000	.000
	Wilks' Lambda	.373	79.223[a]	10.000	1242.000	.000
	Hotelling's Trace	1.661	102.952	10.000	1240.000	.000
	Roy's Largest Root	1.647	204.900[b]	5.000	622.000	.000

a. Exact statistic
b. The statistic is an upper bound on F that yields a lower bound on the significance level.
c. Design: Intercept + business_cycle

FIGURE 54.7 Multivariate tests of statistical significance.

of multivariate variance not explained by the effect and subtracting that value from 1.00 gives a sense of the strength of the **business_cycle** effect (a multivariate analog of eta square). In this analysis, $1.00 - .373 = .627$, and so it appears that **business_cycle** explained approximately 63% of the variance of the discriminant score variable. With the multivariate effect of **business_cycle** statistically significant, we can examine the univariate results.

Levene's Test of Equality of Error Variances[a]

	F	df1	df2	Sig.
neoneuro	1.649	2	625	.193
neoextra	1.593	2	625	.204
neoopen	4.345	2	625	.013
neoagree	2.901	2	625	.056
neoconsc	5.678	2	625	.004

Tests the null hypothesis that the error variance of the dependent variable is equal across groups.

a. Design: Intercept + business_cycle

FIGURE 54.8
Levene's test results for each dependent variable.

Levene's test of homogeneity of variance, presented in Figure 54.8, returned statistically significant results for **neoopen** and **neoconsc**, borderline significance for **neoagree**, and nonsignificant results for **neoneuro** and **neoextra**; we should therefore increase the stringency of our revised alpha level beyond the Bonferroni level when evaluating **neoopen** and **neoconsc**, and even possibly **neoagree**.

The summary table for the univariate ANOVAs can been seen in Figure 54.9. Because we are in the **General Linear Model** module, we obtain the full model solution (see Chapter 47). Our interest is in the reduced model and so we focus on the rows for **business_cycle**, **Error**, and **Corrected Total**. Each dependent variable is given its own row within each of these portions of the summary table; even though they are displayed together, these represent the results of five separate univariate ANOVAs.

We are well advised to impose the Bonferroni correction on our nominal .05 alpha level. Dividing .05 by 5 (the number of dependent variables) brings our corrected alpha level to .01. For variables that violated the homogeneity assumption, we should make the alpha level even more stringent (perhaps .005), but even then the effects of all five dependent variables are statistically significant ($p < .001$).

Eta square values are computed with reference to the **Corrected Total**. For **neoneuro**, we thus divide its sum of squares (**49350.149**) by its **Corrected Total** (**82665.373**). The result is .597. We note that this is also shown in the footnote at the bottom of the table, where the eta square values (called **R Squared** in the footnote) for all of the dependent variables may be found.

Levene's test indicated that **neoneuro** and **neoextra** yielded homogeneous variances. We can therefore examine the results of the **R-E-G-W-Q** post hoc tests for these variables (see Figure 54.10). These tests revealed that all three groups differed from each other on each of these measures. Thus, the **rapid expansion** group was most neurotic and extraverted, the **steady state** group was less neurotic and extraverted, and the **rapid expansion** group was least neurotic and extraverted.

Levene's test further indicated that **neoopen** and **neoconsc** demonstrated heterogeneity of variances and that **neoagree** was suggestive of heterogeneity of variances. We can therefore examine the results of **Tamhane's T2** post hoc tests for these variables (see Figure 54.11). These tests revealed that all three groups differed significantly on **neoagree**. Thus, the **rapid expansion** group was most agreeable, the **steady state** group was less agreeable, and the **rapid expansion** group was least agreeable.

Tests of Between-Subjects Effects

Source	Dependent Variable	Type III Sum of Squares	df	Mean Square	F	Sig.
Corrected Model	neoneuro	49350.149[a]	2	24675.075	462.909	.000
	neoextra	16348.497[b]	2	8174.248	60.986	.000
	neoopen	2408.468[c]	2	1204.234	9.165	.000
	neoagree	9691.101[d]	2	4845.550	36.912	.000
	neoconsc	5309.765[e]	2	2654.882	22.381	.000
Intercept	neoneuro	1752362.84	1	1752362.84	32874.663	.000
	neoextra	1574990.69	1	1574990.69	11750.676	.000
	neoopen	1772487.97	1	1772487.97	13490.426	.000
	neoagree	1233614.95	1	1233614.95	9397.365	.000
	neoconsc	1101784.85	1	1101784.85	9288.241	.000
business_cycle	neoneuro	49350.149	2	24675.075	462.909	.000
	neoextra	16348.497	2	8174.248	60.986	.000
	neoopen	2408.468	2	1204.234	9.165	.000
	neoagree	9691.101	2	4845.550	36.912	.000
	neoconsc	5309.765	2	2654.882	22.381	.000
Error	neoneuro	33315.224	625	53.304		
	neoextra	83771.282	625	134.034		
	neoopen	82117.864	625	131.389		
	neoagree	82045.269	625	131.272		
	neoconsc	74138.419	625	118.621		
Total	neoneuro	1804801.50	628			
	neoextra	1851646.87	628			
	neoopen	1989850.38	628			
	neoagree	1450994.02	628			
	neoconsc	1286832.50	628			
Corrected Total	neoneuro	82665.373	627			
	neoextra	100119.778	627			
	neoopen	84526.332	627			
	neoagree	91736.370	627			
	neoconsc	79448.184	627			

a. R Squared = .597 (Adjusted R Squared = .596)
b. R Squared = .163 (Adjusted R Squared = .161)
c. R Squared = .028 (Adjusted R Squared = .025)
d. R Squared = .106 (Adjusted R Squared = .103)
e. R Squared = .067 (Adjusted R Squared = .064)

FIGURE 54.9 Summary table for the five ANOVAs.

Homogeneous Subsets

neoneuro

	business_cycle	N	Subset		
			1	2	3
Ryan-Einot-Gabriel-Welsch Range[a]	3 rapid expansion	254	43.3233		
	2 steady state	229		53.5585	
	1 bankruptcy	145			66.3253
	Sig.		1.000	1.000	1.000

Means for groups in homogeneous subsets are displayed.
Based on observed means.
The error term is Mean Square(Error) = 53.304.

a. Alpha =

neoagree

	business_cycle	N	Subset		
			1	2	3
Ryan-Einot-Gabriel-Welsch Range[a]	1 bankruptcy	145	40.5809		
	2 steady state	229		45.6679	
	3 rapid expansion	254			50.6869
	Sig.		1.000	1.000	1.000

Means for groups in homogeneous subsets are displayed.
Based on observed means.
The error term is Mean Square(Error) = 131.272.

a. Alpha =

neoextra

	business_cycle	N	Subset		
			1	2	3
Ryan-Einot-Gabriel-Welsch Range[a]	1 bankruptcy	145	44.0397		
	2 steady state	229		53.3882	
	3 rapid expansion	254			57.2991
	Sig.		1.000	1.000	1.000

Means for groups in homogeneous subsets are displayed.
Based on observed means.
The error term is Mean Square(Error) = 134.034.

a. Alpha =

neoconsc

	business_cycle	N	Subset	
			1	2
Ryan-Einot-Gabriel-Welsch Range[a]	1 bankruptcy	145	38.7657	
	2 steady state	229		44.3662
	3 rapid expansion	254		46.2805
	Sig.		1.000	.060

Means for groups in homogeneous subsets are displayed.
Based on observed means.
The error term is Mean Square(Error) = 118.621.

a. Alpha =

neoopen

	business_cycle	N	Subset	
			1	2
Ryan-Einot-Gabriel-Welsch Range[a]	1 bankruptcy	145	52.7634	
	2 steady state	229	53.9871	
	3 rapid expansion	254		57.3912
	Sig.		.364	1.000

Means for groups in homogeneous subsets are displayed.
Based on observed means.
The error term is Mean Square(Error) = 131.389.

a. Alpha =

FIGURE 54.10 The results of the **R-E-G-W-Q** post hoc tests.

Multiple Comparisons

Dependent Variable		(I) business_cycle	(J) business_cycle	Mean Difference (I-J)	Std. Error	Sig.	95% Confidence Interval	
							Lower Bound	Upper Bound
neoneuro	Tamhane	1 bankruptcy	2 steady state	12.7668*	.77923	.000	10.8951	14.6386
			3 rapid expansion	23.0020*	.79153	.000	21.1012	24.9027
		2 steady state	1 bankruptcy	-12.7668*	.77923	.000	-14.6386	-10.8951
			3 rapid expansion	10.2351*	.65200	.000	8.6729	11.7974
		3 rapid expansion	1 bankruptcy	-23.0020*	.79153	.000	-24.9027	-21.1012
			2 steady state	-10.2351*	.65200	.000	-11.7974	-8.6729
neoextra	Tamhane	1 bankruptcy	2 steady state	-9.3486*	1.28832	.000	-12.4419	-6.2552
			3 rapid expansion	-13.2595*	1.22238	.000	-16.1964	-10.3225
		2 steady state	1 bankruptcy	9.3486*	1.28832	.000	6.2552	12.4419
			3 rapid expansion	-3.9109*	1.04144	.001	-6.4067	-1.4152
		3 rapid expansion	1 bankruptcy	13.2595*	1.22238	.000	10.3225	16.1964
			2 steady state	3.9109*	1.04144	.001	1.4152	6.4067
neoopen	Tamhane	1 bankruptcy	2 steady state	-1.2237	1.27255	.709	-4.2806	1.8332
			3 rapid expansion	-4.6278*	1.25567	.001	-7.6446	-1.6110
		2 steady state	1 bankruptcy	1.2237	1.27255	.709	-1.8332	4.2806
			3 rapid expansion	-3.4041*	1.01259	.003	-5.8304	-.9778
		3 rapid expansion	1 bankruptcy	4.6278*	1.25567	.001	1.6110	7.6446
			2 steady state	3.4041*	1.01259	.003	.9778	5.8304
neoagree	Tamhane	1 bankruptcy	2 steady state	-5.0871*	1.30525	.000	-8.2251	-1.9490
			3 rapid expansion	-10.1060*	1.32822	.000	-13.2981	-6.9140
		2 steady state	1 bankruptcy	5.0871*	1.30525	.000	1.9490	8.2251
			3 rapid expansion	-5.0190*	.97542	.000	-7.3562	-2.6818
		3 rapid expansion	1 bankruptcy	10.1060*	1.32822	.000	6.9140	13.2981
			2 steady state	5.0190*	.97542	.000	2.6818	7.3562
neoconsc	Tamhane	1 bankruptcy	2 steady state	-5.6005*	1.16915	.000	-8.4058	-2.7952
			3 rapid expansion	-7.5148*	1.05263	.000	-10.0426	-4.9870
		2 steady state	1 bankruptcy	5.6005*	1.16915	.000	2.7952	8.4058
			3 rapid expansion	-1.9143	1.02135	.174	-4.3623	.5337
		3 rapid expansion	1 bankruptcy	7.5148*	1.05263	.000	4.9870	10.0426
			2 steady state	1.9143	1.02135	.174	-.5337	4.3623

Based on observed means.
The error term is Mean Square(Error) = 118.621.

*. The mean difference is significant at the

FIGURE 54.11 The results of **Tamhane's T2** post hoc tests.

For **neoopen**, the **bankruptcy** and **steady state** groups did not differ but the **rapid expansion** group was significantly more open than either of the other two groups. Thus, the **rapid expansion** group was more open than either the **steady state** or the **rapid expansion** group.

For **neoconsc**, the **steady state** and **rapid expansion** groups did not differ but the **bankruptcy** group was significantly less conscientious than either of the other two groups.

CHAPTER 55

Discriminant Function Analysis

55.1 OVERVIEW

Discriminant function analysis is an alternative (and not incompatible) way to conceptualize a one-way between-subjects MANOVA. As we discussed in Chapter 54, the quantitative dependent variables in the MANOVA analysis are synthesized together to form a variate or weighted linear composite, the value of which is a discriminant score when solved for each case. These variates are discriminant functions. In discriminant function analysis, our interest is less on group differences per se and more on interpreting the discriminant functions that characterize the group differences. Interpreting a discriminant function is a matter of identifying the latent construct or dimension underlying it.

The orientation in discriminant function analysis toward interpreting latent dimensions of variates shifts the nominal roles of the variables in the statistical analysis (but not in the data collection procedure). The quantitative variables that were the dependent variables in the MANOVA are treated as the predictors or independent variables in discriminant function analysis (akin to multiple regression analysis); group membership is the independent variable in the MANOVA but it is the dependent variable in discriminant function analysis. Thus, in discriminant function analysis, we use the quantitative variables to predict group membership. Regardless of such role differences, however, the same dynamics underlie both MANOVA and discriminant function analysis if we think in terms of quantitative variables and a grouping variable.

The focus on the latent dimensions representing the quantitative variables emerges in two aspects of the discriminant function data analysis: description or explanation and prediction or classification. The differences between groups can be relatively complex, falling along several dimensions (e.g., university professors, corporate executives, and politicians differ in a host of ways). In discriminant function analysis, the number of dimensions (discriminant functions) along which we can describe the group differences (and which together comprise the discriminant function model) is the smaller of the following two quantities:

- the number of quantitative predictor variables in the analysis;
- the number of groups − 1 (the degrees of freedom of the group variable)—this quantity is almost always smaller than the number of quantitative variables and

Performing Data Analysis Using IBM SPSS®, First Edition.
Lawrence S. Meyers, Glenn C. Gamst, and A. J. Guarino.

so almost always determines the number of discriminant functions that can be extracted from the data set.

Each discriminant function represents one latent dimension along which the groups differ. The variables in each function have their predictive weights (unstandardized and standardized discriminant coefficients analogous to those in multiple regression), but they also have associated structure coefficients. These structure coefficients are the same as we have seen in multiple regression and factor analysis; they represent the correlation between the variable and weighted linear composite and are used to interpret the latent dimension represented by the variate. If there are three groups in the analysis, for example, then there are two discriminant functions; if both are statistically significant, then the groups may be said to differ along both the latent dimensions. That said, because the discriminant functions are extracted sequentially and are orthogonal to each other, as we have seen in principal components analysis (see Chapter 38), the first function typically accounts for the bulk of the explained variance and should therefore be given appropriate emphasis in the interpretation.

Another major aspect of discriminant function analysis deals with prediction or classification. In addition to the discriminant weights and structure coefficients produced by the analysis, IBM SPSS® also produces a set of classification coefficients or weights. These classification weights are applied to the quantitative variables to predict group membership; that is, they are used to classify cases based on their quantitative variable scores into the group to which they most likely belong. Each group is associated with a unique set of classification weights. When these weights are applied to a given case, the case is classified into the group for which the total classification score is highest. Prediction will ordinarily not be perfect, and the percentage of correct predictions can be used to gauge the effectiveness of the model. Although no test is provided in the IBM SPSS output of whether cases are classified significantly better than chance, such as Press' *Q* statistic (Press, 1972), readers can consult other sources (e.g., Meyers et al., 2013) to perform such tests on their own.

IBM SPSS provides two different procedures to perform the classification analysis. The default analysis uses all cases to derive the discriminant model and then applies the model to each case. This procedure maximizes the rate of successful classification of the cases in the data set but would not be as successful if applied to a new set of cases.

IBM SPSS also provides us with a *jackknife* (Quenouille, 1956; Tukey, 1958) classification procedure called the **Leave-one-out** method (see Meyers et al. (2013)). In this method, one case is omitted from the analysis and the discriminant function model is generated and applied to classify that case. The case is then reabsorbed into the sample, another case is selected to be left out of the analysis, and so on until all cases have been treated in this way. This attempt at cross-validation provides a more realistic picture of classification effectiveness and has much greater external validity than the default method.

55.2 NUMERICAL EXAMPLE

We use the same **business climate** data file used in Chapter 54 and will also use **business_cycle** as our group variable with its levels of **bankruptcy** (coded as **1** in the data file), **steady state** (coded as **2** in the data file), and **rapid expansion** (coded as **3** in the data file). For this analysis, we use the following quantitative variables as predictors of group membership: **beckdep** (depression), **regard** (self-regard), **selfcon** (self-control), **neoneuro** (neuroticism), **neoextra** (extraversion), **neoopen** (openness), **neoagree** (agreeableness), **neoconsc** (conscientiousness), **posafect** (positive affect), **negafect** (negative affect), **sanx** (state anxiety), **tanx** (trait anxiety), and **acceptnc** (self-acceptance).

55.3 ANALYSIS SETUP

We open the data file **business climate** and from the main menu, we select **Analyze ➔ Classify ➔ Discriminant**. This opens the main **Discriminant Analysis** window shown in Figure 55.1. We move the quantitative variables of **beckdep**, **regard**, **selfcon**, **neoneuro**, **neoextra**, **neoopen**, **neoagree**, **neoconsc**, **posafect**, **negafect**, **sanx**, **tanx**, and **acceptnc** into the **Independents** panel,

We next move **business_cycle** into the **Grouping Variable** panel. Upon doing so, we see two question marks in parentheses next to the variable name; this is because IBM SPSS must be informed of the group codes we have used. To accomplish this, we select the **Define Range** pushbutton to reach the **Define Range** dialog window (see Figure 55.2), type **1** in the **Minimum** panel and **3** in the **Maximum** panel, and select **Continue** to return to the main dialog window where our codes have been specified in the parentheses next to **business_cycle** as shown in Figure 55.3.

Selecting the **Statistics** pushbutton opens the **Statistics** dialog window shown in Figure 55.4. In the **Descriptives** area, check **Means** and **Univariate ANOVAs** (to obtain tests of statistical significance for each predictor). Under **Function Coefficients**, check both **Fisher's** (to obtain the classification coefficients) and **Unstandardized** (to obtain the discriminant coefficients for each predictor). Selecting **Continue** returns us to the main dialog window.

Selecting the **Classify** pushbutton opens the **Classification** dialog window shown in Figure 55.5. In the **Prior Probabilities** area, select **All groups equal**. This requires the prediction of group membership to be based exclusively on the discriminant model (the alternative option **Compute from group sizes** allows the software to take advantage of differences in group sizes to "improve" prediction). Retain the default of **Within-groups** in the **Use Covariance Matrix** area.

In the **Display** area, we select **Summary table** and **Leave-one-out classification** to obtain the standard classification results as well as the cross-validation results. In

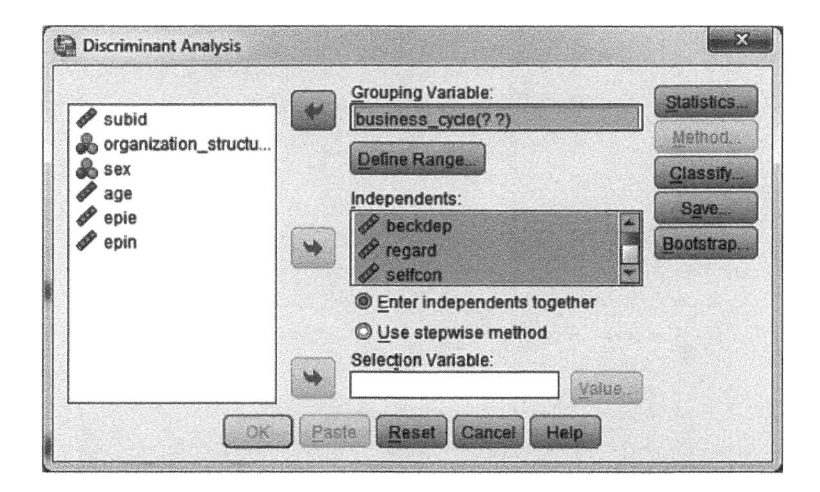

FIGURE 55.1
The main **Discriminant Analysis** window.

FIGURE 55.2
The **Define Range** dialog window.

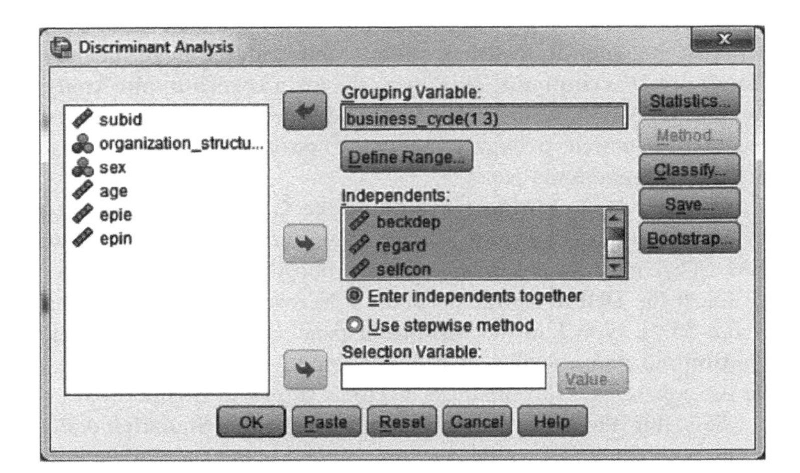

FIGURE 55.3

The main **Discriminant Analysis** window with the ranges defined.

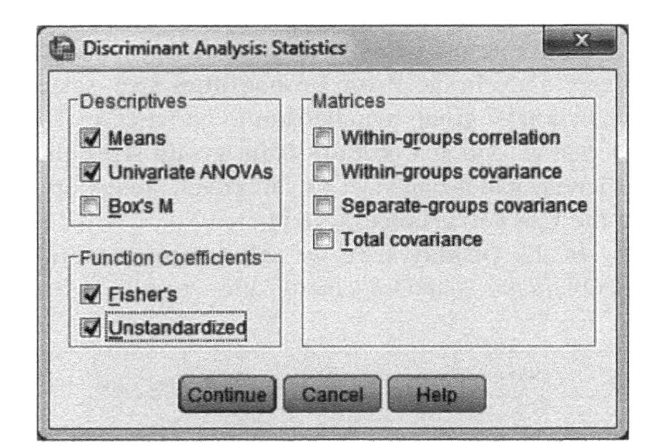

FIGURE 55.4

The **Statistics** dialog window of **Discriminant Analysis**.

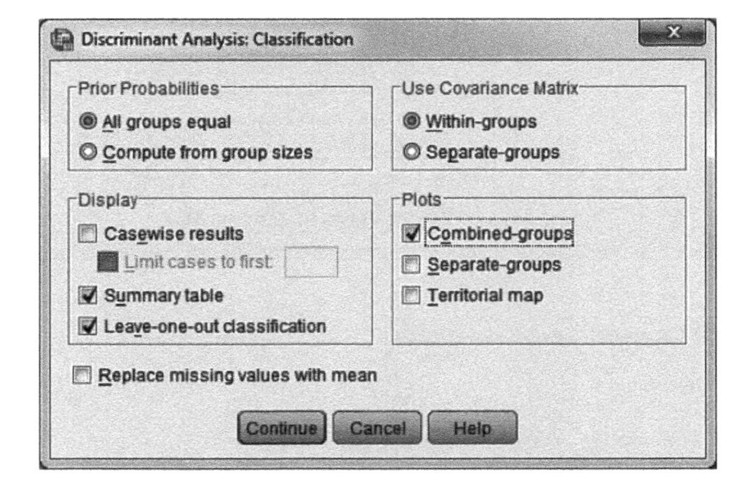

FIGURE 55.5

The **Classification** dialog window of **Discriminant Analysis**.

the **Plots** field, we select **Combined-groups** to obtain a pictorial representation of the multivariate group differences. Click **Continue** to return to the main dialog window and click **OK** to perform the analysis.

55.4 ANALYSIS OUTPUT

The **Descriptive Statistics** with the mean, standard deviation, and sample sizes is shown in Figure 55.6. As all groups are equally weighted (we specified **All groups equal** in the **Prior Probabilities** area), the **Unweighted** and the **Weighted Valid N**s are equal.

Figure 55.7 presents the results of the univariate ANOVAs for each quantitative predictor. Group differences are tested using Wilks' lambda statistic. As may be seen in the table, the three groups (that is why **df1** is given as **2**) were significantly different on all of the predictors ($p < .001$).

The results of the overall (omnibus) multivariate analysis are shown in Figure 55.8. The bottom table provides the **Wilks' lambda** output. Had we performed MANOVA using these variables, the **Wilks' lambda** test of multivariate statistical significance would have corresponded to the first row of the table labeled **1 through 2**.

The **Wilks' lambda** table is a *dimension reduction analysis*. The first row evaluates the entire discriminant model, which in our example consists of two discriminant functions (because there are three groups). Functions **1 through 2** generated a **Wilks' lambda** of **.145**. Evaluated against a chi-square distribution with **26** degrees of freedom (13 predictors in each of functions), functions **1 through 2** as a set are statistically significant ($p < .001$). Subtracting the value of **Wilks' lambda** (the amount of unexplained multivariate variance) from 1.00 yields .855. Thus, the discriminant model with its two functions accounted for almost 86% of the multivariate group difference variance.

The dimension reduction analysis continues to completion. Having examined all of the functions as a set, the first function is removed and the remaining functions as a set are evaluated. Here, the remaining set contains only discriminant function **2**. It is statistically significant as well.

To achieve a sense of the effectiveness of each function in differentiating the groups, we examine the **Eigenvalues** table. As we have seen in factor analysis, eigenvalues are interpretable in terms of explained variance and, although the numerical calculations are a bit different here, the meaning is essentially the same. The values in the **Eigenvalues** table are taken with respect to the total amount of *explained* variance (not the total variance). Of the explained variance, the first function accounts for the vast majority (97%), with an **Eigenvalue** of **4.968**. The canonical correlation associated with the first function is **.912**; squaring this value yields approximately .83. We can thus say the first discriminant function accounts for about 83% of the variance in the discriminant score differences of the groups.

The second function is much less potent than the first. Its **Eigenvalue** is **.152**, and it explains the remaining 3% of the *explained* variance. With a canonical correlation of **.364**, it appears that it accounts for about 13% of the variance in the discriminant score differences of the groups.

Figure 55.9 presents the standardized and unstandardized discriminant function coefficients. These weights are akin to those in multiple regression; the raw score coefficients are associated with a constant to be included in the equation.

The **Structure Matrix** is presented in Figure 55.10 and is analogous to what we have seen in principal components and factor analysis (already sorted by magnitude per function) in Chapter 38. We interpret these structure coefficients as though they were factors, although it is not unusual that the magnitudes of the values are often lower in

Group Statistics

business_cycle		Mean	Std. Deviation	Valid N (listwise)	
				Unweighted	Weighted
1 bankruptcy	beckdep	18.4621	8.37574	145	145.000
	regard	37.2517	7.22773	145	145.000
	selfcon	3.6255	.50544	145	145.000
	neoneuro	66.3253	7.65660	145	145.000
	neoextra	44.0397	12.24871	145	145.000
	neoopen	52.7634	12.54515	145	145.000
	neoagree	40.5809	13.50701	145	145.000
	neoconsc	38.7657	10.18847	145	145.000
	posafect	4.5103	2.83607	145	145.000
	negafect	9.7724	2.66861	145	145.000
	sanx	49.5931	14.23049	145	145.000
	tanx	54.4759	8.79084	145	145.000
	acceptnc	42.5207	10.30993	145	145.000
2 steady state	beckdep	8.6769	5.17843	229	229.000
	regard	51.4170	7.02156	229	229.000
	selfcon	3.9753	.42116	229	229.000
	neoneuro	53.5585	6.81634	229	229.000
	neoextra	53.3882	11.96425	229	229.000
	neoopen	53.9871	11.05835	229	229.000
	neoagree	45.6679	10.10016	229	229.000
	neoconsc	44.3662	12.21000	229	229.000
	posafect	6.8777	2.87196	229	229.000
	negafect	7.0175	3.04493	229	229.000
	sanx	38.1965	11.95829	229	229.000
	tanx	41.2402	7.26255	229	229.000
	acceptnc	47.0742	8.22990	229	229.000
3 rapid expansion	beckdep	4.2008	3.46568	254	254.000
	regard	62.1220	6.33680	254	254.000
	selfcon	4.2298	.47570	254	254.000
	neoneuro	43.3233	7.51283	254	254.000
	neoextra	57.2991	10.80371	254	254.000
	neoopen	57.3912	11.17124	254	254.000
	neoagree	50.6869	11.33657	254	254.000
	neoconsc	46.2805	9.98008	254	254.000
	posafect	8.6220	2.28392	254	254.000
	negafect	3.6693	2.90170	254	254.000
	sanx	29.4803	8.13848	254	254.000
	tanx	31.7362	6.46282	254	254.000
	acceptnc	61.6949	7.98643	254	254.000
Total	beckdep	9.1258	7.79533	628	628.000

FIGURE 55.6 Descriptive statistics.

Tests of Equality of Group Means

	Wilks' Lambda	F	df1	df2	Sig.
beckdep	.505	305.867	2	625	.000
regard	.335	621.251	2	625	.000
selfcon	.799	78.738	2	625	.000
neoneuro	.403	462.909	2	625	.000
neoextra	.837	60.986	2	625	.000
neoopen	.972	9.165	2	625	.000
neoagree	.894	36.912	2	625	.000
neoconsc	.933	22.381	2	625	.000
posafect	.735	112.540	2	625	.000
negafect	.593	214.915	2	625	.000
sanx	.676	149.750	2	625	.000
tanx	.413	444.399	2	625	.000
acceptnc	.525	282.976	2	625	.000

FIGURE 55.7 Results of the univariate ANOVAs for each quantitative predictor.

Eigenvalues

Function	Eigenvalue	% of Variance	Cumulative %	Canonical Correlation
1	4.968[a]	97.0	97.0	.912
2	.152[a]	3.0	100.0	.364

a. First 2 canonical discriminant functions were used in the analysis.

Wilks' Lambda

Test of Function(s)	Wilks' Lambda	Chi-square	df	Sig.
1 through 2	.145	1193.529	26	.000
2	.868	87.772	12	.000

FIGURE 55.8 **Eigenvalues** and **Wilks' lambda** results.

discriminant function analysis. The latent dimensions represented here describe the way the groups differ. The first function appears to represent higher levels of self-regard and lower levels of neuroticism and trait anxiety; this could be interpreted as representing emotional stability and positive feelings toward oneself. The second function appears to represent higher levels of self-acceptance and depression; this could be interpreted as representing an acceptance of feelings of melancholy.

Figure 55.11 presents the centroids for each function in table (**Functions at Group Centroids**) and graphic (**Canonical Discriminant Functions**) forms. The centroids—think of them as multivariate means—are in units of the Mahalanobis distance and are standardized akin to *z* scores (see Chapter 19).

Standardized Canonical Discriminant Function Coefficients

	Function	
	1	2
beckdep	-.213	.522
regard	.600	-.163
selfcon	.104	.008
neoneuro	-.419	.149
neoextra	.094	.017
neoopen	-.104	.062
neoagree	.058	.065
neoconsc	-.051	.049
posafect	-.041	.031
negafect	-.072	-.435
sanx	-.088	-.202
tanx	-.004	.274
acceptnc	.537	.732

Canonical Discriminant Function Coefficients

	Function	
	1	2
beckdep	-.038	.094
regard	.088	-.024
selfcon	.224	.017
neoneuro	-.057	.020
neoextra	.008	.001
neoopen	-.009	.005
neoagree	.005	.006
neoconsc	-.005	.004
posafect	-.016	.012
negafect	-.025	-.150
sanx	-.008	-.018
tanx	-.001	.037
acceptnc	.062	.084
(Constant)	-4.765	-5.944

Unstandardized coefficients

FIGURE 55.9 Standardized and unstandardized discriminant function coefficients.

Structure Matrix

	Function	
	1	2
regard	.631*	-.272
neoneuro	-.545*	.179
tanx	-.533*	.278
negafect	-.371*	-.147
sanx	-.310*	.127
posafect	.268*	-.128
selfcon	.224*	-.112
neoagree	.154*	.006
acceptnc	.407	.732*
beckdep	-.433	.557*
neoextra	.192	-.275*
neoconsc	.115	-.199*
neoopen	.074	.120*

Pooled within-groups correlations between discriminating variables and standardized canonical discriminant functions
Variables ordered by absolute size of correlation within function.

*. Largest absolute correlation between each variable and any discriminant function

FIGURE 55.10
The structure coefficients.

Functions at Group Centroids

business_cycle	Function 1	Function 2
1 bankruptcy	-3.378	.394
2 steady state	-.466	-.507
3 rapid expansion	2.348	.233

Unstandardized canonical discriminant functions evaluated at group means

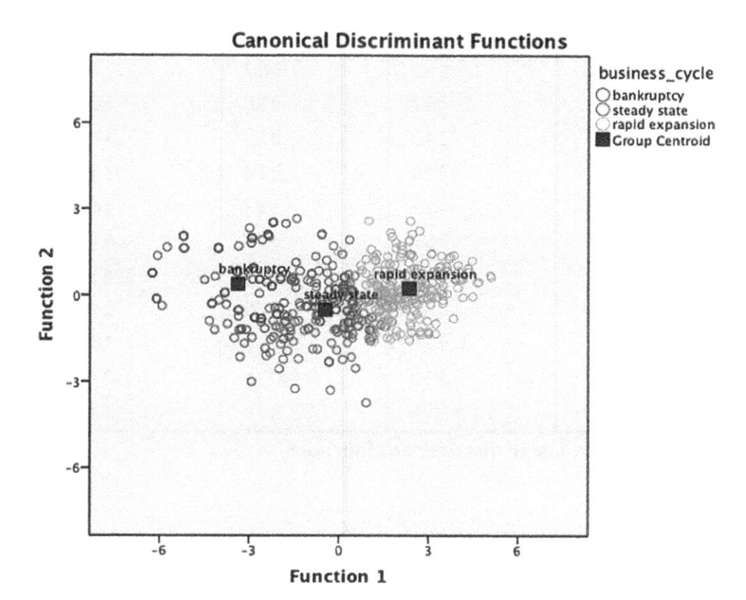

FIGURE 55.11

Group centroids in the table and plot forms.

We determined from the eigenvalues that the first function was overwhelmingly more potent than the second in terms of explaining differences between the groups, and the potency of the first function can be seen in both the tabled and the plotted displays. In the table labeled **Functions at Group Centroids**, the groups are very much differentiated in the first function, spanning a wide range −**3.378** to **2.348**, whereas their centroids are much closer (less differentiated) in the second function where they span the range −.**507** to .**394**. This differentiation can also be seen in the **Combined-groups** plot. The groups are substantially spaced along the horizontal axis representing **Function 1** but are not that far apart vertically (representing **Function 2**).

The **Classification Function Coefficients** are shown in Figure 55.12. These are used to predict the group membership of each case (we use them to classify the cases into predicted groups). The scores of the quantitative variables for each case would be entered into a classification function using the weights in this table. Thus, in our example ,we would compute three classification functions for each case. These are unstandardized weights and so the appropriate constant would be included in the solution for each function. The case would be classified into (predicted to be a member of) the group whose value of the function is highest. Because the true group membership of the case is also known, it is possible to track the success rate of classification across the sample.

The results of the success rate of the classification procedure are presented in Figure 55.13 in the **Classification Results** table. Rows represent actual group membership, and the columns represent the group membership predictions based on the discriminant model. Within each major portion of the table (**Original** and **Cross-validated**),

Classification Function Coefficients

	business_cycle		
	1 bankruptcy	2 steady state	3 rapid expansion
beckdep	-.181	-.378	-.416
regard	.781	1.059	1.290
selfcon	21.862	22.498	23.140
neoneuro	1.552	1.367	1.220
neoextra	.357	.380	.404
neoopen	.242	.210	.189
neoagree	.304	.314	.332
neoconsc	.415	.397	.387
posafect	-.481	-.537	-.572
negafect	-1.298	-1.235	-1.416
sanx	.086	.079	.044
tanx	1.174	1.138	1.164
acceptnc	.931	1.035	1.272
(Constant)	-179.996	-182.970	-203.325

Fisher's linear discriminant functions

FIGURE 55.12
The classification coefficients.

Classification Results[b,c]

		business_cycle	Predicted Group Membership			Total
			1 bankruptcy	2 steady state	3 rapid expansion	
Original	Count	1 bankruptcy	131	14	0	145
		2 steady state	4	220	5	229
		3 rapid expansion	0	8	246	254
	%	1 bankruptcy	90.3	9.7	.0	100.0
		2 steady state	1.7	96.1	2.2	100.0
		3 rapid expansion	.0	3.1	96.9	100.0
Cross-validated[a]	Count	1 bankruptcy	130	15	0	145
		2 steady state	5	217	7	229
		3 rapid expansion	0	8	246	254
	%	1 bankruptcy	89.7	10.3	.0	100.0
		2 steady state	2.2	94.8	3.1	100.0
		3 rapid expansion	.0	3.1	96.9	100.0

a. Cross validation is done only for those cases in the analysis. In cross validation, each case is classified by the functions derived from all cases other than that case.
b. 95.1% of original grouped cases correctly classified.
c. 94.4% of cross-validated grouped cases correctly classified.

FIGURE 55.13 The classification table for **Original** and **Leave-one-out** (**Cross-validated**) methods.

correct classifications occupy the upper left to lower right diagonal. The **Original** classifications are based on all of the cases, whereas the **Cross-validated** classifications are based on the **Leave-one-out** method. With our huge sample size and the substantial differences between the groups, both methods achieved remarkable overall classification success as shown in the footnotes to the table (95.1% for the **Original** method and 94.4% for the **Cross-validated** method).

CHAPTER 56

Two-Way Between-Subjects MANOVA

56.1 OVERVIEW

A two-way between-subjects MANOVA is an extension of the one-way MANOVA design discussed in Chapter 54, and everything we discussed in that chapter applies here as well. The difference, and it is an important one, is that we have the addition of a second independent variable, and by including it, we can evaluate not only the main effects of the independent variables but also their interaction.

The two-way between-subjects MANOVA design is analogous to the univariate two-way design but is considered from a multivariate standpoint. The dependent variables are still combined into a weighted linear composite but there are three multivariate effects of interest (the two main effects and the interaction). If our focus is on group differences, and it almost always is if we structure the design as a two-way, then after we assess the multivariate interaction and the main effects, we end up interpreting the univariate effects.

For each multivariate effect, there are v univariate effects, where v is the number of dependent variables. We saw this in Chapter 54 where we evaluated each of the five univariate effects after obtaining a statistically significant multivariate effect. The same general strategy works for the two-way design, but we must also remember that interaction effects supercede main effects. Given this context, we process the results of the analysis using the strategy that is presented as a flowchart in Figure 56.1.

As shown in the top portion of Figure 56.1, we first examine the multivariate interaction effect. If the multivariate interaction is not statistically significant, we go on to the multivariate main effects of both independent variables for all dependent variables. If the multivariate interaction effect is statistically significant, then we examine the univariate interaction for each dependent variable using the Bonferroni-corrected alpha level (or a more stringent one if homogeneity of variance has been violated). For any dependent variables whose univariate interaction effect has reached statistical significance, we graph the estimated marginal means, perform simple effects analyses, and interpret the results. For any dependent variables whose interaction effect has not reached statistical significance, we examine (for those dependent variables only) each of the main effects.

For those dependent variables not involved in a statistically significant interaction, we start with the multivariate main effect of one of the independent variables and

Performing Data Analysis Using IBM SPSS®, First Edition.
Lawrence S. Meyers, Glenn C. Gamst, and A. J. Guarino.

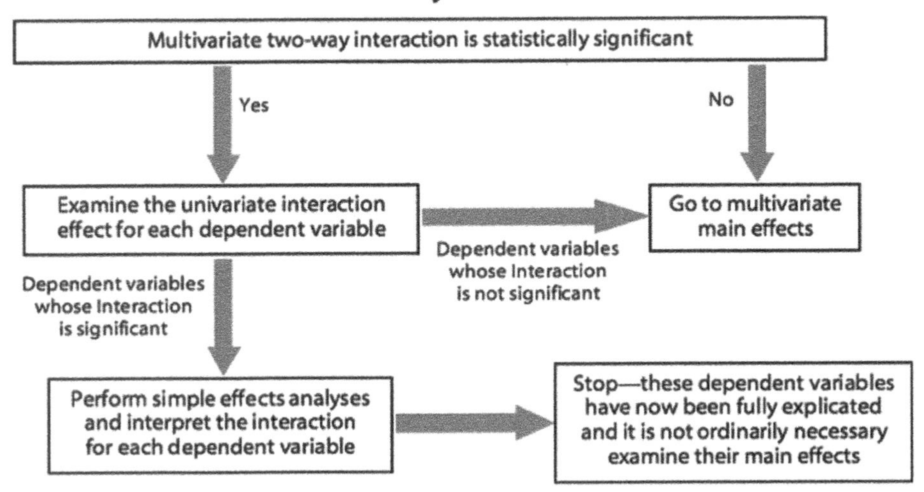

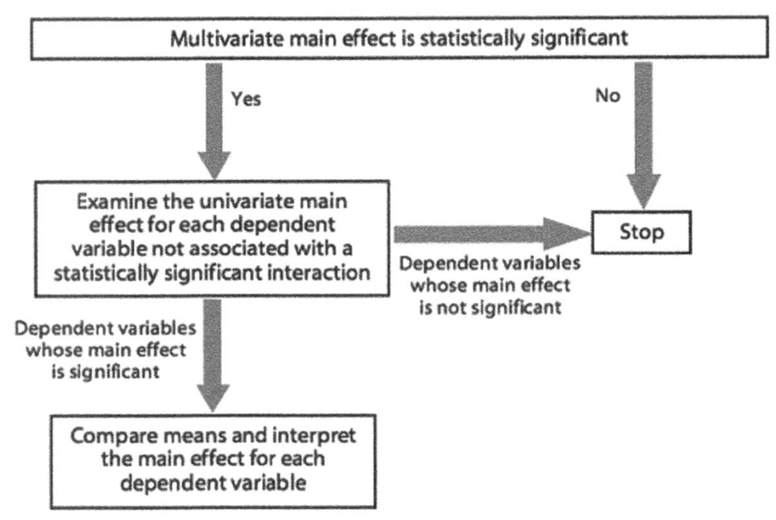

FIGURE 56.1 Flowchart for processing multivariate and univariate interactions and main effects. *Source*: Modified from Meyers et al. (2013).

then examine the other multivariate main effect; this is shown in the bottom portion of Figure 56.1. If the multivariate main effect is not statistically significant, we go on to the other multivariate main effect, repeating the process we describe here (because the main effects are independent of each other). If the multivariate main effect is statistically significant, then we examine the univariate main effects for each dependent variable using the Bonferroni-corrected alpha level (or a more stringent one if homogeneity of variance has been violated). We evaluate only those dependent variables (if any) that were not involved in a statistically significant univariate interaction. If there are more than two levels of the independent variable, we perform multiple comparison tests and interpret the results; if there are only two levels, we interpret the mean difference directly.

56.2 NUMERICAL EXAMPLE

We use the same **business climate** data file used in Chapter 54 and will also use **business_cycle** as one of our independent variables; this factor has the following three levels: **bankruptcy** (coded as **1** in the data file), **steady state** (coded as **2** in the data file), and **rapid expansion** (coded as **3** in the data file). The other independent variable we use here is **organization_structure** with its two levels **hierarchical** (coded as **1** in the data file) and **egalitarian** (coded as **2** in the data file). The three dependent variables we use in this example are extraversion (**neoextra**), openness (**neoopen**), and the level of self-regard the employees feel for themselves (**regard**).

56.3 ANALYSIS SETUP

We open the **business climate** data file and from the main menu select **Analyze ➔ General Linear Model ➔ Multivariate**. This opens the main **Multivariate** dialog window as shown in Figure 56.2. We move **business_cycle** and **organization_structure** into the **Fixed Factor(s)** panel and move **neoextra**, **neoopen**, and **regard** into the **Dependent Variables** panel.

Selecting the **Options** pushbutton opens the **Options** dialog window shown in Figure 56.3. In the **Display** area, check **Descriptive statistics, Residual SSCP Matrix** (to obtain Bartlett's test of sphericity evaluating whether or not there is sufficient correlation between the dependent variables), and **Homogeneity tests** (to obtain Box's M and the Levene tests). On the possibility that we will obtain a statistically significant interaction effect, in the **Estimated Marginal Means** area of the window, we move **business_cycle*organization_structure** from the **Factor(s) and Factor interactions** panel to the **Display Means for** panel.

We will anticipate the possibility of a statistically significant main effect and thus select the **Post Hoc** pushbutton. In the **Post Hoc** window (see Figure 56.4), we move **business_cycle** (because it has more than two levels) into the panel labeled **Post Hoc Tests for** and select **R-E-G-W-Q** under **Equal Variances Assumed** (the tests for violation of homogeneity of variance are not available in the two-way design).

Continuing to anticipate the possibility of a statistically significant interaction effect, we will generate a plot of the interaction for each dependent variable. Selecting the

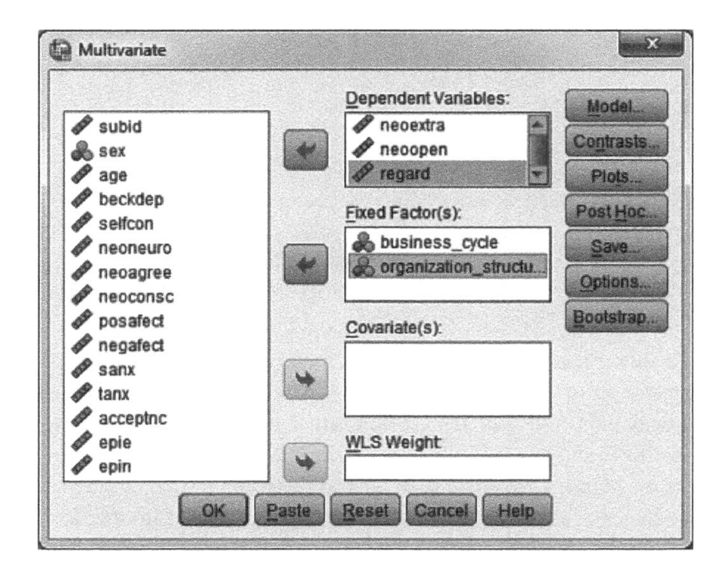

FIGURE 56.2

The main dialog window of the **Multivariate** module of **General Linear Model**.

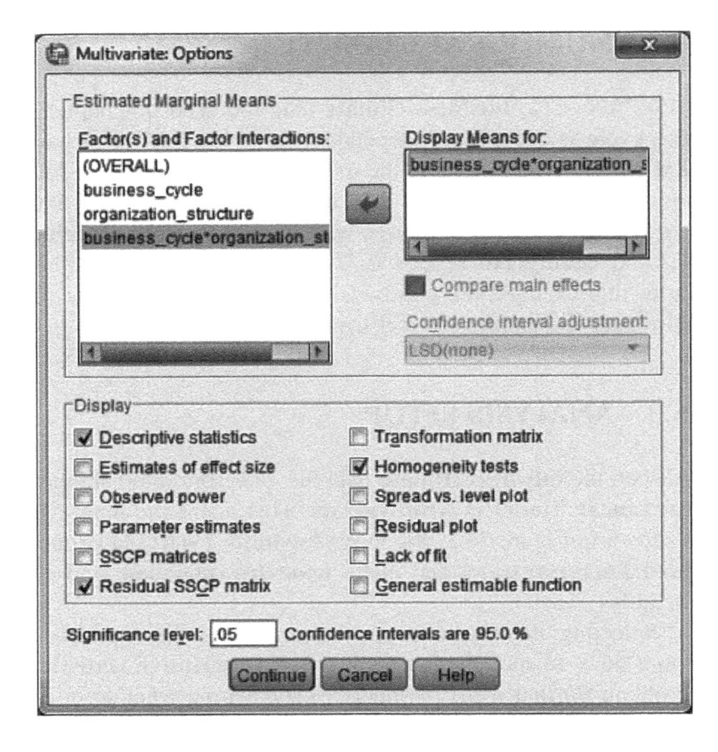

FIGURE 56.3
The **Options** dialog window of the **Multivariate** module of **General Linear Model**.

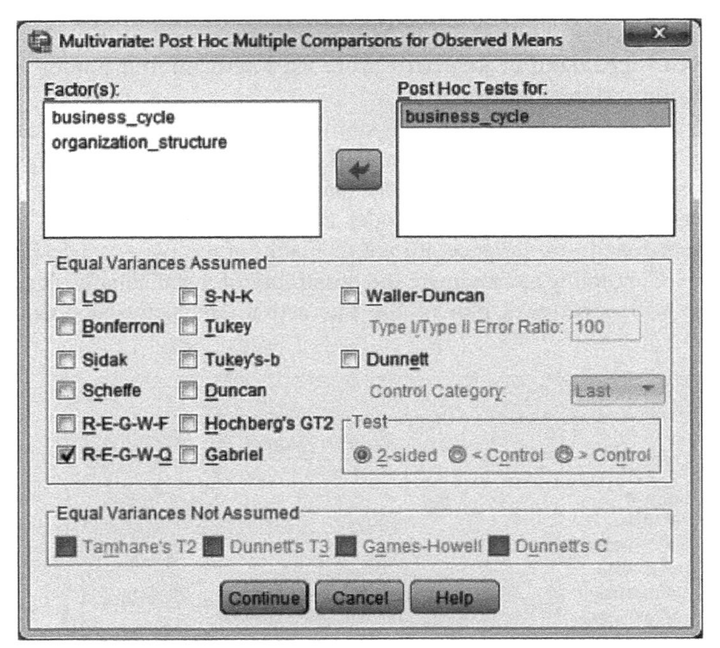

FIGURE 56.4
The **Post Hoc** dialog window of the **Multivariate** module of **General Linear Model**.

Plots pushbutton opens the **Profile Plots** dialog window. With both independent variables as categorical, it is arbitrary as to which goes on the *X*-axis. We therefore place **business_cycle** on the **Horizontal axis** and will have **Separate Lines** for each level of **organization_structure**; this setup is shown in Figure 56.5. Click the **Add** pushbutton to place the plot in the **Plots** panel and click **Continue** to return to the main dialog window.

Given that we wish to anticipate the possibility of a statistically significant interaction effect, we continue to configure the simple effects analysis. We select **Paste** to obtain the

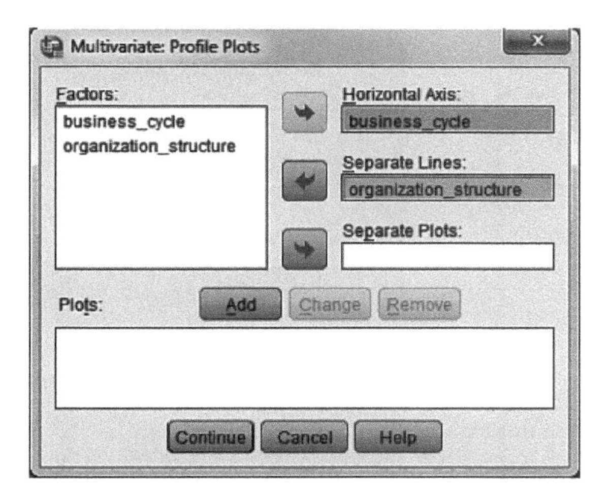

FIGURE 56.5
The **Profile Plots** dialog window of the **Multivariate** module of **General Linear Model**.

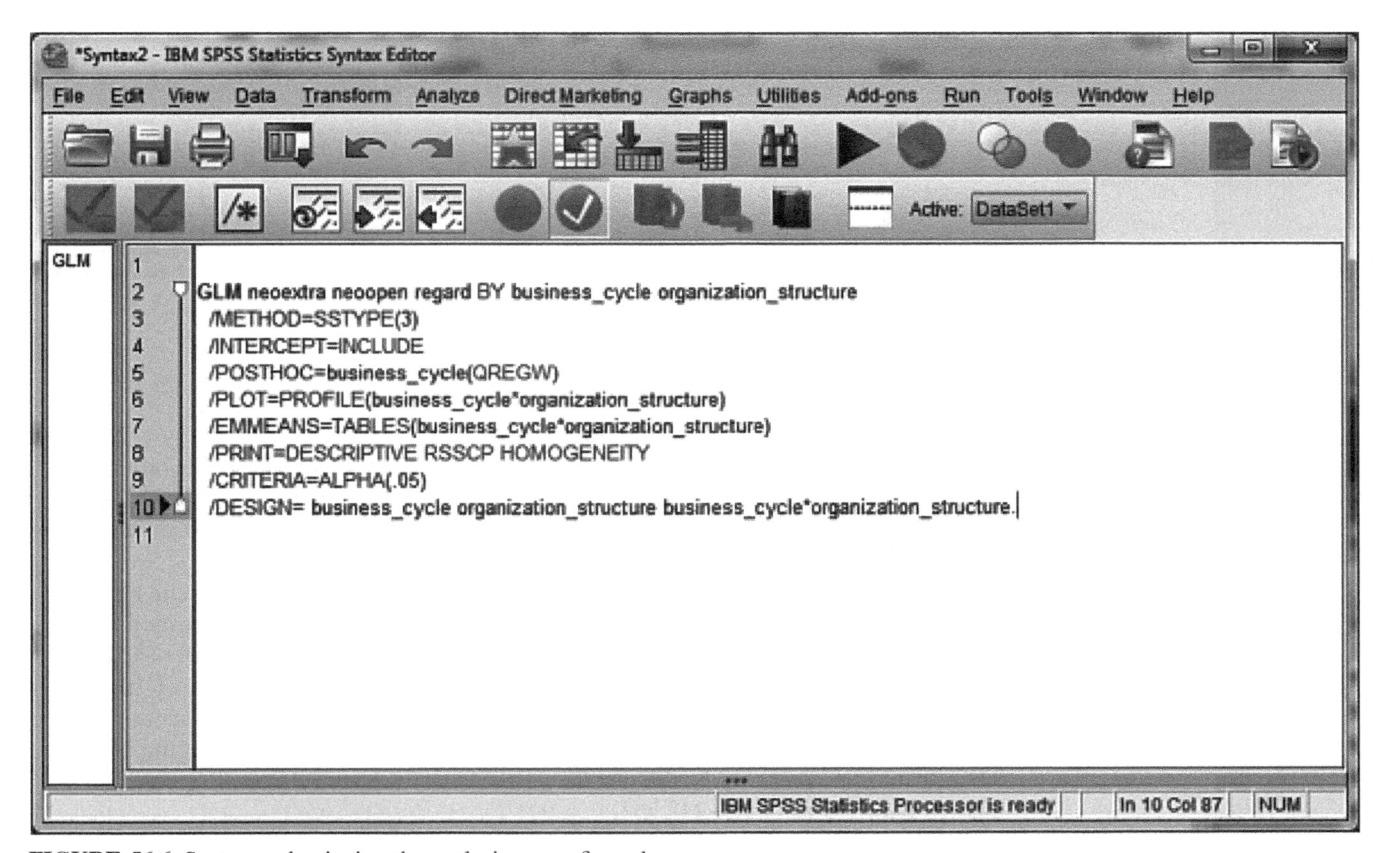

FIGURE 56.6 Syntax underpinning the analysis as configured.

window containing the syntax that represents our analysis setup shown in Figure 56.6. We configure our simple effects in a manner analogous to that described in Chapter 50; this is shown in Figure 56.7. From the main menu select **Run ➔ All** to perform the analysis.

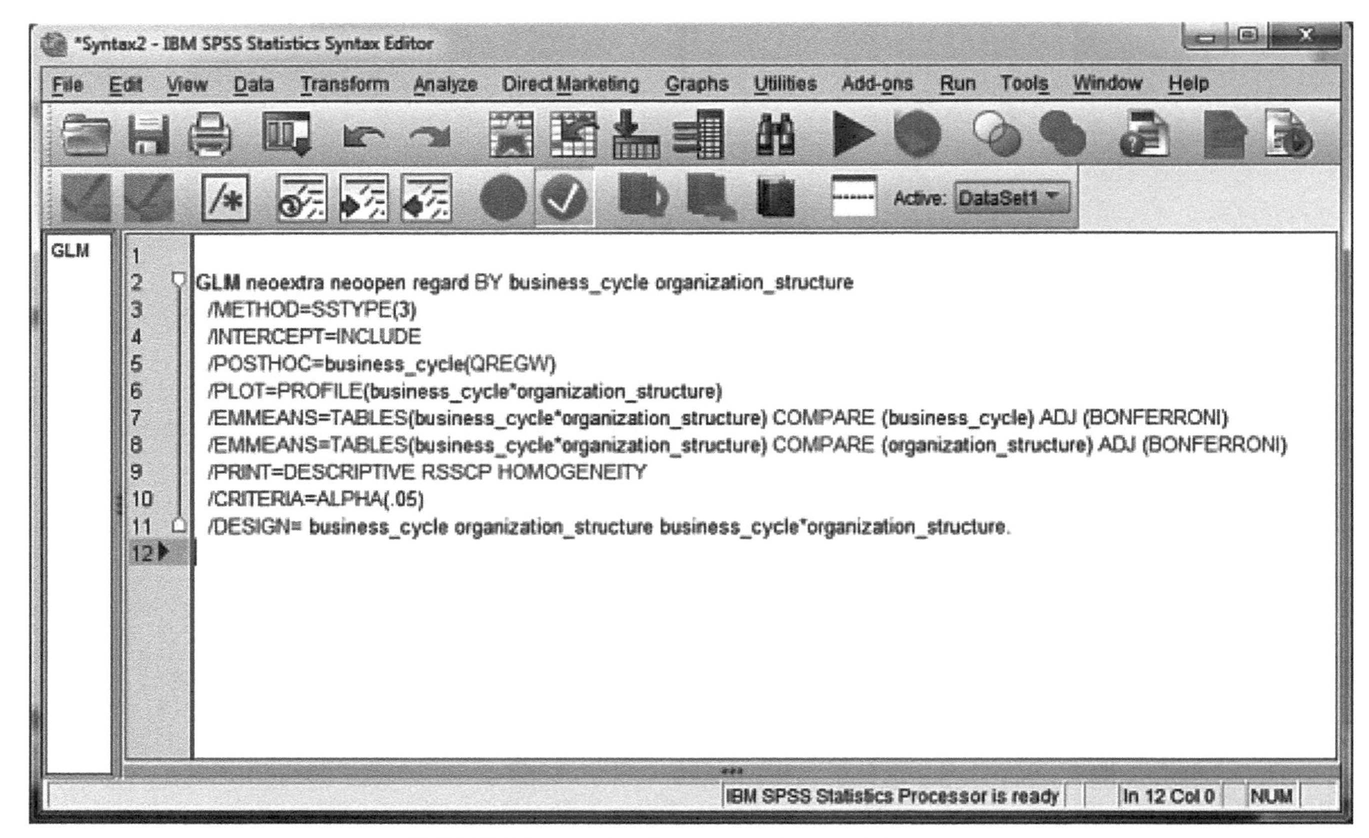

FIGURE 56.7 Simple effects comparisons added to the syntax.

56.4 ANALYSIS OUTPUT

The **Descriptive Statistics** and sample sizes are shown in Figure 56.8. It is worth noting that our very large sample size has provided us with more statistical power than is usual for analyses of group differences. Thus, to be statistically significant, an effect need not be especially potent.

Figure 56.9 shows the results of **Box's** test and **Bartlett's** test. The results from the **Box's Test of Equality of Covariance Matrices** suggests that the assumption of equal dependent variable covariance matrices has apparently not been met (Box's $M = \mathbf{154.479}$, $p < .001$). However, Box's M statistic has a good deal of statistical power when the sample size is large (Norman & Streiner, 2008), and so we will still evaluate our results (using Pillai's Trace) but exercise caution. The statistically significant outcome of **Bartlett's Test of Sphericity** indicates that there the correlations are sufficient for our analysis (approximate chi-square = **240.982**, $p < .001$).

Figure 56.10 displays the results of the four multivariate tests of statistical significance; Pillai's Trace (as well as the other multivariate tests) yielded all three multivariate effects as statistically significant ($p < .001$). Wilks' lambda represents the amount of multivariate variance not explained by the effect and subtracting that value from 1.00 gives a sense of the strength of each multivariate effect. For **business_cycle**, **organization_structure**, and **business_cycle*organization_structure**, the subtraction yields .614, .104, and .078, respectively. Even with the strong effect of **business_cycle**, it is still worthwhile to evaluate the interaction, as the main effect of **business_cycle** will still be evident at a richer information level.

Between-Subjects Factors

		Value Label	N
business_cycle	1	bankruptcy	145
	2	steady state	229
	3	rapid expansion	254
organization_structure	1	hierarchical	369
	2	egalitarian	259

Descriptive Statistics

	business_cycle	organization_structure	Mean	Std. Deviation	N
neoextra	1 bankruptcy	1 hierarchical	51.5567	10.98037	43
		2 egalitarian	40.8707	11.38283	102
		Total	44.0397	12.24871	145
	2 steady state	1 hierarchical	55.4152	10.34242	121
		2 egalitarian	51.1172	13.23640	108
		Total	53.3882	11.96425	229
	3 rapid expansion	1 hierarchical	59.1626	9.89426	205
		2 egalitarian	49.5031	11.05791	49
		Total	57.2991	10.80371	254
	Total	1 hierarchical	57.0475	10.47856	369
		2 egalitarian	46.7765	13.00853	259
		Total	52.8115	12.63648	628
neoopen	1 bankruptcy	1 hierarchical	50.0674	11.24969	43
		2 egalitarian	53.8999	12.93672	102
		Total	52.7634	12.54515	145
	2 steady state	1 hierarchical	53.5845	10.84479	121
		2 egalitarian	54.4381	11.32640	108
		Total	53.9871	11.05835	229
	3 rapid expansion	1 hierarchical	58.0120	9.70449	205
		2 egalitarian	54.7939	15.77965	49
		Total	57.3912	11.17124	254
	Total	1 hierarchical	55.6344	10.63737	369
		2 egalitarian	54.2934	12.85159	259
		Total	55.0814	11.61080	628
regard	1 bankruptcy	1 hierarchical	40.12	7.288	43
		2 egalitarian	36.04	6.887	102
		Total	37.25	7.228	145
	2 steady state	1 hierarchical	53.55	7.315	121
		2 egalitarian	49.03	5.846	108
		Total	51.42	7.022	229
	3 rapid expansion	1 hierarchical	61.70	6.241	205
		2 egalitarian	63.88	6.496	49
		Total	62.12	6.337	254
	Total	1 hierarchical	56.51	9.717	369
		2 egalitarian	46.73	11.992	259
		Total	52.48	11.740	628

FIGURE 56.8 The group sizes and **Descriptive Statistics**.

Box's Test of Equality of Covariance Matrices[a]

Box's M	154.479
F	5.064
df1	30
df2	202517.901
Sig.	.000

Tests the null hypothesis that the observed covariance matrices of the dependent variables are equal across groups.

a. Design: Intercept + business_cycle + organization_structure + business_cycle * organization_structure

Bartlett's Test of Sphericity[a]

Likelihood Ratio	.000
Approx. Chi-Square	240.982
df	5
Sig.	.000

Tests the null hypothesis that the residual covariance matrix is proportional to an identity matrix.

a. Design: Intercept + business_cycle + organization_structure + business_cycle * organization_structure

FIGURE 56.9

Results of Box's and Bartlett's tests.

Levene's test of homogeneity of variance, presented in Figure 56.11, returned statistically significant results for all three dependent variables. Thus, we will increase the stringency of our revised alpha level beyond the Bonferroni level when evaluating the univariate effects.

The summary table for the univariate ANOVAs showing the full model solution can be seen in Figure 56.12. Because we are using the **General Linear Model** module, we obtain the results of the full general linear model (see Chapter 47). Our interest is in the reduced model, and so we focus on the rows for **business_cycle**, **organization_structure**, **business_cycle*organization_structure**, **Error**, and **Corrected Total**. Each dependent variable is given its own row within each of these portions of the summary table, representing the results of three separate univariate ANOVAs.

We choose to impose the Bonferroni correction on our nominal .05 alpha level. Dividing .05 by 3 (the number of dependent variables) brings our corrected alpha level to .017. Because all three dependent variables violated the homogeneity assumption, we should make the alpha level even more stringent; we will use an alpha level of .005 in this example.

When we first examine the univariate interaction effects against our stringent .005 alpha level, we determine that **regard** is statistically significant ($p < .001$) but **neoextra**

Multivariate Tests[c]

Effect		Value	F	Hypothesis df	Error df	Sig.
Intercept	Pillai's Trace	.984	12920.690[a]	3.000	620.000	.000
	Wilks' Lambda	.016	12920.690[a]	3.000	620.000	.000
	Hotelling's Trace	62.519	12920.690[a]	3.000	620.000	.000
	Roy's Largest Root	62.519	12920.690[a]	3.000	620.000	.000
business_cycle	Pillai's Trace	.622	93.462	6.000	1242.000	.000
	Wilks' Lambda	.386	126.169[a]	6.000	1240.000	.000
	Hotelling's Trace	1.574	162.362	6.000	1238.000	.000
	Roy's Largest Root	1.561	323.135[b]	3.000	621.000	.000
organization_structure	Pillai's Trace	.104	24.059[a]	3.000	620.000	.000
	Wilks' Lambda	.896	24.059[a]	3.000	620.000	.000
	Hotelling's Trace	.116	24.059[a]	3.000	620.000	.000
	Roy's Largest Root	.116	24.059[a]	3.000	620.000	.000
business_cycle * organization_structure	Pillai's Trace	.078	8.453	6.000	1242.000	.000
	Wilks' Lambda	.922	8.521[a]	6.000	1240.000	.000
	Hotelling's Trace	.083	8.589	6.000	1238.000	.000
	Roy's Largest Root	.070	14.545[b]	3.000	621.000	.000

a. Exact statistic
b. The statistic is an upper bound on F that yields a lower bound on the significance level.
c. Design: Intercept + business_cycle + organization_structure + business_cycle * organization_structure

FIGURE 56.10 Multivariate tests of statistical significance.

Levene's Test of Equality of Error Variances[a]

	F	df1	df2	Sig.
neoextra	2.253	5	622	.048
neoopen	9.098	5	622	.000
regard	3.526	5	622	.004

Tests the null hypothesis that the error variance of the dependent variable is equal across groups.

a. Design: Intercept + business_cycle + organization_structure + business_cycle * organization_structure

FIGURE 56.11

Levene's test results for each dependent variable.

and **neoopen** were not (p = **.012** and **.034**, respectively). Thus, we examine the plot and the simple effects for **regard** and then examine the main effects for **neoextra** and **neoopen**. Eta square values are computed with reference to the **Corrected Total**. For **regard**, we divide its interaction sum of squares (**1166.410**) by its **Corrected Total** (**86421.142**). The result is .014.

The plot of the interaction is shown in Figure 56.13. Self-regard appears to be greater for **steady state** than for **bankruptcy** and greater still for **rapid expansion** than for **steady state** (this is the main effect of **business_cycle** showing), but employees of hierarchical organizations who seemed to have higher self-regard if their businesses were in the **bankruptcy** or **steady state** cycle appear to lose that advantage in businesses that are in **rapid expansion** (this crossover accounts for the interaction effect).

The tests of simple effects can fine-tune our visual examination, and these are contained in two separate portions of the output. Figure 56.14 presents the **Pairwise**

Tests of Between-Subjects Effects

Source	Dependent Variable	Type III Sum of Squares	df	Mean Square	F	Sig.
Corrected Model	neoextra	24546.766[a]	5	4909.353	40.406	.000
	neoopen	3303.882[b]	5	660.776	5.060	.000
	regard	59349.421[c]	5	11869.884	272.723	.000
Intercept	neoextra	1247315.98	1	1247315.98	10265.974	.000
	neoopen	1390441.35	1	1390441.35	10647.974	.000
	regard	1220640.56	1	1220640.56	28045.444	.000
business_cycle	neoextra	5274.344	2	2637.172	21.705	.000
	neoopen	1367.407	2	683.703	5.236	.006
	regard	41937.703	2	20968.851	481.780	.000
organization_structure	neoextra	8004.587	1	8004.587	65.881	.000
	neoopen	28.398	1	28.398	.217	.641
	regard	541.578	1	541.578	12.443	.000
business_cycle * organization_structure	neoextra	1076.560	2	538.280	4.430	.012
	neoopen	884.409	2	442.205	3.386	.034
	regard	1166.410	2	583.205	13.400	.000
Error	neoextra	75573.012	622	121.500		
	neoopen	81222.450	622	130.583		
	regard	27071.721	622	43.524		
Total	neoextra	1851646.87	628			
	neoopen	1989850.38	628			
	regard	1815771.50	628			
Corrected Total	neoextra	100119.778	627			
	neoopen	84526.332	627			
	regard	86421.142	627			

a. R Squared = .245 (Adjusted R Squared = .239)
b. R Squared = .039 (Adjusted R Squared = .031)
c. R Squared = .687 (Adjusted R Squared = .684)

FIGURE 56.12 Summary table for the univariate effects.

Comparisons (see Chapters 49 and 50 for guidance on reading the table) for the three levels of **business_cycle** for each type of **organization_structure**. The estimated marginal mean differences here are equal to the observed mean differences that can be gleaned from Figure 56.8. For example, the first mean difference value shown in the **Pairwise Comparisons** table is −**13.429** and represents the difference between **bankruptcy** and **steady state** for **hierarchical** structures. The **regard** means for these two groups are **40.12** and **53.55**, respectively, and subtracting the second from the first gives us a value of −**13.43**.

We learn from these **Pairwise Comparisons** that there are statistically significant differences between all three **business_cycle** groups for both types of **organization_structure**. Translated to the plot in Figure 56.13, employees in both **hierarchical** (solid line) and **egalitarian** (dashed–dotted line) organizations exhibited the highest level of self-regard in the **rapid expansion** cycle, next highest level of self-regard in the **steady state** cycle, and the lowest level of self-regard in the **bankruptcy** cycle.

The other half of the simple effects results is shown in Figure 56.15. Here we have the **Pairwise Comparisons** for the two levels of **organization_structure** for each type

regard

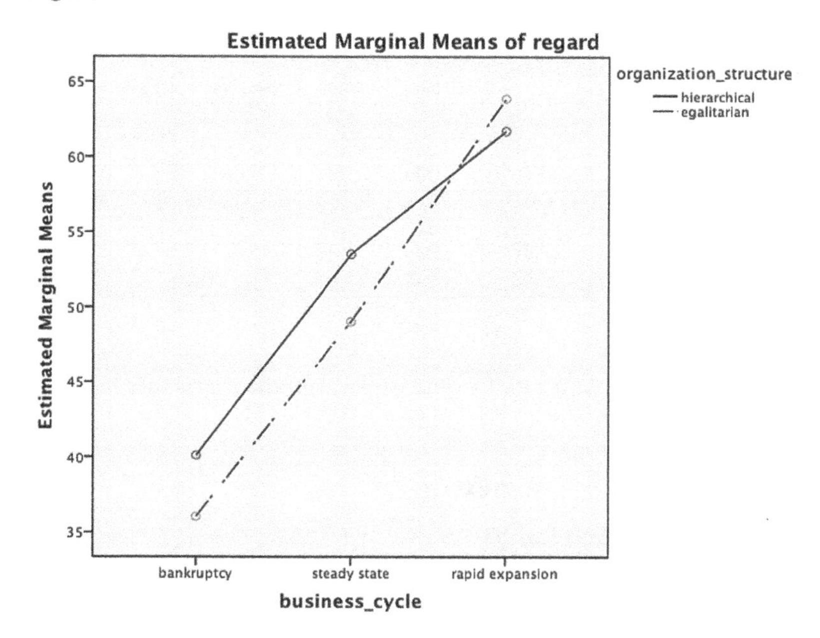

FIGURE 56.13
Plot of the interaction effect for **regard**.

Pairwise Comparisons

Dependent Variable	organization_structure	(I) business_cycle	(J) business_cycle	Mean Difference (I-J)	Std. Error	Sig.[a]	95% Confidence Interval for Difference[a] Lower Bound	Upper Bound
regard	1 hierarchical	1 bankruptcy	2 steady state	-13.429*	1.171	.000	-16.241	-10.618
			3 rapid expansion	-21.586*	1.107	.000	-24.242	-18.930
		2 steady state	1 bankruptcy	13.429*	1.171	.000	10.618	16.241
			3 rapid expansion	-8.157*	.756	.000	-9.972	-6.341
		3 rapid expansion	1 bankruptcy	21.586*	1.107	.000	18.930	24.242
			2 steady state	8.157*	.756	.000	6.341	9.972
	2 egalitarian	1 bankruptcy	2 steady state	-12.988*	.911	.000	-15.175	-10.802
			3 rapid expansion	-27.833*	1.147	.000	-30.586	-25.081
		2 steady state	1 bankruptcy	12.988*	.911	.000	10.802	15.175
			3 rapid expansion	-14.845*	1.136	.000	-17.573	-12.117
		3 rapid expansion	1 bankruptcy	27.833*	1.147	.000	25.081	30.586
			2 steady state	14.845*	1.136	.000	12.117	17.573

Based on estimated marginal means

a. Adjustment for multiple comparisons: Bonferroni.
*. The mean difference is significant at the

FIGURE 56.14 Simple effects tests comparing the three levels of **business_cycle** for each type of **organization_structure**.

of **business_cycle**. Although the Bonferroni correction is taken into account in reporting the probability (**Sig.**) levels, the variances are not homogeneous and we should probably make the alpha level even more stringent. We will use .01 for our evaluation.

We learn from these **Pairwise Comparisons** that there are statistically significant differences between **hierarchical** and **egalitarian** organizations under **bankruptcy** and **steady state** but not under **rapid expansion** cycles. Translated to the plot in Figure 56.13, employees in both **hierarchical** (solid line) organizations exhibited higher levels of self-regard under **bankruptcy** and **steady state** cycles but (even though the **egalitarian**

Pairwise Comparisons

Dependent Variable	business_cycle	(I) organization_structure	(J) organization_structure	Mean Difference (I-J)	Std. Error	Sig.[a]	95% Confidence Interval for Difference[a] Lower Bound	Upper Bound
regard	1 bankruptcy	1 hierarchical	2 egalitarian	4.072*	1.200	.001	1.717	6.428
		2 egalitarian	1 hierarchical	-4.072*	1.200	.001	-6.428	-1.717
	2 steady state	1 hierarchical	2 egalitarian	4.513*	.873	.000	2.798	6.228
		2 egalitarian	1 hierarchical	-4.513*	.873	.000	-6.228	-2.798
	3 rapid expansion	1 hierarchical	2 egalitarian	-2.175*	1.049	.039	-4.235	-.115
		2 egalitarian	1 hierarchical	2.175*	1.049	.039	.115	4.235

Based on estimated marginal means

*. The mean difference is significant at the

a. Adjustment for multiple comparisons: Bonferroni.

FIGURE 56.15 Simple effects tests comparing the two levels of **organization_structure** for each type of **business_cycle**.

neoextra

Ryan-Einot-Gabriel-Welsch Range[a]

business_cycle	N	Subset 1	2	3
1 bankruptcy	145	44.0397		
2 steady state	229		53.3882	
3 rapid expansion	254			57.2991
Sig.		1.000	1.000	1.000

Means for groups in homogeneous subsets are displayed.
Based on observed means.
The error term is Mean Square(Error) = 121.500.

a. Alpha = .005

FIGURE 56.16
The results of the **R-E-G-W-Q** post hoc tests for **neoextra**.

employees seem to have a higher average score) were not significantly different in their feelings of self-regard in businesses that were under **rapid expansion**.

Having finished with the interaction effect associated with **regard**, we can look into the main effects for the other two dependent variables (in the summary table shown in Figure 56.12). Against our stringent alpha level of .005, only **neoextra** yields a statistically significant main effect of **business_cycle**. The eta square associated with this effect is computed by dividing its sum of squares (**5272.344**) by its **Corrected Total** (**100119.778**), which is .053. The results of the **R-E-G-W-Q** post hoc tests for **neoextra** are presented in Figure 56.16. All groups appear to be significantly different from each other.

For the main effect of **organization_structure** (see Figure 56.12), only one (**neoextra**) of the two dependent variables not subsumed in the interaction yielded a statistically significant effect. With only two levels of **organization_structure**, we can directly interpret the mean difference. The **Descriptive Statistics** table shown in Figure 56.8 provides us with these two means in the fourth main row labeled **Total** in the **neoextra** (top third) of the table. On average, employees in **hierarchical** structures (M = **57.2991**, SD = **10.47856**) exhibited higher levels of self-regard than did those in **egalitarian** structures (M = **52.8115**, SD = **12.63648**).

PART 17

MULTIDIMENSIONAL SCALING

CHAPTER 57

Multidimensional Scaling: Classical Metric

57.1 OVERVIEW

Chapters 57 and 59 cover multidimensional scaling (MDS), a statistical technique that is used to describe the dimensional structure underlying a set of objects or stimuli. The technique was introduced, developed, and popularized by Kruskal (1964), Shepard (1962), and Torgerson (1952, 1958). Readable descriptions of MDS can be found in Davison (1992), Giguère (2007), Kruskal and Wish (1978), Meyers et al. (2013), and Young and Harris (2012).

The primary focus of MDS is to assess the ways in which objects or stimuli are dissimilar to each other. Dissimilarity can be thought of in terms of the relative distance (the proximity) of objects to each other. Objects that are further apart are more dissimilar. In MDS, objects are arranged in multidimensional space as a function of their dissimilarity. This space is defined by the number of dimensions it is presumed by the researchers to contain (usually two or three).

The MDS procedure is typically performed on (very roughly) about a dozen or so objects or stimuli that are evaluated in pairs. For each pair, the distance between the objects is determined. In some contexts, distance can be assessed on some objective metric (e.g., mileage between cities). In other contexts, and the one we illustrate in this chapter, ratings are made on a summative response scale indicating the degree of perceived dissimilarity between pairs of stimuli. For example, respondents could be asked to rate the similarity or dissimilarity of a group of automobile brands.

In practice, MDS places more similar stimuli closer together in a multidimensional space configuration. MDS maps each object as a point in two (or three) multidimensional space. These dimensionality data are typically arrayed in the form of a *proximity* or *dissimilarity matrix*. Once these distances between objects are determined, IBM SPSS® uses a special MDS algorithm called *ALSCAL* (Alternating Least Squares Scaling) developed by Takane, Young, and de Leeuw (1977). This procedure locates the coordinates for each object and depicts them in the multidimensional space such that the distance between points resembles respondents' original dissimilarity judgments.

We focus here on *classical metric MDS* (CMDS). This approach uses a single matrix of metric dissimilarities (or a single matrix that reflects averages across participants) to perform the analysis. Using this single matrix, IBM SPSS initiates the ALSCAL algorithm to create a multidimensional space based on the number of dimensions specified by the researchers.

Performing Data Analysis Using IBM SPSS®, First Edition.
Lawrence S. Meyers, Glenn C. Gamst, and A. J. Guarino.

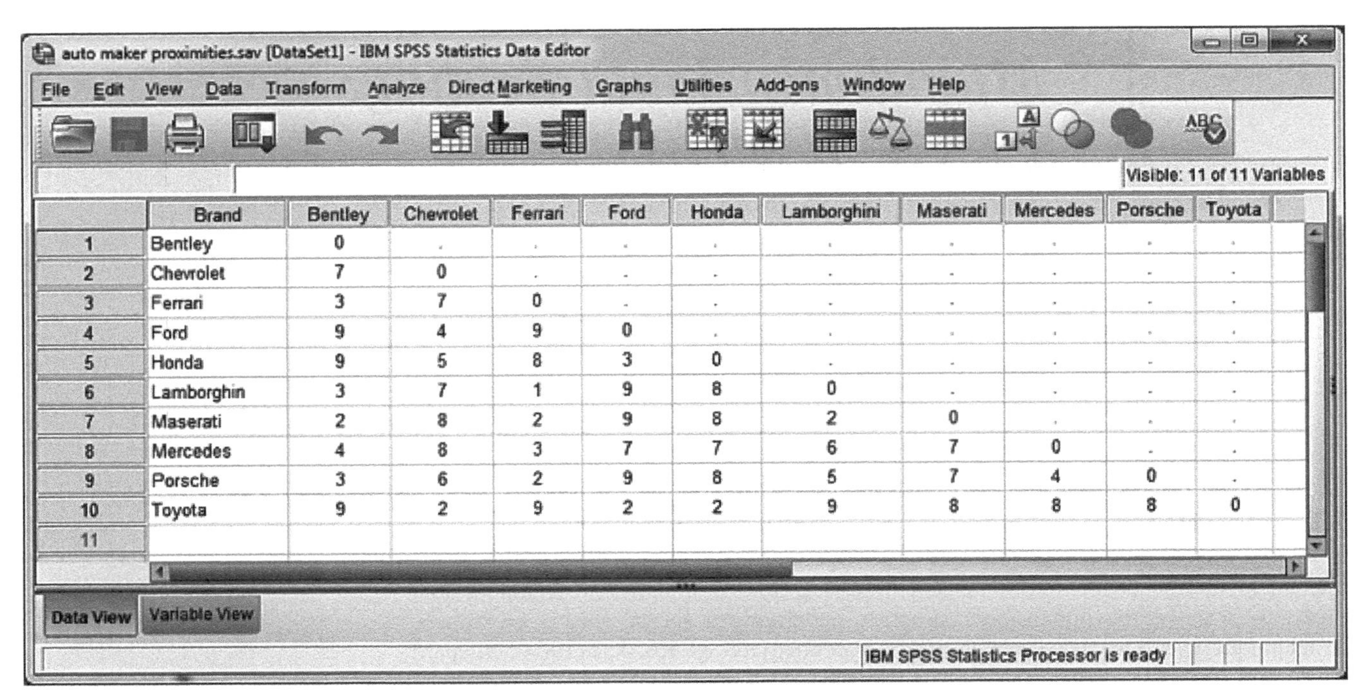

	Brand	Bentley	Chevrolet	Ferrari	Ford	Honda	Lamborghini	Maserati	Mercedes	Porsche	Toyota
1	Bentley	0	.	.	.	.	.	.	.	.	.
2	Chevrolet	7	0	.	.	.	.	.	.	.	.
3	Ferrari	3	7	0	.	.	.	.	.	.	.
4	Ford	9	4	9	0	.	.	.	.	.	.
5	Honda	9	5	8	3	0	.	.	.	.	.
6	Lamborghin	3	7	1	9	8	0	.	.	.	.
7	Maserati	2	8	2	9	8	2	0	.	.	.
8	Mercedes	4	8	3	7	7	6	7	0	.	.
9	Porsche	3	6	2	9	8	5	7	4	0	.
10	Toyota	9	2	9	2	2	9	8	8	8	0
11											

FIGURE 57.1 The proximity data.

57.2 NUMERICAL EXAMPLE

The present data set represents the average ratings of 15 respondents who compared the similarity/dissimilarity of 10 automobile brands on a 1 (*very similar*) to 9 (*very dissimilar*) summative response scale. The data can be found in the file named **auto maker proximities** and can be seen in Figure 57.1.

Note that the data are structured quite differently from what we have seen in the previous chapters. Because MDS operates on proximity data between pairs of objects, the data must be entered in that form. Thus, we directly enter a proximity matrix into the data file. For example, the average rating for the pair **Chevrolet–Bentley** was **7**; thus, respondents viewed these auto brands as relatively dissimilar on average. On the other hand, the pair **Toyota–Honda** yielded an average of **2**; thus, respondents viewed these auto brands as relatively similar. Because the data are depicted by means of a symmetric matrix, ratings above the matrix diagonal do not need to be entered, as they are mirror images of the lower diagonal counterparts.

57.3 ANALYSIS SETUP

Open the file named **auto maker proximities** and from the main menu select **Analyze → Scale → Multidimensional Scaling (ALSCAL)**. This produces the **Multidimensional Scaling** main dialog window shown in Figure 57.2.

We have moved the automobile brands to the **Variables** panel. In the **Distances** panel, we have kept the default **Data are distances** with its default setting at **Square symmetric** for **Shape** because our data are distances between objects (auto brands) and the shape of the distances matrix is square and symmetric (see Meyers et al. (2013) for more details on this topic).

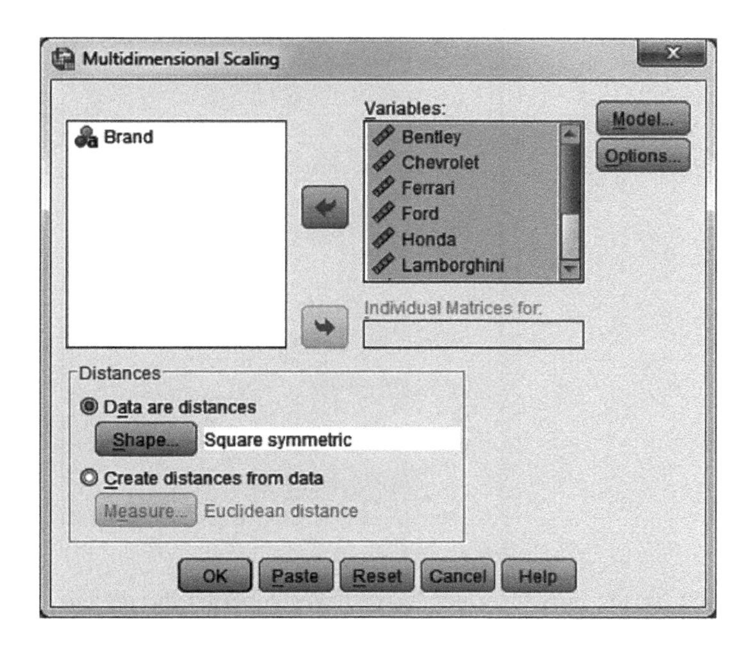

FIGURE 57.2

The main **Multidimensional Scaling** dialog window.

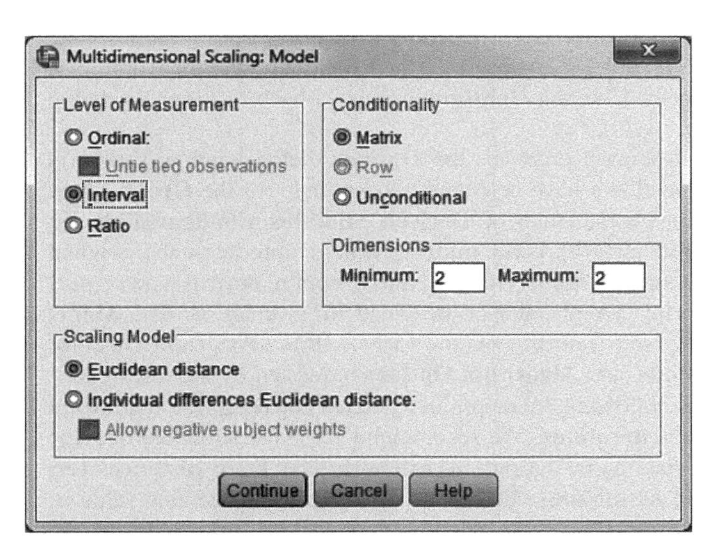

FIGURE 57.3

The **Model** window of **Multidimensional Scaling**.

Selecting the **Model** pushbutton produces the **Model** dialog window shown in Figure 57.3. This dialog screen contains four separate panels. The **Level of Measurement** panel (upper left of screen) indicates the measurement level of the data: **Ordinal, Interval**, or **Ratio**. For the present example, we have activated the **Interval** level for our summative response measure, as we will presume that our ratings approximate this level of measurement.

The **Conditionality** panel provides three options: **Matrix** (for data measured on the same measurement scale), **Row** (for rectangular data matrices where the data cannot be compared with data in other matrices), and **Unconditional** (unconditional data matrices can be compared with each other). The default **Matrix** option is selected in the present example.

The **Dimensions** panel allows researchers to specify the **Minimum** and **Maximum** number of dimensions (solutions) they wish to obtain. IBM SPSS allows for one to

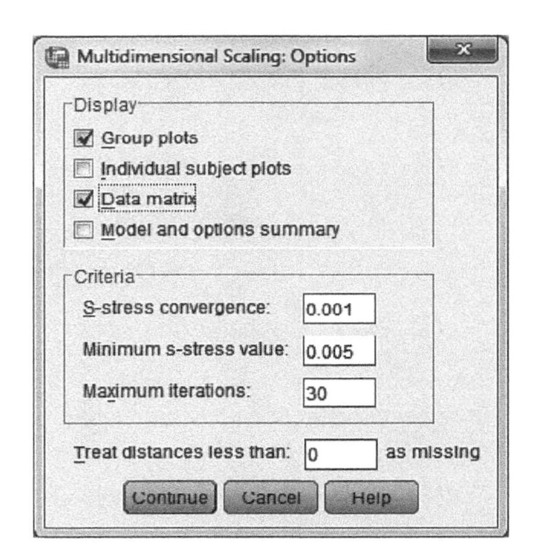

FIGURE 57.4

The **Options** window of **Multidimensional Scaling**.

six dimensions to be specified. In the present example, we have requested the default **Minimum** and **Maximum** of **2** dimensions.

Finally, the **Scaling Model** panel offers either a **Euclidean distance** (appropriate for CMDS analyses) or an **Individual differences Euclidean distance** option (appropriate for *replicated MDS* and *weighted MDS* analyses). The **Euclidean distance** option is activated for the present analysis. Clicking **Continue** brings us back to the main dialog window.

Selecting the **Options** pushbutton produces the **Options** dialog window shown in Figure 57.4. In the **Display** panel, we have activated two options: (a) the **Group plots**, which produces the CMDS perceptual map or **Derived Stimulus Configuration** and the **Scatterplot of Linear Fit**, and (b) **Data matrix**, which reproduces the original data matrix. The **Individual subject plots** (not activated) option provides perceptual maps for each participant during a replicated or weighted MDS analysis. The **Model and options summary** (not activated) documents the various IBM SPSS **Data Options**, **Model Options**, **Output Options**, and **Algorithm Options** requested by the researchers.

The **Criteria** panel has the following three options: **S-stress convergence**, **Minimum S-stress value**, and **Maximum iterations**. We recommend for most situations leaving these values in their default state as in the present example. The **Treat distances less than [a to-be-filled-in value] as missing** option was also kept at its default value of **0**. Clicking **Continue** returns us to the main window and clicking **OK** performs the analysis.

57.4 ANALYSIS OUTPUT

Figure 57.5 displays the **Raw (unscaled) Data for Subject 1**. Note that these data are the original dissimilarity data averaged across our 15 respondents, and so **Subject 1** here (in the title of the table) is the only proximity matrix in the data file.

Figure 57.6 provides the **Iteration history for the 2 dimensional solution (in squared distances)**. This table provides two measures of model fit known as Young's **S-stress** and Kruskal's **Stress Index**. **Stress**, as developed by Kruskal (1964), is the difference between the input raw disparities and the output distances in the multidimensional map and is an index of the fit of solution to the data. **Stress** varies from a minimum value of 0 (the dimensional structure perfectly fits the data) to a maximum value of 1 (the dimensional structure does not fit the data). According to Kruskal and

Raw (unscaled) Data for Subject 1

	1	2	3	4	5	6	7	8	9	10
1	.000									
2	7.000	.000								
3	3.000	7.000	.000							
4	9.000	4.000	9.000	.000						
5	9.000	5.000	8.000	3.000	.000					
6	3.000	7.000	1.000	9.000	8.000	.000				
7	2.000	8.000	2.000	9.000	8.000	2.000	.000			
8	4.000	8.000	3.000	7.000	7.000	6.000	7.000	.000		
9	3.000	6.000	2.000	9.000	8.000	5.000	7.000	4.000	.000	
10	9.000	2.000	9.000	2.000	2.000	9.000	8.000	8.000	8.000	.000

FIGURE 57.5 The proximity matrix from the data file.

```
Iteration history for the 2 dimensional solution (in squared distances)

                    Young's S-stress formula 1 is used.

                  Iteration      S-stress       Improvement

                      1          .17068
                      2          .15548           .01520
                      3          .15445           .00103
                      4          .15434           .00012

                          Iterations stopped because
                  S-stress improvement is less than    .001000

                                        Stress and squared correlation (RSQ) in distances

                              RSQ values are the proportion of variance of the scaled data (disparities)
                                         in the partition (row, matrix, or entire data) which
                                          is accounted for by their corresponding distances.
                                            Stress values are Kruskal's stress formula 1.


                  For  matrix
    Stress   =    .15922         RSQ =   .88048
```

FIGURE 57.6 Iteration history for the 2 dimensional solution and **Stress** levels.

Wish (1978) and Giguère (2007), **Stress** values less than .05 are considered *excellent*, .05 to less than.10 are considered *good*, .10 to less than .20 are *fair*, and values greater than .20 are considered a *poor* fit. The **Stress** value found at the bottom of Figure 57.6 is **Stress** = **.15922** indicating a fair model fit to the original proximity data.

At the top of Figure 57.6 is a second fit index known as **S-stress** (developed by Takane et al. (1977)). **S-stress** is derived from the **Stress** measure and differs only in that it is defined by squared distances and disparities. IBM SPSS provides an **Iteration history** that depicts for each iteration (up to a maximum of 30) the **S-stress** value and its improvement over the previous iteration. From Figure 57.6, we note **S-stress** began with a value of **.17068** and improved slightly after **4** iterations to a value of **.15434**, indicating a fair or modest model fit (it is interpreted in the same way as Kruskal's **Stress Index**).

One last measure of model fit, the **squared correlation index** (labeled by IBM SPSS as **RSQ**) is also provided at the bottom of Figure 57.6. **RSQ** indicates the amount of input data variance accounted for by the MDS model. An acceptable fit is typically indicated by RSQ values of .60 or greater. The present RSQ value of **.88048** indicates a reasonable fit to our data.

Figure 57.7 displays the **Optimally scaled data (disparities) for subject 1**. Recall that **subject 1** refers here to the data file consisting of only one matrix. This matrix of

Optimally scaled data (disparities) for subject 1

	1	2	3	4	5	6	7	8	9	10
1	.000									
2	2.271	.000								
3	.905	2.271	.000							
4	2.954	1.246	2.954	.000						
5	2.954	1.588	2.612	.905	.000					
6	.905	2.271	.222	2.954	2.612	.000				
7	.563	2.612	.563	2.954	2.612	.563	.000			
8	1.246	2.612	.905	2.271	2.271	1.929	2.271	.000		
9	.905	1.929	.563	2.954	2.612	1.588	2.271	1.246	.000	
10	2.954	.563	2.954	.563	.563	2.954	2.612	2.612	2.612	.000

FIGURE 57.7 **Optimally scaled** data matrix.

Stimulus Coordinates

Stimulus Number	Stimulus Name	Dimension 1	Dimension 2
1	Bentley	1.2992	.0007
2	Chevrole	-1.1196	.4061
3	Ferrari	1.1947	-.0900
4	Ford	-1.7656	-.1546
5	Honda	-1.5217	.0377
6	Lamborgh	1.1654	.6136
7	Maserati	1.0283	1.0539
8	Mercedes	.5274	-1.1987
9	Porsche	.8480	-.9246
10	Toyota	-1.6561	.2560

FIGURE 57.8
Stimulus Coordinates used for the plot for the perceptual map.

data represents the raw data that was transformed by means of a least-squares algorithm employed by the IBM SPSS MDS procedure.

Figure 57.8 provides the **Stimulus Coordinates** for **Dimensions 1** and **2** for each automobile manufacturer. These coordinates are used by the IBM SPSS MDS procedure to locate or position each object (auto brand) in a two-dimensional space configuration.

The **Derived Stimulus Configuration** or *perceptual map* of the auto manufacturers/models can be seen in Figure 57.9. Objects that are closer to each other are perceived as more similar. For example, we can see that Maserati and Lamborghini vehicles and likewise Toyota and Honda vehicles are both plotted relatively close together, indicating a substantial amount of perceived similarity between Maserati and Lamborghini and also between Toyota and Honda. Conversely, there is considerable distance between Maserati and Mercedes, indicating a considerable perceived dissimilarity between these two makes of cars.

Researchers are tasked with the challenge of imposing meaning to the dimensional configuration by naming the separate dimensions. **Dimension 1**, the horizontal dimension, separates Toyota, Honda, Ford, and Chevrolet (on the left side of the map) from a grouping of Maserati, Lamborghini, Bentley, Ferrari, Porsche, and Mercedes. **Dimension 1** appears to represent a dimension of *value* or the level of exoticness, with relatively inexpensive (or less exotic) vehicles at the negative (left) end of **Dimension** 1 and expensive (or more exotic) autos at the positive (right) end.

The interpretation of **Dimension 2**, the vertical dimension, is less clear. One possibility is that it may represent a *performance* dimension, with expensive high performance auto brands at the positive end of the dimension and other manufacturers clustered around the low or negative end of the dimension. Another possibility is that this dimension may represent the perceived level of excitement involved in driving such vehicles. For

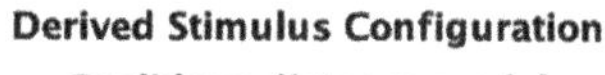

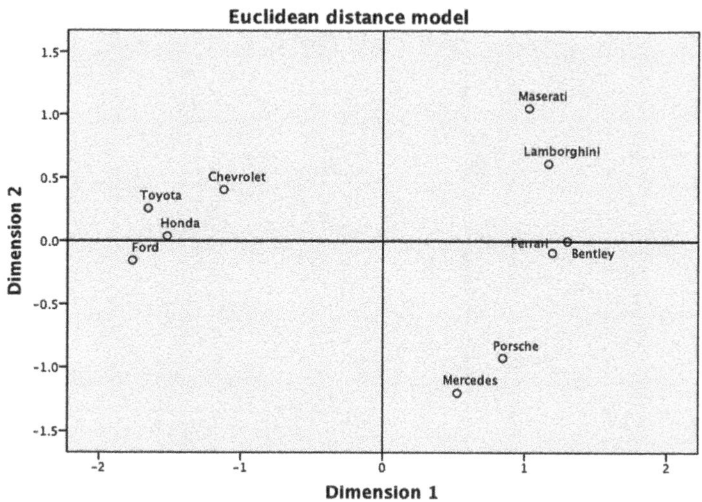

FIGURE 57.9

The perceptual map showing the dimensions representing the solution.

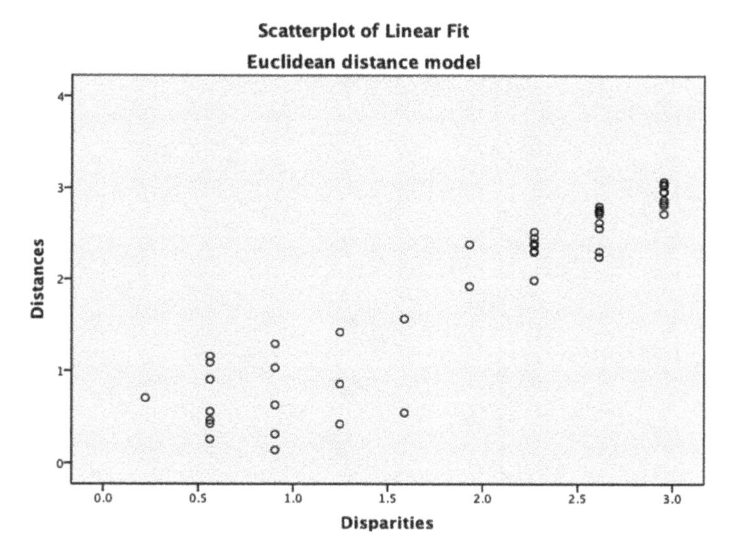

FIGURE 57.10

The **Scatterplot of Linear Fit**.

example, Maserati and Lamborghini brands may be seen as more exciting vehicles or that driving them may result in a more exciting experience, whereas Mercedes and Porches are much less exciting (with our apologies to Porsche owners).

Figure 57.10 presents the **Scatterplot of Linear Fit**. As indicated in its title, it is based on **Euclidean distance** (as we specified in the **Model** dialog window). This figure is sometimes referred to as the *Shepard diagram*. It has on the horizontal axis the original raw distance data (labeled as **Disparities**) and on the vertical axis the Euclidean distances between pairs of objects. Ideally, we would like to see a strong linear relationship (from lower left to upper right of the figure). The present scatterplot indicates some degree of error for this model (and is to be expected, given our **Stress** level of **.15922** indicated that our model represented only a fair fit to the original proximity data).

CHAPTER 58

Multidimensional Scaling: Metric Weighted

58.1 OVERVIEW

This chapter provides a brief overview of weighted multidimensional scaling (WMDS), also known as individual differences scaling (INDSCAL). WMDS extends MDS to multiple matrices of dissimilarity data (i.e., one for each respondent) under the assumption that the multidimensional space (perceptual map) is *different* for each matrix (or respondent). WMDS allows for differences (typically cognitive or perceptual) among the respondents (exhibited in their dissimilarity matrices) through the IBM SPSS® ALSCAL algorithm (see Carroll & Chang (1970) and Takane et al. (1977)), which builds the usual (unweighted) stimulus space configuration as well as a weighted stimulus space.

The *unweighted* space is shared or is common across individuals or matrices (Young & Harris, 2012). The *weighted* or *personal stimulus space* contains information that is *unique* to the individual. Weights vary from 0.0 to 1.0 for each individual on each dimension of the stimulus; larger weights indicate greater importance of the dimension, and smaller weights indicate less importance to the individual (Meyers et al., 2013; Young & Harris, 2012).

58.2 NUMERICAL EXAMPLE

The present data set represents 12 students from one of our universities who were asked to make dissimilarity judgments on 10 pairs of the following automobile manufacturers: Chevrolet, Ferrari, Ford, Honda, Lamborghini, Maserati, Mercedes, Porsche, Rolls Royce, and Toyota. Judgments were made on a summative response scale with anchors of 1 (*very similar*) to 9 (*very dissimilar*). Respondents were given separate randomized lists of 45 (n $(n-1)/2$ or $10\ (10-1)/2 = 90/2 = 45$) paired automobile manufacturers on which to make dissimilarity judgments.

The data can be found in the file named **auto maker proximities 12 raters** and can be partially seen in Figure 58.1. This data file includes 12 individual symmetric matrices embedded in one file (only the first of which is fully displayed in Figure 58.1). Data must be submitted to IBM SPSS in matrix form to run a WMDS. Only the bottom portion of each matrix and the diagonal elements are entered.

Performing Data Analysis Using IBM SPSS®, First Edition.
Lawrence S. Meyers, Glenn C. Gamst, and A. J. Guarino.

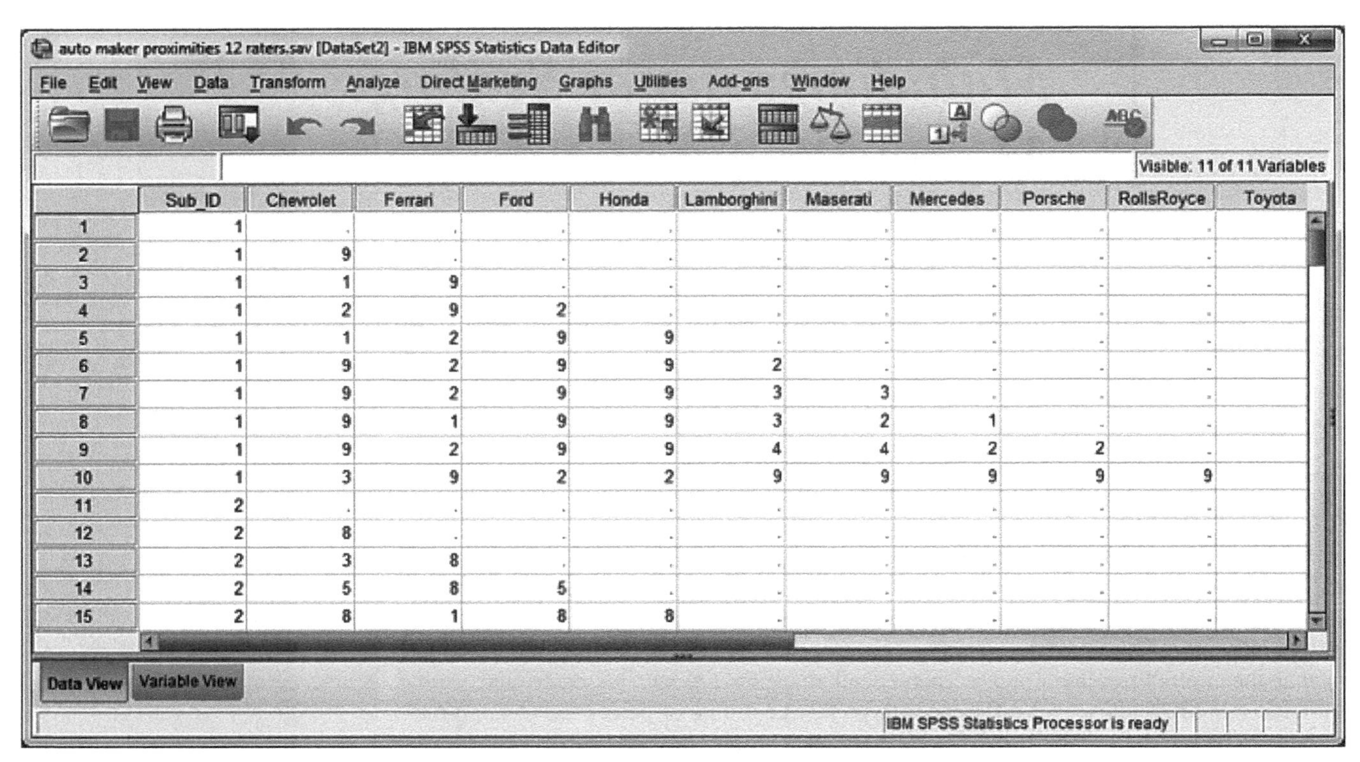

	Sub_ID	Chevrolet	Ferrari	Ford	Honda	Lamborghini	Maserati	Mercedes	Porsche	RollsRoyce	Toyota
1	1	.	.	.	.	.	.	.	.	.	
2	1	9	.	.	.	.	.	.	.	.	
3	1	1	9	.	.	.	.	.	.	.	
4	1	2	9	2	.	.	.	.	.	.	
5	1	1	2	9	9	.	.	.	.	.	
6	1	9	2	9	9	2	.	.	.	.	
7	1	9	2	9	9	3	3	.	.	.	
8	1	9	1	9	9	3	2	1	.	.	
9	1	9	2	9	9	4	4	2	2	.	
10	1	3	9	2	2	9	9	9	9	9	
11	2	.	.	.	.	.	.	.	.	.	
12	2	8	.	.	.	.	.	.	.	.	
13	2	3	8	.	.	.	.	.	.	.	
14	2	5	8	5	.	.	.	.	.	.	
15	2	8	1	8	8	.	.	.	.	.	

FIGURE 58.1 A portion of the data file.

Sub_ID, the first variable in the file, provides a subject number identifier for each line of each dissimilarity matrix. Thus, **Subject 1** has a **1** at the beginning of the 10 lines of data that comprise his or her symmetric matrix. This **Sub_ID** variable serves no functional purpose because IBM SPSS already "knows" the beginning and end points of each of the 12 symmetric matrices (Giguère, 2007), but it does serve a useful purpose in terms of data entry verification.

The remaining 10 variables in the columns of the file represent the names of the auto manufacturers. The data represent paired comparisons. For example, for **Subject 1**, the first comparison is left blank because it represents a comparison between **Chevrolet** and itself. The next comparison is between **Chevrolet** and **Ferrari** and received a **9** (indicating *very dissimilar*). The next ratings were between **Chevrolet** and **Ford** and **Ferrari** and **Ford** and were rated **1** and **9**, respectively. This indicates an appraisal of *very similar* in the first comparison and *very dissimilar* in the second comparison.

58.3 ANALYSIS SETUP

Open the file named **auto maker proximities 12 raters** and from the main menu select **Analyze ➔ Scale ➔ Multidimensional Scaling (ALSCAL)**. This produces the **Multidimensional Scaling** main dialog window seen in Figure 58.2.

We have moved the automobile brands to the **Variables** panel. In the **Distances** panel, we have kept the default **Data are distances** with its default setting at **Square symmetric** for **Shape** because our data are distances between objects (auto brands) and the shape of the distances matrix is square and symmetric.

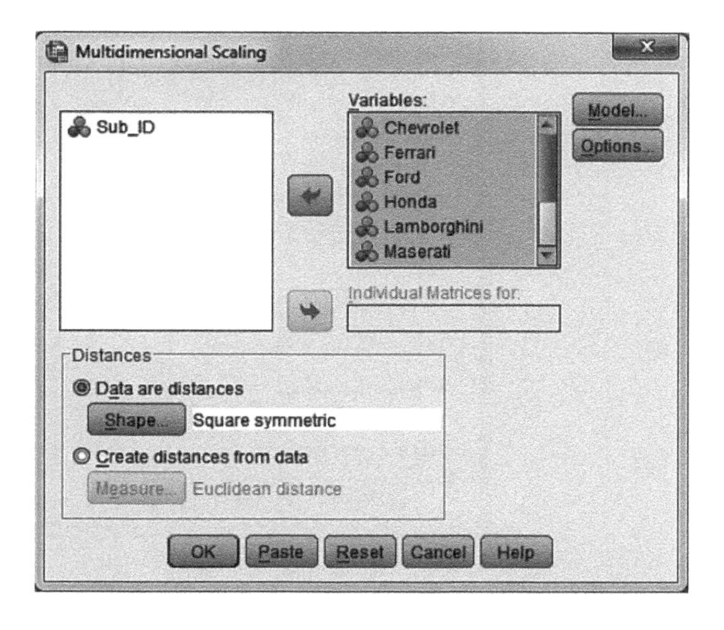

FIGURE 58.2

The main **Multidimensional Scaling** dialog window.

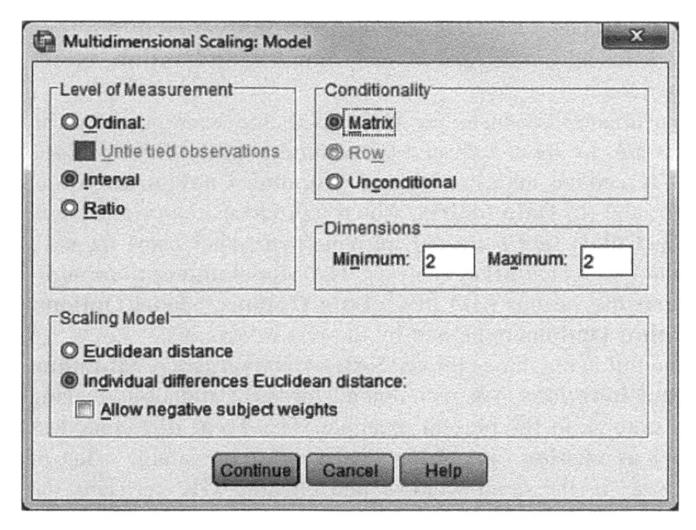

FIGURE 58.3

The **Model** window of **Multidimensional Scaling**.

Selecting the **Model** pushbutton produces the **Model** dialog window shown in Figure 58.3. This dialog screen contains four separate panels. The **Level of Measurement** panel (upper left of screen) indicates the measurement level of the data: **Ordinal, Interval**, or **Ratio**. For the present example, we have activated the **Interval** level for our summative response measure based on the presumption that our ratings approximate this level of measurement.

The **Conditionality** panel provides three options: **Matrix** (for data measured on the same measurement scale), **Row** (for rectangular data matrices where the data cannot be compared with data in other matrices), and **Unconditional** (unconditional data matrices can be compared with each other). The default **Matrix** option is selected in the present example.

The **Dimensions** panel allows researchers to specify the **Minimum** and **Maximum** number of dimensions (solutions) they wish to obtain. IBM SPSS allows for one to six dimensions to be specified. In the present example, we have requested the default **Minimum** and **Maximum** of **2** dimensions.

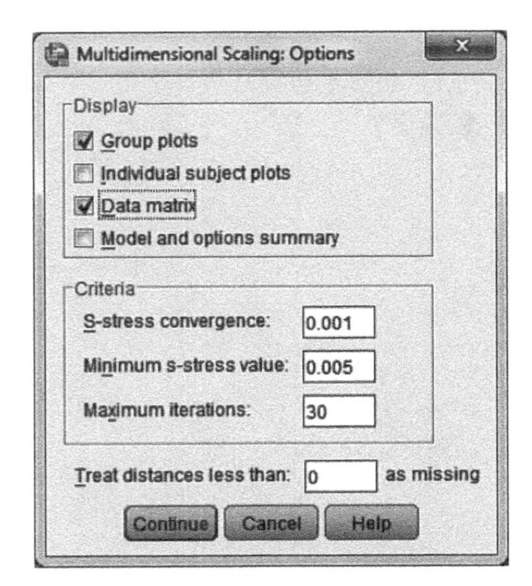

FIGURE 58.4
The **Options** window of **Multidimensional Scaling**.

Finally, the **Scaling Model** panel offers either a **Euclidean distance** (appropriate for CMDS analyses) or an option for **Individual differences Euclidean distance** (appropriate for *replicated MDS* and *weighted MDS* analyses. The **Individual differences Euclidean distance** option is activated for the present analysis. Clicking **Continue** brings us back to the main dialog window.

Selecting the **Options** pushbutton produces the **Options** dialog window shown in Figure 58.4. In the **Display** panel, we have activated two options: (a) the **Group plots**, which produces the WMDS perceptual map or **Derived Stimulus Configuration** and the **Scatterplot of Linear Fit**, and (b) **Data matrix**, which reproduces the original data matrix. The **Individual subject plots** (not activated) provide perceptual maps for each participant during a replicated or weighted MDS analysis. The **Model and options summary** (not activated) documents the various IBM SPSS **Data Options**, **Model Options**, **Output Options**, and **Algorithm Options** requested by the researchers.

The **Criteria** panel has the following three options: **S-stress convergence**, **Minimum S-stress value**, and **Maximum iterations**. We recommend for most situations leaving these values in their default state as in the present example. The **Treat distances less than [a to-be-filled-in value] as missing** option was also kept at its default value of **0**. Clicking **Continue** returns us to the main window and clicking **OK** performs the analysis.

58.4 ANALYSIS OUTPUT

The output begins with the **Raw (unscaled) Data for Subject 1-12** (not shown as it is a transcription of the values in the data file). Figure 58.5 provides the **Iteration history** where convergence occurred after five iterations. The analysis yielded a **Young's S-stress** value of **.29889**, indicating a relatively poor fit. The result is followed toward the bottom of the figure by separate **Stress** and **RSQ** values for each of the 12 respondents or matrices (see Chapter 57 for details on these fit indexes). For example, **Subject 1** (represented in the output as **Matrix 1**) had a **Stress** value of **.240** and an **RSQ** value of **.865**, etc. The **Stress** values range from **.156** to **.381** and the **RSQ** values range from **.162** to **.917**. The average Kruskal's **Stress** value (collapsed across all 12 matrices) was **.25572** and the average **RSQ** was **.69504**, both of which indicate a fair to poor model fit.

Figure 58.6 presents the **Subject Weights** table and the plot of those weights. The values in the table indicate the relative importance of each dimension to each respondent.

```
Iteration history for the 2 dimensional solution (in squared distances)

              Young's S-stress formula 1 is used.

            Iteration     S-stress      Improvement

                0          .32581
                1          .32456
                2          .30556          .01901
                3          .30101          .00454
                4          .29950          .00152
                5          .29889          .00061

                 Iterations stopped because
            S-stress improvement is less than   .001000
```

```
                          Stress and squared correlation (RSQ) in distances

                  RSQ values are the proportion of variance of the scaled data (disparities)
                        in the partition (row, matrix, or entire data) which
                         is accounted for by their corresponding distances.
                          Stress values are Kruskal's stress formula 1.
```

Matrix	Stress	RSQ	Matrix	Stress	RSQ	Matrix	Stress	RSQ	Matrix	Stress	RSQ
1	.240	.865	2	.185	.863	3	.381	.162	4	.362	.255
5	.216	.787	6	.156	.917	7	.218	.794	8	.150	.910
9	.224	.790	10	.245	.798	11	.249	.764	12	.321	.434

```
    Averaged (rms) over  matrices
  Stress  =   .25572      RSQ =  .69504
```

FIGURE 58.5 **Iteration history for the 2 dimensional solution** and **Stress** levels.

Also displayed is the **Weirdness** index for each respondent. The **Overall importance of each dimension** across the 12 respondents is shown at the bottom of the table. In the present example, the overall importance values are **.6509** and **.0441** for Dimensions 1 and 2, respectively. Thus, in terms of the dimensionality, the respondents gave considerably more subjective importance to Dimension 1 than they did to Dimension 2. When the individual **Subject Weights** are fairly proportional to the overall weights for each dimension, the **Weirdness** index is closer to 0.00. Larger **Weirdness** index values suggest that respondents are investing somewhat different importance weights to the dimensions, and **Weirdness** index values greater than .50 may be considered atypical or outliers rendering them as possible candidates for elimination from subsequent analyses. In the present example, **Subjects 3**, **4**, and **6** meet this consideration criterion; for illustration purposes, we will retain them in the analysis.

The **Derived Subject Weights** plot depicts how each respondent's assessments of the auto brand pairings were weighted on the two dimensions of the Euclidean distance model. These are simply the coordinates for the two dimensions shown in the table above the plot. We note a small grouping of respondents (e.g., 3, 4, and 12) who weight Dimension 1 fairly low and Dimension 2 relatively high.

The **Optimally scaled data (disparities)** for each respondent are not shown due to space considerations. These disparities represent the transformed raw data based on the ALSCAL algorithm used by IBM SPSS.

Figure 58.7 presents the **Scatterplot of Linear Fit Individual differences (weighted) Euclidean distance model**. This plot shows some scatter and departure from linearity, indicating a modest but not an ideal model fit.

Figure 58.8 provides the **Flattened Subject Weights** and a plot of the **Flattened Subject Weights Individual differences (weighted) Euclidean distance model**. IBM SPSS calculates *flattened weights*. These coefficients transform raw subject weights into distance coordinates. The flattening algorithm reduces d dimensions to $d - 1$ dimensions; in the present case, this computes to $2 - 1$, or 1 dimension. It has been labeled as **Variable 1** on the Y-axis in Figure 58.8. From this plot, we note that respondents with relatively low **Weirdness** scores and symmetrical subject weights on the two dimensions

```
Subject weights measure the importance of each dimension to each subject.
Squared weights sum to RSQ.

A subject with weights proportional to the average weights has a weirdness of
zero, the minimum value.
A subject with one large weight and many low weights has a weirdness near one.
A subject with exactly one positive weight has a weirdness of one,
the maximum value for nonnegative weights.
```

Subject Weights

Subject Number	Weird-ness	Dimension 1	2
1	1.0000	.9301	.0000
2	.1480	.9121	.1757
3	.6224	.3145	.2507
4	.5346	.4264	.2714
5	.2861	.8272	.3208
6	.5549	.9540	.0847
7	.0238	.8677	.2036
8	.1893	.9388	.1691
9	.0831	.8690	.1857
10	.3416	.8848	.1226
11	.1025	.8562	.1774
12	.4694	.5775	.3177
Overall importance of each dimension:		.6509	.0441

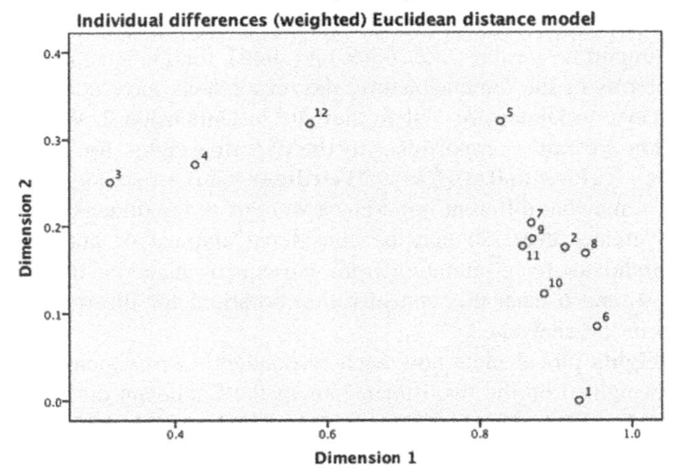

FIGURE 58.6 Importance weights for the two dimensions.

will appear in the middle of the plot (e.g., **Subjects 8**, **2**, **9**, and **7**) and subjects with relatively high **Weirdness** scores and high weights on one or the other dimensions will appear near the top or bottom of the *Y*-axis (e.g., **Subjects 1**, **6**, **4**, and **3**).

Figure 58.9 provides the **Stimulus Coordinates** for **Dimensions 1** and **2** for each automobile manufacturer. These coordinates are used by the IBM SPSS MDS procedure to locate or position each object (manufacturer) in a two-dimensional space configuration.

Figure 58.10 displays the perceptual map or **Derived Stimulus Configuration Individual differences (weighted) Euclidean distance model**. Four obvious sets of auto manufacturer clusters (one in each quadrant) can be discerned in this two-dimensional configuration. **Dimension 1** (horizontal axis) appears to represent a "Cost or Value" dimension; lower perceived cost/value is toward the right and higher perceived cost/value

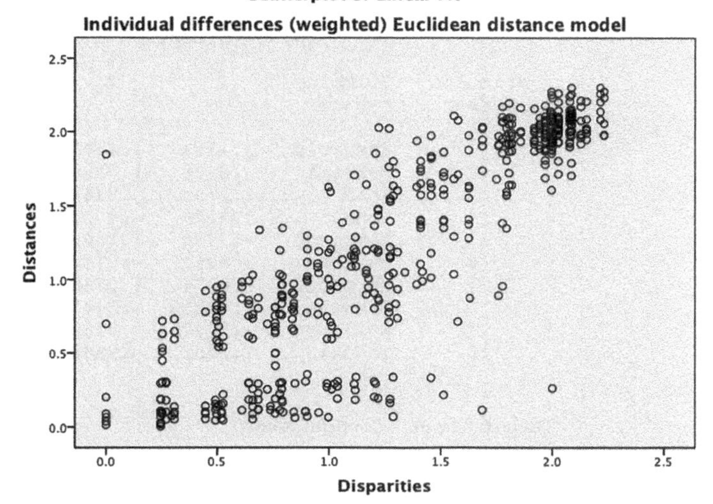

FIGURE 58.7
Scatterplot of Linear Fit Individual differences (weighted) Euclidean distance model.

Flattened Subject Weights

Subject Number	Plot Symbol	Variable 1
1	1	1.6726
2	2	.3871
3	3	-1.8575
4	4	-1.4226
5	5	-.5516
6	6	1.0236
7	7	.1602
8	8	.4580
9	9	.2714
10	A	.7043
11	B	.3067
12	C	-1.1521

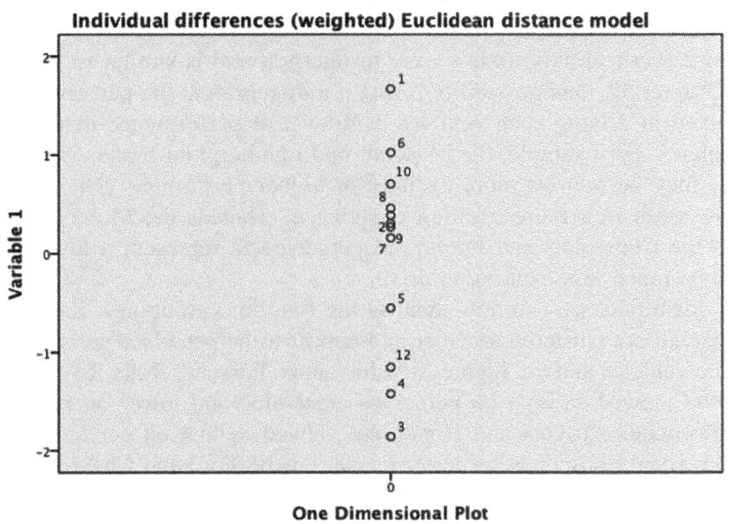

FIGURE 58.8
The **Flattened Subject Weights** and a plot of the **Flattened Subject Weights Individual differences (weighted) Euclidean distance model**.

Stimulus Coordinates

Stimulus Number	Stimulus Name	Dimension 1	Dimension 2
1	Chevrole	1.1169	1.4543
2	Ferrari	-.8306	-1.0028
3	Ford	1.2100	.8761
4	Honda	1.3148	-.8054
5	Lamborgh	-.7986	-1.2006
6	Maserati	-.8251	-.8763
7	Mercedes	-.6374	1.0748
8	Porsche	-.8455	.5441
9	RollsRoy	-.9447	.9168
10	Toyota	1.2402	-.9808

FIGURE 58.9
Stimulus Coordinates used for the plot for the perceptual map.

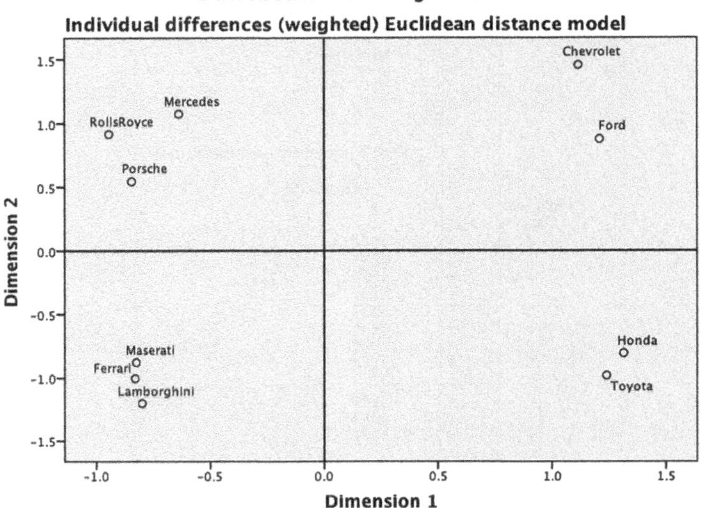

FIGURE 58.10
The perceptual map showing the dimensions representing the solution.

is on the left. **Dimension 2** (vertical axis) is less easy to interpret and is similar to the solution we obtained in Chapter 57. One possibility is that it may represent the perceived level of excitement involved in driving such vehicles or a level of performance that is associated with those vehicles. For example, the Maserati and Lamborghini brands (and the Toyotas and Hondas) may be seen as more exciting or higher performing vehicles or that driving them may result in a more exciting experience, whereas the Mercedes and Porsches (as well as the Chevrolets and Fords) are perceived to represent a lower performance level and to be much less exciting to drive.

With the dimensions identified, we can now identify the four auto groupings. Lamborghini, Ferrari, and Maserati are clustered together and appear to be perceived as high excitement or performance vehicles and are high cost/value autos. Porsche, Rolls Royce, and Mercedes appear to be viewed as high on perceived cost/value and lower on perceived excitement or performance. Toyota and Honda are viewed as low on perceived cost/value and high on perceived excitement or performance. Lastly, Ford and Chevrolet are seen as low on both the cost/value and excitement or performance dimensions. We should put in a disclaimer here that this small sample of psychology students is not necessarily representative of the general population of marketplace consumers.

PART 18

CLUSTER ANALYSIS

CHAPTER 59

Hierarchical Cluster Analysis

59.1 OVERVIEW

As is true for MDS covered in Chapters 57 and 58, cluster analysis also uses as its base the proximities or distances between cases in the process of organizing them. But unlike MDS where we attempt to align the cases along two or sometimes more dimensions, in hierarchical cluster analysis, we place the cases into groupings or clusters based on their proximities. Although hierarchical cluster analysis can be applied to large sets of cases as part of larger analyses (e.g., IBM SPSS® has a **TwoStep Cluster** procedure that uses a hierarchical analysis as its first stage), when used as a stand-alone analysis, it is almost always applied to a relatively small set of cases (e.g., one or maybe two dozen).

Cases are measured on a set of quantitative variables, and the proximities of the cases are computed with respect to those variables; if a different set of variables was used, the distances between the cases, and thus the groupings or clusters, could be quite different. Cases can be any entities that have characteristics that can be measured. Examples include people belonging to a given group, such as military personnel in a training program or female college students in a sorority; manufacturers of certain products, such as automobiles or refrigerators; and geopolitical entities, such as cities or countries.

A hierarchical clustering solution starts with a given number of cases and establishes one joining or combining of entities (called *linkage*) at each step until all cases are ultimately combined into a single cluster. Consider an example in which we cluster 12 large US cities. Each city is associated with several quantitative variables that have been gleaned from public records; these variables might include such measures as median household income, population density, and the percentage of residents with a graduate or professional degree. At the start of the process, we have 12 cities each comprising a cluster. Proximities (similarities) between each pair of cities, assessed in terms of the measured variables, are computed. In the first pass or step through the cases, the two cities that are most similar (most proximal) on the measured variables are joined or linked together to form a cluster. The clustering process is hierarchical in that once joined, these cities remain joined for the duration of the analysis.

After the first two cities have been placed into a cluster, we have 11 entities remaining; that is, we have 10 stand-alone cities and one cluster composed of two cities. The next step in the hierarchical cluster analysis evaluates the proximities (similarities) of these 11 entities, and the two entities most similar are joined together. Thus, the two-city

Performing Data Analysis Using IBM SPSS®, First Edition.
Lawrence S. Meyers, Glenn C. Gamst, and A. J. Guarino.

cluster may be joined with some other city or two stand-alone cities may be joined into their own cluster. At the end of this step, there are 10 entities, and the process is repeated, one step at a time, until all of the cities are part of a single cluster.

By virtue of the clustering algorithm, regardless of how dissimilar the entities may be, they are ultimately going to be linked together by the end of the analysis. The job of the researchers is to determine how many "viable" clusters there are. As there is no statistical test of any null hypothesis in cluster analysis, researchers must be guided by the pattern of clustering over the course of the analysis displayed in both tabular and visual forms, showing the joining process at each step as well as their knowledge of the entities in the analysis.

The clustering process is computationally complex, involving two separate types of calculations, one of *distance* and another of *linkage*, with several choices of method available for each; these are more fully described in Meyers et al. (2013). Proximities are measures of distance, and in hierarchical cluster analysis, there are several such distance measures available (the squared Euclidean distance, the Euclidean distance, city-block or the Manhattan distance, the Chebyshev distance, the Minkowski distance, and power metric distance). Distance is assessed in terms of the difference between all corresponding measures between all of the entities (e.g., calculating the difference between each pair on median household income, population density), although each method makes this determination somewhat differently.

There are several methods available to link the entities together into clusters; these include the unweighted pair-group method with arithmetic averages (also called the average linkage between-groups) method, the average linkage within-groups method, the nearest neighbor (also called the single linkage) method, the furthest neighbor (also called the complete linkage) method, the unweighted pair-group centroid method (often called the centroid method), the weighted pair-group centroid method (often called the median method), and Ward's method.

Because we calculate differences and perform other arithmetical operations on the variables, it is desirable for all of the variables to be set to the same metric before data analysis, and the most convenient way to do that is to transform the values to z scores. The **Hierarchical Clustering** procedure in IBM SPSS has a built-in procedure to accomplish this standardization if we request so.

59.2 NUMERICAL EXAMPLE

For this example, we use the **demographics 12 US cities** data file. In addition to the **City** variable containing the name of the city, the data file contains the following quantitative variables for each city: **age** (median resident age), **income** (median household income in thousands of dollars), **house_price** (mean detached home price in thousands of dollars), **rent** (median cost of rental), **cost** (cost of living index), **African_Amer** (percentage of African-Americans residing in a city), **Latino_Amer** (percentage of Latino-Americans residing in a city), **density** (population density), **higher_degree** (percentage with a graduate or professional degree), **never_marry** (percentage who have never been married), **divorced** (percentage who have been divorced), **foreign_born** (percentage who were born outside of the United States), **same_sex** (percentage of same sex households), **obesity** (adult obesity rate), and **offenders** (number of registered sex offenders).

59.3 ANALYSIS SETUP

We open the **demographics 12 US cities** data file and from the main menu select **Analyze → Classify → Hierarchical Cluster**. This opens the main dialog window shown in

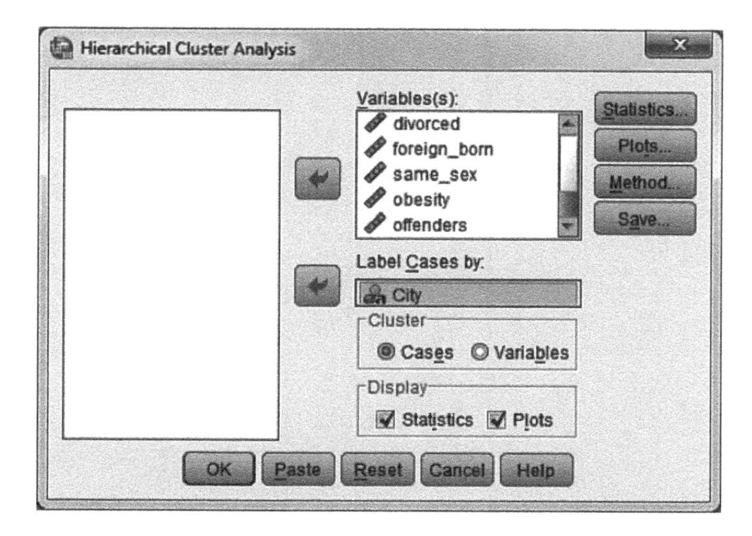

FIGURE 59.1

The main dialog window of **Hierarchical Cluster Analysis**.

Figure 59.1. We move **City** into the **Label Cases by** panel so that we will see the city names in our output. We then move all of the other variables into the **Variable(s)** panel and retain the defaults of **Cases** under **Cluster** and of **Statistics** and **Plots** under **Display**.

Selecting the **Method** pushbutton opens the **Method** dialog window shown in Figure 59.2. It is common practice to perform a hierarchical cluster analysis using various combinations of clustering methods and distance measures (to determine which one will be used to represent the clustering solution). One of the solutions that yielded a relatively parsimonious and interpretable outcome is illustrated here. For the **Cluster Method**, we select **Ward's method** from the drop-down menu, and for our distance **Measure**, we select the default of **Squared Euclidean distance**. In the **Transform Values** area of the window, we choose **Z scores** from the **Standardize** drop-down menu and retain the default **By variable**. Clicking **Continue** returns us to the main dialog window.

In the **Statistics** window (see Figure 59.3), we select **Agglomeration schedule** (to show progression of the clustering process in tabular form) and **Proximity matrix** to

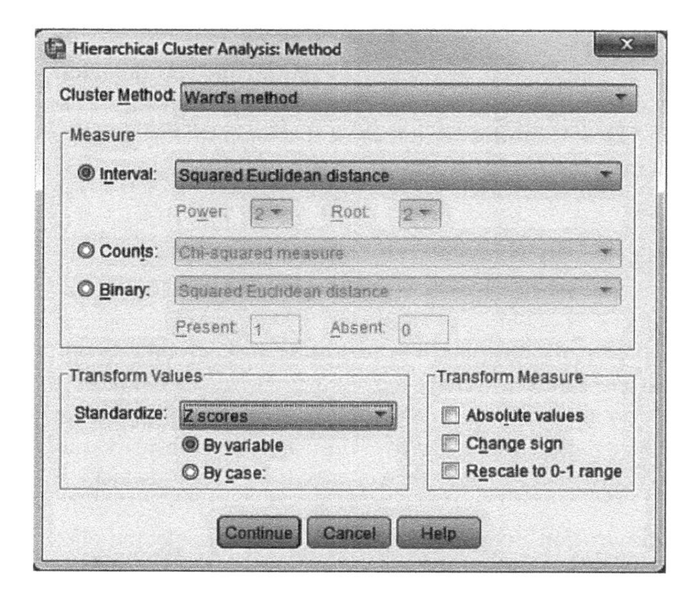

FIGURE 59.2

The **Method** dialog window of **Hierarchical Cluster Analysis**.

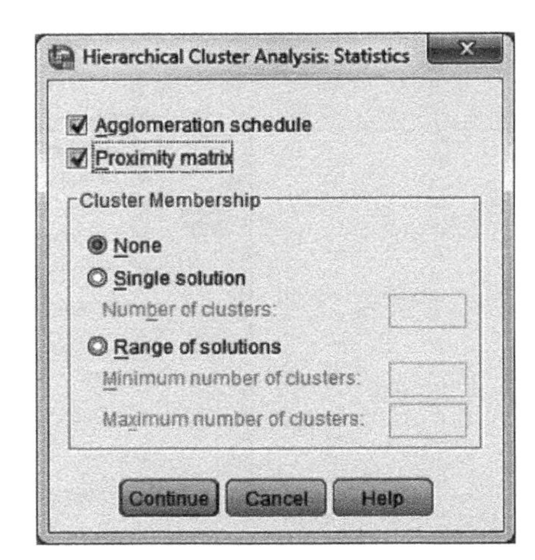

FIGURE 59.3

The **Statistics** dialog window of **Hierarchical Cluster Analysis**.

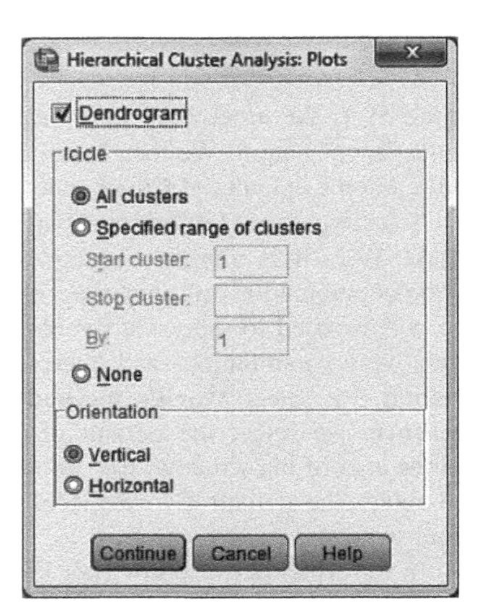

FIGURE 59.4

The **Plots** dialog window of **Hierarchical Cluster Analysis**.

see the initial distances between the cities. Click **Continue** to return to the main dialog window.

In the **Plots** window shown in Figure 59.4, we select **Dendrogram** and an **Icicle** plot for **All clusters**. In the **Orientation** area, we select **Vertical** to present the icicle plot in its most traditional form. Click **Continue** to return to the main dialog window and select **OK** to perform the analysis.

59.4 ANALYSIS OUTPUT

The **Proximity Matrix** for the cities is shown in Figure 59.5. These are the squared Euclidean distances between each pair of cities, and the two cities that are closest will be joined together in a cluster at the first clustering step. Larger values are indicative of greater distances (hence the footnote indicating that this is really a **dissimilarity matrix**). The pair of cities with the smallest value (the two closest cities as measured by the variables in our analysis) is Austin and Phoenix; their squared Euclidean distance is **6.053**.

Figure 59.6 shows the **Agglomeration Schedule** for the clusters. It tracks the clustering process using shorthand that is not necessarily intuitive, but the most important

Proximity Matrix

Case	Squared Euclidean Distance											
	1:New York	2:Chicago	3:Boston	4:Los Angeles	5:San Francisco	6:Miami	7:Atlanta	8:Seattle	9:Phoenix	10: Jacksonville	11:Austin	12:Memphis
1:New York	.000	21.162	16.019	18.173	23.588	29.643	34.814	23.994	33.213	47.121	30.104	58.948
2:Chicago	21.162	.000	19.839	11.106	50.396	26.210	12.861	21.460	11.371	17.636	8.940	24.736
3:Boston	16.019	19.839	.000	25.429	25.017	50.316	18.769	16.059	33.096	47.859	18.505	47.125
4:Los Angeles	18.173	11.106	25.429	.000	35.683	22.165	32.111	25.590	18.891	33.802	19.763	55.557
5:San Francisco	23.588	50.396	25.017	35.683	.000	62.121	46.269	18.204	62.071	76.913	50.648	98.310
6:Miami	29.643	26.210	50.316	22.165	62.121	.000	43.737	39.751	19.922	31.448	29.738	46.280
7:Atlanta	34.814	12.861	18.769	32.111	46.269	43.737	.000	12.220	20.006	19.981	11.642	17.535
8:Seattle	23.994	21.460	16.059	25.590	18.204	39.751	12.220	.000	23.125	30.165	13.718	43.786
9:Phoenix	33.213	11.371	33.096	18.891	62.071	19.922	20.006	23.125	.000	9.114	6.053	23.058
10:Jacksonville	47.121	17.636	47.859	33.802	76.913	31.448	19.981	30.165	9.114	.000	18.112	12.447
11:Austin	30.104	8.940	18.505	19.763	50.648	29.738	11.642	13.718	6.053	18.112	.000	26.557
12:Memphis	58.948	24.736	47.125	55.557	98.310	46.280	17.535	43.786	23.058	12.447	26.557	.000

This is a dissimilarity matrix

FIGURE 59.5 **Proximity Matrix** for the cities.

Ward Linkage

Agglomeration Schedule

Stage	Cluster Combined		Coefficients	Stage Cluster First Appears		Next Stage
	Cluster 1	Cluster 2		Cluster 1	Cluster 2	
1	9	11	3.026	0	0	6
2	2	4	8.579	0	0	6
3	7	8	14.689	0	0	9
4	10	12	20.913	0	0	10
5	1	3	28.922	0	0	7
6	2	9	39.374	2	1	8
7	1	5	52.906	5	0	9
8	2	6	68.707	6	0	10
9	1	7	88.046	7	3	11
10	2	10	114.548	8	4	11
11	1	2	165.000	9	10	0

FIGURE 59.6 **Agglomeration Schedule** for the clusters.

information in the table is contained under **Coefficients**. This column represents the distance at which clusters were linked (with Ward's method, it is the within-cluster sum of squares) at each step in the analysis.

It may be computationally complex but we interpret the **Coefficients** relatively simply and subjectively: larger jumps in the value of the coefficients across steps (**Stages**) suggest that, in overly simplified terms, the joining was rather "forced" (clusters that are relatively dissimilar are being joined) and that we might want to identify the number of clusters in the solution we wish to accept as falling just before a relatively large jump. In our results, we have a jump of 26.502 between **Stage 9** and **Stage 10** (**114.548** – **88.046**) and a much larger jump of 50.452 between **Stage 10** and **Stage 11** (**165.000** – **114.548**). With 12 cities in the analysis, all 12 will be joined into one cluster at **Stage 12**. Based on the increments in the coefficient values (and without peeking at which cities are linked together at each stage), it would appear that we probably would be inclined to accept the solution at the end of **Stage 11** with an outside possibility of accepting the solution at the end of **Stage 10**.

The classical output of a hierarchical cluster analysis is the vertical icicle plot as presented in Figure 59.7. It derives its name from the appearance of icicles having formed on the roof of a house, and it is read from bottom to top. A full (dark) vertical bar from the top (analysis complete) to the bottom (11 clusters identified) represents each city. For example, in the 11-cluster step (the location where the vertical bars begin at the bottom of the plot), we have 10 stand-alone cities and one cluster where two cities have been linked.

We know from the **Proximity Matrix** that the first joining is between Austin and Phoenix. Note that Austin and Phoenix occupy adjacent bars, as IBM SPSS used the information of how these cities will be sequentially linked to align them across the top of the icicle plot. Also note that to signify their joining, a vertical bar has been drawn between these two cities down to the 11-cluster mark.

Moving up to the 10-cluster position, we see that Los Angeles and Chicago have been linked. Thus, we have eight stand-alone cities and two clusters each containing two cities. Again moving up one notch to the nine-cluster position, we see that Seattle and Atlanta have been joined together.

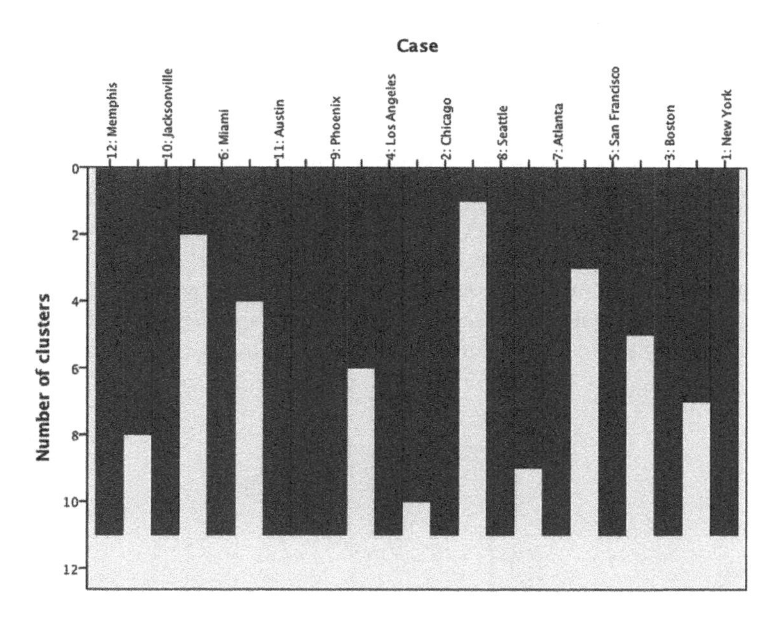

FIGURE 59.7
The vertical icicle plot.

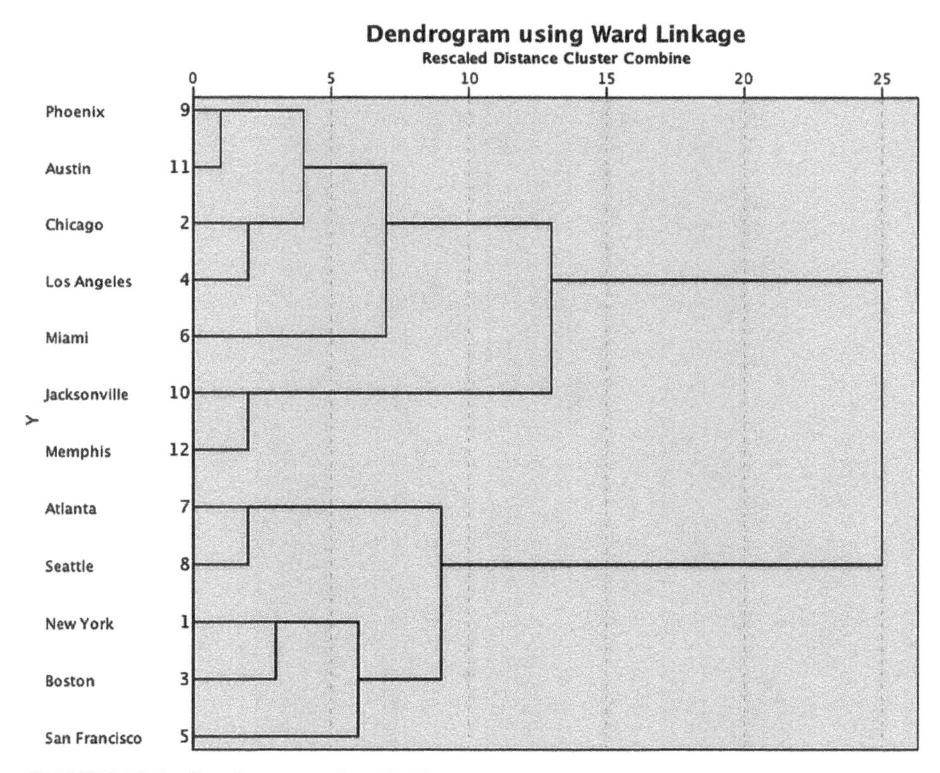

FIGURE 59.8 **Dendrogram** plot of cities.

The light spacing in the icicle plot helps us to distinguish the clusters as they grow. Just before we reach the top row, we can distinguish the jump in coefficient values that caught our attention in the **Coefficients** column of the **Agglomeration Schedule**. It appears to present us with two major clusters. One cluster (on the far right of the plot) appears to contain Seattle, Atlanta, San Francisco, Boston, and New York; the other contains the remaining cities.

Examining the next lowest position allows us to discern the alternative and less appealing solution that we highlighted in the **Agglomeration Schedule**. Here, we appear to have a set of three clusters. Again, Seattle, Atlanta, San Francisco, Boston, and New York represent one of the clusters. Another cluster appears to be on the far left of the plot and consists of Memphis and Jacksonville. The third cluster lies between these two in the plot and consists of Miami, Austin, Phoenix, Los Angeles, and Chicago.

An alternative way to view the clustering process is the **Dendrogram** plot shown in Figure 59.8. It may be somewhat more difficult to discern the three-cluster solution in this diagram but the two-cluster solution stands out strikingly clear, being almost pointed to by the rightmost pair of horizontal lines. Given the results of the **Agglomeration Schedule**, together with the **Icicle** and **Dendrogram** plots, we would conclude that the two-cluster solution best characterizes the similarities among the studied cities.

CHAPTER 60

k-Means Cluster Analysis

60.1 OVERVIEW

While hierarchical clustering discussed in Chapter 59 is most often used to group a relatively few entities on the basis of several variables, *k*-means clustering is used to form a small set of groups or clusters from relatively many entities based on a small subset of variables. As was true for hierarchical clustering, the variables should be in *z* score form at the start of the analysis; however, this must be done in advance for *k*-means clustering in the **Descriptives** procedure of IBM SPSS®. The algorithm used for the *k*-means clustering procedure was named by J. MacQueen (1967), but the version of the algorithm that is currently used by most software programs was put forward by Hartigan and Wong (1979).

The *k*-means procedure begins with the researchers specifying the number of clusters they wish to have formed; the *k* in *k*-means is the number of clusters. Because we do not know the "best" (most interpretable, most sensible, most appealing) solution at the outset, researchers almost always perform a set of analyses to generate a small range of solutions (e.g., three to five clusters); they will then choose the one that works best for their needs.

The *k*-means analysis is an *iterative* process. The first step in the analysis is the identification of *k* cases in the data file that have quite different values (are far apart) on the clustering variables. These cases will serve as *seed points* or initial cluster centers (outliers must have been screened out, as they are otherwise strong candidates to be selected as seed points and thus distort the results). In this first phase, clusters are built with a case at a time being assigned to the cluster to which it is closest. A modified centroid method uses the Euclidean distance between the centroid or center of the cluster (an average of the clustering variables) and the nonclustered entities. The smallest distance determines which entity is to be assigned to which cluster.

Once all of the cases have been assigned cluster membership, the center of the cluster (its centroid) is determined. This centroid is, simplistically, the multivariate mean of the cases assigned to the cluster based on the variables that were used in the clustering process. These centroids take the role of new seed points, the cases are "freed" from their prior cluster assignment, and a *reassignment* phase is initiated akin to what was done at the start of the procedure.

After all of the cases have been reassigned to the new clusters, centroids based on this reconfiguration are computed. IBM SPSS then determines the largest change in

Performing Data Analysis Using IBM SPSS®, First Edition.
Lawrence S. Meyers, Glenn C. Gamst, and A. J. Guarino.

distance between clusters that has taken place. If that change is greater than a preset threshold, it starts again (iterates) using the current cluster centers as the initial centers (seed points) of another reclassification phase. Iteration continues until either the change in distance reaches its criterion threshold or IBM SPSS reaches the number of iterations that was specified by the researchers. Cluster membership can be saved to the data file and can then be used as a categorical grouping variable in subsequent analyses.

60.2 NUMERICAL EXAMPLE: *k*-MEANS CLUSTERING

For this example, we use the **personality** data file with 425 cases. We will cluster the cases based on three variables: **neoconsc** (conscientiousness), **negafect** (amount of negative affect, such as distress, hostility, and fear respondent have reported as having been recently experienced), and **regard** (a measure of self-regard).

60.3 ANALYSIS STRATEGY

We first perform a *z* score transformation of the three clustering variables in the **Descriptives** procedure. We then use those transformed variables in a *k*-means procedure. For illustration purposes, we obtain the four-cluster solution, as we believe it presents a viable and interpretable solution, although a three-cluster solution is not unreasonable. Finally, we will use the groups generated by the *k*-means procedure as an independent variable in a one-way between-subjects ANOVA to illustrate the use of the cluster variable in a subsequent (and very simple) analysis.

60.4 TRANSFORMING CLUSTER VARIABLES TO *Z* SCORES

We open the **personality** data file and from the main menu select **Analyze ➔ Descriptive Statistics ➔ Descriptives**. This opens the main **Descriptives** dialog window shown in Figure 60.1. We move **neoconsc**, **negafect**, and **regard** into the **Variable(s)** panel, check the box for **Save standardized values as variables**, and select **OK** to perform the analysis.

The result of our procedure can be seen in Figure 60.2. The newly transformed variables, named with an upper case **Z** in front of the original name, appear at the end of the data file.

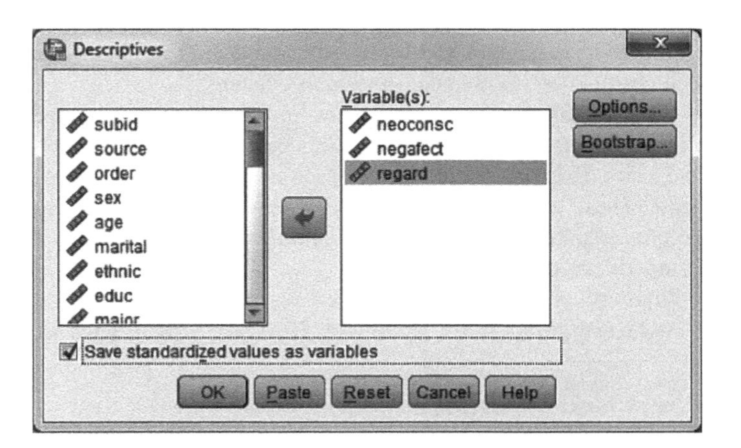

FIGURE 60.1

Using **Descriptives** to create *z* score transformations of the three variables for the *k*-means cluster analysis.

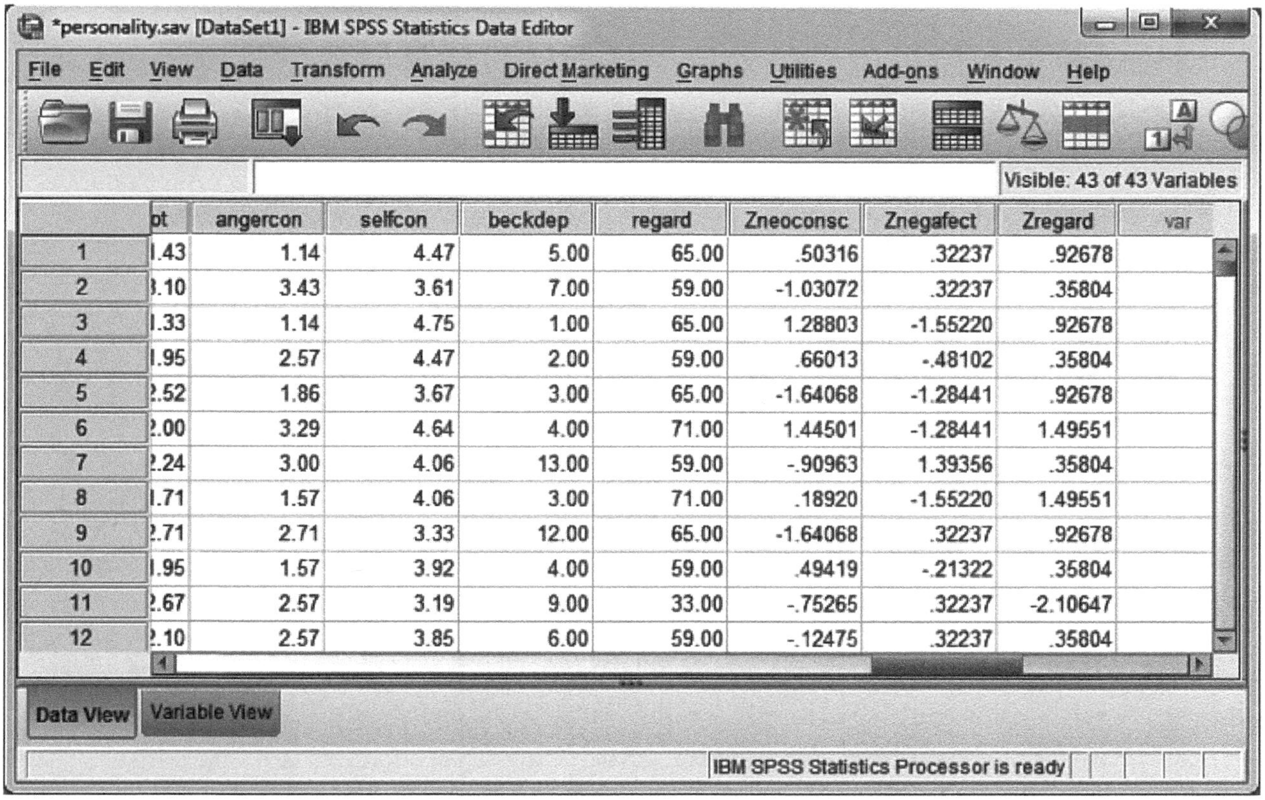

*personality.sav [DataSet1] - IBM SPSS Statistics Data Editor

File Edit View Data Transform Analyze Direct Marketing Graphs Utilities Add-ons Window Help

Visible: 43 of 43 Variables

	pt	angercon	selfcon	beckdep	regard	Zneoconsc	Znegafect	Zregard	var
1	1.43	1.14	4.47	5.00	65.00	.50316	.32237	.92678	
2	3.10	3.43	3.61	7.00	59.00	-1.03072	.32237	.35804	
3	1.33	1.14	4.75	1.00	65.00	1.28803	-1.55220	.92678	
4	1.95	2.57	4.47	2.00	59.00	.66013	-.48102	.35804	
5	2.52	1.86	3.67	3.00	65.00	-1.64068	-1.28441	.92678	
6	2.00	3.29	4.64	4.00	71.00	1.44501	-1.28441	1.49551	
7	2.24	3.00	4.06	13.00	59.00	-.90963	1.39356	.35804	
8	1.71	1.57	4.06	3.00	71.00	.18920	-1.55220	1.49551	
9	2.71	2.71	3.33	12.00	65.00	-1.64068	.32237	.92678	
10	1.95	1.57	3.92	4.00	59.00	.49419	-.21322	.35804	
11	2.67	2.57	3.19	9.00	33.00	-.75265	.32237	-2.10647	
12	2.10	2.57	3.85	6.00	59.00	-.12475	.32237	.35804	

Data View Variable View

IBM SPSS Statistics Processor is ready

FIGURE 60.2 A portion of the data file showing the transformed variables.

60.5 ANALYSIS SETUP: *k*-MEANS CLUSTERING

From the main menu select **Analyze ➔ Classify ➔ *k*-Means Cluster**. This opens the main dialog window shown in Figure 60.3. We move **Zneoconsc**, **Znegafect**, and **Zregard** into the **Variable(s)** panel. We specify **4** in the **Number of Clusters** panel and retain the default of **Iterate and classify** under **Method**.

Selecting the **Iterate** pushbutton opens the **Iterate** dialog window shown in Figure 60.4. The default number of iterations established by IBM SPSS is **10**, a carryover from long-ago days when too many iterations would stress the limited processing power of the ancient machines. We bump the iterations to 100 to ensure that our analysis will go to completion. Clicking **Continue** returns us to the main dialog window.

In the **Options** window shown in Figure 60.5, we select in the **Statistics** area **Initial cluster centers** (to show the initial seed points) and **ANOVA table** to compare the groups on the clustering variables. We also retain the option of **Exclude cases listwise** in the **Missing Values** portion of the window. Click **Continue** to return to the main dialog window.

The **Save** window is shown in Figure 60.6. We select **Cluster membership**. This will create a new categorical variable at the end of the data file, with numeric codes indicating the cluster to which each case has been assigned in the final solution. Click **Continue** to return to the main dialog window and select **OK** to perform the analysis.

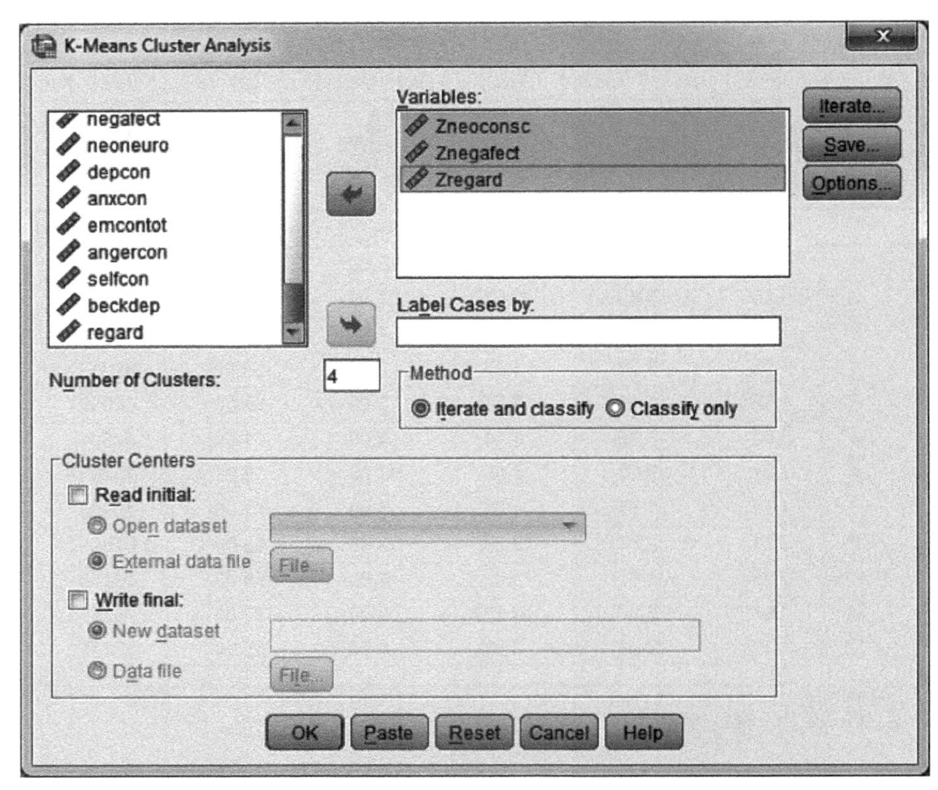

FIGURE 60.3 The main dialog window of ***k*-Means Cluster Analysis**.

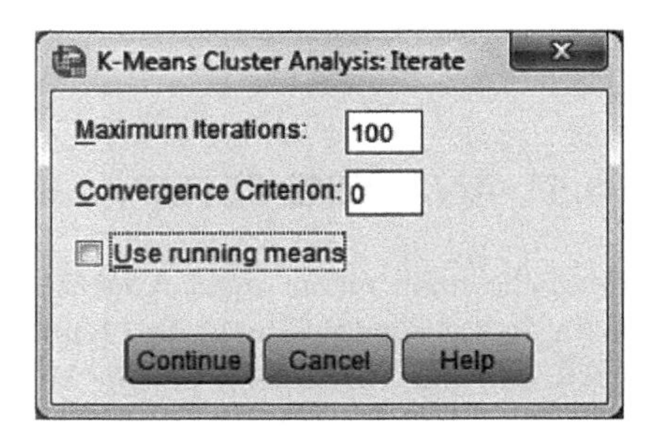

FIGURE 60.4
The **Iterate** dialog window of ***k*-Means Cluster Analysis**.

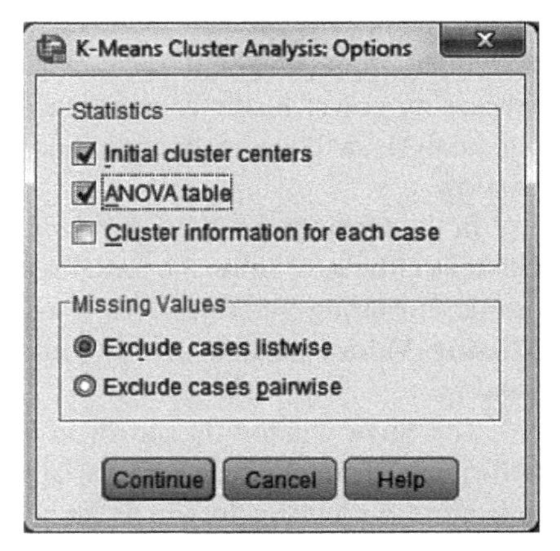

FIGURE 60.5
The **Options** dialog window of ***k*-Means Cluster Analysis**.

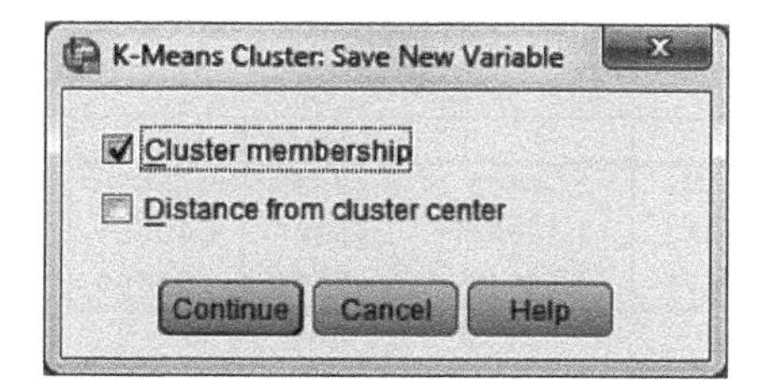

FIGURE 60.6 The **Save** dialog window of *k*-**Means Cluster Analysis**.

60.6 ANALYSIS OUTPUT: k-MEANS CLUSTERING

Figure 60.7 shows a screenshot of a portion of the data file, with the cluster membership of each case designated under the variable **QCL_1**. The name of the variable is shorthand for "Quick Cluster, First Save Command of this Session." We can see that the case in the first row is assigned to Cluster **3**, the case in the second row is assigned to Cluster **1**, and so on.

The initial seed points for the analysis are shown in the **Initial Cluster Centers** table in Figure 60.8. The values are in z score form, with a sample mean of zero and standard deviation of 1.00. Most of the seed point z scores are well in excess of one standard deviation distance from their respective means, and many are somewhat over the distance of two standard deviations. We note two points. First, we do not see a true outlier (a score beyond three standard deviation units), and that is reassuring. Second, these seed points are very distant from the sample mean, but the iterative process will abate these distances to a large extent by the time we obtain the final cluster centers.

Figure 60.9 displays the **Iteration History**. Our modification of the default appeared to be unnecessary here, as the convergence threshold was achieved by the tenth iteration.

One concern of researchers in performing a k-means cluster analysis is making sure that the number of cases assigned to each cluster in the final solution is acceptable. This information is shown in Figure 60.10, where we see that sample sizes ranged from 66

*personality.sav [DataSet1] - IBM SPSS Statistics Data Editor

File Edit View Data Transform Analyze Direct Marketing Graphs Utilities Add-ons Window Help

Visible: 44 of 44 Variables

	emcontot	angercon	selfcon	beckdep	regard	Zneoconsc	Znegafect	Zregard	QCL_1	var	var
1	1.43	1.14	4.47	5.00	65.00	.50316	.32237	.92678	3		
2	3.10	3.43	3.61	7.00	59.00	-1.03072	.32237	.35804	1		
3	1.33	1.14	4.75	1.00	65.00	1.28803	-1.55220	.92678	3		
4	1.95	2.57	4.47	2.00	59.00	.66013	-.48102	.35804	3		
5	2.52	1.86	3.67	3.00	65.00	-1.64068	-1.28441	.92678	1		
6	2.00	3.29	4.64	4.00	71.00	1.44501	-1.28441	1.49551	3		
7	2.24	3.00	4.06	13.00	59.00	-.90963	1.39356	.35804	4		
8	1.71	1.57	4.06	3.00	71.00	.18920	-1.55220	1.49551	3		
9	2.71	2.71	3.33	12.00	65.00	-1.64068	.32237	.92678	1		
10	1.95	1.57	3.92	4.00	59.00	.49419	-.21322	.35804	3		
11	2.67	2.57	3.19	9.00	33.00	-.75265	.32237	-2.10647	2		
12	2.10	2.57	3.85	6.00	59.00	-.12475	.32237	.35804	1		

Data View Variable View

IBM SPSS Statistics Processor is ready

FIGURE 60.7 A portion of the data file showing the saved cluster membership variable.

Initial Cluster Centers

	Cluster			
	1	2	3	4
Zneoconsc	-2.16543	-2.16543	2.22989	1.75896
Znegafect	-.74881	2.19694	-1.55220	2.19694
Zregard	.92678	-2.58042	1.49551	-.96900

FIGURE 60.8
The table of **Initial Cluster Centers**.

Iteration History[a]

Iteration	Change in Cluster Centers			
	1	2	3	4
1	1.687	1.652	1.673	1.875
2	.093	.155	.070	.116
3	.065	.045	.016	.063
4	.010	.034	.009	.023
5	.000	.040	.009	.028
6	.000	.022	.012	.015
7	.000	.036	.000	.021
8	.000	.017	.000	.010
9	.000	.021	.000	.013
10	.000	.000	.000	.000

a. Convergence achieved due to no or small change in cluster centers. The maximum absolute coordinate change for any center is . 000. The current iteration is 10. The minimum distance between initial centers is 4.242.

FIGURE 60.9
The **Iteration History** table.

Number of Cases in each Cluster

Cluster	1	125.000
	2	66.000
	3	119.000
	4	110.000
Valid		420.000
Missing		5.000

FIGURE 60.10
The sample size falling into each of the four clusters.

(**Cluster 2**) to 125 (**Cluster 1**). We judge this as quite good. To be avoided is the situation where one cluster has very few cases (maybe a dozen or fewer cases), rendering such a group as not viable in subsequent analyses.

The ANOVA summary table is shown in Figure 60.11. The footnote to the table is important, which informs us that the analysis needs to be treated as strictly exploratory. Here, there are statistically significant group differences in all of the clustering variables. The primary use of this table to researchers is to recognize when a clustering variable does not yield a statistically significant effect; this would indicate that the variable is not effective for its chosen purpose and might suggest to researchers that they may wish to repeat the analysis swapping it out for another variable if one is reasonable and available.

The table of **Final Cluster Centers** is shown in Figure 60.12. This table shows the heart of the results and is used to characterize the clusters. Note that the relatively extreme values of the seed points have been substantially moderated. These z scores are based on the sample statistics and so must be interpreted relative to this sample rather than in any population sense.

ANOVA

	Cluster		Error		F	Sig.
	Mean Square	df	Mean Square	df		
Zneoconsc	82.103	3	.415	416	197.782	.000
Znegafect	82.524	3	.410	416	201.285	.000
Zregard	79.953	3	.420	416	190.213	.000

The F tests should be used only for descriptive purposes because the clusters have been chosen to maximize the differences among cases in different clusters. The observed significance levels are not corrected for this and thus cannot be interpreted as tests of the hypothesis that the cluster means are equal.

FIGURE 60.11 Results of an exploratory ANOVA examining group differences for each of the three clustering variables.

Final Cluster Centers

	Cluster			
	1	2	3	4
Zneoconsc	-.70895	-1.02073	.86603	.48118
Znegafect	-.39747	1.09736	-.87484	.72163
Zregard	.39823	-1.44798	.75074	-.36451

FIGURE 60.12
The table of **Final Cluster Centers**.

Because the standardization is based on a relatively large sample size ($N = 420$), we can interpret the magnitude of the z scores in a functional way. Thus, we can judge a z score in the range about $\pm.50$ to about $\pm.75$ as substantial, a z score of approximately ±1.00 as relatively considerable, and a z score of anything greater than about ±1.30 as fairly extreme with respect to the sample mean. In our characterization, we look for these higher z scores to guide our interpretation. We characterize these clusters (groups) as follows:

- *Cluster 1.* The only distinguishing feature of Cluster 1 is a substantial (negative) conscientiousness mean, suggesting that the cases in this cluster can on average be generally characterized as being relatively more careless and irresponsible than average for the sample.
- *Cluster 2.* The cases on average in this cluster appear to be characterized by a relative lack of conscientiousness and self-regard but appear to experience considerable negative affect; they may be characterized as being relatively more hopeless and lost than average for the sample.
- *Cluster 3.* The cases on average in this cluster appear to be characterized by substantial conscientiousness and self-regard while experiencing very little negative affect; they may be characterized as being relatively more psychologically healthy than average for the sample.
- *Cluster 4.* The cases on average in this cluster appear to be relatively conscientiousness but experience substantial negative affect; they may be characterized as being more likely than average for the sample to brood while completing the tasks of life without taking much pleasure in their actions.

60.7 FOLLOW-UP ONE-WAY BETWEEN-SUBJECTS ANALYSIS

To illustrate the use of the cluster membership variable in a subsequent analysis, we perform a one-way between-subjects ANOVA using the **One-Way ANOVA** procedure in IBM SPSS. We discussed this procedure in Chapter 48 and opt for it here because it is not unusual for groups formed by a clustering procedure to differ quite a bit in variability in the dependent variable, and in the **One-Way ANOVA** procedure, we can request specialized tests of significance of the omnibus effect (an overall group difference) that take into account a violation of the homogeneity of variance assumption.

60.8 NUMERICAL EXAMPLE: ONE-WAY ANOVA

In the **personality** data file now containing our **QCL_1** variable that will be used as the independent variable representing our groups, we will use **selfcon** (self-control) as our dependent variable. This dependent variable is the score on the Self-Control Schedule (Rosenbaum, 1980) and represents ways of coping with life by taking positive control of one's thoughts and feelings.

60.9 ANALYSIS SETUP: ONE-WAY ANOVA

From the main menu select **Analyze ➔ Compare Means One-Way ANOVA**. This opens the main dialog window shown in Figure 60.13. We move **selfcon** into the **Dependent List** panel and **QCL_1** into the **Factor** panel.

In the **Options** dialog window presented in Figure 60.14, we check **Descriptive** (to obtain several descriptive statistics for the dependent variable) and **Homogeneity**

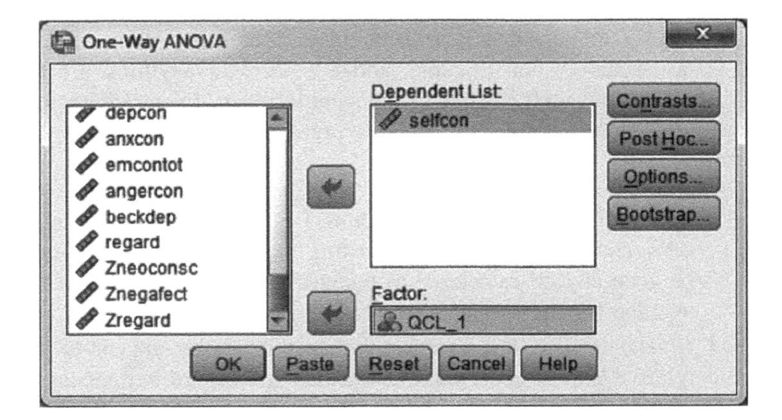

FIGURE 60.13
The main dialog window of **One-Way ANOVA**.

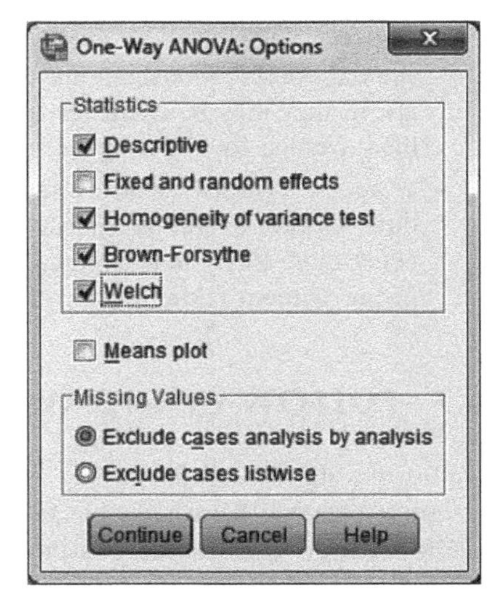

FIGURE 60.14
The **Options** dialog window of **One-Way ANOVA**.

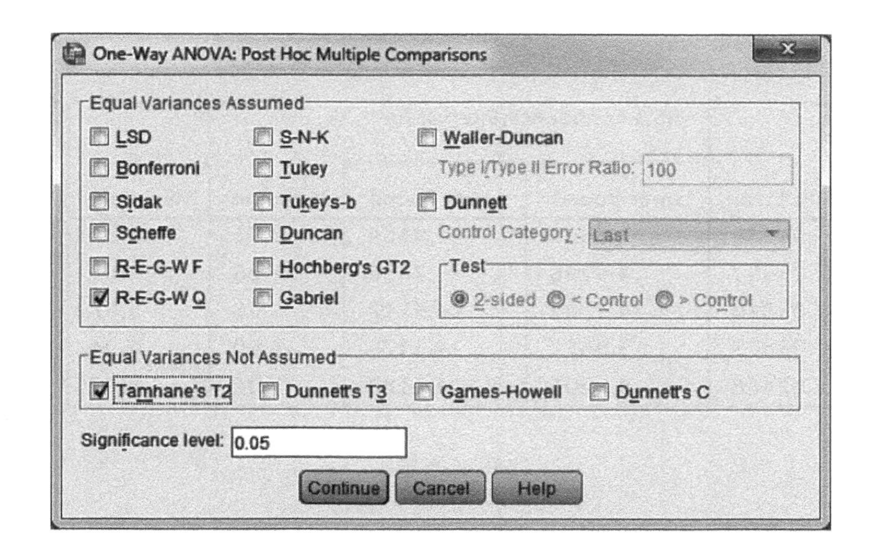

FIGURE 60.15
The **Post Hoc** dialog window of **One-Way ANOVA**.

of variance test (to obtain Levene's test). To obtain statistical significance tests of the between-subjects effect if the assumption of homogeneity of variance has not been met, we also check **Brown-Forsythe** and **Welch**. Click **Continue** to return to the main dialog window.

We will optimistically anticipate a statistically significant omnibus group difference and request in the **Post Hoc** dialog window (see Figure 60.15) the **R-E-G-W-Q** test (to be used if we have homogeneity of variance) and **Tamhane's T2** test (to be used if we have violated the homogeneity of variance assumption). Click **Continue** to return to the main dialog window and click **OK** to perform the analysis.

60.10 ANALYSIS OUTPUT: ONE-WAY ANOVA

The descriptive statistics shown in the top table of Figure 60.16 provide relatively complete information describing each group. Based on visual inspection of the differences in means with respect to their standard errors, it appears likely that the analysis will yield a statistically significant omnibus effect.

The bottom table in Figure 60.16 displays the results of the Levene test. With **3** and **416** degrees of freedom, the value of the test statistic (**1.205**) is not statistically significant ($p =$ **.307**). This outcome indicates that we have not violated the assumption of equal group variances and can use the ordinary F ratio as our test of statistical significance for the overall effect.

The ANOVA summary table is presented in the top table of Figure 60.17. The omnibus effect of the independent variable (**Between Groups** in the table) yielded an F ratio of **54.490**. With **3** and **416** degrees of freedom, the effect is statistically significant ($p < .001$). The value of eta square is computed as the ratio of **Between Groups Sum of Squares** to **Total Sum of Squares**; here, it is **30.685/108.772**, or **.282**. Thus, approximately 28% of the total variance of **selfcon** is explained by **QCL_1** (the cluster group membership).

The bottom table of Figure 60.17 shows the results for the **Robust Tests of Equality of Means**. Had we violated the assumption of equal variance, we would have used one of these tests to evaluate the statistical significance of the omnibus effect.

With the omnibus F ratio statistically significant, we can examine the results for our post hoc tests. We focus on the **R-E-G-W-Q** test in that we have observed comparable variances across our four groups, and these results are presented in Figure 60.18. We

Descriptives

selfcon self-control: Rosenbaum sched: rscs

	N	Mean	Std. Deviation	Std. Error	95% Confidence Interval for Mean		Minimum	Maximum
					Lower Bound	Upper Bound		
1	125	3.9646	.47357	.04236	3.8807	4.0484	2.61	5.25
2	66	3.6366	.43524	.05357	3.5296	3.7436	2.06	4.61
3	119	4.4458	.41563	.03810	4.3703	4.5212	3.58	5.67
4	110	4.0411	.40160	.03829	3.9652	4.1170	3.00	4.94
Total	420	4.0694	.50951	.02486	4.0205	4.1183	2.06	5.67

Test of Homogeneity of Variances

selfcon self-control: Rosenbaum sched: rscs

Levene Statistic	df1	df2	Sig.
1.205	3	416	.307

FIGURE 60.16 The **Descriptives** and **Test of Homogeneity of Variances** output.

ANOVA

selfcon self-control: Rosenbaum sched: rscs

	Sum of Squares	df	Mean Square	F	Sig.
Between Groups	30.685	3	10.228	54.490	.000
Within Groups	78.087	416	.188		
Total	108.772	419			

Robust Tests of Equality of Means

selfcon self-control: Rosenbaum sched: rscs

	Statistic[a]	df1	df2	Sig.
Welch	56.010	3	206.115	.000
Brown-Forsythe	54.791	3	368.026	.000

a. Asymptotically F distributed.

FIGURE 60.17 The ANOVA summary table and the **Robust Tests of Equality of Means** (used if homogeneity of variance has been violated).

described how to read this output in Section 47.5; briefly, means within a subset do not statistically differ, whereas means that are in different subsets do significantly differ. Thus, cases in Cluster 2 (relatively more hopeless and lost than average) reported the lowest level of self-control, those in Clusters 1 (relatively more careless and irresponsible than average) and 4 (somewhat conscientious but more likely than average to brood) reported the second highest degree of self-control, and those in Cluster 3 (relatively more psychologically healthy than average) reported the first highest level of self-control.

We can also evaluate the effect size using Cohen's d statistic. For example, the largest mean difference was between Cluster 3 and Cluster 2. From the descriptive statistics output in Figure 60.16, we can compute the (weighted) average standard deviation of

selfcon self-control: Rosenbaum sched: rscs

	QCL_1 Cluster Number of Case	N	Subset for alpha = 0.05		
			1	2	3
Ryan-Einot-Gabriel-Welsch Range[a]	2	66	3.6366		
	1	125		3.9646	
	4	110		4.0411	
	3	119			4.4458
	Sig.		1.000	.345	1.000

Means for groups in homogeneous subsets are displayed.

a. Critical values are not monotonic for these data. Substitutions have been made to ensure monotonicity. Type I error is therefore smaller.

FIGURE 60.18 The results of the **R-E-G-W-Q** post hoc test.

these two groups as ((**119** * **.41563**) + (**66** * **.43524**))/(**119** + **66**) = 78.1858/185 = .423. Cohen's *d* is the ratio of the mean difference to the average standard deviation; for the difference between Cluster 3 and Cluster 2, it can be computed as **4.4458** – **3.6366**/.423 = .8092/.423 = 1.913. This would be interpreted as a large effect size. However, because these groupings were created statistically with an algorithm biased toward separating the clusters on at least the clustering variables, and because **selfcon** is likely correlated with the clustering variables to a certain extent, we would use Cohen's *d* effect size index in this instance with considerable caution.

PART 19

NONPARAMETRIC PROCEDURES FOR ANALYZING FREQUENCY DATA

CHAPTER 61

Single-Sample Binomial and Chi-Square Tests: Binary Categories

61.1 OVERVIEW

The procedures we have described in most parts of this book are applicable when we are working with variables that are assessed on quantitative scales of measurement, but it is not uncommon for researchers to also collect frequency data. Frequency data can be collected on nominal or categorical variables where we count the number of occurrences observed for each mutually exclusive category of a qualitative variable. Examples of such categorical variables are the number of students enrolled in a particular university who are on financial aid or not; the number of patrons of the local coffee house purchasing a café latte or a café mocha on a given day; the number of red, silver, or black vehicles that pull in for gas at a particular service station; and the number of patients in a particular health provider facility given a flu shot during past October.

In the most general sense, our research focus in such qualitative studies is on whether the pattern of observed frequencies meets our expectations. To address this focus, we apply a single-sample nonparametric test, either the *binomial test* or the *single-sample chi-square test* developed by Karl Pearson (1900b). The binomial test is based on the binomial distribution and is applied only to variables that have binary (exactly two) categories; we use a z statistic to compare the observed and expected frequencies. The single-sample chi-square can be applied to two or more categories; it will provide the same result as the binomial test for a binary variable.

To qualify as a single-sample design, we process only one piece of information regarding the cases—the category into which each case is classified. Using binary or dichotomous categorical variables, cases are associated with either one arbitrary numerical code (e.g., 1 representing "yes") or another (e.g., 0 representing "no"). For each case, a student is on financial aid or is not, a patron purchased either a café latte or a café mocha, and so on. If we had collected frequency data on a variable that had three or more categories, cases are still associated with only one of the arbitrarily assigned multiple codes (e.g., 1, 2, and 3 represent red, silver, and black vehicles, respectively), but this schema does not meet the requirement of a binary categorical variable; how to handle such variables is treated in Chapter 62.

The frequencies we expect to observe in the study must be made explicit in setting up these nonparametric tests. Often, our expectations will be based entirely on chance, that is, a 50-50 split for dichotomous variables. For example, we might suppose that half of the students sampled will indicate that they are on financial aid and half will not, or

Performing Data Analysis Using IBM SPSS®, First Edition.
Lawrence S. Meyers, Glenn C. Gamst, and A. J. Guarino.

that half of the patrons who buy one of the two drinks will purchase a café latte and that the other half will purchase a café mocha. In other circumstances, on either a theoretical or an actuarial basis, we might expect some different breakdown. For example, we might know that on a national basis, about 67% of all college students received some financial aid during the previous year and so we might expect a two-thirds to one-third showing in the frequencies on the particular campus where the study is being conducted.

61.2 NUMERICAL EXAMPLE

We use the data file named **financial aid** containing the variable **financial_aid**. This is a **Nominal** variable (defined as such in the **Measure** column of the **Variable View**) assessing whether the student who was sampled was on financial aid or not. **Financial_aid** is a binary variable with the two categories of **No** (coded as **0** in the data file) and **Yes** (coded as **1** in the data file). A screenshot of a portion of the data file is presented in Figure 61.1 to illustrate the data structure of a one-sample design.

61.3 ANALYSIS STRATEGY

IBM SPSS® has revamped the **Nonparametric Tests** module in recent versions but has retained the older procedures under **Legacy Dialogs** within the **Nonparametric Tests**

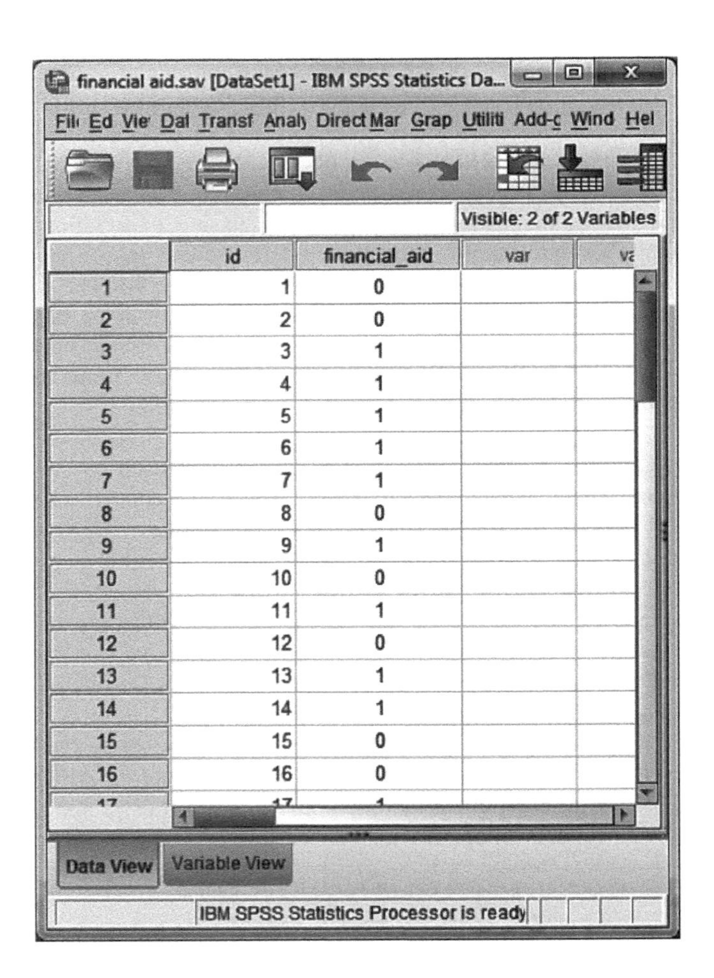

	id	financial_aid	var
1	1	0	
2	2	0	
3	3	1	
4	4	1	
5	5	1	
6	6	1	
7	7	1	
8	8	0	
9	9	1	
10	10	0	
11	11	1	
12	12	0	
13	13	1	
14	14	1	
15	15	0	
16	16	0	

FIGURE 61.1
A portion of the data file.

submenu. The **Legacy** procedures have some features that, in our opinion, make user errors more likely to occur. For example, if the proportions for the categories need to be user-defined rather than having equal probabilities in the **Binomial test**, the specified proportion will be applied to the first value in the data file, thus forcing users to either carefully select the first case in data entry or to sort the file before performing the analysis. These features are much improved in the revamped **Nonparametric Tests** module, and so we describe this module in this chapter. Both the binomial and chi-square tests can be used for a binary coded variable, and we perform both in a single data analysis run to illustrate how each works.

61.4 FREQUENCIES ANALYSIS

We have performed a **Frequencies** analysis on the variable **financial_aid** to determine its distribution and to provide some context, the results of which are shown in Figure 61.2. As can be seen, the breakdown of students in the sample is as evenly divided as possible, given an *N* of 35.

61.5 ANALYSIS SETUP

We open **financial aid** and from the main menu select **Analyze ➔ Nonparametric Tests ➔ One Sample**. This opens the **Objective** screen in the initial **One-Sample Nonparametric Tests** window shown in Figure 61.3. We select **Customize analysis** to gain full control over the specifications for our analysis and click the **Fields** tab.

Despite our selection of **Customize analysis** in the previous screen, the **Fields** window opens with the **Use predefined roles** button selected and all of our variables already placed in the **Test Fields** panel (see Figure 61.4). We want only **financial_aid** in the **Test Fields** panel, and the easiest way to accomplish this is to highlight **id** and move it into the **Fields** panel by clicking the horizontal arrow. This action not only moves **id** across the panels but also switches the choice at the top left portion of the screen to **Use custom field assignments** as shown in Figure 61.5.

Selecting the **Settings** tab opens the **Settings** window (see Figure 61.6). The choices in this window allow us to specify our statistical analysis, but again we must wrest control from IBM SPSS by selecting in the **Choose Tests** screen the option of **Customize tests** and thereby negate the default of **Automatically choose the tests based on the data** (an option that allows users to blindly perform analyses without necessarily knowing what they are doing). The only two tests applicable to our frequency data are the **Binomial test** and the **Chi-Square test** (the first two checkboxes). In Figure 61.6, we have selected **Customize tests** and have checked **Compare observed binary probability to hypothesized (Binomial test)**.

financial_aid

		Frequency	Percent	Valid Percent	Cumulative Percent
Valid	0 no financial aid	18	51.4	51.4	51.4
	1 yes financial aid	17	48.6	48.6	100.0
	Total	35	100.0	100.0	

FIGURE 61.2 Output from **Frequencies** analysis.

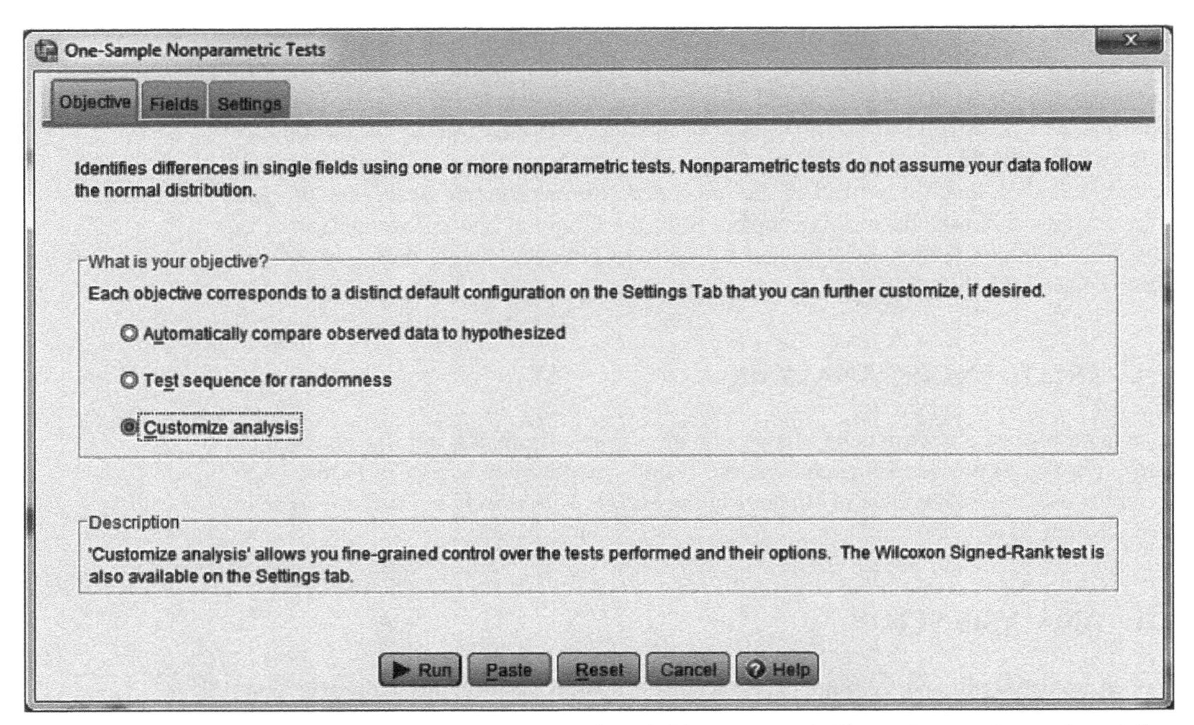

FIGURE 61.3 Initial screen of **One-Sample Nonparametric Tests** where we have selected **Customize analysis**.

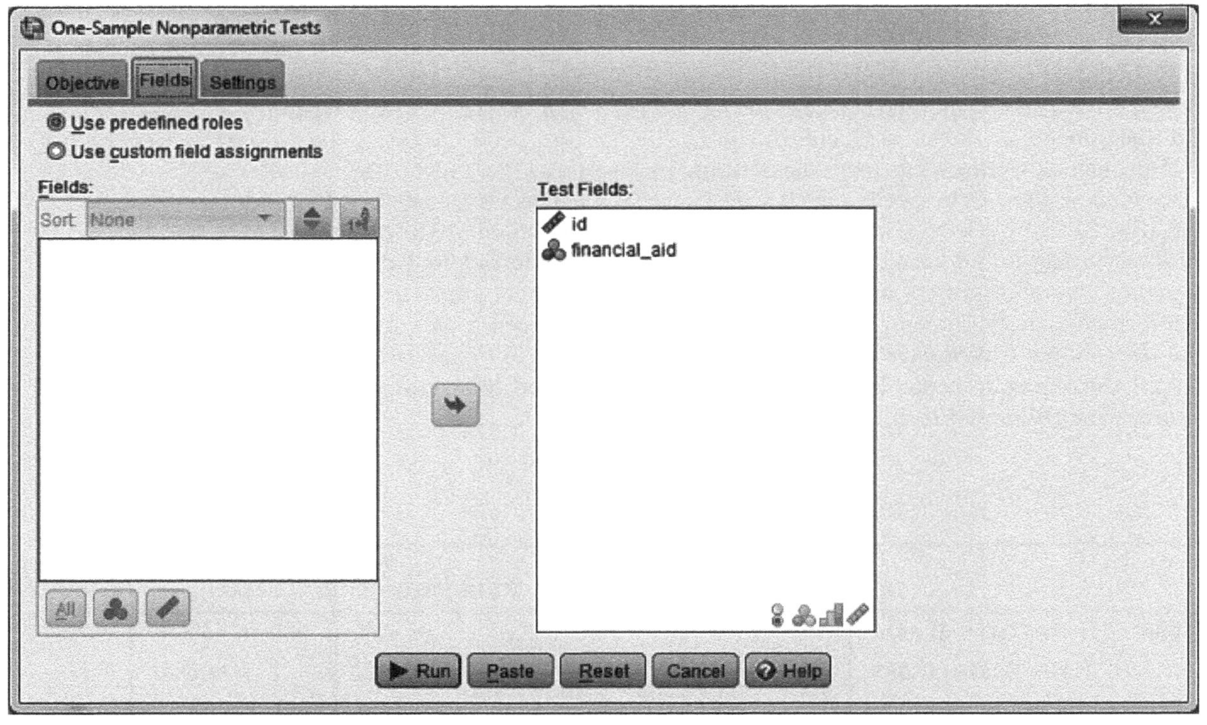

FIGURE 61.4 The initial window on the **Fields** tab.

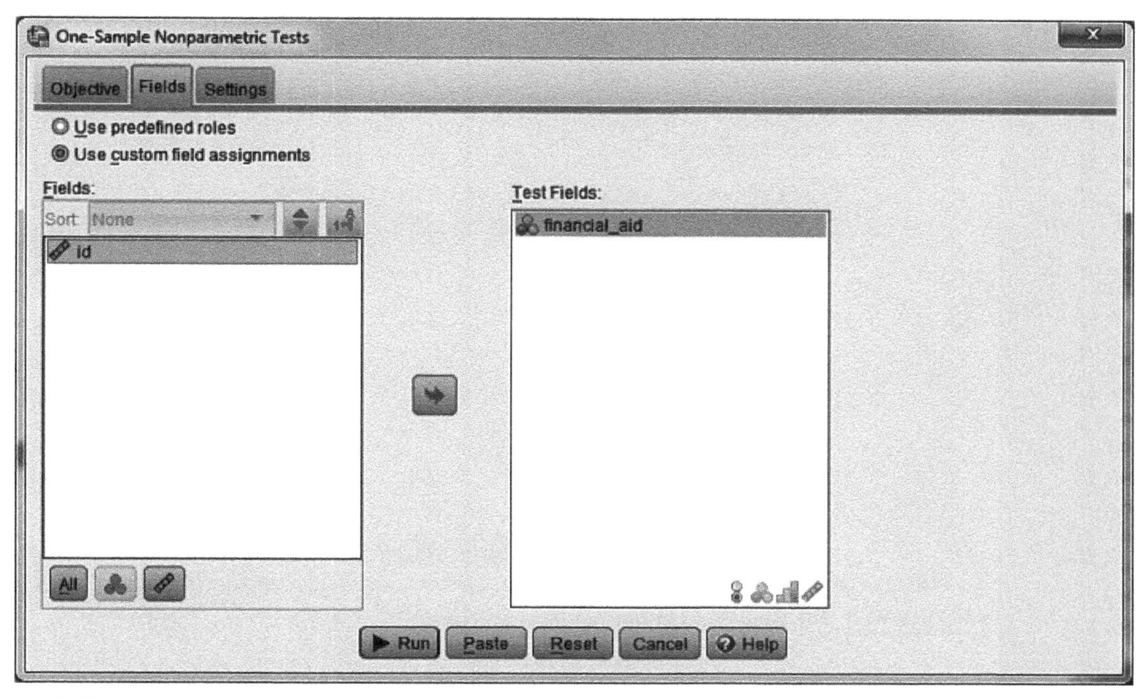

FIGURE 61.5 The **Fields** window configured with **financial_aid** in the **Test Fields** panel.

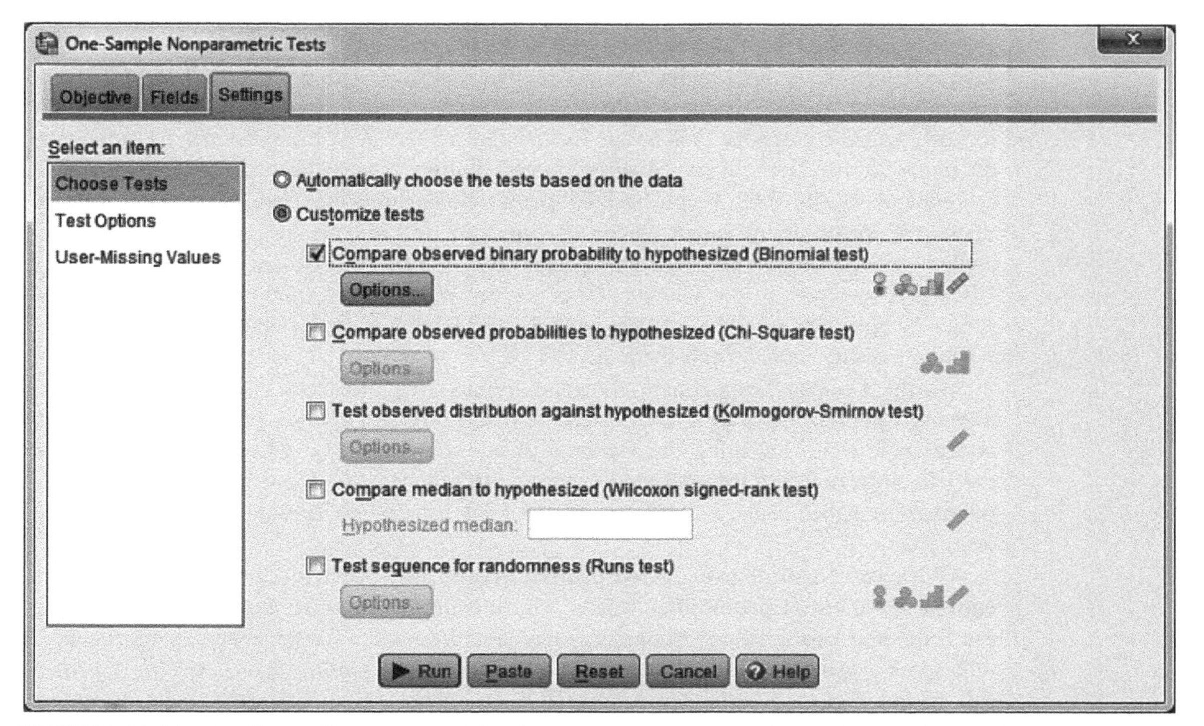

FIGURE 61.6 In the **Choose Tests** screen of **Settings**, we have chosen to **Customize** the **Binomial test**.

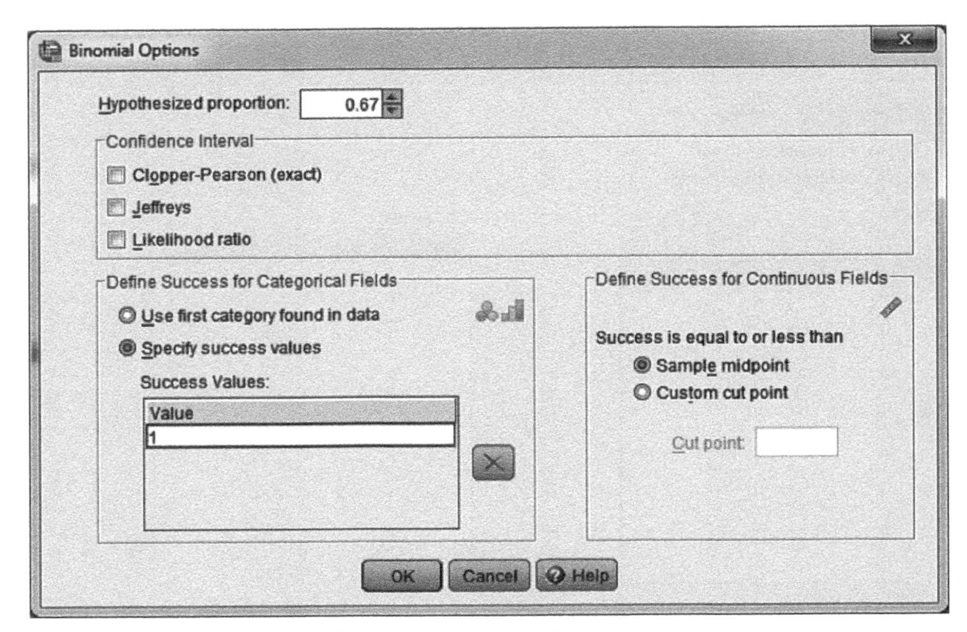

FIGURE 61.7 The **Binomial Options** are now specified.

Selecting the **Options** pushbutton under the **Binomial test** opens the **Binomial Options** window shown in Figure 61.7. In the **Hypothesized proportion** panel, we replace the default of **0.5** with **0.67** by typing it in (we could also use the arrow buttons to modify the value), as we wish to evaluate the proportion of students in our sample who are receiving financial aid against the national proportion of 67%.

In the **Define Success for Categorical Fields**, we wish to reject the default option (carried over from the **Legacy** procedures) of **Use first category found in data** and instead select **Specify success values**. The term **success** is used in a generic manner to designate the category that is our focal interest (here it is the proportion of students on financial aid). Opting for **Specify success values** places the user in control of the assignment of the **0.67 Hypothesized proportion**. Selecting **Specify success values** opens the **Value** dialog panel where we type in **1** to indicate the category assigned to students in the sample who are on financial aid. All this is also shown in Figure 61.7. Click **OK** to return to the **Choose Tests** screen. We need not modify the **Test Options** screen (it specifies an alpha of .05) or the **User-Missing Values** (which specifies that we exclude missing values from the analysis).

In the **Choose Tests** screen shown in Figure 61.8, we now also check **Compare observed probabilities to hypothesized (Chi-Square test)**, as we can perform this test at the same time. Selecting the **Options** pushbutton under the **Chi-Square test** opens the **Chi Square Test Options** window shown in Figure 61.9. There is no **Hypothesized proportion** panel because this test can be (and usually is) applied to more than two categories.

We select **Customize expected probability**, thereby deactivating the default **All categories have equal probability** option. On selecting **Customize expected probability**, the **Expected probabilities** dialog area becomes available. We have two categories and will specify probabilities for both in whatever order we prefer. Thus, we type **1** in the cell in the first row under **Category**, click the panel in the first row under **Relative Frequency**, and type **67**. Then we click the second row under **Category** that has now become available, type **0**, click the panel under **Relative Frequency**, and type **33**. We strongly recommend that these relative frequencies add to 100 because IBM SPSS will

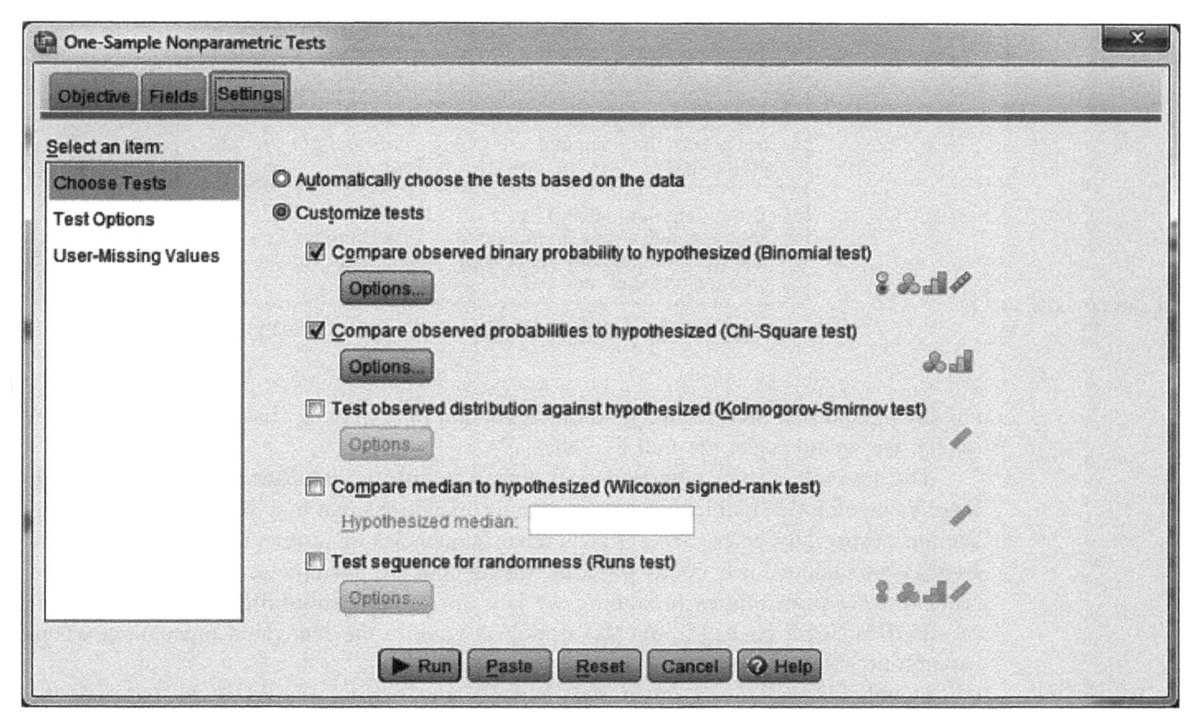

FIGURE 61.8 In the **Choose Tests** screen of **Settings**, we have chosen to **Customize** the **Chi-Square test**.

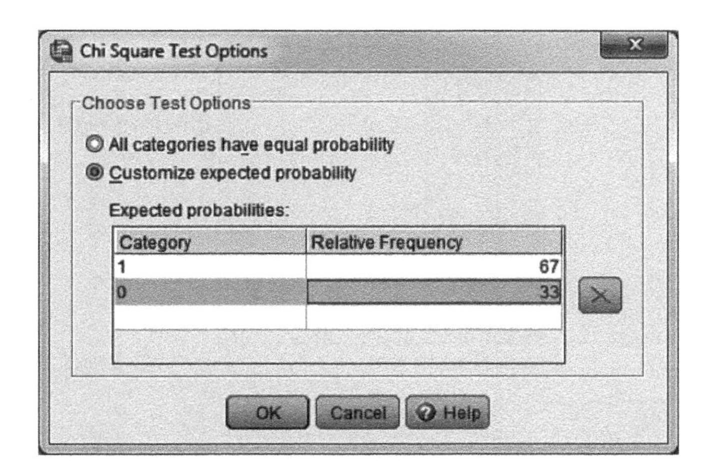

FIGURE 61.9

Chi Square Options are now specified.

total them to determine the percent of total represented by each of the relative frequencies. The finished specifications are shown in Figure 61.9. Click **OK** to return to the **Choose Tests** screen, and select **Run** to perform the analyses.

61.6 ANALYSIS OUTPUT

An overview of the results is shown in the summary table in Figure 61.10. The top row provides output from the chi-square test and informs us (in the first column) that we had specified the probabilities; the bottom row addresses the binomial test and provides us with the proportions we specified. Both tests are statistically significant (p = **.020** for

Hypothesis Test Summary

	Null Hypothesis	Test	Sig.	Decision
1	The categories of financial_aid occur with the specified probabilities.	One-Sample Chi-Square Test	.020	Reject the null hypothesis.
2	The categories defined by financial_aid = (yes financial aid) and (no financial aid) occur with probabilities 0.67 and 0.33.	One-Sample Binomial Test	.016	Reject the null hypothesis.

Asymptotic significances are displayed. The significance level is .05.

FIGURE 61.10

The basic output that can be expanded by double-clicking.

chi-square and $p = \mathbf{.016}$ for the binomial test), and in both cases, based on an alpha level of .05, we would reject the null hypothesis.

The footnote to the table indicates that **Asymptotic significances are displayed**. Briefly, asymptotic significance levels are approximations that may falter with very small samples, especially where several cells have frequencies of fewer than five cases. In the **Legacy** procedures, we could override the default of asymptotic results and instead request exact probabilities (assuming we had the **Exact Probabilities** software as part of our IBM SPSS package), but this option is gone in the renovated procedures where asymptotic results are always shown.

Double-clicking on a given row expands the output. Figure 61.11 presents the expanded view of the binomial test. The bar chart visually displays the observed and

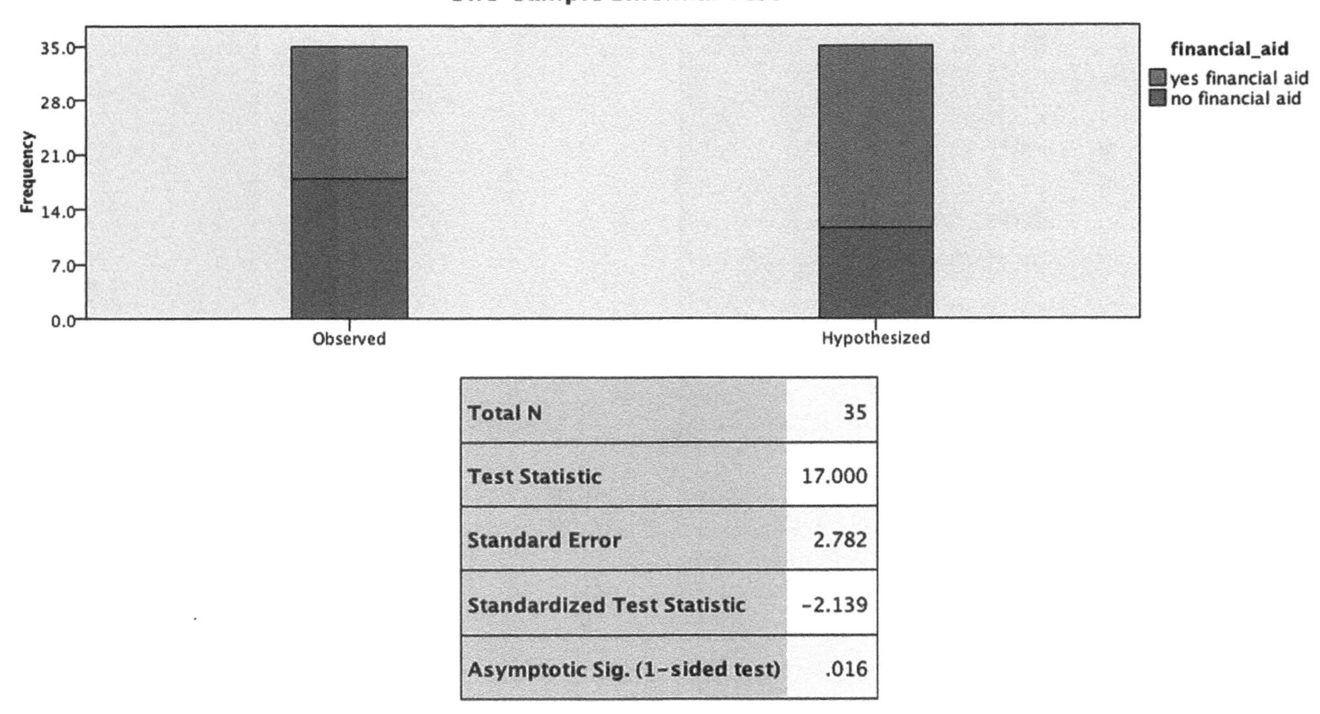

Total N	35
Test Statistic	17.000
Standard Error	2.782
Standardized Test Statistic	-2.139
Asymptotic Sig. (1-sided test)	.016

1. The alternative hypothesis is that the proportion of records in the success group is less than the hypothesized success probability.

FIGURE 61.11 Expanded view of binomial test results.

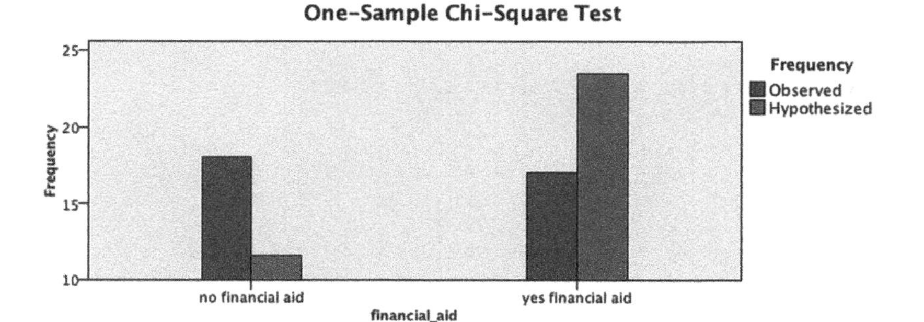

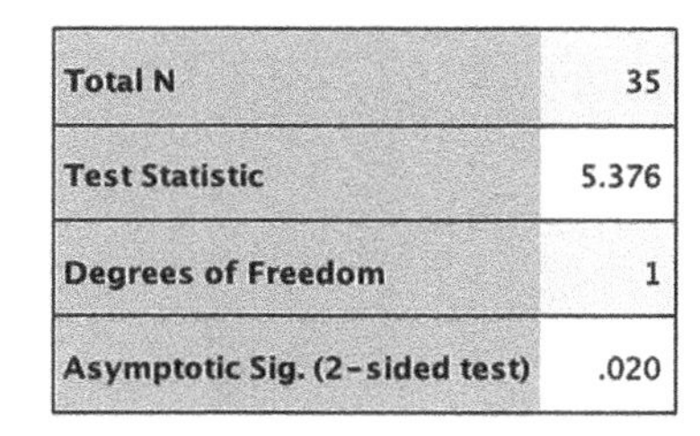

Total N	35
Test Statistic	5.376
Degrees of Freedom	1
Asymptotic Sig. (2-sided test)	.020

1. There are 0 cells (0%) with expected values less than 5. The minimum expected value is 11.550.

FIGURE 61.12
Expanded view of chi-square results.

hypothesized frequencies. In the statistical table below the graph, we see that the *z* value is **−2.139** and that the chances of obtaining that value or one displaced even further from the mean of a normal curve is **.016**. We therefore conclude that the frequencies were not distributed as hypothesized; rather, there were proportionally fewer students on financial aid in this sample than was true for the national average.

Figure 61.12 presents the expanded view of the chi-square test. The chi-square value is shown in the table as **5.376**. Evaluated against a chi-square distribution with one degree of freedom, the chances of obtaining that value or greater are **.020**. The conclusion we draw is identical to the outcome of the binomial test.

CHAPTER 62

Single-Sample (One-Way) Multinominal Chi-Square Tests

62.1 OVERVIEW

The one-way chi-square test described in Chapter 61 is applicable not only to binary variables but also to multinomial (more than two categories) categorical variables. The analysis setup is a simple extension of the two-category design. It is called a one-way chi-square analysis, in that there is a single variable under study. Each category within that variable represents a different group of cases, and our question is whether the sample sizes of the groups (the frequencies) are in accord with our expectations.

62.2 NUMERICAL EXAMPLE

The fictional data we use for this example are derived from clients of a stress-reduction clinic serving many companies in its metropolitan area. The 110 clients in this study worked in high stress jobs for a large technology firm and were referred to the clinic during the past fiscal year by the Human Resources manager based on recommendations from supervisors. These clients were presenting distress symptoms or signals and were categorized by the clinic according to the strongest set of signals they presented: symptoms of **cognitive decline** included lessened ability to make decisions, lowered concentration, and less likelihood of remembering important things (coded a **1**); symptoms of **emotional deterioration** included general irritability, loss of trust in people, and procrastination (coded a **2**); and symptoms of **physical changes** included increase in alcohol consumption, use of prescription or illegal drugs, and weight change (coded a **3**) under the variable of **distress_signal**. The data file is named **distress signals**.

62.3 ANALYSIS STRATEGY

We use the revamped **Nonparametric Tests** module as described in Chapter 61 to perform the analysis. Assume for our example that it was determined, based on clients from other types of industries seen by the clinic, that the strongest set of symptoms presented by clients would be equally likely across the three categories. We therefore perform the omnibus (three-category) analysis under the expectation that the categories should have equal frequencies. If we obtain a statistically significant chi-square (indicating that the

Performing Data Analysis Using IBM SPSS®, First Edition.
Lawrence S. Meyers, Glenn C. Gamst, and A. J. Guarino.

frequencies are not equally divided among the three categories), we will then perform post hoc comparisons analogous to what we would do in a one-way ANOVA design. As there is no official post hoc option in the module, we perform additional chi-square analyses on the pairs of the categories in order to pinpoint the group differences.

62.4 FREQUENCIES ANALYSIS

We have performed a **Frequencies** analysis on the **distress_signal** variable to determine its distribution, the results of which are shown in Figure 62.1. As can be seen, the most prevalent category is **cognitive decline** with **48** (**43.6**%) of the cases, the next most prevalent category is **emotional deterioration** with **38** (**34.5**%) of the cases, and the least frequently seen category is **physical changes** with **24** (**21.8**%) of the cases.

62.5 ANALYSIS SETUP: OMNIBUS ANALYSIS

We open **distress signals** and from the main menu select **Analyze ➔ Nonparametric Tests ➔ One Sample**. In the **Objective** screen in the initial **One-Sample Nonparametric Tests** window as shown in Figure 62.2, we select **Customize analysis**.

In the **Fields** screen, we highlight **id** and move it into the **Fields** panel by clicking the horizontal arrow. This action leaves **distress_signal** in the **Test Fields** panel and switches the choice at the top left portion of the screen to **Use custom field assignments** as shown in Figure 62.3.

Selecting the **Settings** tab allows us to specify our statistical analysis. We select **Customize tests** in the **Choose Tests** screen and check **Compare observed probabilities to hypothesized (Chi-Square test)** as shown in Figure 62.4.

Selecting the **Options** pushbutton under the **Chi-Square test** opens the **Chi Square Test Options** window shown in Figure 62.5. In the **Choose Test Options** panel, we select **All categories have equal probability** for this omnibus analysis, with all three categories included. Click **OK** to return to the **Settings** window and click **Run** to perform the analysis.

62.6 ANALYSIS OUTPUT: OMNIBUS ANALYSIS

The summary output for the omnibus analysis is shown in Figure 62.6. Under **Null Hypothesis**, we are reminded in the first column that we opted for expecting equal frequencies. The chi-square value is statistically significant (p = **.019**), indicating that we should reject the null hypothesis.

distress_signal

		Frequency	Percent	Valid Percent	Cumulative Percent
Valid	1 cognitive decline	48	43.6	43.6	43.6
	2 emotional deterioration	38	34.5	34.5	78.2
	3 physical changes	24	21.8	21.8	100.0
	Total	110	100.0	100.0	

FIGURE 62.1 Output from **Frequencies** analysis.

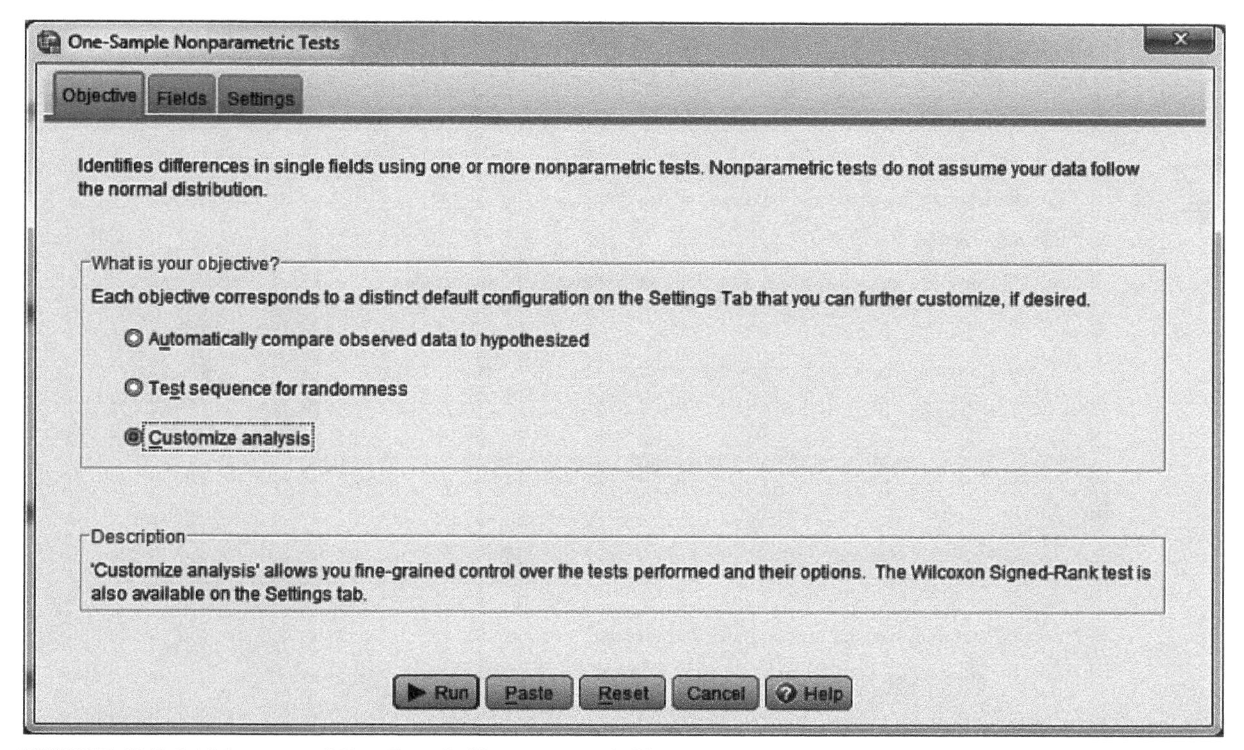

FIGURE 62.2 Initial screen of **One-Sample Nonparametric Tests** where we have selected **Customize analysis**.

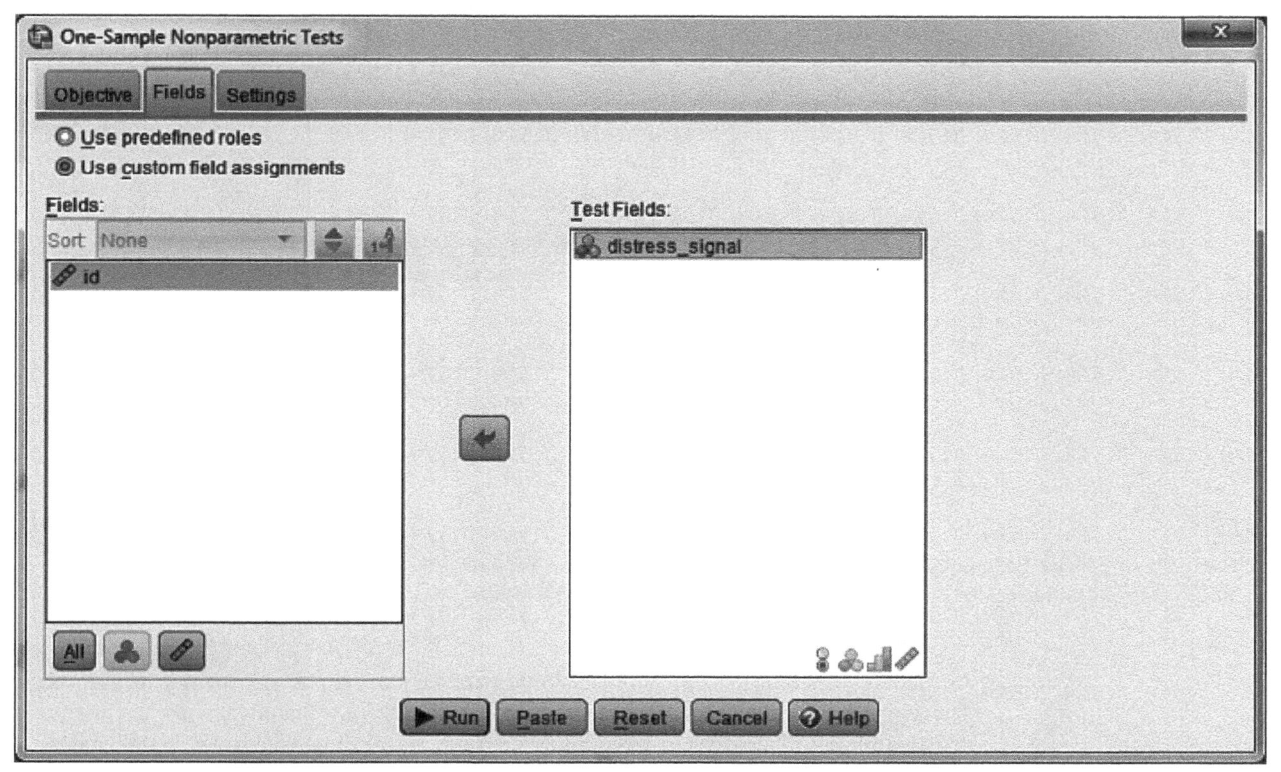

FIGURE 62.3 The **Fields** window configured with **distress_signal** in the **Test Fields** panel.

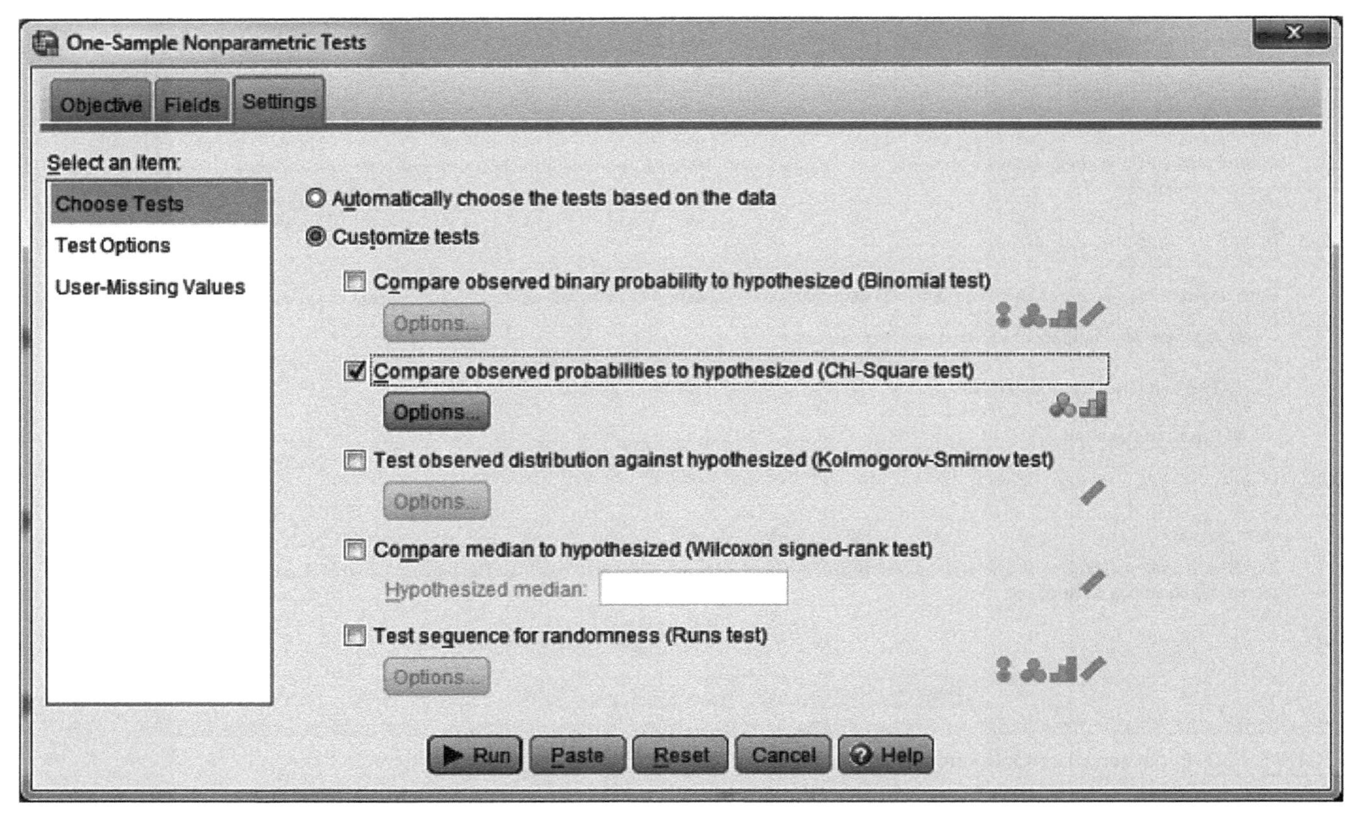

FIGURE 62.4 In the **Choose Tests** screen of **Settings**, we have chosen to **Customize** the **Chi-Square test**.

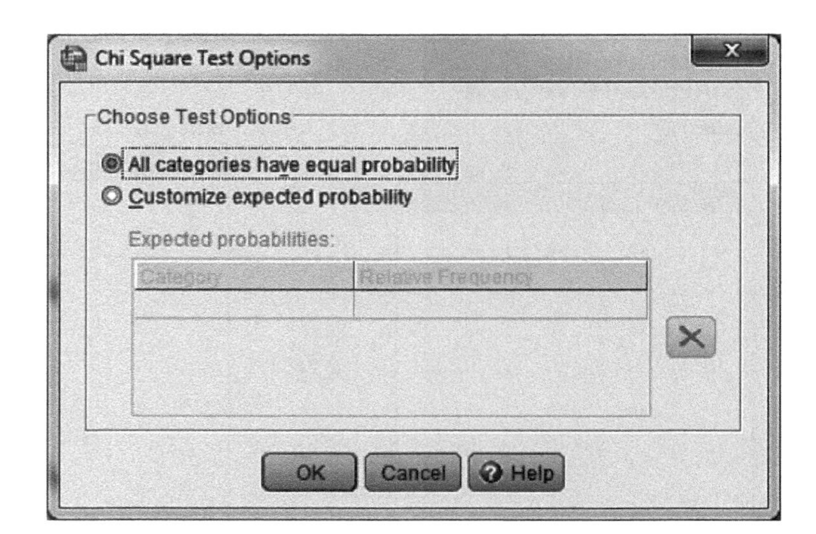

FIGURE 62.5
We have selected **All categories have equal probability** in the **Options** window.

Double-clicking on the summary table expands the output as shown in Figure 62.7. The chi-square value is shown in the table as **7.927**. Evaluated against a chi-square distribution with **2** degrees of freedom (because there are three categories in the analysis), the chances of obtaining that value or greater are **.019**. The omnibus analysis has thus informed us that the frequencies across the categories are not distributed equally. While

Hypothesis Test Summary

	Null Hypothesis	Test	Sig.	Decision
1	The categories of distress_signal occur with equal probabilities.	One-Sample Chi-Square Test	.019	Reject the null hypothesis.

Asymptotic significances are displayed. The significance level is .05.

FIGURE 62.6 The summary of the results for the omnibus analysis.

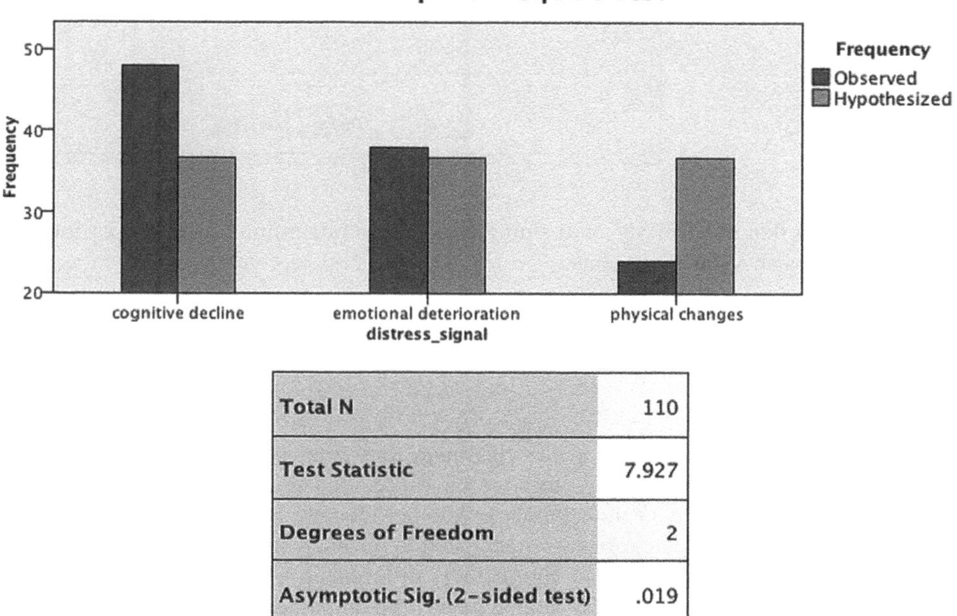

Total N	110
Test Statistic	7.927
Degrees of Freedom	2
Asymptotic Sig. (2-sided test)	.019

1. There are 0 cells (0%) with expected values less than 5. The minimum expected value is 36.667.

FIGURE 62.7 Detailed view of the results for the omnibus analysis.

this result may be sufficient for some research purposes, it is often useful to engage in finer-grain analyses by comparing pairs of categories or groups.

62.7 ANALYSIS SETUP: COMPARISON OF CATEGORIES 1 AND 2

We set up the analysis exactly as described earlier up to the point where we configure the **Chi-Square Options** window. As shown in Figure 62.8, we select **Customize expected probability**. In the **Expected probabilities** panel, we click the cell under **Category** in the first row, type **1**, click the cell next to it under **Relative Frequency**, and type **50**. We then do the same for the second row but now specify **Category 2**. This configuration accomplishes two goals at once:

- It restricts the chi-square analysis to only Categories **1** and **2**, thereby automatically excluding Category **3**. The total frequency is the sum of Categories **1** and **2**.
- It sets the expectation that the two categories in the analysis should have equal frequencies. We make sure that the values in the **Relative Frequency** column sum

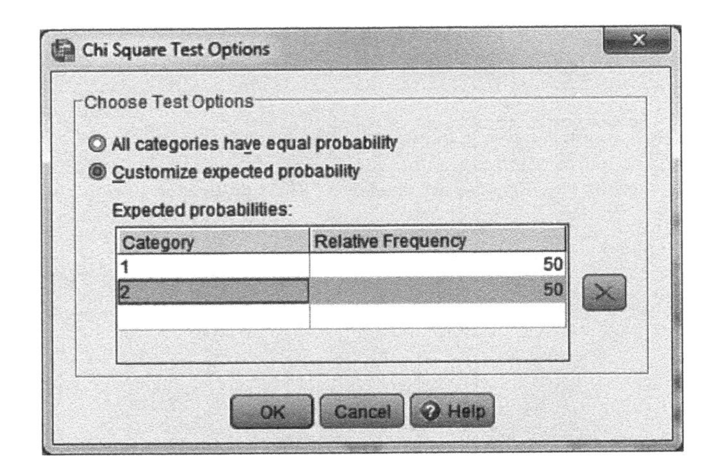

FIGURE 62.8
We have configured the **Options** to compare only the first and second categories.

to 100, in that IBM SPSS® will sum the values and determine what proportion of the total each value represents.

62.8 ANALYSIS OUTPUT: COMPARISON OF CATEGORIES 1 AND 2

Figure 62.9 presents the summary of the chi-square test comparing Categories **1** and **2**. Under the **Null Hypothesis**, we see that we have specified the probabilities but those specifications do not appear in the summary. Another element not included in the summary table is exactly what categories were included in the analysis. However, we do learn that the included categories were not significantly different in frequency (p = **.281**).

Double-clicking on the summary table expands the output as shown in Figure 62.10. The bar graph above the statistical table shows us that the categories in the analysis were **cognitive decline** and **emotional deterioration**, but the choice of scale on the Y-axis, starting at **38**, is unable to clearly display the observed frequency for **emotional deterioration** (it is 38). The chi-square value is shown in the table as **1.163**. Evaluated against a chi-square distribution with **1** degree of freedom (because there are two categories in the analysis), the chances of obtaining that value or greater are **.281**. We therefore conclude that the frequencies of cognitive and emotional distress signals in the sample were not statistically different.

62.9 ANALYSIS SETUP: COMPARISON OF CATEGORIES 1 AND 3

We set up the analysis exactly as described earlier up to comparing the first two categories, except that in the **Chi-Square Options** window, we specify **1** and **3** in the **Category** column, each still at a **Relative Frequency** of **50**. This is shown in Figure 62.11.

Hypothesis Test Summary

	Null Hypothesis	Test	Sig.	Decision
1	The categories of distress_signal occur with the specified probabilities.	One-Sample Chi-Square Test	.281	Retain the null hypothesis.

Asymptotic significances are displayed. The significance level is .05.

FIGURE 62.9
The summary of the results comparing Categories 1 and 2.

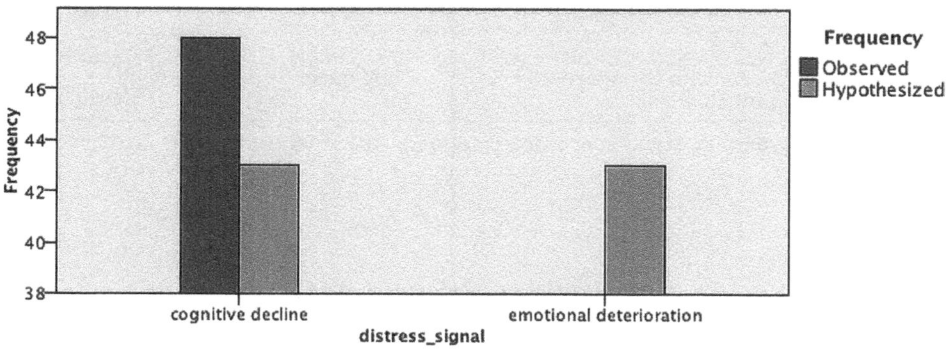

Total N	86
Test Statistic	1.163
Degrees of Freedom	1
Asymptotic Sig. (2-sided test)	.281

1. There are 0 cells (0%) with expected values less than 5. The minimum expected value is 43.

FIGURE 62.10 Detailed view of the results comparing Categories 1 and 2.

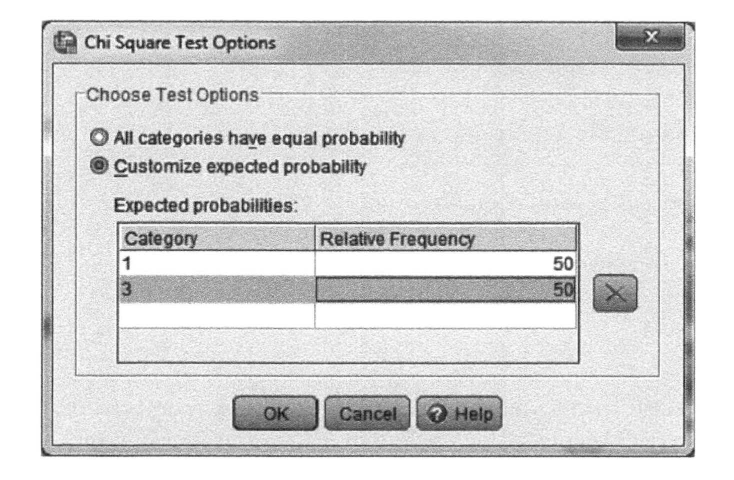

FIGURE 62.11

We have configured the **Options** to compare only the first and third categories.

62.10 ANALYSIS OUTPUT: COMPARISON OF CATEGORIES 1 AND 3

Figure 62.12 presents the summary of the chi-square test comparing Categories **1** and **3**. The results indicate that the included categories were significantly different in frequency ($p =$ **.005**).

Double-clicking on the summary table expands the output as shown in Figure 62.13. The bar graph above the statistical table identifies the categories in the analysis as **cognitive decline** and **physical changes**. The chi-square value is shown in the table as **8.000**.

FIGURE 62.12

The summary of the results comparing Categories 1 and 3.

Hypothesis Test Summary

	Null Hypothesis	Test	Sig.	Decision
1	The categories of distress_signal occur with the specified probabilities.	One-Sample Chi-Square Test	.005	Reject the null hypothesis.

Asymptotic significances are displayed. The significance level is .05.

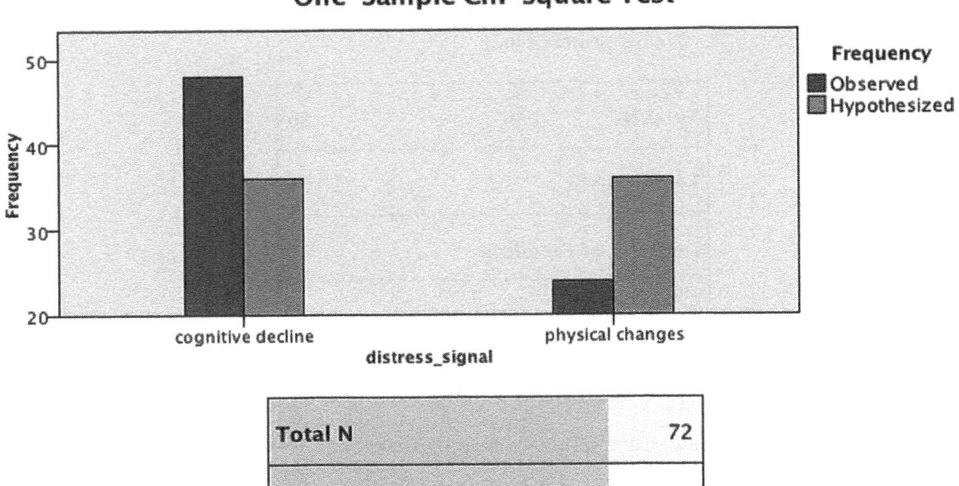

Total N	72
Test Statistic	8.000
Degrees of Freedom	1
Asymptotic Sig. (2-sided test)	.005

1. There are 0 cells (0%) with expected values less than 5. The minimum expected value is 36.

FIGURE 62.13 Detailed view of the results comparing Categories 1 and 3.

Evaluated against a chi-square distribution with **1** degree of freedom (because there are two categories in the analysis), the chances of obtaining that value or greater are **.005**. On the basis of this analysis, we therefore conclude that there were significantly more clients presenting cognitive signals of distress than those presenting physical signals of distress.

62.11 ANALYSIS SETUP: COMPARISON OF CATEGORIES 2 AND 3

We set up the analysis exactly as described earlier up to comparing the first two categories, except that in the **Chi-Square Options** window, we specify **2** and **3** in the **Category** column. This is shown in Figure 62.14.

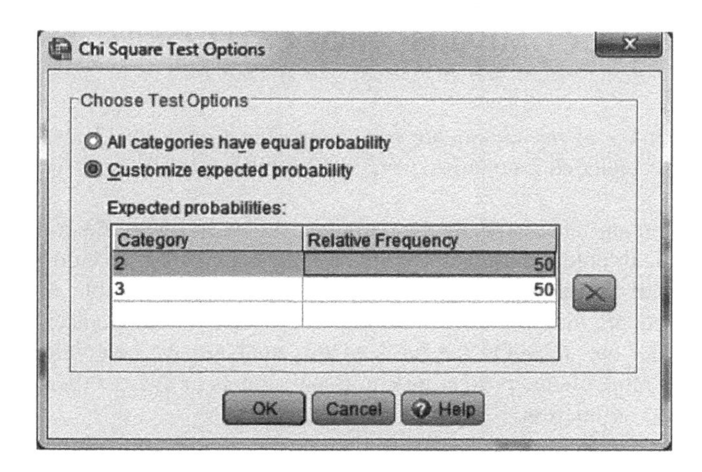

FIGURE 62.14

We have configured the **Options** to compare only the second and third categories.

Hypothesis Test Summary

	Null Hypothesis	Test	Sig.	Decision
1	The categories of distress_signal occur with the specified probabilities.	One-Sample Chi-Square Test	.075	Retain the null hypothesis.

Asymptotic significances are displayed. The significance level is .05.

FIGURE 62.15

The summary of the results comparing Categories 2 and 3.

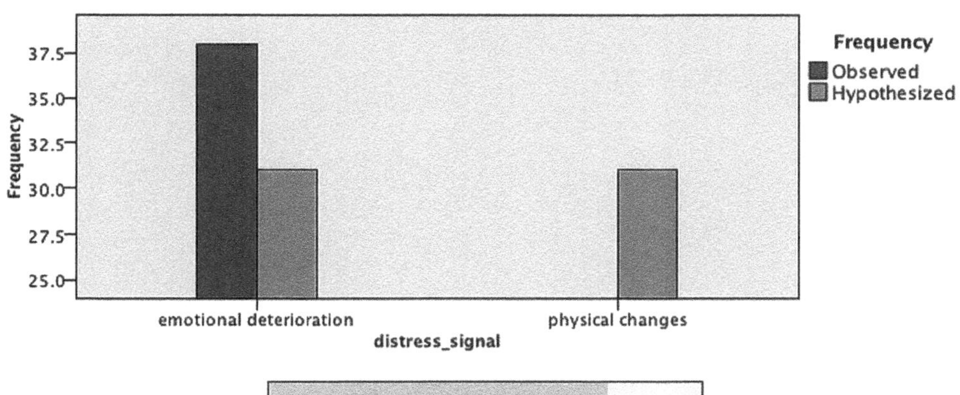

Total N	62
Test Statistic	3.161
Degrees of Freedom	1
Asymptotic Sig. (2-sided test)	.075

1. There are 0 cells (0%) with expected values less than 5. The minimum expected value is 31.

FIGURE 62.16 Detailed view of the results comparing Categories 2 and 3.

62.12 ANALYSIS OUTPUT: COMPARISON OF CATEGORIES 2 AND 3

Figure 62.15 presents the summary of the chi-square test comparing Categories **2** and **3**. The results indicate that the included categories were not significantly different in frequency ($p =$ **.075**).

Double-clicking on the summary table expands the output as shown in Figure 62.16. The bar graph above the statistical table identifies the categories in the analysis as **emotional deterioration** and **physical changes**. The chi-square value is shown in the table as **3.161**. Evaluated against a chi-square distribution with **1** degree of freedom, the chances of obtaining that value or greater are **.075**. On the basis of this analysis, we conclude that there were not significantly more clients presenting emotional signals of distress than those presenting physical signals of distress.

In conclusion, the post hoc analyses indicated that there were significantly more clients presenting symptoms of **cognitive decline** than symptoms of **physical changes** or symptoms of **emotional deterioration**.

CHAPTER 63

Two-Way Chi-Square Test of Independence

63.1 OVERVIEW

The two-way chi-square design is an extension of the one-way design covered in Chapters 61 and 62; the difference, and it is a substantial one, is that we include a second categorical variable in the analysis. The frequencies in a two-way design are placed in a table where one variable is represented by the rows (e.g., females and males) and another by the columns (e.g., whether the patron ordered a café latte or a café mocha at a local coffee house). Such a configuration is called a *two-way contingency table* in the sense that the frequencies are contingent on the levels of two categorical variables (e.g., females who ordered a café latte).

To distinguish between the two variables in the research study, some researchers place the variable representing the groups of interest on the rows and have the other variable (analogous to an outcome or focus of the study) represented by the columns, but such placement is arbitrary and other researchers suggest the reverse arrangement. For example, if we were studying the drink preferences of the two sexes in a local coffee house, we would envision the rows representing the different sexes and each column representing one of the coffee drinks. Such a configuration is less relevant when we are just comparing the observed with the expected frequencies as we do in this chapter but becomes somewhat more relevant in the analysis of risk that we address in Chapter 64.

The null hypothesis in a two-way design is that the two variables are *independent*. Consider the coffee house study where females and males are ordering drinks. If the two variables are independent, then roughly the same proportion of females as males would order each drink. For example, if 25% of each sex ordered a café latte (and thus 75% of each ordered a café mocha), then we would obtain a nonsignificant chi-square value and we would conclude that the two variables are independent; that is, if the two variables are independent, the same pattern of drink ordering would apply to each sex, and knowing the sex of the patron would not enhance our prediction of what drink is ordered.

If the two variables are related (if they are not independent), then the opposite of what we just described will occur. For example, if 10% of the females but 75% of the males ordered a café latte, then we would obtain a statistically significant chi-square value and we would conclude that the two variables are not independent (i.e., they are related). In this situation, sex of patron is related to the ordering preferences for the drinks, and sex becomes a predictor of ordering behavior; given the above scenario, males would be more likely than females to order a café latte.

Performing Data Analysis Using IBM SPSS®, First Edition.
Lawrence S. Meyers, Glenn C. Gamst, and A. J. Guarino.

Another difference between the one-way and two-way chi-square design is the manner in which the expected frequencies are generated. As we have seen in Chapter 62, in the one-way design, they are generated theoretically, actuarially, or on the basis of chance; in any case, it is the researchers who must decide on the strategy to generate the values of the expected frequencies. In the two-way design, the expected frequencies are generated *empirically* based on the observed frequencies, and they represent the null hypothesis (that there is no relationship between the variables).

In our coffee house study, with patron sex as the rows and the type of drink as the columns, we can establish the expected frequencies by first determining the column totals, that is, how many patrons overall ordered a café latte and how many overall ordered a café mocha. Suppose that 40 patrons ordered café latte and 60 ordered café mocha. The total, N, is thus 100, and of those 100 patrons, 40% overall ordered a café latte and 60% overall ordered a café mocha.

The expected frequencies for each sex are based on these observed overall (marginal) frequencies; thus, we would hypothesize that if the variables are independent (i.e., the null hypothesis), then 40% of the female patrons and 40% of male patrons would order a café latte and 60% of the female patrons and 60% of male patrons would order a café mocha. These expected frequencies are compared to the observed frequencies, and a chi-square test is used to determine if these two sets of frequencies are comparable (yielding nonsignificant chi-square value) or not (yielding statistically significant chi-square value).

To say that two variables are related is to say that they are correlated, and there are several indexes that assess the degree of relationship between categorical variables. Two indexes that are not applicable here are the tetrachoric (introduced by Pearson (1900a)) and polychoric correlations; these are used when underlying continuous or latent variables have been dichotomized or divided into three or more ordered categories, respectively. The three indexes that are applicable to a two-way contingency table are as follows:

- *Phi Coefficient.* Phi is a correlation coefficient describing the relationship between two truly categorical variables when each variable has only two levels (it can be applied only to 2×2 contingency tables). Karl Pearson (1900a) derived it from his Pearson r formula. Phi is computed by taking the square root of (chi-square/N), where N is the total frequency. We typically square phi and interpret it analogously to r^2 as the amount of variance shared by the two variables.
- *Contingency Coefficient.* This is a variant of phi coefficient used when the number of rows and columns is equal and exceeds 2×2. It is computed similarly to phi except that the denominator is $N +$ chi-square.
- *Cramer's V.* This is another variant of phi used when the number of rows or columns exceeds two levels. It is computed similarly to phi except that the denominator is $N * (k - 1)$, where k is the smaller value of the number of rows or columns.

Out of a concern that the chi-square test is less appropriate when applied to 2×2 contingency tables containing cells with low frequencies (because the chi-square distribution is continuous but frequencies are discontinuous and small frequencies accentuate the discontinuity), Yates (1934) offered a correction for this lack of continuity. One step in computing chi-square is to determine the difference between the observed and expected frequencies in each cell. Yates suggested reducing the difference by 0.5 to correct for the accentuated discontinuity in low observed frequencies (this correction has increasingly less effect for increasingly larger frequencies). For many years following the suggestion by Yates, this correction was routinely recommended for analyses, with observed cell frequencies ranging between 5 and 10 or so (see Meyers et al. (2009)). However, using this correction has become the subject of controversy over the past several decades, with several authors suggesting that it should not be used (e.g., Haviland, 1990; Jaccard & Becker, 1990) because it is too conservative a correction to the computed chi-square value.

63.2 ANALYSIS STRATEGY

A chi-square analysis can be performed on any (reasonable) sized two-way contingency table. However, each of the cells must have an expected frequency greater than 1 and most (e.g., 80% or better) of the cells should have expected frequencies greater than 5. We will illustrate an analysis of a 2 × 2 table first, and then extend that to an analysis of a 4 × 2 table.

63.3 NUMERICAL EXAMPLE: 2 x 2 CHI-SQUARE

The data we use for this example (selected from IBM SPSS® sample data files) are risk factors for 1048 patients at a large HMO. We deal with two of the variables here. For the variable **physically_active**, clients were dichotomized into those who were **not active** (coded as **0** in the data file) and those who were **active** (coded as **1** in the data file). For the variable **obesity**, clients were dichotomized into those whose were not diagnosed as obese (**no**, coded as **0** in the data file) and those who were (**yes**, coded as **1** in the data file). The data file is named **lifestyle medical study**. A screenshot of a portion of the data file is shown in Figure 63.1.

*lifestyle medical study.sav [DataSet1] - IBM SPSS Statistics Data Editor

File Edit View Data Transform Analyze Direct Marketing Graphs Utilities Add-ons Window Help

Visible: 11 of 11 Variables

	sex	physically_active	obesity	smoker	cholesterol	hist_angina	hist_myocardial_infarction	prescribed_nitroglycerin	hist_transient_ischemic	age_category	blood_pressure
1	0	1	0	0	0	0	0	0	0	3	0
2	0	1	0	0	0	0	0	0	0	2	1
3	1	1	0	1	0	1	0	0	0	3	2
4	1	1	1	0	0	0	0	0	0	1	1
5	0	0	1	0	0	1	0	0	0	1	0
6	0	1	0	0	1	0	0	0	0	2	2
7	1	1	0	0	1	0	0	0	0	1	1
8	0	0	0	0	0	0	0	0	0	1	2
9	1	0	0	0	0	0	0	0	0	1	0
10	0	1	0	1	1	1	0	0	0	3	1
11	1	0	0	0	1	1	0	0	0	2	2
12	1	0	0	1	0	0	0	0	0	3	1
13	0	0	0	1	1	0	1	0	1	3	0
14	1	1	0	0	0	0	0	1	1	3	2
15	0	0	0	0	0	0	0	0	0	2	2

Data View Variable View

IBM SPSS Statistics Processor is ready

FIGURE 63.1 A portion of the data file.

63.4 ANALYSIS SETUP: 2 x 2 CHI-SQUARE

We open **lifestyle medical study** and from the main menu select **Analyze ➔ Descriptive Statistics ➔ Crosstabs**. This produces the main **Crosstabs** dialog window shown in Figure 63.2. We move **physically_active** into the **Row(s)** panel to represent our groups for the purpose of this analysis, and we move **obesity** into the **Column(s)** panel to represent our outcome variable for the purpose of this analysis.

Selecting the **Exact** tab (some readers may not have this optional module, in which case the analysis defaults to the **Asymptotic** solution) opens the **Exact Tests** screen (see Figure 63.3) where we have selected **Exact**. The default time limit to perform these more memory-intensive computations (compared to the **Asymptotic** estimate) is set at **5** minutes, but it would take a huge data file for most personal computers to exceed this default. Click **Continue** to return to the main dialog window.

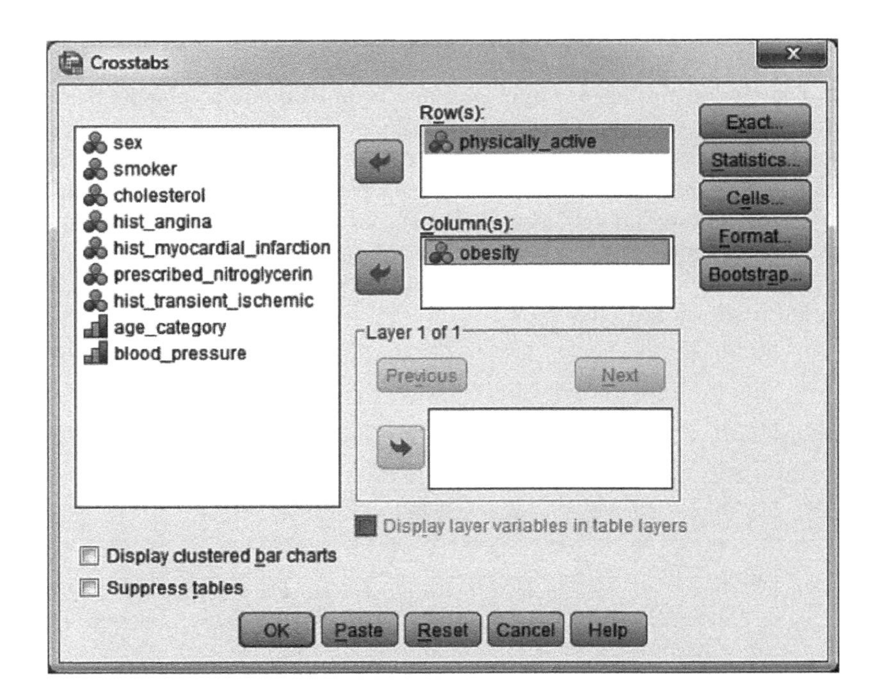

FIGURE 63.2
The main dialog window of **Crosstabs**.

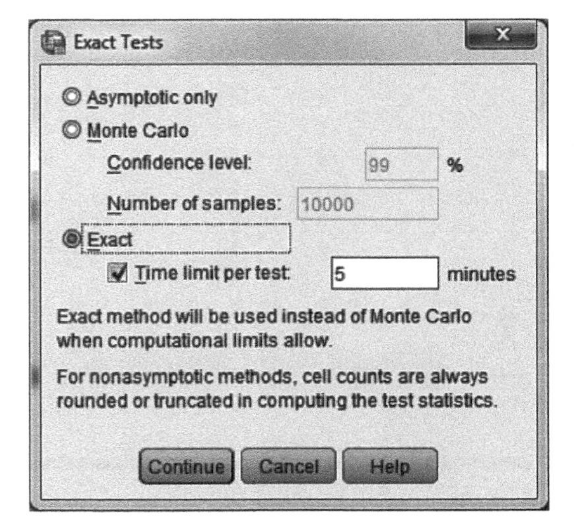

FIGURE 63.3
The **Exact** dialog window of **Crosstabs**.

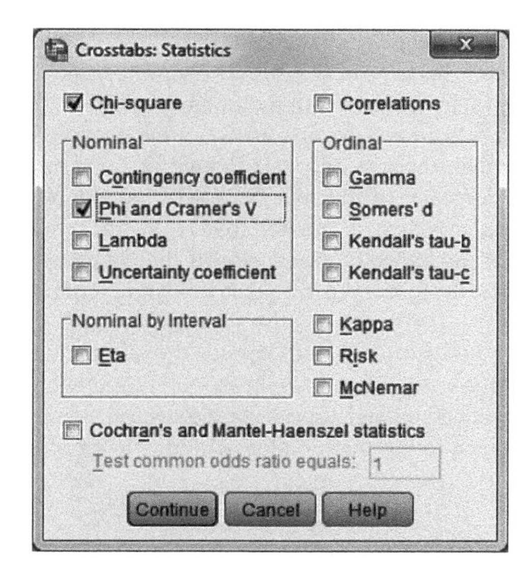

FIGURE 63.4
The **Statistics** dialog window of **Crosstabs**.

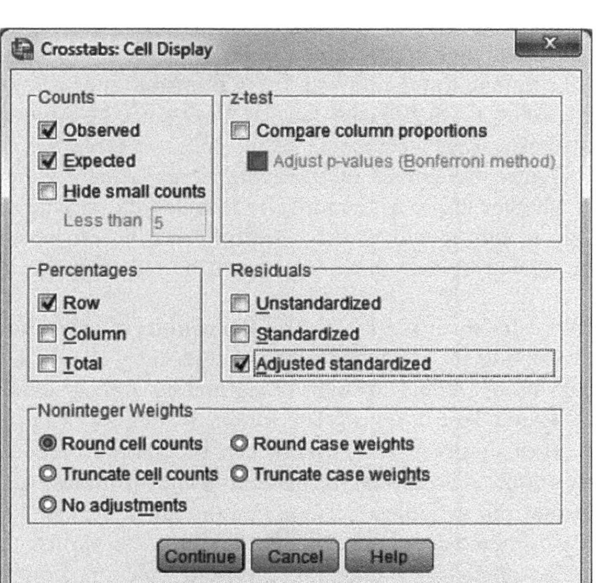

FIGURE 63.5
The **Cell Display** dialog window of **Crosstabs**.

Selecting the **Statistics** pushbutton produces the **Statistics** dialog window shown in Figure 63.4. We select **Chi-square** in the upper left area of the screen, and **Phi and Cramer's V** in the **Nominal** area (for use with categorical variables). Click **Continue** to return to the main dialog window.

Selecting the **Cells** pushbutton produces the **Cell Display** dialog window shown in Figure 63.5. In the **Counts** area, **Observed** is already checked; we also select **Expected**. In the **Percentages** area, we select **Row** (this will facilitate in understanding how the expected frequencies were generated).

In the **Residuals** area, we select **Adjusted standardized**. Residuals are the differences between the observed and expected values. The **Unstandardized** residuals are the differences based on the original metric and may not be interpretable if the metrics are different (e.g., weight in pounds and heart beats in beats per minute). To correct this problem, the residuals can be **Standardized** and interpreted as any standard score on the normal curve with a mean of 0 and a standard deviation of 1. If the **Standardized** residual for a particular cell exceeds the absolute value of 1.96, then that cell is associated with a statistically significant difference between the expected and the observed count.

The **Adjusted Standardized** residual takes into account both the number of comparisons made and sample size and reports a more accurate difference between the observed and expected counts. In a 2 × 2 contingency table, there is only one degree of freedom, and so the **Adjusted Standardized** residuals for each of the four cells will have equal absolute values. If the **Adjusted Standardized** residual is greater than 1.96, then the observed frequency is significantly greater than the expected frequency; if the **Adjusted Standardized** residual is less than −1.96, then the observed frequency is significantly less than the expected frequency.

In designs with more cells, the chi-square test is an omnibus test, and examination of the **Adjusted Standardized** residual can inform us of which cells are associated with statistically significant differences between the observed and expected frequencies. Select **Adjusted Standardized**, click **Continue** to return to the main dialog window, and click **OK** to perform the analysis.

63.5 ANALYSIS OUTPUT: 2 × 2 CHI-SQUARE

The overall results of the analysis are shown in the **Chi-Square Tests** table in Figure 63.6. The **Pearson Chi-Square** was the original test employed for **Crosstabs** and is still commonly reported. The chi-square value is **26.398** and labeled **Pearson Chi-Square** by IBM SPSS in honor of its developer, Karl Pearson (1900b). A 2 × 2 table has **1** degree of freedom and, evaluated against its degrees of freedom, the chi-square value is statistically significant ($p < .001$) whether tested asymptotically (the default output) or exactly (as we requested). Thus, we would conclude that being physically active and presenting symptoms of obesity are related to each other (we reject the null hypothesis of independence).

The results of other tests are also shown in the table. The **Continuity Correction** results from applying the Yates correction to the frequencies. A more recently developed test is the **Likelihood Ratio**, which for more complex designs (e.g., three-way analyses) is almost exclusively applied. **Fisher's Exact Test** is appropriate for small samples or large samples where some cells have small or empty counts; here it can be ignored because none of the cells had small or empty counts. The **Linear-by-Linear Association** test can also be ignored here because it assumes the variables are on a continuous scale.

The estimates of the strength of relationship between the variables are shown in Figure 63.7. With a 2 × 2 design and a statistically significant chi-square value indicating

Chi-Square Tests

	Value	df	Asymp. Sig. (2-sided)	Exact Sig. (2-sided)	Exact Sig. (1-sided)	Point Probability
Pearson Chi-Square	26.398[a]	1	.000	.000	.000	
Continuity Correction[b]	25.657	1	.000			
Likelihood Ratio	26.738	1	.000	.000	.000	
Fisher's Exact Test				.000	.000	
Linear-by-Linear Association	26.373[c]	1	.000	.000	.000	.000
N of Valid Cases	1048					

a. 0 cells (.0%) have expected count less than 5. The minimum expected count is 122.34.
b. Computed only for a 2x2 table
c. The standardized statistic is -5.135.

FIGURE 63.6 Chi-square values from the analysis.

Symmetric Measures

		Value	Approx. Sig.	Exact Sig.
Nominal by Nominal	Phi	-.159	.000	.000
	Cramer's V	.159	.000	.000
N of Valid Cases		1048		

FIGURE 63.7 Estimates of the strength of relationship between the variables.

that the two variables are related, we focus on the **Phi** value of **−.159**, ignoring the negative sign, to address the strength of this relationship. This is the correlation between the two variables, and squaring this value yields .025. We would therefore infer that the two variables share about 2.5% common variance (or that physical activity accounts for approximately 2.5% of the variance of obesity). This may be judged to be a weak relationship in many academic laboratory studies, but it may be clinically important in applied medical research as the amount of increased risk of obesity from a lack of physical activity may be worthy of attention (see Chapter 64).

Figure 63.8 presents the observed and expected frequencies together with row percentages and adjusted residual values. The bottom row labeled as **Total** gives us the strategy to compute our expected frequencies. Of the 1048 clients, 800 (76.3%) of them were not obese and 248 (23.7%) of them were evaluated as obese. These overall or marginal percentages were then applied to the observed frequencies to generate the expectations. For example, of the 531 **not active** clients, it was expected based on the null hypothesis that 76.3% of them would not be obese; this translates to 531 * .763, or approximately 405, clients.

physically_active * obesity Crosstabulation

			obesity		
			0 no	1 yes	Total
physically_active	0 not active	Count	370	161	531
		Expected Count	405.3	125.7	531.0
		% within physically_active	69.7%	30.3%	100.0%
		Adjusted Residual	-5.1	5.1	
	1 active	Count	430	87	517
		Expected Count	394.7	122.3	517.0
		% within physically_active	83.2%	16.8%	100.0%
		Adjusted Residual	5.1	-5.1	
Total		Count	800	248	1048
		Expected Count	800.0	248.0	1048.0
		% within physically_active	76.3%	23.7%	100.0%

FIGURE 63.8 The observed and expected frequencies together with row percentages and adjusted residual values.

With a statistically significant chi-square, we anticipate that the **Adjusted Standardized Residuals** for each cell (they will be equal in a 2 × 2 table) are statistically significant. In this example, the absolute value of **5.1** can be interpreted as a *z* score and, in exceeding an absolute value of 1.96, would be considered statistically significant; thus, (as is true by definition in a 2 × 2 table when the chi-square is significant), the observed frequency for each cell significantly departs from its expected cell frequency. Thus, those who were not active were more frequently diagnosed as obese than expected, whereas those who were active were less frequently diagnosed as obese.

63.6 NUMERICAL EXAMPLE: 4 x 2 CHI-SQUARE

We use the same **lifestyle medical study** data file but select two different variables for the present analysis. The variable of **age_category** codes clients into four categories based on their age: **45–54** years (coded as **1** in the data file), **55–64** years (coded as **2** in the data file), **65–74** years (coded as **3** in the data file), and **75+** years (coded as **4** in the data file). The variable of **myocardial_infarction** indicates whether clients have not (**no**, coded as **0** in the data file) or have (**yes**, coded as **1** in the data file) had a history of a heart attack.

63.7 ANALYSIS SETUP: 4 x 2 CHI-SQUARE

We open **lifestyle medical study** and from the main menu select **Analyze ➔ Descriptive Statistics ➔ Crosstabs**. This produces the main **Crosstabs** dialog window shown in Figure 63.9. We move **age_category** into the **Row(s)** panel to represent our groups for the purpose of this analysis, and we move **myocardial_infarction** into the **Column(s)** panel to represent our outcome variable for the purpose of this analysis. The analysis is otherwise configured in exactly the same way as described in Section 63.4.

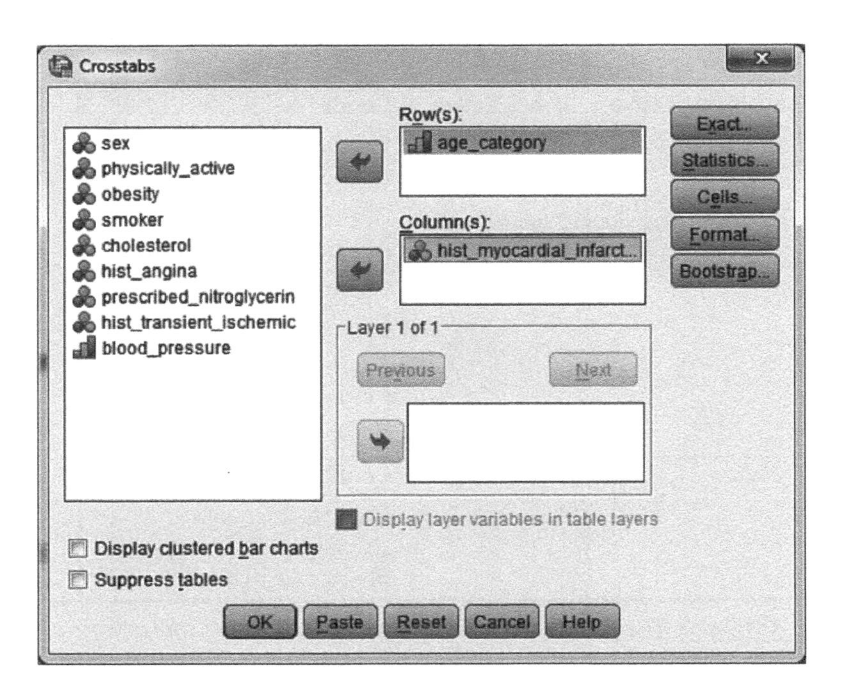

FIGURE 63.9
The main dialog window of **Crosstabs**.

Chi-Square Tests

	Value	df	Asymp. Sig. (2-sided)	Exact Sig. (2-sided)	Exact Sig. (1-sided)	Point Probability
Pearson Chi-Square	36.277[a]	3	.000	.000		
Likelihood Ratio	36.743	3	.000	.000		
Fisher's Exact Test	36.611			1.000		
Linear-by-Linear Association	35.757[b]	1	.000	.000	.000	.000
N of Valid Cases	1048					

a. 0 cells (.0%) have expected count less than 5. The minimum expected count is 34.83.
b. The standardized statistic is 5.980.

Symmetric Measures

		Value	Approx. Sig.	Exact Sig.
Nominal by Nominal	Phi	.186	.000	.000
	Cramer's V	.186	.000	.000
N of Valid Cases		1048		

FIGURE 63.10 Omnibus chi-square output.

63.8 ANALYSIS OUTPUT: 4 x 2 CHI-SQUARE

The results of the omnibus analysis are shown in the **Chi-Square Tests** table in Figure 63.10. The **Pearson Chi-Square** value is **36.277**. Degrees of freedom for a two-way contingency table are equal to (number of rows − 1) * (number of columns − 1); here that is (4 − 1) * (2 − 1), or 3 * 1 or 3. Evaluated against **3** degrees of freedom, the chi-square value is statistically significant ($p < .001$) whether tested asymptotically (the default output) or exactly (as we requested). Thus, we would conclude that having a history of a heart attack is related to age. **Cramer's V** is **.186**. Squaring **.186** yields a value of approximately .035; it therefore appears that the two variables share approximately 3.5% of their variance.

Figure 63.11 presents the observed and expected frequencies together with row percentages and adjusted residual values. The bottom row labeled as **Total** gives us the strategy to compute our expected frequencies. Of the 1048 clients, 765 (73%) of them had no heart attack history and 283 (27%) of them did have a history of heart attack. This was then applied to the observed frequencies to generate the expectations. For example, of the 240 clients who were aged between 45 and 54 years, it was expected based on the null hypothesis that 73% of them would not have had a heart attack; this translates to 240 * .73, or approximately 175, clients.

With a statistically significant chi-square, we can examine the **Adjusted Standardized Residuals** for each cell (to "simplify" the omnibus analysis analogous to multiple comparison tests following an omnibus ANOVA). As these are interpreted as z scores, we are looking for absolute residuals values of 1.96 or greater to meet an alpha level of .05. Only two cells do not meet this threshold, those for the age group of 55–64 years; thus, there were no statistically significant differences between the observed and expected frequencies for these cases. For the other groups, all of the standardized residuals are greater than 1.96 in absolute value and indicate that the observed and expected frequencies differ significantly. Thus, those in the 45–54 years age group had significantly fewer heart attacks than expected, whereas those in the 65–74 and 75+ years age groups had significantly more heart attacks than expected.

age_category * hist_myocardial_infarction Crosstabulation

			hist_myocardial_infarction		Total
			0 no	1 yes	
age_category	1 45-54	Count	203	37	240
		Expected Count	175.2	64.8	240.0
		% within age_category	84.6%	15.4%	100.0%
		Adjusted Residual	4.6	-4.6	
	2 55-64	Count	286	97	383
		Expected Count	279.6	103.4	383.0
		% within age_category	74.7%	25.3%	100.0%
		Adjusted Residual	.9	-.9	
	3 65-74	Count	202	94	296
		Expected Count	216.1	79.9	296.0
		% within age_category	68.2%	31.8%	100.0%
		Adjusted Residual	-2.2	2.2	
	4 75+	Count	74	55	129
		Expected Count	94.2	34.8	129.0
		% within age_category	57.4%	42.6%	100.0%
		Adjusted Residual	-4.3	4.3	
Total		Count	765	283	1048
		Expected Count	765.0	283.0	1048.0
		% within age_category	73.0%	27.0%	100.0%

FIGURE 63.11 The observed and expected frequencies together with row percentages and adjusted residual values.

Note that in our example, our expected frequencies were based on the observed frequencies of the sample as a whole. This served to illustrate how we would perform the analysis on a contingency table larger than 2 × 2. However, if we were interested in comparing our sample of those in the 75+ years age group (for example) to the percentage based on a national average, we would perform a one-way chi-square analysis as described in Chapter 62. Such a restructuring implies that we would be examining a different research question than that was dealt with in our present example.

CHAPTER 64

Risk Analysis

64.1 OVERVIEW

Used extensively in medical and health behavior research but applicable in a variety of fields, the 2 × 2 contingency tables provide us with the opportunity to compute indexes of risk assessment. The two indexes we address here are the *risk ratio* (also called *relative risk*) and the *odds ratio*. These ratios provide a comparison of two groups in terms of risk. Because they are both ratios, a risk ratio of 1.00 and an odds ratio of 1.00 both indicate that there is no difference in risk between the groups involved in the comparison. However, these two indexes have different bases and are interpreted differently (e.g., Cohen, 2000; Holcomb, Chaiworapongsa, Luke, & Burgdorf, 2001).

Risk represents the likelihood or probability that a condition or event of interest (positive or negative) central to the main issue of the study will happen or will be observed (e.g., a patient will experience a heart attack or become free of a certain debilitating symptom). Risk is a rate of occurrence that is assessed on two groups of cases. One of the groups can generally be thought of as representing a factor of interest in the study; for example, the group can represent a "risk" condition or a treatment group. The other group can be thought of as the comparison, base, or reference condition representing, for example, a "less risky" condition or a no-treatment group.

A risk ratio is the ratio of two probabilities: (a) the probability of displaying the condition of interest for the (higher) risk condition and (b) the probability of not displaying the condition of interest for the lower risk condition. We often use the probability associated with the high risk group in the numerator, but this decision depends on the context of the research study.

Odds themselves are ratios that are not translatable into probabilities. They are determined by dividing the number of occurrences of an event by the number of times the event is not observed. For example, if the event occurs thrice and does not occur twice out of five possibilities, then the odds are 3 to 2 (a ratio of 1.5) in favor of the event occurring. We determine the odds for the same two groups that we described for the risk ratio.

An odds ratio (discussed in Chapters 30 and 32) is the ratio of the odds for each group (which makes it a ratio of two ratios and rendering it a little more difficult to conceptualize). We often use the odds associated with the risk group in the numerator, but again this depends on the context of the research study.

Performing Data Analysis Using IBM SPSS®, First Edition.
Lawrence S. Meyers, Glenn C. Gamst, and A. J. Guarino.

An odds ratio greater than 1.00 indicates that the odds for the condition of interest occurring in the risk group are higher relative to the odds for the base or comparison group. An odds ratio less than 1.00 indicates that the odds for the condition of interest occurring in the risk group are lower relative to the odds for the base or comparison group (Harris & Taylor, 2009).

To illustrate how these ratios are generated, consider the table of frequencies shown in Figure 63.8 (and reproduced in Figure 64.3) where we have crossed physical activity with obesity. Let us assume that our interest is in the risk of becoming obese. Let us further assume that one possible factor responsible for obesity is lack of physical activity. Our most basic question concerns the obesity consequences of physical activity (the risk of obesity for each physical activity group), given the frequency data collected from the sample.

The risk ratio is a ratio of the obesity rate or probability of occurrence of obesity for each physical activity group. The rate of obesity is computed by dividing the number of observed obese cases by the number of cases in total. Of the 531 cases in the **not active** group, 161 were diagnosed as obese. The rate of occurrence (the probability) of obesity for this group is 161/531, or approximately .303. Of the 517 cases in the **active** group, 87 were defined as obese. The rate of occurrence of obesity for this group is 87/517, or approximately .168. Dividing the risk for the "higher risk" (**not active**) condition by the risk for the "lower risk" (**active**) condition yields a risk ratio of .303/.168, or approximately 1.80. Thus, the risk of becoming obese is approximately 1.8 times higher for those who are not physically active relative to those who are physically active.

IBM SPSS® will actually calculate the risk for the other outcome as well, which is in this case, the outcome of not being diagnosed as obese. The analogous reasoning is as follows. Of the 531 cases in the **not active** group, 370 were diagnosed as not obese. The rate of occurrence (the probability) of not being obese for this group is 370/531, or approximately .697. Of the 517 cases in the **active** group, 430 were defined as not obese. The rate of occurrence of not being obese for this group is 430/517, or approximately .832. Dividing the risk for the "higher risk" (**not active**) condition by the risk for the "lower risk" (**active**) condition yields a risk ratio of .697/.832, or approximately .83. Thus, the "risk" of not being obese is approximately .83 times as great (it is lower) for those who are not physically active than for those who are physically active. This outcome, with its verbalization as a double negative, is probably a less elegant and clear but an equally valid way to communicate the results of the study.

The odds ratio is a ratio of the odds of being diagnosed as obese for each physical activity group. Odds are computed by dividing the number of cases with the target outcome (those who are obese) by the number of cases with the alternative outcome. For the **not active** group, 161 were diagnosed as obese and 370 were diagnosed as not being obese. The odds of obesity in this **not active** group are 161/370, or .435. For the **active** group, 87 were diagnosed as obese and 430 were diagnosed as not being obese. The odds of obesity in this **active** group are 87/430, or .202.

Dividing the odds for the "higher risk" condition by the odds for the "lower risk" condition yields an odds ratio of 2.153. Thus, the odds of being diagnosed as obese for those who are not physically active is over twice as great as the odds of being diagnosed as obese for those who are physically active. IBM SPSS actually calculates the other side of this coin by dividing the group coded as **0** (our "lower risk" condition) by the group coded as **1** (our "higher risk" condition). This computation yields an odds ratio of 0.464 (it is the inverse of 2.153 in that 1/2.153 = .464) and indicates that the odds of being diagnosed as obese for those who are physically active are about half as great as the odds for those who are not physically active being diagnosed as obese.

The differences between how risk ratios and odds ratios are computed link to their application in different types of research designs (MacDonald, 2012; Schlesselman, 1992). Generally, studies that are *prospectively* (future) oriented are conducive to the use of the risk ratio in that the risk ratio (as an index of incidence or probability of an event occurring) makes sense when we are projecting what the future might hold in store. In a cohort study, for example, a sample of cases is selected and measured or followed over time often to determine if a certain medical condition surfaces, and a risk ratio would be an appropriate index of risk in such a study.

On the other hand, studies that are *retrospective* (past) oriented are not conducive to the use of the risk ratio, and we must revert to the odds ratio instead. For example, in a case control study, cases with the condition of interest are selected and compared to those who do not have the condition. Often, such a study is carried out on archival records (an existing data file). The incidence rate of the condition is not applicable here in that the cases were selected *because* they either had the condition or not. With the incidence rate not applicable in such studies because the cases were selected on the basis of having or not having the particular condition, we cannot meaningfully interpret the risk ratio; instead, we use the odds ratio as our gauge of risk.

The risk ratio and the odds ratio calculated on the same 2×2 contingency table will both be either greater than or less than 1.00, although their absolute values and interpretation will differ. As pointed out by Merrill (2013), the odds ratio will generally overestimate the strength of the relationship between the variables. More specifically, the odds ratio will approximate the value of the risk ratio when the condition of interest occurs relatively infrequently (in the range 10% or less), but the values of the two ratios will tend to be different for relatively more frequent events (Harris & Taylor, 2009).

64.2 NUMERICAL EXAMPLE

We use two of the variables here as we did in the first analysis presented in Chapter 63. For the variable **physically_active**, clients were dichotomized into those who were **not active** (coded as **0** in the data file) and those who were **active** (coded as **1** in the data file). For the variable **obesity**, clients were dichotomized into those who were not obese (**no**, coded as **0** in the data file) and those whose were (**yes**, coded as **1** in the data file). The data file is named **lifestyle medical study**.

64.3 ANALYSIS SETUP

We open **lifestyle medical study** and from the main menu select **Analyze → Descriptive Statistics → Crosstabs**. This produces the main **Crosstabs** dialog window shown in Figure 64.1. We configure this analysis exactly as we did in Chapter 63 but with one exception: in the **Statistics** window, in addition to selecting **Chi-square** and **Phi and Cramer's V**, we select **Risk**. This is shown in Figure 64.2.

64.4 ANALYSIS OUTPUT

The overall results of the analysis are identical to those shown in Chapter 63, and so we show only the table of observed and expected frequencies in Figure 64.3. What is new with this analysis is the **Risk Estimate** table shown in Figure 64.4.

The two risk ratios are provided in the second and third rows of the table. We have already described how these values were calculated and how they are interpreted.

- The row labeled **For cohort obesity = 0 no** focuses on the absence of obesity. Dividing the risk for the "higher risk" (**not active**) condition by the risk for the "lower risk" (**active**) condition yields a risk ratio of .697/.832, or approximately **.838**.

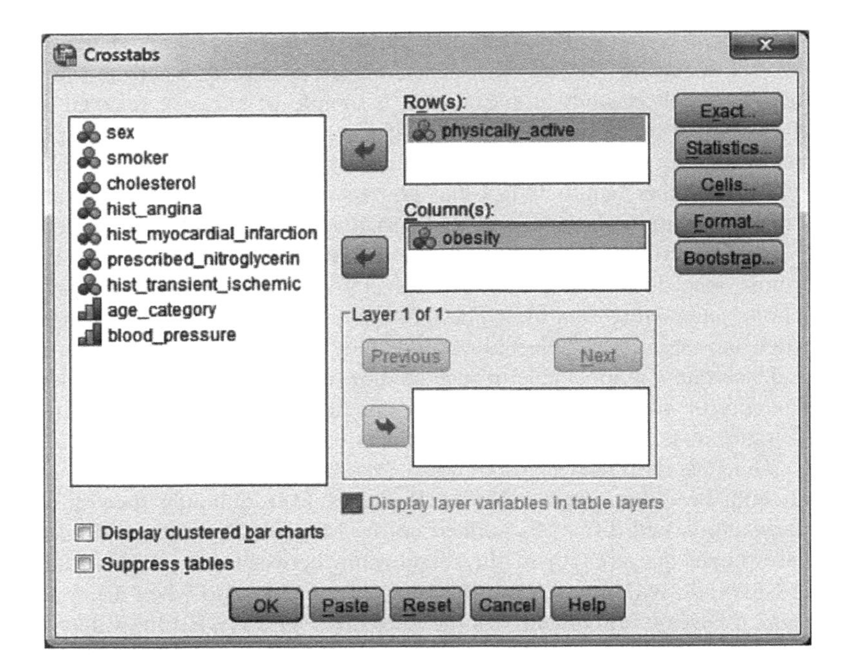

FIGURE 64.1
The main dialog window of **Crosstabs**.

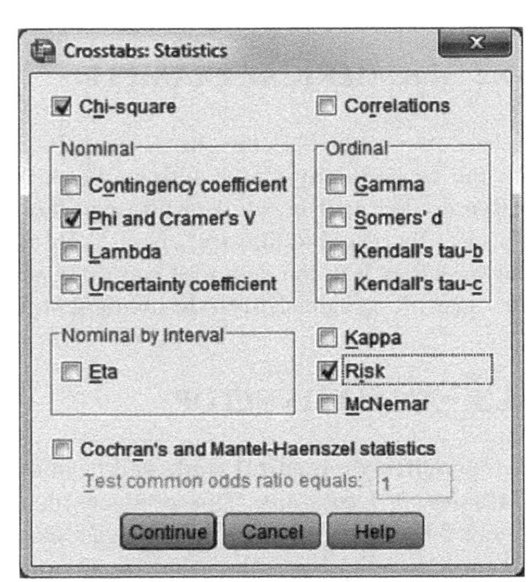

FIGURE 64.2
The **Statistics** dialog window of **Crosstabs**.

- The row labeled **For cohort obesity = 1 yes** focuses on the occurrence of obesity. Dividing the risk for the "higher risk" (**not active**) condition by the risk for the "lower risk" (**active**) condition yields a risk ratio of .303/.168, or approximately **1.802**.

The output also provides the lower and upper values for the 95% confidence interval for each ratio, as it is appropriate for researchers to provide these when reporting the risk ratio.

Only one odds ratio is computed with the row label of **Odds Ratio for physically_active (0 not active / 1 active)**. Given the label, it may be clear that this computation uses odds for the **not active** group in the numerator, focusing on the lack of obesity (the **0** code) rather than the occurrence of obesity. We can calculate this odds ratio in two ways.

physically_active * obesity Crosstabulation

			obesity		
			0 no	1 yes	Total
physically_active	0 not active	Count	370	161	531
		Expected Count	405.3	125.7	531.0
		% within physically_active	69.7%	30.3%	100.0%
		Adjusted Residual	-5.1	5.1	
	1 active	Count	430	87	517
		Expected Count	394.7	122.3	517.0
		% within physically_active	83.2%	16.8%	100.0%
		Adjusted Residual	5.1	-5.1	
Total		Count	800	248	1048
		Expected Count	800.0	248.0	1048.0
		% within physically_active	76.3%	23.7%	100.0%

FIGURE 64.3 The observed and expected frequencies together with row percentages and adjusted residual values.

Risk Estimate

		95% Confidence Interval	
	Value	Lower	Upper
Odds Ratio for physically_active (0 not active / 1 active)	.465	.346	.625
For cohort obesity = 0 no	.838	.783	.897
For cohort obesity = 1 yes	1.802	1.430	2.270
N of Valid Cases	1048		

FIGURE 64.4 Risk analysis results.

- We can say that the odds of not being obese for the **not active** group are 370/161, or approximately 2.298, and the odds for not being obese for the **active** group are 430/87, or approximately 4.943. Dividing 2.298 by 4.943 yields an odds ratio of approximately .465.
- We can say that the odds of being obese for the **not active** group are 161/370, or approximately .435, and the odds for being obese for the **active** group are 87/430, or approximately .202. Dividing .202 by .435 yields an odds ratio of approximately .465.

Whichever way we view it, the odds of developing obesity for those who are physically active are a little less than half the odds of developing obesity for those who are not physically active. Again, the table provides the lower and upper values for the 95% confidence interval, as it is appropriate for researchers to provide these when reporting the odds ratio.

CHAPTER 65

Chi-Square Layers

65.1 OVERVIEW

One way to enrich the information we obtain in a two-way chi-square design is to superimpose the presence of a third categorical variable, referred to as a **layer** in IBM SPSS®. This third variable serves as a stratifying variable; that is, we examine our ordinary two-way contingency table for each category of the **layer** variable. For example, we could use a variable indicating whether the person was a smoker or not in our obesity and physical activity study. In such a situation, we would obtain separate chi-square analyses (physical activity by obesity) for smokers, nonsmokers, and the combined sample.

65.2 NUMERICAL EXAMPLE

We use the same two variables here as we did in the analysis presented in Chapter 64. For the variable **physically_active**, clients were dichotomized into those who were **not active** (coded as **0** in the data file) and those who were **active** (coded as **1** in the data file). For the variable **obesity**, clients were dichotomized into those who were not obese (**no**, coded as **0** in the data file) and those who were (**yes**, coded as **1** in the data file). For this analysis, we now add the binary coded variable of **smoker**, coded as **0** for **no** and **1** for **yes**. The data file is named **lifestyle medical study**.

65.3 ANALYSIS SETUP

We open **lifestyle medical study** and from the main menu select **Analyze ➔ Descriptive Statistics ➔ Crosstabs**. This produces the main **Crosstabs** dialog window shown in Figure 65.1. We move **physically_active** into the **Row(s)** panel and **obesity** into the **Column(s)** panel as we have done in previous analyses. However, we also move **smoker** into the **Layer** panel as shown in Figure 65.1.

The remaining setup is the same as we described in Chapter 64. Briefly, choose the **Exact** pushbutton (if it is available) and select **Exact** (see Figure 65.2). In the **Statistics**

Performing Data Analysis Using IBM SPSS®, First Edition.
Lawrence S. Meyers, Glenn C. Gamst, and A. J. Guarino.

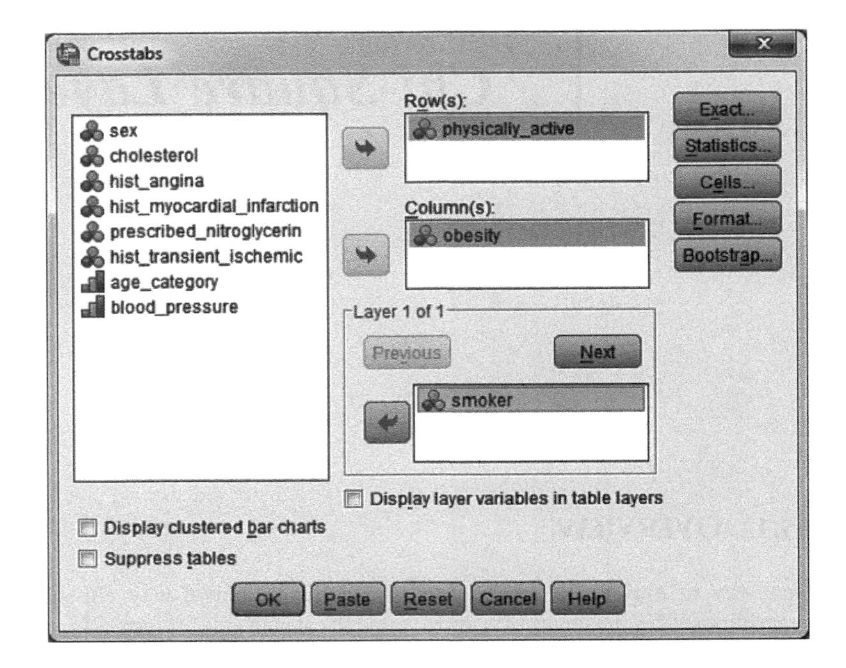

FIGURE 65.1

The main dialog window of **Crosstabs** including a **Layer** variable.

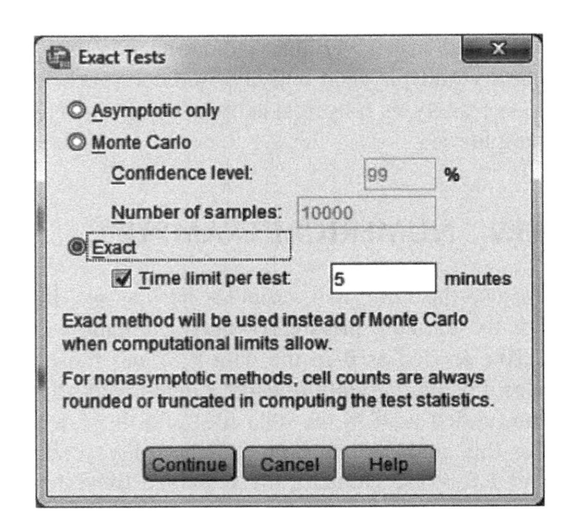

FIGURE 65.2

The **Exact** dialog window of **Crosstabs**.

dialog window shown in Figure 65.3, we select **Chi-square** and **Phi and Cramer's V** in the **Nominal** area, and **Risk**. In the **Cell Display** dialog window shown in Figure 65.4, we select **Observed**, **Expected**, **Row Percentages**, and **Adjusted standardized**. Click **OK** to perform the analysis.

65.4 ANALYSIS OUTPUT

Figure 65.5 presents the chi-square values for the layer analysis. The bottom major row of the **Chi-Square Tests** table is a full-sample result (labeled as **Total**) and is identical to the results described in Chapter 63; the **Pearson Chi-Square** is **26.398** and is statistically significant ($p < .001$). But in the layer analysis, this overall analysis is partitioned into the two levels of **smoker**, and these results are shown in the first two major rows of the table.

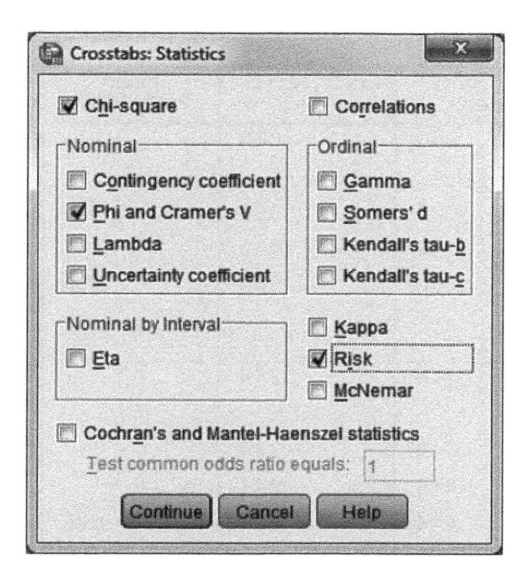

FIGURE 65.3

The **Statistics** dialog window of **Crosstabs**.

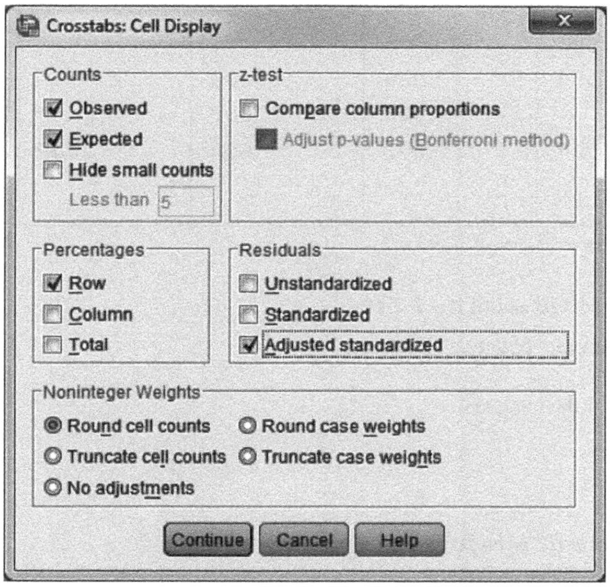

FIGURE 65.4

The **Cell Display** dialog window of **Crosstabs**.

The first major row provides the results for those cases (N = **826**) who indicated that they did not smoke (coded as **0** in the data file). For this group, the **Pearson Chi-Square** was **29.051** and with one degree of freedom (this is still a 2×2 contingency table) it is statistically significant ($p < .001$) based on either an exact two-tail test of significance or an asymptotic test. Thus, there is a relationship between the variables of **physically_active** and **obesity** for nonsmokers.

The second major row provides the results for those cases (N = **222**) who indicated that they did smoke (coded as **1** in the data file). For this group, the **Pearson Chi-Square** was **0.718** and with one degree of freedom it is not statistically significant based on either an exact two-tail test of significance (p = **.444**) or an asymptotic test (p = **.397**). Thus, the variables of **physically_active** and **obesity** appear to be independent of each other for nonsmokers.

Estimates of the strength of the relationship between the variables are shown in Figure 65.6, with the same major partitions in the table. The phi coefficient for the

Chi-Square Tests

smoker		Value	df	Asymp. Sig. (2-sided)	Exact Sig. (2-sided)	Exact Sig. (1-sided)	Point Probability
0 no	Pearson Chi-Square	29.051[a]	1	.000	.000	.000	
	Continuity Correction[b]	28.168	1	.000			
	Likelihood Ratio	29.618	1	.000	.000	.000	
	Fisher's Exact Test				.000	.000	
	Linear-by-Linear Association	29.016[c]	1	.000	.000	.000	.000
	N of Valid Cases	826					
1 yes	Pearson Chi-Square	.718[d]	1	.397	.444	.244	
	Continuity Correction[b]	.481	1	.488			
	Likelihood Ratio	.718	1	.397	.444	.244	
	Fisher's Exact Test				.444	.244	
	Linear-by-Linear Association	.714[e]	1	.398	.444	.244	.086
	N of Valid Cases	222					
Total	Pearson Chi-Square	26.398[f]	1	.000	.000	.000	
	Continuity Correction[b]	25.657	1	.000			
	Likelihood Ratio	26.738	1	.000	.000	.000	
	Fisher's Exact Test				.000	.000	
	Linear-by-Linear Association	26.373[g]	1	.000	.000	.000	.000
	N of Valid Cases	1048					

a. 0 cells (.0%) have expected count less than 5. The minimum expected count is 93.65.
b. Computed only for a 2x2 table
c. The standardized statistic is -5.387.
d. 0 cells (.0%) have expected count less than 5. The minimum expected count is 28.24.
e. The standardized statistic is -.845.
f. 0 cells (.0%) have expected count less than 5. The minimum expected count is 122.34.
g. The standardized statistic is -5.135.

FIGURE 65.5 Chi-square values from the analysis.

Symmetric Measures

smoker			Value	Approx. Sig.	Exact Sig.
0 no	Nominal by Nominal	Phi	-.188	.000	.000
		Cramer's V	.188	.000	.000
	N of Valid Cases		826		
1 yes	Nominal by Nominal	Phi	-.057	.397	.444
		Cramer's V	.057	.397	.444
	N of Valid Cases		222		
Total	Nominal by Nominal	Phi	-.159	.000	.000
		Cramer's V	.159	.000	.000
	N of Valid Cases		1048		

FIGURE 65.6 Estimates of the strength of relationship between the variables.

physically_active * obesity * smoker Crosstabulation

smoker				obesity 0 no	obesity 1 yes	Total
0 no	physically_active	0 not active	Count	291	130	421
			Expected Count	323.7	97.3	421.0
			% within physically_active	69.1%	30.9%	100.0%
			Adjusted Residual	-5.4	5.4	
		1 active	Count	344	61	405
			Expected Count	311.3	93.7	405.0
			% within physically_active	84.9%	15.1%	100.0%
			Adjusted Residual	5.4	-5.4	
	Total		Count	635	191	826
			Expected Count	635.0	191.0	826.0
			% within physically_active	76.9%	23.1%	100.0%
1 yes	physically_active	0 not active	Count	79	31	110
			Expected Count	81.8	28.2	110.0
			% within physically_active	71.8%	28.2%	100.0%
			Adjusted Residual	-.8	.8	
		1 active	Count	86	26	112
			Expected Count	83.2	28.8	112.0
			% within physically_active	76.8%	23.2%	100.0%
			Adjusted Residual	.8	-.8	
	Total		Count	165	57	222
			Expected Count	165.0	57.0	222.0
			% within physically_active	74.3%	25.7%	100.0%
Total	physically_active	0 not active	Count	370	161	531
			Expected Count	405.3	125.7	531.0
			% within physically_active	69.7%	30.3%	100.0%
			Adjusted Residual	-5.1	5.1	
		1 active	Count	430	87	517
			Expected Count	394.7	122.3	517.0
			% within physically_active	83.2%	16.8%	100.0%
			Adjusted Residual	5.1	-5.1	
	Total		Count	800	248	1048
			Expected Count	800.0	248.0	1048.0
			% within physically_active	76.3%	23.7%	100.0%

FIGURE 65.7 The observed and expected frequencies together with row percentages and adjusted residual values.

overall (**Total**) analysis is again identical to what we presented in Chapter 63, but we also obtain separate phi coefficients for nonsmokers and smokers. For nonsmokers, we know that the chi-square value was statistically significant and so we can move forward with these results. Squaring the value of **.188** (and ignoring the negative valence) yields a phi square value of approximately .035. We may thus conclude that for nonsmokers, physical activity accounts for approximately 3.5% of the variance of obesity. For smokers, we learned that the chi-square value was not statistically significant, and so we know that the phi coefficient (computed to be −**.057**) is not statistically different from zero (there is no relationship between the variables).

The observed and expected frequencies together with row percentages and adjusted residual values in Figure 65.7, again structured by the **Total** analysis (duplicating what we have seen in Chapter 63) and by levels of **smoker**. All expected frequencies are based on the **Total** percentages, as they represent a common frame of reference.

The nonsmoker frequencies were associated with a statistically significant chi-square, and we can see in those cells in what way the two variables are related for this group. For those who are not active, approximately two-thirds are not obese and one-third are obese, but for the cases who are physically active, approximately 85% are not obese, whereas only approximately 15% are obese.

We obtained a nonsignificant chi-square for the smokers, indicating that physical activity was not related to obesity for these cases. The cell frequencies show us that, regardless of whether the cases are or are not physically active, about three-quarters of the cases are not obese and about one-quarter of the cases are obese.

Because each of the partitions is still based on a 2 × 2 contingency table, we are able to perform a risk analysis, the results of which are presented in Figure 65.8. Again,

Risk Estimate

smoker		Value	95% Confidence Interval	
			Lower	Upper
0 no	Odds Ratio for physically_active (0 not active / 1 active)	.397	.282	.559
	For cohort obesity = 0 no	.814	.754	.878
	For cohort obesity = 1 yes	2.050	1.562	2.691
	N of Valid Cases	826		
1 yes	Odds Ratio for physically_active (0 not active / 1 active)	.770	.421	1.410
	For cohort obesity = 0 no	.935	.801	1.092
	For cohort obesity = 1 yes	1.214	.774	1.904
	N of Valid Cases	222		
Total	Odds Ratio for physically_active (0 not active / 1 active)	.465	.346	.625
	For cohort obesity = 0 no	.838	.783	.897
	For cohort obesity = 1 yes	1.802	1.430	2.270
	N of Valid Cases	1048		

FIGURE 65.8
Risk analysis results.

the outcome for the **Total** sample is exactly what we obtained earlier. But in this layer analysis, we also obtain risk estimates for each level of **smoker**.

The chi-square analysis indicated that there was no relationship between the variables for smokers. Thus, with respect to obesity, it did not matter whether cases were physically active or not. It then follows that both the risk ratio and the odds ratio should be effectively 1.00. As can be seen from the **Risk Estimate** table, these values are not exactly 1.00, and there is no test of statistical significance for these indexes. However, IBM SPSS has provided us with an invaluable piece of information, namely, the lower and upper limits of the 95% confidence interval. Note that the confidence interval for all three of the smoker risk estimates subsumes the value 1.00, and so we can indirectly assert that the obtained ratios could reasonably have been 1.00.

The results for nonsmokers were statistically significant, and so the risk ratios and odds ratio are viable (they are not effectively 1.00). The values shown in the table are not markedly different from those representing the **Total** sample and suggest that the risk of becoming obese is better than two times higher for those who are not physically active than for those who are physically active.

CHAPTER 66

Hierarchical Loglinear Analysis

66.1 OVERVIEW

Loglinear analysis is an extension of a two-way **Crosstabs** analysis in that we cross three (or more) categorical variables together (e.g., Variables A, B, and C), but the focus is still on the differences between the observed and expected frequencies. In the two-way **Crosstabs** analysis, a statistically significant chi-square value indicates that the two variables are not independent of each other; that is, they are related in the sense that it is the particular combinations of their levels that are associated with (predictive of) different frequencies. This is tantamount to saying that there is a statistically significant interaction between the two variables. We tend not to verbalize the relationship in this way because there are just two variables in the analysis and it is ordinarily sufficient to say that they are related.

With three or more variables, however, there are many combinations of (relationships among) the variables that could be associated with differences between the observed and expected frequencies, and so it becomes convenient to use the ANOVA terminology and refer to these combinations as interactions. With three variables in our study, we therefore can speak of a three-way interaction or of two-way interactions. We use loglinear analysis to assess these relationships among the variables (Fienberg, 1994; Knoke, & Burke, 1980). Loglinear analysis thus bears a resemblance to factorial ANOVA; it can also be (under certain circumstances) transformed into a logistic regression design. Briefly, the ties to these other designs are as follows:

- Loglinear analysis resembles factorial ANOVA designs in that the various combinations of variables (e.g., AB, ABC) are thought of as interactions. In a three-way loglinear design, for example, we can have one three-way interaction, three two-way interactions, and three main effects. In a loglinear design, a statistically significant interaction indicates that the observed frequencies result from particular combinations of the levels of the variables involved. But unlike ANOVA, there is no dependent variable per se (in an ANOVA sense), as it is the frequency of occurrence of various combinations of the levels of the variables that is the target of the analysis.
- Logistic regression can be invoked if one of the variables does indeed appear to better assume the role of dependent variable. As would be true for any regression

Performing Data Analysis Using IBM SPSS®, First Edition.
Lawrence S. Meyers, Glenn C. Gamst, and A. J. Guarino.

analysis, we could evaluate the effectiveness of the main effects and interactions of the other variables to predict the occurrence of the dependent variable (e.g., coded as 0 for absent and 1 for observed).

The general approach of a hierarchical loglinear analysis is to evaluate the fit of models containing all or some of the effects (interactions and main effects) to predict the observed frequencies. Assessing model differences focuses on the deviations (differences) between the observed frequencies and the expected frequencies. Because of the complexity of the calculations in dealing with the increased number of relationships over the two-way classification design, loglinear analysis transforms the frequency value in the cells (count data) to its natural log.

A natural logarithm uses the base e (with a value of approximately 2.72) in its computation, and the natural log of a number is the power to which 2.72 needs to be raised to achieve it. For example, the natural log of 7.398 is 2 because 2.72 raised to the second power produces a value of 7.398. If a cell had a frequency of 44, then its natural log is 3.78 (computed as $2.72^{3.78} = 44$). This transformation from frequency to natural logs produces interval-level data that are appropriate for linear regression analysis (thus "log" and "linear" are joined in the name "loglinear" analysis).

As a simplified description, the full loglinear model is an equation (analogous to a multiple regression equation) that predicts expected cell frequencies as a function of terms representing the observed frequencies associated with the interactions and main effect terms. The algorithm attempts to calculate expected scores that are as close to the observed scores as possible; that is, the residuals or differences between the predicted (expected) score and the actual (observed) score are as small as possible. The smaller the residuals, the better the model fits the data.

Model fit is assessed by the chi-square test. If a model fits the data well, the results of the chi-square will be nonsignificant. In the context of hierarchical loglinear analysis, it is recommended that the p values exceed .10 to be acceptable (Knoke & Burke, 1980). The important point is that a nonsignificant chi-square supports the plausibility of the model. The proposed model should predict scores that are *not* statistically significantly different from the observed data, thus fitting the model to the data.

The hierarchical strategy is the most popular variation of loglinear analysis (Fienberg, 1994). The process begins with the full or *saturated* model containing all of the effects. In a three-way design, which is what we cover here, this model subsumes the three-way interaction, all of the two-way interactions, and all of the main effects. We then proceed in an exploratory fashion to search for the best fitting model. The analysis is hierarchical in the sense that if an interaction term is included in a model, then all of the lower order interactions and main effects involving those variables must also be included. For example, if the model contains the A × B × C term, then all two-way terms (A × B, A × C, and B × C) and all main effects (A, B, and C) must also remain in the model. Including these lower order associations is necessary because in most applications, it is not meaningful to include higher order interactions without including the lower order interactions (Agresti, 2002).

To identify which of the hierarchical associations are statistically related, a systematic backward deletion process is applied to the saturated model with the highest order association removed first. The saturated model will predict the cell frequencies perfectly (there will be no residuals), and so the issue becomes one of determining if a simplified model can do virtually as well, that is, determining if a simplified model can fit the data as indicated by a statistically nonsignificant chi-square value. The strategy used in a hierarchical loglinear analysis is one of systematically deleting the highest order association and proceeding to the lower order effects; such a strategy is referred to as **Generating Class**.

The backward deletion process starts by removing the three-way interaction from the model and comparing that model to the saturated model using a chi-square statistic to assess if the two models are statistically significantly different. If the result of the chi-square is statistically significant, then the three-way interaction is necessary in fitting the model and the analysis is completed. However, if the result of the chi-square is not significant (i.e., the reduced model fits the data virtually as well as the saturated model), then the three-way interaction is not necessary in fitting the model and can be removed.

If the three-way interaction is removed, then the model remaining contains all of the two-way interactions along with the main effects. Because there are three two-way interactions, each of the two-way interactions is considered for deletion. The two-way interactions that achieve statistical significance remain in the model (because their deletion would cause a significant drop in fit), as well as the main effects of individual variables comprising the interactions. The two-way interactions that are not significant are deleted from the model, as well as the main effects of the individual variables.

66.2 NUMERICAL EXAMPLE

The data file is named **lifestyle medical study**, a data file we have used in previous chapters. For the variable **physically_active**, clients were dichotomized into those who were **not active** (coded as **0** in the data file) and those who were **active** (coded as **1** in the data file). For the variable **obesity**, clients were dichotomized into those who were not obese (**no**, coded as **0** in the data file) and those who were (**yes**, coded as **1** in the data file). The variable **hist_myocardial_infarction** represents clients who did not have (**no**, coded as **0** in the data file) and did have (**yes**, coded as **1** in the data file) a history of heart attack. Thus, we have a $2 \times 2 \times 2$ design.

66.3 ANALYSIS SETUP

We open **lifestyle medical study** and from the main menu select **Analyze ➔ Loglinear ➔ Model Selection**. This produces the main **Loglinear Model Selection** dialog window shown in Figure 66.1. We select the three variables to be included in the analysis

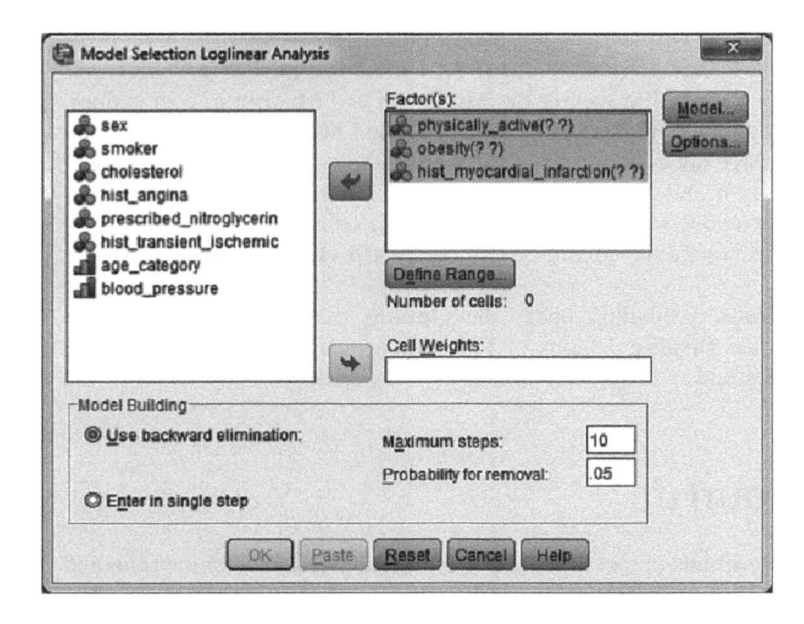

FIGURE 66.1
The main **Loglinear Model Selection** dialog window.

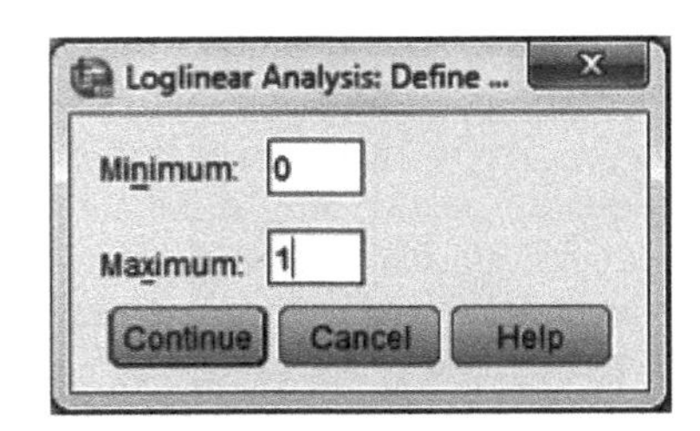

FIGURE 66.2

The **Define Range** window of **Loglinear Model Selection**.

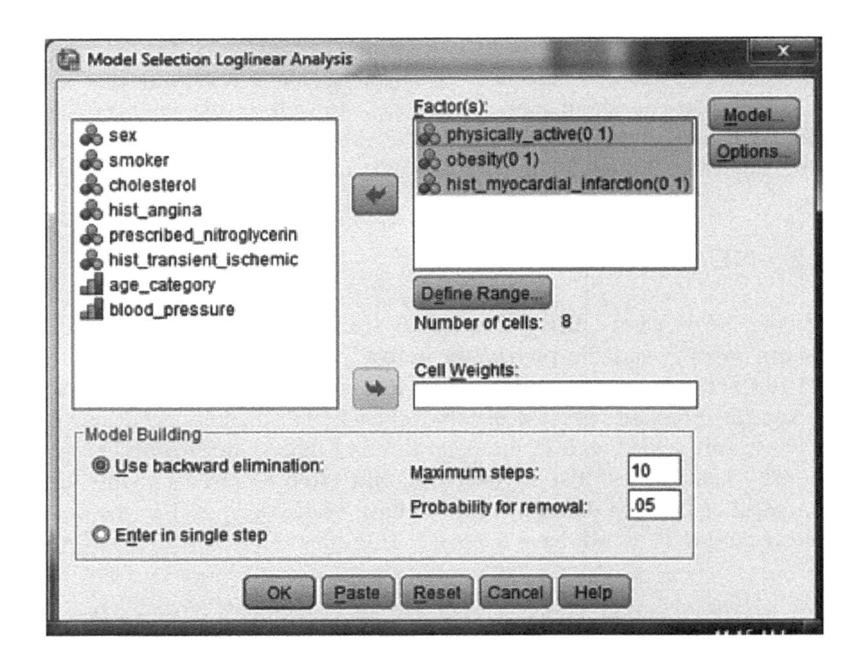

FIGURE 66.3

The main **Loglinear Model Selection** dialog window with the ranges now defined.

(**physically_active**, **obesity**, and **hist_myocardial_infarction**) and move them to the **Factor(s)** panel. When they are placed in the **Factor(s)** panel, the expression **(? ?)** appears after each variable; this serves as a visual reminder that we must define the range of codes for each. To facilitate the defining of range for our variables, we have highlighted all three (because they are coded identically).

Selecting the **Define Range** pushbutton opens the **Define Range** dialog window. As seen in Figure 66.2, we are asked to specify the **Minimum** and **Maximum** code values. What we type in will apply to all highlighted variables in the **Factor(s)** panel of the main dialog window. As all three are coded in the same way, and as all three are highlighted, we can supply the codes **0** and **1** as seen in Figure 66.2. Selecting **Continue** returns us to the main dialog window where the codes are now associated with each variable (see Figure 66.3). We also retain the default of **Use backward elimination** under **Model Building**.

Selecting the **Options** pushbutton opens the **Options** dialog window shown in Figure 66.4. We retain the **Display** defaults of **Frequencies** and **Residuals** but change the **Delta** value (a value added to the cell frequencies for saturated models) from the default of **.5–0**.

66.4 ANALYSIS OUTPUT

The **Cell Counts and Residuals** output in Figure 66.5 presents the results of the saturated model. This model contains all of the effects (the three-way interaction, all two-way

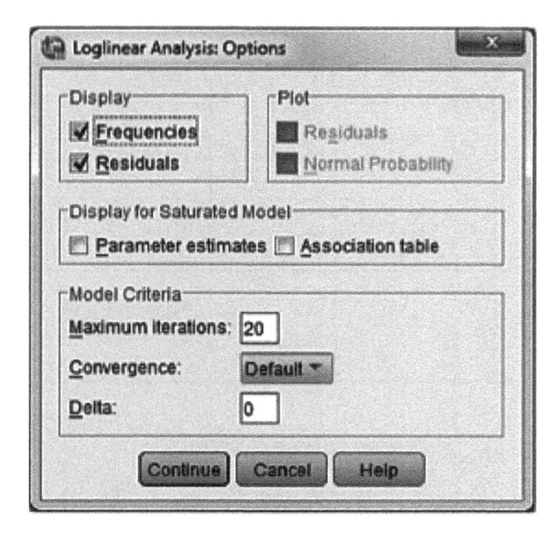

FIGURE 66.4
The **Options** window of **Loglinear Model Selection**.

Cell Counts and Residuals

physically_active	obesity	hist_myocardial_infarction	Observed Count[a]	Observed %	Expected Count	Expected %	Residuals	Std. Residuals
0 not active	0 no	0 no	262.000	25.0%	262.000	25.0%	.000	.000
		1 yes	108.000	10.3%	108.000	10.3%	.000	.000
	1 yes	0 no	80.000	7.6%	80.000	7.6%	.000	.000
		1 yes	81.000	7.7%	81.000	7.7%	.000	.000
1 active	0 no	0 no	365.000	34.8%	365.000	34.8%	.000	.000
		1 yes	65.000	6.2%	65.000	6.2%	.000	.000
	1 yes	0 no	58.000	5.5%	58.000	5.5%	.000	.000
		1 yes	29.000	2.8%	29.000	2.8%	.000	.000

a. For saturated models, .000 has been added to all observed cells.

FIGURE 66.5 **Observed** and **Expected** cell frequencies and **Residuals**.

Goodness-of-Fit Tests

	Chi-Square	df	Sig.
Likelihood Ratio	.000	0	.
Pearson	.000	0	.

FIGURE 66.6
The **Goodness-of-Fit Tests** output.

interactions, and the main effects). It perfectly fits the data, as shown by the fact that the **Residuals** (the differences between expectations based on the model and the observed frequencies), are equal to zero. It is this model that will be subjected to backward elimination to detect any significant hierarchical associations.

Figure 66.6 shows the **Goodness-of-Fit Tests** for the saturated model. Because there are no residuals, the **Likelihood Ratio** and the chi-square value (shown as **Pearson**) are both zero; neither can be meaningfully evaluated for statistical significance.

The **K-Way and Higher-Order Effects** table is shown in Figure 66.7 and provides a global view of the results of the analysis. The value of **K** in the rows of the table represents main effects (shown as **1**), two-way interactions (shown as **2**), and three-way interactions (shown as **3**), and the degrees of freedom (**df**) provides a count of the number of effects involved.

K-Way and Higher-Order Effects

	K	df	Likelihood Ratio		Pearson		Number of Iterations
			Chi-Square	Sig.	Chi-Square	Sig.	
K-way and Higher Order Effects[a]	1	7	639.674	.000	745.313	.000	0
	2	4	103.288	.000	118.908	.000	2
	3	1	.165	.684	.166	.684	4
K-way Effects[b]	1	3	536.386	.000	626.405	.000	0
	2	3	103.123	.000	118.742	.000	0
	3	1	.165	.684	.166	.684	0

a. Tests that k-way and higher order effects are zero.
b. Tests that k-way effects are zero.

FIGURE 66.7 The **K-Way and Higher-Order Effects** table.

The first major row carries the same name as provided for the entire table and is divided into a set of three rows. The following is the message conveyed in the upper half of the table: *If these effects and all higher order effects are removed from the model, then here are the consequences.* For example, the first row labeled as **1** shows the consequences of removing all of the main effects and all higher order effects (i.e., everything) from the model; as seen by the **df**, we are talking about removing **7** effects (three main effects + three two-way interactions + one three-way interaction). Not surprisingly, the consequences are a statistically significant chi-square (**745.313**), informing us that the degree of predicting the cell frequencies is significantly worse than that in the saturated model. This suggests that at least one of those seven effects is sufficiently important that it needs to be included in the model.

As we can see from the table, we would draw the same conclusion regarding **K** = **2**; removing all two-way interactions and the three-way interaction (removing **4** effects) from the model also significantly deteriorates our prediction prowess. Thus, at least one of these four effects is needed for prediction.

Finally, the results for row **3** indicate that removing the single three-way interaction from the model yields a nonsignificant chi-square (p = **.684**); thus, removing this single effect does not significantly lower our prediction of cell frequencies and it therefore becomes expendable. We may therefore infer that at least one of the two-way interactions was driving the results in **K** = **2**, as it is now clear that the three-way effect is nonessential.

The second major row of the **K-Way and Higher-Order Effects** table examines the same global scenarios but without packaging the higher order effects. Thus, the row for **K** = **1** deals with removing the main effects only, the row for **K** = **2** deals with removing the two-way interaction effects only, and row for **K** = **3** deals with removing the single three-way interaction effect from the model. The results here are similar to those in the first row. Specifically, removing from the model the main effects and the two-way interactions significantly decreases our prediction ability but removing the three-way interaction does not significantly decrease prediction (this third finding duplicates the information we obtained in the first major row because the three-way interaction is the highest order effect in the model).

With the global results now understood, we are in a position to examine the consequences of removing the specific effects. Even with the three-way interaction able to be removed, retaining one or more of the two-way interactions in this hierarchical approach will automatically retain the main effects associated with them as well. The detailed analysis dealing with this is shown in the **Step Summary** table in Figure 66.8. The **Generating Class** in each step is the set of effects under consideration for deletion.

Step Summary

Step[a]			Effects	Chi-Square[c]	df	Sig.	Number of Iterations
0	Generating Class[b]		physically_active*obesity*hist_myocardial_infarction	.000	0	.	
	Deleted Effect	1	physically_active*obesity*hist_myocardial_infarction	.165	1	.684	4
1	Generating Class[b]		physically_active*obesity, physically_active*hist_myocardial_infarction, obesity*hist_myocardial_infarction	.165	1	.684	
	Deleted Effect	1	physically_active*obesity	15.635	1	.000	2
		2	physically_active*hist_myocardial_infarction	29.821	1	.000	2
		3	obesity*hist_myocardial_infarction	35.461	1	.000	2
2	Generating Class[b]		physically_active*obesity, physically_active*hist_myocardial_infarction, obesity*hist_myocardial_infarction	.165	1	.684	

a. At each step, the effect with the largest significance level for the Likelihood Ratio Change is deleted, provided the significance level is larger than .050.
b. Statistics are displayed for the best model at each step after step 0.
c. For 'Deleted Effect', this is the change in the Chi-Square after the effect is deleted from the model.

FIGURE 66.8 The **Step Summary** table.

Convergence Information[a]

Generating Class	physically_active*obesity, physically_active*hist_myocardial_infarction, obesity*hist_myocardial_infarction
Number of Iterations	0
Max. Difference between Observed and Fitted Marginals	.070
Convergence Criterion	.365

a. Statistics for the final model after Backward Elimination.

FIGURE 66.9

The **Convergence Information** table.

Cell Counts and Residuals

			Observed		Expected			
physically_active	obesity	hist_myocardial_infarction	Count	%	Count	%	Residuals	Std. Residuals
0 not active	0 no	0 no	262.000	25.0%	263.239	25.1%	-1.239	-.076
		1 yes	108.000	10.3%	106.765	10.2%	1.235	.120
	1 yes	0 no	80.000	7.6%	78.763	7.5%	1.237	.139
		1 yes	81.000	7.7%	82.234	7.8%	-1.234	-.136
1 active	0 no	0 no	365.000	34.8%	363.761	34.7%	1.239	.065
		1 yes	65.000	6.2%	66.235	6.3%	-1.235	-.152
	1 yes	0 no	58.000	5.5%	59.237	5.7%	-1.237	-.161
		1 yes	29.000	2.8%	27.766	2.6%	1.234	.234

FIGURE 66.10 **Cell Counts and Residuals** output.

At the start of the process (**Step 0**), we consider only the three-way interaction. As we have already seen, removing that effect from the model does not significantly reduce prediction ($p = .684$), and so it is deleted.

The **Generating Class** for the next step (**Step 1**) is the set of two-way interactions. We begin this step with them all in the model (**Generating Class** in the top portion of the **Step 1** row), and the **Sig**. column just repeats the information from the row above it, thus letting us know that prediction is not significantly reduced ($p = .684$). Now each individual two-way effect is evaluated. As can be seen, removing any of them results in a statistically significant chi-square ($p < .001$), indicating that they are all making a statistically significant contribution. Thus, there is no basis to remove them from the model.

Step 2 would ordinarily consider the main effects not subsumed in the significant interactions. But with all two-way effects in the model, and with the model being structured in a hierarchical manner, the main effects are all subsumed in the interactions and so are not eligible for removal. Therefore, the backward elimination process ends with just the three-way effect removed.

In summary, the final model includes all the main effects and all the two-way interactions. This means that the combinations of (a) **physically_active** and **obesity**, (b) **physically_active** and **myocardial_infarction**, and (c) **obesity** and **myocardial_infarction** are significant predictors of the frequencies.

Figure 66.9 displays the **Convergence Information**. This is a summary of the **Generating Class** consisting of the three two-way interaction effects.

The **Cell Counts and Residuals** are shown in Figure 66.10. A residual is the difference between the observed and expected frequencies. In the table, the residuals are relatively small because the model fits the data.

Goodness-of-Fit Tests

	Chi-Square	df	Sig.
Likelihood Ratio	.165	1	.684
Pearson	.166	1	.684

FIGURE 66.11
The **Goodness-of-Fit Tests** output.

The **Goodness-of-Fit Tests** table presented in Figure 66.11 informs us whether there are significant differences between the expected and the observed frequencies. **IBM SPSS®** generates both the **Likelihood Ratio** and the **Pearson**. If the model fits the data, the **Goodness-of-Fit Tests** will be nonsignificant. Because of the complicated calculations involved with assessing model fit, researchers generally prefer to report the **Likelihood Ratio** rather than the traditional Pearson chi-square (Read & Cressie, 1988). In our example, both tests indicated nonsignificance ($p =$ **.684**), suggesting that final model adequately fits the data.

66.5 THE NEXT STEPS

If the three-way interaction significantly contributed to prediction, the table of frequencies shown in Figure 66.10 would represent the final solution. However, with the three-way interaction unnecessary but with all of the two-way interactions as significant predictors, it would be a reasonable next step to perform a series of two-way chi-square analyses on each interaction effect. These would mirror the analyses described in Chapter 63. Furthermore, if any of the two-way interactions represented a 2×2 array and if the research context and variables warranted, then an analysis of risk as described in Chapter 64 should also be performed.

APPENDIX

Statistics Tables

TABLE A.1 Selected Critical Values of Chi-Square Distribution with Degrees of Freedom

df	p = .10	p = .05	p = .025	p = .01	p = .001
1	2.706	3.841	5.024	6.635	10.828
2	4.605	5.991	7.378	9.210	13.816
3	6.251	7.815	9.348	11.345	16.266
4	7.779	9.488	11.143	13.277	18.467
5	9.236	11.070	12.833	15.086	20.515
6	10.645	12.592	14.449	16.812	22.458
7	12.017	14.067	16.013	18.475	24.322
8	13.362	15.507	17.535	20.090	26.125
9	14.684	16.919	19.023	21.666	27.877
10	15.987	18.307	20.483	23.209	29.588
11	17.275	19.675	21.920	24.725	31.264
12	18.549	21.026	23.337	26.217	32.910
13	19.812	22.362	24.736	27.688	34.528
14	21.064	23.685	26.119	29.141	36.123
15	22.307	24.996	27.488	30.578	37.697
16	23.542	26.296	28.845	32.000	39.252
17	24.769	27.587	30.191	33.409	40.790
18	25.989	28.869	31.526	34.805	42.312
19	27.204	30.144	32.852	36.191	43.820
20	28.412	31.410	34.170	37.566	45.315
21	29.615	32.671	35.479	38.932	46.797
22	30.813	33.924	36.781	40.289	48.268
23	32.007	35.172	38.076	41.638	49.728
24	33.196	36.415	39.364	42.980	51.179
25	34.382	37.652	40.646	44.314	52.620
26	35.563	38.885	41.923	45.642	54.052
27	36.741	40.113	43.195	46.963	55.476
28	37.916	41.337	44.461	48.278	56.892
29	39.087	42.557	45.722	49.588	58.301
30	40.256	43.773	46.979	50.892	59.703

(*continued*)

Performing Data Analysis Using IBM SPSS®, First Edition.
Lawrence S. Meyers, Glenn C. Gamst, and A. J. Guarino.

TABLE A.1 (Continued)

df	$p = .10$	$p = .05$	$p = .025$	$p = .01$	$p = .001$
31	41.422	44.985	48.232	52.191	61.098
32	42.585	46.194	49.480	53.486	62.487
33	43.745	47.400	50.725	54.776	63.870
34	44.903	48.602	51.966	56.061	65.247
35	46.059	49.802	53.203	57.342	66.619
36	47.212	50.998	54.437	58.619	67.985
37	48.363	52.192	55.668	59.893	69.347
38	49.513	53.384	56.896	61.162	70.703
39	50.660	54.572	58.120	62.428	72.055
40	51.805	55.758	59.342	63.691	73.402
41	52.949	56.942	60.561	64.950	74.745
42	54.090	58.124	61.777	66.206	76.084
43	55.230	59.304	62.990	67.459	77.419
44	56.369	60.481	64.201	68.710	78.750
45	57.505	61.656	65.410	69.957	80.077
46	58.641	62.830	66.617	71.201	81.400
47	59.774	64.001	67.821	72.443	82.720
48	60.907	65.171	69.023	73.683	84.037
49	62.038	66.339	70.222	74.919	85.351
50	63.167	67.505	71.420	76.154	86.661
51	64.295	68.669	72.616	77.386	87.968
52	65.422	69.832	73.810	78.616	89.272
53	66.548	70.993	75.002	79.843	90.573
54	67.673	72.153	76.192	81.069	91.872
55	68.796	73.311	77.380	82.292	93.168
56	69.919	74.468	78.567	83.513	94.461
57	71.040	75.624	79.752	84.733	95.751
65	79.973	84.821	89.177	94.422	105.988
66	81.085	85.965	90.349	95.626	107.258
67	82.197	87.108	91.519	96.828	108.526
68	83.308	88.250	92.689	98.028	109.791
69	84.418	89.391	93.856	99.228	111.055
70	85.527	90.531	95.023	100.425	112.317
71	86.635	91.670	96.189	101.621	113.577
72	87.743	92.808	97.353	102.816	114.835
73	88.850	93.945	98.516	104.010	116.092
74	89.956	95.081	99.678	105.202	117.346
75	91.061	96.217	100.839	106.393	118.599
76	92.166	97.351	101.999	107.583	119.850
77	93.270	98.484	103.158	108.771	121.100
78	94.374	99.617	104.316	109.958	122.348
79	95.476	100.749	105.473	111.144	123.594
80	96.578	101.879	106.629	112.329	124.839
81	97.680	103.010	107.783	113.512	126.083
82	98.780	104.139	108.937	114.695	127.324
83	99.880	105.267	110.090	115.876	128.565
84	100.980	106.395	111.242	117.057	129.804
85	102.079	107.522	112.393	118.236	131.041
86	103.177	108.648	113.544	119.414	132.277
87	104.275	109.773	114.693	120.591	133.512
88	105.372	110.898	115.841	121.767	134.746
89	106.469	112.022	116.989	122.942	135.978
90	107.565	113.145	118.136	124.116	137.208

TABLE A.1 (Continued)

df	p = .10	p = .05	p = .025	p = .01	p = .001
91	108.661	114.268	119.282	125.289	138.438
92	109.756	115.390	120.427	126.462	139.666
93	110.850	116.511	121.571	127.633	140.893
94	111.944	117.632	122.715	128.803	142.119
95	113.038	118.752	123.858	129.973	143.344
96	114.131	119.871	125.000	131.141	144.567
97	115.223	120.990	126.141	132.309	145.789
98	116.315	122.108	127.282	133.476	147.010
99	117.407	123.225	128.422	134.642	148.230
100	118.498	124.342	129.561	135.807	149.449

http://www.itl.nist.gov/div898/handbook/eda/section3/eda3674.htm.

National Institute of Standards and Technology. The NIST/SEMATECH Engineering Statistics Handbook, Web Based statistics handbook.

Developed as a joint partnership between the Statistical Engineering Division of NIST and the Statistical Methods Group of SEMATECH.

http://www.itl.nist.gov/div898/handbook/ for complete handbook.

REFERENCES

Agresti, A. (2002). *Categorical data analysis* (2nd ed.). New York, NY: John Wiley & Sons, Inc.

American Psychiatric Association. (2000). *Diagnostic and statistical manual of mental disorders* (4th ed., Text Revision). Washington, DC: Author.

American Psychological Association. (2009). *Publication manual* (6th ed.). Washington, DC: Author.

Aroian, L. A. (1947). The probability function of the product of two normally distributed variables. *Annals of Mathematical Statistics*, *18*, 265–271.

Baron, R. M., & Kenny, D. A. (1986). The moderator-mediator variable distinction in social psychological research: Conceptual, strategic, and statistical considerations. *Journal of Personality and Social Psychology*, *51*, 1173–1182.

Bentler, P. M. (1990). Comparative fit indexes in structural models. *Psychological Bulletin*, *107*, 238–246.

Berkman, E. T., & Reise, S. P. (2012). *A conceptual guide to statistics using SPSS*. Thousand Oaks, CA: Sage.

Bickel, R. (2007). *Multilevel analysis for applied research*. New York, NY: Guilford Press.

Blau, P., & Duncan, O. (1967). *The American occupational structure*. New York, NY: John Wiley & Sons, Inc.

Bollen, K. A. (1989). *Structural equations with latent variables*. New York, NY: John Wiley & Sons, Inc.

Bollen, K. A. (2002). Latent variables in psychology and the social sciences. *Annual Review of Psychology*, *53*, 605–634.

Brown, T. A. (2006). *Confirmatory factor analysis for applied research*. New York, NY: Guilford Press.

Brown, T. A., & Moore, M. T. (2012). Confirmatory factor analysis. In R. H. Hoyle (Ed.), *Handbook of structural equation modeling* (pp. 361–379). New York, NY: Guilford Press.

Brown, W. (1910). Some experimental results in the correlation of mental abilities. *British Journal of Psychology*, *3*, 296–322.

Browne, M. W., & Cudeck, R. (1989). Single sample cross-validation indices for covariation structures. *Multivariate Behavioral Research*, *24*, 445–455.

Browne, M. W., & Cudeck, R. (1993). Alternative ways of assessing model fit. In K. A. Bollen & J. S. Long (Eds.), *Testing structural equation models* (pp. 136–162). Newbury Park, CA: Sage.

Budescu, D. V. (1996). Dominance analysis: A new approach to the problem of relative importance of predictors in multiple regression. *Psychological Bulletin*, *114*, 542–551.

Performing Data Analysis Using IBM SPSS®, First Edition.
Lawrence S. Meyers, Glenn C. Gamst, and A. J. Guarino.

Byrne, B. M. (2010). *Structural equation modeling with AMOS: Basic concepts, applications and programming* (2nd ed.). New York, NY: Routledge.

Carroll, J. D., & Chang, J. J. (1970). Analysis of individual differences in multidimensional scaling via an N-way generalization of "Eckart-Young" decomposition. *Psychometrika*, *35*, 238–319.

Cattell, R. B. (1966). The scree test for the number of factors. *Multivariate Behavioral Research*, *1*, 245–276.

Clark, L. A., & Watson, D. (1995). Construction validity: Basic issues in objective scale development. *Psychological Assessment*, *7*(3), 309–319.

Cohen, J. (1960). A coefficient of agreement for nominal scales. *Educational and Psychological Measurement*, *20*, 37–46.

Cohen, J. (1988). *Statistical power analysis for the behavioral sciences* (2nd ed.). Hillsdale, NJ: Lawrence Erlbaum Associates.

Cohen, J., Cohen, P., West, S. G., & Aiken, L. (2003). *Applied multiple regression/correlation analysis for the behavioral sciences* (3rd ed.). Hillsdale, NJ: Lawrence Erlbaum Associates.

Cohen, M. P. (2000). Note on the odds ratio and the probability ratio. *Journal of Educational and Behavioral Statistics*, *25*, 249–252.

Cortina, J. M. (1993). What is coefficient alpha? An examination of theory and application. *Journal of Applied Psychology*, *78*, 98–104.

Costa, P. T., Jr., & McCrae, R. R. (1992). *The NEO PI-R professional manual*. Odessa, FL: Psychological Assessment Resources.

Courville, T., & Thompson, B. (2001). Use of structure coefficients in published multiple regression articles: x is not enough. *Educational and Psychological Measurement*, *61*, 229–248.

Cox, D. R. (1972). Regression models and life tables. *Journal of the Royal Statistical Society, Series B: Methodological*, *34*, 187–220.

Cox, D. R., & Oakes, D. (1984). *Analysis of survival data*. New York, NY: Chapman & Hall.

Cronbach, L. J. (1951). Coefficient alpha and the internal structure of tests. *Psychometrika*, *16*, 297–334.

Cudeck, R., & MacCallum, R. C. (Eds.) (2007). *Factor analysis at 100: Historical developments and future directions*. Mahwah, NJ: Lawrence Erlbaum Associates.

Darlington, R. B. (1968). Multiple regression in psychological research and practice. *Psychological Bulletin*, *69*, 161–182.

Darlington, R. B. (1990). *Regression and linear models*. New York, NY: McGraw-Hill.

Davison, M. L. (1992). *Multidimensional scaling*. New York, NY: Krieger.

Di Eugenio, B., & Glass, M. (2004). The kappa statistic: A second look. *Computational Linguistics*, *30*, 95–101.

Enders, C. K. (2010). *Applied missing data analysis*. New York, NY: Guilford Press.

Eknoyan, G. (2008). Adolphe Quetelet (1796–1874)—The average man and indices of obesity. *Nephrology, dialysis, transplantation: Official publication of the European Dialysis and Transplant Association—European Renal Association*, *23*, 47–51.

Edwards, J. R., & Bagozzi, R. P. (2000). On the nature and direction of relationships between constructs and measures. *Psychological Methods*, *5*, 155–174.

Enders, C. K., & Tofighi, D. (2007). Centering predictor variables in cross-sectional multilevel models: A new look at an old issue. *Psychological Methods*, *12*, 121–138.

Feldt, L. S., & Qualls, A. L. (1996). Estimation of measurement error variance at specific score levels. *Journal of Educational Measurement*, *33*, 141–156.

Fienberg, S. E. (1994). *The analysis of cross-classified categorical data* (2nd ed.). Cambridge, MA: MIT Press.

Fisher, R. A. (1921a). Some remarks on the methods formulated in a recent article on the qualitative analysis of plant growth. *Annals of Applied Biology*, *7*, 367–372.

Fisher, R. A. (1921b). Studies in crop variation. I. An examination of the yield of dressed grain from broadbalk. *Journal of Agricultural Science*, *11*, 107–135.

Fisher, R. A. (1925). *Statistical methods for research workers*. Edinburgh, England: Oliver & Boyd.

Fisher, R. A. (1935a). *The design of experiments*. Edinburgh, England: Oliver & Boyd.

Fisher, R. A. (1935b). The logic of inductive inference. *Journal of the Royal Statistical Society*, *98*, 39–54.

Fisher, R. A. (1950). *Statistical methods for research workers* (11th ed.). New York, NY: Hafner.

Fisher, R. A., & Eden, T. (1927). Studies in crop variation. IV. The experimental determination of the value of top dressings with cereals. *Journal of Agricultural Science*, *17*, 548–562.

Fisher, R. A., & Mackenzie, W. A. (1923). Studies in crop variation. II. The manorial responses of different potato varieties. *Journal of Agricultural Science*, *13*, 311–320.

Fitzmaurice, G., Laird, N. M., & Ware, J. H. (2011). *Applied longitudinal analysis* (2nd ed.). Hoboken, NJ: John Wiley & Sons, Inc.

Fleiss, J. L. (1971). Measuring nominal scale agreement among many raters. *Psychological Bulletin*, *76*, 378–382.

Fleiss, J. L., & Cohen, J. (1973). The equivalence of weighted kappa and the intraclass correlation coefficient as measures of reliability. *Educational and Psychological Measurement*, *33*, 613–619.

Freedman, L. S., & Schatzkin, A. (1992). Sample size for studying intermediate endpoints within intervention trials of observational studies. *American Journal of Epidemiology*, *136*, 1148–1159.

Galton, F. (1886). Heredity stature. *Journal of the Anthropological Institute*, *15*, 489–499.

Galton, F. (December 13, 1888). Co-relations and their measurement, chiefly from anthropometric data. *Proceedings of the Royal Society*, *45*, 135–145.

Gamst, G., Meyers, L. S., & Guarino, A. J. (2008). *Analysis of variance designs: A conceptual and computational approach with SPSS and SAS*. New York, NY: Cambridge University Press.

Gay, L. R. (2010). *Educational research: Competencies for analysis and applications* (8th ed.). Upper Saddle River, NJ: Pearson Prentice Hall.

Giguère, G. (2007). Collecting and analyzing data in multidimensional scaling experiments: A guide for psychologists using SPSS. *Tutorials in Quantitative Methods for Psychology*, *2*, 26–37.

Goldstein, H. (2011). *Multilevel statistical models* (4th ed.). Chichester, West Sussex: John Wiley & Sons, Ltd.

Goodman, L. A. (1960). On the exact variance of products. *Journal of the American Statistical Association*, *55*, 708–713.

Gonzalez, R., & Griffin, D. (2004). Measuring individuals in a social environment: Conceptualizing dyadic and group interaction. In C. Sansone, C. C. Morf, & A. T. Panter (Eds.), *The Sage handbook of methods in social psychology* (pp. 313–334). Thousand Oaks, CA: Sage.

Gorsuch, R. L. (2003). Factor analysis. In I. B. Weiner, J. A. Schinka, & W. F. Velicer (Eds.), *Handbook of psychology: Volume 2 Research methods in psychology* (pp. 143–164). Hoboken, NJ: John Wiley & Sons, Inc.

Graham, J. W. (2009). Missing data analysis: Making it work in the real world. *Annual Review of Psychology*, *60*, 549–576.

Guilford, J. P., & Fruchter, B. (1978). *Fundamental statistics in psychology and education* (6th ed.). New York, NY: McGraw-Hill.

Hallgren, K. A. (2012). Computing inter-rater reliability for observational data: An overview and tutorial. *Tutorial in Quantitative Methods for Psychology*, *8*, 23–34.

Harris, M. J., & Taylor, G. (2009). *Medical and health science statistics made easy* (2nd ed.). Sudbury, MA: Jones & Bartlett.

Hartigan, J. A., & Wong, M. A. (1979). A K-means clustering algorithm. *Applied Statistics*, *28*, 100–108.

Haviland, M. G. (1990). Yates's correction for continuity and the analysis of 2 x 2 contingency tables. *Statistics in Medicine*, *9*, 363–367.

Hayes, A. F. (2009). Beyond Baron and Kenny: Statistical mediation analysis in the new millennium. *Communication Monographs*, *76*, 408–420.

Hays, W. L. (1981). *Statistics* (3rd ed.). New York, NY: Holt, Rinehart & Winston.

Holcomb, W. L., Chaiworapongsa, T., Luke, D. A., & Burgdorf, K. D. (2001). An odd measure of risk: Use and misuse of the odds ratio. *Obstetrics and Gynecology*, *84*, 685–688.

Hotelling, H. (1933). Analysis of a complex of statistical variables into principal components. *Journal of Educational Psychology*, *24*, 498–520.

Hotelling, H. (1936). Simplified calculation of principal components. *Psychometrika*, *1*, 27–35.

Howell, D. (2010). *Statistical methods for psychology* (7th ed.). Belmont, CA: Wadsworth.

Hox, J. J. (2010). *Multilevel analysis: Techniques and applications* (2nd ed.). New York, NY: Routledge.

Hox, J. J., & Roberts, J. K. (2010). *Handbook of advanced multilevel analysis*. New York, NY: Psychology Press.

Hu, L.-T., & Bentler, P. M. (1999). Cutoff criteria for fit indices in covariance structure analysis: Conventional criteria versus new alternatives. *Structural Equation Modeling*, *6*, 1–55.

Huberty, C. J. (1989). Problems with stepwise methods—better alternatives. In B. Thompson (Ed.), *Advances in social science methodology* (Vol. 1, pp. 43–70). Greenwich, CT: JAI Press.

Jaccard, J., & Becker, M. A. (1990). *Statistics for the behavioral sciences* (2nd ed.). Belmont, CA: Wadsworth.

Jöreskog, K. G., & Sörbom, D. (1996). *LISREL 8 user's reference guide*. Chicago, IL: Scientific Software International.

Johnson, J. W. (2000). A heuristic method for estimating the relative weight of predictor variables in multiple regression. *Multivariate Behavioral Research*, *35*, 1–19.

Johnson, J. W., & LeBreton, J. M. (2004). History and relative use of importance indices in organizational research. *Organizational Research Methods*, *7*, 238–257.

Kaiser, H. F. (1970). A second-generation Little Jiffy. *Psychometrika*, *35*, 401–415.

Kaiser, H. F. (1974). An index of factorial simplicity. *Psychometrika*, *39*, 31–36.

Kaplan, E. L., & Meier, P. (1958). Nonparametric estimation from incomplete observations. *Journal of the American Statistical Association*, *53*, 457–481.

Kasser, T., & Ryan, R. M. (1993). A dark side of the American dream: Correlates of financial success as a central life aspiration. *Journal of Personality and Social Psychology*, *65*, 410–422.

Kasser, T., & Ryan, R. M. (1996). Further examining the American dream: Differential correlates of intrinsic and extrinsic goals. *Personality and Social Psychology Bulletin*, *22*, 280–287.

Katz, M. H. (2006). *Multivariable analysis: A practical guide for clinicians* (2nd ed.). New York, NY: Cambridge University Press.

Kleinbaum, D. G. (1996). *Survival analysis*. New York, NY: Springer-Verlag.

Keppel, G., & Wickens, T. D. (2004). *Design and analysis: A researcher's handbook* (4th ed.). Upper Saddle River, NJ: Pearson Prentice Hall.

Kendall, M. G. (1938). A new measure of rank correlation. *Biometrika*, *30*, 81–89.

Kendall, M. G. (1948). *Rank correlation methods*. London, England: Charles Griffin & Company.

Kim, K., & Neil, T. (2007). *Univariate and multivariate general linear models* (2nd ed.). Boca Raton, FL: Chapman & Hall/CRC.

Kline, R. B. (2011). *Principles and practice of structural equation modeling* (3rd ed.). New York, NY: Guilford Press.

Knoke, D., & Burke, P. J. (1980). *Log-linear models*. Thousand Oaks, CA: Sage.

Krippendorff, K. (1980). *Content analysis: An introduction to its methodology*. Beverley Hills, CA: Sage.

Kruskal, J. B. (1964). Multidimensional scaling by optimizing goodness of fit to a non-metric hypothesis. *Psychometrika*, *29*, 1–28.

Kruskal, J. B., & Wish, M. (1978). *Multidimensional scaling*. Newbury Park, CA: Sage.

Kuder, G. F., & Richardson, M. W. (1937). The theory of the estimation of test reliability. *Psychometrika*, *2*, 151–160.

Lee, E. T., & Wang, J. W. (2003). *Statistical methods for survival data analysis* (3rd ed.). Hoboken, NJ: John Wiley & Sons, Inc.

Lee, V. E. (2000). Using hierarchical linear modeling to study social contexts: The case of school effects. *Educational Psychologist*, *35*, 125–141.

Lleras, C. (2005). Path analysis. In K. Kempf-Leonard (Ed.), *Encyclopedia of social measurement*, (Vol. 3, pp. 25–30). Amsterdam, The Netherlands: Elsevier.

Lord, F. M. (1984). Standard errors of measurement at different ability levels. *Journal of Educational Measurement*, *21*, 239–243.

Lord, F. M., & Norick, M. R. (1968). *Statistical theories of mental test scores*. Reading, MA: Addison-Wesley.

MacDonald, P. D. M. (2012). *Methods in field epidemiology*. Burlington, MA: Jones & Bartlett.

MacKinnon, D. P. (2008). *Introduction to statistical mediation analysis*. Mahwah, NJ: Lawrence Erlbaum Associates.

MacKinnon, D. P., Fairchild, A. J., & Fritz, M. S. (2007). Mediation analysis. *Annual Review of Psychology*, *58*, 593–614.

MacKinnon, D. P., Lockwood, C. M., Hoffman, J. M., West, S. G., & Sheets, V. (2002). A comparison of methods to test mediation and other intervening variable effects. *Psychological Methods*, *7*, 83–104.

MacQueen, J. (1967). Some methods for classification and analysis of multivariate observations. In L. M. LeCam, & J. Neyman (Eds.), *Proceedings of the 5th Berkeley Symposium on Mathematical Statistics and Probability* (Vol. 5, pp. 281–297). Berkeley, CA: University of California Press.

Marascuilo, L. A., & McSweeney, M. (1977). *Nonparametric and distribution-free methods for the social sciences*. Monterey, CA: Brooks/Cole.

McKnight, P. E., McKnight, K. M., Sidani, S., & Figuerdo, A. J. (2007). *Missing data: A gentle introduction*. New York, NY: Guilford Press.

Menard, S. (2010). *Logistic regression: From introductory to advanced concepts and applications*. Thousand Oaks, CA: Sage.

Merrell, M. (1947). Time-specific life tables contrasted with observed survivorship. *Biometrics*, *3*, 129–136.

Merrill, R. M., (2013). *Fundamentals of epidemiology and biostatistics: Combining the basics*. Burlington, MA: Jones & Bartlett.

Meyers, L. S., Gamst, G., & Guarino, A. J. (2009). *Data analysis using SAS enterprise guide*. New York, NY: Cambridge University Press.

Meyers, L. S., Gamst, G., & Guarino, A. J. (2013). *Applied multivariate research: Design and interpretation* (2nd ed.). Thousand Oaks, CA: Sage.

Moser, B. K., & Stevens, G. R. (1992). Homogeneity of variance in the two-sample means test. *The American Statistician*, *46*, 19–21.

Norman, G. R., & Streiner, D. L. (2008). *Biostatistics: The bare essentials* (3rd ed.). Hamilton, Ontario: BC Decker.

Norusis, M. J. (2008). *SPSS statistics 17.0: Advanced statistical procedures companion*. Upper Saddle River, NJ: Pearson Prentice Hall.

Norusis, M. J. (2012). *IBM SPSS statistics 19 advanced statistical procedures companion*. Upper Saddle River, NJ: Pearson Prentice Hall.

Nunnally, J. C., & Bernstein, I. H. (1994). *Psychometric theory* (3rd ed.). New York, NY: McGraw-Hill.

Osborne, J. W. (2008). Best practices in data transformation: The overlooked effect of minimum values. In J. Osborne (Ed.), *Best practices in quantitative methods* (pp. 197–204). Thousand Oaks, CA: Sage.

Pearson, K. (1896). Mathematical contributions to the mathematical theory of evolution. III Regression, heredity, and panmixia. *Philosophical Transactions of the Royal Society of London*, *187*, 253–318.

Pearson, K. (1900a). Mathematical contributions to the theory of evolution. VII. On the correlation of characters not quantitatively measurable. *Philosophical Transactions of the Royal Society of London, Series A*, *195*, 1–47.

Pearson, K. (1900b). On the criterion that a given system of deviations from the probable in the case of correlated system of variables is such that it can be reasonably supposed to have arisen from random sampling. *Philosophical Magazine*, *50*, 157–175.

Pearson, K. (1901). Mathematical distributions to the theory of evolution IX. On the principle of homotyposis and its relation to heredity, to the variability of the individual, and to that of the race. Part I: Homotyposis in the vegetable kingdom. *Philosophical Transactions of the Royal Society of London: Series A*, *197*, 285–379.

Pearson, K. (1901). On lines and planes of closest fit to systems of points in space. *Philosophical Magazine*, 2, 559–572.

Pedhazur, E. J. (1997). *Multiple regression in behavioral research: Explanation and prediction* (3rd ed.). Orlando, FL: Harcourt Brace.

Pedhazur, E. J., & Schmelkin, L. P. (1991). *Measurement, design, and analysis: An integrated approach*. Hillsdale, NJ: Lawrence Erlbaum Associates.

Preacher, K. J., & Hayes, A. F. (2004). SPSS and SAS procedures for estimating indirect effects in simple mediation models. *Behavior Research Methods, Instruments, and Computers*, *36*, 717–731.

Preacher, K. J., & Hayes, A. F. (2008). Asymptotic and resampling strategies for assessing and comparing indirect effects in multiple mediator models. *Behavior Research Methods*, *40*, 879–891.

Prentice, R. L., & Marek, P. (1979). A qualitative discrepancy between censored data rank tests. *Biometrics*, *35*, 861–867.

Press, S. J. (1972). *Applied multivariate analysis*. New York, NY: Holt, Rinehart & Winston.

Quenouille, M. H. (1956). Note on bias in estimation. *Biometrika*, *43*, 353–360.

Quinn, G. P., & Keough, M. J. (2002). *Experimental design and data analysis for biologists*. New York, NY: Cambridge University Press.

Raudenbush, S. W., & Bryk, A. S. (2002). *Hierarchical linear models: Applications and data analysis methods* (2nd ed.). Thousand Oaks, CA: Sage.

Raykov, T., & Marcoulides, G. A. (2008). *An introduction to applied multivariate analysis*. New York, NY: Routledge.

Read, T. R. C., & Cressie, N. A. C. (1988). *Goodness-of-fit statistics for discrete multivariate data*. New York, NY: Springer-Verlag.

Reise, S. F., & Duan, N. (2003). *Multilevel modeling: Methodological advances, issues, and applications*. Mahwah, NJ: Lawrence Erlbaum Associates.

Robinson, J. P., Shaver, P. R., & Wrightsman, L. S. (1991). *Measures of personality and social psychological attitudes*. San Diego, CA: Academic Press.

Rosenbaum, M. (1980). A schedule for assessing self-control behaviors: Preliminary findings. *Behavior Therapy*, *11*, 109–121.

Salsburg, D. (2001). *The lady tasting tea: How statistics revolutionized science in the twentieth century*. New York, NY: W. H. Freeman.

Satterthwaite, F. W. (1946). An approximate distribution of estimates of variance components. *Biometrics Bulletin*, *65*, 110–114.

Schlesselman, J. J. (1992). *Case-control studies: Design, conduct, analysis*. New York, NY: Oxford University Press.

Schreiber, J. B., Nora, A., Stage, F. K., Barlow, E. A., & King, J. (2006). Reporting structural equation modeling and confirmatory factor analysis results: A review. *Journal of Educational Research*, *99*, 323–337.

Shadish, W. R., Cook, T. D., & Campbell, D. T. (2002). *Experimental and quasi-experimental designs for generalized causal inference*. Boston, MA: Houghton-Mifflin.

Shepard, R. N. (1962). The analysis of proximities: Multidimensional scaling with an unknown distance function. I. *Psychometrika*, *27*, 125–140.

Shrout, P. E., & Fleiss, J. L. (1979). Intraclass correlations: Uses in assessing rater reliability. *Psychological Bulletin*, *86*, 420–428.

Singer, J. D., & Willett, J. B. (1993). It's about time: Using discrete-time survival analysis to study duration and the timing of events. *Journal of Educational Statistics*, *18*, 155–195.

Singer, J. D., & Willett, J. B. (2003). *Applied longitudinal data analysis*. New York, NY: Oxford University Press.

Snedecor, G. W. (1934). *Analysis of variance and covariance*. Ames, IA: Collegiate Press.

Snijders, T. A. B., & Bosker, R. J. (2012). *Multilevel analysis: An introduction to basic and advanced multilevel modeling* (2nd ed.). Thousand Oaks, CA: Sage.

Sobel, M. E. (1982). Aysmptotic confidence intervals for indirect effects and their standard errors in structural equation models. In N. Tuma (Ed.), *Sociological methodology* (pp. 159–186). San Francisco, CA: Jossey-Bass.

Sobel, M. E. (1986). Some new results on indirect effects in structural equation models. In S. Leinhardt (Ed.), *Sociological methodology* (pp. 290–312). Washington, DC: American Sociological Association.

Spearman, C. (1904a). "General intelligence," objectively determined and measured. *American Journal of Psychology*, *15*, 201–293.

Spearman, C. (1904b). The proof and measurement of association between two things. *American Journal of Psychology*, *15*, 72–101.

Spearman, C. (1907). Demonstration of formulae for true measurement of correlation. *American Journal of Psychology*, *18*, 161–169.

Spearman, C. (1910). Correlation calculated from faulty data. *British Journal of Psychology*, *3*, 271–295.

Steinberg, L., & Thissen, D. (1996). Uses of item response theory and the testlet concept in the measurement of psychopathology. *Psychological Methods*, *1*, 81–97.

Stigler, S. M. (1986). *The history of statistics: The measurement of uncertainty before 1900*. Cambridge, MA: Harvard University Press.

Stevens, J. P. (2009). *Applied multivariate statistics: A modern approach* (3rd ed.). Mahwah, NJ: Lawrence Erlbaum Associates.

Student . (1908). The probable error of a mean. *Biometrika*, *6*, 1–25.

Tabachnick, B. G., & Fidell, L. S. (2007).*Using multivariate statistics* (5th ed.). Boston, MA: Pearson.

Takane, Y. Young, F. W., & de Leeuw, J. (1977). Nonmetric individual differences multidimensional scaling: An alternating least squares method with optimal scaling features. *Psychometrika*, *42*, 7–67.

Tekle, F. B., & Vermunt, J. K. (2012). Event history analysis. In H. Cooper, P. M. Camic, D. L. Long, A. T. Panter, D. Rindskopf, & K. J. Sher (Eds.), *APA handbook of research methods in psychology: Data analysis and research publication* (Vol. 3, pp. 267–290). Washington, DC: American Psychological Association.

Thompson, B. (1995). Stepwise regression and stepwise discriminant analysis need not apply here: A guidelines editorial. *Educational and Psychological Measurement*, *55*, 525–534.

Thompson, B. (2006). *Foundations of behavioral statistics: An insight-based approach*. New York, NY: Guilford Press.

Thompson, B., & Borrello, G. M. (1985). The importance of structure coefficients in regression research. *Educational and Psychological Measurement*, *56*, 203–209.

Thurstone, L. L. (1931). Multiple factor analysis. *Psychological Review*, *38*, 406–427.

Thurstone, L. L. (1935). *Vectors of the mind*. Chicago, IL: University of Chicago Press.

Thurstone, L. L. (1947). *Multiple factor analysis*. Chicago, IL: University of Chicago Press.

Tonidandel, S., & LeBreton, J. M. (2011). Relative importance analysis: A useful supplement to regression analysis. *Journal of Business Psychology*, *26*, 1–9.

Torgerson, W. S. (1952). Multidimensional scaling: I. Theory and method. *Psychometrika*, *17*, 401–419.

Torgerson, W. S. (1958), *Theory and method of scaling*. New York, NY: John Wiley & Sons, Inc.

Traub, R. E. (1997). Classical test theory in historical perspective. *Educational Measurement: Issues and Practice*, *16*, 8–14.

Tukey, J. W. (1958). Bias and confidence in not quite large samples (Abstract). *Annals of Mathematical Statistics*, *29*, 614.

Tukey, J. W. (1977). *Exploratory data analysis*. Reading, MA: Addison-Wesley.

Varmuza, K., & Filzmoser, P. (2009). *Introduction to multivariate statistical analysis in chemometrics*. Boca Raton, FL: CRC Press.

Vogt, P. W. (2007). *Quantitative research methods for professionals*. Boston, MA: Allyn and Bacon.

Wang, J., & Wang, X. (2012). *Structural equation modeling: Applications using Mplus*. Chichester, West Sussex: John Wiley & Sons, Ltd.

Welch, B. L. (1937). The significance of the difference between two means when the population variances are unequal. *Biometrika*, *29*, 350–362.

West, S. G., Taylor, A. B., & Wu, W. (2012). Model fit and model selection in structural equation modeling. In R. H. Hoyle, (Ed.), *Handbook of structural equation modeling* (pp. 209–231). New York, NY: Guilford Press.

Wilcox, R. (2012). *Introduction to robust estimation & hypothesis testing* (3rd ed.). Waltham, MA: Elsevier.

Wright, S. (1920). The relative importance of heredity and environment in determining the piebald pattern of guinea-pigs. *Proceedings of the National Academy of sciences*, *6*, 320–332.

Yates, F. (1934). Contingency tables involving small numbers and the χ^2 test. *Supplement to the Journal of the Royal Statistical Society*, *1*, 217–235.

Yen, W. M. (1993). Scaling performance assessments: Strategies for managing local item dependence. *Journal of Educational Measurement*, *30*, 187–213.

Yerkes, R. M., & Dodson, J. D. (1908). The relation of strength of stimulus to rapidity of habit-formation. *Journal of Comparative Neurology and Psychology*, *18*, 459–482.

Young, F. W., & Harris, D. F. (2012). Multidimensional scaling. In M. J. Norusis (Ed.), *SPSS Statistics 19.0: Advanced statistics procedures companion* (pp. 361–430). Upper Saddle River, NJ: Pearson Prentice Hall.

AUTHOR INDEX

Performing Data Analysis Using IBM SPSS®, First Edition.
Lawrence S. Meyers, Glenn C. Gamst, and A. J. Guarino.

SUBJECT INDEX

Performing Data Analysis Using IBM SPSS®, First Edition.
Lawrence S. Meyers, Glenn C. Gamst, and A. J. Guarino.